Sustainable Biotechnological Remedial Frameworks for the Rejuvenation of Heavily Polluted Environments

This book discusses advanced research in mitigation of heavily polluted environments through a sustainable biotechnological remedial framework along with their role in betterment of conventional treatment methods. It includes a sustainable biotechnology-based approach to rejuvenate the heavily polluted environment, highlighting the technical, scientific, regulatory, safety, and societal impacts. It discusses current sustainable solutions and contemporary challenges of bioremediation of environmental pollutants using biotechnology-based interventions.

Features:

- Provides understanding of a biotechnology-based sustainable framework and its potential in rejuvenation of heavily polluted environments.
- Discusses nature-based solutions as a sustainable tool for net-zero emission transition and emerging economies.
- Provides the latest information and innovative scaffold as a support tool for fixing the heavily polluted environmental issues in a sustainable way.
- Sheds light on future research needs on advances of biotechnological intervention in managing the complicated environmental issues.
- Highlights the importance of multi-disciplinary research and integrated bioengineering systems.

This book is aimed at graduate students and researchers in waste management, environmental engineering, biotechnology and clean technologies.

Sustainable Biotechnological Remedial Frameworks for the Rejuvenation of Heavily Polluted Environments

Edited by
Naveen Dwivedi
Shubha Dwivedi
Lucas Meili

CRC Press
Taylor & Francis Group
Boca Raton London New York

CRC Press is an imprint of the
Taylor & Francis Group, an **informa** business

Designed cover image: © Shutterstock

First edition published 2026
by CRC Press
2385 NW Executive Center Drive, Suite 320, Boca Raton FL 33431

and by CRC Press
4 Park Square, Milton Park, Abingdon, Oxon, OX14 4RN

CRC Press is an imprint of Taylor & Francis Group, LLC

ISBN: 978-1-032-68945-6 (hbk)
ISBN: 978-1-041-13272-1 (pbk)
ISBN: 978-1-003-66896-1 (ebk)

DOI: 10.1201/9781003668961

Typeset in Times
by SPi Technologies India Pvt Ltd (Straive)

Contents

About the Editors xi
Contributors xii
Preface xvii

Chapter 1 Biotechnological Framework for the Treatment of Environmental Pollution and Challenges Thereof 1

Ayush Madan and Palak Sangal

1.1 Introduction 1
1.2 Biotechnological Approaches 5
1.3 Microbial Remediation: The Role of Microorganisms in Breaking Down Pollutants 9
1.4 Phytoremediation: Using Plants to Absorb and Detoxify Pollutants 13
1.5 Enzymatic Degradation: Applications of Enzymes in Breaking Down Pollutants 17
1.6 Challenges in Biotechnological Pollution Treatment 21
1.7 Complexity of Pollutants and Associated Challenges 22
1.8 Challenges in Scaling Up Biotechnological Processes 24
1.9 Strategies to Overcome Scale-Up Challenges 25
1.10 Integration of Biotechnological Solutions with Conventional Methods 27
1.11 Potential Synergies and Challenges in Integrating Biotechnological Solutions with Conventional Methods 28
1.12 Emerging Technologies and Future Prospects in Environmental Biotechnology 29
1.13 Nanobiotechnology in Pollution Treatment 30
1.14 Synthetic Biology in Pollution Remediation 31
1.15 Ethical Considerations and Regulatory Challenges in Synthetic Biology for Pollution Remediation 32
References 33

Chapter 2 Integration of Bioremediation Planning and Execution for Restoring the Ecosystem 39

Latifa EL Bardiy, Abdelmjid Bouazizi, Mounia Baidou, and Mohammed Zeriouh

2.1 Introduction 39
2.2 Bioremediation: Concept and Approach 40
2.3 Environmental Challenges of Pollutants 43
2.4 Environmental Revolution: Bioremediation and Pollution Management 45
2.5 Challenges of Bioremediation 50
2.6 Planning and Execution of the Bioremediation Approach 51
2.7 Conclusion, Recommendation, and Future Challenges 54
References 55

Chapter 3 Water Resources Management Through Nature-Based Solutions 60

Rohit Sharma, Himanshu Mishra, Yashaswani Sharma, and Vinod Kumar

3.1 Introduction 60
3.2 Historical Perspective on Water Management 64
3.3 Principles and Types of Nature-Based Solutions 67
3.4 Benefits of Nature-Based Solutions for Water Management 70
3.5 Challenges and Limitations of Nature-Based Solutions 71
3.6 Policy and Governance for Promoting NBS 72
3.7 Future Directions and Research Needs 74
3.8 Conclusion 75
References 76

Chapter 4 Nano-Biomaterials Used for the Elimination of Harmful Contaminants 80

Abdelmjid Bouazizi, Mounia Baidou, Latifa EL Bardiy, Doha El Machtani Idrissi, Mohamed Douma, Mohammed Zeriouh, Najib Tijani, and Mohamed Ouammou

4.1 Introduction 80
4.2 Contaminants in Water Environments 82
4.3 Utilization Methods for Eliminating Water Contaminants 85
4.4 Extracting Heavy Metal Ions from Polluted Water 88
4.5 Clearing Wastewater of Hazardous Organic Substances Through Adsorption 88
4.6 Enhanced Nano-biomaterials for Eco-friendly Pollution Remediation 89
4.7 Nano-Biomaterials for Environmental Cleanup 91
4.8 Recommendation and Future Challenges 96
References 96

Chapter 5 Production of Carbon with High Specific Surface Area from West African Agricultural Waste: Evaluation of Product Quality and potential Application 101

Joshua O. Ighalo and Chinenye Adaobi Igwegbe

5.1 Introduction 101
5.2 Materials and Methods 103
5.3 Results and Discussion 104
5.4 Conclusions 112
References 113

Chapter 6 Global Situation of Environmental Pollution and Its Sustainable Remedial Solutions 117

João Paulo Tenório da Silva Santos, Luiz Daniel da Silva Neto, Adeildo Junior de Oliveira, Thamyres Fernandes Messa Moreira, Vinicius Del Colle, Lucas Meili, and Danilo Henrique da Silva Santos

6.1 Introduction 117
6.2 Environmental Pollution, Health, and Development 118
6.3 Conventional Technologies for Mitigation and Elimination of Pollutants Around the World 121
6.4 Innovation in Sustainable Approaches to Environmental Pollution 126
6.5 Conclusions 131
References 132

Chapter 7 Microbial Rejuvenation of Heavily Polluted Soil 136

Hira Singh Gariya, Himanshi Rawat, and Arun Bhatt

7.1 Introduction 136
7.2 Pollutants 137
7.3 Mechanism of Microbial Degradation 148
7.4 Challenges and Limitations in Microbial Rejuvenation of Heavily Polluted Soil 170
7.5 Advancements in Bioremediation Techniques 171
7.6 Future Directions 172
7.7 Conclusion 172
References 173

Chapter 8 Sustainable Energy Production Through Microbial Fuel Cells and Waste Management 176

Namrata Bist, Anirbid Sircar, and Tithee Bhavsar

8.1 Introduction 176
8.2 Literature Study 177
8.3 Challenges in MFC Bioelectricity Generation 189
8.4 Utilization of MFC in Power Generation and Wastewater Treatment 190
8.5 Conclusion 191
References 191

Chapter 9 Insights into Omics Approaches in Plant Phytoremediation of Heavy Metals 195

Yuh Nien Chow and Keng Yuen Foo

9.1 Introduction 195
9.2 Phytoremediation 196
9.3 Governing Factors of Successive Phytoremediation 200
9.4 Omics Approaches for the Phytoremediation of Heavy Metals 204
9.5 Conclusion and Future Perspectives 213
References 214

Chapter 10 Resources Recovery Potential from Solid Waste and Bioeconomy Model Toward Sustainable Production and Responsible Consumption 220

Abiola Folakemi Olaniran, Adekunbi Adetola Malomo, Bolanle Akinsanola, and Olubukola David Olaniran

10.1 Introduction 220
10.2 Sustainable Solid Waste Management Practices 221
10.3 The Sociological Perspective of Solid Waste Resource Recovery and Its Sustainability 222
10.4 Resource Recovery Technologies and the Bioeconomy Model 224
10.5 Recent Research and Innovation in Resource Recovery and Bioeconomy ... 227
10.6 Future Trends and Outlook Toward Technological Advancements and Sustainability 230
References 230

Chapter 11 Microbial Electrolysis: An Emerging Biotechnology for Energy Production and Wastewater Treatment 234

Sajjad Hussain, Hammad Khan, Arslan Maqbool, Muhammad Usama, and Saima Gul

11.1 Introduction 234
11.2 Fundamentals of Bio Electrochemical System 236
11.3 Applications Microbial Electrolysis 240
11.4 Integration with Renewables 244
11.5 Technological Advancements 247
11.6 Challenges and Outlook 248
References 251

Chapter 12 Prediction of Heavy Metal Adsorption by Biomass Ash Using Artificial Neural Network 256

T.J. Kalanja, K. Chigayo, J. Kurasha, B. Mapani, H. Musiyarira, T.N. Sithole, and T. Falayi

12.1 Introduction 256
12.2 Materials and Methods 258
12.3 Literature Review 259
12.4 Results and Discussions 260
12.5 Conclusions 267
References 267

Chapter 13 Bio-Inspired Materials for Reducing Ecological Footprint and Life Cycle Assessment 269

Shubha Dwivedi, Naveen Dwivedi, and Aline Viancelli

13.1 Introduction 269
13.2 Ecological Footprint 269
13.3 Carbon Footprint 271
13.4 Bioengineering-Inspired Materials 275
13.5 Characteristics of Biomaterials 276
13.6 Life Cycle Assessment (LCA) 278
13.7 Bio-Inspired Materials for Reducing Ecological Footprint and Life Cycle Assessment 280
13.8 Bioengineering-Inspired Materials: Concept and Principles 281
13.9 Benefits of Bioengineering-Inspired Materials 282
13.10 Conclusion 283
References 284

Chapter 14 Enzymes Based Technologies for Bioremediation of Environmental Pollution 285

Josimayra Almeida Medeiros, Carlos Eduardo de Farias Silva, Gabrielly Karla Silva Santos, João Victor Ferro, João Victor Oliveira Nascimento da Silva, Micaela Almeida Alves do Nascimentos, Yasmin Matos Cardoso dos Santos, and Brígida Maria Villar da Gama

14.1 Introduction 285
14.2 Microbial Enzyme Production 286

14.3 Operational Factor Affecting the Enzymatic Degradation Process ... 287
14.4 Mechanisms of Enzymatic Pollutant Degradation ... 287
14.5 Bioremediation Through Enzyme Action ... 293
14.6 Enzyme-Based Technologies ... 294
14.7 Conclusion and Future Perspectives ... 299
References ... 299

Chapter 15 Bio-refinery Approach for a Sustainable Eco-system and Circular Bio-economy ... 304

Divya Bisen, Ashish Pratap Singh Chouhan, and Manish Pant

15.1 Introduction to Biorefinery Concepts ... 304
15.2 Advancements in the Research and Analysis of Biorefinery Initiatives ... 306
15.3 Conversion Technology ... 308
15.4 Sustainability Assessment of Biorefineries ... 315
15.5 Strengths and Weaknesses of Biorefinery ... 315
15.6 Environmental Impact and Footprint Assessment ... 318
15.7 Waste Bio-Refinery Promoting a Circular Economy ... 318
15.8 Challenges and Opportunities in Biorefinery ... 319
15.9 Conclusion ... 320
References ... 320

Chapter 16 Metabolic Engineering in Waste Prevention ... 325

Prashant Saxena and Madhvi Saxena

16.1 Introduction ... 325
16.2 Fundamentals of Metabolic Engineering ... 326
16.3 Microbial Conversion of Waste ... 328
16.4 Biodegradation and Bioremediation ... 331
16.5 Metabolic Engineering in the Circular Economy ... 335
16.6 Challenges and Future Directions ... 338
16.7 Conclusion ... 339
References ... 340

Chapter 17 IoT/ AI Based Solutions in Environmental Waste Management and Monitoring ... 344

Madhura Bodkhe, Gauri Chavan, Sunil Kulkarni, and Ajaygiri Goswami

17.1 Introduction ... 344
17.2 IoT for Waste Treatment, Monitoring and Management ... 353
17.3 AI for Waste Treatment, Monitoring and Management ... 357
17.4 Discussion ... 359
17.5 Conclusion ... 360
References ... 360

Chapter 18 Nano-Biosensors for Detecting Environmental Pollutants ... 366

Bhaskar Garg

18.1 Introduction ... 366
18.2 Nano-Biosensors ... 368
18.3 Optical Nano-Biosensors ... 372

18.4 Electrochemical Nano-Biosensors 376
18.5 Field Effect Transistors 382
18.6 Conclusive Remarks and Future Directions 382
References 383

Chapter 19 Industrial Framework for Net Zero Emission Transition in Emerging Economies 387

Riti Kumari, Naveen Dwivedi and Harmanpreet Meehnian

19.1 Introduction 387
19.2 Global Net Zero Emission Goals and Industrial Responsibility 389
19.3 Framework for Net Zero Transition in Industry 394
19.4 Artificial Intelligence and Digital Innovations for Net Zero 400
19.5 Challenges for Emerging Economies 401
19.6 Case Studies of Emerging Economies in Transition 404
19.7 Future Directions and Multilateral Cooperation 407
19.8 Conclusion 409
References 409

Chapter 20 A Vision on the Development of Carbon Industry, Enactment of Adaptation Policies, Guidelines, Ethical Considerations and Public Perception 413

Kah Yee Lim and Keng Yuen Foo

20.1 Introduction 413
20.2 Recent Development of the Carbon Industry 414
20.3 Overcapacity Issues in the Carbon Industry 417
20.4 Enactment of Adaptation Policy and Guidelines 418
20.5 Ethical Considerations and Public Perception 421
20.6 Conclusion 430
References 431

Index 437

About the Editors

Naveen Dwivedi is Professor and Head in the Department of Biotechnology Engineering and Food Technology, University Institute of Engineering, Chandigarh University, Mohali, Punjab, India. He obtained his M. Tech. degree in Biotechnology Engineering from Institute of Engineering & Technology, Lucknow. Dr. Dwivedi has completed his M. Tech. dissertation work from Fermentation Division of Central Drug Research Institute (CSIR-CDRI), Lucknow and Ph.D. research work from Bioenergy & Environmental Engineering Research Lab, Indian Institute of Technology Roorkee. Dr. Dwivedi is broadly interested in the fields of environmental biotechnology, bioprocess engineering, bioenergy, biological remediation of wastewater and waste valorization. Dr. Dwivedi has more than 21 years of teaching and research experience. He has a good number of research publications in peer-reviewed journals and conferences, has edited books and written more than two dozen chapters in books published by reputed international publishers.

Shubha Dwivedi is Associate Professor and Head in the Department of Biotechnology, IIMT University, Meerut, India. She received her postgraduate degree in biochemistry from Jiwaji University Gwalior and Ph.D. research work from Sustainable Processing and Water Treatment Research Lab, Indian Institute of Technology Roorkee. She was associated with the Institute of Genomics and Integrative Biology (CSIR-IGIB) Delhi during her masters. Her working area has been in the development of eco-friendly technology for the removal of hazardous contaminants from wastewater. Dr. Dwivedi has more than 19 years of teaching and research experience. Her outstanding research work is evidenced by a good number of research publications, books and book chapters, patents and so on.

Lucas Meili is Full Professor at the Center for Technology of the Federal University of Alagoas (UFAL) in Maceió, Alagoas, Brazil. Professor Meili graduated in chemical engineering from the Federal University of Rio Grande (Brazil) and obtained his Ph.D. in chemical engineering from the Federal University of São Carlos (Brazil) in 2009. His areas of interest are focused on separation processes, water and wastewater treatment, and materials synthesis. He has several publications in peer-reviewed high-impact scientific journals, covering various topics related to pollutant removal, adsorption, materials synthesis and water treatment. His publications demonstrate a particular focus on the application of materials as adsorbents in pollutant removal processes, as well as the evaluation of synthesis methods and water treatment processes.

Contributors

Abdelmjid Bouazizi
Materials, Membranes and Nanotechnology Team (EMaMNaT)
Moulay Ismail University of Meknès
Faculty of Sciences
BP: 11201 Zitoune Meknès, Morocco

Jadara University Research Center
Jadara University
Jordan

Interdisciplinary Laboratory for Modeling
Simulation and Scientific Innovation (LIMSIS)
École

Abiola Folakemi Olaniran
Landmark University SDG 12 (Responsible Consumption and Production research Group) and Department of Food Science and Microbiology
College of Pure and Applied Sciences, Landmark University
Omu-Aran, Nigeria

Adeildo Junior de Oliveira
Department of Chemistry, Federal University of Alagoas-Campus Arapiraca
Av. Manoel Severino Barbosa s/n
Arapiraca, Brazil

Adekunbi Adetola Malomo
Department of Food Science and Technology, Faculty of Technology
Obafemi Awolowo University
Ile-Ife, Nigeria

Ajaygiri Goswami
University Institute of Chemical Technology
Kavayitri Bahinabai Chaudhari North Maharashtra University
Jalgaon, India

Aline Viancelli
Universidade Do Contestado
Concordia, SC, Brazil

Anirbid Sircar
Pandit Deendayal Energy University
Gandhinagar, India

Arslan Maqbool
Faculty of Materials and Chemical Engineering
GIK Institute of Engineering Sciences and Technology
Topi, Pakistan

Arun Bhatt
Biotechnology Department
G.B.P.I.E.T, Ghurdauri
Pauri Garhwal

Ashish Pratap Singh Chouhan
School of Chemical Engineering and Physical Sciences
Lovely the Professional University
Phagwara, India

Ayush Madan
Department of Biotechnology, School of Research & Technology
People's University
Bhanpur, India

B. Mapani
Department of Civil, Mining and Process Engineering
Namibia University of Science and Technology
Windhoek, Namibia

Bhaskar Garg
Department of Chemistry, School of Basic Sciences & Technology
IIMT University
Meerut, India

Bolanle Akinsanola
Landmark University SDG 12 (Responsible Consumption and Production research Group) and Department of Food Science and Microbiology, College of Pure and Applied Sciences
Landmark University
Omu-Aran, Nigeria

Brígida Maria Villar da Gama
Technology Center
Federal University of Alagoas
Maceió, Brazil

Carlos Eduardo de Farias Silva
Technology Center
Federal University of Alagoas
Maceió, Brazil

Chinenye Adaobi Igwegbe
Department of Chemical Engineering
Nnamdi Azikiwe University
Awka, Nigeria

and

Department of Applied Bioeconomy
Wroclaw University of Environmental and Life Sciences
Wroclaw, Poland

Clinton Emeka Okonkwo
Department of Food Science and Technology, College of Agriculture and Veterinary Medicine
United Arab Emirates University
AI ain, United Arab Emirates

Danilo Henrique da Silva Santos
Technology Center of Federal Universityof Alagoas
Av. Lourival Melo Mota, s/n, Campus A.C. Simões, Tabuleiro do Martins
Maceió-AL, Brazil

Divya Bisen
School of Chemical Engineering and Physical Sciences
Lovely the Professional University
Phagwara, India

Doha El Machtani Idrissi
Laboratory of Materials, Membranes and Environment
Faculty of Sciences and Technologies of Mohammedia
Hassan II University of Casablanca
Morocco

Gabrielly Karla Silva Santos
Technology Center
Federal University of Alagoas
Maceió, Brazil

Gauri Chavan
University Institute of Chemical Technology
Kavayitri Bahinabai Chaudhari North Maharashtra University
Jalgaon, India

H. Musiyarira
Department of Civil, Mining and Process Engineering
Namibia University of Science and Technology
Windhoek, Namibia

Hammad Khan
Faculty of Materials and Chemical Engineering
GIK Institute of Engineering Sciences and Technology
Topi, Pakistan

Harmanpreet Meehnian
Department of Biotechnology Engineering and Food Technology
University Institute of Engineering, Chandigarh University
Mohali, India

Himanshi Rawat
Biotechnology
Shri Guru Ram Rai Public School
Patel Nagar, Dehradun

Himanshu Mishra
Department of Biotechnology Engineering and Food Technology
UIE, Chandigarh University
Mohali, India

Hira Singh Gariya
Research Fellow
CDRI
Lucknow

J. Kurasha
Department of Civil, Mining and Process Engineering
Namibia University of Science and Technology
P. Bag, Windhoek, Namibia

João Paulo Tenório da Silva Santos
Postgraduate Program in Agriculture and Environment
Federal University of Alagoas-Campus Arapiraca
Av. Manoel Severino Barbosa s/n
Arapiraca, Brazil

João Victor Ferro
Technology Center
Federal University of Alagoas
Maceió, Brazil

João Victor Oliveira Nascimento da Silva
Technology Center
Federal University of Alagoas
Maceió, Brazil

Joshua O. Ighalo
Department of Chemical Engineering
Nnamdi Azikiwe University
P. M. B. 5025, Awka, Nigeria

Josimayra Almeida Medeiros
Technology Center
Federal University of Alagoas
Maceió, Brazil

K. Chigayo
Department of Civil, Mining and Process Engineering
Namibia University of Science and Technology
Windhoek, Namibia

Kah Yee Lim
River Engineering and Urban Drainage Research Centre (REDAC)
UniversitiSains Malaysia
Nibong Tebal, Malaysia

Keng Yuen Foo
River Engineering and Urban Drainage Research Centre (REDAC)
Universiti Sains Malaysia
Nibong Tebal, Malaysia

Latifa EL Bardiy
Research team "Data Science, innovation and Development"; affiliated with the Faculty of Legal
Economic and Social Sciences of Tangier
Morocco

Research Group in Territorial and Organizational Economics, affiliated with the Faculty of Legal
Economic and Social Sciences of Tangier
Morocco

Lucas Meili
Technology Center of Federal Universityof Alagoas
Maceió-AL, Brazil

Luiz Daniel da Silva Neto
Drying Center of Pastes, Suspensions, and Seeds, Department of Chemical Engineering
Federal University of São Carlos (UFSCar)
São Carlos, Brazil

Madhura Bodkhe
University Institute of Chemical Technology
Kavayitri Bahinabai Chaudhari North Maharashtra University
Jalgaon, India

Madhvi Saxena
Department of Artificial Intelligence
Vishwakarma University
Pune, Maharastra

Manish Pant
Department of Chemistry-School of Physical Sciences
DIT University
Dehradun, India

Micaela Almeida Alves do Nascimentos
Technology Center
Federal University of Alagoas
Maceió, Brazil

Mohamed Douma
Materials, Membranes and Nanotechnology Team (EMaMNaT)
Moulay Ismail University of Meknès, Faculty of Sciences
BP: 11201 Zitoune Meknès, Morocco

Mohamed Ouammou
Laboratory of Materials, Membranes and Environment
Faculty of Sciences and Technologies of Mohammedia
Hassan II University of Casablanca
Morocco

Mohammed Zeriouh
Education Sciences, Philosophy, and Humanities
UMI
ENS, Meknes

Mounia Baidou
Materials, Membranes and Nanotechnology Team (EMaMNaT)
Moulay Ismail University of Meknès, Faculty of Sciences
BP: 11201 Zitoune Meknès, Morocco

Muhammad Usama
Faculty of Materials and Chemical Engineering

Najib Tijani
Laboratory of Materials, Membranes and Nanotechnology, Faculty of Sciences
Moulay Ismail University
Zitoune, Meknes, Morocco

Namrata Bist
Pandit Deendayal Energy University
Gandhinagar, India

Naveen Dwivedi
Department of Biotechnology Engineering and Food Technology
University Institute of Engineering, Chandigarh University
Mohali, India

Olubukola David Olaniran
Department of Sociology and Anthropology
Obafemi Awolowo University
Ile-Ife, Nigeria

Palak Sangal
Department of Biological Sciences
Birla Institute of Technology & Science
Pilani, India

Prashant Saxena
Department of Biotechnology Engineering
University Institute of Engineering, Chandigarh University
Ghuran, India

Riti Kumari
Department of Biotechnology Engineering and Food Technology
University Institute of Engineering, Chandigarh University
Mohali, India

Rohit Sharma
Department of Biotechnology Engineering and Food Technology, UIE
Chandigarh University
Mohali, India

Saima Gul
Department of Chemistry
Islamia College Peshawar
Pakistan

Sajjad Hussain
Faculty of Materials and Chemical Engineering
GIK Institute of Engineering Sciences and Technology
Topi, Pakistan

and

Department of Chemistry
Federal University of Amazonas
Manaus, AM, CEP, Brazil

Shubha Dwivedi
Department of Biotechnology
IIMT University
Meerut, India

Sunil Kulkarni
Chemical Engineering Department
Gharda Institute of Technology
Lavel, India

T. Falayi
Department of Chemical Engineering
University of Johannesburg
Doornfontein, South Africa

T.J. Kalanja
Department of Civil, Mining and Process Engineering
Namibia University of Science and Technology
Windhoek, Namibia

T.N. Sithole
Department of Chemical Engineering
University of Johannesburg
Doornfontein, South Africa

Thamyres Fernandes Messa Moreira
Paulista State University Julio de Mesquita Filho
Chemistry Instituteof Araraquara, Departament of Physical-Chemistry
Araraquara, São Paulo, Brazil

Tithee Bhavsar
Pandit Deendayal Energy University
Gandhinagar, India

Vinicius Del Colle
Department of Chemistry
Federal University of Alagoas-Campus Arapiraca
Arapiraca, Brazil

Vinod Kumar
Peoples' Friendship,
University of Russia (RUDN University),
Moscow, The Russian Federation

Yashaswani Sharma
Department of Biotechnology Engineering and Food Technology, UIE
Chandigarh University
Mohali, India

Yasmin Matos Cardoso dos Santos
Technology Center
Federal University of Alagoas
Maceió, Brazil

Yuh Nien Chow
River Engineering and Urban Drainage Research Centre (REDAC)
Universiti Sains Malaysia, Engineering Campus
Nibong Tebal, Malaysia

Preface

Waste production and its management is presently one of the major environmental challenges around the world due to an increasingly urbanized population, industrial growth and unsustainable agricultural practices, which have led to the accumulation of hazardous pollutants in the water, soil and air, jeopardizing ecosystems and human health. Therefore, there is a pressing need to develop a universal, flexible and sustainable solution that leverages biological processes to detoxify and restore heavily polluted environments. Sustainable biotechnological remedial frameworks represent a vital pathway for addressing the urgent challenge of environmental pollution. Sustainable bio-based solutions employed by harnessing the power of living organisms and natural processes not only treat polluted environments but also contribute to the restoration of ecological integrity and promote long-term sustainability. By espousing sustainable practices we can envision a world that is healthy and resilient by remediating heavily polluted environments for coming generations. The successful implementation of these frameworks will require a collaborative effort among scientists, researchers, policymakers and communities, paving the way for a healthier and more sustainable future and forging a path toward the United Nation's Sustainable Development Goals (SDGs).

Therefore, this book provides a broad impression of the advanced research in the mitigation of heavily polluted environments through a sustainable biotechnological-based remedial framework along with their prospective role in revolutionizing conventional treatment methods. The realization of smart and sustainable treatment technology will substantially provide significant economic and social benefits. The book is also focused on a global outlook of its challenges, trends and policies in relation to hazardous waste prevention, minimization and management. In this book, an interdisciplinary team of researchers presents a sustainable biotechnology-based approach to rejuvenating the heavily polluted environment with their advancements and highlights the technical, scientific, regulatory, safety and societal impacts. Readers will learn all about the recent and sustainable progress in both theoretical and practical aspects and future potential applications of biotechnology in the environment. We are thankful to the publisher (Engineering) team, CRC Press and all our contributors whose great efforts made this book a proper appearance.

Naveen Dwivedi
Chandigarh University, Punjab, India
Shubha Dwivedi
Meerut, Uttar Pradesh, India

1 Biotechnological Framework for the Treatment of Environmental Pollution and Challenges Thereof

Ayush Madan and Palak Sangal

1.1 INTRODUCTION

1.1.1 Overview of Environmental Pollution

Environmental pollution is one of the most critical issues facing the global community today. It encompasses the contamination of air, water, and soil by harmful substances, leading to adverse effects on human health, ecosystems, and the climate. The severity of environmental pollution is underscored by its widespread impact, affecting both developed and developing nations. This chapter delves into the different facets of environmental pollution, highlighting its sources, effects, and the urgent need for sustainable solutions.

1.1.2 Sources of Environmental Pollution

Environmental pollution originates from a variety of sources, broadly categorized into industrial, agricultural, and domestic activities. Industrial pollution is primarily driven by the emission of pollutants from factories and power plants. These pollutants include heavy metals, particulate matter, and greenhouse gases such as carbon dioxide (CO_2) and methane (CH_4). Agricultural activities contribute to pollution through the use of pesticides, fertilizers, and the improper disposal of animal waste, which contaminate water bodies and soil. Domestic pollution arises from household waste, sewage, and the use of non-biodegradable products, adding to the growing burden on the environment.

1.1.3 Air Pollution

Air pollution is one of the most pervasive forms of environmental pollution. According to the World Health Organization (WHO), around 91% of the world's population lives in places where air quality levels exceed WHO limits. Air pollutants, such as particulate matter (PM2.5 and PM10), nitrogen oxides (NOx), sulfur dioxide (SO_2), and volatile organic compounds (VOCs), pose significant health risks. They are linked to respiratory and cardiovascular diseases, lung cancer, and adverse pregnancy outcomes (WHO, 2021). Most common causes of air pollution are illustrated in Figure 1.1.

A study by Landrigan et al. (2018) highlighted the global health burden of air pollution, estimating that it contributed to 8.8 million deaths in 2015. The economic costs associated with air pollution are also staggering, amounting to trillions of dollars in healthcare costs and lost labor productivity. Furthermore, air pollution exacerbates climate change by increasing the concentration of greenhouse gases in the atmosphere, leading to global warming and extreme weather events.

DOI: 10.1201/9781003668961-1

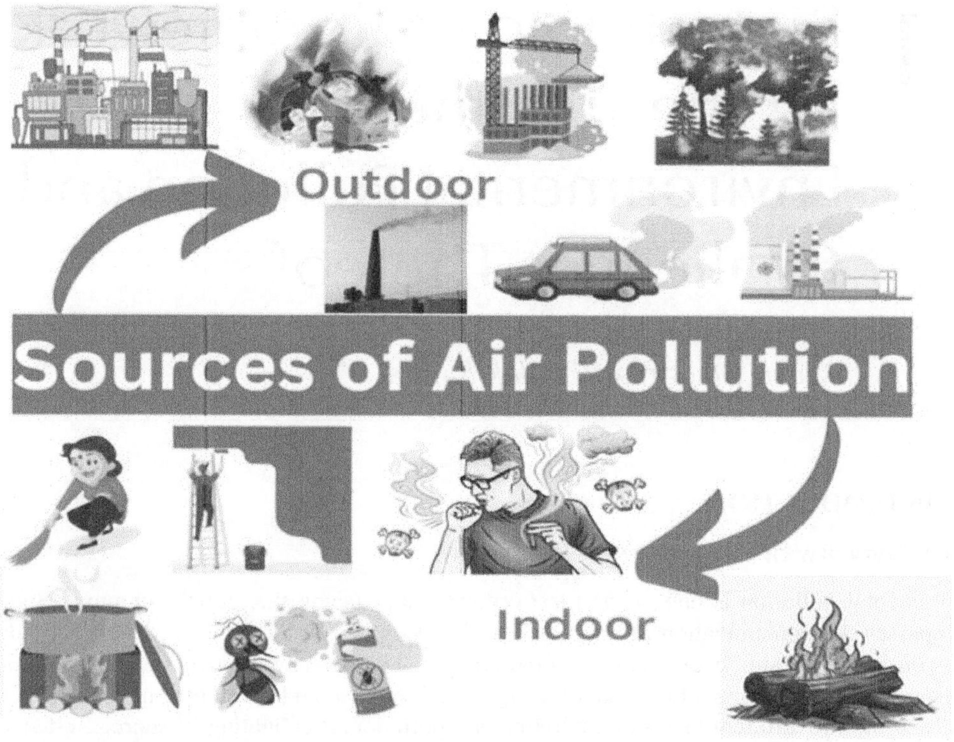

FIGURE 1.1 Most common sources of air pollution.

1.1.4 Water Pollution

Water pollution affects rivers, lakes, oceans, and groundwater, posing a threat to both human health and aquatic ecosystems. Industrial discharge, agricultural runoff, and untreated sewage are primary contributors to water pollution. Contaminants such as heavy metals, nitrates, phosphates, and pathogens degrade water quality, making it unsafe for drinking, irrigation, and recreational activities.

The United Nations (UN) has identified water pollution as a significant barrier to achieving Sustainable Development Goal 6, which aims to ensure the availability and sustainable management of water and sanitation for all. Research by Vörösmarty et al. (2010) indicates that more than 80% of the world's wastewater is discharged into the environment without adequate treatment, leading to widespread water contamination. This not only affects human health but also disrupts aquatic ecosystems, resulting in the loss of biodiversity and the collapse of fisheries.

1.1.5 Soil Pollution

Soil pollution is another critical issue, largely driven by industrial activities, mining, improper waste disposal, and excessive use of agrochemicals. Contaminants such as heavy metals, pesticides, hydrocarbons, and persistent organic pollutants (POPs) accumulate in the soil, reducing its fertility and posing risks to food safety. Soil pollution affects crop yield and quality, leading to economic losses for farmers and food insecurity for communities.

A study by Zhang et al. (2016) emphasizes the global extent of soil pollution, noting that industrial and agricultural activities have significantly contaminated soils worldwide. The researchers call for urgent action to address soil pollution, highlighting the need for stricter regulations, sustainable agricultural practices, and effective remediation techniques.

1.1.6 Global Impact and the Need for Action

The global impact of environmental pollution is profound, affecting the health and well-being of billions of people (Table 1.1). It contributes to a range of health problems, including respiratory diseases, cardiovascular conditions, cancers, and neurological disorders. Moreover, pollution exacerbates socioeconomic inequalities, as marginalized communities often bear the brunt of its adverse effects. Addressing environmental pollution requires a multifaceted approach that involves international cooperation, stringent regulations, and the adoption of clean technologies. Governments, industries, and individuals must work together to reduce emissions, manage waste responsibly, and promote sustainable practices. Policies aimed at improving air and water quality, regulating industrial emissions, and encouraging the use of renewable energy sources are crucial in mitigating the impact of pollution. Environmental pollution poses a severe threat to the planet, with far-reaching

TABLE 1.1
Different Types of Pollution, Problems they Cause, and Biotechnological Interventions to Treat Various Types of Pollution

Type of Pollution	Problems Caused	Biotechnological Interventions
Air pollution	Respiratory diseases, cardiovascular problems, climate change, acid rain	Biofilters and bioscrubbers using microorganisms to remove pollutants; algal biofilters for CO_2 absorption
Water pollution	Waterborne diseases, toxic effects from heavy metals and chemicals, eutrophication	Bioremediation using bacteria and fungi; phytoremediation with aquatic plants; engineered wetlands
Soil pollution	Loss of soil fertility, contamination of food crops, health hazards from heavy metals and pesticides	Bioremediation using microbes; phytoremediation with hyperaccumulator plants; mycoremediation with fungi
Noise pollution	Hearing loss, stress, sleep disturbances, cardiovascular issues	Noise mitigation through biotechnological designs in urban planning, such as green walls and roofs
Radioactive pollution	Cancer, genetic mutations, radiation sickness	Bioremediation using radioresistant bacteria and fungi; phytoremediation with plants that can accumulate radioactive elements
Thermal pollution	Disruption of aquatic ecosystems, decreased oxygen levels in water bodies	Bioremediation using thermophilic microorganisms to degrade organic pollutants in thermal effluents
Plastic pollution	Marine life harm, food chain disruption, human health risks through ingestion of microplastics	Enzymatic degradation of plastics by engineered microorganisms; bioplastics development using microbial processes
Pesticide pollution	Acute and chronic health effects, endocrine disruption, biodiversity loss	Biodegradation of pesticides using genetically modified microorganisms; integrated pest management with biocontrol agents
Heavy metal pollution	Toxicity leading to organ damage, developmental issues, and cancer	Biosorption using algae, fungi, and bacteria; phytoremediation with metal-accumulating plants
Oil pollution	Damage to marine and coastal ecosystems, economic losses in fisheries and tourism	Bioremediation using hydrocarbon-degrading bacteria and fungi; biostimulation and bioaugmentation techniques
E-waste pollution	Toxic effects from heavy metals and chemicals in electronics, environmental contamination	Biorecovery of metals using microbes; biotechnological recycling processes to break down electronic waste safely
Pharmaceutical pollution	Development of antibiotic-resistant bacteria, disruption of aquatic life	Biodegradation using specialized microbes and fungi; constructed wetlands to treat pharmaceutical-laden wastewater

consequences for human health, ecosystems, and the climate. It is imperative to recognize the urgency of this issue and implement effective solutions to combat pollution. By fostering a global commitment to environmental sustainability, we can protect the health of our planet and ensure a better future for generations to come.

1.1.7 The Need for Effective and Sustainable Solutions to Environmental Pollution

Environmental pollution poses a significant threat to the health of ecosystems, human well-being, and the stability of the planet's climate systems. The necessity for effective and sustainable solutions to mitigate and prevent pollution has never been more critical. This study highlights the pressing need for such solutions, supported by extensive research and scientific literature. Environmental pollution encompasses a wide range of contaminants, including air, water, soil pollution, and hazardous waste. These pollutants have far-reaching impacts on both natural ecosystems and human health.

Air pollution, primarily caused by industrial activities, transportation, and energy production, has severe health implications. It contributes to respiratory and cardiovascular diseases and premature mortality. Studies indicate that fine particulate matter (PM2.5) and nitrogen oxides (NOx) are particularly harmful (Smith et al., 2014). Water pollution arises from agricultural runoff, industrial discharges, and improper waste disposal. Contaminants like heavy metals, pesticides, and plastics degrade water quality, affecting aquatic life and human health (Hoekstra et al., 2011). Soil pollution results from industrial activities, agricultural chemicals, and improper waste disposal. It leads to reduced soil fertility, disrupted ecosystems, and health risks from exposure to toxic substances (Galloway et al., 2003). The improper handling and disposal of hazardous waste, including electronic waste and chemical byproducts, lead to environmental contamination and health hazards (Rockström et al., 2009).

1.1.7.1 The Urgency for Sustainable Solutions

Climate Change and Environmental Stability

Climate change exacerbates the effects of environmental pollution. The interplay between climate change and pollution creates a feedback loop, where pollutants contribute to global warming, and climate change, in turn, worsens pollution levels (Steffen et al., 2015).

Biodiversity Loss

Pollution is a major driver of biodiversity loss. The degradation of habitats due to pollution disrupts ecosystems, leading to the decline of species and loss of biodiversity (Newbold et al., 2016).

Human Health Impacts

The health impacts of pollution are profound. Exposure to pollutants is linked to a range of diseases, including respiratory illnesses, cardiovascular diseases, and cancers. The World Health Organization estimates millions of deaths annually due to air pollution alone (Whitmee et al., 2015).

1.1.7.2 Effective and Sustainable Solutions

Regulatory Measures and Policies

Implementing stringent regulatory measures and policies is crucial. Regulations should enforce limits on emissions, promote the use of cleaner technologies, and ensure proper waste management (Sachs, 2012).

Technological Innovations

Technological advancements offer promising solutions to pollution. Innovations in renewable energy, waste recycling, and pollution control technologies can significantly reduce environmental impacts (Foley et al., 2005).

Sustainable agricultural practices

Adopting sustainable agricultural practices, such as precision farming and organic agriculture, can reduce the runoff of pesticides and fertilizers into water bodies, thereby mitigating water pollution (Tilman & Clark, 2014).

Community Engagement and Education

Engaging communities and raising awareness about the impacts of pollution and the importance of sustainable practices is vital. Education can drive behavioral changes and support the implementation of local solutions (Moser & Dilling, 2007).

1.1.7.3 Case Studies

Success in Reducing Air Pollution

Countries like Sweden and Germany have successfully reduced air pollution through comprehensive policies, including promoting renewable energy, improving public transportation, and enforcing emission standards (Griggs et al., 2013).

Water Pollution Management

The Clean Water Act in the United States has significantly improved water quality by regulating discharges of pollutants into water bodies and funding sewage treatment plants (Parry et al., 2007).

Soil Remediation Efforts

China's efforts in soil remediation, including the use of phytoremediation and bioremediation techniques, have shown promising results in restoring contaminated soils (Diaz et al., 2019).

The need for effective and sustainable solutions to environmental pollution is undeniable. Addressing this global challenge requires a multifaceted approach, combining regulatory measures, technological innovations, sustainable practices, and community engagement. By drawing on scientific research and successful case studies, we can develop strategies that protect the environment, promote public health, and ensure a sustainable future.

1.2 BIOTECHNOLOGICAL APPROACHES

Biotechnology involves using living organisms, systems, or processes to develop products and technologies that help improve human life and the environment. Biotechnological methods have shown immense potential in addressing various forms of environmental pollution, including air, water, and soil pollution. Here we explore these methods and their applications, supported by valid references.

1.2.1 Biotechnological Methods

1.2.1.1 Bioremediation

Bioremediation utilizes microorganisms, fungi, green plants, or their enzymes to detoxify polluted environments. It includes various approaches such as bioaugmentation, biostimulation, and phytoremediation.

1.2.1.2 Bioaugmentation

Bioaugmentation involves the introduction of specific strains of microorganisms to accelerate the degradation of pollutants. For example, *Pseudomonas putida* has been used to degrade toluene and other aromatic hydrocarbons effectively (Megharaj et al., 2011).

1.2.1.3 Biostimulation

Biostimulation enhances the activity of indigenous microorganisms by adding nutrients or electron donors. This method has been successfully applied to clean up oil spills and contaminated groundwater (Mulligan et al., 2001).

1.2.1.4 Phytoremediation

Phytoremediation uses plants to absorb, sequester, and degrade pollutants from soil and water. Plants such as willow and poplar have been used to remediate sites contaminated with heavy metals and organic pollutants (Pilon-Smith et al., 2006).

1.2.1.5 Biofiltration

Biofiltration involves the use of living material to capture and biologically degrade pollutants. Biofilters are commonly used to treat air emissions containing VOCs and odors from industrial processes.

1.2.1.6 Air Biofiltration

Biofiltration systems use microorganisms immobilized on a filter bed to degrade pollutants in the air. These systems are effective for removing VOCs, hydrogen sulfide, and other malodorous compounds (Delhoménie & Heitz, 2005).

1.2.1.7 Genetic Engineering

Genetic engineering involves the modification of an organism's DNA to enhance its capabilities to degrade pollutants. Genetically engineered microorganisms can be designed to degrade specific contaminants more efficiently than their natural counterparts.

1.2.1.8 Engineered Microorganisms

For example, bacteria have been genetically engineered to degrade polychlorinated biphenyls and other persistent organic pollutants (Sayler & Ripp, 2000).

1.2.1.9 Enzymatic Treatment

Enzymatic treatment employs isolated enzymes to catalyze the breakdown of pollutants. Enzymes such as laccases and peroxidases have been used to degrade dyes, pesticides, and pharmaceuticals in wastewater.

1.2.1.10 Enzymatic Degradation

Laccase enzymes, derived from fungi, have been used to degrade synthetic dyes and other recalcitrant organic pollutants in wastewater treatment processes (Rodríguez Couto & Toca Herrera, 2006).

1.2.1.11 Biosorption

Biosorption involves using biological materials, such as algae, bacteria, and fungi, to adsorb heavy metals and other pollutants from aqueous solutions.

1.2.1.12 Algal biosorption

Marine algae have been shown to effectively remove heavy metals like cadmium, lead, and mercury from wastewater (Volesky, 2007).

1.2.2 Potential and Applications

1.2.2.1 Wastewater Treatment

Biotechnological methods are extensively used in wastewater treatment plants to remove organic pollutants, nutrients, and heavy metals. Activated sludge processes, biofilters, and bioreactors are common applications.

1.2.2.2 Activated Sludge Process

The activated sludge process relies on microbial consortia to degrade organic matter in sewage. Enhanced biological phosphorus removal is a biotechnological advancement in this field, reducing phosphorus levels in wastewater (Henze et al., 2008).

1.2.2.3 Soil Remediation

Bioremediation techniques, such as bioventing and biosparging, are used to treat contaminated soils. These methods enhance the microbial degradation of hydrocarbons and other pollutants in the soil.

1.2.2.4 Bioventing

Bioventing supplies oxygen to contaminated soils to stimulate the aerobic degradation of hydrocarbons. It has been effective in cleaning up sites contaminated with petroleum hydrocarbons (Leeson & Hinchee, 1997).

1.2.2.5 Air Pollution Control

Biofiltration and bioscrubbers are used to treat industrial air emissions. These methods are particularly effective for removing VOCs, sulfur compounds, and ammonia from the air.

1.2.2.6 Bioscrubbers

Bioscrubbers combine a physical scrubbing process with biological degradation. Pollutants are absorbed into a liquid phase and then degraded by microorganisms in a separate bioreactor (Kennes & Veiga, 2004).

1.2.2.7 Sustainable Agriculture

Biotechnology plays a crucial role in sustainable agriculture by reducing the need for chemical fertilizers and pesticides. Biofertilizers and biopesticides help in maintaining soil health and reducing environmental pollution.

1.2.2.8 Biofertilizers

Biofertilizers, such as rhizobium and mycorrhiza, enhance nutrient availability to plants, promoting sustainable agriculture and reducing the environmental impact of chemical fertilizers (Vessey, 2003).

Biotechnological methods offer effective and sustainable solutions to various forms of environmental pollution. These methods harness the natural capabilities of organisms and their metabolic processes to degrade, detoxify, and remove pollutants from the environment. The potential of biotechnology in pollution control is vast, and continued research and development will further enhance its applications and efficacy.

Biotechnological frameworks are increasingly recognized for their eco-friendly and resource-efficient characteristics. These methods utilize natural biological processes and organisms to mitigate environmental pollution, reduce resource consumption, and promote sustainability. Here we emphasize the advantages of biotechnological approaches in environmental management.

1.2.3 Eco-Friendly Characteristics of Biotechnological Frameworks

1.2.3.1 Reduced Chemical Usage

Biotechnological methods often replace or reduce the need for harmful chemicals in pollution control and agricultural practices.

Example: Biopesticides, derived from natural materials such as animals, plants, bacteria, and certain minerals, are less harmful to the environment compared to conventional chemical pesticides. They target specific pests without affecting nontarget species, reducing ecological disruption (Chandler et al., 2011).

1.2.3.2 Biodegradability

Biotechnological products are often biodegradable, breaking down into harmless substances that do not accumulate in the environment.

Example: Biodegradable Plastics. Bioplastics produced from renewable biomass sources, such as starch and cellulose, degrade naturally, reducing the burden of plastic pollution (Shah et al., 2008).

1.2.3.3 Minimal Environmental Impact

Biotechnological processes are designed to have minimal environmental impact, often using less energy and generating fewer by-products than traditional methods.

Example: Microbial Fuel Cells generate electricity from organic compounds in wastewater through the metabolic activities of bacteria. This technology not only treats wastewater but also produces clean energy with minimal environmental footprint (Logan et al., 2006).

1.2.4 Resource-Efficient Nature of Biotechnological Frameworks

1.2.4.1 Utilization of Renewable Resources

Biotechnological processes often use renewable resources, reducing dependence on nonrenewable materials and promoting sustainability.

Example: Biofuels produced from biomass, such as bioethanol from sugarcane or corn, are renewable alternatives to fossil fuels. They contribute to reducing greenhouse gas emissions and reliance on oil (Demirbas, 2009).

1.2.4.2 Energy Efficiency

Biotechnological methods can be more energy-efficient compared to conventional techniques, leading to lower energy consumption and reduced greenhouse gas emissions.

Example: Anaerobic digestion of organic waste produces biogas, which can be used as a renewable energy source. This process reduces waste volume and generates energy with lower emissions compared to landfill or incineration (Weiland, 2010).

1.2.4.3 Waste Reduction

Biotechnological processes can convert waste materials into valuable products, thus reducing waste and promoting resource efficiency.

Example: Composting uses microorganisms to decompose organic waste into nutrient-rich compost, which can be used as a natural fertilizer. This process reduces landfill waste and recycles nutrients back into the soil (Haug, 2018).

1.2.5 Eco-Friendly and Resource-Efficient Biotechnological Applications

1.2.5.1 Bioremediation

Bioremediation uses microorganisms to degrade environmental contaminants, providing an eco-friendly solution to pollution.

Example: Oil Spill Cleanup. Microbial bioremediation has been effectively used to clean up oil spills. Bacteria such as *Alcanivorax borkumensis* degrade hydrocarbons, reducing the environmental impact of oil spills (Yakimov et al., 2007).

1.2.5.2 Phytoremediation

Phytoremediation uses plants to absorb, sequester, and detoxify pollutants from soil and water, providing a green and sustainable method for environmental cleanup.

Example: Heavy Metal Removal. Plants such as Indian mustard (*Brassica juncea*) have been used to remove heavy metals like lead and cadmium from contaminated soils, offering a cost-effective and eco-friendly remediation strategy (Salt et al., 1995).

1.2.5.3 Biosorption

Biosorption involves using biological materials to adsorb heavy metals and other pollutants from aqueous solutions, offering a low-cost and effective method for wastewater treatment.

Example: Algal Biosorption. Marine algae have been shown to effectively remove heavy metals like cadmium, lead, and mercury from wastewater, providing an eco-friendly solution for water pollution (Volesky, 2007).

Biotechnological frameworks are inherently eco-friendly and resource-efficient, providing sustainable solutions to environmental pollution. By leveraging natural processes and renewable resources, biotechnology minimizes environmental impact, reduces chemical usage, and promotes energy and resource efficiency. The continued development and application of biotechnological methods hold great promise for creating a more sustainable and less polluted world.

1.3 MICROBIAL REMEDIATION: THE ROLE OF MICROORGANISMS IN BREAKING DOWN POLLUTANTS

Microbial remediation, also known as bioremediation, utilizes microorganisms to degrade, detoxify, or transform pollutants into less harmful forms. This process leverages the natural metabolic pathways of bacteria, fungi, and other microbes to clean up contaminated environments, including soil, water, and air. This chapter explores the role of microorganisms in breaking down various pollutants, supported by references for each key aspect.

1.3.1 Mechanisms of Microbial Remediation

Microorganisms utilize several mechanisms to break down pollutants, including enzymatic degradation, cometabolism, and biosorption.

1.3.1.1 Enzymatic Degradation

Microorganisms produce enzymes that catalyze the breakdown of complex pollutants into simpler, less toxic compounds.

Example: Hydrocarbon Degradation. Certain bacteria, such as *Pseudomonas* and *Mycobacterium*, produce enzymes like monooxygenases and dioxygenases, which initiate the degradation of hydrocarbons in oil spills (Head et al., 2006). These enzymes add oxygen to the hydrocarbon molecules, making them more soluble and easier to further degrade.

1.3.1.2 Cometabolism

Cometabolism occurs when microorganisms degrade pollutants incidentally while metabolizing another substrate. This process often requires the presence of a co-substrate that induces the production of necessary enzymes.

Example: Trichloroethylene (TCE) degradation. The bacteria *Burkholderia cepacia* can degrade TCE, a common groundwater contaminant, through cometabolism. When grown on methane or phenol, the bacteria produce enzymes that also degrade TCE (Leahy et al., 2003).

1.3.1.3 Biosorption

Biosorption involves the binding of pollutants to the cell surface of microorganisms, effectively removing them from the environment.

Example: Heavy Metal Biosorption. Fungi such as *Aspergillus niger* and algae like *Spirulina platensis* have been shown to biosorb heavy metals, including lead, cadmium, and chromium, from wastewater. The cell walls of these microorganisms contain functional groups that bind metal ions, facilitating their removal from aqueous solutions (Volesky, 2001).

1.3.2 Applications of Microbial Remediation

1.3.2.1 Soil Remediation

Microorganisms play a crucial role in the remediation of contaminated soils by degrading organic pollutants and immobilizing heavy metals.

Example: Polycyclic Aromatic Hydrocarbons (PAH) Degradation. PAHs are persistent organic pollutants commonly found in contaminated soils. Bacteria such as *Mycobacterium* and *Rhodococcus* are known to degrade PAHs through enzymatic oxidation, transforming them into less harmful compounds (Cerniglia, 1992).

1.3.2.2 Groundwater Remediation

Microbial remediation is effective in treating contaminated groundwater, particularly for pollutants such as chlorinated solvents and nitrates.

Example: Denitrification. Denitrifying bacteria, such as *Pseudomonas stutzeri*, reduce nitrates to nitrogen gas, thereby removing excess nitrates from groundwater. This process is essential in preventing the eutrophication of aquatic ecosystems (Zumft, 1997).

1.3.2.3 Wastewater Treatment

Microbial remediation is widely used in wastewater treatment plants to degrade organic pollutants and remove nutrients. Example: Activated Sludge Process. The activated sludge process employs a mixed microbial community to degrade organic matter in sewage. Aerobic bacteria play a central role in converting organic pollutants into carbon dioxide, water, and biomass (Seviour & Nielsen, 2010).

1.3.2.4 Air Pollution Control

Microorganisms can also be used to treat air emissions by degrading VOCs and other pollutants.

Example: Biofiltration of VOCs. Biofiltration systems use microorganisms immobilized on a filter medium to degrade VOCs from industrial exhaust streams. Bacteria such as *Pseudomonas putida* can metabolize VOCs like toluene, effectively reducing air pollution (Devinny et al., 1998).

1.3.3 Advantages of Microbial Remediation

1.3.3.1 Eco-Friendliness

Microbial remediation is environmentally friendly as it utilizes natural processes to degrade pollutants without the need for harmful chemicals (EPA, 2020).

1.3.3.2 Cost-Effectiveness

Microbial remediation is often more cost-effective compared to conventional remediation methods, particularly for large-scale environmental cleanup projects (Vidali, 2001).

1.3.3.3 Sustainability

Microbial remediation promotes sustainability by converting pollutants into harmless substances, thereby preventing secondary pollution and conserving natural resources (Juwarkar et al., 2010).

Microbial remediation leverages the natural capabilities of microorganisms to degrade, detoxify, and transform pollutants, offering an eco-friendly and sustainable approach to environmental cleanup. The diverse mechanisms employed by microbes, including enzymatic degradation, cometabolism, and biosorption, enable the effective remediation of various pollutants in soil, water, and air. The continued research and development in microbial remediation holds great promise for addressing global pollution challenges.

1.3.4 Case Studies Showcasing Successful Microbial Remediation

Microbial remediation has been successfully implemented in various environmental cleanup projects worldwide. These case studies highlight the practical applications and effectiveness of microbial remediation in addressing different types of pollution, supported by relevant references.

1.3.4.1 Exxon Valdez Oil Spill, Alaska

In March 1989, the Exxon Valdez oil tanker spilled approximately 11 million gallons of crude oil into Prince William Sound, Alaska. This disaster resulted in extensive environmental damage, affecting marine life, coastal ecosystems, and local communities.

Microbial Remediation Approach – Bioremediation was employed as a key strategy to clean up the spilled oil. Nutrient amendments, such as nitrogen and phosphorus fertilizers, were applied to stimulate the growth of indigenous oil-degrading bacteria. These bacteria, including species of *Pseudomonas* and *Alcanivorax*, used the hydrocarbons in the oil as a carbon and energy source, breaking them down into less harmful substances.

Outcome – The bioremediation efforts significantly accelerated the natural degradation of the oil, reducing its impact on the environment. Studies showed that treated shorelines had a higher rate of hydrocarbon degradation compared to untreated areas (Bragg et al., 1994).

1.3.4.2 Love Canal, New York

Love Canal, a neighborhood in Niagara Falls, New York, became infamous in the late 1970s due to widespread contamination from buried chemical waste. The site was found to contain numerous hazardous substances, including dioxins, benzene, and other toxic compounds, posing serious health risks to residents.

Microbial Remediation Approach – In situ bioremediation was implemented to address the contamination at Love Canal. The strategy involved the use of specialized bacteria capable of degrading aromatic hydrocarbons and chlorinated compounds. Techniques such as bioventing, where oxygen is supplied to enhance microbial activity, were used to stimulate the breakdown of contaminants in the soil.

Outcome – The bioremediation efforts successfully reduced the concentration of hazardous chemicals in the soil, contributing to the overall cleanup of the site. The project demonstrated the potential of microbial remediation to address complex and heavily contaminated environments (Demuth and Walsh, 1989).

1.3.4.3 Kuwaiti Oil Fields, Kuwait

During the Gulf War in 1991, retreating Iraqi forces set fire to hundreds of oil wells in Kuwait, resulting in extensive oil spills and soot deposition. The environmental disaster led to the contamination of large areas of desert soil with crude oil.

Microbial Remediation Approach – Bioremediation was employed to treat the oil-contaminated soil in Kuwait. The approach involved the use of indigenous hydrocarbon-degrading bacteria, combined with nutrient amendments to enhance microbial activity. Techniques such as landfarming, where contaminated soil is periodically tilled to promote aeration and microbial growth, were used.

Outcome – The bioremediation efforts significantly reduced the concentration of oil pollutants in the soil, restoring the affected areas over time. The project highlighted the effectiveness of microbial remediation in large-scale environmental disasters (Al-Mutairi et al., 2008).

1.3.4.4 Groundwater Contamination at a Chemical Plant, California

A chemical manufacturing plant in California was found to have contaminated the groundwater with TCE, a common industrial solvent and a potential carcinogen. The contamination posed a risk to the local water supply and required effective remediation.

Microbial Remediation Approach – A bioremediation approach involving bioaugmentation and biostimulation was used to treat the TCE–contaminated groundwater. Bioaugmentation involved the introduction of a mixed microbial culture known for its ability to degrade TCE. Biostimulation involved adding nutrients and electron donors, such as lactate, to enhance the microbial activity.

Outcome – The bioremediation approach successfully reduced TCE concentrations in the groundwater to below regulatory limits. The project demonstrated the potential of microbial remediation to effectively treat chlorinated solvent contamination in groundwater (Gossett, 2010).

1.3.4.5 Uranium Contamination at the Hanford Site, Washington

The Hanford Site in Washington State, a former nuclear production complex, has been plagued with uranium contamination in the groundwater. The contamination posed significant environmental and health risks due to the radioactive nature of uranium.

Microbial Remediation Approach – Researchers implemented a bioremediation strategy that involved stimulating indigenous microbial communities capable of reducing soluble uranium to insoluble uranium, thereby immobilizing it and preventing its spread in groundwater. The process involved adding electron donors, such as acetate, to the groundwater to enhance microbial activity.

Outcome – The bioremediation efforts successfully reduced the concentration of soluble uranium in the groundwater, demonstrating the feasibility of microbial remediation for treating radioactive contaminants. The project provided valuable insights into the potential for bioremediation in complex and high-risk environments (Anderson et al., 2003).

These case studies highlight the successful application of microbial remediation in diverse environmental contexts, including oil spills, industrial contamination, and radioactive waste. The effectiveness of microbial remediation in these scenarios underscores its potential as a sustainable and eco-friendly solution for environmental cleanup.

Microbial remediation has also been successfully implemented in various environmental cleanup projects across India. These case studies highlight the practical applications and effectiveness of microbial remediation in addressing different types of pollution in the Indian context.

1.3.4.6 Oil Spill in the Sundarbans, West Bengal

In 2014, an oil spill occurred in the Sundarbans, one of the largest mangrove forests in the world and a UNESCO World Heritage Site. The spill posed a significant threat to the delicate ecosystem, affecting both flora and fauna.

Microbial Remediation Approach – Bioremediation strategies involved the use of indigenous oil-degrading bacteria isolated from the Sundarbans' sediment. These bacteria, including strains of *Pseudomonas* and *Bacillus*, were used to treat the contaminated areas. Nutrient amendments were added to enhance microbial activity and accelerate the degradation process.

Outcome – The microbial remediation efforts led to a significant reduction in oil pollutants, demonstrating the potential of using native microorganisms for environmental cleanup in sensitive ecosystems like the Sundarbans (Sarkar and Sarma, 2017).

1.3.4.7 Chromium Contamination in Kanpur, Uttar Pradesh

Kanpur, an industrial city in Uttar Pradesh, has faced severe pollution issues due to the discharge of untreated tannery effluents containing high levels of hexavalent chromium (Cr(VI)) into the Ganga River. This contamination posed serious health risks to the local population and aquatic life.

Microbial Remediation Approach – Bioremediation strategies involved the use of chromium-reducing bacteria, such as *Pseudomonas putida* and *Bacillus subtilis*, to reduce toxic Cr(VI) to the less toxic trivalent chromium (Cr(III)). These bacteria were applied to contaminated sites, along with organic substrates to enhance microbial activity.

Outcome – The bioremediation approach successfully reduced Cr(VI) levels in the contaminated water and soil, making the environment safer for local communities and wildlife. This case study highlighted the potential of microbial remediation in addressing heavy metal pollution (Dhal et al., 2013).

1.3.4.8 Oil Sludge Contamination in Gujarat

Gujarat, a state with significant oil refining and petrochemical industries, has faced issues with oil sludge contamination in its industrial zones. These sludges contain a mixture of hydrocarbons and heavy metals, posing environmental and health risks.

Microbial Remediation Approach – Bioremediation strategies included the use of hydrocarbon-degrading bacteria and fungi, such as *Acinetobacter* sp., and *Aspergillus* sp., to treat oil sludge. Landfarming techniques were employed, where contaminated soil was regularly tilled and aerated to promote microbial degradation.

Outcome – The bioremediation efforts led to a significant reduction in hydrocarbon content in the treated soils, demonstrating the effectiveness of microbial remediation for managing industrial waste in Gujarat (Parmar and Singh, 2015).

1.3.4.9 Pesticide Contamination in Punjab

Punjab, an agriculturally intensive state, has faced significant pesticide contamination in its soils and water bodies due to the extensive use of agrochemicals. This contamination has had adverse effects on the environment and human health.

Microbial Remediation Approach – Bioremediation efforts involved the use of pesticide-degrading bacteria and fungi, such as *Pseudomonas fluorescens* and *Phanerochaetechrysosporium*. These microorganisms were applied to contaminated fields, and biostimulation techniques, such as the addition of organic matter, were used to enhance microbial activity.

Outcome – The bioremediation strategies led to a significant reduction in pesticide residues in the soil and water, helping to restore the agricultural productivity and environmental health of the affected areas (Cycon and Piotrowska-Seget, 2016).

1.3.4.10 Heavy Metal Contamination in Vellore, Tamil Nadu

The Vellore district in Tamil Nadu has experienced significant heavy metal contamination in its soil and water due to industrial effluents from tanneries and electroplating industries. The presence of heavy metals like lead, cadmium, and nickel posed severe environmental and health risks.

Microbial Remediation Approach – Bioremediation efforts included the use of heavy metal-accumulating bacteria, such as *Bacillus cereus* and *Rhodococcus* sp., to treat contaminated sites. Techniques like bioaugmentation and biosorption were employed to enhance the removal of heavy metals from the environment.

Outcome – The bioremediation strategies successfully reduced the concentrations of heavy metals in the affected areas, demonstrating the potential of microbial remediation in mitigating industrial pollution in Vellore (Prasad and Freitas, 2003).

These case studies highlight the successful application of microbial remediation in various environmental contexts across India. The effectiveness of microbial remediation in these scenarios underscores its potential as a sustainable and eco-friendly solution for environmental cleanup.

1.4 PHYTOREMEDIATION: USING PLANTS TO ABSORB AND DETOXIFY POLLUTANTS

Phytoremediation is an environmentally friendly and cost-effective method that uses plants to absorb, accumulate, and detoxify pollutants from soil, water, and air. This technique leverages the natural processes of plants and their associated microorganisms to remove contaminants, including heavy metals, organic pollutants, and radionuclides, thereby restoring polluted environments. The mechanisms of phytoremediation for key aspects of this technology is discussed next.

1.4.1 Mechanisms of Phytoremediation

Phytoremediation encompasses several mechanisms, each suited to different types of contaminants and environmental conditions:

1.4.1.1 Phytoextraction

Phytoextraction, or phytoaccumulation, involves the uptake of contaminants by plant roots and their translocation to above-ground parts (shoots and leaves). This process is particularly effective for heavy metals and certain organic pollutants. Example: Cadmium and Lead Phytoextraction.

Plants such as Indian mustard (*Brassica juncea*) and sunflower (*Helianthus annuus*) have been shown to effectively accumulate heavy metals like cadmium (Cd) and lead (Pb) from contaminated

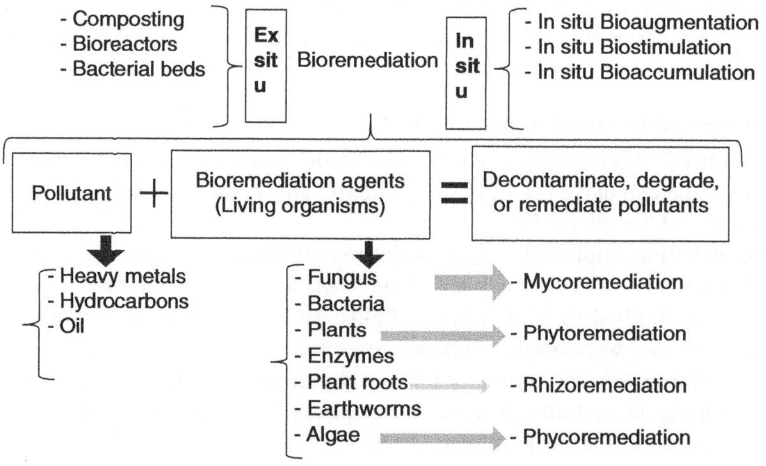

FIGURE 1.2 The process of phytoremediation to get rid of heavy metals from soil using sunflower plant.

soils (Figure 1.2). These plants concentrate the metals in their tissues, which can then be harvested and safely disposed of or processed (Salt et al., 1995).

1.4.1.2 Phytodegradation

Phytodegradation involves the breakdown of contaminants within the plant through metabolic processes. Plants produce enzymes that degrade organic pollutants into less harmful substances. Example: Trichloroethylene (TCE) Degradation

Poplar trees (*Populus* spp.) have been used to degrade TCE, a common groundwater contaminant. The trees take up TCE through their roots, and enzymes within the plant, such as dehalogenases, transform TCE into nontoxic compounds (Newman and Reynolds, 2004).

1.4.1.3 Rhizofiltration

Rhizofiltration involves the use of plant roots to absorb, concentrate, and precipitate heavy metals and other pollutants from aqueous environments, including wastewater and groundwater. Example: Uranium Removal

Sunflower (*Helianthus annuus*) roots have been used to remove uranium from contaminated water. The roots absorb and concentrate the uranium, which can then be harvested and treated (Dushenkov et al., 1995).

1.4.1.4 Phytostabilization

Phytostabilization uses plants to immobilize contaminants in the soil, preventing their migration to groundwater or air. Plants can achieve this by absorbing contaminants into their roots, adsorbing them onto root surfaces, or precipitating them within the rhizosphere.

Example: Arsenic Stabilization

Plants such as Pterisvittata (Chinese brake fern) can stabilize arsenic-contaminated soils by adsorbing arsenic onto root surfaces and reducing its mobility (Vamerali et al., 2010).

1.4.1.5 Phytovolatilization

Phytovolatilization involves the uptake of volatile contaminants by plants and their subsequent release into the atmosphere through transpiration. This process can convert contaminants into less harmful forms.

Example: Selenium Volatilization

Indian mustard (*Brassica juncea*) and other plants have been used to volatilize selenium from contaminated soils, transforming it into gaseous methylselenides, which are less toxic (Terry et al., 2000).

1.4.2 Applications of Phytoremediation

1.4.2.1 Soil Remediation

Phytoremediation is widely used to remediate contaminated soils, particularly for heavy metals and organic pollutants.

Example: Heavy Metal Contamination in Mining Areas

In mining regions, plants like willow (*Salix* spp.) and Indian mustard (*Brassica juncea*) have been used to extract heavy metals such as zinc, cadmium, and lead from contaminated soils (Robinson et al., 2000).

1.4.2.2 Groundwater Remediation

Phytoremediation can be used to treat contaminated groundwater, especially for organic solvents and heavy metals.

Example: TCE Remediation Using Poplar Trees

Poplar trees have been planted at contaminated sites to absorb and degrade TCE from groundwater, reducing its concentration and preventing its spread (Vroblesky and Casey, 2001).

1.4.2.3 Wastewater Treatment

Phytoremediation is employed in constructed wetlands and other systems to treat industrial and municipal wastewater.

Example: Constructed Wetlands for Nutrient Removal

Constructed wetlands with plants such as *Typha* (cattails) and *Phragmites* (reeds) are used to remove nutrients, heavy metals, and organic pollutants from wastewater, improving water quality (Vymazal, 2011).

1.4.2.4 Air Pollution Control

Plants can also be used to mitigate air pollution by absorbing VOCs and other airborne pollutants.

Example: Indoor Air Quality Improvement

Plants like spider plants (*Chlorophytum comosum*) and peace lilies (*Spathiphyllum* spp.) are known to absorb VOCs, such as formaldehyde and benzene, improving indoor air quality (Wolverton et al., 1989).

1.4.3 Advantages and Limitations of Phytoremediation

Phytoremediation, the use of plants to remediate contaminated environments, offers numerous advantages and some limitations.

1.4.3.1 Advantages of Phytoremediation

Eco-Friendliness

Phytoremediation is an environmentally friendly approach that leverages natural plant processes to detoxify pollutants without introducing harmful chemicals into the environment. This makes it a sustainable option for long-term environmental management (Pilon-Smits, 2005).

Cost-Effectiveness

Compared to traditional remediation methods, phytoremediation is often more cost-effective, especially for large-scale sites with diffuse contamination. The costs associated with planting, maintaining, and harvesting plants are typically lower than those for mechanical or chemical remediation techniques (Glass, 1999).

1.4.3.2 Aesthetic and Ecological Benefits

Phytoremediation can enhance the aesthetic value of a site by creating green spaces. Additionally, plants used in phytoremediation can provide habitats for wildlife, improve soil structure, and contribute to the ecological health of an area (Doty, 2008).

1.4.3.3 Soil Health Improvement

Certain plants used in phytoremediation can enhance soil fertility and structure through root growth and organic matter addition. This can improve soil quality and support subsequent agricultural or ecological uses of the site (Robinson et al., 2000).

1.4.3.4 Minimal Disturbance

Phytoremediation involves minimal disturbance to the site compared to more invasive methods like excavation or soil washing. This makes it suitable for areas where maintaining the integrity of the landscape is important (Salt et al., 1998).

1.4.3.5 Applicability to a Wide Range of Contaminants

Phytoremediation can address a broad spectrum of pollutants, including heavy metals, radionuclides, organic compounds, and emerging contaminants. Different plant species and mechanisms can be tailored to target specific types of contamination (Vamerali et al., 2010).

1.4.3.6 Limitations of Phytoremediation

Limited Depth of Effectiveness

Phytoremediation is primarily effective in the root zone (rhizosphere), limiting its application to shallow contaminants. Deeply buried pollutants or those in groundwater far below the surface are less accessible to plant roots (Vangronsveld and Cunningham, 1998).

Time Consuming

Phytoremediation is generally slower than conventional methods, often requiring several growing seasons to achieve significant contaminant reduction. This may not be suitable for sites requiring immediate remediation (Chaney et al., 1997).

Plant Biomass Disposal

The accumulation of contaminants in plant biomass poses a challenge for disposal. Harvested plants may contain high levels of pollutants, requiring safe and environmentally sound disposal or processing methods (McCutcheon & Schnoor, 2003).

Variability in Plant Performance

The effectiveness of phytoremediation can vary based on plant species, soil conditions, climate, and the type of contaminant. This variability can make it difficult to predict outcomes and optimize the process for specific sites (Pilon-Smits, 2005).

Risk of Contaminant Release

In some cases, phytoremediation can lead to the volatilization of contaminants or the release of secondary pollutants. For example, phytovolatilization can transfer contaminants from soil to air, potentially creating new exposure pathways (Doty, 2008).

Site-Specific Constraints
Certain environmental conditions, such as extreme pH, salinity, or arid climates, may limit the growth and survival of phytoremediation plants. This can restrict the applicability of the method to certain geographic regions or contaminated sites (Arthur et al., 2005).

Phytoremediation offers numerous advantages, including eco-friendliness, cost-effectiveness, and minimal site disturbance. However, its limitations, such as the depth of effectiveness, time requirements, and variability in plant performance, must be carefully considered when evaluating its suitability for a specific site. Ongoing research and technological advancements continue to improve the efficacy and applicability of phytoremediation, making it a valuable tool in the suite of environmental remediation strategies.

1.5 ENZYMATIC DEGRADATION: APPLICATIONS OF ENZYMES IN BREAKING DOWN POLLUTANTS

Enzymatic degradation involves using specific enzymes to break down pollutants into less harmful substances. This biotechnological approach harnesses the catalytic power of enzymes, which can degrade a wide range of contaminants, including organic compounds, heavy metals, and persistent pollutants. The mechanisms of enzymatic degradation and its applications in environmental remediation are discussed in the following.

1.5.1 Mechanisms of Enzymatic Degradation

Enzymatic degradation relies on the catalytic properties of enzymes, which are biological molecules that accelerate chemical reactions. Several classes of enzymes are involved in the degradation of pollutants, each with specific substrates and mechanisms of action.

1.5.1.1 Oxidoreductases

Oxidoreductases, such as peroxidases, laccases, and cytochrome P_{450} enzymes, catalyze oxidation-reduction reactions, breaking down complex organic pollutants into simpler, less toxic compounds.

Example: Laccase in Lignin Degradation

Laccases, found in fungi and plants, are capable of degrading lignin, a complex polymer found in plant cell walls. This property makes laccases useful in breaking down lignin-derived pollutants in industrial effluents (Strong & Claus, 2011).

1.5.1.2 Hydrolases

Hydrolases, including esterases, lipases, and proteases, catalyze the hydrolysis of chemical bonds, such as ester, amide, and glycosidic bonds. These enzymes are effective in breaking down a variety of organic pollutants.

Example: Esterase in Pesticide Degradation

Esterases can hydrolyze ester bonds in organophosphate pesticides, rendering them nontoxic. This makes esterases valuable for bioremediation of pesticide-contaminated soils and water (Singh and Walker, 2006).

1.5.1.3 Lyases

Lyases, such as ammonia lyases and carbon-carbon lyases, catalyze the cleavage of bonds without hydrolysis or oxidation, often forming double bonds or rings. These enzymes can degrade a range of organic pollutants.

Example: Ammonia Lyase in Nitrile Degradation

Nitrile-degrading enzymes, including nitrilases and nitrile hydratases, can break down nitriles, which are common in industrial waste. This results in less toxic products, such as carboxylic acids and ammonia (Banerjee et al., 2002).

1.5.1.4 Transferases

Transferases, such as methyltransferases and glycosyltransferases, catalyze the transfer of functional groups between molecules. These enzymes can modify pollutants, making them more amenable to further degradation.

Example: Methyltransferase in Mercury Detoxification

Methyltransferases play a role in the detoxification of mercury by converting toxic methylmercury to less harmful forms, facilitating its removal from the environment (Parks et al., 2013).

1.5.2 Applications of Enzymatic Degradation

1.5.2.1 Soil Remediation

Enzymatic degradation is widely used to remediate contaminated soils, particularly those polluted with organic compounds, pesticides, and heavy metals.

Example: Laccase in Soil Bioremediation

Laccases have been used to degrade PAHs in contaminated soils. These enzymes oxidize PAHs, breaking them down into less toxic compounds (Asif and Hai, 2019).

1.5.2.2 Water Treatment

Enzymes are employed in treating industrial and municipal wastewater to remove organic pollutants, heavy metals, and emerging contaminants.

Example: Peroxidase in Wastewater Treatment

Peroxidases, such as horseradish peroxidase, have been used to treat phenolic compounds in wastewater. These enzymes catalyze the oxidation of phenols, resulting in their polymerization and precipitation (Husain, 2010).

1.5.2.3 Air Pollution Control

Enzymatic degradation is also applied in air pollution control, particularly for the removal of VOCs and other airborne pollutants.

Example: Laccase in VOC Degradation

Laccases have been used in biofilters to degrade VOCs emitted from industrial processes. These enzymes oxidize VOCs, transforming them into less volatile and less toxic compounds (Baldrian, 2006).

1.5.2.4 Industrial Waste Management

Enzymatic degradation is utilized in managing industrial waste, particularly for the treatment of hazardous chemicals and byproducts.

Example: Esterase in Plastic Degradation

Esterases have been used to degrade polyethylene terephthalate (PET), a common plastic. These enzymes hydrolyze the ester bonds in PET, breaking it down into its monomers, which can be further processed or recycled (Tournier et al., 2020).

1.5.2.5 Bioremediation of Oil Spills

Enzymatic degradation plays a crucial role in bioremediation strategies for oil spills, breaking down complex hydrocarbons into simpler, less harmful compounds.

Example: Lipase in Oil Spill Remediation

Lipases, which hydrolyze fats into glycerol and fatty acids, have been used to degrade petroleum hydrocarbons in oil spill–affected areas. This enzymatic action enhances the breakdown and dispersion of oil (Wang and Cai, 2006).

1.5.3 Advantages of Enzymatic Degradation

1.5.3.1 Specificity

Enzymes exhibit high specificity for their substrates, allowing targeted degradation of specific pollutants without affecting nontarget compounds. This precision reduces the risk of collateral damage to the environment (Ahuja et al., 2004).

1.5.3.2 Efficiency

Enzymes catalyze reactions at high rates, significantly speeding up the degradation process compared to natural microbial degradation alone. This efficiency can lead to faster remediation times (Sheldon and van Pelt, 2013).

1.5.3.3 Environmentally Friendly

Enzymatic degradation is a green technology that operates under mild conditions (ambient temperature and pH) and does not require harmful chemicals or extreme physical conditions. This minimizes the environmental footprint of remediation activities (Turner et al., 2007).

1.5.3.4 Versatility

Enzymes can be produced from a variety of sources, including bacteria, fungi, plants, and animals, making them versatile tools for bioremediation across different environments and pollutants (Van Beilen and Li, 2002).

1.5.4 Limitations of Enzymatic Degradation

1.5.4.1 Stability

Enzymes can be sensitive to environmental conditions, such as temperature, pH, and the presence of inhibitors, which can affect their stability and activity. Stabilizing enzymes for practical applications can be challenging (Guisán, 2006).

1.5.4.2 Cost

The production, purification, and immobilization of enzymes can be expensive, which may limit the economic feasibility of enzymatic degradation for large-scale applications. However, advances in biotechnology are reducing these costs (Sheldon and Woodley, 2018).

1.5.4.3 Limited Range of Pollutants

While enzymes are effective against a wide range of organic pollutants, their effectiveness against certain types of contaminants, such as heavy metals, may be limited. Combining enzymatic degradation with other remediation techniques can address this limitation (Gianfreda & Rao, 2004).

Enzymatic degradation offers a promising approach for the remediation of contaminated environments, leveraging the specificity, efficiency, and eco-friendliness of enzymes. While there are challenges related to enzyme stability, cost, and the range of effective pollutants, ongoing research and technological advancements continue to enhance the feasibility and application of this biotechnological tool.

1.5.5 Specificity of Enzymatic Degradation

Enzymatic degradation exhibits high specificity, allowing enzymes to selectively target and degrade particular substrates or pollutants while leaving other compounds untouched. This specificity is crucial for minimizing collateral damage to the environment and optimizing the efficiency of remediation processes.

1.5.5.1 Mechanisms of Specificity

Enzymes achieve specificity through several mechanisms:

1. Active Site Configuration: Enzymes have a specific three-dimensional structure in their active site that complements the shape and chemical properties of their substrate. This "lock-and-key" fit ensures that only the correct substrate can bind and undergo catalysis (Bugg et al., 2011).
2. Substrate Recognition: Enzymes often recognize specific functional groups or chemical bonds within a substrate. This recognition allows them to selectively cleave or modify targeted molecules, such as hydrolyzing ester bonds in pesticides or oxidizing aromatic compounds (Liu et al., 2014).
3. Regulation and Specificity: Enzymatic specificity can also be regulated through environmental factors, co-factors, or co-substrates that influence enzyme activity and substrate binding. This regulatory control enhances the precision of enzymatic reactions in complex environments (Hibi et al., 2007).

1.5.6 Efficiency of Enzymatic Degradation

Enzymatic degradation processes are highly efficient due to the catalytic nature of enzymes, which accelerate reaction rates significantly compared to nonenzymatic processes. This efficiency is beneficial for rapid pollutant removal and reducing the overall remediation time.

1.5.6.1 Factors Contributing to Efficiency

1. Catalytic Turnover: Enzymes can catalyze multiple substrate molecules per unit time, often undergoing thousands to millions of turnovers per second. This rapid turnover rate amplifies the degradation efficiency, even at low enzyme concentrations (Gotor-Fernández and Alfonso, 2011).
2. Reaction Specificity: Enzymes exhibit high specificity not only in substrate recognition but also in reaction pathways, ensuring that degradation proceeds via predictable and controlled pathways. This specificity minimizes side reactions and enhances overall efficiency (Schomburg & Schomburg, 2006).
3. Environmental Adaptation: Many enzymes are adaptable to a wide range of environmental conditions, including temperature, pH, and substrate concentration. This adaptability allows enzymatic degradation processes to function optimally in diverse environmental settings (Bornscheuer & Kazlauskas, 2004).

1.5.7 Applications and Implications

The combination of specificity and efficiency makes enzymatic degradation a valuable tool in environmental remediation, industrial processes, and medical applications. Understanding these characteristics is essential for optimizing enzyme-based technologies and advancing sustainable biotechnological solutions.

1.6 CHALLENGES IN BIOTECHNOLOGICAL POLLUTION TREATMENT

Biotechnological approaches for pollution treatment offer promising solutions but also face several challenges that need to be addressed for effective implementation (Table 1.2). This discussion explores key challenges which are as follows:

1.6.1 Complexity of Pollutants

Pollutants in natural environments are often complex mixtures of chemicals, including organic compounds, heavy metals, and emerging contaminants. Biotechnological methods must contend with this complexity, requiring versatile approaches capable of addressing diverse pollutants simultaneously (Mohan et al., 2014).

1.6.2 Substrate Availability and Accessibility

The availability and accessibility of pollutants as substrates for biodegradation or transformation by biotechnological agents can vary significantly. This variability influences the effectiveness and efficiency of bioremediation processes (Cycon and Piotrowska-Seget, 2016).

TABLE 1.2
Challenges Faced by Biotechnology in Environmental Pollution

Challenge	Challenge	Examples
Complexity of Pollutants	Diverse and complex nature of pollutants makes degradation challenging.	Mixed waste streams, POPs
Scalability	Difficulty in scaling up lab-scale processes to industrial-scale applications.	Bioreactors, field-scale bioremediation projects
Environmental Variability	Changes in environmental conditions affect microbial activity and efficiency.	Temperature, pH, nutrient availability
Regulatory Hurdles	Complex and varying regulatory requirements across regions.	Approval processes for GMOs, environmental impact assessments
Economic Viability	High costs of biotechnological treatments compared to conventional methods.	Cost of microbial cultures, bioreactor setup and maintenance
Public Acceptance	Public concerns about safety and ethical issues of genetically modified organisms (GMOs) and synthetic biology.	Community resistance, lack of awareness
Gene Transfer Risks	Risk of horizontal gene transfer to nontarget organisms in the environment.	Spread of engineered genes, antibiotic resistance
Monitoring and Control	Challenges in monitoring and controlling biological processes in the field.	Real-time tracking of microbial activity, pollutant levels
Microbial Adaptation	Risk of microorganisms adapting and losing efficiency over time.	Development of resistance, mutations
Integration with Conventional Methods	Difficulty in integrating biotechnological solutions with existing methods.	Combining bioremediation with physical/chemical treatments
Environmental Impact	Potential negative impacts of biotechnological interventions on ecosystems.	Disruption of native microbial communities, bioaccumulation
Technical Expertise	Need for specialized knowledge and skills to design and implement treatments.	Training personnel, interdisciplinary collaboration

1.6.3 Environmental Condition

Biotechnological processes are often sensitive to environmental conditions such as temperature, pH, and oxygen levels. Variations in these parameters can affect the activity and stability of enzymes, microorganisms, or plants involved in pollution treatment (Sarkar et al., 2012).

1.6.4 Scale-up and Engineering Challenges

Moving from laboratory-scale experiments to large-scale applications poses significant challenges in terms of engineering, process optimization, and cost-effectiveness. Scaling up biotechnological processes while maintaining efficiency and reducing operational costs remains a critical hurdle (Behera et al., 2017).

1.6.5 Long-Term Effectiveness and Monitoring

Ensuring the long-term effectiveness of biotechnological treatments requires continuous monitoring of pollutant levels, microbial activity, and environmental parameters. Predicting and managing potential rebound effects or secondary pollution sources are essential considerations (Pattapulavar and Christopher, 2025).

1.6.6 Public Acceptance and Regulatory Compliance

The acceptance of biotechnological solutions for pollution treatment by stakeholders, including communities and regulatory bodies, plays a crucial role in their implementation. Addressing concerns regarding safety, efficacy, and long-term environmental impacts is essential for gaining public trust and regulatory approval (Torres-Duarte et al., 2009).

1.6.7 Cost-Efficiency and Economic Viability

The economic feasibility of biotechnological pollution treatment methods depends on factors such as the cost of biocatalysts, substrate availability, energy requirements, and maintenance. Achieving cost-efficiency relative to traditional remediation methods is a significant challenge (Margesin and Schinner, 2001).

Addressing these challenges requires interdisciplinary approaches integrating biotechnology, environmental science, engineering, and policymaking. Ongoing research and technological advancements are essential to overcome these hurdles and realize the full potential of biotechnological pollution treatment methods.

1.7 COMPLEXITY OF POLLUTANTS AND ASSOCIATED CHALLENGES

1.7.1 Complexity of Pollutants

Pollutants in natural and industrial environments exhibit considerable complexity due to their diverse chemical compositions and physical properties. These pollutants encompass a wide range of organic and inorganic compounds, including POPs, heavy metals, pesticides, pharmaceuticals, and emerging contaminants such as microplastics and nanoparticles. The complex nature of these pollutants presents significant challenges for effective environmental remediation strategies.

1.7.2 Challenges Associated with Diverse and Complex Pollutants

1.7.2.1 Diversity in Chemical Structures

Pollutants vary widely in their chemical structures, which influences their behavior in the environment and their susceptibility to degradation processes. Biotechnological methods must be versatile enough to address different classes of pollutants, each requiring specific enzymatic or microbial degradation pathways (Mohan et al., 2014).

1.7.2.2 Persistence and Bioaccumulation

Some pollutants are highly persistent in the environment, resisting degradation over extended periods. Persistent pollutants can accumulate in organisms through bioaccumulation and biomagnification, posing long-term risks to ecosystem health and human health (Tchounwou et al., 2012).

1.7.2.3 Interactions and Synergistic Effects

Pollutants often coexist in complex mixtures in environmental matrices, leading to interactions that can amplify or mitigate their individual effects. Synergistic effects among pollutants may result in enhanced toxicity or persistence, complicating remediation efforts (Lagunas-Rangel et al., 2022).

1.7.2.4 Multiple Sources and Pathways

Pollutants can enter the environment through diverse pathways, including industrial discharges, agricultural runoff, atmospheric deposition, and improper waste disposal. Identifying and mitigating multiple pollution sources is crucial for effective pollution control and remediation (Cycon & Piotrowska-Seget, 2016).

1.7.2.5 Complexity of Remediation Technologies

Developing remediation technologies capable of addressing diverse pollutants requires integration of multiple approaches, including bioremediation, phytoremediation, chemical treatments, and physical methods. Each technology has specific limitations and effectiveness, depending on the pollutant and environmental conditions (Sarkar et al., 2012).

1.7.2.6 Regulatory and Public Perception Challenges

Regulating and managing diverse pollutants involves navigating complex regulatory frameworks and addressing public concerns about environmental risks and health impacts. Public perception of pollution risks can influence policy decisions and the acceptance of remediation strategies (Torres-Duarte et al., 2009).

The complexity of pollutants presents multifaceted challenges for environmental scientists, engineers, and policymakers striving to develop effective remediation strategies. Addressing these challenges requires interdisciplinary collaboration, innovative technologies, and a nuanced understanding of pollutant behavior and interactions in the environment. By integrating insights from referenced studies, advancements can be made toward more sustainable and comprehensive approaches to pollution management and remediation.

Biotechnological frameworks encompass a diverse array of approaches that can adapt to various types of pollutants present in different environmental matrices. This adaptability is crucial for effectively addressing pollution challenges. Following is an exploration of how biotechnological frameworks can adapt to different types of pollutants.

1. Organic Pollutants
   Biotechnological frameworks such as enzymatic degradation and microbial bioremediation are particularly effective against organic pollutants. Enzymes and microorganisms can break down organic compounds such as petroleum hydrocarbons, pesticides, pharmaceuticals, and solvents into simpler, less harmful substances through biochemical processes.
   Enzymatic Degradation: Specific enzymes like oxidases, peroxidases, and hydrolases catalyze the degradation of organic pollutants by breaking down chemical bonds or modifying functional groups (Husain, 2010).
   Microbial Bioremediation: Certain bacteria, fungi, and algae have metabolic pathways capable of metabolizing and utilizing organic pollutants as carbon sources, thereby facilitating their degradation (Pattapulavar and Christopher, 2025).

2. Heavy Metals
   Biotechnological approaches for heavy metal remediation often involve bioaccumulation, biosorption, or biotransformation processes. Certain plants (phytoremediation), algae, fungi, and bacteria can sequester or transform heavy metals through mechanisms such as ion exchange, precipitation, or enzymatic reduction.
   Phytoremediation: Plants like sunflowers, willows, and aquatic plants uptake heavy metals from soil or water into their tissues, where they can be concentrated, immobilized, or transformed (Behera et al., 2017).
   Microbial Biomineralization: Certain bacteria can transform soluble toxic metals into insoluble, less bioavailable forms through biomineralization processes, reducing their environmental impact (Gadd, 2010).
3. Emerging Contaminants
   Biotechnological frameworks are also being developed to address emerging contaminants such as pharmaceutical residues, microplastics, and nanoparticles, which pose novel environmental challenges.
   Enzymatic Degradation of Pharmaceuticals: Enzymes capable of degrading pharmaceutical compounds are being explored for wastewater treatment to mitigate their environmental impact (Feng et al., 2025).
   Microbial Degradation of Microplastics: Certain microbial species have been identified that can degrade and metabolize components of microplastics, offering potential solutions to mitigate their accumulation in aquatic ecosystems (Syranidou et al., 2019).
4. Mixed Pollutants and Complex Mixtures
   Biotechnological frameworks are increasingly being adapted to address mixed pollutants and complex contaminant mixtures found in industrial effluents, urban runoff, and contaminated soils. Integrated approaches combining multiple biotechnological strategies are often necessary to achieve comprehensive remediation.
   Combined Bioremediation Strategies: Integrating microbial consortia, enzymatic treatments, and phytoremediation techniques can synergistically enhance the degradation of complex mixtures of pollutants (Mohan et al., 2014).

Biotechnological frameworks demonstrate versatility in adapting to different types of pollutants by leveraging enzymatic activities, microbial metabolisms, plant physiological processes, and innovative biotechnological tools. Continued research and development are essential to optimize these frameworks for diverse environmental applications and to address emerging pollution challenges effectively.

1.8 CHALLENGES IN SCALING UP BIOTECHNOLOGICAL PROCESSES

Scaling up biotechnological processes from laboratory-scale experiments to industrial or environmental applications poses significant challenges that need to be carefully addressed to ensure efficiency, effectiveness, and economic viability. Here are some key considerations and challenges associated with scaling up biotechnological processes:

1.8.1 Process Optimization and Reproducibility

Laboratory-scale processes often operate under controlled conditions that may not reflect real-world environments. Scaling up requires optimizing parameters such as substrate concentrations, pH, temperature, and oxygen levels to ensure reproducibility and consistent performance at larger scales (Bell et al., 2021).

1.8.2 Mass Transfer Limitations

Effective mass transfer of substrates, nutrients, and gases becomes more challenging as process volumes increase. In large-scale bioreactors or environmental remediation sites, ensuring adequate mixing and distribution of components throughout the system is critical for maintaining microbial activity and enzyme effectiveness (Pakshirajan et al., 2014).

1.8.3 Biocatalyst Stability and Longevity

Biocatalysts such as enzymes or microbial consortia may exhibit reduced stability or activity under large-scale operational conditions, leading to decreased process efficiency over time. Strategies for enhancing biocatalyst stability, such as immobilization techniques or genetic engineering, are crucial for maintaining performance (Sheldon and van Pelt, 2013).

1.8.4 Economic Feasibility

Scaling up biotechnological processes often involves substantial capital investment and operational costs. Balancing these costs against the benefits of pollutant removal or product yield is essential for ensuring economic feasibility and market competitiveness (Torres-Duarte et al., 2009).

1.8.5 Monitoring and Control

Monitoring large-scale biotechnological processes requires robust analytical methods to assess key parameters such as pollutant concentrations, microbial activity, and biocatalyst performance. Implementing real-time monitoring and control systems helps optimize process conditions and detect potential issues early (Mohan et al., 2014).

1.8.6 Regulatory and Safety Considerations

Scaling up biotechnological processes for environmental applications requires compliance with regulatory standards governing pollutant discharge, waste management, and health and safety protocols. Addressing these considerations ensures legal compliance and public acceptance of biotechnological solutions (Pattapulavar and Christopher, 2025).

1.9 STRATEGIES TO OVERCOME SCALE-UP CHALLENGES

Pilot-Scale Testing: Conducting pilot-scale studies allows for optimization of process parameters and assessment of scalability before full-scale implementation.

Engineering Solutions: Implementing advanced bioreactor designs, such as fed-batch or continuous-flow systems; enhances process control and efficiency.

Biocatalyst Engineering: Developing genetically modified organisms or immobilizing enzymes can improve stability and activity under large-scale conditions.

Process Monitoring and Automation: Integrating sensor technologies and automated control systems facilitates real-time monitoring and adjustment of process variables.

Collaboration and Interdisciplinary Approaches: Engaging multidisciplinary teams can leverage expertise in biotechnology, engineering, economics, and regulatory affairs to address complex scale-up challenges holistically.

Addressing scale-up challenges in biotechnological processes requires a systematic approach that integrates scientific innovation, engineering principles, economic analysis, and regulatory compliance. By overcoming these challenges, biotechnological solutions can be effectively deployed to tackle environmental pollution, enhance industrial processes, and contribute to sustainable development goals. Continued research and collaboration are essential to advance scalable biotechnological solutions for global environmental and industrial challenges.

Overcoming scalability issues in biotechnological processes requires a combination of strategic planning, innovative technologies, and rigorous implementation. Here are key strategies and approaches to address scalability challenges effectively:

1. Pilot-Scale Testing and Optimization
   Strategy: Conducting pilot-scale studies allows for systematic optimization of process parameters and validation of scalability before full-scale implementation. This phase helps identify potential bottlenecks and refine operational procedures.
   Implementation: Start with small-scale reactors or experimental setups that mimic operational conditions. Gradually increase system complexity and size while monitoring performance metrics such as substrate conversion rates, product yields, and biocatalyst stability.
2. Advanced Bioreactor Design and Engineering
   Strategy: Adopting advanced bioreactor designs, such as fed-batch, continuous-flow, or membrane bioreactors; enhances process control, mass transfer efficiency, and scalability. These designs optimize nutrient supply, gas exchange, and microbial growth conditions.
   Implementation: Integrate sensors for real-time monitoring of key process parameters (e.g., pH, temperature, dissolved oxygen) and automated control systems to adjust conditions dynamically. This approach ensures stable operation and maximizes biocatalyst performance (Pakshirajan et al., 2014).
3. Biocatalyst Engineering and Immobilization
   Strategy: Enhancing biocatalyst stability and activity through genetic engineering, protein engineering, or immobilization techniques improves performance under large-scale conditions. Immobilization on solid supports or within matrices prolongs biocatalyst lifespan and facilitates reuse.
   Implementation: Develop genetically modified microorganisms or enzymes with enhanced tolerance to process conditions (e.g., temperature extremes, high substrate concentrations). Implement immobilization strategies using carriers or matrices compatible with bioreactor operations (Sheldon and van Pelt, 2013).
4. Process Monitoring and Control Systems
   Strategy: Implementing robust monitoring systems and automated control strategies optimizes process performance, detects deviations early, and ensures consistency across different scales. Real-time data acquisition facilitates timely adjustments to maintain desired operational conditions.
   Implementation: Deploy sensors, probes, and analytical instruments to monitor critical parameters continuously. Integrate data acquisition systems with control algorithms for automated feedback and feedforward control of nutrient addition, agitation rates, and aeration levels (Mohan et al., 2014).
5. Collaboration and Interdisciplinary Approaches
   Strategy: Engaging multidisciplinary teams comprising biotechnologists, engineers, chemists, economists, and regulatory experts fosters comprehensive problem-solving and innovation. Collaboration facilitates holistic understanding of scalability challenges and accelerates technology transfer.
   Implementation: Foster partnerships between academia, industry, and governmental agencies to leverage diverse expertise and resources. Establish collaborative research initiatives

to address specific scalability issues, share best practices, and navigate regulatory requirements (Pattapulavar and Christopher, 2025).

Addressing scalability issues in biotechnological processes demands a strategic approach encompassing pilot-scale testing, advanced engineering solutions, biocatalyst optimization, robust monitoring and control systems, and collaborative interdisciplinary efforts. By implementing these strategies, researchers and practitioners can effectively transition innovative biotechnological solutions from laboratory settings to practical applications, thereby advancing sustainable practices and addressing global challenges in environmental remediation, industrial production, and healthcare. Continued research and development are essential to refine and optimize these strategies for future scalability and impact.

1.10 INTEGRATION OF BIOTECHNOLOGICAL SOLUTIONS WITH CONVENTIONAL METHODS

Integrating biotechnological solutions with traditional pollution control methods offers a comprehensive approach to addressing environmental challenges. This integration leverages the strengths of both biotechnology and conventional methods to enhance efficiency, sustainability, and cost-effectiveness in pollution management. Following is an exploration of the need for integration this approach.

1.10.1 Need for Integration

1. Complementary Strengths
   Biotechnological solutions, such as bioremediation and enzymatic degradation, often offer environmentally friendly and specific approaches to pollutant removal. However, they may face limitations in scalability or effectiveness under certain conditions. Integrating these with conventional methods, such as physical-chemical treatments or filtration, can provide synergistic effects and broaden applicability (Cycon and Piotrowska-Seget, 2016).
2. Enhanced Efficiency and Versatility
   Conventional methods like activated carbon adsorption or chemical oxidation are effective for treating specific pollutants or achieving rapid remediation. By integrating biotechnological approaches, these methods can target residual or complex contaminants that may not be effectively treated by traditional means alone (Sarkar et al., 2012).
3. Environmental and Economic Benefits
   Biotechnological solutions are often more sustainable and less energy-intensive compared to conventional methods, contributing to reduced carbon footprints and lower operational costs over time. Integration allows for optimizing resource use and minimizing environmental impacts while achieving regulatory compliance (Margesin and Schinner, 2001).
4. Risk Mitigation and Long-Term Sustainability
   Combining biotechnological and conventional methods provides redundancy in treatment processes, enhancing reliability and resilience against operational failures or unforeseen challenges. This integrated approach supports long-term sustainability by addressing pollutant persistence and potential rebound effects (Pattapulavar and Christopher, 2025).
   Integrating biotechnological solutions with conventional pollution control methods represents a forward-thinking approach to tackling diverse environmental challenges. By harnessing the complementary strengths of both approaches, stakeholders can achieve more effective, sustainable, and economically viable solutions for pollution management. Continued research,

technological innovation, and collaborative efforts across disciplines are essential to optimizing integrated strategies and advancing environmental stewardship globally.

1.11 POTENTIAL SYNERGIES AND CHALLENGES IN INTEGRATING BIOTECHNOLOGICAL SOLUTIONS WITH CONVENTIONAL METHODS

Integrating biotechnological solutions with conventional pollution control methods offers potential synergies that enhance overall effectiveness in environmental remediation. However, this integration also presents challenges that need to be carefully addressed to maximize synergistic benefits. Following is a detailed discussion on both aspects:

1.11.1 Potential Synergies

1. Complementary Mechanisms
Biotechnological solutions, such as bioremediation and enzymatic degradation, complement conventional methods by targeting different pollutants or stages of treatment. Bioremediation can be effective for organic contaminants and some heavy metals, whereas conventional methods like adsorption or chemical oxidation may excel in removing specific pollutants or achieving rapid treatment (Cycon and Piotrowska-Seget, 2016).
2. Enhanced Efficiency and Effectiveness
Integrating biotechnological and conventional methods can lead to synergistic effects, where combined treatments achieve higher removal efficiencies or reduce residual contaminant levels more effectively than either method alone. This synergy can optimize treatment outcomes across varying environmental conditions and pollutant profiles (Sarkar et al., 2012).
3. Environmental Sustainability
Biotechnological solutions often offer environmentally friendly approaches, such as natural attenuation or biological degradation, which minimize chemical usage and energy consumption. Integrating these methods with conventional treatments supports sustainable practices by reducing environmental footprints and enhancing long-term remediation effectiveness (Margesin and Schinner, 2001).
4. Cost Efficiency and Operational Flexibility
Combining biotechnological and conventional methods can optimize resource use and operational costs. Biotechnological approaches may offer lower operational costs over time due to reduced chemical inputs and energy requirements, while conventional methods provide rapid treatment capabilities that complement biotechnological processes (Pattapulavar and Christopher, 2025).

1.11.2 Challenges

1. Compatibility and Integration Issues
Biotechnological and conventional methods may require different operational conditions, such as pH, temperature, and substrate availability, which can complicate integration efforts. Ensuring compatibility and optimizing process parameters for both approaches to work synergistically can be challenging (Bell et al., 2021).
2. Monitoring and Control
Integrating diverse treatment methods necessitates robust monitoring and control systems to optimize performance and detect deviations promptly. Ensuring real-time data acquisition and adaptive control strategies across integrated processes is crucial for maintaining treatment efficacy (Pakshirajan et al., 2014).

3. Regulatory Compliance
   Compliance with regulatory standards and environmental permits can pose challenges when integrating biotechnological and conventional methods. Demonstrating the safety, effectiveness, and reliability of integrated treatments requires navigating complex regulatory frameworks and obtaining necessary approvals (Mohan et al., 2014).
4. Scale-Up and Implementation Costs
   Scaling up integrated biotechnological and conventional treatments from laboratory or pilot-scale to industrial or environmental applications involves significant capital investment and operational costs. Balancing these costs against the benefits of enhanced treatment efficacy and sustainability is essential for feasibility and long-term success (Torres-Duarte et al., 2009).

Integrating biotechnological solutions with conventional pollution control methods holds promise for enhancing environmental remediation effectiveness, sustainability, and cost efficiency. By leveraging synergies between biotechnological and conventional approaches, stakeholders can develop holistic strategies to address diverse environmental challenges. Addressing integration challenges through innovation, collaboration, and adaptive management approaches is essential for realizing the full potential of integrated pollution control solutions in practice. Continued research, technological advancements, and interdisciplinary cooperation will drive the evolution and adoption of integrated approaches for sustainable environmental stewardship.

1.12 EMERGING TECHNOLOGIES AND FUTURE PROSPECTS IN ENVIRONMENTAL BIOTECHNOLOGY

Emerging technologies in environmental biotechnology are shaping the future of pollution control, remediation, and sustainable resource management. These innovative approaches hold promise for addressing complex environmental challenges while advancing toward a more sustainable and resilient future. Following is a discussion on key emerging technologies and their future prospects.

1. Nanotechnology in Environmental Remediation
   Nanotechnology offers novel materials and approaches for environmental remediation. Nanomaterials such as nanoparticles and nanocomposites exhibit unique properties that enhance pollutant adsorption, catalysis, and degradation efficiency. Nano-enabled sensors for real-time environmental monitoring, nanostructured membranes for water purification, and nanocatalysts for pollutant degradation are examples of how nanotechnology is revolutionizing environmental biotechnology. Continued research into safe synthesis, environmental fate, and regulatory aspects will drive the integration of nanotechnology into mainstream environmental remediation strategies (Theron et al., 2008).
2. Genomic and Metagenomic Approaches
   Advances in genomics and metagenomics are enabling comprehensive analyses of microbial communities and their functional potentials in environmental processes. This includes understanding genetic adaptations, metabolic pathways, and microbial interactions relevant to bioremediation and ecosystem restoration. Metagenomic tools are used to discover novel enzymes for pollutant degradation, engineer microbial consortia for enhanced remediation efficiency, and assess microbial community responses to environmental stressors. Integration of high-throughput sequencing, bioinformatics, and synthetic biology will accelerate the development of tailored biotechnological solutions for diverse environmental contaminants and conditions (Riaz Ejaz et al., 2024).

3. Bioelectrochemical Systems (BES)
   BES harness microbial electrochemical activities to convert organic pollutants into electricity or valuable chemicals. These systems utilize microbial metabolism at electrode surfaces for pollutant degradation, metal recovery, and energy generation. BES are applied in wastewater treatment for organic matter removal, microbial fuel cells for renewable energy production, and bioremediation of contaminated soils and sediments. Optimization of electrode materials, microbial electrode interactions, and system scalability will expand the application of BES in decentralized wastewater treatment and resource recovery (Saratale et al., 2022).
4. Synthetic Biology and Engineered Microorganisms
   Synthetic biology integrates principles from biology, engineering, and computational sciences to design and construct novel biological systems or redesign existing organisms for specific applications in environmental biotechnology. Engineered microorganisms are engineered for enhanced pollutant degradation, biosynthesis of biofuels from waste materials, and biosensing of environmental contaminants. Advances in genome editing techniques (e.g., CRISPR-Cas) and synthetic DNA synthesis will enable precise manipulation of microbial genomes for customized environmental solutions, paving the way for sustainable bioproduction and pollution control (Dehkordi et al., 2023).
5. Artificial Intelligence (AI) and Machine Learning (ML)
   AI and ML algorithms are increasingly applied in environmental monitoring, predictive modeling of pollutant fate and transport, and optimization of biotechnological processes. These technologies enhance decision-making, resource allocation, and adaptive management in environmental management. AI–driven platforms analyze big data sets from environmental sensors, satellite imagery, and modeling simulations to optimize remediation strategies, predict ecological impacts, and support policymaking. Integration of AI with real-time monitoring systems and autonomous environmental platforms will enable adaptive and responsive pollution control strategies in dynamic environmental contexts (Okechukwuyem Ojji, 2024).

1.13 NANOBIOTECHNOLOGY IN POLLUTION TREATMENT

Nanobiotechnology, an interdisciplinary field combining nanotechnology with biological sciences, holds great promise for revolutionizing pollution treatment strategies. It leverages nanomaterials and biological systems to develop efficient, targeted, and sustainable solutions for environmental remediation. Nanomaterials such as nanoparticles and nanocomposites exhibit high surface area–to-volume ratios and unique physicochemical properties. These properties enhance their capacity for adsorption of pollutants and catalytic degradation of contaminants in air, water, and soil (Theron et al., 2008). Functionalized nanomaterials can be engineered to selectively target specific pollutants or contaminants. This targeted delivery enhances treatment efficiency while minimizing unintended environmental impacts and resource use (Anjum et al., 2019).

Nanozymes, nanomaterials with enzyme-like catalytic activities, have emerged as efficient tools for pollutant degradation. They offer advantages such as stability, reusability, and tunable catalytic properties, making them promising for wastewater treatment and soil remediation (Nagendran et al. 2024). Engineered nanomaterials, including carbon nanotubes, graphene oxide, and metal-organic frameworks, are being explored for their exceptional adsorption capacities and filtration efficiencies. These materials can selectively remove pollutants like heavy metals, dyes, and organic compounds from contaminated water and air (Hashmi et al. 2025).

1.13.1 Potential Applications of Nanobiotechnology

1. Water Purification and Desalination
Nanotechnology-enabled membranes and filtration systems improve water purification efficiency by removing pathogens, microplastics, and chemical contaminants. Nanomaterial-based desalination technologies offer sustainable solutions for freshwater production from seawater and brackish water sources (Nishu, 2023). In India, researchers have developed nanomaterial-based filters and membranes for efficient water purification. For instance, graphene oxide and carbon nanotube membranes have been studied for their ability to remove heavy metals and organic pollutants from contaminated water sources (Oluwafemi Abioye et al., 2024).
2. Soil Remediation and Agriculture
Nano-enabled agrochemicals and soil amendments enhance nutrient uptake, soil fertility, and pest management in agriculture. Nanoscale delivery systems for nutrients and bioactive compounds improve crop productivity while minimizing environmental impacts (Mim et al., 2025). Nanoscale zero-valent iron and other nanoparticles have been tested in India for soil remediation purposes, targeting contaminants like heavy metals and pesticides. These nanoparticles facilitate the degradation and immobilization of pollutants, improving soil quality and agricultural productivity (Mohanapragash et al., 2025).
Nanobiotechnology represents a transformative approach in pollution treatment, offering innovative solutions to complex environmental challenges. From enhancing catalytic degradation and adsorption capabilities to enabling targeted delivery systems and sustainable resource management, nanotechnology is poised to drive advancements in environmental remediation. Continued research, development of safe nanomaterial synthesis, understanding of environmental impacts, and regulatory oversight are crucial for harnessing the full potential of nanobiotechnology in achieving sustainable and resilient environmental solutions globally.
3. Nanozymes for Environmental Catalysis
Indian scientists have explored nanozymes, such as metal nanoparticles and metal-oxide nanomaterials, for catalytic degradation of pollutants in wastewater. These nanozymes mimic the enzymatic activity and have shown promising results in breaking down organic pollutants and dyes (Nagendran et al., 2024).
4. Nanoparticles in Air Pollution Control
Indian researchers have investigated the use of nanoparticles for air pollution control, focusing on particulate matter removal and gas-phase pollutant oxidation. Nanomaterial-based filters and catalysts have demonstrated potential in reducing air pollutants emitted from industrial sources and vehicular emissions (Roy et al., 2021).

These examples illustrate how Indian researchers are leveraging nanobiotechnology to address various environmental challenges, including water and soil pollution, air quality management, and sustainable agriculture. By advancing nanomaterial synthesis, application techniques, and understanding their environmental impacts, India is contributing to the global effort in developing innovative and sustainable solutions for pollution treatment using nanotechnology.

1.14 SYNTHETIC BIOLOGY IN POLLUTION REMEDIATION

Synthetic biology represents a cutting-edge field that applies engineering principles to design and construct biological systems or organisms with novel functions. In the context of pollution remediation, synthetic biology offers innovative approaches to tailor organisms for efficient and targeted degradation of pollutants. Following is an exploration of synthetic biology's role and its applications in designing customized organisms for pollution remediation.

1.14.1 Role of Synthetic Biology in Pollution Remediation

1. Customized Pathway Design
   Synthetic biology enables the design and optimization of metabolic pathways within microorganisms to degrade specific pollutants. This approach involves engineering genetic circuits, regulatory elements, and enzyme systems to enhance pollutant uptake, transformation, and detoxification (Rafeeq et al., 2023).
2. Bioremediation of Complex Pollutants
   Engineered microorganisms can be designed to metabolize complex pollutants such as PAHs, chlorinated compounds, and pharmaceutical residues. Synthetic biology facilitates the creation of tailored biocatalysts with improved substrate specificity and degradation efficiency (Rafeeq et al., 2023).

1.14.2 Use of Synthetic Biology in Designing Customized Organisms for Pollution Remediation

1. Engineering Microbial Consortia
   Synthetic biology allows the construction of synthetic microbial consortia with complementary metabolic capabilities for synergistic pollutant degradation. By assembling consortia of engineered microorganisms, complex environmental contaminants can be targeted more effectively (Rafeeq et al., 2023).
2. Genome Editing and Genetic Manipulation
   Techniques such as CRISPR-Cas9 enable precise genome editing and genetic manipulation of microorganisms. This capability facilitates the introduction of pollutant-degrading enzymes, pathways, or regulatory elements into microbial hosts, enhancing their remediation potential (Rafeeq et al., 2023).

Synthetic biology holds tremendous potential in designing customized organisms tailored for pollution remediation through engineered metabolic pathways, microbial consortia, and genome editing techniques.

1.15 ETHICAL CONSIDERATIONS AND REGULATORY CHALLENGES IN SYNTHETIC BIOLOGY FOR POLLUTION REMEDIATION

Synthetic biology offers promising solutions for pollution remediation through engineered organisms and pathways. However, its application raises important ethical considerations and regulatory challenges that must be addressed to ensure responsible and safe deployment. Following is a discussion on these issues.

1.15.1 Ethical Considerations

1. Unintended Consequences
   Engineering organisms for pollutant degradation may introduce unintended ecological consequences. Modified organisms could potentially proliferate or interact unpredictably with natural ecosystems, impacting biodiversity and ecosystem services (Wang et al., 2024).
2. Risk Assessment and Mitigation
   Assessing the risks associated with GMOs is crucial. Ethical considerations include evaluating the potential for horizontal gene transfer, persistence in the environment, and effects on nontarget organisms (Dhan Prakash et al., 2011).

3. Public Perception and Engagement
Public acceptance of synthetic biology applications in pollution remediation is essential. Addressing public concerns about safety, environmental impact, and ethical implications through transparent communication and stakeholder engagement is crucial for fostering trust and acceptance (Anguzu et al., 2025).

1.15.2 Regulatory Challenges

1. Regulatory Frameworks
Existing regulatory frameworks may not fully encompass the unique characteristics and risks associated with synthetic biology applications. Developing adaptive regulatory frameworks that balance innovation with environmental safety is essential (Bohua et al., 2023).
2. International Standards and Collaboration
Harmonizing international standards and regulations for synthetic biology applications in pollution remediation is challenging. Collaboration among regulatory agencies, scientists, industry, and policymakers is necessary to establish guidelines that ensure responsible innovation and global safety (Sun et al., 2022).
3. Ethical and Social Implications
Addressing ethical and social implications of synthetic biology, such as equity in access to benefits, intellectual property rights, and potential disparities in environmental impacts, requires interdisciplinary dialogue and ethical frameworks (Bohua et al., 2023).

Synthetic biology holds immense potential for pollution remediation but requires careful consideration of ethical implications and robust regulatory oversight. By addressing ethical concerns, engaging stakeholders, and developing adaptive regulatory frameworks, synthetic biology can contribute responsibly to environmental sustainability and human well-being.

REFERENCES

Ahuja, S. K., Ferreira, G. M., & Moreira, A. R. (2004). Application of enzymes as biocatalysts in the synthesis of pharmaceuticals and fine chemicals. *Catalysis Today*, 99(1–2), 457–471.

Al-Mutairi, N., Bufarsan, A. A., & Al-Rukaibi, F. (2008). Ecorisk evaluation and treatability potential of bioremediated soils contaminated with oil sludge. *Soil and Sediment Contamination: An International Journal*, 17(2), 195–208.

Anderson, R. T., Vrionis, H. A., Ortiz-Bernad, I., Resch, C. T., Long, P. E., Dayvault, R., ... & Lovley, D. R. (2003). Stimulating the in situ activity of Geobacter species to remove uranium from the groundwater of a uranium-contaminated aquifer. *Applied and Environmental Microbiology*, 69(10), 5884–5891.

Anguzu, S., Pineda, I. T., Muñoz-Solórzano, L. & Vats, I. (2025). Preprint. *Risk Assessment of Synthetic Cell Technology in Environmental Remediation: A stakeholder policy perception analysis*. https://doi.org/10.21203/rs.3.rs-6601051/v1

Anjum, M., Miandad, R., Waqas, M., Gehany, F., & Barakat, M.A. (2019). Remediation of wastewater using various nano-materials. *Arabian Journal of Chemistry*, 12(8), 4897–4919.

Arthur, E. L., Rice, P. J., Rice, P. J., Anderson, T. A., Baladi, S. M., Henderson, K. L. D., & Coats, J. R. (2005). Phytoremediation—An overview. *Critical Reviews in Plant Sciences*, 24(2), 109–122.

Asif, M. B., & Hai, F. I. (2019). Laccase: Microbial sources, production, purification, and potential biotechnological applications. *Biocatalysis and Agricultural Biotechnology*, 20, 101027.

Baldrian, P. (2006). Fungal laccases – occurrence and properties. *FEMS Microbiology Reviews*, 30(2), 215–242.

Banerjee, A., Sharma, R., & Banerjee, U. C. (2002). The nitrile-degrading enzymes: Current status and future prospects. *Applied Microbiology and Biotechnology*, 60(1–2), 33–44.

Behera, B. C., Sethi, B. K., Mishra, R. R., Dutta, S. K., & Thatoi, H. N. (2017). Microbial cellulases – Diversity & biotechnology with reference to mangrove environment: A review. *Journal of Genetic Engineering and Biotechnology*, 15(1), 197–210.

Bell, E.L., Finnigan, W., France, S.P., et al. (2021). Biocatalysis. *Nat Rev Methods Primers*, 1(46). https://doi.org/10.1038/s43586-021-00044-z

Bohua, L., Yuexin, W., Yakun, O., Kunlan, Z., Huan, L., & Ruipeng, L. (2023). Ethical framework on risk governance of synthetic biology. *Journal of Biosafety and Biosecurity*, 5(2), 45–56.

Bornscheuer, U. T., & Kazlauskas, R. J. (Eds.). (2004). *Hydrolases in Organic Synthesis: Regio- and StereoselectiveBiotransformations*. Wiley-VCH.

Bragg, J. R., Prince, R. C., Harner, E. J., & Atlas, R. M. (1994). Effectiveness of bioremediation for the Exxon Valdez oil spill. *Nature*, 368(6470), 413–418.

Bugg, T. D., Ahmad, M., Hardiman, E. M., & Rahmanpour, R. (2011). Pathways for degradation of lignin in bacteria and fungi. *Natural Product Reports*, 28(12), 1883–1896.

Cerniglia, C. E. (1992). Biodegradation of polycyclic aromatic hydrocarbons. *Biodegradation*, 3(2–3), 351–368.

Chandler, D., Bailey, A. S., Tatchell, G. M., Davidson, G., Greaves, J., & Grant, W. P. (2011). The development, regulation, and use of biopesticides for integrated pest management. *Philosophical Transactions of the Royal Society B: Biological Sciences*, 366(1573), 1987–1998.

Chaney, R. L., Malik, M., Li, Y. M., Brown, S. L., Brewer, E. P., Angle, J. S., & Baker, A. J. M. (1997). Phytoremediation of soil metals. *Current Opinion in Biotechnology*, 8(3), 279–284.

Cycon, M., & Piotrowska-Seget, Z. (2016). Bioremediation of pesticide-contaminated soils: A review. *Microbial Research*, 173, 71–83.

Dehkordi, J. A., Rahimi, S., Golzar-Ahmadi, M., Singh, A., Lopez, J., Ledesma-Amaro, R., & Mijakovic, I. (2023). Synthetic biology tools for environmental protection. *Biotechnology Advances*, 68, 108239.

Delhoménie, M. C., & Heitz, M. (2005). Biofiltration of air: A review. *Critical Reviews in Biotechnology*, 25(1–2), 53–72.

Demirbas, A. (2009). Biofuels securing the planet's future energy needs. *Energy Conversion and Management*, 50(9), 2239–2249.

Demuth, S., & Walsh, M. E. (1989). Bioremediation of chlorinated solvents at Love Canal. *Hazardous Waste Consultant*, 7(6), 1–23.

Devinny, J. S., Deshusses, M. A., & Webster, T. S. (1998). *Biofiltration for air pollution control*. CRC Press.

Dhal, B., Thatoi, H. N., Das, N. N., & Pandey, B. D. (2013). Chemical and microbial remediation of hexavalent chromium from contaminated soil and mining/metallurgical solid waste: A review. *Journal of Hazardous Materials*, 250–251, 272–291.

Dhan Prakash, D., Verma, S., Bhatia, R., & Tiwary, B. N. (2011). *Risks and Precautions of Genetically Modified Organisms*. Wiley.

Diaz, S., Settele, J., Brondizio, E. S., Ngo, H. T., Agard, J., Arneth, A., ... & Zayas, C. N. (2019). Pervasive human-driven decline of life on Earth points to the need for transformative change. *Science*, 366(6471).

Doty, S. L. (2008). Enhancing phytoremediation through the use of transgenics and endophytes. *New Phytologist*, 179(2), 318–333.

Dushenkov, V., Kumar, P. B. A. N., Motto, H., & Raskin, I. (1995). Rhizofiltration: The use of plants to remove heavy metals from aqueous streams. *Environmental Science & Technology*, 29(5), 1239–1245.

EPA. (2020). Bioremediation: Using microbes to clean up contaminated sites. Environmental Protection Agency. Retrieved from https://www.epa.gov/land-research/bioremediation-using-microbes-clean-contaminated-sites

Feng, S., Guo, W., Ding, A., Parsa, S. M., Pan, J., Cheng, D., Van Tung, T., Ngo, H. H., (2025). Enzyme sources in wastewater treatment: Their influence on enzymatic bioremediation and large-scale applications. *Chemical Engineering Journal*, 510, 161891.

Foley, J. A., DeFries, R., Asner, G. P., Barford, C., Bonan, G., Carpenter, S. R., ... & Snyder, P. K. (2005). Global consequences of land use. *Science*, 309(5734), 570–574.

Gadd G. M. (2010). Metals, minerals and microbes: geomicrobiology and bioremediation. *Microbiology (Reading, England)*, *156*(Pt 3), 609–643. https://doi.org/10.1099/mic.0.037143-0

Galloway, J. N., Aber, J. D., Erisman, J. W., Seitzinger, S. P., Howarth, R. W., Cowling, E. B., & Cosby, B. J. (2003). The nitrogen cascade. *BioScience*, 53(4), 341–356.

Gianfreda, L., & Rao, M. A. (2004). Potential of extra cellular enzymes in remediation of polluted soils: A review. *Enzyme and Microbial Technology*, 35(4), 339–354.

Glass, D. J. (1999). Economic potential of phytoremediation. In *Phytoremediation: Past promises and future practices* (pp. 15–31).

Gossett, J. M. (2010). Sustained aerobic oxidation of vinyl chloride at low oxygen concentrations. *Environmental Science & Technology*, 44(4), 1405–1411.

Gotor-Fernández, V., & Alfonso, I. (2011). Enzyme catalysis in organic synthesis: A perspective from reaction interconversions in aqueous media. *Topics in Catalysis*, 54(14–15), 965–970.

Griggs, D., Stafford-Smith, M., Gaffney, O., Rockström, J., Öhman, M. C., Shyamsundar, P., ... & Noble, I. (2013). Sustainable development goals for people and planet. *Nature*, 495(7441), 305–307.

Guisán, J. M. (2006). Immobilization of enzymes as the 21st century begins. *Methods in Biotechnology*, 22, 1–50.

Hashmi, Z., Idriss, I. M., Gapsari, F., Samsuddin, N., & Roil Bilad, M. (2025). The role of nanomaterials in enhancing membrane-based treatment for emerging contaminants: A review. *Sustainable Chemistry for Climate Action*, 7, 100092.

Haug, R. T. (2018). *The practical handbook of compost engineering*. Routledge.

Head, I. M., Jones, D. M., & Röling, W. F. (2006). Marine microorganisms make a meal of oil. *Nature Reviews Microbiology*, 4(3), 173–182.

Henze, M., van Loosdrecht, M. C., Ekama, G. A., & Brdjanovic, D. (Eds.). (2008). *Biological wastewater treatment: Principles, modelling and design*. IWA Publishing.

Hibi, M., Ogawa, J., & Yamada, M. (2007). Oxidation of methanol to formaldehyde catalyzed by hydroxylamine oxidoreductase from an aerobic methanol-utilizing bacterium. *Archives of Microbiology*, 187(3), 287–294.

Hoekstra, A. Y., Chapagain, A. K., Aldaya, M. M., & Mekonnen, M. M. (2011). *The water footprint assessment manual: Setting the global standard*. Earthscan.

Husain, Q. (2010). Peroxidase mediated decolorization and remediation of wastewater containing industrial dyes: A review. *Reviews in Environmental Science and Bio/Technology*, 9(2), 117–140.

Juwarkar, A. A., Singh, S. K., & Mudhoo, A. (2010). A comprehensive overview of elements in bioremediation. *Reviews in Environmental Science and Bio/Technology*, 9(3), 215–288.

Kennes, C., & Veiga, M. C. (2004). Fungal biocatalysts in the biofiltration of VOC-polluted air. *Journal of Biotechnology*, 113(1–3), 305–319.

Lagunas-Rangel, F. A., Linnea-Niemi, J.V., Kudłak, B., Williams, M.J., Jönsson, J., & Schiöth, H.B. (2022). Role of the synergistic interactions of environmental pollutants in the development of cancer. *Geohealth*, 6(4), e2021GH000552. https://doi.org/10.1029/2021GH000552. PMID: 35493962; PMCID: PMC9036628.

Landrigan, P. J., Fuller, R., Acosta, N. J., Adeyi, O., Arnold, R., Basu, N., ... & Zhong, M. (2018). The Lancet Commission on pollution and health. *The Lancet*, 391(10119), 462–512.

Leahy, J. G., Byrne, A. M., & Olsen, R. H. (2003). Comparison of factors influencing trichloroethylene degradation by phenol-oxidizing bacteria. *Applied and Environmental Microbiology*, 69(8), 4742–4750.

Leeson, A., & Hinchee, R. E. (Eds.). (1997). *Bioventing Principles and Practice*. CRC Press.

Liu, Z., Dang, Z., & Wang, D. (2014). Microbial degradation of sulfamethoxazole in aqueous solution by a novel bacterium Enterobacter cloacae DT-1: Kinetics, pathways, and mechanisms. *Applied Microbiology and Biotechnology*, 98(19), 8307–8317.

Logan, B. E., Hamelers, B., Rozendal, R., Schröder, U., Keller, J., Freguia, S., ... & Rabaey, K. (2006). Microbial fuel cells: Methodology and technology. *Environmental Science & Technology*, 40(17), 5181–5192.

Margesin R., Schinner F. (2001). Biodegradation and bioremediation of hydrocarbons in extreme environments. *Applied Microbiology Biotechnology*. *Sep*; 56(5–6):650–663. doi: 10.1007/s002530100701. PMID: 11601610

McCutcheon, S. C., & Schnoor, J. L. (2003). *Phytoremediation: Transformation and Control of Contaminants*. Wiley-Interscience.

Megharaj, M., Ramakrishnan, B., Venkateswarlu, K., Sethunathan, N., & Naidu, R. (2011). Bioremediation approaches for organic pollutants: A critical perspective. *Environment International*, 37(8), 1362–1375.

Mim, J.J., Maksudur Rahman, S.M., Khan, F., Paul, D., Sikder, S., Pollab Das, H., Khan, S., Tabassum Orny, N., Rifat Hossain Shuvo, Md. & Hossain, N. (2025). Towards smart agriculture through nano-fertilizer – A review. *Materials Today Sustainability*, 30, 101100.

Mohan, D., S. Ankur, Ok, Y. S., Pittman Jr., C. U. (2014). Organic and inorganic contaminants removal from water with biochar, a renewable, low cost and sustainable adsorbent – A critical review. *Bioresource Technology*. 160, 191–202.

Mohanapragash, A. G., Kaleeswari, R. K., Madhupriyaa, D, Salma Santhosh, S., Meena, S., Baskar, M., Umamaheswari, T., Selvamurugan, M. & Ramesh, T., (2025). Innovative nanoremediation techniques for soil contamination: exploring metal based nanomaterial approaches – A review. *Soil and Sediment Contamination: An International Journal*, 1–69. https://doi.org/10.1080/15320383.2025.2477818

Moser, S. C., & Dilling, L. (Eds.). (2007). *Creating a Climate for Change: Communicating Climate Change and Facilitating Social Change*. Cambridge University Press.

Mulligan, C. N., Yong, R. N., & Gibbs, B. F. (2001). Remediation technologies for metal-contaminated soils and groundwater: An evaluation. *Engineering Geology*, 60(1–4), 193–207.

Nagendran, V., Goveas, L. C., Vinayagam, R., Varadavenkatesan, T., & Selvaraj, R. (2024). Nanozymes in environmental remediation: A bibliometric and comprehensive review of their oxidoreductase-mimicking capabilities. *Microchemical Journal*, 207, 111748.

Newbold, T., Hudson, L. N., Arnell, A. P., Contu, S., De Palma, A., Ferrier, S., ... & Purvis, A. (2016). Has land use pushed terrestrial biodiversity beyond the planetary boundary? A global assessment *Science*, 353(6296), 288–291.

Newman, L. A., & Reynolds, C. M. (2004). Phytodegradation of organic compounds. *Current Opinion in Biotechnology*, 15(3), 225–230.

Nishu, S. K. (2023). Smart and innovative nanotechnology applications for water purification. *Hybrid Advances*, 3, 100044.

Okechukwuyem Ojji, S. (2024). Emerging technology integration – Artificial intelligence (AI) and Machine learning (ML) for predictive analysis for safety and toxicity assessment in environmental toxicology. *International Journal of Scientific Research and Management (IJSRM)*, 12(05):1182–1195.

Oluwafemi Abioye, S., Majooni, Y., Moayedi, M., Rezvani, H., Kapadia, M., & Yousefi, N. (2024). Graphene-based nanomaterials for the removal of emerging contaminants of concern from water and their potential adaptation for point-of-use applications. *Chemosphere*, 355, 141728.

Pakshirajan, K., Rene, E.R., & Ramesh, A. (2014). Biotechnology in environmental monitoring and pollution abatement. *Biomed Research International*, 2014, 235472. https://doi.org/10.1155/2014/235472. Epub 2014 Apr 23. PMID: 24864232; PMCID: PMC4017724.

Parks, J. M., Johs, A., Podar, M., Bridou, R., Hurt, R. A., Smith, S. D., ... & Liang, L. (2013). The genetic basis for bacterial mercury methylation. *Science*, 339(6125), 1332–1335.

Parmar, N., & Singh, A. (2015). Bioremediation of petroleum hydrocarbon polluted sites: A review. *Indian Journal of Environmental Protection*, 35(7), 586–595.

Parry, M. L., Canziani, O. F., Palutikof, J. P., Van der Linden, P. J., & Hanson, C. E. (Eds.). (2007). *Climate change 2007: Impacts, adaptation and vulnerability*. Cambridge University Press.

Pattapulavar, V., Christopher, J. (2025). The role of actinomycetes in a cleaner environment. In: Jeyaseelan, A., Murugasen, K. (Eds.). *Sustainable Environmental Remediation: Avenues in Nano and Biotechnology. Environmental Science and Engineering*. Springer. https://doi.org/10.1007/978-3-031-78483-5_1

Pilon-Smith, E. A., Kelly, J. O., Larkin, M. J., Day, M. J., & Head, I. M. (2006). The potential for accelerated in situ bioremediation of petroleum hydrocarbons in contaminated groundwater using denitrifying bacteria from a low-temperature environment. *Journal of Applied Microbiology*, 101(2), 255–263.

Pilon-Smits, E. (2005). Phytoremediation. *Annual Review of Plant Biology*, 56, 15–39.

Prasad, M. N. V., & Freitas, H. (2003). Metal hyperaccumulation in plants: Biodiversity prospecting for phytoremediation technology. *Electronic Journal of Biotechnology*, 6(3), 285–321.

Rafeeq, H., Afsheen, N., Rafique, S., Arshad, A., Intisar, M., Hussain, A., Bilal, M., & Iqbal, M. N. (2023). Genetically engineered microorganisms for environmental remediation. *Chemosphere*, 310, 136751.

Riaz Ejaz, M.Badr, K., Ul Hassan, Z., Al-Thani, R., & Jaoua, S. (2024). Metagenomic approaches and opportunities in arid soil research. *Science of the Total Environment*, 953, 176173.

Robinson, B. H., Mills, T. M., Petit, D., Fung, L. E., Green, S. R., & Clothier, B. E. (2000). Natural and induced cadmium-accumulation in poplar and willow: Implications for phytoremediation. *Plant and Soil*, 227(1), 301–306.

Rockström, J., Steffen, W., Noone, K., Persson, Å., Chapin, F. S., Lambin, E. F., ... & Foley, J. A. (2009). A safe operating space for humanity. *Nature*, 461(7263), 472–475.

Rodríguez Couto, S., & Toca Herrera, J. L. (2006). Industrial and biotechnological applications of laccases: A review. *Biotechnology Advances*, 24(5), 500–513.

Roy, A., Sharma, A., Yadav, S., Jule, L. T., & Krishnaraj, R. (2021). Nanomaterials for remediation of environmental pollutants. *Bioinorganic Chemistry and Applications*, 28:1764647. https://doi.org/10.1155/2021/1764647. PMID: 34992641; PMCID: PMC8727162.

Sachs, J. D. (2012). From millennium development goals to sustainable development goals. *The Lancet*, 379(9832), 2206–2211.

Salt, D. E., Blaylock, M., Kumar, N. P., Dushenkov, V., Ensley, B. D., Chet, I., & Raskin, I. (1995). Phytoremediation: A novel strategy for the removal of toxic metals from the environment using plants. *Nature Biotechnology*, 13(5), 468–474.

Salt, D. E., Smith, R. D., & Raskin, I. (1998). Phytoremediation. *Annual Review of Plant Physiology and Plant Molecular Biology*, 49, 643–668.

Saratale, G. D., Rajesh Banu, J., Nastro, R. A., Kadier, A., Ashokkumar, V., Lay, C.-H., Jung, J.-H., Shin, H. S., & Chandrasekhar, K. (2022). Bioelectrochemical systems in aid of sustainable biorefineries for the production of value-added products and resource recovery from wastewater: A critical review and future perspectives. *Bioresource Technology*, 359, 127435.

Sarkar, D., & Sarma, P. (2017). Oil spill and bioremediation: A study of the Sundarbans. *Environmental Science and Pollution Research*, 24(34), 26744–26755.

Sarkar, N., Kumar Ghosh, S., Bannerjee, S., Aikat K. (2012). Bioethanol production from agricultural wastes: An overview. *Renewable Energy*, 37(1) 19–27.

Sayler, G. S., & Ripp, S. (2000). Field applications of genetically engineered microorganisms for bioremediation processes. *Current Opinion in Biotechnology*, 11(3), 286–289.

Schomburg, I., & Schomburg, D. (Eds.). (2006). *Springer Handbook of Enzymes*. Springer.

Seviour, R., & Nielsen, P. H. (2010). *Microbial ecology of activated sludge*. IWA Publishing.

Shah, A. A., Hasan, F., Hameed, A., & Ahmed, S. (2008). Biological degradation of plastics: A comprehensive review. *Biotechnology Advances*, 26(3), 246–265.

Sheldon, R. A., & van Pelt, S. (2013). Enzyme immobilization in biocatalysis: Why, what, and how. *Chemical Society Reviews*, 42(15), 6223–6235.

Sheldon, R. A., & Woodley, J. M. (2018). Role of biocatalysis in sustainable chemistry. *Chemical Reviews*, 118(2), 801–838.

Singh, B. K., & Walker, A. (2006). Microbial degradation of organophosphorus compounds. *FEMS Microbiology Reviews*, 30(3), 428–471.

Smith, K. R., Woodward, A., Campbell-Lendrum, D., Chadee, D. D., Honda, Y., Liu, Q., … & Sauerborn, R. (2014). Human health: Impacts, adaptation, and co-benefits. In *Climate change 2014: Impacts, adaptation, and vulnerability. Part A: Global and sectoral aspects. Contribution of Working Group II to the Fifth Assessment Report of the Intergovernmental Panel on Climate Change* (pp. 709–754). Cambridge University Press.

Steffen, W., Richardson, K., Rockström, J., Cornell, S. E., Fetzer, I., Bennett, E. M., … & Sörlin, S. (2015). Planetary boundaries: Guiding human development on a changing planet. *Science*, 347(6223).

Strong, P. J., & Claus, H. (2011). Laccase: A review of its past and its future in bioremediation. *Critical Reviews in Environmental Science and Technology*, 41(4), 373–434.

Sun, T., Song, J., Wang, M., Zhao, C., & Zhang, W. (2022). Challenges and recent progress in the governance of biosecurity risks in the era of synthetic biology. *Journal of Biosafety and Biosecurity*, 4(1), 59–67.

Syranidou, E., Karkanorachaki, K., Amorotti, F., Avgeropoulos, A., Kolvenbach, B., Zhou, N. Y., Fava, F., Corvini, P. F.-X., Kalogerakis, N. (2019). Biodegradation of mixture of plastic films by tailored marine consortia, *Journal of Hazardous Material*, 375, 33–42.

Tchounwou, P. B., Yedjou, C.G., Patlolla, A. K., & Sutton D.J. (2012). Heavy metal toxicity and the environment. *Experientia Supplementum*, 101:133–164. https://doi.org/10.1007/978-3-7643-8340-4_6. PMID: 22945569; PMCID: PMC4144270.

Terry, N., Zayed, A. M., De Souza, M. P., & Tarun, A. S. (2000). Selenium in higher plants. *Annual Review of Plant Biology*, 51(1), 401–432.

Theron, J., Walker, J. A., & Cloete, T. E. (2008). Nanotechnology and water treatment: Applications and emerging. *Critical Reviews in Microbiology*, 34:43–69.

Tilman, D., & Clark, M. (2014). Global diets link environmental sustainability and human health. *Nature*, 515(7528), 518–522.

Torres-Duarte, C., Roman, R., Tinoco, R., Vazquez-Duhalt, R. (2009). Halogenated pesticide transformation by a laccase–mediator system. *Chemosphere*, 77(5) 687–692.

Tournier, V., Topham, C. M., Gilles, A., David, B., Folgoas, C., Moya-Leclair, E., … & Marty, A. (2020). An engineered PET depolymerase to break down and recycle plastic bottles. *Nature*, 580(7802), 216–219.

Turner, P., Mamo, G., & Karlsson, E. N. (2007). Potential and utilization of thermophiles and thermostable enzymes in biorefining. *Microbial Cell Factories*, 6(1), 9.

Vamerali, T.,Bandiera, M., & Mosca, G. (2010). Field crops for phytoremediation of metal-contaminated land. A review. *Environmental Chemistry Letters*, 8(1), 1–17.

Van Beilen, J. B., & Li, Z. (2002). Enzyme technology: An overview. *Current Opinion in Biotechnology*, 13(4), 338–344.

Vangronsveld, J., & Cunningham, S. D. (Eds.). (1998). *Metal-Contaminated Soils: In Situ Inactivation and Phytorestoration*. Springer.

Vessey, J. K. (2003). Plant growth promoting rhizobacteria as biofertilizers. *Plant and Soil*, 255(2), 571–586.

Vidali, M. (2001). Bioremediation. An overview. *Pure and Applied Chemistry*, 73(7), 1163–1172.

Volesky, B. (2001). Detoxification of metal-bearing effluents: Biosorption for the next century. *Hydrometallurgy*, 59(2–3), 203–216.

Volesky, B. (2007). Biosorption and me. *Water Research*, 41(18), 4017–4029.

Vörösmarty, C. J., McIntyre, P. B., Gessner, M. O., Dudgeon, D., Prusevich, A., Green, P., ... & Davies, P. M. (2010). Global threats to human water security and river biodiversity. *Nature*, 467(7315), 555–561.

Vroblesky, D. A., & Casey, C. C. (2001). Use of trees for assessing groundwater contamination and remediation. US Geological Survey Fact Sheet, FS-084-01.

Vymazal, J. (2011). Constructed wetlands for wastewater treatment: Five decades of experience. *Environmental Science & Technology*, 45(1), 61–69.

Wang, F., Xiang, L., Sze-Yin Leung, K. S.-Y. ... & Tiedje, J. M. (2024). Emerging contaminants: A one health perspective. *The Innovation*, 5(4):100612.

Wang, L., & Cai, Y. (2006). The role of enzymes in the bioremediation of oil-contaminated environments: A review. *Journal of Environmental Science and Health, Part C*, 24(2), 155–181.

Weiland, P. (2010). Biogas production: Current state and perspectives. *Applied Microbiology and Biotechnology*, 85(4), 849–860.

Whitmee, S., Haines, A., Beyrer, C., Boltz, F., Capon, A. G., Ferreira de Souza Dias, B., ... & Yach, D. (2015). Safeguarding human health in the Anthropocene epoch: Report of The Rockefeller Foundation–Lancet Commission on planetary health. *The Lancet*, 386(10007), 1973–2028.

Wolverton, B. C., Johnson, A., & Bounds, K. (1989). Interior landscape plants for indoor air pollution abatement. NASA Stennis Space Center Report.

World Health Organization (WHO). (2021). Ambient (outdoor) air pollution. Retrieved from https://www.who.int/news-room/fact-sheets/detail/ambient-(outdoor)-air-quality-and-health

Yakimov, M. M., Timmis, K. N., & Golyshin, P. N. (2007). Obligate oil-degrading marine bacteria. *Current Opinion in Biotechnology*, 18(3), 257–266.

Zhang, H., Lombi, E., Smolders, E., Zhao, F. J., & McGrath, S. P. (2016). Multielement screening of worldwide soils reveals effects of soil properties on bioavailability. *Environmental Science & Technology*, 50(14), 8423–8432.

Zumft, W. G. (1997). Cell biology and molecular basis of denitrification. *Microbiology and Molecular Biology Reviews*, 61(4), 533–616.

2 Integration of Bioremediation Planning and Execution for Restoring the Ecosystem

Latifa EL Bardiy, Abdelmjid Bouazizi, Mounia Baidou, and Mohammed Zeriouh

2.1 INTRODUCTION

Undoubtedly, the core of economic and social challenges lies in the limitless needs of human beings, faced with limited resources. Economists have ceaselessly devised methods to use our resources rationally while maximizing the utility of economic actors with minimal resources. From an economic perspective, in addition to capital and labor, natural resources represent a valuable, scarce, and finite factor of production. It is crucial to manage them wisely and ensure their preservation. However, the technological advancements and industrialization that have surged in recent years have caused damages and compromised the quality of this wealth.

The literature unanimously agrees on the observation that the quantities of chemical compounds released into the environment due to the increasing human activities have resulted in the contamination of soils, surface waters, and groundwater, thereby posing a serious threat to human health (Murthy et al., 2023; Roy et al., 2023; Demarco et al., 2023). Subsequently, the accumulation of these toxic compounds beyond permitted levels jeopardizes the overall quality of the environment and the quality of life on Earth. For this reason, the issue of pollution is not solely the concern of biologists and field specialists; it is a problem that interests researchers across various disciplines, including physicists, legal experts, and economists. Researchers from all fields must collaborate synergistically to advance and safeguard our survival as well as that of future generations.

The researchers thus agree that solutions can often be found to reduce the level of environmental pollution and remediate contaminated sites. Among the proposed solutions, bioremediation is considered an increasingly popular alternative to conventional methods for treating contaminated compounds and environments. Researchers continue to seek naturally occurring decontaminating organisms capable of cleaning contaminated soils. This chapter aims to detail these biological processes that complement chemical or physical remediation methods and explore the progress made in this field. It is known that environmental issues are constantly evolving, bringing new challenges, and creating the need to develop more suitable technologies. The study of literature addressing topics related to bioremediation will enable us to meet these challenges, explore the latest strategies developed for healthy remediation of contaminated sites, and consider methods deemed to be the least costly. The researchers thus agree that solutions can often be found to reduce the level of environmental pollution and remediate contaminated sites. Among the proposed solutions, bioremediation is considered an increasingly popular alternative to conventional methods for treating contaminated compounds and environments. Researchers continue to seek naturally occurring decontaminating organisms capable of cleaning contaminated soils. This chapter aims to detail these biological processes that complement chemical or physical remediation methods and explore the progress made in this field. It is known that environmental issues are constantly evolving, bringing new challenges, and creating the need to develop more suitable technologies. The study of literature

DOI: 10.1201/9781003668961-2

addressing topics related to bioremediation will enable us to meet these challenges, explore the latest strategies developed for healthy remediation of contaminated sites, and consider methods deemed to be the least costly.

The current chapter endeavors to explore this inquiry by conducting an extensive literature review. To achieve this, a careful examination of select articles published in 2023 will be undertaken to scrutinize the research directions, study locations, bioremediation methodologies employed for degrading various pollutants, and the challenges encountered by researchers. Through the identification of these elements, this chapter offers crucial insights to enhance practices and propel research in this vital realm of environmental conservation. Additionally, suggestions are put forth for charting a course toward the effective integration of planning and execution in bioremediation efforts aimed at restoring degraded ecosystems. A comprehensive analysis was carried out wherein the research directions, pollutants, bioremediation strategies utilized for their degradation, and the principal outcomes reported by each author were pinpointed and organized into nodes using Nvivo software. These nodes were then presented based on the recurrence frequency of each idea. This collective investigation thus furnishes a comprehensive overview of the current landscape in bioremediation, serving as a robust foundation for future developments and concrete actions to effectively address contemporary environmental challenges.

2.2 BIOREMEDIATION: CONCEPT AND APPROACH

It is important to note that conventional remediation methods typically involve excavating contaminated soil and transporting it to a landfill or covering contaminated areas on-site. These approaches carry the risk of relocating contamination to other locations and may pose significant hazards during the transportation of hazardous substances. Bioremediation is an alternative that allows for the neutralization or rendering harmless of various contaminants by harnessing natural biological activity (Hou et al., 2023).

History indicates that environmental biotechnology is not a new field; wastewater treatments are familiar examples of age-old environmental biotechnologies. For centuries, civilizations have employed natural bioremediation in wastewater treatment, but its deliberate use for hazardous waste reduction is a more recent development. Thus, the use of microorganisms in modern bioremediation can be traced, in part, to George Robinson, who utilized microbes to clean up an oil spill along the coast of Santa Barbara, California, in the late 1960s (Roy et al., 2023). Since the 1980s, bioremediation of oil spills and other hazardous wastes has garnered increased attention (El-Saadony et al., 2023).

2.2.1 Bioremediation: Definition and Related Terms

It has been noted that the literature examined in this study discusses various concepts related to bioremediation that necessitate clarification. While the majority of the research utilizes the term "bioremediation", it also has been observed that authors use other terms interchangeably, such as bio-restoration, biodegradation, or biotransformation, to refer to similar processes involving the breakdown or transformation of compounds, particularly pollutants, facilitated by living organisms. Although some authors use these terms interchangeably, there are nuanced differences among them. The concept of bioremediation, as presented by Narayanan et al. (2023), is defined as an efficient process in which harmful pollutants are removed from the environment using microorganisms or their enzymes. Additionally, according to Sudarshan et al. (2023), the bioremediation approach focuses on the biological cleaning of toxic contaminants from industrial effluents. Furthermore, Gupta and Gandhi (2023) emphasized that this approach represents an environmentally friendly technology. Bioremediation, as defined by Narayanan et al. (2023), is described as an efficient process wherein harmful pollutants are eliminated from the environment using microorganisms or their enzymes. Sudarshan et al. (2023) emphasize that the bioremediation approach is centered on the biological purification of toxic contaminants from industrial effluents. Furthermore, Gupta and Gandhi (2023) underscore that this approach embodies an environmentally friendly technology.

Moreover, Vaksmaa et al. (2023) have proposed a more generic definition of bioremediation, presenting it as a term encompassing biologically mediated processes, during which an undesirable compound is transformed, degraded, sequestered, and/or eliminated from the ecosystem. On their part, Rana et al. (2023) highlighted that bioremediation offers the possibility of either completely eliminating pollutants or transforming them into nonhazardous compounds using natural biological agents. Thus, these various perspectives converge toward a comprehensive view of bioremediation as an ecological and versatile method for treating environmental contaminants. Therefore, bioremediation entails employing microorganisms, plants, or enzymes to degrade or remove pollutants found in soil, water, or air, with the objective of rejuvenating a contaminated environment. Conversely, biodegradation refers to the natural breakdown of a compound through biological processes facilitated by living organisms, resulting in the production of simpler and less harmful substances (Kumar & Singh, 2023; Dash & Osborne, 2023; Xu et al., 2023; Bhatt et al., 2023; Sun et al., 2023). According to Davis (2023), the biodegradation and biotransformation of contaminants in groundwater generally occur naturally. However, natural rates of biodegradation can be slow, leading to the formation of extensive plumes of contaminants and presenting prolonged risks, thus requiring a more significant remediation intervention (Davis, 2023; Eckel et al., 2023). Improving the biodegradation of contaminants in groundwater can be initiated by adding amendments to alter the geochemical conditions, making them more conducive to native or added biota (Davis, 2023). In his approach, bio-restoration involves using living organisms or biological processes to rehabilitate a polluted environment contaminated by toxic substances, with the goal of breaking down or removing pollutants. Thus, bio-restoration encompasses a diverse array of techniques focused on restoring a contaminated environment (Murthy et al., 2023; Davis, 2023; Narayanan & Ma, 2023; An et al., 2023).

In summary, research has cited numerous approaches and forms of bioremediation, alongside these related terms that can be used to refer to the same phenomenon.

2.2.2 Bioremediation: Approaches and Forms

Bioremediation technologies can be broadly classified as ex situ or in situ. In situ bioremediation involves treating the contaminated material on-site, while ex situ involves removing the contaminated material to treat it elsewhere (Pathania et al., 2023; Xu & Wu, 2023). According to Xu and Wu (2023), in situ treatments take place directly at the site of contamination, without requiring the removal of sediments, while ex situ remediation involves removing contaminated sediments from the riverbed for treatment outside the original site. Figure 2.1 highlights the elements of bioremediation that have been addressed and discussed in the reviewed articles.

Certain research highlights various methods of bioremediation, including mycoremediation, phytoremediation, and phycoremediation. Mycoremediation is a bioremediation method that uses fungi to decontaminate an area. The term mycoremediation was coined by Paul Stamets and specifically refers to the use of fungal mycelium in bioremediation (Vaksmaa et al., 2023; Rana et al., 2023; Zainab et al., 2023). Phytoremediation, which involves the use of plants, is also addressed in several articles (Demarco et al., 2023; Gupta & Gandhi, 2023; Dash & Osborne, 2023; Azuazu et al., 2023). Rhizoremediation is a technique that uses plant roots to degrade, absorb, or immobilize contaminants (Ansari et al., 2023). Furthermore, phycoremediation, a phytoremediation technique that utilizes algae, is mentioned in the study by Kankanamalage et al. (2023) and Rambabu et al. (2023). Other authors have discussed both phyco- and mycoremediation technologies to detoxify sites polluted with polychlorinated biphenyls (PCBs). These two technologies represent algal and fungal biosystems, respectively (Kaleem et al., 2023). The studies included in this review have also examined several bioremediation technologies, including nanotechnology (Narayanan et al., 2023; Zainab et al., 2023; Saravanan et al., 2023; Geetha et al., 2023), biotechnology (Narayanan & Ma, 2023;

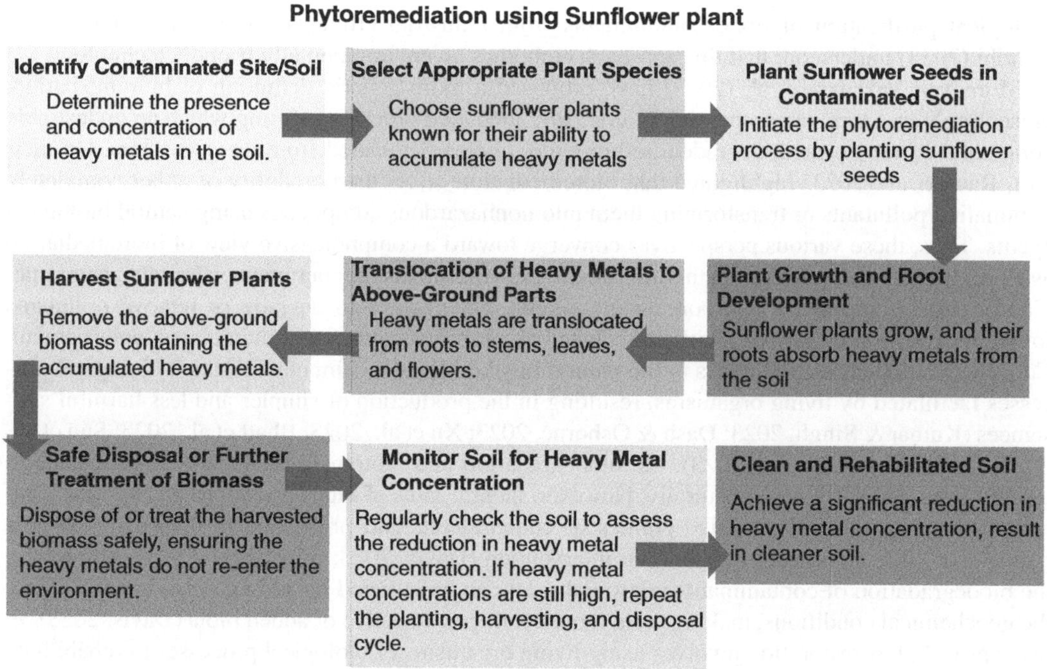

FIGURE 2.1 Overview of the mechanism of bioremediation.

Mali et al., 2023), bioinformatics (Dash & Osborne, 2023; Borella et al., 2023), structural biology (Dash & Osborne, 2023), machine learning (Xu et al., 2023), nanofiltration (Rendón-Castrillón et al., 2023), and artificial intelligence (Patowary et al., 2023). This indicates that bioremediation is an interdisciplinary technology involving microbiology, engineering, ecology, geology, and chemistry. According to Xu et al. (2023), machine learning, quantum chemistry, and biomolecular modeling are promising tools for predicting the environmental fate of chemicals and exploring functional enzymes. However, their application in bioremediation has only recently emerged. Some writings have suggested the implementation of mixed approaches by combining different bioremediation strategies. For example, Dash and Osborne (2023) compared various bioremediation techniques to recommend the simultaneous implementation of more than one remediation strategy. According to these authors, a combinatorial approach creates an advantageous hybrid technique for bioremediation. Similarly, Rendón-Castrillón et al. (2023) developed a hybrid approach combining bioremediation and nanofiltration. Radhakrishnan et al. (2023) have suggested that an integrated approach to bioremediation, including biostimulation and bioaugmentation, is preferably advocated for the cost-effective removal of toxic petroleum hydrocarbons and their derivatives. Furthermore, Hidangmayum et al. (2023) and Patowary et al. (2023) have emphasized that new technologies, such as the combined process of nanotechnology and bioremediation, are urgently needed to expedite the cost-effective remediation process in order to mitigate toxic contaminants compared to conventional remediation methods. In the long term, this combination of the two technologies, known as nanobioremediation, could significantly alter the field of environmental remediation. Several authors have delved into emerging approaches to bioremediation, particularly bioelectroremediation, which integrates principles from both biotechnology and electrochemistry (Syed et al., 2023). Discussions also extend to bioelectrochemical systems. These new systems integrate biological methods and electrochemistry to reduce the environmental impacts of hazardous and recalcitrant molecules (Gebregiorgis Ambaye et al., 2023; Saadaoui et al., 2023). Additionally, microbial fuel cells (MFCs) have been mentioned in several studies (Idris et al., 2023; Hamdan & Salam, 2023; Bhattacharya et al., 2023;

Ma et al., 2023). According to Ma et al. (2023), MFCs are a promising sustainable bioelectrochemical energy technology that enables wastewater recycling while generating bioelectricity. These MFCs have attracted remarkable attention worldwide in recent decades.

2.3 ENVIRONMENTAL CHALLENGES OF POLLUTANTS

Recent research demonstrates a significant focus on the bioremediation of chemical compounds and petroleum products. The majority of the studies examined in this research specifically tackle this subject matter.

2.3.1 Environmental Hazards of Chemical and Petroleum Compounds

Indeed, authors have dedicated significant attention to organophosphates (OP), chemical compounds commonly employed as insecticides and herbicides. Mali et al. (2023) suggested that OP, essential in contemporary agriculture, accumulate in soil, thus contaminating terrestrial and aquatic ecosystems via groundwater. Acute exposure to OP can induce toxicity in insects, plants, animals, and humans. This is illustrated in Figure 2.2, which outlines the topics covered by previous studies (Mali et al., 2023).

The topics represented are:

- Bioremediation of Heavy Metals
- Bioremediation of Various Pollutants
- Bioremediation of Chemical Compounds and Petroleum Products

The chart shows that "Bioremediation of Chemical Compounds and Petroleum Products" is the most studied topic, followed by "Bioremediation of Various Pollutants," and lastly "Bioremediation of Heavy Metals".

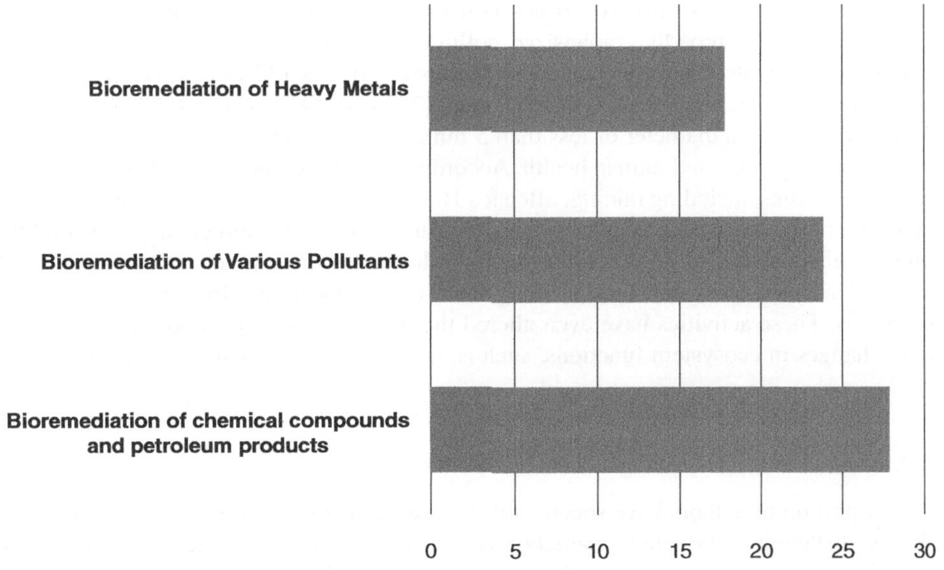

FIGURE 2.2 Topics covered by some previous studies.

Organo-halogenated chemicals have also been the subject of studies by several authors. Xu et al. (2023) emphasized that organo-halogenated chemicals such as PCBs, polybrominated diphenyl ethers (PBDEs), and perfluoroalkyl substances (PFAS) are common contaminants in the environment. In addition, other chemical substances present in the environment as pollutants have been extensively addressed in the literature. These include, among others, formaldehyde (FA), persistent organic pollutants (POP), polycyclic aromatic hydrocarbons (PAH), hydrocarbon pollutants, trichloroethylene (TCE), PCB, and p-nitrophenol (PNP) (Radhakrishnan et al., 2023; Idris et al., 2023; Zhou et al., 2023; Thacharodi et al., 2023). As emphasized by Hamdan and Salam (2023), the global pollution of nearly all natural environments by these toxic chemical compounds is a major issue for public health and the environment due to their carcinogenic, genotoxic, mutagenic, and cytotoxic properties. The bioremediation of petroleum hydrocarbons, including crude oil, diesel, gasoline, and other petroleum products, also remains a subject that captures the attention of several researchers (Gebregiorgis Ambaye et al., 2023; Bhadra et al., 2023). According to Patowary et al. (2023), crude oil spills have become a global concern for environmental pollution and pose a serious threat to flora and fauna.

2.3.2 Environmental Hazards of Various Pollutants

While the majority of researchers have concentrated primarily on chemical compounds and petroleum products, as mentioned earlier, a notable subset of authors have also addressed various other effluents. For instance, Masotti et al. (2023) directed their attention to glyphosate. According to these authors, the widespread use of this herbicide has led to environmental issues that need to be addressed. On their part, Narayanan et al. (2023) targeted various harmful pollutants, including dyes, polycyclic aromatic hydrocarbons, plastics, heavy metals, and pesticides. Gupta and Gandhi (2023) added that these pollutants cause severe soil and water pollution, leading to toxicity. For Sudarshan et al. (2023), environmental contamination resulting from the discharge of wastewater by textile industries is a growing global concern. Textile industries produce a significant volume of wastewater containing a diverse range of chemicals, primarily composed of dyes. Discharging such effluents into the aquatic environment leads to pollution that negatively impacts aquatic organisms (Sudarshan et al., 2023). These concerns have also been echoed by Ullah Khan et al. (2023), who assert that wastewater polluted with dyes poses a serious threat to the health of humans, animals, and plants. Hou et al. (2023) broadly emphasized pollution from industrial activities, estimating that rehabilitating approximately 5 million abandoned industrial sites worldwide is necessary to enhance urban ecosystem services. Furthermore, for Sun et al. (2023) and Meng et al. (2023), microplastics, defined as particles with a diameter of less than 5 mm, are emerging environmental pollutants that pose a threat to ecosystems and human health. According to Anekwe and Isa (2023), anthropogenic and industrial activities, including mining, often lead to the generation of acid mine drainage (AMD), considered one of the most catastrophic ecological issues. Hidangmayum et al. (2023) highlighted those anthropogenic activities such as industry, agriculture, and other practices can result in excessive levels of hazardous toxic substances, along with sewage discharges from residential and commercial sources. These activities have even altered the nitrogen and phosphorus cycles, leading to substantial changes in ecosystem functions, such as the provision of clean air and water (Filippini et al., 2023).

2.3.3 Environmental Hazards of Heavy Metals

A significant portion of authors have specifically focused their interest on a hazardous type of pollutant that is challenging to degrade, namely heavy metals. According to Saravanan et al. (2023), heavy metals pose critical exposure risks to humans and animals, even at low concentrations. Similarly, Narayanan and Ma (2023) confirmed that heavy metal contamination has toxic effects on plants and constitutes a significant constraint on plant quality and yield. This observation was shared by Saberi Riseh et al. (2023), who confirmed that heavy metals rapidly accumulate in the soil due to industrial activities, significantly impacting the morphology, physiology, and yield of crops, which seriously affects human and animal health. Furthermore, Jeyakumar et al. (2023) emphasized that contamination of the system by heavy metals remains one of the most serious global threats. Even at minimal concentrations, heavy metals and metalloids such as chromium, lead, mercury, cadmium, nickel, and cobalt are toxic and carcinogenic, posing a serious threat to human life. Syed et al. (2023) also delved into heavy metal pollution caused by various industrial and mining activities. According to them, heavy metal pollution poses a serious threat to all forms of life in the environment due to the persistence and toxicity of metal ions. It is worth noting that Demarco et al. (2023) confirmed that no well-defined mechanism has been found for the specific absorption of heavy metals by plants; however, some results may indicate that, in the case of sporadic contamination by certain pollutants, plants can serve as indicators with some adsorption. Murthy et al. (2023) focused on the bioremediation of chromium (Cr). According to these authors, the use of chromium in tanneries makes it one of the most hazardous environmental pollutants. It is toxic and has detrimental effects on human health and the environment. On their part, Cui et al. (2023) investigated uranium, which serves as a major pollutant, posing significant risks to human health. These authors specified the sources of uranium contamination based on 789 soil sampling sites collected from 76 studies (Cui et al., 2023).

On the other hand, Zhang et al. (2023) delved into the bioremediation of vanadium. In fact, the authors asserted that vanadium is a highly toxic multivalent element sensitive to redox reactions. It is widely distributed in the environment and used in various industrial applications. Tambat et al. (2023) also explored vanadium, presenting it as a dangerous heavy metal that poses health problems at an exposure concentration of around 2 ppm. The authors postulated that vanadium is regularly discharged by certain industries, representing an environmental challenge. They also emphasized that there is currently no sustainable and eco-friendly treatment method for the discharged effluents to mitigate the risks associated with vanadium for humans and the environment (Tambat et al., 2023). While the previously mentioned authors focused on a specific metal, Roy et al. (2023) emphasized 20 different metal ions [arsenic, mercury, lead, chromium, iron, copper, cadmium, palladium, zinc, cobalt, antimony, gold, silver, platinum, selenium, manganese, molybdenum, nickel, tungsten, and uranium]. According to the authors, these toxic metalloids are a primary source of water system pollution (Roy et al., 2023)

Conversely, Kumar and Singh (2023) gave particular focus to lead, highlighting its status as a naturally occurring metal with a grayish-blue color, existing in small amounts within the Earth's crust. The authors asserted that lead's nonbiodegradable nature and its continued use result in an accumulation of lead concentration in the environment, causing various adverse effects such as neurotoxicity and changes in the psychological and behavioral development of different organisms. Similarly, Wang et al. (2023) investigated lead discharged by rural industries. According to them, this metal poses a significant threat to the environment and human health.

2.4 ENVIRONMENTAL REVOLUTION: BIOREMEDIATION AND POLLUTION MANAGEMENT

2.4.1 Superiority of Bioremediation Compared to Conventional Methods

According to numerous authors, bioremediation has proven effective for the remediation of sites contaminated with vanadium, lead, chromium, AMD, and harmful environmental pollutants (Murthy et al., 2023; Roy et al., 2023; Narayanan et al., 2023; Sudarshan et al., 2023; Gupta & Gandhi, 2023; Kumar & Singh, 2023; Xu et al., 2023; Davis, 2023; Pathania et al., 2023; Mali et al., 2023; Idris et al., 2023; Masotti et al., 2023; Anekwe & Isa, 2023; Zhang et al., 2023; Wang et al., 2023; Li et al., 2023; Madison et al., 2023). For Saravanan et al. (2023), microbial bioremediation emerges as a promising method to mitigate pollutants in contaminated water, leveraging the flexibility of microorganisms to eliminate toxic pollutants. Furthermore, Jeyakumar et al. (2023) and Gao et al. (2023) reinforce this concept by asserting the superiority of bioremediation, supporting that this method significantly surpasses other approaches in the removal of heavy metals and micro plastics. Syed et al. (2023) highlight that the potential of bioremediation materializes through bacterial biofilms, demonstrating substantial capacity for the bioremediation of heavy metals, offering a sustainable and efficient approach for the removal of these pollutants in the environment. Similarly, the study by Sun et al. (2023) presents the degradation of microplastics by biofilms as an ecologically favorable strategy for treating these pollutants. Thakur et al. (2023) proposes the integration of biofilm-based bioreactors as a potential solution to eliminate ammoniacal nitrogen residues, contributing to an effective bioremediation of residual water in the system. As per the findings of Hou et al. (2023), sustainable remediation technologies, including enduring immobilization, low-impact bioremediation, novel in situ chemical treatment methods, and innovative passive barriers, are proposed as effective means to substantially diminish the environmental impact of remediation while maximizing overall net benefits. Conversely, Zainab et al. (2023) highlight that the effectiveness of bioremediation varies depending on the type of petroleum hydrocarbons. Innovative approaches such as genetically modified microorganisms, nanotechnology, and bioaugmentation are identified as having high potential to enhance the remediation of PAH contamination.

In addition to the following listed advantages, a significant portion of the conducted research has converged toward the conclusion that bioremediation represents a more cost-effective approach compared to other decontamination methods. This finding is justified, on one hand, by the fact that the microorganisms used in the bioremediation process are often naturally present in the environment, thereby reducing costs associated with the introduction of external decontaminating agents. On the other hand, bioremediation leverages existing biological processes to degrade contaminants, eliminating the need for massive investments in costly equipment. Table 2.1 summarizes various findings from the literature, providing evidence that bioremediation is an economical and cost-effective approach.

TABLE 2.1
The Benefits of Bioremediation

Element	Author
Bioremediation of chromium using bacteria yields environmentally friendly and cost-effective results.	(Murthy et al., 2023)
Bioremediation outperforms other approaches due to its environmentally benign nature, economic viability, and low requirements in terms of labor and effort.	(Jeyakumar et al., 2023)
Bioremediation: The most effective technique, environmentally friendly, and cost-effective.	(Kumar & Singh, 2023)
Bioremediation: A promising alternative for wastewater treatment due to its high efficiency.	(Wang et al., 2023)
Bioremediation of OP pesticides in polluted environments: Safe, reliable, cost-effective, and environmentally friendly methods.	(Mali et al., 2023)
Bioremediation strategies aim to minimize the costs associated with the pollution removal process.	(Narayanan et al., 2023)
The techno-economic analysis demonstrated that the hybrid biological treatment approach had an average unit cost of 0.97 USD/m^3, indicating economic feasibility and competitiveness compared to commercial purification processes for sustainable water reuse.	(Rendón-Castrillón et al., 2023)
Bioremediation approach: Generates minimal or no sludge and is cost-effective, efficient, and environmentally friendly.	(Sudarshan et al., 2023)
Bioremediation: Enhances socio-economic benefits while providing carbon sequestration or green energy.	(Hou et al., 2023)
Bioremediation: An environmentally friendly technology. A low-cost and self-sustaining process.	(Gupta & Gandhi, 2023)

(Continued)

TABLE 2.1 ***(Continued)***
The Benefits of Bioremediation

Element	Author
Bioremediation using microalgae: A cost-effective and environmentally friendly system that can be applied to treat aquaculture wastewater and simultaneously produce valuable microalgal biomass.	(Kashem et al., 2023)
Microalgae-based bioremediation: An environmentally friendly alternative to traditional methods, providing operational cost-effectiveness and process optimization through the adaptability of microalgae and the production of valuable biomass.	(Kashem et al., 2023)
Bioremediation as a green method: Reduces overall costs for vanadium treatment.	(Tambat et al., 2023)
Biorestoration has proven effective and can serve as a cost-effective, environmentally friendly approach for remediating sites contaminated by AMD.	(Anekwe & Isa, 2023)
Bioelectrochemical systems (BES): Single-chamber air-cathode reactors based on biochar are preferred due to their lower cost compared to other conventional configurations.	(Gebregiorgis Ambaye et al., 2023)
Microbe-based bioremediation is an effective, economical, and environmentally friendly strategy.	(Dell'Anno et al., 2023)
Bioremediation and phytoremediation are low-carbon remediation technologies that are particularly promising due to their cost-effectiveness and adaptability to local conditions.	(Azuazu et al., 2023)
Various physical and chemical approaches are employed to remove contaminants. However, these technologies have limitations, such as high costs and low efficiency. Therefore, bioremediation is considered one of the most promising solutions to these issues.	(Rana et al., 2023)
Phycoremediation is gaining attention not only as a pollutant mitigation approach but also as one of the most cost-effective paths to achieve carbon neutrality.	(Rambabu et al., 2023)
Bioremediation provides a sustainable and cost-effective approach to remove or transform inorganic contaminants.	(Zaib et al., 2023)

2.4.2 Potential of Remediation Agents

Numerous previous studies have focused on algae, considering them as remediation agents with high potential (Kankanamalage et al., 2023; Kashem et al., 2023; Kashem et al., 2023; Dubey et al., 2023; Satya et al., 2023). For instance, Dubey et al. (2023) proposed that algae possess captivating sustainable characteristics, addressing odors and toxicity, co-remediating various common and emerging inorganic and organic pollutants from gaseous and aqueous environments, and generating biomass for a range of valuable products. Additionally, Dubey et al. (2023) confirmed that algae provide a better option than any other non-algal processes for several high CO_2–emitting industries. Satya et al. (2023) stated that microalgae are well-suited to contribute to the circular economy due to their rapid growth and adaptability, supported by their biological robustness in nutrient consumption. This makes them ideal candidates for bioremediation efforts. Kankanamalage et al. (2023) also explored the potential of microalgae, demonstrating that microalgae-based bioremediation generates valuable biomass with metabolites, namely lipids, proteins, and carbohydrates, which can be used in the value-added production of biofuels, biofertilizers, and so on. On their part, Kharel et al. (2023) confirmed that cyanidiales, belonging to red algae, have the potential for heavy metal remediation as they can resist and tolerate extreme stresses such as heat, acidic salts, and heavy metals.

Indeed, organisms from all walks of life can participate in bioremediation; however, Vaksmaa et al. (2023) believe that fungi are particularly promising candidates. They possess metabolic capabilities to break down complex molecules, making fungi the ultimate degraders of refractory organic matter in nature. In line with these ideas, Rana et al. (2023) present yeast as agents characterized by high biodegradability and ease of cultivation. For Geetha et al. (2023), fungal symbiosis with the plant root system is widely recognized as plant growth-promoting fungi (PGPF). Thus, PGPF play a key role in improving agricultural quality, increasing agricultural production, and paving the way for sustainable agriculture (Geetha et al., 2023). Finally, Mahmood et al. 2023, presented *F. pinicola* IEBL-4 as a fungus suitable for the bioremediation of industrial textile effluents (Table 2.2).

TABLE 2.2
Pollutants and Their Bioremediation Agents

Pollutants	Bioremediation Agent	Author
*Polycyclic Aromatic Hydrocarbons (PAHs)	*Bacteria *Fungi *Algae	(Thacharodi et al., 2023)
*Many inorganic and organic pollutants	*Various species of algae (Chlorella species)	(Dubey et al., 2023)
*Microplastics *Antibiotics *Heavy metal	*Microalgae	(Satya et al., 2023)
*Organophosphorus pesticides (OPP)	*Microorganisms *Microbial enzymes *Rhizospheric bacteria (PGPR)	(Dash & Osborne, 2023)
*Dyes *Polycyclic aromatic hydrocarbons *Plastics, *Heavy metals *Pesticides	Microbial enzymes: *Hydrolases, *Lipases, *Oxidoreductases, *Oxygenases *Laccases.	(Narayanan et al., 2023)
*Petroleum hydrocarbons (PHC)	*Biosurfactants *Biochar	(Gebregiorgis Ambaye et al., 2023)
*Heavy metals	*Plant Growth-Promoting Rhizobacteria (PGPR)	(Saberi Riseh et al., 2023)
*Polychlorinated biphenyls (PCBs) *Polybrominated diphenyl ethers (PBDEs) *Polyfluoroalkyl substances (PFAS).	*Microbes *Potent microorganisms	(Xu et al., 2023)
*Organic and inorganic contaminants *Heavy metals	*Macrophytes (plants)	(Demarco et al., 2023)
*Pesticides *Pharmaceuticals and personal care products *Plastics *Heavy metals	*Fungi	(Vaksmaa et al., 2023)
*Organophosphates (OP)	*Microbes *Microbial biocatalysts.	(Mali et al., 2023)
*Total Petroleum Hydrocarbons (TPH): Long-chain Hydrocarbons C21-C40	Tensioactive Agents (SEBR): *Sodium Dodecylbenzenesulfonate (SDBS) *Alpha-Olefin Sulfonate (AOS) *Triton X-100 (TX-100) *Tween80 *Rhamnolipid	(Ling et al., 2023)
*Textile industrial effluents	*Fomitopsis pinicola IEBL-4 (brown rot fungus)	(Mahmood et al., 2023)
*Xenobiotic substances	*Yeasts (Candida)	(Rana et al., 2023)
Microplastics: *Low-Density Polyethylene (LDPE), *Polylactic Acid (PLA), *Polybutylene Adipate Terephthalate (PBAT)	*Earthworm (Oligochaeta)	(Meng et al., 2023)

(Continued)

TABLE 2.2 *(Continued)*
Pollutants and Their Bioremediation Agents

Pollutants	Bioremediation Agent	Author
*Glyphosate *Isoproturon *Ligninpolymer *Parathion	*The laccase	(Bhatt et al., 2023)
*Polychlorinated biphenyls (PCBs)	*The resuscitation-promoting factor (Rpf) derived from *Micrococcus luteus*	(Zhou et al., 2023)
*Aquaculture wastewater (generated by a tilapia fish farm in Qatar)	Microalgae: **Haematococcus* sp., **Neochloris* sp., **Monoraphidium* sp.	(Kashem et al., 2023)
*Ammonia (NH_4^+) *Phosphate (PO_4^{3-})	*The shells	(Filippini et al., 2023)
*Heavy metals and metalloids: chromium, lead, mercury, cadmium, nickel, and cobalt	*Microbes: bacteria and fungi *Plants *Other agents: algae, biosurfactants, biofilms, genetically modified microorganisms	(Jeyakumar et al., 2023)
*Heavy metals	*Microbes: Microbial biofilms	(Syed et al., 2023)
*Nutrients from wastewater streams *Sludge generated during wastewater treatment *Residual pollutants, nutrients, and pathogens	* Microalgae	(Kankanamalage et al., 2023)
*Metal ions:Arsenic (As); Mercury (Hg); Lead (Pb); Chromium (Cr); Iron (Fe); Copper (Cu); Cadmium (Cd); Palladium (Pd); Zinc (Zn); Cobalt (Co); Antimony (Sb); Gold (Au); Silver (Ag); Platinum (Pt); Selenium (Se); Manganese (Mn); Molybdenum (Mo); Nickel (Ni); Tungsten (W); Uranium (U)	*Metal-resistant bacteria	(Roy et al., 2023)
*Petroleum hydrocarbons (PHC) such as crude oil, diesel, gasoline, and other petroleum products. *Polycyclic aromatic hydrocarbons (PAHs).	*Genetically modified microorganisms *Algae *Fungi	(Zainab et al., 2023)
*Glyphosate (GLY)	*Microorganisms	(Masotti et al., 2023)
*Lead (Pb)	*Bacterial strains *Algal species *Fungal strains *Various microorganisms	(Kumar & Singh, 2023)
*Microplastics	*Biofilms	(Sun et al., 2023)
*Synthetic dyes	*Microorganisms: Microalgae and bacteria, fungi, yeasts, and enzymes.	(Sudarshan et al., 2023)
*Radionuclides (from nuclear contamination)	*Plants or microorganisms	(Li et al., 2023)
*Ammonia (NH_3) *Nitrite (NO_2^-) *Nitrate (NO_3^-)	*Biofilm-based bioreactors	(Thakur et al., 2023)
*Aromatic and aliphatic hydrocarbon fractions	* The strain of *Lysinibacillus sphaericus* IITR51	(Gaur et al., 2023)
*Chrome (Cr (III)/Cr (VI))	*Bacteria	(Murthy et al., 2023)

(Continued)

TABLE 2.2 *(Continued)*
Pollutants and Their Bioremediation Agents

Pollutants	Bioremediation Agent	Author
*Black liquor effluents from the paper industry	*Thermophilic bacteria (*Serratia* sp. AXJ-M)	(An et al., 2023)
*Crude oil	*Compound microbial agent containing *Pseudomonas aeruginosa* and *Bacillus licheniformis*	(Liu et al., 2023)
*The brown azoic textile dye 703	*The bacterial strain *Pseudomonas aeruginosa*	(Ullah Khan et al., 2023)
*Crude oil *Aromatic hydrocarbons *Alkanes *Aliphatic hydrocarbons *Asphaltenes *Naphthenes *Other petroleum products	*The biosurfactants	(Bhadra et al., 2023)
*Vanadium	Microalgae: *Chlorellasorokiniana SU1 *Picochlorumoklahomensis	(Tambat et al., 2023)
*Trichloroethylene (TCE)	*Bacteria Dehalococcoides	(Madison et al., 2023)
*Polycyclic Aromatic Hydrocarbons (PAHs) *Heavy Metals (HM)	Innovative bacterial strains: *Alcanivorax xenomutans strain-SRM1 *Halomonas alkaliantarctica strain-SRM2)	(Dell'Anno et al., 2023)
*Lead (Pb)	*Aerobic Algae-Bacteria Granules (A-B AGS)	(Wang et al., 2023)
*Radionuclides	*Microbes *Plants	(Singh et al., 2023)
*Uranium (U)	*Microbes *Plants	(Cui et al., 2023)
*Heavy metals	*Seaweed	(Varabih & Nofirman, 2023)
*Detergents and dyes *Cyanide and sulfuric acid *Cement and metals	*Microorganisms: bacteria, fungi, and algae *Plants	(Ayilara & Babalola, 2023)
*Toxic pollutants in the environment	*Zeolite nanoparticles, *Biopolymers	(Varabih & Nofirman, 2023)
*Heavy metals	*Cyanidiales (red algae: the genera *Cyanidioschyzon*, *Cyanidium*, and *Galdieria*)	(Kharel et al., 2023)
*Hydrocarbons (crude oil)	*Animal manure (poultry droppings, cow dung, goat manure, and pig manure)	(Adewoyin & Arimoro, 2023)
*Glyphosate	*Bacterial consortium (*Pseudomonas stutzeri*, *Comamonas odontotermitis*, *Sinomonas atrocyanea*) *Microalgae (*Chlorella protothecoides*)	(Borella et al., 2023)

2.5 CHALLENGES OF BIOREMEDIATION

As previously mentioned, the literature has suggested that bioremediation techniques have superiority over other conventional methods. However, bioremediation still encounters some significant challenges that restrict its large-scale implementation. Firstly, it appears that the effectiveness of bioremediation depends on various parameters, namely the selected substrates, the type of microbe, site conditions, and other physical parameters (Radhakrishnan et al., 2023). According to Das et al. (2023), in the natural environment, several biotic and abiotic factors, such as temperature, snow, precipitation, soil pH, and low nutrient content, influence the bioremediation process. Furthermore, the complexity of transitioning from controlled laboratory conditions to real environments has also been highlighted. According to Radhakrishnan et al. (2023), bioremediation experiments conducted in the laboratory with limited volumes need to be extrapolated to larger volumes and validated in the field. In this context, Hellal et al. (2023) argued that the challenges of bioremediation stem from current gaps in understanding biotransformation processes, whether in controlled laboratory conditions or relevant environmental settings. These gaps encourage microbial ecotoxicologists to innovate at the experimental level. While simplified microbial experimental systems are useful for addressing ecological questions and allowing experimental controls, they do not fully capture the complexity of the real world. Therefore, the authors emphasized the need to go beyond these relatively simple models to better understand and account for this complexity (Hellal et al., 2023). According to Haas and Nikel (2023), the restriction of chemical elements in the microbial "biochemical palate" can pose challenges for bioremediation, as certain compounds require specific biological processes for degradation or transformation. The limited diversity of elements may impede the effectiveness of biotechnology in pollutant treatment (Haas & Nikel, 2023). The performance of bioremediation also depends on the tools utilized. In fact, researchers are encouraged to employ numerical simulations and other methods to identify the potential effectiveness of the process (Radhakrishnan et al., 2023). Additionally, the existence of multi-omic tools, such as genomics, proteomics, transcriptomics, and metabolomics, provides significant data, enhancing the understanding of microbial behavior in contaminated environments and thus promoting more effective bioremediation (Das et al., 2023). Another challenge mentioned in the literature lies in the variability of remediation costs. For instance, bioaugmentation is typically emphasized as more costly than biostimulation. Consequently, the cost of remediation varies depending on various factors such as the site location, the extent of contamination, and the type of treatment method employed (Radhakrishnan et al., 2023).

Moreover, in their research, Gao et al. (2023) validated that utilizing microalgae-based bioremediation techniques is not optimal for eliminating microplastics from wastewater due to extended retention times. The authors noted that the hydraulic retention time in standard biological reactors typically ranges from 4 to 12 hours, which fails to meet the criteria for swiftly treating microplastics in wastewater (Gao et al., 2023). Furthermore, despite the availability of numerous bioremediation strategies aimed at tackling challenges associated with contaminant removal, it has been established that a solitary bioremediation pathway is inadequate for degrading all components simultaneously (Sudarshan et al., 2023). In their study, Crecca et al. (2023) highlighted international competition in the development of bioremediation solutions, as evidenced by the protection of biotechnology-related inventions being primarily concentrated in China, the United States, Japan, and the European Union. Despite this, the authors argued that the biological processes required for bioremediation are still challenging to manipulate and control optimally. They emphasized that the current state of knowledge in the field of bioremediation is still evolving, indicating that there is much more to be discovered and improved (Crecca et al., 2023) (Table 2.3).

2.6 PLANNING AND EXECUTION OF THE BIOREMEDIATION APPROACH

In order to address the various challenges of bioremediation, it is essential to adopt a plan that guides researchers throughout the process, in terms of both planning and execution. Drawing inspiration from the literature, key steps to better structure the bioremediation approach have been identified. Indeed, the importance of planning and executing bioremediation has been emphasized by numerous authors. For instance, Rana et al. (2023) underscored the importance of discovering appropriate long-term treatment methods. This highlights the need for careful planning and methodical execution to effectively integrate bioremediation into ecosystem restoration. Similarly, Zhou et al. (2023)

TABLE 2.3
Some Challenges of Bioremediation

Challenges	Authors
The effectiveness of bioremediation depends on various parameters	(Radhakrishnan et al., 2023; Ayilara & Babalola, 2023; Das et al., 2023)
Complexity of transitioning from controlled laboratory conditions to real environments	(Radhakrishnan et al., 2023; Hellal et al., 2023)
Variability of bioremediation costs	(Radhakrishnan et al., 2023)
Prolonged retention time	(Gao et al., 2023)
Insufficiency of a single bioremediation pathway	(Das et al., 2023)
Limitation of the spectrum of chemical elements	(Haas & Nikel, 2023)
Emphasis on patent protection and ongoing complexities of biological processes	(Crecca et al., 2023)

emphasized the necessity of adopting an integrated strategy involving both planning (selection of bioremediation agents) and execution (application on the contaminated site) to restore affected ecosystems. The review indicates that researchers employing experimental methods follow a plan when conducting bioremediation. For instance, Zhao et al. (2023) initially selected promoters, built strains, and optimized the degradation pathway. With this plan, they developed a salt-tolerant bioremediation platform from the halophilic strain *Halomonas cupida* J9. They also simulated their experiments on a contaminated site with high salinity conditions, replicating highly saline environments. Similarly, Dell'Anno et al. (2023) started by simultaneously selecting bacteria resistant to heavy metals. They then subjected the enriched mixed bacterial culture to complete DNA sequencing. Subsequently, they annotated the genomes assembled through metagenomics, followed by subcultures to obtain the main bacterial species in the form of pure strains. Finally, they tested the ability of the isolated strains to degrade pollutants (Dell'Anno et al., 2023).

Thus, as shown in Figure 2.3, it appears that the initial step to undertake lies in the identification of the problem and the assessment of the nature and degree of danger. The goal is to gain a thorough understanding of the problem by involving personnel with expertise in microbiology, chemistry, geology, soil science, environmental engineering, chemical engineering, and other specialties. This multidisciplinary approach is necessary to comprehend the underlying biogeochemical mechanisms in order to effectively plan bioremediation strategies. The study of the impact of this identified problem on human, animal, and environmental health is also necessary to confirm the urgent need for corrective solutions. During this stage, mastery of the geographical distribution of the problem is recommended. According to Zhang et al. (2023), recognizing the geographical distributions and environmental impacts of contaminants is crucial for integrated planning and successful execution of bioremediation on a regional or global scale. Many authors have supported this step (Xu et al., 2023; Narayanan & Ma, 2023; Hidangmayum et al., 2023; Gebregiorgis Ambaye et al., 2023; Saberi Riseh et al., 2023; Zhang et al., 2023).

Once the problem is clearly identified, researchers will be able to identify emerging pollutants that require specific management. In fact, it has been observed that the majority of recently published articles have focused their attention on emerging pollutants such as microplastics, antibiotics, various types of wastewaters, and heavy metals. It has also been noted that these emerging pollutants require the adoption of a specific protocol. The focus on a specific pollutant and the relevant choice has been advocated by several studies (Narayanan & Ma, 2023; Thacharodi et al., 2023; Saberi Riseh et al., 2023; Jeyakumar et al., 2023; Satya et al., 2023; Zhao et al., 2023).

Before choosing the right bioremediation approach to adopt for degrading the identified pollutant, it is necessary to rely on a timely review of bioremediation progress and new tools. This allows staying informed about recent developments for effective planning and execution (Xu et al., 2023).

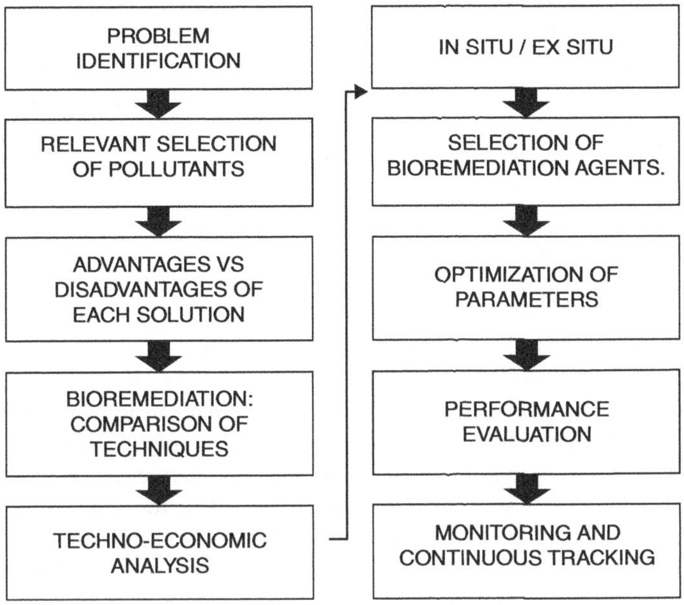

FIGURE 2.3 Planning and execution of bioremediation.

In fact, the need for in-depth studies on bioremediation mechanisms has been emphasized by Vaksmaa et al. (2023). The goal is to stay updated on existing techniques and advancements in the field (Gupta & Gandhi, 2023). This step could be integrated into ecosystem restoration plans to make them more effective. Through this review, the advantages and disadvantages of each approach can be identified, allowing for the maximization of results while minimizing drawbacks (Davis, 2023; Dubey et al., 2023).

After relevantly specifying the pollutant, its impacts, and identifying possible pathways for its detoxification and bioremediation, the choice of a degradation technique becomes necessary. Thus, the literature has unanimously agreed on the superiority of bioremediation compared to other conventional methods. In addition to its ecologically benign nature, bioremediation is characterized by economic viability and low labor requirements. This strengthens the idea of integrating this approach into the planning of ecosystem restoration (Narayanan et al., 2023; Narayanan & Ma, 2023; Azuazu et al., 2023; Saravanan et al., 2023; Saberi Riseh et al., 2023; Jeyakumar et al., 2023). Moreover, the bioremediation approach involves a range of techniques and mechanisms. The comparison of these techniques is essential to select those most suitable for the restoration of contaminated sites. It is worth noting that the literature strongly suggests the simultaneous implementation of more than one remediation strategy or a combinatorial approach to benefit from synergistic advantages (Demarco et al., 2023; Vaksmaa et al., 2023; Dash & Osborne, 2023; Radhakrishnan et al., 2023; Hidangmayum et al., 2023; Jeyakumar et al., 2023).

As emphasized by Rendón-Castrillón et al. (2023), a techno-economic analysis must be conducted before the implementation of the bioremediation approach. The objective is to highlight the economic feasibility of the bioremediation approach and its various techniques. Additionally, the economic analysis allows for the selection of cost-effective and environmentally sustainable technologies (Anekwe & Isa, 2023). This study should also anticipate the necessary equipment to ensure the success of the bioremediation process (Kashem et al., 2023; Mahmood et al., 2023).

Based on the feasibility analysis conducted, a decision can be made regarding whether the intervention will take place in a laboratory or at a specific site. Zhao et al. (2023) emphasized the importance of evaluating the effectiveness of field interventions during the planning process. On the other hand, Li et al. (2023) highlighted the significance of a practical approach and the necessity to integrate these techniques in real-world environments. This same perspective was shared by Madison et al. (2023), who prefer a practical and operational application of bioremediation. However, the choice between in situ/ex situ depends on the technical and economic feasibility of the experiment (Madison et al., 2023).

The need to integrate various organisms into strategies for restoring contaminated ecosystems is crucial (Dash & Osborne, 2023; Thacharodi et al., 2023; Saberi Riseh et al., 2023). Indeed, microorganisms play a key role in bioremediation, producing enzymes capable of degrading various contaminants (Narayanan et al., 2023). Therefore, during this stage, it is essential to select the most suitable organisms based on the specific conditions of the restoration sites (Dubey et al., 2023). For example, the specific use of fungi in mycoremediation suggests a targeted and strategic implementation of bioremediation (Vaksmaa et al., 2023). Additionally, the types of enzymes to be integrated depend on the types of contaminants present in the ecosystem to be restored (Narayanan et al., 2023).

The literature identifies numerous environmental factors that can influence the effectiveness of bioremediation, including nutrients, grain size, seasonality, microbial competition, environmental conditions, pH, temperature, internal resistance, and substrate nature, among others. According to Filippini et al. (2023), these parameters are crucial as they allow for anticipating the responses of bioremediation agents in changing environmental conditions. Planning must take these factors into account to ensure the success of ecosystem restoration (Zainab et al., 2023; Ullah Khan et al., 2023). Additionally, these parameters need to be optimized to enhance the efficiency of microorganisms in their natural habitat, maximize the impact of bioremediation, and design tailor-made solutions (Thacharodi et al., 2023; Wang et al., 2023; Zhao et al., 2023).

The literature also recommends using indicators to assess the quality of bioremediation. For instance, Mahmood et al. (2023) emphasized that after treatment, evaluating biochemical oxygen demand and chemical oxygen demand provides indicators of the quality of biodegradation, thus offering a comprehensive assessment of the process. Additionally, Narayanan et al. (2023) addressed immobilization strategies, genetic engineering, and nanotechnology applications to enhance enzyme performance. These strategies can be considered in integrated planning to maximize the efficiency of bioremediation.

According to several references, the importance of continuous monitoring is a crucial aspect for the successful integration of planning and execution in bioremediation (Zainab et al., 2023). In fact, the ability to monitor degraders during bioremediation provides tools to assess the effectiveness of interventions in real-time. This can contribute to adjusting plans based on observed results, thus incorporating an optimization process into overall planning (Zhao et al., 2023).

2.7 CONCLUSION, RECOMMENDATION, AND FUTURE CHALLENGES

In conclusion, this study aimed to provide an updated overview of bioremediation, exploring recent advancements and emerging trends in the field. The findings offer a comprehensive understanding of key concepts such as biorestoration and biodegradation. It's worth noting that while a substantial portion of the reviewed literature consisted of theoretical studies, others involved practical experiments utilizing various methodologies. For instance, some researchers utilized molecular docking and simulation techniques to investigate enzyme-compound interactions, while others conducted field experiments in specific locations. The variety of experimental approaches underscores the versatility and adaptability of bioremediation strategies to different environmental contexts and contaminants. Furthermore, the inclusion of studies from various geographic locations highlights the global relevance and applicability of bioremediation techniques in addressing environmental pollution. Based on the findings, there's a clear recommendation to prioritize experimental research in bioremediation. Additionally, suggestions from scholars include the need for more field trials, optimization of biorestoration technologies for complex contamination scenarios, exploration of fungal species and their metabolic pathways, continued investigation into the efficacy of microorganisms like Candida, understanding polymer depolymerization mechanisms in earthworms for microplastic remediation, and the crucial importance of policy frameworks supporting sustainable remediation practices. Furthermore, it is imperative to incorporate more recent literature, extending to the years 2020, 2021, and 2022, in future endeavors to ensure a comprehensive understanding of the latest developments in bioremediation research. This would facilitate staying abreast of cutting-edge technologies and methodologies, fostering innovation and advancement in the field of environmental remediation.

REFERENCES

Adewoyin, J. A., & Arimoro, F. O. (2023). Animal manure as a biostimulant in bioremediation of oil-contaminated soil: The role of earthworms. *Environmental Monitoring and Assessment, 195*(293). https://doi.org/10.1007/s10661-022-10884-1

Anekwe, I. M. S., & Isa, Y. M. (2023). Bioremediation of acid mine drainage – Review. *Alexandria Engineering Journal, 65*, 1047–1075. https://doi.org/10.1016/j.aej.2022.09.053

An, X., Li, N., Zhang, S., Han, Y., & Zhang, Q. (2023). Integration of proteome and metabolome profiling to reveal heat stress response and tolerance mechanisms of Serra-tia sp. AXJ-M for the bioremediation of papermaking black liquor. *Journal of Hazardous Materials, 450*, 131092. https://doi.org/10.1016/j.jhazmat.2023.131092

Ansari, F., Momina, A., Ahmad, A., & Rafatullah, M. (2023). Review on bioremediation technologies of polycyclic aromatic hydrocarbons (PAHs) from soil: Mechanisms and future perspective. *International Biodeterioration & Biodegradation, 179*, 105582. https://doi.org/10.1016/j.ibiod.2023.105582

Azuazu, I. N., Sam, K., Campo, P., & Coulon, F. (2023). Challenges and opportunities for low-carbon remediation in the Niger Delta: Towards sustainable environmental management. *Science of the Total Environment, 900*, 165739. https://doi.org/10.1016/j.scitotenv.2023.165739

Ayilara, M. S., & Babalola, O. O. (2023). Bioremediation of environmental wastes: The role of microorganisms. *Frontiers in Agronomy, 5*, 1183691. https://doi.org/10.3389/fagro.2023.1183691

Bhadra, S., Chettri, D., & Kumar Verma, A. (2023). Biosurfactants: Secondary metabolites involved in the process of bioremediation and biofilm removal. *Applied Biochemistry and Biotechnology*, 195(15), 5541–5567. https://doi.org/10.1007/s12010-022-03951-3

Bhattacharya, R., Bose, D., Yadav, J., Sharma, B., Sangli, E., Patel, A., Mukherjee, A., & Singh, A. A. (2023). Bioremediation and bioelectricity from Himalayan rock soil in sediment-microbial fuel cell using carbon-rich substrates. *Fuel*, 341, 127019. https://doi.org/10.1016/j.fuel.2022.127019

Bhatt, P., Bhatt, K., Chen, W.-J., Huang, Y., Xiao, Y., Wu, S., Lei, Q., Zhong, J., Zhu, X., & Chen, S. (2023). Bioremediation potential of laccase for catalysis of glyphosate, iso-proturon, lignin, and parathion: Molecular docking, dynamics, and simulation. *Journal of Hazardous Materials*, 443(Part B), 130319. https://doi.org/10.1016/j.jhazmat.2022.130319

Borella, L., Novello, G., Gasparotto, M., Renella, G., Roverso, M., Bogialli, S., Filippini, F., & Sforza, E. (2023). Design and experimental validation of an optimized micro-algae-bacteria consortium for the bioremediation of glyphosate in continuous photobioreactors. *Journal of Hazardous Materials*, 441, 129921. https://doi.org/10.1016/j.jhazmat.2022.129921

Crecca, V. M. T., da Silva, J. M., & de Souza, P. A. R. (2023). Technological prospecting: Patent mapping of bioremediation of soil contaminated with agrochemicals using fungi. *World Patent Information*, 73, 102196. https://doi.org/10.1016/j.wpi.2023.102196

Cui, Q., Zhang, Z., Beiyuan, J., Cui, Y., Chen, L., Chen, H., & Fang, L. (2023). A critical review of uranium in the soil-plant system: Distribution, bioavailability, toxicity, and bioremediation strategies. *Critical Reviews in Environmental Science and Technology*, 53(3), 340–365. https://doi.org/10.1080/10643389.2022.2054246

Dash, D. M., & Osborne, W. J. (2023). A systematic review on the implementation of advanced and evolutionary biotechnological tools for efficient bioremediation of organophosphorus pesticides. *Chemosphere*, 313, 137506. https://doi.org/10.1016/j.chemosphere.2022.137506

Das, N., Das, A., Das, S., Bhatawadekar, V., Pandey, P., Choure, K., Damare, S., & Pandey, P. (2023). Petroleum hydrocarbon catabolic pathways as targets for metabolic engineering strategies for enhanced bioremediation of crude-oil-contaminated environments. *Fermentation*, 9(2), 196. https://doi.org/10.3390/fermentation9020196

Davis, G. B. (2023). Reviewing the bioremediation of contaminants in groundwater: Investigations over 40 years provide insights into what's achievable. *Frontiers in Bioscience (Elite Edition)*, 15(3), 16. https://doi.org/10.31083/j.fbe1503016

Dell'Anno, F., van Zyl, L. J., Trindade, M., Buschi, E., Cannavacciuolo, A., Pepi, M., Sansone, C., Brunet, C., Ianora, A., de Pascale, D., Golyshin, P. N., Dell'Anno, A., & Rastelli, E. (2023). Microbiome enrichment from contaminated marine sediments unveils novel bacterial strains for petroleum hydrocarbon and heavy metal bioremediation. *Environmental Pollution*, 317, 120772. https://doi.org/10.1016/j.envpol.2022.120772

Demarco, C. F., Quadro, M. S., Selau Carlos, F., Pieniz, S., Morselli, L. B. G. A., & Andreazza, R. (2023). Bioremediation of aquatic environments contaminated with heavy metals: A review of mechanisms, solutions and perspectives. *Sustainability*, 15, 1411. https://doi.org/10.3390/su15021411

Dubey, S., Chen, C.-W., Haldar, D., Tambat, V. S., Kumar, P., Tiwari, A., Singhania, R. R., Dong, C.-D., & Patel, A. K. (2023). Advancement in algal bioremediation for organic, inorganic, and emerging pollutants. *Environmental Pollution*, 317, 120840. https://doi.org/10.1016/j.envpol.2022.120840

Eckel, F., Sinzinger, K., Van Opdenbosch, D., Schieder, D., Sieber, V., & Zollfrank, C. (2023). Influence of microbial biomass content on biodegradation and mechanical properties of poly(3-hydroxybutyrate) composites. *Biodegradation*, 35, 209–224.

El-Saadony, M. T., Saad, A. M., El-Wafai, N. A., Abou-Aly, H. E., Salem, H. M., Soliman, S. M., Abd El-Mageed, T. A., Elrys, A. S., Selim, S., Abd El-Hack, M. E., Kappachery, S., El-Tarabily, K. A., & AbuQamar, S. F. (2023). Hazardous wastes and management strategies of landfill leachates: A comprehensive review. *Environmental Technology & Innovation*, 31, 103150. https://doi.org/10.1016/j.eti.2023.103150

Filippini, G., Dafforn, K. A., & Bugnot, A. B. (2023). Shellfish as a bioremediation tool: A review and meta-analysis. *Environmental Pollution*, 316(Part 2), 120614. https://doi.org/10.1016/j.envpol.2022.120614

Gao, N., Ning, R., & Deng, X. (2023). Feasibility, challenges, and future prospects of microalgae-based bioremediation technique for removing microplastics from wastewater. *Frontiers in Bioengineering and Biotechnology*, 11, 1288439. https://doi.org/10.3389/fbioe.2023.1288439

Gaur, V. K., Tripathi, V., Gupta, P., Thakur, R. S., Kaur, I., Regar, R. K., Srivastava, P. K., & Manickam, N. (2023). Holistic approach to waste mobil oil bioremediation: Valorizing waste through biosurfactant production for soil restoration. *Journal of Environmental Management*, *348*, 119207. https://doi.org/10.1016/j.jenvman.2023.119207

Gebregiorgis Ambaye, T., Vaccari, M., Franzetti, A., Prasad, S., Formicola, F., Rosatelli, A., Hassani, A., Aminabhavi, T. M., & Rtimi, S. (2023). Microbial electrochemical bioremediation of petroleum hydrocarbons (PHCs) pollution: Recent advances and outlook. *Chemical Engineering Journal*, 452(3), 139372. https://doi.org/10.1016/j.cej.2022.139372

Geetha, N., Sunilkumar, C. R., Bhavya, G., Nandini, B., Abhijith, P., Satapute, P., Shetty, H. S., Govarthanan, M., & Jogaiah, S. (2023). Warhorses in soil bioremediation: Seed biopriming with PGPF secretome to phytostimulate crop health under heavy metal stress. *Environmental Research*, 216(Part 1), 114498. https://doi.org/10.1016/j.envres.2022.114498

Gupta, P. K., & Gandhi, M. (2023). Bioremediation of organic pollutants in soil–water system: A review. *BioTech*, 12, 36. https://doi.org/10.3390/biotech12020036

Haas, R., & Nikel, P. I. (2023). Challenges and opportunities in bringing nonbiological atoms to life with synthetic metabolism. *Trends in Biotechnology*, 41(1). https://doi.org/10.1016/j.tibtech.2022.06.004

Hamdan, H. Z., & Salam, D. A. (2023). Sediment microbial fuel cells for bioremediation of pollutants and power generation: A review. *Environmental Chemistry Letters*, 21(8), 2761–2787. https://doi.org/10.1007/s10311-023-01625-y

Hellal, J., Barthelmebs, L., Bérard, A., Cébron, A., Cheloni, G., Colas, S., Cravo-Laureau, C., De Clerck, C., Gallois, N., Hery, M., Martin-Laurent, F., Martins, J., Morin, S., Palacios, C., Pesce, S., Richaume, A., & Vuilleumier, S. (2023). Unlocking secrets of microbial ecotoxicology: Recent achievements and future challenges. *FEMS Microbiology Ecology*, 99(10), fiad102. https://doi.org/10.1093/femsec/fiad102

Hidangmayum, A., Debnath, A., Guru, A., Singh, B. N., Upadhyay, S. K., & Dwivedi, P. (2023). Mechanistic and recent updates in nano-bioremediation for developing green technology to alleviate agricultural contaminants. *International Journal of Environmental Science and Technology*, 20(1), 11693–11718. https://doi.org/10.1007/s13762-022-04560-7

Hou, D., Al-Tabbaa, A., O'Connor, D., Hu, Q., Zhu, Y.-G., Wang, L., Kirkwood, N., Ok, Y. S., Tsang, D. C. W., Bolan, N. S., & Rinklebe, J. (2023). Sustainable remediation and redevelopment of brownfield sites. *Nature Reviews Earth & Environment*, 4(1), 271–286. https://doi.org/10.1038/s43017-023-00404-1

Idris, M. O., Noh, N. A. M., Ibrahim, M. N. M., & Yaqoob, A. A. (2023). Sustainable microbial fuel cell functionalized with a bio-waste: A feasible route to formaldehyde bioremediation along with bioelectricity generation. *Chemical Engineering Journal*, 455, 140781. https://doi.org/10.1016/j.cej.2022.140781

Jeyakumar, P., Debnath, C., Vijayaraghavan, R., & Muthuraj, M. (2023). Trends in bioremediation of heavy metal contaminations. *Environmental Engineering Research*, 28(4), 220631. https://doi.org/10.4491/eer.2021.631

Kaleem, M., Mumtaz, A. S., Hashmi, M. Z., Saeed, A., Inam, F., Waqar, R., & Jabeen, A. (2023). Myco- and phyco-remediation of polychlorinated biphenyls in the environment: A review. *Environmental Science and Pollution Research*, 30(1), 13994–14007. https://doi.org/10.1007/s11356-022-24902-9

Kankanamalage, G., Nishshanka, H. S., Thevarajah, B., Nimarshana, P. H. V., Prajapati, S. K., & Ariyadasa, T. U. (2023). Real-time integration of microalgae-based bioremediation in conventional wastewater treatment plants: Current status and prospects. *Journal of Water Process Engineering*, 56, 104248. https://doi.org/10.1016/j.jwpe.2023.104248

Kashem, A. H. M., Das, P., AbdulQuadir, M., Khan, S., Thaher, M. I., Alghasal, G., Hawari, A. H., & Al-Jabri, H. (2023). Microalgal bioremediation of brackish aquaculture wastewater. *Science of the Total Environment*, 873, 162384. https://doi.org/10.1016/j.scitotenv.2023.162384

Kharel, H. L., Shrestha, I., Tan, M., Nikookar, M., Saraei, N., & Selvaratnam, T. (2023). Cyanidiales-based bioremediation of heavy metals. *BioTech*, 12, 29. https://doi.org/10.3390/biotech12020029

Kumar, K., & Singh, D. (2023). Toxicity and bioremediation of lead: A critical review. *International Journal of Environmental Health Research*. https://doi.org/10.1080/09603123.2023.2165047

Li, Z., He, Y., Sonne, C., Lam, S. S., Kirkham, M. B., Bolan, N., Rinklebe, J., Chen, X., & Peng, W. (2023). A strategy for bioremediation of nuclear contaminants in the environment. *Environmental Pollution*, 319, 120964. https://doi.org/10.1016/j.envpol.2022.120964

Ling, H., Hou, J., Du, M., Zhang, Y., Liu, W., Christie, P., & Luo, Y. (2023). Surfactant-enhanced bioremediation of petroleum-contaminated soil and microbial community response: A field study. *Chemosphere*, 322, 138225. https://doi.org/10.1016/j.chemosphere.2023.138225

Liu, Q., Wang, Y., Sun, S., Tang, F., Chen, H., Chen, S., Zhao, C., & Li, L. (2023). A novel chitosan-biochar immobilized microorganism strategy to enhance bioremediation of crude oil in soil. *Chemosphere*, 313, 137367. https://doi.org/10.1016/j.chemosphere.2022.137367

Ma, J., Zhang, J., Zhang, Y., Guo, Q., Hu, T., Xiao, H., Lu, W., & Jia, J. (2023). Progress on anodic modification materials and future development directions in microbial fuel cells. *Journal of Power Sources*, 556, 232486. https://doi.org/10.1016/j.jpowsour.2022.232486

Madison, A. S., Sorsby, S. J., Wang, Y., & Key, T. A. (2023). Increasing in situ bioremediation effectiveness through field-scale application of molecular biological tools. *Frontiers in Microbiology*, 13, 1005871. https://doi.org/10.3389/fmicb.2022.1005871

Mahmood, R. T., Asad, M. J., Hadri, S. H., El-Shorbagy, M. A., Mousa, A. A. A., Dara, R. N., Awais, M., & Tlili, I. (2023). Bioremediation of textile industrial effluents by *Fomitopsispinicola* IEBL-4 for environmental sustainability. *Human and Ecological Risk Assessment: An International Journal, 29*(2), 285–302. https://doi.org/10.1080/10807039.2022.2057277

Mali, H., Shah, C., Raghunandan, B. H., Prajapati, A. S., Patel, D. H., Trivedi, U., & Subramanian, R. B. (2023). Organophosphate pesticides: An emerging environmental contaminant - Pollution, toxicity, bioremediation progress, and remaining challenges. *Journal of Environmental Sciences*, *127*, 234–250. https://doi.org/10.1016/j.jes.2022.04.023

Masotti, F., Garavaglia, B. S., Gottig, N., & Ottado, J. (2023). Bioremediation of the herbicide glyphosate in polluted soils by plant-associated microbes. *Current Opinion in Microbiology*, *73*, 102290. https://doi.org/10.1016/j.mib.2023.102290

Meng, K., Huerta Lwanga, E., van der Zee, M., Munhoz, D. R., & Geissen, V. (2023). Fragmentation and depolymerization of microplastics in the earthworm gut: A potential for microplastic bioremediation? *Journal of Hazardous Materials*, *447*, 130765. https://doi.org/10.1016/j.jhazmat.2023.130765

Murthy, M. K., Khandayataray, P., Padhiary, S., & Samal, D. (2023). A review on chromium health hazards and molecular mechanism of chromium bioremediation. *Reviews on Environmental Health*, *38*(3), 461–478. https://doi.org/10.1515/reveh-2021-0139

Narayanan, M., & Ali, S. S., & El-Sheekh, M. (2023). A comprehensive review on the potential of microbial enzymes in multipollutant bioremediation: Mechanisms, challenges, and future prospects. *Journal of Environmental Management*, *334*, 117532. https://doi.org/10.1016/j.jenvman.2023.117532

Narayanan, M., & Ma, Y. (2023). Metal tolerance mechanisms in plants and microbe-mediated bioremediation. *Environmental Research*, *222*, 115413. https://doi.org/10.1016/j.envres.2023.115413

Patowary, R., Devi, A., & Mukherjee, A. K. (2023). Advanced bioremediation by an amalgamation of nanotechnology and modern artificial intelligence for efficient restoration of crude petroleum oil-contaminated sites: A prospective study. *Environmental Science and Pollution Research*, *30*, 74459–74484. https://doi.org/10.1007/s11356-023-27698-4

Pathania, T., Bottacin-Busolin, A., & Eldho, T. I. (2023). Evaluating the effect of aquifer heterogeneity on multiobjective optimization of in-situ groundwater bioremediation. *Engineering Analysis with Boundary Elements*, *148*, 336–350. https://doi.org/10.1016/j.enganabound.2023.01.005

Radhakrishnan, A., Balaganesh, P., Vasudevan, M., Natarajan, N., Chauhan, A., Arora, J., Ranjan, A., Rajput, V. D., Sushkova, S., Minkina, T., & Srivastav, R. (2023). Bioremediation of hydrocarbon pollutants: Recent promising sustainable approaches, scope, and challenges. *Sustainability*, *15*(1), 5847. https://doi.org/10.3390/su15075847

Rambabu, K., Avornyo, A., Gomathi, T., Thanigaivelan, A., Show, P. L., & Banat, F. (2023). Phycoremediation for carbon neutrality and circular economy: Potential, trends, and challenges. *Bioresource Technology*, *367*, 128257. https://doi.org/10.1016/j.biortech.2022.128257

Rana, S., Handa, S., Aggarwal, Y., Puri, S., & Chatterjee, M. (2023). Role of *Candida* in the bioremediation of pollutants: A review. *Letters in Applied Microbiology*, *76*(9), ovad103. https://doi.org/10.1093/lambio/ovad103

Rendón-Castrillón, L., Ramírez-Carmona, M., Ocampo-López, C., González-López, F., Cuartas-Uribe, B., & Mendoza-Roca, J. A. (2023). Treatment of water from the textile industry contaminated with indigo dye: A hybrid approach combining bioremediation and nanofiltration for sustainable reuse. *Case Studies in Chemical and Environmental Engineering*, *8*, 100498. https://doi.org/10.1016/j.cscee.2023.100498

Saberi Riseh, R., Gholizadeh Vazvani, M., Hajabdollahi, N., & Thakur, V. K. (2023). Bioremediation of heavy metals by rhizobacteria. *Applied Biochemistry and Biotechnology*, *195*(1), 4689–4711. https://doi.org/10.1007/s12010-022-04177-z

Roy, R., Samanta, S., Pandit, S., Naaz, T., Banerjee, S., Rawat, J. M., Chaubey, K. K., & Saha, R. P. (2023). An overview of bacteria-mediated heavy metal bioremediation strategies. *Applied Biochemistry and Biotechnology*, *196*(6), 1712–1751. https://doi.org/10.1007/s12010-023-04614-7

Saadaoui, S., Erable, B., Saidi, N., Etcheverry, L., Neifar, M., Masmoudi, A. S., Driouech, R., Cherif, A., & Chouchane, H. (2023). Saline sediments as a suitable source for halophilic inoculums to degrade azo dyes in synthetic and real textile wastewaters by microbial electrochemical systems. *Applied Sciences*, *13*, 5581. https://doi.org/10.3390/app13095581

Saravanan, A., Senthil Kumar, P., Duc, P. A., & Rangasamy, G. (2023). Strategies for microbial bioremediation of environmental pollutants from industrial wastewater: A sustainable approach. *Chemosphere*, *313*, 137323. https://doi.org/10.1016/j.chemosphere.2022.137323

Satya, A. D. M., Cheah, W. Y., Yazdi, S. K., Cheng, Y.-S., Khoo, K. S., Vo, D.-V. N., Bui, X. D., Vithanage, M., & Show, P. L. (2023). Progress on microalgae cultivation in wastewater for bioremediation and circular bioeconomy. *Environmental Research*, *218*, 114948. https://doi.org/10.1016/j.envres.2022.114948

Singh, G., Bhadange, S., Bhawna, F., Shewale, P., Dahiya, R., Aggarwal, A., Manju, F., & Arya, S. K. (2023). Phytoremediation of radioactive elements, possibilities and challenges: Special focus on agricultural aspects. *International Journal of Phytoremediation*, *25*(1), 1–8. https://doi.org/10.1080/15226514.2022.2043239

Sudarshan, S., Harikrishnan, S., RathiBhuvaneswari, G., Alamelu, V., Aanand, S., Rajasekar, A., & Govarthanan, M. (2023). Impact of textile dyes on human health and bioremediation of textile industry effluent using microorganisms: Current status and future prospects. *Journal of Applied Microbiology*, *134*(2), lxac064. https://doi.org/10.1093/jambio/lxac064

Sun, X.-L., Xiang, H., Xiong, H.-Q., Fang, Y.-C., & Wang, Y. (2023). Bioremediation of microplastics in freshwater environments: A systematic review of biofilm culture, degradation mechanisms, and analytical methods. *Science of the Total Environment*, *863*, 160953. https://doi.org/10.1016/j.scitotenv.2022.160953

Syed, Z., Sogani, M., Rajvanshi, J., & Sonu, K. (2023). Microbial biofilms for environmental bioremediation of heavy metals: A review. *Applied Biochemistry and Biotechnology*, *195*(6), 5693–5711. https://doi.org/10.1007/s12010-022-04276-x

Tambat, V. S., Patel, A. K., Chen, C.-W., Raj, T., Chang, J.-S., Singhania, R. R., & Dong, C.-D. (2023). A sustainable vanadium bioremediation strategy from aqueous media by two potential green microalgae. *Environmental Pollution*, *323*, 121247. https://doi.org/10.1016/j.envpol.2023.121247

Thacharodi, A., Hassan, S., Singh, T., Mandal, R., Chinnadurai, J., Khan, H. A., Hussain, M. A., Brindhadevi, K., & Pugazhendhi, A. (2023). Bioremediation of polycyclic aromatic hydrocarbons: An updated microbiological review. *Chemosphere*, *328*, 138498. https://doi.org/10.1016/j.chemosphere.2023.138498

Thakur, K., Kuthiala, T., Singh, G., Arya, S. K., Boonthai Iwai, C., Ravindran, B., ... Awasthi, M. K. (2023). An alternative approach towards nitrification and bioremediation of wastewater from aquaponics using biofilm-based bioreactors: A review. *Chemosphere*, *316*, 137849. https://doi.org/10.1016/j.chemosphere.2023.137849

Ullah Khan, A., Zahoor, M., Ur Rehman, M., Ikram, M., Zhu, D., Naveed Umar, M., Ullah, R., & Ali, E.A. (2023). Bioremediation of Azo Dye Brown 703 by *Pseudomonas aeruginosa*: An effective treatment technique for dye-polluted wastewater. *Microbiology Research*, *14*, 1049–1066. https://doi.org/10.3390/microbiolres14030070

Varabih, C. A., & Nofirman, N. (2023). The importance of seaweed as bioremediation natural agent. *Journal of Oceanography and Aquatic Science*, *1*(2). https://doi.org/10.56855/joane.v1i2.343

Vaksmaa, A., Guerrero-Cruz, S., Ghosh, P., Zeghal, E., Hernando-Morales, V., & Niemann, H. (2023). Role of fungi in bioremediation of emerging pollutants. *Frontiers in Marine Science*, *10*, 1070905. https://doi.org/10.3389/fmars.2023.1070905

Wang, Z., Wang, H., Nie, Q., Ding, Y., Lei, Z., Zhang, Z., Shimizu, K., & Yuan, T. (2023). Pb(II) bioremediation using fresh algal-bacterial aerobic granular sludge and its underlying mechanisms highlighting the role of extracellular polymeric substances. *Journal of Hazardous Materials*, *444*(Part B), 130452. https://doi.org/10.1016/j.jhazmat.2022.130452

Xu, G., Zhao, S., Liu, J., & He, J. (2023). Bioremediation of organohalide pollutants: Progress, microbial ecology, and emerging computational tools. *Current Opinion in Environmental Science & Health*, *32*, 100452. https://doi.org/10.1016/j.coesh.2023.100452

Xu, Q., & Wu, B. (2023). Recent progress on ex situ remediation technology and resource utilization for heavy metal contaminated sediment. *Toxics*, *11*(3), 207. https://doi.org/10.3390/toxics11030207

Zaib, M., Aryan, M., & Ullah, H. (2023). Innovative inorganic pollutant bioremediation approaches for industrial wastewater treatment: A review. *International Journal of Scientific Research and Engineering Development*, *6*(4), 1294. Available at www.ijsred.com

Zainab, R., Hasnain, M., Ali, F., Dias, D. A., El-Keblawy, A., & Abideen, Z. (2023). Exploring the bioremediation capability of petroleum-contaminated soils for enhanced environmental sustainability and minimization of ecotoxicological concerns. *Environmental Science and Pollution Research*, *30*, 104933–104957. https://doi.org/10.1007/s11356-023-29801-1

Zhang, B., Zhang, H., He, J., Zhou, S., Dong, H., Rinklebe, J., & Ok, Y. S. (2023). Vanadium in the environment: Biogeochemistry and bioremediation. *Environmental Science & Technology*, *57*(39), 14770–14786. https://doi.org/10.1021/acs.est.3c04508

Zhao, W., Xiong, W., Liu, Y., Guo, H., Wang, S., Chen, Y., Liu, R., Li, B., & Yang, C. (2023). Establishment of a halotolerant bioremediation platform from *Halomonascupida* using synthetic biology approaches. *Chemical Engineering Journal*, *473*, 145285. https://doi.org/10.1016/j.cej.2023.145285

Zhou, X., Zhang, S., Wang, R., An, Z., Sun, F., Shen, C., Lin, H., & Su, X. (2023). A novel strategy for enhancing bioremediation of polychlorinated biphenyl-contaminated soil with resuscitation promoting factor and resuscitated strain. *Journal of Hazardous Materials*, *447*, 130781. https://doi.org/10.1016/j.jhazmat.2023.130781

3 Water Resources Management Through Nature-Based Solutions

Rohit Sharma, Himanshu Mishra, Yashaswani Sharma, and Vinod Kumar

3.1 INTRODUCTION

In the face of increasing global water challenges, the need for innovative and sustainable management practices has never been more critical. Traditional approaches, often based on planned infrastructure, have proven insufficient to solve the difficult and multifaceted challenges of today's water challenges. This chapter examines the transformative potential of nature-based solutions (NBS) in water resource management, providing a paradigm shift toward more sustainable, cost-effective, and ecologically harmonious practices. The availability of sufficient and high-quality water is essential for sustainable development. Water, a key component of the planet's life support system, is essential to sustain all life forms and almost all human endeavors. Nature-based solutions harness the power of natural processes and ecosystems for water-related issues and challenges such as flood control, water quality improvement, and drought mitigation. These solutions include a variety of practices from restoring wetlands and riparian buffers to implementing green infrastructure in urban environments. By integrating these strategies simultaneously, communities can improve water security, biodiversity, and resilience to climate change. Implementing NBS is not just a technical choice, but a holistic approach that requires a deep understanding of ecological, social, and economic dynamics. This chapter discusses the principles of NBS, their various applications, and the advantages they offer over traditional methods. Through case studies and practical examples, we illustrate how NBS can be effectively designed, implemented, and managed to suffice the demands of several stakeholders, from rural farmers to urban planners. As we move into the 21st century, it is becoming increasingly clear that our survival and prosperity depend on our ability to live in harmony with nature. Adopting nature-based water resource management solutions is a crucial step in this direction as it promotes a sustainable future where human activities and natural systems thrive together. This chapter will help to deliver an in-depth outline and practical understanding of the integration of NBS into water management practices, encouraging readers to imagine and implement sustainable solutions in their contexts. Fresh water is a key resource and tool in many economic sectors and households, but its use often comes at the expense of natural ecosystems. Water management must change to address pressing issues and protect water ecosystems and their services while meeting the competing demands of cities, agriculture, industry, energy, and transportation (Table 3.1).

3.1.1 Importance of Water Resources Management

Water resources management is the strategic implementation, distribution, and regulation of water resources. It aims to optimize water use to fulfill present and future human needs while ensuring the sustainability of aquatic ecosystems. The multifaceted nature of water resources management includes several economic sectors, such as agriculture, industry, households, and environmental

DOI: 10.1201/9781003668961-3

TABLE 3.1
Key Elements of Water Resources Management and Their Importance

Key Element	Importance in Water Resources Management
Assessment and Monitoring	Provides data for informed decision-making, ensures sustainable resource use
Planning and Development	Establishes frameworks for long-term sustainability, infrastructure growth
Allocation and Regulation	Ensures fair distribution, prevents overuse, and pollution
Conservation and Restoration	Maintains ecological balance, protects biodiversity

Source: Author's compilation based on Gleick (2003), UN-Water (2020), Postel & Richter (2003), and WHO (2019).

protection. Water resources management involves a comprehensive framework that includes several key elements:

- Assessment and monitoring: Regular monitoring of water resources to collect data on quality, quantity, and distribution.
- Planning and development: Development of long-term plans for sustainable water use and development of infrastructure for storage, treatment, and distribution of water.
- Allocation and Regulation: Implement policies to efficiently allocate water between different users and enforce regulations to prevent overuse and pollution.
- Conservation and restoration: Promote water conservation practices and restore degraded water bodies to maintain ecological balance.

Effective management of water resources is crucial for several reasons:

- Sustainability: Due to global population growth and the effects of climate change, sustainable management ensures the availability of water resources for future generations. Proper management helps maintain a balance between water supply and demand (Gleick, 2003).
- Economic development: Water is an essential part of various economic activities, including agriculture, industry, and energy production. Effective management supports economic growth by ensuring reliable water supply to these sectors (UN-Water, 2020).
- Environmental protection: Healthy aquatic ecosystems are essential for biodiversity. Management practices that prevent pollution and overharvesting help protect the aquatic environment and the species that depend on it (Postel and Richter, 2003).
- Public Health: Access to clean and safe water is essential for human health. Proper management ensures that communities have access to drinking water, which reduces the incidence of waterborne diseases (WHO, 2019).
- Climate Resilience: Managing water resources is critical to adapting to climate change. By improving water resources, flood management, and drought mitigation measures, communities can better withstand climate impacts (IPCC, 2014).

3.1.2 Challenges in Water Resources Management

In spite of being crucial, the management of water resources faces several challenges:

- Water crisis: Many regions experience water scarcity due to excessive use, pollution, and climate change, which requires a more efficient use and management of available resources (Falkenmark and Rockström, 2004).

- Conflicting interests: The allocation of water resources often involves conflicting interests of different users, making it difficult to develop fair and sustainable management plans (Gleick, 2000).
- Infrastructure needs: The development of the facilities required to hold, treat, and deliver water requires significant investments, which can be an obstacle in case of resource constraints (World Bank, 2019).
- Policy and governance: Effective water management requires sound policies and governance structures. However, political and administrative challenges often prevent the implementation of comprehensive water management strategies (Rogers and Hall, 2003). Water resources management faces many challenges that complicate efforts to ensure sustainable and equitable water availability and use. Some of the most pressing challenges are:
- Water scarcity and over-abstraction: Many regions of the world are experiencing water scarcity due to excessive extraction of water resources for agricultural, industrial, and domestic needs. Such unsustainable use of water resources threatens the availability of fresh water in the long term (Gleick, 2014).
- Pollution: Contamination of water sources by industrial discharges, agricultural drainage, and raw sewage significantly degrades water quality. These pollutants, including heavy metals, pesticides, and disease-causing organisms, severely harm aquatic ecosystems, making them unsafe for human use (UNEP, 2016).
- Climate change: Climate change acts as a multiplier of water supply challenges, disrupting rainfall patterns, making droughts and floods more frequent and intense, and ultimately impacting the accessibility and distribution of freshwater. These changes require adaptive management strategies to mitigate and address climate impacts (IPCC, 2014).
- Infrastructure gaps: Many areas lack the necessary infrastructure to store, treat, and distribute water. Aging infrastructure in developed countries and inadequate infrastructure in developing countries hinder effective water supply and service delivery.
- Management and policy issues: Effective water management requires a sound management framework and policies. However, many regions struggle with fragmented governance, inadequate enforcement of regulations, and a lack of coherent water management methods. These problems complicate the coordination and implementation of sustainable water management (Rogers and Hall, 2003).
- Conflicting water use: Competition between agricultural, industrial, and domestic water use often leads to conflicts over water sharing. Balancing these competing demands is a significant challenge for water managers, especially in areas of water scarcity (Gleick, 2003) (Figure 3.1).

3.1.3 Traditional Approaches in Water Resources Management

Traditional methods of water resource management have been the basis for responding to water challenges. Some of the main traditional approaches are:

- Surface water management: Traditionally, water management has focused on the use of surface water through the construction of dams, reservoirs, and canals. These infrastructures store and deliver water utilized for diverse applications, such as crop nourishment, potable water provision, and hydroelectric power generation (May 2006).
- Groundwater management: Groundwater has been an important source of water in many regions, especially in arid and semi-arid regions. Traditional management practices include

FIGURE 3.1 Various challenges faced in water resource management.

drilling wells and installing pumps to extract groundwater for agricultural, industrial, and domestic use (Foster and Chilton, 2003).

- Irrigation systems: Traditional irrigation systems such as surface irrigation have been widely used to support agriculture. These systems involve channeling water from rivers or reservoirs to fields through canals, enabling crop production in water-scarce areas (Burt et al., 1997).
- Water harvesting: Traditional water harvesting techniques such as rainwater harvesting and construction of protective dams and reservoirs were used to collect rainwater and store it for later use. These methods are particularly valuable in regions with variable rainfall (Agarwal and Narain, 1997).
- Riparian rights and customs: In many regions, traditional water management was based on riparian rights and customs. These systems share water use rights based on land ownership adjacent to water bodies and established community practices (Meinzen-Dick and Pradhan, 2002).

3.1.4 Limitations of Traditional Approaches

Although traditional approaches have helped manage water resources, they have limitations to meet today's challenges:

- Environmental impacts: Massive construction endeavors like dams and water storage facilities can severely disrupt the natural environment, including habitat destruction and natural change in flow regimes and community migration (WCD, 2000).
- Efficiency and equity: Traditional irrigation systems often suffer from inefficiencies such as water loss due to percolation and evaporation. In addition, these systems may not distribute water fairly, leading to differences in water availability between users (Burt et al., 1997).

TABLE 3.2
Challenges Faced in Water Resource Management and Their Impacts with Relevant Examples

Challenge	Description	Impact	Examples
Water Scarcity	Unequal distribution of freshwater resources, coupled with increasing demand due to population growth and climate change.	Droughts, competition for water, limitations on food production, ecosystem degradation.	Colorado River Basin water shortages, conflict over Nile River water rights.
Water Quality Deterioration	Pollution from agriculture, industry, and untreated wastewater contaminates freshwater sources.	Public health risks, harm to aquatic ecosystems, limitations on water use.	Nitrates from agricultural runoff, industrial waste spills, and raw sewage discharge into rivers.
Climate Change	Rising temperatures, changes in precipitation patterns, and increased frequency of extreme weather events disrupt water cycles.	More frequent and severe droughts, floods, changes in river flow, and saltwater intrusion into coastal aquifers.	Glacier melt in the Himalayas impacting regional water supplies, increased flooding in Southeast Asia.
Infrastructure Deficiencies	Aging and inadequate water treatment and distribution systems lead to leaks, losses, and inefficiencies.	Water scarcity, wasted resources, and potential for contamination.	Leakage in urban water pipes, and limited access to clean water in rural areas.
Governance Issues	Lack of effective water policies, fragmented management structures, and limited enforcement mechanisms hinder sustainable use.	Overexploitation of resources, inequitable water allocation, and difficulty in responding to crises.	Ineffective water pricing structures, and lack of stakeholder participation in decision-making.
Social Equity	Unequal access to clean water and sanitation disproportionately impacts marginalized communities.	Health problems, economic hardship, and social conflict.	Water scarcity in developing countries, and lack of sanitation facilities in slums.

Source: Author's compilation based on Gleick (2000, 2003, 2014), UNEP (2016), IPCC (2014), and World Bank (2019).

- Adaptability: Traditional approaches may not have the necessary flexibility to adapt to changing conditions such as climate change and population growth. Modern water management requires adaptive strategies that can respond to dynamic and uncertain future scenarios (IPCC, 2014).

Water resources management faces major challenges in today's world, from water scarcity and pollution to climate change and governance. Although traditional approaches have provided basic strategies for managing water resources, they often do not respond to today's complex problems. To overcome these challenges and ensure the long-term availability and quality of water resources, a transition to integrated, adaptive, and sustainable water management practices is essential to overcome these challenges and ensure the long-term availability and quality of water resources (Table 3.2).

3.2 HISTORICAL PERSPECTIVE ON WATER MANAGEMENT

Water management has a long and varied history, reflecting the different ways in which human societies have interacted with water resources over time. Understanding the historical perspective of water management provides valuable information about how past practices have shaped today's approaches and challenges.

3.2.1 Ancient Civilizations and Early Water Supply

Water supply has been an important part of human civilization since ancient times. Early societies recognized the importance of water for agriculture, drinking, sanitation, and transportation, leading to the development of various water management technologies.

- Mesopotamia: Some of the earliest irrigation systems were developed in the cradle of civilization in the Tigris-Euphrates basin. The Sumerians built extensive networks of canals, dams, and reservoirs to control and distribute water for agriculture. These systems allowed them to grow plants in otherwise arid regions (Jacobsen & Adams, 1958).
- Egypt: The Nile was central to the development of ancient Egyptian civilization. The annual floods of the Nile deposited fertile mud on the floodplain, enriching the soil. The Egyptians harnessed this fertility through basin irrigation, building canals, dams, and reservoirs to control water flow and support agriculture. (Butzer, 1976).
- Indus Valley: The Indus Valley Civilization, located in what is now Pakistan and northwestern India, developed complex urban water supply systems. Cities such as Mohenjo-Daro and Harappa developed drainage systems, public baths, and reservoirs, emphasizing their importance in sanitation and water supply (Kenoyer, 1998).
- China: Ancient China developed extensive water supply systems to support agriculture and control floods. The construction of the Dujiangyan irrigation system in the 3rd century BC is an example of their engineering skills. This system diverted water from the Mini River to irrigate the Chengdu Plain, greatly increasing agricultural productivity (Needham, 1971).

3.2.2 Medieval and Early Modern Periods

In the Middle Ages and early modern times, the development of water supply continued thanks to technological innovations and the growing demands of the population.

- Golden Age of Islam: During the Golden Age of Islam (8th to 13th century), scientists and engineers made significant contributions to water management. They developed qanats (underground canals) to transport water in arid regions, built complex hydraulic structures, and improved irrigation techniques (Hill, 1996).
- Europe: In medieval Europe, water supply was closely related to agriculture and urbanization. The construction of water mills to grind grain and saw wood became more and more common. In addition, the development of complex systems of canals and ditches supported agriculture and facilitated trade and transportation (Hoffmann, 2014).
- Colonial period: European colonization brought new water management practices to different parts of the world. In the Americas, the Spanish adopted the irrigation techniques of the Iberian Peninsula to support agriculture in arid regions. Similarly, the British and Dutch built extensive canal systems in their colonies to promote trade and agriculture (Worster, 1985).

3.2.3 Modern Water Management

The Industrial Revolution and the consolidation of nation-states as the dominant political entity in the 19th and 20th centuries marked a new era in water supply. Advances in design, technology, and science have revolutionized the management of water resources.

- Large-scale infrastructure projects: The construction of large dams, reservoirs, and irrigation systems has become a characteristic of modern water management. Projects such as the Hoover Dam in the United States and the Aswan High Dam in Egypt are examples of the scope and ambition of these efforts. These infrastructures provided water for irrigation, hydropower, flood control, and urban water supply (McCully, 2001).

- Public health and sanitation: The development of modern sewage systems and the treatment of drinking water have greatly improved public health. The construction of sewage systems and sewage treatment plants helped control waterborne diseases and improve living conditions in cities (Tarr, 1996).
- Ecological Awareness: In the late 20th century, awareness of the environmental effects of water management practices grew. Driven by anxieties about environmental well-being, the richness of life on Earth, and the long-term viability of water resources, the integrated water resources management (IWRM) approach emerged.

These approaches emphasize the coordinated water resource development and management to take into account societal, financial, and ecological considerations (GWP, 2000). A historical perspective on water management reveals a rich tapestry of human ingenuity, adaptation, and innovation. From the irrigation systems of ancient Mesopotamia and Egypt to today's large infrastructure projects, water management practices have evolved to meet the changing needs of society. Understanding this history provides valuable lessons for today's water management and emphasizes the importance of sustainable and integrated approaches to solving today's complex challenges.

3.2.4 Emerging Challenges in Water Resources Management

Water resource management is becoming increasingly complex due to many emerging challenges. These challenges are caused by factors such as environmental shifts, demographic expansion, urban sprawl, contamination, and social disparities. Addressing these challenges requires innovative and integrated approaches to ensure a water system that meets future needs without compromising the environment.

- Climate change
  Climate change poses a major threat to water resources worldwide, affecting both water availability and quality.
  - Altered hydrological cycles: Climate change affects precipitation, leading to stronger and increasing occurrence of severe weather phenomena, including prolonged periods of aridity and inundations, throwing traditional water supply systems into disarray and increasing variability in water availability (IPCC, 2014).
  - Melting of glaciers: Accelerated melting of glaciers due to rising temperatures contributes to sea level rise and affects river flows, especially in areas that depend on glacial meltwater for freshwater (Huss and Hock, 2018).
- Population growth and urbanization
  Rapid population growth and urbanization are putting enormous pressure on water resources.
  - Increased demand: The growing population requires more water for domestic, agricultural, and industrial use. This increased demand often exceeds the capacity of existing water supply systems, leading to excessive withdrawal of groundwater and depletion of surface water sources (UN, 2018).
  - Urban Water Management: Urban areas face unique challenges in managing water resources, including the need for comprehensive water supply, wastewater, and stormwater management infrastructure. Rapid urbanization often outpaces infrastructure development, leading to inadequate water services and increased pollution (UN-Habitat, 2016).
- Water Pollution
  Water pollution remains a critical challenge that threatens water quality and endangers human health and ecosystems.
  - Agricultural runoff: The heavy application of fertilizers and pest control chemicals in agriculture causes nutrient flow into waterways, causing eutrophication and harmful algal blooms. These conditions degrade water quality and harm aquatic organisms (Carpenter et al., 1998).

  - Industrial discharges: Industrial processes discharge contaminants into waterways, including heavy metals, chemicals, and hazardous waste. These pollutants pose serious health risks and degrade ecosystems (Schwarzenbach et al., 2010).
  - Plastic pollution: More and more plastic waste, especially microplastics, enters water bodies. This pollution affects marine life and can easily make its way into the human food chain (Jambeck et al., 2015).
- Socio-economic differences
  Socio-economic differences exacerbate water supply problems and affect the availability and distribution of water resources.
  - Inequitable access: In many areas, marginalized and poor communities lack safe and reliable water supplies. This gap is often due to inadequate infrastructure, political marginalization, and financial barriers (WHO/UNICEF, 2019).
  - Price and distribution of water: Economic differences can affect the price and distribution of water, which can lead to an unequal distribution of water resources. Wealthier individuals and industries can ensure access to water at the expense of poorer communities (UN-Water, 2018).
- Administrative and institutional challenges
  Effective water management requires a solid administrative and institutional framework, which is often lacking.
  - Fragmented management: Water management tasks are often divided between multiple agencies and levels of government, resulting in fragmented and sometimes conflicting policies. Such fragmentation hinders the development and implementation of coherent water management strategies (OECD, 2015).
  - Corruption and abuse: Corruption and abuse of water resources undermine efforts to achieve sustainable and equitable water use. This problem is particularly prevalent in regions with weak governance structures and accountability mechanisms (Transparency International, 2008).
- Technology and information gaps
  Advances in technology and information management are essential for effective water resources management, but gaps still exist.
  - Monitoring and data collection: Precise and timely updated data on water availability, quality, and use is critical to making informed decisions. However, many regions lack sufficient monitoring systems and data management capacity (UNESCO, 2019).
  - Technology integration: Incorporating cutting-edge technologies like earth observation, spatial data analysis tools, and smart water control systems can optimize and streamline water management practices. However, adoption of such technologies is often limited by financial, technical, and institutional barriers (Gleick et al., 2011).

Meeting the emerging challenges of water resource management needs a diverse approach that combines scientific, technological, social, and political dimensions. Sustainable water management practices must adapt to changing conditions, be equitably distributed, and withstand future uncertainty. Cooperation at the local, national, and international levels is necessary for developing and implementing strategies that ensure the sustainable and equitable use of water resources for both present and upcoming generations.

3.3 PRINCIPLES AND TYPES OF NATURE-BASED SOLUTIONS

NBS use natural processes and services to solve social problems, including water management, climate change adaptation, and biodiversity conservation. These solutions are increasingly recognized for their potential to provide sustainable and cost-effective alternatives to traditional design methods. Understanding the principles and types of NBS is important for their successful implementation and integration into water resources management.

3.3.1 Principles of Nature-Based Solutions

NBS are guided by several core principles that ensure their effectiveness and sustainability:

- Sustainability and resilience: NBS enhances the strength and flexibility of ecosystems and communities by promoting sustainable management practices that maintain ecosystem health and function (Cohen-Shacham et al., 2016).
- Integration with natural processes: NBS works with, rather than against, natural processes. They harness the ability of natural systems to regulate water flows, sequester carbon, and support biodiversity (Eggermont et al., 2015).
- Multi-functionality: NBS provides multiple benefits simultaneously, addressing environmental, social, and economic objectives. For example, wetland restoration can improve water quality, enhance flood protection, and provide recreational opportunities (Maes & Jacobs, 2017).
- Local context and community involvement: Effective NBS are tailored to the specific ecological, cultural, and socio-economic contexts of the areas where they are implemented. Community involvement in planning and decision-making is crucial for the acceptance and success of NBS (Raymond, Frantzeskaki, et al., 2017).
- Adaptive management: NBS requires adaptive management approaches that allow for monitoring, learning, and adjusting practices based on outcomes and changing conditions. This flexibility is vital for responding to uncertainties and ensuring long-term success (IPBES, 2019) (Table 3.3).

3.3.2 Types of Nature-Based Solutions

NBS encompasses various strategies and practices, each tailored to specific environmental and social challenges. The main types of NBS in water resources management are:

- Restoration and creation of wetlands: Wetlands are natural water filters and buffers against floods and droughts. Restoring degraded wetlands or creating new ones can improve water quality, conserve floodwater and provide habitat for wildlife. For example: The restoration of the Danube floodplain in Europe improved flood protection, improved biodiversity, and increased recreational opportunities for local communities (Schneider et al., 2011).
- Riparian buffer zones: These are vegetated areas along waterways that help filter pollutants, stabilize shorelines, and create habitat corridors. Establishing or restoring these zones

TABLE 3.3
Challenges in Water Resources Management and Nature-Based Solutions

Challenges	Nature-Based Solutions
Water Scarcity	Rainwater harvesting, groundwater recharge through green infrastructure
Conflicting Interests	Stakeholder engagement, participatory approaches in decision-making
Infrastructure Needs	Natural flood management, wetland restoration for water storage
Policy and Governance	Integrated watershed management, adaptive governance frameworks

Source: Author's compilation based on Cohen-Shacham et al. (2016), Eggermont et al. (2015), Maes & Jacobs (2017), and Raymond, Frantzeskaki et al. (2017).

can significantly improve water quality and ecosystem health. Example: The Chesapeake Bay Program in the United States promotes riparian buffers to decrease nutrient washoff from fertile fields, elevate water quality, and support aquatic life (Palone & Todd, 1998).

- Green infrastructure: It includes practices such as landscaped rooftops, bioretention basins, and infiltration surfaces that control stormwater runoff and mitigate urban heat islands. These solutions mimic natural processes to collect, filter, and percolate rainwater. For example: New York's Green Infrastructure Plan combines green rooftops and biowalls to manage water, reduce sewer overflows, and improve green spaces (NYC Environmental Protection, 2010).
- Afforestation and reforestation: This implies the planting of trees in deforested or degraded areas and the creation of new forests. These practices increase water infiltration, reduce erosion, and sequester carbon. For example: China's Grain for Green program encourages farmers to convert farmland to forest, which reduces soil erosion and improves water body health (Liu et al., 2008).
- Agricultural Best Management Practices (BMPs): Agricultural BMPs include techniques such as cover cropping, conservation tillage, and agroforestry that improve soil health, reduce runoff, and improve water retention. These practices reconcile agricultural productivity with environmental sustainability. Example: The use of cover crops in the Mississippi River Basin reduces nutrient runoff and soil erosion, contributing to improved water quality in the Gulf of Mexico (Gentry et al., 2009).
- Management of marine and coastal ecosystems: NBS for marine and coastal areas includes restoration of mangroves, coral reefs, and sea grasses. These ecosystems protect the coast from erosion, support fisheries, and sequester carbon. Example: Restoration of mangroves in Vietnam's Mekong Delta protects the coast, increases fisheries, and sequesters carbon, benefiting local communities and biodiversity (Primavera, 2006) (Figure 3.2).

Nature-based solutions offer a promising path to sustainable water resource management by exploiting the inherent properties of natural systems. Following the principles of sustainable development, integration, multifunctionality, local context, and adaptive management, NBS can offer several advantages that meet environmental, social, and economic challenges. As such, they are an integral part of modern water management strategies.

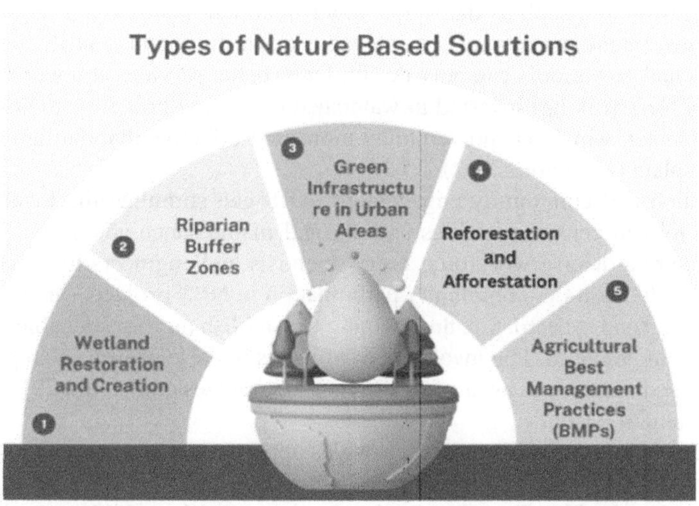

FIGURE 3.2 The figure lists the various types of nature-based solutions.

3.4 BENEFITS OF NATURE-BASED SOLUTIONS FOR WATER MANAGEMENT

NBS use natural processes to solve social problems and provide sustainable and universal benefits. In the area of water management, NBS includes different approaches such as wetland restoration, forest regeneration, and the creation of green infrastructures like bioswales and rain gardens. These solutions provide many environmental, economic, social, and cultural benefits, which are described:

- Environmental benefits
  - Increase biodiversity: NBS play a crucial role in increasing biodiversity. By restoring natural habitats such as wetlands, rivers, and forests, NBS creates environments where different species can thrive. For example, wetland restoration projects can provide critical habitat for birds, fish, and amphibians that are often lost to urbanized or agricultural landscapes (Dale and Beyeler, 2001). In addition, NBS can connect fragmented habitats, allowing for greater genetic diversity and healthier ecosystems.
  - Better water quality: NBS significantly improves water quality. For example, wetlands act as natural filters, trapping sediments and pollutants before they reach larger bodies of water (Mitsch and Gosselink, 2000). Riparian buffers, which are vegetated areas along watercourses, can reduce nutrient runoff from agricultural lands by preventing the eutrophication of lakes and rivers (Osborne and Kovacic, 1993). These natural processes eliminate the requirement for money-intensive water treatment plants and promote healthier aquatic ecosystems.
  - Climate resilience and adaptation: NBS improves climate resilience and adaptation by preventing extreme weather events. For example, reforestation and afforestation can reduce flood risk by increasing soil infiltration and stabilizing riverbanks (Bradshaw et al., 2007). In addition, urban green spaces such as parks and green roofs can lower temperatures in cities, preventing the urban heat island phenomenon and reducing energy consumption for cooling (Gill et al., 2007). These solutions not only help communities adapt to climate change but also reduce its impact through carbon sequestration.
- Economic benefits
  - Cost efficiency compared to gray infrastructure: One of the most important economic advantages of NBS is its cost efficiency compared to traditional gray infrastructure. Gray infrastructure such as dams and water treatment plants often require significant capital investment and ongoing maintenance costs. In contrast, NBS such as restored wetlands and green roofs can provide similar or better services at lower total costs. For example, New York has invested in watershed protection programs to ensure access to drinking water, which has proven to be more cost-effective than building a new water filtration plant (Appleton, 2002).
  - Job creation and community engagement: NBS can stimulate the local economy by creating jobs in conservation, restoration, and maintenance activities. These projects often require a diverse workforce, from scientists and engineers to local workers and volunteers. In addition, community participation in NBS projects can promote a sense of ownership and stewardship, thus helping to establish more sustainable and long-term management practices. The involvement of locals in these projects also promotes environmental education and awareness (Raymond, Frantzeskaki, et al., 2017).
- Social and cultural benefits
  - Improved recreational spaces: NBS provides opportunities to create recreational spaces that enhance the overall lifestyle of communities. Parks, green spaces, and restored natural areas provide places for outdoor activities such as hiking, cycling, and bird watching, which promote physical and mental well-being. These spaces also serve as community gathering places that promote social interaction and cohesion (Chiesura,

2004). In addition, access to natural areas has been associated with reduced stress and improved general health (Ulrich et al., 1991).
  - Cultural and aesthetic values: Incorporating NBS into water management can enhance cultural and aesthetic values. Natural landscapes often have cultural significance and provide local communities with a sense of place and identity. By preserving and restoring these landscapes, NBS helps preserve cultural heritage and traditions (Plieninger et al., 2015). In addition, aesthetically pleasing environments created through NBS can increase property values and attract tourism, further stimulating the local economy.

3.5 CHALLENGES AND LIMITATIONS OF NATURE-BASED SOLUTIONS

Although NBS offer several advantages in water resource management, their implementation is not without challenges and limitations. These can be broadly categorized into implementation issues, socioeconomic constraints, and monitoring and evaluation issues.

- Implementation challenges
  - Technical and scientific challenges: Implementation of NBS often requires overcoming various technical and scientific challenges. These include understanding complex ecological processes, predicting outcomes of interventions, and integrating NBS into existing gray infrastructure systems. For example, restoration of a wetland to its natural state requires detailed hydrological and ecological assessments to ensure that it functions as intended and provides the desired benefits (Mitsch & Gosselink, 2000). In addition, the lack of standardized methodologies and metrics for designing and evaluating NBS may hinder their widespread adoption (Raymond, Frantzeskaki et al., 2017).
  - Political and regulatory barriers. Political and regulatory barriers can significantly impede the implementation of NBS. In many areas, current policies and regulations are heavily biased toward traditional gray infrastructure solutions, making it difficult to include NBS in planning and development processes. Regulatory frameworks may lack provisions to recognize and assess the benefits of NBS, which reduces their use (Kabisch et al., 2016). In addition, obtaining permits and approvals for NBS projects can be time-consuming and difficult, especially when many stakeholders and jurisdictions are involved.
- Socio-economic constraints
  - Finance and financial constraints: Ensuring adequate funding for NBS projects is a common challenge. Although NBS can be cost-effective in the long term, the initial investment required for planning, design, and implementation can be significant (Dunwiddie et al., 2009). In addition, traditional financing mechanisms and financing models are often not suitable to support NBS, as they prefer gray infrastructure projects with more predictable results and established financial flows. This may limit the availability of NBS's financial resources and prevent their deployment.
  - Community acceptance and participation: Community acceptance and participation are critical to the success of NBS but can be difficult to achieve. Public perceptions of NBS can vary, with some communities viewing them as less reliable or efficient compared to traditional gray infrastructure (Andersson et al., 2014). Building trust and understanding through education and outreach is essential to gaining community support. In addition, involving local communities in the implementation of NBS can promote a sense of ownership and ensure that solutions meet local needs and preferences (Raymond, Frantzeskaki, et al., 2017).
- Monitoring and evaluation
  - Measuring effectiveness and long-term effects: Measuring the effectiveness and long-term effects of NBS are major challenges. Unlike gray infrastructure, which can often

be assessed using simple technical metrics, NBS performance encompasses complex ecological, social, and economic variables (Bennett et al., 2015). It is critical to develop robust monitoring and evaluation frameworks that can capture these multifaceted outcomes. This includes establishing baselines, setting clear goals, and using appropriate indicators to track progress over time.

– Adaptive management and continuous improvement: Adaptive management is a key component of a successful NBS, requiring continuous monitoring, evaluation, and adjustment of management strategies based on observed results (Holling, 1978). However, implementing adaptive management can be difficult due to the need for continuous data collection, analysis, and stakeholder engagement. Moreover, the flexibility of management practices and the willingness to revise approaches based on new knowledge can be energy and resource draining in the long run (Allen and Gunderson, 2011).

3.6 POLICY AND GOVERNANCE FOR PROMOTING NBS

Effective policy and management are critical to the successful implementation and sustainability of NBS for water resource management. This section examines the various political frameworks and strategies, management and institutional arrangements, and financial mechanisms and incentives that support the promotion of NBS.

3.6.1 Policy Frameworks and Strategies

- International Policies and Agreements Supporting NBS International policies and agreements play a key role in promoting NBS. Global frameworks such as the UN Sustainable Development Goals (SDGs) specifically recognize the importance of NBS in achieving sustainable water management (UN, 2015). SDG 6 mainly aims to ensure access to water and sanitation for all and sustainable management, while SDG 15 focuses on the protection, restoration, and promotion of sustainable use of the country's ecosystems.
  In addition, the Paris Agreement on Climate Change recognizes the role of ecosystems in climate adaptation and mitigation and encourages parties to integrate NBS into their national strategies (UNFCCC, 2015). The Convention on Biological Diversity (CBD) also promotes ecosystem-based approaches consistent with the principles of the NBS (CBD, 2010).
- National and regional policy initiatives: At national and regional levels, policy initiatives are increasingly incorporating NBS into water management strategies. For example, the European Union's Green Infrastructure Strategy aims to promote natural processes and green infrastructure across Europe, supporting biodiversity and ecosystem services (European Commission, 2013). National policies such as the Wetlands Policy in Canada and the Natural Flood Management Strategy in Great Britain emphasize the restoration and conservation of natural habitats to ensure the sustainable management of water resources (Environment and Climate Change Canada, 1991; Scottish Government, 2015) (Table 3.4).

3.6.2 Governance and Institutional Arrangements

- Multi-stakeholder participation and collaboration: Effective NBS management requires multi-stakeholder participation and collaboration, including government bodies, nongovernmental organizations (NGOs), the private sector, and various local bodies. Multi-stakeholder platforms help in sharing information, resources, and responsibility, ensuring that a variety of opinions are taken into consideration in decision-making processes

TABLE 3.4
Various Policies and Strategies Adopted by Some Developed Nations to Manage Their Water Resources Using Nature-Based Systems

Policy/Strategy	Region/Country	Year	Focus
Green Infrastructure Strategy	European Union	2013	Enhance natural processes and green infrastructure, support biodiversity and ecosystem services
Wetlands Policy	Canada	1991	Restoration and preservation of natural habitats for sustainable water resource management
Natural Flood Management Strategy	United Kingdom	2015	Restoration and preservation of natural habitats for sustainable water resource management

Source: Author's compilation based on European Commission (2013), Environment and Climate Change Canada (1991), and Scottish Government (2015).

(Frantzeskaki et al., 2017). Collaborative management models such as river basin management councils and river basin organizations are examples of successful multi-stakeholder participation. These units bring together stakeholders from different sectors and jurisdiction to jointly manage water resources and implement NBS (Margerum and Robinson, 2015).

- Role of governments, NGOs, and the private sector: The government plays a key role in creating the operating environments for NBS through policy development, regulation, and funding. They can establish standards and guidelines, provide technical support, and facilitate stakeholder participation. For example, the US Environmental Protection Agency (EPA) supports green infrastructure initiatives with grants and technical assistance (EPA, 2014).

NGOs contribute to NBS by promoting political change, raising public awareness, and implementing on-the-ground projects. Organizations such as The Nature Conservancy and WWF have promoted NBS worldwide (WWF, 2016).

The private sector also plays a key role in investing in NBS, particularly in sectors such as real estate, insurance, and water utilities. Public-private partnerships can leverage the strengths of both sectors to increase NBS implementation (Browder et al., 2019).

3.6.3. Funding Mechanisms and Incentives

1. Public and private sources of funding: Ensuring adequate funding is essential for the implementation and maintenance of NBS. National financial sources are the state budget, environmental subsidies, and international development aid. For example, the LIFE program of the European Union finances environmental and climate action projects, including projects focused on NBS (European Commission, 2020).
   Private funding can come from corporate investments, philanthropy, and impact investment funds. Innovative financing mechanisms such as green bonds and environmental bonds are increasingly used to finance NBS projects (OECD, 2018).
2. Incentives for NBS adoption and retention: Incentive structures can encourage NBS adoption and retention. Governments can offer tax incentives, grants, and subsidies to landowners and developers who implement NBS. For example, the United States offers tax incentives for the installation of green roofs, which encourages the integration of NBS in urban areas (EPA, 2014).

Payment for Ecosystem Services rewards landowners for managing their land in ways that provide ecological benefits. These systems can be particularly effective in promoting NBS in agricultural landscapes (Engel et al., 2008).

3.7 FUTURE DIRECTIONS AND RESEARCH NEEDS

The future of NBS in water resources management is promising but requires innovative approaches, addressing research gaps, and creating a vision for widespread adoption and sustainability. This section discusses these aspects in detail.

- Innovative approaches and technologies
  - Emerging technologies supporting NBS: The integration of new technologies can improve the implementation and effectiveness of NBS. Central to NBS monitoring and management are remote sensing and geographic information systems (GIS), which enable real-time data collection and analysis of large-scale environmental changes (Lang et al., 2021). Drones and unmanned aerial vehicles provide high-resolution images to assess ecosystem health and identify areas for intervention (Getzin et al., 2012). In addition, advances in biotechnology and ecological engineering are creating new opportunities for NBS. Technologies such as phytoremediation, which uses plants to remove pollutants from soil and water, and the development of hybrid systems combining green and gray infrastructure expand the toolbox of sustainable water management (Vymazal, 2010; Brown et al., 2020).
  - Combining traditional knowledge and modern science: Combining traditional ecological knowledge with modern scientific approaches can improve the design and implementation of NBS. Indigenous and local communities often have valuable knowledge about ecosystem functions and sustainable practices developed over centuries (Berkes, 2012). Combining this wisdom with modern scientific methods can lead to more culturally appropriate and effective NBS. For example, traditional water harvesting techniques in arid regions can be revived and integrated into modern hydrological models to improve water security (Reij et al., 2005).
- Disadvantages and opportunities of research
  - Areas in need of further research: Despite progress, there are still several research gaps in the field of NBS. One of the key areas is the need for more empirical evidence on the long-term performance and cost-effectiveness of NBS compared to traditional gray infrastructure (Kabisch et al., 2016). Understanding the socio-economic impacts of NBS, especially in terms of equity and inclusion, is also crucial (Geneletti and Zardo, 2016). Research is also needed to determine the effectiveness of NBS in different climates and extreme weather conditions. This includes investigating how the NBS can be optimized to withstand the effects of climate change, such as the increase and intensity of floods and droughts (Narayan et al., 2016).
  - Multidisciplinary science: Promoting NBS requires interdisciplinary research that spans the gap between ecology, engineering, social sciences, and economics. Collaborative research across multiple sectors, such as agriculture, urban planning, and public health, can lead to holistic solutions that address different aspects of water management (Raymond, Frantzeskaki, et al., 2017). Cross-sector partnerships between universities, government agencies, NGOs, and the private sector can facilitate the exchange of information and the co-creation of innovative solutions. Such collaboration can also help align research directions with policy needs and practical applications and ensure that research findings are relevant and actionable (Pahl-Wostl et al., 2011).

- Future vision
  - Possible future scenarios with widespread NBS: Seeing future scenarios with widespread adoption of NBS requires envisioning cities and landscapes where natural processes are an integral part of water management. Urban areas could have extensive networks of green roofs, permeable pavements, and restored waterways to manage stormwater, reduce heat islands, and increase biodiversity (Benedict and McMahon, 2006). Rural landscapes can enhance agroforestry, wetland restoration, and the use of riparian buffers that improve water quality, sequester carbon, and support agricultural productivity (Bentrup, 2008). Communities around the world could benefit from sustainable water systems that provide clean water, protect against flooding, and sustain livelihoods even in the face of climate change.
  - Long-term sustainability and goals: Achieving long-term sustainability through NBS requires setting ambitious goals and making continuous improvements. This includes the development of comprehensive policies that link NBS to all aspects of water management, from urban planning to agriculture and disaster risk reduction (Eggermont et al., 2015). Monitoring and adaptive management frameworks should be established to ensure the effective contribution of NBS to the achievement of the SDGs and resilience to changing conditions (Allen and Gunderson, 2011). Training and building initiatives are also important to equip stakeholders with the knowledge and skills needed to implement and sustain NBS (Pahl-Wostl et al., 2007).

3.8 CONCLUSION

- Summary of key points

  NBS offer a transformative approach to water resource management by harnessing natural processes to solve water-related problems. In this chapter, we highlighted the many environmental, economic, social, and cultural benefits of NBS, including increased biodiversity, improved water quality, climate resilience, cost-effectiveness, job creation, improved recreational opportunities, and cultural values (Eggermont et al., 2015; Browder et al., 2019).

  Successful case studies from around the world demonstrate the effectiveness of NBS in various contexts. For example, wetland restoration in the Mississippi Basin has greatly improved water quality, reduced flood risk, and provided critical habitat for wildlife (Mitsch and Gosselink, 2000). Likewise, the implementation of green infrastructure in urban areas, such as New York's Green Infrastructure Plan, effectively managed stormwater, reduced urban heat islands, and created green spaces for residents (NYC Environmental Protection, 2010).

  These examples highlight the potential of NBS to provide sustainable and versatile water supply solutions. Best practices emerging from these cases highlight the importance of integrated planning, stakeholder engagement, and adaptive management to ensure the long-term success and sustainability of NBS (Frantzeskaki et al., 2017).
- Call to action

  Realizing the full potential of NBS requires joint action by policymakers, actors, and researchers. Policymakers are encouraged to create enabling environments for NBS through supportive policies, regulations, and funding mechanisms. Integrating NBS into national and regional water management strategies can encourage widespread adoption and ensure that natural solutions are preferred over traditional infrastructure (European Commission, 2013).

  Participants, including urban planners, engineers, and environmental managers, must use innovative approaches and techniques that improve the design and implementation of NBS. The use of new tools such as remote sensing, GIS, and biotechnology can improve the efficiency and scalability of NBS projects (Lang et al., 2021).

Researchers have an important role in developing the knowledge base and correcting existing gaps. Interdisciplinary and multidisciplinary studies are needed to collect reliable evidence on the effectiveness, cost-effectiveness, and socioeconomic impact of NBS. Research collaboration can also promote the joint creation of scientifically justified, socially acceptable, and economically viable solutions (Frantzeskaki et al., 2017).

Continued collaboration and innovation are paramount to the success of NBS. By fostering partnerships between governments, NGOs, the private sector, and local communities, we can ensure that diverse perspectives and resources are used to achieve common goals. This joint effort contributes to the long-term resilience and sustainability of our watersheds, benefiting both people and nature (Kabisch et al., 2016).

REFERENCES

Agarwal, A., & Narain, S. (1997). *Dying Wisdom: Rise, Fall, and Potential of India's Traditional Water Harvesting Systems*. Centre for Science and Environment.

Allen, C. R., & Gunderson, L. H. (2011). Pathology and Failure in the Design and Implementation of Adaptive Management. *Journal of Environmental Management, 92*(5), 1379–1384.

Andersson, E., Tengö, M., McPhearson, T., & Kremer, P. (2014). Cultural Ecosystem Services as a Gateway for Improving Urban Sustainability. *Ecosystem Services, 12*, 165–168.

Appleton, A. F. (2002). *How New York City Used an Ecosystem Services Strategy Carried Out Through an Urban-Rural Partnership to Preserve the Pristine Quality of its Drinking Water and Save Billions of Dollars*. Forest Trends.

Benedict, M. A., & McMahon, E. T. (2006). *Green Infrastructure: Linking Landscapes and Communities*. Island Press.

Bennett, E. M., Peterson, G. D., & Gordon, L. J. (2015). Understanding Relationships Among Multiple Ecosystem Services. *Ecology Letters, 12*(12), 1394–1404.

Bentrup, G. (2008). Conservation Buffers: Design Guidelines for Buffers, Corridors, and Greenways. General Technical Report SRS-109. Asheville, NC: Department of Agriculture, Forest Service, Southern Research Station.

Berkes, F. (2012). *Sacred Ecology*. Routledge.

Bradshaw, C. J., Sodhi, N. S., & Brook, B. W. (2007). Tropical Turmoil: A Biodiversity Tragedy in Progress. *Frontiers in Ecology and the Environment, 5*(2), 79–87.

Browder, G., Ozment, S., Rehberger Bescos, I., Gartner, T., & Lange, G. M. (2019). *Integrating Green and Gray: Creating Next Generation Infrastructure*. World Bank and World Resources Institute.

Brown, R. R., Rogers, B. C., & Werbeloff, L. (2020). Advancing the Uptake of Nature-based Solutions. *Urban Water Journal, 17*(6), 497–507.

Burt, C. M., Clemmens, A. J., Strelkoff, T. S., Solomon, K. H., Bliesner, R. D., Hardy, L. A., Howell, T. A., ASCE Members, & Eisenhauer, D. E. (1997). Irrigation Performance Measures: Efficiency and Uniformity. *Journal of Irrigation and Drainage Engineering, 123*(6), 423–442.

Butzer, K. W. (1976). *Early Hydraulic Civilization in Egypt: A Study in Cultural Ecology*. University of Chicago Press.

Carpenter, S. R., Caraco, N. F., Correll, D. L., Howarth, R. W., Sharpley, A. N., & Smith, V. H. (1998). Nonpoint Pollution of Surface Waters with Phosphorus and Nitrogen. *Ecological Applications, 8*(3), 559–568.

CBD. (2010). *Strategic Plan for Biodiversity 2011–2020 and the Aichi Targets*. Convention on Biological Diversity.

Chiesura, A. (2004). The Role of Urban Parks for the Sustainable City. *Landscape and Urban Planning, 68*(1), 129–138.

Cohen-Shacham, E., Walters, G., Janzen, C., & Maginnis, S. (2016). *Nature-based Solutions to Address Global Societal Challenges*. IUCN.

Dale, V. H., & Beyeler, S. C. (2001). Challenges in the Development and Use of Ecological Indicators. *Ecological Indicators, 1*(1), 3–10.

Dunwiddie, P. W., Hall, S. A., Ingraham, M. W., Bakker, J. D., Nelson, K. S., Fuller, R. N., & Gray, E. (2009). Rethinking Conservation Practice: Applying the Science of Ecological Restoration in the Twenty-First Century. *Frontiers in Ecology and the Environment, 7*(10), 527–534.

Eggermont, H., Balian, E., Azevedo, J. M. N., Beumer, V., Brodin, T., Claudet, J., Fady, B., Grube, M., Keune, H., Lamarque, P., Reuter, K., Smith, M., van Ham, C., Weisser, W. W., & Le Roux, X. (2015). Nature-based Solutions: New Influence for Environmental Management and Research in Europe. *GAIA-Ecological Perspectives for Science and Society*, *24*(4), 243–248.

Engel, S., Pagiola, S., & Wunder, S. (2008). Designing Payments for Environmental Services in Theory and Practice: An Overview of the Issues. *Ecological Economics*, *65*(4), 663–674.

Environment and Climate Change Canada. (1991). Federal Policy on Wetland Conservation.

EPA. (2014). *Enhancing Sustainable Communities with Green Infrastructure*. U.S. Environmental Protection Agency.

European Commission. (2013). Green Infrastructure (GI) — Enhancing Europe's Natural Capital.

European Commission. (2020). LIFE Programme.

Falkenmark, M., & Rockström, J. (2004). *Balancing Water for Humans and Nature: The New Approach in Ecohydrology*. Earthscan.

Foster, S., & Chilton, J. (2003). Groundwater: The Processes and Global Significance of Aquifer Degradation. *Philosophical Transactions of the Royal Society of London. Series B: Biological Sciences*, *358*(1440), 1957–1972.

Frantzeskaki, N., Kabisch, N., & McPhearson, T. (2017). Advancing Urban Environmental Governance: Understanding Theories, Practices and Processes Shaping Urban Sustainability and Resilience. *Environmental Science & Policy*, *77*, 1–6.

Geneletti, D., & Zardo, L. (2016). Ecosystem-based Adaptation in Cities: An Analysis of European Urban Climate Adaptation Plans. *Land Use Policy*, *50*, 38–47.

Gentry, L. E., David, M. B., Smith, K. M., & Kovacic, D. A. (2009). Nitrogen Cycling and Tile Drainage Nitrate Loss in a Corn/Soybean Watershed. *Agriculture, Ecosystems & Environment*, *131*(3–4), 255–260.

Getzin, S., Wiegand, K., & Schöning, I. (2012). Assessing Biodiversity in Forests Using Very High-resolution Images and Unmanned Aerial Vehicles. *Methods in Ecology and Evolution*, 3(2), 397–404.

Gill, S. E., Handley, J. F., Ennos, A. R., & Pauleit, S. (2007). Adapting Cities for Climate Change: The Role of the Green Infrastructure. *Built Environment*, *33*(1), 115–133.

Gleick, P. H. (2000). *The World's Water 2000–2001: The Biennial Report on Freshwater Resources*. Island Press.

Gleick, P. H. (2003). Water Use. *Annual Review of Environment and Resources*, *28*(1), 275–314.

Gleick, P. H. (2014). *The World's Water Volume 8: The Biennial Report on Freshwater Resources*. Island Press.

Gleick, P. H., Cooley, H., & Christian-Smith, J. (2011). *The World's Water Volume 7: The Biennial Report on Freshwater Resources*. Island Press.

GWP (Global Water Partnership). (2000). Integrated Water Resources Management. TAC Background Papers No. 4.

Hill, D. R. (1996). *Islamic Science and Engineering*. Edinburgh University Press.

Hoffmann, R. C. (2014). *An Environmental History of Medieval Europe*. Cambridge University Press.

Holling, C. S. (1978). *Adaptive Environmental Assessment and Management*. Wiley.

Huss, M., & Hock, R. (2018). Global-scale Hydrological Response to Future Glacier Mass Loss. *Nature Climate Change*, *8*(2), 135–140.

IPBES. (2019). *Global Assessment Report on Biodiversity and Ecosystem Services*. Intergovernmental Science-Policy Platform on Biodiversity and Ecosystem Services.

IPCC. (2014). *Climate Change 2014: Impacts, Adaptation, and Vulnerability*. Cambridge University Press.

Jacobsen, T., & Adams, R. M. (1958). Salt and Silt in Ancient Mesopotamian Agriculture. *Science*, *128*(3334), 1251–1258.

Jambeck, J. R., Geyer, R., Wilcox, C., Siegler, T. R., Perryman, M., Andrady, A., Narayan, R., & Law, K. L. (2015). Plastic Waste Inputs from Land into the Ocean. *Science*, *347*(6223), 768–771.

Kabisch, N., Frantzeskaki, N., Pauleit, S., Naumann, S., Davis, M., Artmann, M., Haase, D., Knapp, S., Korn, H., Stadler, J., Zaunberger, K., & Bonn, A. (2016). Nature-based Solutions to Climate Change Mitigation and Adaptation in Urban Areas: Perspectives on Indicators, Knowledge Gaps, Barriers, and Opportunities for Action. *Ecology and Society*, *21*(2), 39.

Kenoyer, J. M. (1998). *Ancient Cities of the Indus Valley Civilization*. Oxford University Press.

Lang, M. W., McCarty, G. W., Oesterling, R., & Yeo, I. Y. (2021). Topographic Metrics for Improved Mapping of Forested Wetlands. *Remote Sensing*, *13*(2), 325.

Liu, J., Li, S., Ouyang, Z., Tam, C., & Chen, X. (2008). Ecological and Socioeconomic Effects of China's Policies for Ecosystem Services. *Proceedings of the National Academy of Sciences*, *105*(28), 9477–9482.

Maes, J., & Jacobs, S. (2017). Nature-based Solutions for Europe's Sustainable Development. *Conservation Letters*, *10*(1), 121–124.

Margerum, R. D., & Robinson, C. J. (2015). *The Challenges of Collaboration in Environmental Governance: Barriers and Responses*. Edward Elgar Publishing.

McCully, P. (2001). *Silenced Rivers: The Ecology and Politics of Large Dams*. Zed Books.

Meinzen-Dick, R., & Pradhan, R. (2002). Legal Pluralism and Dynamic Property Rights. CAPRi Working Paper No. 22.

Mitsch, W. J., & Gosselink, J. G. (2000). The Value of Wetlands: The Importance of Scale and Landscape Setting. *Ecological Economics*, *35*(1), 25–33.

Narayan, S., Beck, M. W., Wilson, P., Thomas, C. J., Guerrero, A., Shepard, C. C., Reguero, B. G., Franco, G., Ingram, C. J., & Trespalacios, D. (2016). *Coastal Wetlands and Flood Damage Reduction: Using Risk Industry-based Models to Assess Natural Defenses in the Northeastern USA*. Lloyd's Tercentenary Research Foundation.

Needham, J. (1971). *Science and Civilisation in China, Volume 4: Physics and Physical Technology, Part 3: Civil Engineering and Nautics*. Cambridge University Press.

NYC Environmental Protection. (2010). *New York City Green Infrastructure Plan: A Sustainable Strategy for Clean Waterways*. City of New York.

OECD. (2015). *Water Resources Governance in Brazil*. OECD Publishing.

OECD. (2018). Financing Water: Investing in Sustainable Growth.

Osborne, L. L., & Kovacic, D. A. (1993). Riparian Vegetated Buffer Strips in Water-quality Restoration and Stream Management. *Freshwater Biology*, *29*(2), 243–258.

Pahl-Wostl, C., Jeffrey, P., Isendahl, N., & Brugnach, M. (2011). Maturing the New Water Management Paradigm: Progressing from Aspiration to Practice. *Water Resources Management*, *25*(3), 837–856.

Pahl-Wostl, C., Möltgen, J., Sendzimir, J., & Kabat, P. (2007). Managing Change Toward Adaptive Water Management Through Social Learning. *Ecology and Society*, *12*(2), 30.

Palone, R. S., & Todd, A. H. (1998). *Chesapeake Bay Riparian Handbook: A Guide for Establishing and Maintaining Riparian Forest Buffers*. USDA Forest Service.

Plieninger, T., Dijks, S., Oteros-Rozas, E., & Bieling, C. (2015). Assessing, Mapping, and Quantifying Cultural Ecosystem Services at the Community Level. *Land Use Policy*, *33*, 118–129.

Postel, S., & Richter, B. (2003). *Rivers for Life: Managing Water for People and Nature*. Island Press.

Primavera, J. H. (2006). Overcoming the Impacts of Aquaculture on the Coastal Zone. *Ocean & Coastal Management*, *49*(9–10), 531–545.

Raymond, C. M., Frantzeskaki, N., Kabisch, N., Berry, P., Breil, M., Razvan Popescu, V., Geneletti, D., & Calfapietra, C. (2017). A Framework for Assessing and Implementing the Co-benefits of Nature-based Solutions in Urban Areas. *Environmental Science & Policy*, *77*, 15–24.

Reij, C., Tappan, G., & Smale, M. (2005). *Agroenvironmental Transformation in the Sahel: Another Kind of "Green Revolution"*. International Food Policy Research Institute.

Rogers, P., & Hall, A. W. (2003). Effective Water Governance. Global Water Partnership Technical Committee (TEC) Background Papers, No. 7.

Schneider, J., Meyer, V., Scheuer, S., & Haase, D. (2011). The Potential of Restored Floodplains for Sustainable Flood Risk Management – A Case Study on the River Elbe, Germany. *Ecohydrology & Hydrobiology*, *11*(2), 123–135.

Schwarzenbach, R. P., Egli, T., Hofstetter, T. B., von Gunten, U., & Wehrli, B. (2010). Global Water Pollution and Human Health. *Annual Review of Environment and Resources*, *35*, 109–136.

Scottish Government. (2015). Natural Flood Management Handbook.

Tarr, J. A. (1996). *The Search for the Ultimate Sink: Urban Pollution in Historical Perspective*. University of Akron Press.

Transparency International. (2008). *Global Corruption Report 2008: Corruption in the Water Sector*. Cambridge University Press.

Ulrich, R. S., Simons, R. F., Losito, B. D., Fiorito, E., Miles, M. A., & Zelson, M. (1991). Stress Recovery During Exposure to Natural and Urban Environments. *Journal of Environmental Psychology*, *11*(3), 201–230.

UN. (2015). Transforming Our World: The 2030 Agenda for Sustainable Development.

UNEP. (2016). *A Snapshot of the World's Water Quality: Towards a Global Assessment*. United Nations Environment Programme.

UNESCO. (2019). *United Nations World Water Development Report 2019: Leaving No One Behind*. UNESCO.

UNFCCC. (2015). Paris Agreement.

UN-Habitat. (2016). *World Cities Report 2016: Urbanization and Development – Emerging Futures*. UN-Habitat.

United Nations. (2018). *World Population Prospects: The 2018 Revision*. United Nations.

UN-Water. (2018). *Sustainable Development Goal 6 Synthesis Report 2018 on Water and Sanitation*. UN-Water.
UN-Water. (2020). *United Nations World Water Development Report 2020: Water and Climate Change*. UNESCO.
Vymazal, J. (2010). Constructed Wetlands for Wastewater Treatment. *Water*, *2*(3), 530–549.
WCD. (2000). *Dams and Development: A New Framework for Decision-making*. World Commission on Dams.
WHO. (2019). Drinking-Water. Retrieved from https://www.who.int/news-room/fact-sheets/detail/drinking-water
WHO/UNICEF. (2019). *Progress on Household Drinking Water, Sanitation and Hygiene 2000–2017: Special Focus on Inequalities*. UNICEF and World Health Organization.
World Bank. (2019). World Bank Group Water Strategy. Retrieved from https://www.worldbank.org/en/topic/water/overview
Worster, D. (1985). *Rivers of Empire: Water, Aridity, and the Growth of the American West*. Oxford University Press.
WWF. (2016). Nature-based Solutions for Climate Change Adaptation and Disaster Risk Reduction.

4 Nano-Biomaterials Used for the Elimination of Harmful Contaminants

Abdelmjid Bouazizi, Mounia Baidou, Latifa EL Bardiy, Doha El Machtani Idrissi, Mohamed Douma, Mohammed Zeriouh, Najib Tijani, and Mohamed Ouammou

4.1 INTRODUCTION

Water, an indispensable resource for life, undergoes intricate changes in its physical, chemical, and biological properties when exposed to a myriad of toxic substances. These pollutants, originating from industrial, pesticide, domestic, and agricultural sources, not only jeopardize the delicate ecological balance but also pose grave threats to human health. With water being vital for drinking, irrigation, cleaning, and food production, its escalating global contamination, primarily fueled by the discharge of industrial waste laden with heavy metals and toxins, has emerged as a critical concern. The repercussions of water pollution transcend environmental issues, manifesting in a spectrum of health complications such as liver problems, gastrointestinal disorders, nervous system complications, kidney and renal issues, and blood diseases. The urgency to combat this burgeoning global water crisis has prompted a comprehensive reevaluation of international water resource policies. To counteract the contamination inflicted by heavy metals, diverse removal methods, including adsorption (El-Kordy et al., 2022), ion exchange, electro-dialysis, and membrane filtration (Bouazizi et al., 2016; Breida et al., 2018), have been developed. An innovative and efficient approach involves the utilization of biomaterials, sourced from living and nonliving entities like algae, fungi, bacteria, and agricultural by-products, employing biosorption for wastewater treatment. These biomaterials not only offer an environmentally friendly solution but also prove to be economically viable for pollutant removal. The issue extends beyond heavy metals, as pesticides, employed in agriculture to control pests, significantly contribute to water pollution (El Abbadi et al., 2024). The excessive and often unnecessary use of these chemical compounds, reaching approximately 2 million tons annually, heightens toxicity levels in water bodies. While nonpersistent pesticides break down relatively easily and are less harmful, the widespread use of synthetic halogenated pesticides globally perpetuates water contamination. Agricultural waste and industrial effluents are identified as prime culprits in water contamination, altering water acidity and inducing bioaccumulation in the food chain. The persistence of certain pesticide residues poses serious health risks due to their genotoxic, neurotoxic, and mutagenic effects. These challenging-to-degrade substances accumulate in nature, driven by their high molecular mass, low polarity, and lipophilic nature. Biomaterials, comprising both naturally occurring and synthetic materials, play a pivotal role in wastewater treatment. Notably, cellulose, a versatile biomaterial, finds application in wastewater filtration membranes, exhibiting effectiveness in adsorbing heavy metals. The coagulation and flocculation properties of biomaterials assist in removing metals released from pesticides, contributing to the decontamination of wastewater. In regions like Pakistan, where pesticides are extensively used for agriculture, the uncontrolled release of pesticides and heavy metals into the environment underscores the imperative for effective degradation methods. Microorganisms and biomaterials derived from them, including bacteria and

DOI: 10.1201/9781003668961-4

fungi, have shown promise in pesticide degradation; fungi, leveraging their mycelia and extracellular enzymes, prove particularly effective compared to bacteria. Moreover, plant species known as macrophytes play a vital role in phytoremediation, proficiently removing both inorganic and organic contaminants from water. The pressing need to address water pollution demands comprehensive strategies that integrate various approaches, including nano-biomaterials, ensuring the protection of human health and the environment. Transitioning to a different realm, biopolymer-based biomaterials have emerged as pivotal entities in the biomedical field due to their inherent biodegradability, biocompatibility, nontoxicity, and nonallergenic properties. Focused on proteins, polysaccharides, and polyhydroxyalkanoates, this chapter delves into the realm of polysaccharides, specifically chitosan, alginate, and cellulose. These biopolymers, renowned for their ability to interact seamlessly with biological systems without causing cytotoxicity, offer a promising avenue for replacing or enhancing various bodily tissues or functions. Despite their advantages, nano-biopolymer-based nano-biomaterials grapple with challenges such as insufficient mechanical properties and stability in physiological environments. Commonly employed in various biomedical applications, including scaffolds, films, fibers, hydrogels, and nanoparticles, these materials often undergo crosslinking to enhance mechanical properties, albeit with associated drawbacks. Specific applications, such as wound dressing, involve the incorporation of silver nanoparticles for improved wound healing and antimicrobial activity. This chapter provides a detailed exploration of the preparation methods for chitosan, alginate, and cellulose-based nano-biomaterials, offering insights into their application in the biomedical realm. Figure 4.1 depicts a schematic diagram illustrating the pathway of chemical mutagens in the environment. Nitrosamines, a notable category of mutagens found in tobacco, can permeate the human body via processed meats and fish. This process occurs through the interaction

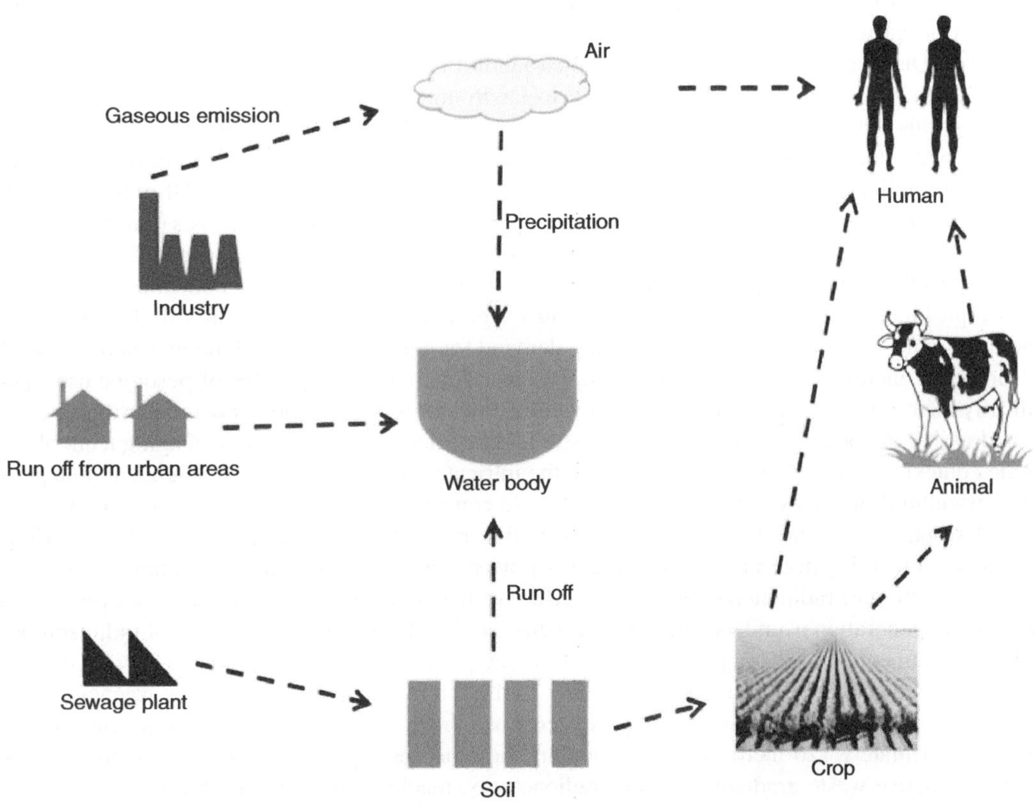

FIGURE 4.1 Schematic diagram illustrating the pathway of chemical mutagens in the environment.

of amines naturally present in food with nitrites added as preservatives (Jakszyn & González, 2006). The primary contributors to soil pollution encompass various sources, including agricultural wastes, industrial by-products, household and municipal waste, and urbanization-related debris. These waste categories contain a range of trace elements such as arsenic (As), cadmium (Cd), chromium (Cr), copper (Cu), mercury (Hg), lead (Pb), manganese (Mn), nickel (Ni), zinc (Zn), and radionuclides, as well as organic contaminants like hydrocarbons. Furthermore, the impact of these pollutants on the environment and human health is substantial. Soil pollution can lead to reduced crop yields, water contamination, and adverse effects on ecosystems. Additionally, exposure to trace elements and organic contaminants poses risks to human health, potentially causing various health problems. Addressing this global issue requires concerted efforts in waste management, sustainable practices, and the development of environmentally friendly technologies. It is crucial to raise awareness and implement policies that promote responsible waste disposal and reduce the generation of harmful pollutants. This chapter explores the application of innovative nano-biomaterials and composites for the removal of harmful pollutants from water. The chapter provides a detailed exploration of nano-biomaterial synthesis and it reviews their use in water treatment, focusing on the adsorptive removal of pollutants like dyes, heavy metal ions, and toxic organic compounds. The synergistic integration of adsorption, photocatalysis, and membrane processes with various nano-biomaterials is thoroughly examined for efficient contaminant removal.

4.2 CONTAMINANTS IN WATER ENVIRONMENTS

4.2.1 Origins of Contaminants

In the realm of mining waste, Turkish Statistical Institute (TURKSTAT) data reveals that 811 million tons of waste were generated by mining operations, with overburden and tailing constituting 99% of it. Disposal methods include controlled landfill sites, backfilling, and reclamation (Ghosh et al., 2022). India, a major coal producer, faces environmental challenges due to opencast mining practices; this leads to mining waste containing harmful elements like chemicals, heavy metals, radionuclides, and polluted water, contributing to environmental degradation. Additionally, global plastic production has dramatically increased from 2 million tons in 1950 to 360 million tons in 2021. The plastic sector's significant growth raises specific concerns about the costs of recycling and processing both pre- and post-consumer plastic waste. It is critical to note that this rapid expansion emphasizes the importance of long-term solutions to reduce the environmental impact of global plastic proliferation (Liu et al., 2021). The third segment delves into agricultural chemicals and chemical mutagens, highlighting the growing demand for agrochemicals and the widespread use of pesticides. Concerns have been raised about the environmental consequences of pesticide use, specifically their impact on air, water, soil, and overall ecosystems. This section underscores the urgency of addressing the potential ecological risks associated with artificially accelerated aging technology in the context of microplastics. As we explore the intricate interactions between chemical mutagens and agricultural practices, it becomes imperative to consider the broader implications for environmental sustainability and the delicate balance within ecosystems (Shahwar et al., 2023). Various operations, including nuclear medicine, nuclear power generation, and uranium mining, contribute to the generation of radionuclide waste. The effective management of nuclear waste not associated with power generation is underscored, with a focus on identifying potential sources of radioisotopes in water treatment plants. Compared to other toxic industrial wastes, nuclear waste is not exceptionally hazardous or difficult to handle. According to the International Atomic Energy Agency, out of the 370,000 metric tons of heavy metal produced since the inception of civil nuclear power production, approximately 120 metric tons have undergone reprocessing. Moreover, the remaining high-level radioactive waste gradually loses its radioactivity, reaching the radiation level of the original ore within approximately 9,000 years—a significantly shorter period compared to the much longer 300,000 years often cited for certain isotopes. Recently, both the radiological protection and radioactive waste management communities have exhibited a heightened interest in effectively addressing non–power-related nuclear waste (Berthiaume, 2023). Another focus is on polycyclic aromatic hydrocarbons (PAHs), naturally present in coal, crude oil, and gasoline. Anthropogenic sources include various industrial activities, and PAHs can enter the environment through both natural processes and human activities, such as the contamination of soils. Motor vehicles are significant emitters of polyarenes, releasing them in substantial quantities, and PAHs can infiltrate soils when contaminated water is used for irrigation. These PAHs, organic compounds with 2–7 benzene rings, pose health risks, causing respiratory issues and an elevated cancer risk. Major sources include food, industrial pollution, and car exhaust. Entry points into the body include direct contact, seafood, inhalation, and contaminated water. A review of 655 articles highlighted PAHs' association with respiratory diseases and various cancers. Mitigating PAH emissions from activities like cooking and car exhaust is crucial for reducing these health risks (Mosallaei et al., 2023). The majority of historical emissions of perfluorocarboxylates (PFCAs) into the environment arise from the manufacturing of fluoropolymers, with about 67% originating from ammonium perfluorooctanoate. The global estimated production of PFCAs is approximately 4400–8000 tons (Nilsson et al., 2013). In conclusion, these diverse environmental challenges underscore the critical need for sustainable and responsible practices across various industries. From mining waste and plastic proliferation to agricultural chemicals and nuclear waste, each issue demands careful consideration and concerted efforts to mitigate their impact on the environment. Addressing these challenges requires a holistic approach that involves regulatory measures, technological innovations, and public awareness to achieve long-term environmental sustainability.

4.2.2 Categories of Contaminants

Limited freshwater, polluted by industrial waste discharge, poses a severe global challenge. Technology-driven industrialization is the primary culprit, introducing harmful contaminants into water sources. Contaminated water leads to widespread diseases, claiming numerous lives annually and adversely impacting both humans and ecosystems. Marine life suffers from sea pollution due to improper waste disposal. Effluents contain inorganic pollutants with toxic metal ions and organic materials like dyes and pesticides, posing threats to the environment. Eliminating these contaminants is crucial for safeguarding both organisms and the ecosystem. Additionally, the presence of contaminants such as colorants, heavy metals, pharmaceutical drugs, and pesticides/insecticides further exacerbates the environmental crisis. It is imperative to address and mitigate these specific pollutants to ensure the preservation of our precious water resources and the well-being of future generations. Join the collective effort in promoting sustainable practices and responsible industrial behavior to protect our water bodies from further deterioration. Together, let us strive for a cleaner, healthier environment (Saxena et al., 2020).

Colorants, also known as dyes, are produced either naturally or synthetically for various applications such as painting, printing, leather, paper, and more. Each year, over seven types of dyes are manufactured in various industries, with approximately 100 tons being discharged into sewage annually. These dyes constitute a hazardous category of organic contaminants that pose risks to both water and human health. Specifically, cationic dyes used in the dyeing of nylon, wool, acrylic, and silk are considered toxic, leading to skin irritation, allergic dermatitis, mutations, and even cancers. Anionic dyes, which carry negative charges, and non-ionic dyes, mainly employed in coloring plastics, also present potential threats to human health and the environment (Thakur et al., 2024). The scientific community bears the responsibility of developing effective methods for removing these dyes from sewage due to their detrimental effects on water quality and aquatic life. Cationic dyes, such as methylene blue, methylene violet, rhodamine 6G, malachite green, and basal yellow, are not only toxic and mutagenic but also carcinogenic. Their abundance in seawater obstructs light penetration, adversely affecting the photosynthesis process of aquatic plants. Similarly, azoic dyes are deemed toxic due to the presence of toxic metal ions and nonbiodegradable aromatic amines. On the other hand, the anionic dyes, characterized by negative charges, encompass various color classes such as triphenylmethane, anthraquinone, azoic, and nitro. Direct and reactive dyes, like direct red 23, Congo red, reactive black 5, and methyl orange, are commonly used, and their application depends on their structure and binding capacity. Finally, non-ionic dyes, particularly dispersant dyes used mainly for dying plastics, polyamide, and polyester, can also induce allergies. Thus, the imperative to treat colored effluents before their discharge into water is crucial for preserving water quality and overall ecosystem health (Ataei et al., 2024) (Figure 4.2).

This section discussed the presence of heavy metals in the environment, focusing on their natural occurrence as well as the impact of human activities on the geochemical cycle. Agricultural practices, coal-related industries, mining, and metallurgy are all sources of excessive heavy metal release, including lead, mercury, chromium, copper, cadmium, nickel, zinc, and iron (Lau & Le, 2023). Heavy metals have an atomic density greater than 5 g/cm^3 and are classified into three groups on the periodic table: the lanthanide series, the actinide series, and some metallic elements from group P. These metals are typically found in groups 3–16 starting in period 4. Heavy metal ions have been released into the environment more frequently as industries and human activities have evolved, particularly in developing countries. Contaminants are found in wastewater from a variety of industrial processes, such as mining, chemical production, plating, paper manufacturing, battery production, and animal waste. International organizations such as the FAO, EPA, and WHO regularly monitor the effects of these metals on human and animal health. If heavy metal concentrations such as nickel, arsenic, zinc, chromium, mercury, lead, selenium, and cadmium exceed WHO permissible limits, it can cause acute and chronic intoxication, causing damage to various bodily systems such as the central nervous system, cardiovascular and digestive systems, liver, kidneys, lungs, and endocrine system (Zhang et al., 2023). The harmful consequences associated with heavy metal exposure within the general populace result from diverse forms of contact, spanning from brief to prolonged or chronic interactions. The detrimental effects of metal toxicity can manifest in various ways, including but not limited to gastrointestinal and renal impairment, neurological disorders, cancer,

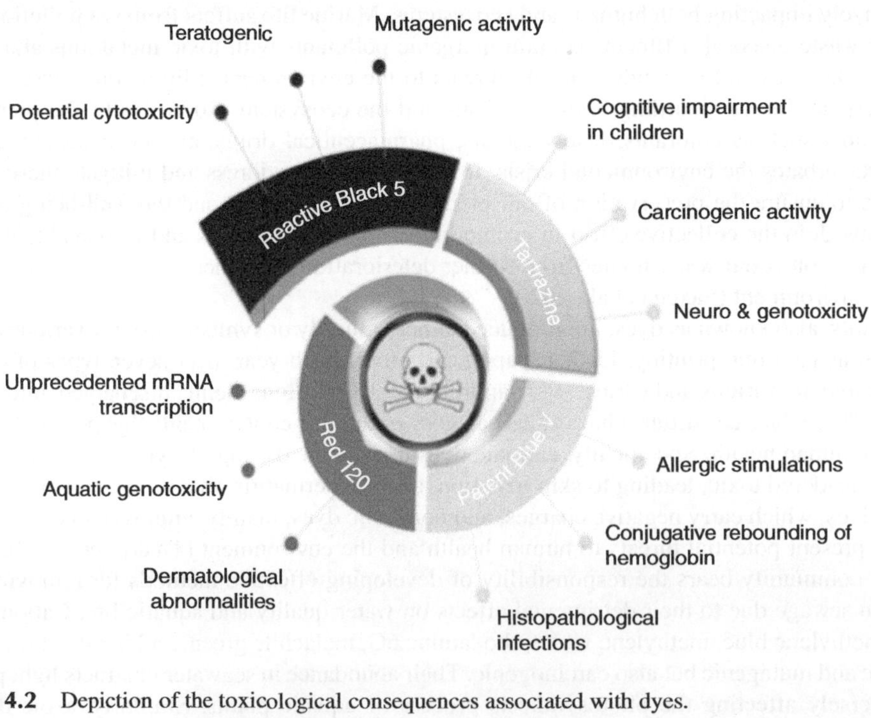

FIGURE 4.2 Depiction of the toxicological consequences associated with dyes.

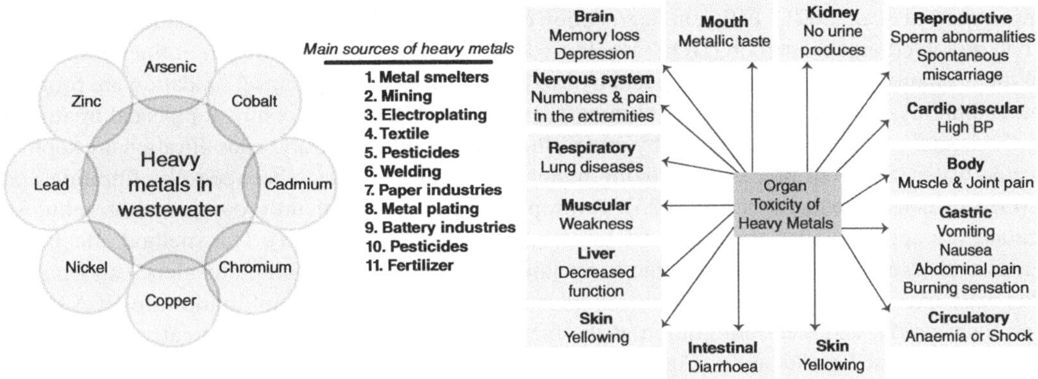

FIGURE 4.3 Diverse heavy metals, their primary origins and the adverse impact.

congenital abnormalities, skin lesions, vascular damage, and complications with the immune system. Figure 4.3. depicts the adverse impact of heavy metals.

4.3 UTILIZATION METHODS FOR ELIMINATING WATER CONTAMINANTS

The current water treatment facilities, designed to efficiently manage pollutants and maximize the benefits of pollution charges, have been constructed in compliance with existing regulations. However, due to the highly polluted nature of sewage, the purification process is more intricate compared to treating regular water. Safe reintroduction of sewage into the environment necessitates thorough treatment. Discharging contaminated water into aquatic environments leads to a significant reduction in the mass density of pollutants, either through decomposition or binding to organic substances or sediments along the banks. This results in decreased downstream concentrations. Nevertheless, the presence of these chemicals in drinking water at concentrations ranging from micrograms per liter to nanograms per liter suggests their ability to evade conventional water treatment processes. When it comes to purifying water, various treatment methods have been developed, each grounded in different assumptions and applications (Patel et al., 2021). Typically, these methods can be categorized into three primary domains: biological, chemical, and physical techniques.

4.3.1 Physical and Chemical Treatment Methods

Physical and chemical treatment methods involve the use of both chemicals and reagents to eliminate contaminants from aqueous media. While these methods exhibit high efficiency in pollutant removal, the associated treatment costs and energy consumption often surpass those of biological techniques. Additionally, the challenge of chemical sludge disposal is prevalent in certain regions, and advanced oxidation processes may lead to the generation of secondary pollutants with potential risks to human health. The electrocoagulation technique, requiring a substantial amount of electricity, can further escalate the overall process costs. On the other hand, chemical treatments such as coagulation-flocculation have proven ineffective in removing pharmaceuticals and endocrine-disrupting compounds (EDCs). Metal salt coagulants like aluminum sulfate ($Al_2(SO_4)_3$) and ferric sulfate ($Fe_2\,(SO_4)\,_3$) failed to eliminate specific elements like carbadox, sulfadimethoxine, trimethoprim, carbamazepine, diclofenac, ketoprofen, and ibuprofen. However, studies indicate that sorption techniques using advanced oxidation and activated carbon, including ozone and UV methods, have shown success in rejecting several pharmaceuticals and EDCs (Bassyouni et al., 2023). Oxidation emerges as a viable elimination method, particularly with chlorine or ozone usage. Nevertheless, the proper combination is crucial, as these substances may react and generate by-products with unclear impacts (Rout et al., 2021). Efficient degradation of EDCs and PPCPs through oxidation is observed when dissolved organic carbon (DOC) levels are low, highlighting the significant influence of DOC on the ozonation method. Both ozone and hydroxyl radicals generated during oxidation are promising and highly reactive strategies for eliminating EDCs and PPCPs. In contrast, physical treatment methods such as membrane filtration by microfiltration, ultrafiltration, and nanofiltration are sophisticated water treatment processes utilizing membranes with pore sizes between ultrafiltration and reverse osmosis (Bouazizi et al., 2023). Hydrophilic membranes, mainly composed of cellulose acetate, are applied under pressures of 5 to 35 bars (Rakcho et al., 2023). This method effectively removes charged and uncharged organic compounds, especially those with multivalency and molecular mass exceeding 300 g/mol. Unlike simpler filtration processes, nanofiltration involves diffusion, convection, and electromigration at the membrane, making it a more intricate and effective technique for water purification (Figure 4.6(a)) (Palaniyappan & Beerappa, 2024). Adsorption and ion exchange separate pollutants from water without inducing chemical or biological changes. Among these, adsorption processes have gained recent attention as a suitable and straightforward technique. However, it's important to note that physical treatments merely transfer pollutants from one medium to another without destroying or degrading them. The adsorption mechanism and isotherm are classified into physical and chemical adsorption based on the forces involved. Physical adsorption relies on van der Waals forces or electrostatic attraction, while chemical adsorption involves chemical bonds. Practical applications often involve a dual mechanism, combining both physical and chemical adsorption due to the porous structure and functional groups of the adsorbent (Figure 4.4(a)).

4.3.2 Biological Treatments

Biological treatments, encompassing both aerobic and anaerobic processes, employ diverse microbial communities such as algae, fungi, and bacteria to break down organic substances in the presence or absence of oxygen (Donzella et al., 2023). Widely applied in conventional wastewater treatment plants, these bioprocesses treat influent wastewater and waste activated sludge with low operational costs and satisfactory performance, sans the use of special chemicals. Although combining aerobic and anaerobic methods is considered effective for eliminating organic substances and nutrients, challenges like lengthy start-up periods and sensitivity to factors like pH, temperature, and nutrients may limit their applicability (Ríos et al., 2024). In their research, Johnson and their team revealed that biological trickling filters and activated sludge can rapidly convert aqueous organic molecules into easily separable biomass. However, not all constituents, such as steroid estrogens, undergo complete dissipation or transformation into biomass, even with advanced technologies. Biological treatment, despite its benefits, removes only a fraction of emerging contaminants, particularly polar pollutants, from the final effluent. Studies emphasize the impact of effluent treatment on the fate of target components in aquatic environments, with factors like pH influencing the removal efficiency of acidic drugs (Rathi et al., 2021). Acidic conditions are found to enhance extraction through adsorption rather than biodegradation in the activated sludge method. Despite the less than 10% expulsion of synthetic and natural estrogens through biological degradation, the majority of constituents remain soluble in polluted water. The research also indicates that extended sludge retention time and hydraulic retention time are crucial for efficient estrogen removal, with specific time-dependent design operational characteristics affecting water treatment plant size and other considerations. Moreover, nitrification plays a key role in the efficiency of biological treatment and estrogen removal, with increased biological diversity and favorable growing conditions facilitating conversion and chemical removal. However, it was observed that nitrifying sludges do not degrade ethinylestradiol, and its breakdown is influenced by factors like pH, dissolved oxygen, and temperature in biological treatment.

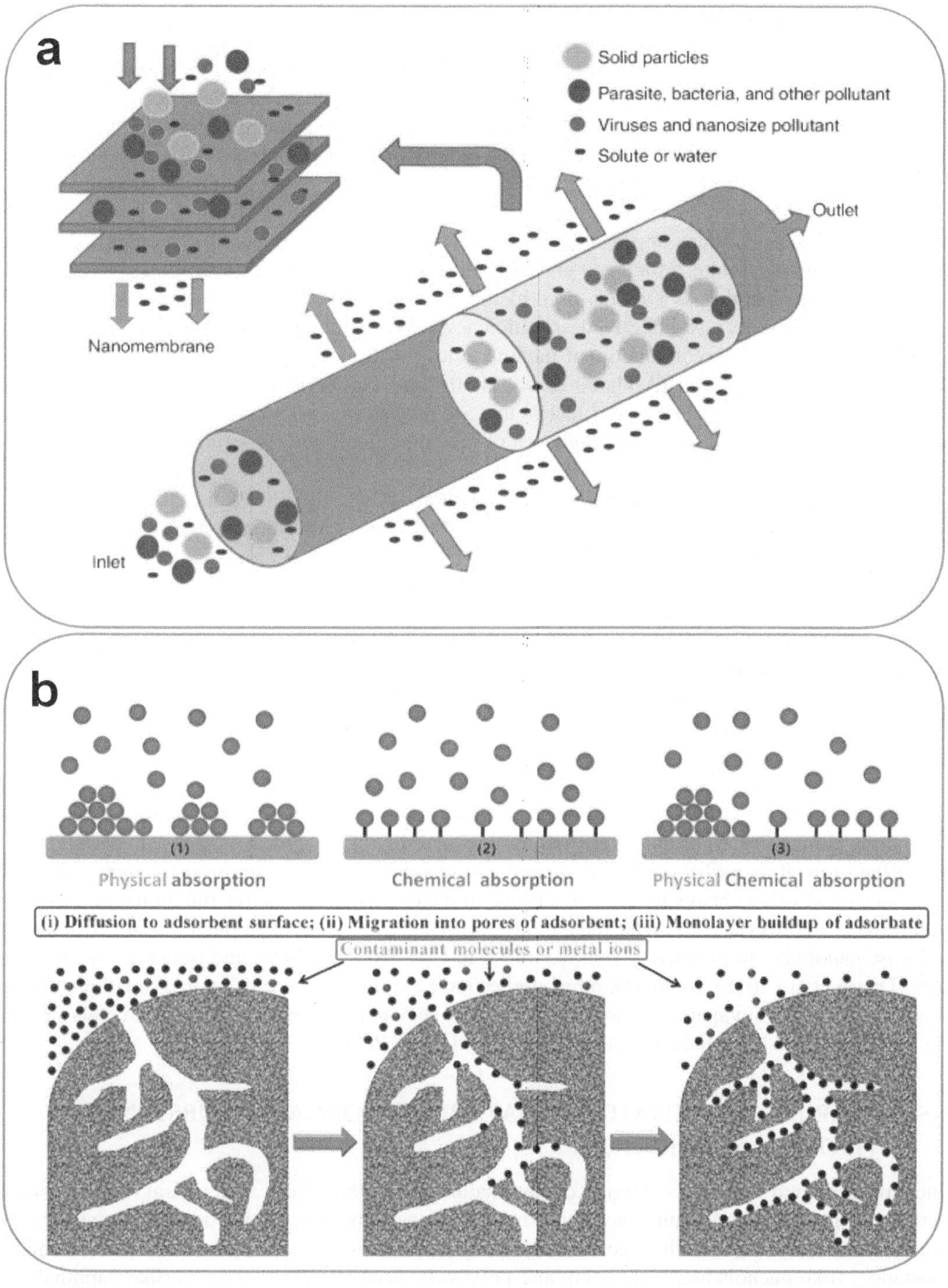

FIGURE 4.4 Integrated Approach: (a) Advanced Nanofiltration process for water contaminant removal and (b) Adsorption Mechanisms on Porous Materials.

(Li & Yu, 2024)

4.4 EXTRACTING HEAVY METAL IONS FROM POLLUTED WATER

Heavy metals, sources of toxic pollution with chronic consequences for the environment and human health, mainly come from agricultural pollutants, industrial effluents, and urban stormwater runoff. Water pollution by these metals is a persistent and irreversible environmental problem, contributing significantly to the long-term deterioration in the quality of water resources. Numerous techniques, including membrane filtration, coagulation-flocculation, chemical precipitation, photocatalysis, and adsorption, are dedicated to the removal of these inorganic pollutants from various wastewater sources. Nevertheless, these approaches have shortcomings such as costly expenses, increased energy requirements, limited efficiency under specific conditions, and the generation of toxic waste.

Adsorption is emerging as a widely used method to effectively remove heavy metals from aqueous solutions, due to its cost-effectiveness, efficiency, and operational flexibility compared to alternatives. Recently, the use of functional nanoparticles, such as SiO_2 (Heidari et al., 2009), TiO_2 (Razzaz et al., 2016), Al_2O_3 (Afkhami et al., 2010), and Fe_3O_4 (Kumari et al., 2015), has been explored as sorbents to extract ions of metals that are present in wastewater. However, their limited practical application results from the high cost of raw ingredients and the stringent preparation conditions associated with nanoparticles.

Some porous sorbents, such as zeolites (Velarde et al., 2023), modified kaolinite (Bhattacharyya & Gupta, 2008), activated carbon (Kobya et al., 2005), and chitosan/graphene nanocomposites (AL-Salman et al., 2023), have shown their potential in heavy metal removal. Adsorption mechanisms, including size and distribution of pore and specific surface area, play an essential role in this process. Aerogels, with their extremely high porosity, minimal density, large specific surface area, and increased capacity for adsorption, are also attracting particular interest. Studies such as those carried out by the efficiency of ethylenediamine-modified calcium alginate (ECAA) aerogels in clearing Pb^{2+} and Cu^{2+} ions from aqueous solutions have been shown by Huang and Wang (2018). The same heavy metals (Pb^{2+}, Cu^{2+}, and Cd^{2+}) were successfully adsorbed using an aerogel composite consisting of calcium alginate and graphene oxide (mp-CA/GO).

Furthermore, bionanomaterials, favored for their environmental friendliness, are increasingly used in wastewater treatment. Biopolymers such as cellulose, chitosan, and keratin stand out as fascinating bionanomaterials thanks to their biodegradability and nontoxicity. These materials, notably in the form of nanofibers, nanofillers, and nanocomposites, are recognized for their enhanced efficiency in adsorbing heavy metals. This is due to the existence of reactive functional groups such as carboxyls, hydroxyls, carbonyls, and amines. Abu-Danso et al., for example, used sulfur-modified cellulose nanofibers to effectively remove Pb^{2+} and Cd^{2+} from synthetic and industrial wastewater (Abu-Danso et al., 2018). Similarly, other studies have explored the use of hydroxyapatite/chitosan nanofibers and hydroxyapatite-incorporated cellulose acetate in the removal of heavy metals such as Pb^{2+}, Fe^{3+}, Co, and Ni (Aliabadi et al., 2014; Hamad et al., 2020).

4.5 CLEARING WASTEWATER OF HAZARDOUS ORGANIC SUBSTANCES THROUGH ADSORPTION

The environmental impacts of organic contaminants, particularly from agricultural run-off, industrial discharges, and other human activities, have been well documented for decades. A wide range of organic substances, including pesticides and hydrocarbons (like petroleum and oils), aromatic substances (like phenols and biphenyls), and polycyclic aromatic hydrocarbons (like naphthalene, phenanthrene, and pyrene), are frequently found in aquatic environments. These persistent pollutants are nonbiodegradable and highly toxic, discharging in significant quantities into marine ecosystems, where they accumulate in the tissues of animals and plants.

Numerous physical, chemical, and biological techniques have been used to treat organic contaminants in wastewater (Anifah et al., 2022). One effective method involves adsorption, a physicochemical process using adsorbents to remove contaminants from fat-containing wastewater. Organic contaminants such as oil and grease are known for their harmful environmental impact, particularly due to the risk of clogging sewers and deteriorating sanitary conditions. To remedy this problem, researchers have explored the effectiveness of adsorption, a technique aimed at reducing the concentration of these contaminants in wastewater. Adsorption offers the advantages of rapid kinetics and high removal efficiency, making a significant contribution to the overall management of organic pollutants in water treatment (Anifah et al., 2022). Several studies have confirmed the effectiveness of adsorption in the removal of organic pollutants, using low-cost adsorbents with enhanced binding capacities, aimed at improving the adsorption efficiency of common wastewater pollutants. Significant research efforts have been made to exploit cost-effective adsorbents in wastewater treatment. These include organic adsorbents and natural clays such as modified kaolinite, modified zeolite, and modified bentonite. In addition, agricultural waste and industrial by-products are also used as adsorbents in this context. Furthermore, the strategic use of nanocomposites as adsorbents represents a significant advance in the quest for increased efficiency in the treatment of contaminated water (Inbaraj et al., 2021). Baskaran Stephen et al. employed a magnetic-activated carbon nanocomposite constructed from residual green tea leaves (MNPs-GTAC) to investigate the adsorption of four priority polycyclic aromatic hydrocarbons (PAHs). Kinetic results demonstrated rapid elimination, reaching 80% in just 5 minutes for PAHs (Inbaraj et al., 2021). Additionally, aerogels have a crucial role in removing organic contaminants from the environment. However, using alpha-cellulose-based aerogels, Toorani et al. (2023) investigated the adsorption of hydrocarbon-based contaminants found in oily wastewater. While water's adsorption capacity decreased from 30.02 to 0.09 g/g, crude oil's increased significantly from 15.43 to 25.53 g/g. The findings of this study underline the particular effectiveness of hydrophobically modified alpha-cellulose aerogel for the selective adsorption of hydrocarbons from wastewater. Biochar from coffee waste and activated carbon are also used as adsorbents to remove the pesticide 2,4-dichlorophenylacetic (2,4-D) and monocyclic aromatic compounds such as aniline, phenol, and pyridine from water (Li et al., 2020; Almahri et al., 2023).

4.6 ENHANCED NANO-BIOMATERIALS FOR ECO-FRIENDLY POLLUTION REMEDIATION

4.6.1 Affordable Biomaterials

This section delves into some of the key biomaterials integral to this discussion. Affordable biomaterials sourced from readily available natural resources, often at minimal or no cost, capable of selectively extracting pollutants from wastewater and binding them to their surfaces, are referred to as low-cost adsorbent biomaterials (Daniell et al., 2023). The development and commercialization of a diverse range of such affordable adsorbents aim to address the pressing issue of wastewater remediation. Given the limited effectiveness and economic viability of traditional wastewater treatment methods, there is a growing emphasis on biological adsorbents (Yao et al., 2024). These materials are considered highly effective in remediating various pollutants, both organic and inorganic, including heavy metals. Low-cost biomaterials, such as activated carbon derived from natural sources like agricultural biomass waste, fruit and vegetable discards, and various industrial by-products, are widely acknowledged for their exceptional adsorption capabilities, as detailed in the preceding text. Biomaterials, in the context of environmental applications, are natural substances derived from living organisms that exhibit unique properties suitable for various purposes, including sorption and remediation of pollutants (Ouyang et al., 2023). These materials, often sourced from organic waste like fruit peels, eggshells, and agricultural by-products, present a sustainable and cost-effective alternative for addressing environmental challenges. Highlighted are advanced (nano) materials, synthesized from agricultural wastes, designed for the reduction/degradation of environmental pollutants (Figure 4.5) (Ouyang et al., 2023). Not only are they readily available, but they also serve as excellent raw materials for the synthesis of potential sorbents. In the realm of biosorption, these biomaterials function as environmentally friendly options due to their renewable nature, biodegradability, and nontoxic characteristics. Their application extends across diverse fields, including wastewater treatment and the removal of contaminants such as heavy metals, dyes, and phenols. Various types of biosorbents, encompassing algal and fungal biomass, plant materials, agricultural waste, magnetic bio-adsorbents, fish scales, and microbial agents, are currently in use or under investigation for their potential and economic viability. Furthermore, the effectiveness of biomaterials in adsorbing pollutants is contingent upon specific characteristics, such as low-cost availability, portability, chemical and mechanical stability, and a high surface area. While finding a single biosorbent possessing all these traits is challenging, the integration of multiple properties enhances their overall potential and reactivity. Synthesizing nano-biomaterials represents a cutting-edge area of research and development in environmental science. Nanotechnology allows for the manipulation of materials at the nanoscale, imparting unique properties that can significantly enhance the performance of biomaterials (Gravina et al., 2022). Nano biomaterials exhibit increased surface area, improved reactivity, and enhanced adsorption capacities, making them promising candidates for advanced environmental remediation applications. The synthesis of nano biomaterials involves employing various techniques, such as sol-gel methods, hydrothermal synthesis, and green synthesis routes utilizing plant extracts or microbial processes. These methods enable the controlled production of nano-sized biomaterials with tailored properties, maximizing their efficiency in pollutant removal. In conclusion, biomaterials offer a sustainable and eco-friendly solution for environmental challenges, particularly in the context of pollutant removal. The synthesis of nano biomaterials represents a frontier in enhancing their capabilities, opening new avenues for advanced environmental applications, and contributing to the development of innovative solutions for a cleaner and healthier planet.

4.6.2 Synthesis of Nano-Biomaterials for Environmental Enhancement

Nano-biomaterials are nanoscale materials with biological and physicochemical properties. Because of their size, high surface-to-volume ratio, atomic nature, and low toxicity, nano-biomaterials have particular physicochemical characteristics that make them useful for several kinds of environmental applications. Nano-biomaterials are attracting a lot of attention in the wastewater treatment industry because of their low production costs, higher adsorption effectiveness, and renewable nature (Annamalai et al., 2022). The synthesis of nano-biomaterials involves the creation of nanoscale materials with biological characteristics or compatibility with biological systems. Various techniques are used in this synthesis, including single-step process and two-step process (Agi et al., 2021). The one-step method to produce nano-biomaterials includes the application of the dry process, nanoprecipitation, desolvation, self-assembly, and gelation. Primarily used for the size reduction of nano-biomaterials, this method presents variations in the size scales of the products obtained. However, the benefit of avoiding particle storage, transfer, drying, and dispersion is that it is restricted to low-pressure base fluids, which helps to maintain product stability. Farrag et al. (2018) used the

FIGURE 4.5 Materials from agri-waste for environmental cleanup.

nanoprecipitation method to elaborate starch nanoparticles from various botanical origins. They incorporated quercetin into the starch nanoparticles to obtain starch-quercetin nanoparticles, offering a biodegradable solution of antioxidants. Similarly, Pandey et al. (2020) developed a simple one-step method for the preparation of chitosan and lignosulfonate (CS/LS) amphiphilic nanospheres of uniform size from 150 to 200 nm. In addition, the two-step method for synthesizing nano-biomaterials involves first solubilizing the biological molecules, then producing the nano-biomaterials by a variety of gelation, precipitation, and polymerization processes. However, nano-biomaterials produced by this method are susceptible to sedimentation and aggregation due to high attractive forces, high surface energy, poor fluid compatibility, and density differences. In order to address these issues, surfactant additions, pH modifications, and intensification techniques like high-shear homogenization and ultrasound can be used (Agi et al., 2021). Chu et al. (2019) used a two-phase method to regulate the size of polysaccharide-based nanospheres, ensuring monodisperse dispersion. Similarly, Le et al. (2013) developed RHA-SNP using the sol-gel method, demonstrating the effectiveness of this approach to produce ultrafine powder particles with a uniform size distribution. Large-scale production is better served by the two-step approach, but stable small-scale manufacturing is better served by the one-step approach.

4.7 NANO-BIOMATERIALS FOR ENVIRONMENTAL CLEANUP

4.7.1 Adsorption and Photocatalytic Oxidation

In response to increasingly stringent environmental regulations, researchers are intensifying their efforts to explore diverse techniques for treating pollutants, aiming to meet ever more rigorous environmental standards (Hitam & Jalil, 2022). This evolving regulatory landscape has prompted a continuous quest for innovative and sustainable solutions to address contaminants in various water sources. Among the investigated approaches, adsorption and photocatalytic oxidation have emerged as promising alternatives (Mishra & Sundaram, 2023). Adsorption is generally recognized as a flexible, efficient, and cost-effective method for eliminating various contaminants from wastewater. Moreover, contaminants adsorbed through this method can be recycled and reused in the future. High-quality adsorbent materials, characterized by large surface areas, low densities, and high porosity, exhibit excellent absorption capacity (Hasanpour & Hatami, 2020). Photocatalysis is also considered a practical choice for wastewater treatment due to its low cost and high efficiency in clearing pollutants, and its ability to minimize the toxicity of organic compounds by converting them into nonhazardous products, thereby preserving the environment (Mishra & Sundaram, 2023). However, the selection of a light source remains a critical obstacle to the widespread industrial application of photocatalysis. Additionally, the type of photocatalyst used in the process influences the decomposition capacity of pollutants. An excellent photocatalyst is expected to be stable, inexpensive, highly photoactive, and environmentally friendly. Furthermore, the size of the photocatalyst plays a crucial role, making photocatalytic nanomaterials promising candidates for various photo-remediation applications. The unique characteristics of nano-biomaterials, such as compressibility, short intraparticle diffusion distance, outstanding durability, reusability, and low toxicity, have garnered significant attention in the wastewater remediation sector. These bio-based materials also serve as advantageous structural components in hybrids that exhibit superior overall performance, particularly in terms of efficiency and environmental friendliness (Valencia et al., 2020). Polysaccharides, as biological materials, can be employed as photocatalysts in photocatalysis procedures for contaminant removal. Modified or functionalized polysaccharides, categorized by variables like hydrophilicity, molecular weight, and origin, exhibit interesting photocatalytic properties. A study by Hitam et al. explored the potential of nanocellulose as a bionanomaterial for environmental photo-remediation. Nanocellulose, with its abundance in the environment, low cost, eco-friendliness, degradability, and excellent mechanical properties, is considered a promising candidate (Hitam & Jalil, 2022). Furthermore, chitosan (CS), as a biodegradable adsorbent, holds significant

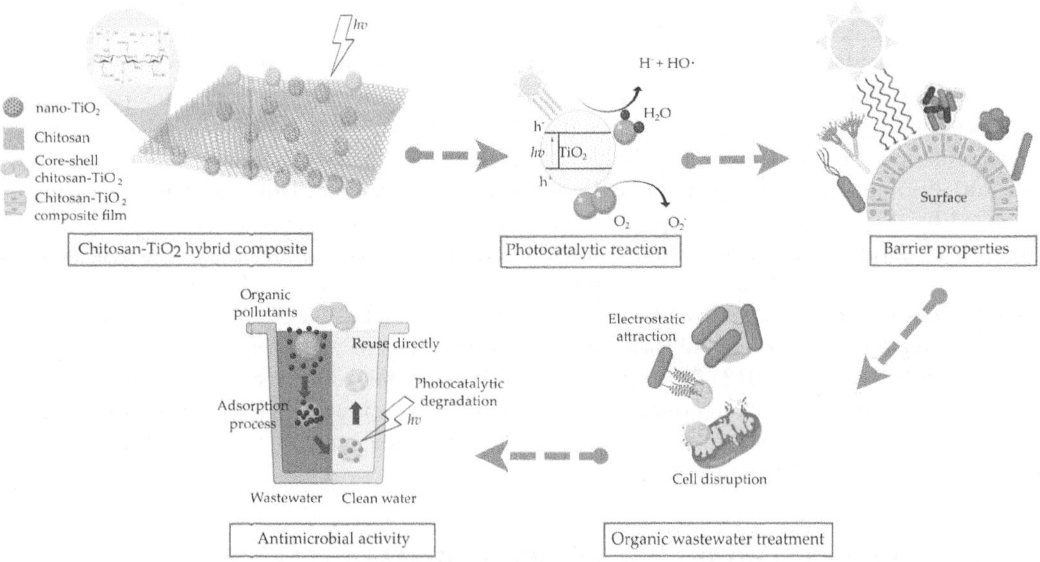

FIGURE 4.6 Chitosan/TiO_2 nano-biomaterial composite designed for the remediation of both organic and inorganic pollutants.

(Anayah-Esparza et al., 2020)

potential in water treatment. The characteristics of CS contribute to pollutant capture, facilitating the degradation process. Recent studies have shown that the semiconductor material TiO_2 stabilizes significantly when incorporated into a polymer matrix, such as chitosan, to create a hybrid composite (CS-TiO_2). The resulting CS-TiO_2 composite enhances treatment efficiency by adding photocatalytic properties to nano-TiO_2, presenting a more effective and viable approach to degrading contaminants in water (Figure 4.6) (Anayah-Esparza et al., 2020).

In the field of environmental remediation, nano-biomaterials play an important role in the advancement of various methods for treating pollutants. Aside from the well-established methods of adsorption and photocatalytic oxidation, nano-biomaterials have proven useful in a variety of novel applications. One notable example is the use of magnetic nano-biomaterials as effective adsorbents for pollutant removal. Magnetic nanoparticles, particularly those based on iron oxides, have excellent adsorption properties and can be easily extracted from water with an external magnetic field. This inherent magnetic property improves their recyclability and reusability, making them promising candidates for the effective removal of heavy metals and organic pollutants from water (Nasiri et al., 2022). In 2020, Chu et al. developed the efficient activated carbon, exhibiting a specific surface area of 1004 m^2/g and a 1.11% quaternary nitrogen content, achieving a significant Cr (VI) adsorption capacity of 3.30 mmol/g (Chu et al., 2020). Yang and co-workers designed a magnetic cobalt-iron modified porous carbon achieved from agricultural wastes to remove Hg^0 in coal-fired flue gas (Yang et al., 2020) (Figure 4.7).

Another intriguing avenue is the application of graphene-based nano-biomaterials in wastewater treatment processes. Graphene, known for its high electrical conductivity and large surface area, acts as an effective adsorbent for a variety of pollutants. Furthermore, graphene-based nano-biomaterials, when combined with semiconductor materials, degrade organic pollutants with remarkable efficiency via photocatalytic oxidation. This synergistic combination takes advantage of graphene's unique properties, opening new possibilities for advanced water treatment technologies. In addition to these examples, the use of nano-biomaterials in nanocomposites containing metal-organic frameworks (MOFs) demonstrates their efficacy in pollutant adsorption. MOFs, with their large surface areas and tunable structures, combine with nano-biomaterials such as cellulose nanofibers or

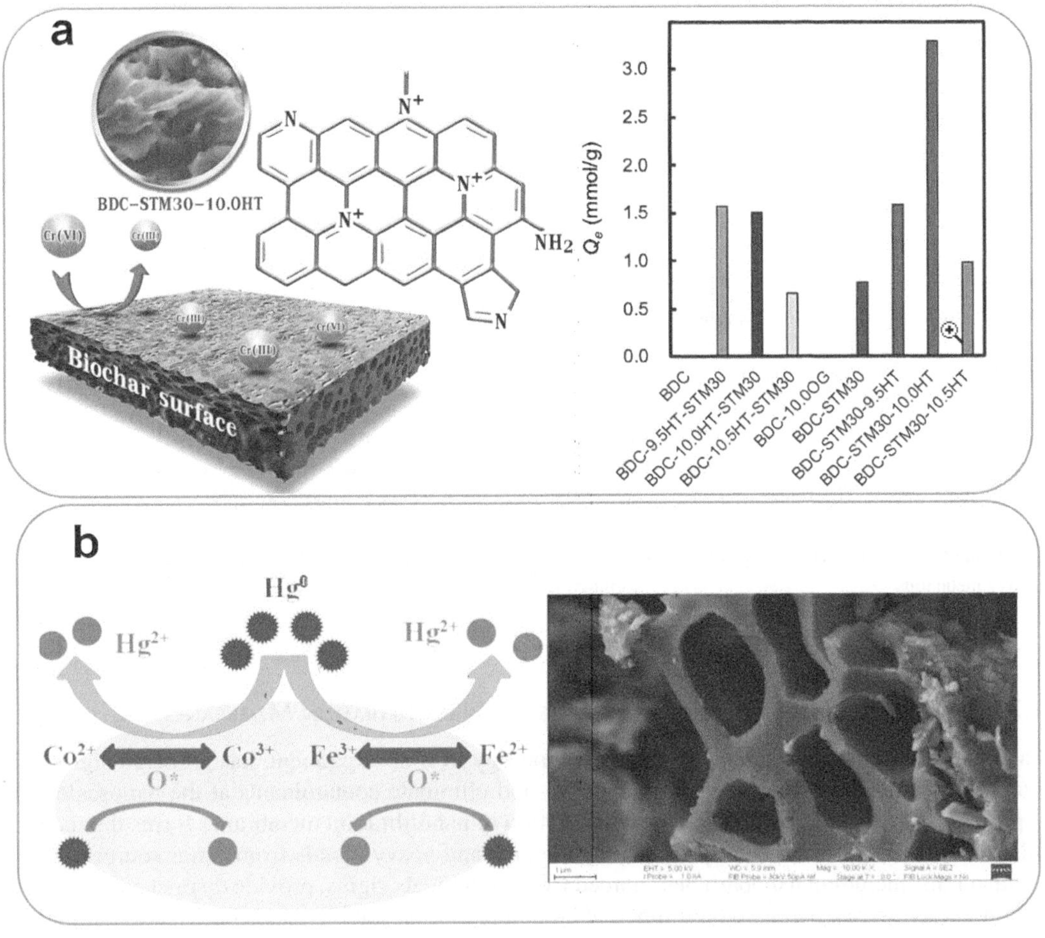

FIGURE 4.7 (a) Depiction of proposed mechanisms for Cr (VI) adsorption onto activated carbon, alongside a comparison of Cr (VI) adsorption capacities among various adsorbents. (b) Synthesis of a magnetic cobalt-iron modified porous carbon for Hg^0 removal.

(Ouyang et al., 2023)

chitosan to significantly improve adsorption properties (Mansoorianfar et al., 2022). This innovation helps to develop efficient and sustainable water systems. In conclusion, nano-biomaterials are invaluable in the field of environmental remediation, with numerous applications beyond traditional adsorption and photocatalytic processes. From magnetic nano-biomaterials for efficient pollutant extraction to graphene-based materials for advanced adsorption and photocatalysis, their versatility propels innovation in addressing the complex challenges of water pollution (Mondal et al., 2023). Other studies have highlighted the practical and economically advantageous role of photocatalysis in pollutant degradation and wastewater treatment. For example, Krishnamurthy et al. (2014) reported on the photocatalytic degradation of wastewater containing methylene blue dye using gold nanosheets obtained from Yucca filament's leaf extract. Similarly, Noman et al. (2020) examined the combined effect of photocatalytic detoxification of textile effluent dyes in the presence of bio-fabricated copper nanoparticles, inducing a change in the physicochemical state of the textile effluent. Table 4.1 presents the photocatalytic performance of a CS–TiO_2 composite in the degradation of various water pollutants, further demonstrating the effectiveness of nanostructured materials in environmental remediation.

TABLE 4.1
The Photocatalytic Performance of a CS–TiO_2 Composite in the Degradation of Various Water Pollutants

Pollutant	Materials	Results	References
Methyl orange	CS–TiO_2 composite	The TiO_2–CS composite (75:25 w/w) exhibited significant efficiency in degrading methyl orange dye, and the recycled material demonstrated effective stability.	Saravanan et al. (2018)
Thymol violet	TiO_2/CS/CMM coating layer	TiO_2/CS/CMMcomposites can adsorb thymol violet for its extraction from water.	Kamal et al. (2016)
Methyl orange/ Congo red	CS/PVA/Na– Titanate/ TiO_2 composites	Complete elimination of methyl orange was achieved under UV irradiation, while 99% removal was observed for Congo red.	Habiba et al. (2017)
Methyl orange/ Congo red	CS/PVA/Na– Titanate/ TiO_2 composites	Full elimination of methyl orange was achieved through UV irradiation, while 99% removal was attained for Congo red.	Zhu et al. (2013)
Cu (II) and Pb (II) heavy metal ions	CS–TiO_2 composite	CS–TiO_2 demonstrated superior potential for sorbing metal ions when compared to CS alone.	Razzaz et al. (2016)

4.7.2 Nano-biomaterials Enhancing Efficient Nanofiltration Membranes

The concept of nanofiltration, a promising technology in water treatment, has garnered significant attention for its capacity to selectively separate and eliminate contaminants at the nanoscale. The exploration of nano-biomaterials in the construction of nanofiltration membranes forms the focus of this section, aiming to efficiently eradicate colorants and heavy metals from water sources. Nano-biomaterials, including nanoparticles sourced from biological origins, provide distinct advantages in membrane synthesis, owing to their inherent biocompatibility, stability, and eco-friendly properties. These qualities position them as ideal candidates for the advancement of sophisticated nanofiltration membranes. The integration of nano-biomaterials into nanofiltration membranes heightens their selectivity in colorant removal. Tailoring the functionalization of these materials allows for a more efficient and customized approach to water treatment. In seawater treatment, the use of eco-friendly membrane modification methods involves blending highly biocompatible biomaterials such as curcumin (Liu et al., 2024), chitosan (El-Saied & Ibrahim, 2020), and various other nanomaterials. This blend enhances membrane antifouling properties, process efficiency, and cost-effectiveness. Additionally, considering membrane characteristics, factors like membrane charge, hydrophilicity or hydrophobicity, roughness, pore morphology, and resistance to fouling influence nanofiltration (NF) performance. NF behavior is predominantly governed by membrane charge (Mallya et al., 2023). Noteworthy challenges in NF include the trade-off between membrane permeability and selectivity, along with fouling issues that hinder stable membrane performance. The choice of membrane material plays a pivotal role in predicting membrane performance in terms of flux, rejection, and antifouling properties. The presence of nano-biomaterials within the polymer matrix can noticeably alter membrane characteristics. From the same viewpoint, efforts to enhance antifouling membrane properties involve investigations into membrane surface modification. Multidisciplinary efforts in membrane engineering, incorporating advanced nano-biomaterials, aim to refine membrane properties such as pore size, hydrophilicity, membrane charge, and fouling resistance. Goh et al. engineered and assessed environmentally sustainable zinc oxide nanoparticles coated with nano-chitosan to efficiently prevent marine fouling. The incorporation of membranes with inherently embedded hydrophilic biomaterials/nanomaterials demonstrates its significance in addressing fouling and related challenges in membrane separation (Goh et al., 2016). In addition, Wu et al. have demonstrated that incorporating functionalized biomaterials and nanomaterials into polymer membranes improves hydrophilicity, permeability, and antifouling properties (Wu et al., 2021). In summary, chitosan is a focal point in biopolymer research, drawing significant attention for its functional groups (hydroxyl and amino), cost-effectiveness, natural occurrence, abundance, and nonhazardous nature. As a result, chitosan demonstrates the ability to efficiently eliminate charged contaminants in water treatment processes.

Nano-biomaterial–based nanofiltration membranes demonstrate remarkable efficiency in removing dyes and heavy metals from water. The unique surface properties and functional groups of these materials facilitate strong adsorption and chelation of heavy metal ions, ensuring an effective and selective separation process. This is particularly valuable in addressing water pollution issues associated with industrial discharges and natural sources of heavy metals.

The use of nano-biomaterials not only enhances membrane performance but also contributes to the overall sustainability of water treatment processes. Biocompatible materials ensure minimal environmental impact and may even offer the potential for bio-based regeneration methods, further reducing the ecological footprint of the membrane technology. In conclusion, while the application of nano-biomaterials in nanofiltration membranes holds great promise, challenges such as scalability, cost-effectiveness, and long-term stability need to be addressed. Continued research is essential to overcome these hurdles and unlock the full potential of nano-biomaterials in water treatment. The integration of nano-biomaterials into nanofiltration membrane fabrication represents a cutting-edge approach to the removal of colorants and heavy metals from water sources. This technology not only demonstrates superior performance but also aligns with the principles of sustainability, opening avenues for innovative solutions to water treatment challenges in the future. On the other hand, Liu et al.'s (2021) review explores the development of air filtration materials (AFMs), which have become a significant research focus due to global concerns about air quality. The challenge lies in balancing high filtration efficiency with low-pressure drop, a dilemma faced by most AFMs. Electrospun nanofibers, with their large surface-to-volume ratio, adjustable porous structure, and straightforward preparation process, emerge as promising candidates for filtration nano-biomaterials. Consequently, electrospun nanofibers have garnered increased attention in air filtration applications. The review examines the preparation methods of high-performance electrospun air filtration membranes and outlines the typical surface structures of electrospun fibers as nano-biomaterials for air filtration. In addition, Table 4.2 presents data on nano-biomaterial–based nanofiltration membranes, focusing on selective contaminant removal and rejection efficiency.

TABLE 4.2
Nano-biomaterial–based Nanofiltration Membranes: Selective Contaminant Removal and Rejection Efficiency

Nano-biomaterial–based Membrane	Filtered Elements	Rejection (%)	Ref.
Functionalized graphene oxide	Organic Dyes (bromothymol blue, methyl orange, and murexide)	94, 93, and 97	Karim et al. (2018)
Cellulose acetate cinnamaldehyde poly (hexamethylene biguanide) ultrafine bimodal nanofibrous membrane	Refers to particulate matter with a diameter of 0.3 micrometers (PM0.3), Bacteriostatic rate (E. coli), Bacteriostatic rate (S. aureus), Mosquito-repellent rate	99.78, 99.65, 99.99, and 92.70	Shao et al. (2023)
Fully bio-based biodegradable nanofiber hydrogel film	Ultrafine suspended solids (SS)	~100%	Jiang et al. (2024)
Electrospun poly (lactic acid) (PLA) nanofibers	Particulate Matter (PM0.3 and PM2.5)	90.68% and 99.82%	Wang et al. (2023)

4.8 RECOMMENDATION AND FUTURE CHALLENGES

In conclusion, the development, utilization, and application of nano-biomaterials for environmental cleanup, particularly in wastewater treatment, represents an important step toward long-term and effective pollutant remediation. These materials' multifaceted properties, including their large surface area, reactivity, and environmental friendliness, make them promising candidates for advanced water treatment technologies. The chapter focused on key applications such as adsorption and photocatalytic oxidation, demonstrating nano-biomaterials' versatility in addressing a wide range of pollutants. Moving forward, it will be critical to encourage additional research into the synthesis of nano-biomaterials using novel techniques such as sol-gel methods, hydrothermal synthesis, and green synthesis routes. This research may result in the development of materials with improved pollutant-removal properties. It is also critical to emphasize the integration of various properties in biosorbents in order to increase their overall potential and reactivity. The combination of properties such as low-cost availability, portability, chemical and mechanical stability, and large surface area can result in more effective pollutant adsorption. Furthermore, research into the use of nanocomposites, particularly those containing MOFs, to improve the adsorption properties of nano-biomaterials is critical. This research could help to develop effective and sustainable water treatment solutions. Addressing challenges related to the scalability and cost-effectiveness of nano-biomaterials in water treatment processes is critical. Finding ways to manufacture these materials on a larger scale while remaining economically viable is critical for widespread adoption. Furthermore, a focus on improving the long-term stability of nano-biomaterials in practical applications is required. Recognizing the durability and performance of these materials over time will be critical for their continued use in environmental cleanup. Furthermore, navigating regulatory pathways and promoting standardization in the use of nano-biomaterials for water treatment is critical. Establishing clear guidelines and ensuring compliance with environmental regulations will help them gain acceptance and deployment on a larger scale. In conclusion, while nano-biomaterials have enormous potential to revolutionize water treatment methodologies, addressing current challenges and implementing recommendations will be critical to their successful integration into environmental cleanup practices. The ongoing commitment to research, innovation, and collaboration among researchers, industry stakeholders, and regulatory bodies will help to create a cleaner, healthier environment for future generations.

REFERENCES

Abu-Danso, E., Peräniemi, S., Leiviskä, T., & Bhatnagar, A. (2018). Synthesis of S-ligand tethered cellulose nanofibers for efficient removal of Pb (II) and Cd (II) ions from synthetic and industrial wastewater. *Environmental Pollution, 242*, 1988–1997. doi:10.1016/j.envpol.2018.07.044

Afkhami, A., Saber-Tehrani, M., & Bagheri, H. (2010). Simultaneous removal of heavy-metal ions in wastewater samples using nano-alumina modified with 2,4-dinitrophenylhydrazine. *Journal of Hazardous Materials, 181*(1–3), 836–844. doi:10.1016/j.jhazmat.2010.05.089

Agi, A., Junin, R., Jaafar, M.Z., Sidek, M.A., Yakasai, F., & Gbadamosi, A., et al. (2021). Formulation of bionanomaterials: A review of particle design towards oil recovery applications. *Journal of Industrial and Engineering Chemistry, 98*, 82–102. doi:10.1016/j.jiec.2021.03.032

Aliabadi, M., Irani, M., Ismaeili, J., & Najafzadeh, S. (2014). Design and evaluation of chitosan/hydroxyapatite composite nanofiber membrane for the removal of heavy metal ions from aqueous solution. *Journal of the Taiwan Institute of Chemical Engineers*, 45 (2), 518–526. doi:10.1016/j.jtice.2013.04.016

Almahri, A., Abou-Melha, K.S., Katouah, H.A., Al-Bonayan, A.M., Saad, F.A., & El-Desouky, M.G. (2023). Adsorption and removal of the harmful pesticide 2,4-dichlorophenylacetic acid from an aqueous environment via coffee waste biochar: Synthesis, characterization, adsorption study and optimization via Box-Behnken design. *Journal of Molecular Structure, 1293*, 136238. doi:10.1016/j.molstruc.2023.136238

Al-Salman, H.N.K., Falih, M., Deab, H.B., Altimari, U.S., Shakier, H.G., Dawood, A.H., et al. (2023). A study in analytical chemistry of adsorption of heavy metal ions using chitosan/graphene nanocomposites. *Case Studies in Chemical and Environmental Engineering, 8*, 100426. doi:10.1016/j.cscee.2023.100426

Anayah-Esparza, L.M., Ruvalcaba-Gómez, J.M., Maytorena-Verdugo, C.I., González-Silva, N., Romero-Toledo, R., Aguilera-Aguirre, S., et al. (2020). Chitosan-TiO_2: A versatile hybrid composite. *Materials*, *13*(4), 811. doi:10.3390/ma13040811

Anifah, E.M., Ariani, I.K., Hayati, R.N., & Nugraha, S.A. (2022). Adsorption of oil and grease in wastewater using activated carbon derived from sewage sludge. *IOP Conference Series: Earth and Environmental Science*, *1098*(1), 012043. doi:10.1088/1755-1315/1098/1/012043

Annamalai, K., Rameshbabu, H., Mahendhran, K., & Ramanathan, M. (2022). Smart Bionanomaterials for the Removal of Contaminants from Wastewater. In: *Advances in Bionanomaterials*, Elsevier Inc. doi:10.1016/B978-0-12-823915-5.00004-6

Ataei, M., Maghsoudi, A.S., & Hassani, S. (2024). Dyes and Colorants. In: *Handbook of Textile and Industrial Dyes*, 4th ed. Elsevier Inc. doi:10.1016/b978-0-12-824315-2.00017-8

Bassyouni, D., Ali, S., Abdel-Aziz, M.H., & Elashtoukhy, E. (2023). Electrocoagulation technique and statistical analysis for treatment of real effluent from the pulp and paper industry. *International Journal of Electrochemical Science*, *18*(12), 100389. doi:10.1016/j.ijoes.2023.100389

Berthiaume, A. (2023). Radionuclide contamination in Canada: A scoping review. *Heliyon*, *9*(6), e16602. doi:10.1016/j.heliyon.2023.e16602

Bhattacharyya, K.G., & Gupta, S.S. (2008). Adsorption of a few heavy metals on natural and modified kaolinite and montmorillonite: A review. *Advances in Colloid and Interface Science*, *140*(2), 114–131. doi:10.1016/j.cis.2007.12.008

Bouazizi, A., Ouammou, M., Aaddane, A., Tijani, N., & Younssi, S.A. (2023). Novel Low-cost Bentonite-based Membranes for Microfiltration, Ultrafiltration, and Nanofiltration Applications. In: *Novel Materials for Environmental Remediation Applications*, Elsevier, pp. 247–275. doi:10.1016/B978-0-323-91894-7.00003-7

Bouazizi, A., Saja, S., Achiou, B., Ouammou, M., Calvo, J.I., Aaddane, A., et al. (2016). Elaboration and characterization of a new flat ceramic MF membrane made from natural Moroccan bentonite. Application to treatment of industrial wastewater. *Applied Clay Science*, *132–133*, 33–40. doi:10.1016/j.clay.2016.05.009

Breida, M., Younssi, S.A., Bouazizi, A., Achiou, B., Ouammou, M., & El Rhazi, M. (2018). Nitrate removal from aqueous solutions by γ-Al2O3 ultrafiltration membranes. *Heliyon*, *4*(1), e00498. doi:10.1016/j.heliyon.2017.e00498

Chu, B., Amano, Y., & Machida, M. (2020). Preparation of bean dreg derived N-doped activated carbon with high adsorption for Cr (VI). *Colloids and Surfaces A: Physicochemical and Engineering Aspects*, *586*, 124262. doi:10.1016/j.colsurfa.2019.124262

Chu, L., Niu, W., Wu, S., Ma, W., Tang, B., Zhang, S., et al. (2019). A two-step approach for size-controlled preparation of monodisperse polysaccharide-based nanospheres. *Materials Research Express*, *6*(5). doi:10.1088/2053-1591/ab0245

Daniell, H., Singh, R., Mangu, V., Nair, S.K., Wakade, G., & Balashova, N. (2023). Corrigendum to affordable oral proinsulin bioencapsulated in plant cells regulates blood sugar levels similar to natural insulin. *Biomaterials*, *298*, 122226. doi:10.1016/j.biomaterials.2023.122226

Donzella, D., Bellis, E., Campisi, P., Crepaldi, G., Data, V., Dapavo, P., et al. (2023). New onset sarcoidosis following biologic treatment in patients with seronegative inflammatory arthritis: A case series and systematic literature review. *Autoimmunity Reviews*, *23*(3), 103481. doi:10.1016/j.autrev.2023.103481

El Abbadi, S., El Moustansiri, H., Douma, M., Bouazizi, A., Arfoy, B., Calvo, J.I., & Tijani, N. (2024). Enhancing the performance of alumina-pillared clay for phenol removal from water solutions and polyphenol removal from olive mill wastewater: Characterization, kinetics, adsorption performance, and mechanism. *Journal of Water Process Engineering*, *63*, 105432. doi:10.1016/j.jwpe.2024.105432

El-Kordy, A., Dehmani, Y., Douma, M., Bouazizi, A., El Moustansiri, H., El Abbadi, S., et al. (2022). Experimental study of phenol removal from aqueous solution by adsorption onto synthesized Faujasite-type Y zeolite. *Desalination and Water Treatment*, *277*, 144–154. doi:10.5004/dwt.2022.28958

El-Saied, H.A.A., & Ibrahim, A.M. (2020). Effective fabrication and characterization of eco-friendly nano chitosan capped zinc oxide nanoparticles for effective marine fouling inhibition. *Journal of Environmental Chemical Engineering*, *8*(4), 103949. doi:10.1016/j.jece.2020.103949

Farrag, Y., Ide, W., Montero, B., Rico, M., Rodríguez-Llamazares, S., Barral, L., et al. (2018). Preparation of starch nanoparticles loaded with quercetin using nanoprecipitation technique. *International Journal of Biological Macromolecules*, *114*, 426–433. doi:10.1016/j.ijbiomac.2018.03.134

Ghosh, A., Manna, M.C., Jha, S., Singh, A.K., Misra, S., Srivastava, R.C., et al. (2022). Impact of Soil-Water Contaminants on Tropical Agriculture, Animal and Societal Environment. In: *Advances in Agronomy*, Vol. 176, Elsevier Inc. doi:10.1016/bs.agron.2022.07.006

Goh, P.S., Matsuura, T., Ismail, A.F., & Hilal, N. (2016). Recent trends in membranes and membrane processes for desalination. *Desalination*, *391*, 43–60. doi:10.1016/j.desal.2015.12.016

Gravina, A.N., D'Elía, N.L., Sartuqui, J., & Messina, P.V. (2022). Bone Tissue Restoration by Nanoscale Features of Biomaterials. In: *Nanostructured Biomaterials for Bone Tissue Engineering*, Elsevier Inc. doi:10.1016/B978-0-323-85148-0.00012-9

Habiba, U., Joo, T.C., Siddique, T.A., Salleh, A., Ang, B.C., & Afifi, A.M. (2017). Effect of degree of deacetylation of chitosan on adsorption capacity and reusability of chitosan/polyvinyl alcohol/TiO_2 nanocomposite. *International Journal of Biological Macromolecules*, *104*, 1133–1142. doi:10.1016/j.ijbiomac.2017.07.007

Hamad, A.A., Hassouna, M.S., Shalaby, T.I., Elkady, M.F., Abd Elkawi, M.A., & Hamad, H.A. (2020). Electrospun cellulose acetate nanofiber incorporated with hydroxyapatite for removal of heavy metals. *International Journal of Biological Macromolecules*, *151*, 1299–1313. doi:10.1016/j.ijbiomac.2019.10.176

Hasanpour, M., & Hatami, M. (2020). Application of three dimensional porous aerogels as adsorbent for removal of heavy metal ions from water/wastewater: A review study. *Advances in Colloid and Interface Science*, *284*, 102247. doi:10.1016/j.cis.2020.102247

Heidari, A., Younesi, H., & Mehraban, Z. (2009). Removal of Ni (II), Cd (II), and Pb (II) from a ternary aqueous solution by amino functionalized mesoporous and nano mesoporous silica. *Chemical Engineering Journal*, *153*(1–3), 70–79. doi:10.1016/j.cej.2009.06.016

Hitam, C.N.C., & Jalil, A.A. (2022). Recent advances on nanocellulose biomaterials for environmental health photoremediation: An overview. *Environmental Research*, *204*(PA), 111964. doi:10.1016/j.envres.2021.111964

Huang, Y., & Wang, Z. (2018). Preparation of composite aerogels based on sodium alginate, and its application in removal of Pb2+ and Cu2+ from water. *International Journal of Biological Macromolecules*, *107*(Part A), 741–747. doi:10.1016/j.ijbiomac.2017.09.057

Inbaraj, B.S., Sridhar, K., & Chen, B.H. (2021). Removal of polycyclic aromatic hydrocarbons from water by magnetic activated carbon nanocomposite from green tea waste. *Journal of Hazardous Materials*, *415*, 125701. doi:10.1016/j.jhazmat.2021.125701

Jakszyn, P., & González, C.A. (2006). Nitrosamine and related food intake and gastric and oesophageal cancer risk: A systematic review of the epidemiological evidence. *World Journal of Gastroenterology*, *12*(27), 4296–4303. doi:10.3748/wjg.v12.i27.4296

Jiang, M., Jing, C., Lei, C., Han, X., Wu, Y., Ling, S., Zhang, Y., et al. (2024). A bio-based nanofibre hydrogel filter for sustainable water purification. *Nature Sustainability*. doi:10.1038/s41893-023-01264-9

Kamal, T., Anwar, Y., Khan, S.B., Chani, M.T.S., & Asiri, A.M. (2016). Dye adsorption and bactericidal properties of TiO2/chitosan coating layer. *Carbohydrate Polymers*, *148*, 153–160. doi:10.1016/j.carbpol.2016.04.042

Karim, A., Achiou, B., Bouazizi, A., Aaddane, A., Ouammou, M., Bouziane, M., et al. (2018). Development of reduced graphene oxide membrane on flat Moroccan ceramic pozzolan support. Application for soluble dyes removal. *Journal of Environmental Chemical Engineering*, *6*(1), 1475–1485. doi:10.1016/j.jece.2018.01.055

Kobya, M., Demirbas, E., Senturk, E., & Ince, M. (2005). Adsorption of heavy metal ions from aqueous solutions by activated carbon prepared from apricot stone. *Bioresource Technology*, *96*(13), 1518–1521. doi:10.1016/j.biortech.2004.12.005

Krishnamurthy, S., Esterle, A., Sharma, N.C., & Sahi, S.V. (2014). Yucca-derived synthesis of gold nanomaterial and their catalytic potential. *Nanoscale Research Letters*, *9*(1), 627. doi:10.1186/1556-276X-9-627

Kumari, M., Pittman, C.U., & Mohan, D. (2015). Heavy metals [chromium (VI) and lead (II)] removal from water using mesoporous magnetite (Fe3O4) nanospheres. *Journal of Colloid and Interface Science*, *442*, 120–132. doi:10.1016/j.jcis.2014.09.012

Lau, C., & Le, X.C. (2023). Cadmium, chromium, copper, iron, lead, mercury, nickel, and zinc in freshwater fish: Assessing trophic transfer using stable isotope ratios of δ15N and δ13C. *Journal of Environmental Sciences (China)*, *128*, 250–257. doi:10.1016/j.jes.2023.01.007

Le, V.H., Thuc, C.N.H., & Thuc, H.H. (2013). Synthesis of silica nanoparticles from Vietnamese rice husk by sol–gel method. *Nanoscale Research Letters*, *8*(1), 1–10. doi:10.1186/1556-276X-8-58

Li, B., Yang, Y., Wu, H., Zhang, C., Zheng, W., & Sun, D. (2020). Adsorptive removal and mechanism of monocyclic aromatics by activated carbons from water: Effects of structure and surface chemistry. *Colloids and Surfaces A: Physicochemical and Engineering Aspects*, *605*, 125346. doi:10.1016/j.colsurfa.2020.125346

Li, Z., & Yu, C. (2024). Adsorption in the Treatment of Three Wastes. In: *Nanostructured Materials for Environmental Applications*, Elsevier, pp. 317–326. doi:10.1016/b978-0-443-19256-2.00013-2

Liu, P., Shi, Y., Wu, X., Wang, H., Huang, H., & Guo, X. (2021). Review of the artificially-accelerated aging technology and ecological risk of microplastics. *Science of the Total Environment*, *768*, 144969. doi:10.1016/j.scitotenv.2021.144969

Liu, X., Miao, X., Bo, S., Deng, X., Zhang, Z., & Zheng, Y. (2024). Macrophage membrane-coated self-assembled curcumin nanoparticle missile for the treatment of colorectal cancer. *Journal of Drug Delivery Science and Technology*, *91*, 105237. doi:10.1016/j.jddst.2023.105237

Mallya, D.S., Abdikheibari, S., Dumée, L.F., Muthukumaran, S., Lei, W., & Baskaran, K. (2023). Removal of natural organic matter from surface water sources by nanofiltration and surface engineering membranes for fouling mitigation – A review. *Chemosphere*, *321*, 138070. doi:10.1016/j.chemosphere.2023.138070

Mansoorianfar, M., Nabipour, H., Pahlevani, F., Zhao, Y., Hussain, Z., Hojjati-Najafabadi, A., et al. (2022). Recent progress on adsorption of cadmium ions from water systems using metal-organic frameworks (MOFs) as an efficient class of porous materials. *Environmental Research*, *214*(P4), 114113. doi:10.1016/j.envres.2022.114113

Mishra, S., & Sundaram, B. (2023). A review of the photocatalysis process used for wastewater treatment. *Materials Today: Proceedings*. doi:10.1016/j.matpr.2023.07.147

Mondal, S., Park, S., Choi, J., Vu, T.T.H., Doan, V.H.M., Vo, T.T., et al. (2023). Hydroxyapatite: A journey from biomaterials to advanced functional materials. *Advances in Colloid and Interface Science*, *321*, 103013. doi:10.1016/j.cis.2023.103013

Mosallaei, S., Hashemi, H., Hoseini, M., & Naz, A. (2023). Polycyclic aromatic hydrocarbons (PAHs) in household dust: The association between PAHs, cancer risk and sick building syndrome. *Building and Environment*, *229*, 109966. doi:10.1016/j.buildenv.2022.109966

Nasiri, A., Rajabi, S., Amiri, A., Fattahizade, M., Hasani, O., Lalehzari, A., et al. (2022). Adsorption of tetracycline using CuCoFe2O4@Chitosan as a new and green magnetic nanohybrid adsorbent from aqueous solutions: Isotherm, kinetic and thermodynamic study. *Arabian Journal of Chemistry*, *15*(8), 104014. doi:10.1016/j.arabjc.2022.104014

Nilsson, H., Kärrman, A., Rotander, A., van Bavel, B., Lindström, G., & Westberg, H. (2013). Biotransformation of fluorotelomer compound to perfluorocarboxylates in humans. *Environment International*, *51*(2013), 8–12. doi:10.1016/j.envint.2012.09.001

Noman, M., Shahid, M., Ahmed, T., Niazi, M.B.K., Hussain, S., Song, F., et al. (2020). Use of biogenic copper nanoparticles synthesized from a native Escherichia sp. as photocatalysts for azo dye degradation and treatment of textile effluents. *Environmental Pollution*, *257*, 113514. doi:10.1016/j.envpol.2019.113514

Ouyang, H., Safaeipour, N., Othman, R.S., Otadi, M., Sheibani, R., Kargaran, F., et al. (2023). Agricultural waste-derived (nano) materials for water and wastewater treatment: Current challenges and future perspectives. *Journal of Cleaner Production*, *421*, 138524. doi:10.1016/j.jclepro.2023.138524

Palaniyappan, J., & Beerappa, R. (2024). Environmental Exposure to Heavy Metals in Ambient Air and Its Human Health Implications. In: *Handbook of Environmental Impact Assessment*, Elsevier Ltd. doi:10.1016/B978-0-323-95282-8.00028-6

Pandey, R.P., Rasool, K., Rasheed, P.A., Gomez, T., Pasha, M., Mansour, S.A., et al. (2020). One-step synthesis of an antimicrobial framework based on covalently cross-linked chitosan/lignosulfonate (CS@LS) nanospheres. *Green Chemistry*, *22*(3), 678–687. doi:10.1039/c9gc03461g

Patel, H.K., Kalaria, R.K., Jokhakar, P.H., Patel, C.R., & Patel, B.Y. (2021). Removal of Emerging Contaminants in Water Treatment by an Application of Nanofiltration and Reverse Osmosis. In: *Development in Wastewater Treatment Research: Process Removal of Emerging Contaminants from Wastewater through Bio-nanotechnology*, Elsevier, pp. 385–400. doi:10.1016/B978-0-323-85583-9.00005-3

Rakcho, Y., Mouiya, M., Bouazizi, A., Abouliatim, Y., Sehaqui, H., Mansouri, S., et al. (2023). Treatment of seawater and wastewater using a novel low-cost ceramic membrane fabricated with red clay and tea waste. *Arabian Journal of Chemistry*, *16*(11), 105277. doi:10.1016/j.arabjc.2023.105277

Rathi, B.S., Kumar, P.S., & Vo, D.V.N. (2021). Critical review on hazardous pollutants in water environment: Occurrence, monitoring, fate, removal technologies and risk assessment. *Science of the Total Environment*, *797*, 149134. doi:10.1016/j.scitotenv.2021.149134

Razzaz, A., Ghorban, S., Hosayni, L., Irani, M., & Aliabadi, M. (2016). Chitosan nanofibers functionalized by TiO2 nanoparticles for the removal of heavy metal ions. *Journal of the Taiwan Institute of Chemical Engineers*, *58*, 333–343. doi:10.1016/j.jtice.2015.06.003

Ríos, F., Caparrós-Salvador, F., Lechuga, M., & Fernández-Serrano, M. (2024). Complete biodegradability assessment of polyoxyethylene glycerol ester non-ionic surfactant: Aerobic, anaerobic, combined biodegradation and inhibitory effects. *Water Research*, 248. doi:10.1016/j.watres.2023.120857

Rout, P.R., Zhang, T.C., Bhunia, P., & Surampalli, R.Y. (2021). Treatment technologies for emerging contaminants in wastewater treatment plants: A review. *Science of the Total Environment, 753*, 141990. doi:10.1016/j.scitotenv.2020.141990

Saravanan, R., Aviles, J., Gracia, F., Mosquera, E., & Gupta, V.K. (2018). Crystallinity and lowering band gap induced visible light photocatalytic activity of TiO2/CS (Chitosan) nanocomposites. *International Journal of Biological Macromolecules, 109*, 1239–1245. doi:10.1016/j.ijbiomac.2017.11.125

Saxena, R., Saxena, M., & Lochab, A. (2020). Recent progress in nanomaterials for adsorptive removal of organic contaminants from wastewater. *ChemistrySelect*, 5 (1), 335–353. doi:10.1002/slct.201903542

Shahwar, D., Khan, Z., Ansari, M.Y.K., & Park, Y. (2023). Relative mutagenic effectiveness and efficiency of chemical mutagens (Caffeine and EMS) and heavy metals [(Pb (NO3) 2 and Cd (NO3) 2)] in developing chlorophyll and morphological mutants in lentil. *Mutation Research - Genetic Toxicology and Environmental Mutagenesis, 890*, 503668. doi:10.1016/j.mrgentox.2023.503668

Shao, Z., Kang, G., Xie, J., Shen, R., Li, H., et al. (2023). Electrospun mutualism-inspired CA/CMA/PHMB ultrafine bimodal nanofibrous membrane for high-performance, antibacterial, and mosquito-repellent air filtration. *Separation and Purification Technology, 327*(15), 124920. doi:10.1016/j.seppur.2023.124920

Thakur, A., Kumar, A., & Singh, A. (2024). Adsorptive removal of heavy metals, dyes, and pharmaceuticals: Carbon-based nanomaterials in focus. *Carbon, 217*, 118621. doi:10.1016/j.carbon.2023.118621

Toorani, F., Aghdasinia, H., Nejatbakhsh, S., & Karimi, A. (2023). Removal of hydrocarbon pollutants from aqueous media using hydrophobic cellulose-based adsorbents. *Chemical Engineering Research and Design, 199*, 413–424. doi:10.1016/j.cherd.2023.10.020

Valencia, L., Kumar, S., Nomena, E.M., Salazar-Alvarez, G., & Mathew, A.P. (2020). In-situ growth of metal oxide nanoparticles on cellulose nanofibrils for dye removal and antimicrobial applications. *ACS Applied Nano Materials, 3*(7), 7172–7181. doi:10.1021/acsanm.0c01511

Velarde, L., Nabavi, M.S., Escalera, E., Antti, M.L., & Akhtar, F. (2023). Adsorption of heavy metals on natural zeolites: A review. *Chemosphere, 328*, 138508. doi:10.1016/j.chemosphere.2023.138508

Wang, C., Song, X., Li, T., Zhu, X., Yang, S., Zhu, J., He, X., Gao, J., & Xu, H. (2023). Biodegradable electroactive nanofibrous air filters for long-term respiratory healthcare and self-powered monitoring. *ACS Applied Materials & Interfaces, 15*(31), 37580–37592. doi:10.1021/acsami.3c08490

Wu, C., Zheng, J., & Hu, J. (2021). Novel antifouling polysulfone matrix membrane modified with zwitterionic polymer. *Journal of the Saudi Chemical Society, 25*(8), 101281. doi:10.1016/j.jscs.2021.101281

Yang, W., Chen, H., Han, X., Ding, S., Shan, Y., & Liu, Y. (2020). Preparation of magnetic Co-Fe modified porous carbon from agricultural wastes by microwave and steam activation for mercury removal. *Journal of Hazardous Materials, 381*, 120981. doi:10.1016/j.jhazmat.2019.120981

Yao, S., Jabeur, F., Pontoni, L., Mechri, S., Jaouadi, B., & Sannino, F. (2024). Sustainable removal of arsenic from waters by adsorption on blue crab, Portunussegnis (Forskål, 1775) chitosan-based adsorbents. *Environmental Technology & Innovation, 33*, 103491. doi:10.1016/j.eti.2023.103491

Zhang, L., Guan, Q., Jiang, J., & Khan, M.S. (2023). Tannin complexation with metal ions and its implication on human health, environment, and industry: An overview. *International Journal of Biological Macromolecules, 253*(P7), 127485. doi:10.1016/j.ijbiomac.2023.127485

Zhu, H., Jiang, R., Xiao, L., Liu, L., Cao, C., & Zeng, G. (2013). CdS nanocrystals/TiO2/crosslinked chitosan composite: Facile preparation, characterization and adsorption-photocatalytic properties. *Applied Surface Science, 273*, 661–669. doi:10.1016/j.apsusc.2013.02.106

5 Production of Carbon with High Specific Surface Area from West African Agricultural Waste

Evaluation of Product Quality and Potential Application

Joshua O. Ighalo and Chinenye Adaobi Igwegbe

5.1 INTRODUCTION

Agricultural wastes are lignocellulosic biomass materials that are obtained from agricultural processes (Dai et al. 2018; Farhadi et al. 2021; Geremew 2017). Agricultural wastes can be considered to be renewable because they will always be produced as long as agricultural processes keep going on (Sud et al. 2008). The availability is assured for the long term because man will always require food for survival. These agricultural processes lead to the generation of a large amount of waste which can have a deleterious effect on various facets of the environment (Bhatnagar et al. 2015; Dungani et al. 2016). The utilisation of these agricultural wastes is an important solid waste management challenge (Dabagh et al. 2021; Iwuozor et al. 2021). Many of such application areas include thermo-chemical and biochemical processes. Carbonisation is one such process that can be used as a solid waste management technique (Ighalo, Igwegbe, and Adeniyi 2022). With proper activation, the technique can yield high quality activated carbon which has a variety of applications (Ali et al. 2022; Kumari et al. 2022).

One major advantage of using agricultural waste for adsorbent production is the low cost involved (Ighalo and Adeniyi 2020; Ighalo, Omoarukhe, et al. 2022). They are considered waste materials and do not have a competitive use (Azeez et al. 2020). Instead of expending energy to incinerate them, they can be used as feedstock for activated carbon production (Mbarki et al. 2022). An important factor is their low moisture content. High moisture content (>10%) would require an extended drying period (which would cost more time and/or energy). High moisture content is only favourable for compositing, bio-gas production and other biochemical processes. It is understood that a good amount of fixed carbon in the biomass is needed to achieve a high yield of activated carbon. Also, the relatively high volatile content of the biomass (>30%) is important as it helps to control the manufacturing process. Some agricultural wastes considered in the current study are *Hevea brasiliensis* (rubber) seed shells, *Mucuna pruriens* seed shells and *Dacryodes edulis* (African pear) seed kernels. They are very suitable for the current application because of their relatively high density, hardness, low initial moisture content and a good amount of volatile matter (Kan et al. 2016; Wanga et al. 2017). The pictorial representations of these agricultural wastes are shown in Figure 5.1(a–c).

DOI: 10.1201/9781003668961-5

FIGURE 5.1 (a–c): Pictorial representations of *Hevea brasiliensis* (rubber) seed shells (a), *Mucuna pruriens* seed shells (b) and *Dacryodes edulis* (African pear) seed kernels (c).

These agricultural wastes can be converted to activated carbon using thermochemical conversion processes (Ighalo 2024). These processes include pyrolysis (Jabar et al. 2022; Liao et al. 2022; Vuppaladadiyam et al. 2022), gasification (Atikah and Harun 2019; Ighalo, Adeniyi, and Igwegbe 2022; Tezer et al. 2022), torrefaction (Abdulyekeen et al. 2021; Lam et al. 2019; Tu et al. 2022) and retort carbonisation (Adeniyi et al. 2023; Lachos-Perez et al. 2022; Saliu et al. 2022). These processes are able to increase the surface area of the biomass due to the development of a superior pore structure (Qu et al. 2021). Increased surface area is very important for adsorption performance since it is a surface phenomenon (Nzediegwu et al. 2021). Many studies have examined the thermochemical conversion of a wide variety of biomass for adsorption applications (Anastopoulos et al. 2022; Ighalo, Igwegbe, and Adeniyi 2022; Kearns et al. 2019; Mokrzycki et al. 2020).

The current chapter examines three case studies of West African agricultural residues. To justify the novelty, the commonly studied rubber seed residues were compared with two rarely studied residues from *Mucuna pruriens* and African pear. Besides the investigation of the yield and physicochemical properties, the study bridges the gap between theory and practice and examines how these properties translate into actual performance in water treatment applications. The aim of the study was to evaluate the optimum process time and temperature for the conversion of some West African agricultural wastes into activated carbon (AC) with high specific surface area. The agricultural wastes considered in this study were: *Hevea brasiliensis* (rubber) seed shells, *Mucuna pruriens* seed shells and *Dacryodes edulis* (African pear) seed kernels. The basis of the optimised condition is the yield and specific surface area. This study sets the blueprint for an economically feasible large-scale activated carbon production process where the goal is to obtain as much yield as possible whilst maintaining quality.

5.2 MATERIALS AND METHODS

5.2.1 Collection and Preparation of Raw Materials

Hevea brasiliensis (rubber) seed shell is termed as RSS, *Mucuna pruriens* seed shell is termed as MSS and *Dacryodes edulis* (African pear) seed kernel is termed as APS. MSS were collected from Mucuna seed processing, Ogbete market, Enugu, Nigeria. MSS was then washed thoroughly with distilled water to remove dirt and oven-dried at 105°C till constant weight. The dried biomass was then ground and sieved to the target particle size of 1–2 mm. It was then stored in an air-tight container for further processing. RSS was obtained from the Rubber Research Institute of Nigeria, Benin City, Nigeria. It was also washed thoroughly with distilled water to make it clean before oven-drying at 105°C till constant weight. Similar to MSS, APS was ground and sieved to the desired particle size of 1–2 mm and stored in an air-tight container for further processing. APS was sourced from Awka environs, Anambra, Nigeria. The cleaning, drying and storage for further processing were the same as in the case of the previous two agricultural wastes. Similar parameters for collection and raw material preparation were used because we did not want them to become extraneous factors that would affect the difference in the yield and quality of the final product obtained.

5.2.2 Production of Activated Carbon

10g of each of the samples were weighed using the analytical balance and impregnated with 60% H_3PO_4 at a 1:1 weight ratio at room temperature (25 ± 2°C) for 24 h. It was placed in the muffle furnace (Model SX-2.5-10) and carbonised at different varying temperatures of 300, 400 and 500°C at different varying times of 1, 2 and 3 h under nitrogen (N_2) gas flow. The heating time was measured from the point when the set carbonisation temperature was achieved. The heating rate from ambient conditions to the set temperature was about 20°C/min. It was cooled and weighed to determine the final weight of the sample. The activated carbons obtained in this study still kept the naming convention (albeit with the AC added to the acronyms, to give MSS-AC, RSS-AC and APS-AC).

5.2.3 Determination of Product Properties

For the characterisation study, samples obtained at 300°C and 3 h were used. The physicochemical characteristics of the activated carbons were determined using the methods described in Nwabanne and Igbokwe (2012). Fourier Transform Infrared (FTIR) spectroscopy was used to elucidate the functional groups on the activated carbons in the wavenumber range of 4600–400 cm^{-1} (Shimadzu S8400, Shimadzu scientific instruments, Japan). X-ray Diffractometry (XRD) was used to determine the mineralogical composition and the diffraction pattern of the activated carbons. The XRD spectra of the activated carbons were obtained on a powder X-ray Diffractometer (Model Philips) using CuKα radiation and operating at 40.0 kV, 30 mA and 8.0 deg/min scanning speed. The spectra were obtained over a 2-theta (2θ) range of $2 - 60°$. The morphology was evaluated using Scanning Electron Microscopy (SEM) (Phenom ProX, Eindhoven Netherlands) and presented after 3D reconstruction. The SEM micrographs were obtained at ×4000 magnification and reconstructed using ImageJ v1.53 (Ighalo, Adeniyi, and Igwegbe 2021). The software was also used to calculate the average roughness (R_a) of the carbons. The original values in pixel (PX) were converted to the conventional values using the factor 1 PX = 264.5833 µm.

5.2.4 Potential Application in Water Treatment

As a preliminary trial, MSS-AC, RSS-AC and APS-AC were tested for Congo red dye removal from aqueous media. The samples used were those obtained at optimum production parameters. A 100 mg/L of the pollutant solution was prepared by dispersing a measured mass of Congo red (0.1 g).

Then, 0.1 g of the activated carbons were introduced separately in three Erlenmeyers containing 100 mL of the solution (to achieve a 1 g/L dosage) each already set at pH 2. The contents were stirred at a constant speed of 120 rpm for 60 min at room temperature (25 ± 2°C). The parameters were not optimised because this is a preliminary test. The activated carbons were separated from the treated solutions with the aid of a centrifuge for 15 min at 3000 rpm. The concentrate was measured for the residual dye concentration through a UV-vis spectrophotometer (Model UV-VIS 754, Yangzhou Wandong Medical Co. Ltd., China) at 419 nm. The percentage Congo red removal was determined via Equation 5.1:

$$\text{Percentage dye removal} = \frac{C_0 - C_e}{C_0} \times 100 \quad (5.1)$$

Where C_0 is the initial concentration of dye in the solution (mg/L); C_e is the final dye concentration of dye in the solution (mg/L).

5.3 RESULTS AND DISCUSSION

5.3.1 Activated Carbon Yield from the Different Biomasses

For the production of activated carbon to be financially feasible and economically sustainable, a high yield is required (Ighalo, Onifade, and Adeniyi 2021). The yield of the prepared activated carbon varied with temperature (300–500°C) and time of activation (1–3 h). The yield of activated carbons obtained at different temperatures and times are shown in Table 5.1 for MSS-AC, RSS-AC and APS-AC respectively. The percentage yield of activated carbons decreased with an increase in temperature and time. This is because the increase of activation temperature and time resulted in the more volatile components being lost, and hence a decreasing percentage yield. The highest yield of activated carbon was achieved at 300°C and 1 h. This yield was 34.2 wt% for MSS-AC, 38.3 wt% for RSS-AC and 35.0 wt% for APS-AC. This reveals that the eventual yield of activated carbon from the process is also dependent on the nature of the feedstock biomass materials used. This is because, different materials have different distributions of lignin, cellulose and hemicelluloses (Dai et al. 2018; Farhadi et al. 2021; Geremew 2017).

TABLE 5.1
Results for the Yield of Activated Carbon from *Mucuna pruriens* Seed Shells (MSS), *Hevea brasiliensis* (Rubber) Seed Shells (RSS), and *Dacryodes edulis* (African Pear) Seed Kernels (APS)

Temperature (°C)	300			400			500		
Time (h)	1	2	3	1	2	3	1	2	3
					MSS-AC				
Activation burn-off (%)	65.8	67.7	69.4	67.9	69.3	73.2	68.9	74.3	77.5
Yield of activated carbon (%)	34.2	32.3	30.6	32.1	30.7	26.8	31.1	25.7	22.5
					RSS-AC				
Activation burn-off (%)	61.7	63.2	67.5	64.4	67.6	72.7	66.8	71.6	76.6
Yield of activated carbon (%)	38.3	36.8	32.5	35.6	32.4	27.3	33.2	28.4	23.4
					APS-AC				
Activation burn-off (%)	65.0	67.5	74.2	66.7	68.9	73.3	69.5	72.3	82.6
Yield of activated carbon (%)	35.0	32.5	25.8	33.3	31.1	26.7	30.5	27.7	17.4

The carbonisation of biomass material is achievable at a reaction temperature above 250°C and can go as high as 600°C (Souza et al. 2021) with the carbonisation time ranging from 1–3 h. As a result of the threshold of biomass thermal disintegration, any modification route that does not reach the reaction temperatures between 250–350°C might not yield carbon (Ighalo, Rangabhashiyam, et al. 2022). Lower temperatures led to a higher yield of activated carbon. This is due to the nature of the carbonisation reactions, which are favoured at higher temperatures. Furthermore, higher temperatures lead to a greater intensity of breakdown of the agricultural wastes. Also, higher process time led to less yield. This is because the longer residence time facilitates the oxidation of the carbon species in the agricultural wastes and their subsequent volatilisation.

5.3.2 Physico-Chemical Properties of the Activated Carbons

The physico-chemical properties of the activated carbons are summarised in Table 5.2. The specific surface area is one of the most important qualities of activated carbon as it is one of the functional properties in many applications (Abdul Shukor et al. 2017; Oemar and Chang 2020). The specific surface area obtained was 841 m^2/g for APS-AC, 847 m^2/g for MSS-AC and 999 m^2/g for RSS-AC. Since the products were obtained at similar conditions, the differences in the specific surface area are due to the nature of the materials. It can be observed that RSS-AC will yield the highest quality activated carbon in this regard. The iodine number obtained was 717 mg/g for APS-AC, 725 mg/g for MSS-AC and 878 mg/g for RSS-AC. The iodine number tells us the amount of micropore present in the activated carbon (Xu et al. 2021). Micropores have pores of <2 nm diameter. The results reveal that RSS has the highest micropore content of the three activated carbons. Different specifications are required based on the intended application. For an intended application in water treatment, a minimum iodine number of 500 mg/g is advised by the American Water Works Association (AWWA). This number is influenced by the carbonisation temperature and time as they significantly affect the micropore formation during the process.

The pH values of the activated carbons were all close to neutral (7). The bulk density is also above the lower 0.25 g/cm^3 limit for good activated carbons. This means the three products meet the quality in this regard. The fixed carbon content of the products is quite high (though RSS has the highest amount). It is usually recommended that the carbon content is above 50% for it to be considered as an activated carbon material or biochar. This requirement is also met by the products. Low ash content is usually desirable for activated carbons because such ash can interfere with the pore structure development. It can be observed that relatively low ash content (5–6%) was achieved for the products for the current investigation.

TABLE 5.2
Physicochemical Properties of Activated Carbon from *Mucuna Pruriens* Seed Shells (MSS), *Hevea Brasiliensis* (Rubber) Seed Shells (RSS), and *Dacryodes Edulis* (African Pear) Seed Kernels (APS)

Samples	MSS-AC	RSS-AC	APS-AC
Moisture content (%)	7.0	3.0	9.0
Ash content (%)	5.38	5.14	5.47
Volatile matter (%)	14.25	12.32	17.40
Fixed carbon (%)	80.37	82.54	77.13
Bulk density (g/cm^3)	0.45	0.55	0.46
pH	7.3	6.6	6.7
Iodine number (mg/g)	725	878	717
Surface area (m^2/g)	847	999	841

5.3.3 Functional Groups and Crystallography of the Activated Carbons

The FTIR spectra for MSS-AC, RSS-AC and APS-ACare shown in the supplementary material (Figures S5.1–S5.3). The specific peaks and assignments are summarised in Table 5.3. Based on the peaks and assignments for MSS-AC, RSS-AC and APS-AC, it can be deduced that similar functional groups are present in the carbons. For materials of biomass origins, it is unsurprising that such functional groups are present (Ighalo, Ogunniyi, et al. 2022). Despite the thermal breakdown of the lignin, cellulose and hemicelluloses in the biomass, the end carbon product usually has these functional groups (Adeniyi et al. 2022; Ighalo, Iwuchukwu, et al. 2022). Such functional groups end up as good sites for the adsorption of pollutants from aqueous media (Lima et al. 2019; Liu et al. 2020; Nguyen et al. 2020). The XRD spectra for MSS-AC, RSS-AC and APS-ACare is shown in the supplementary material (Figures S5.4–S5.6). The specific peaks and assignments are summarised in

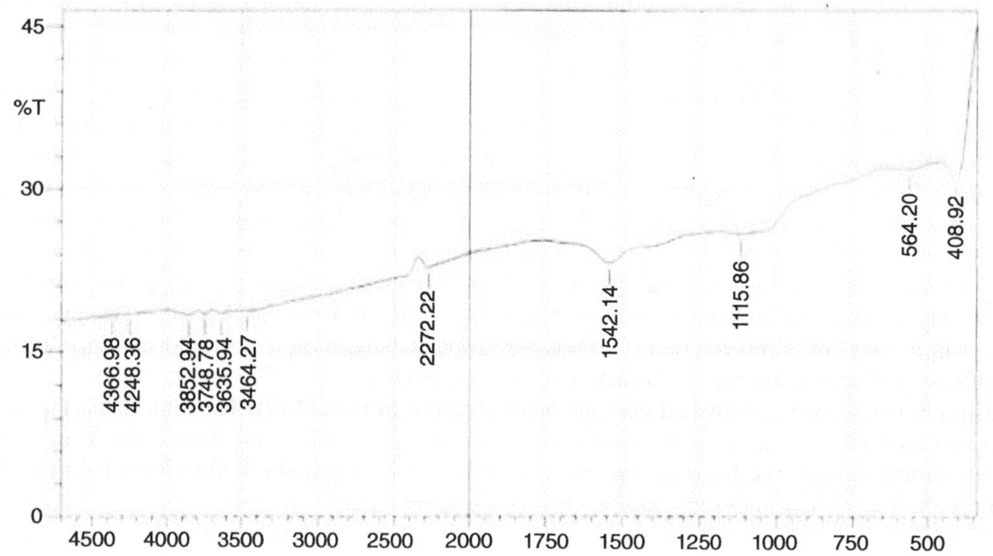

FIGURE S5.1 Supplementary materials.

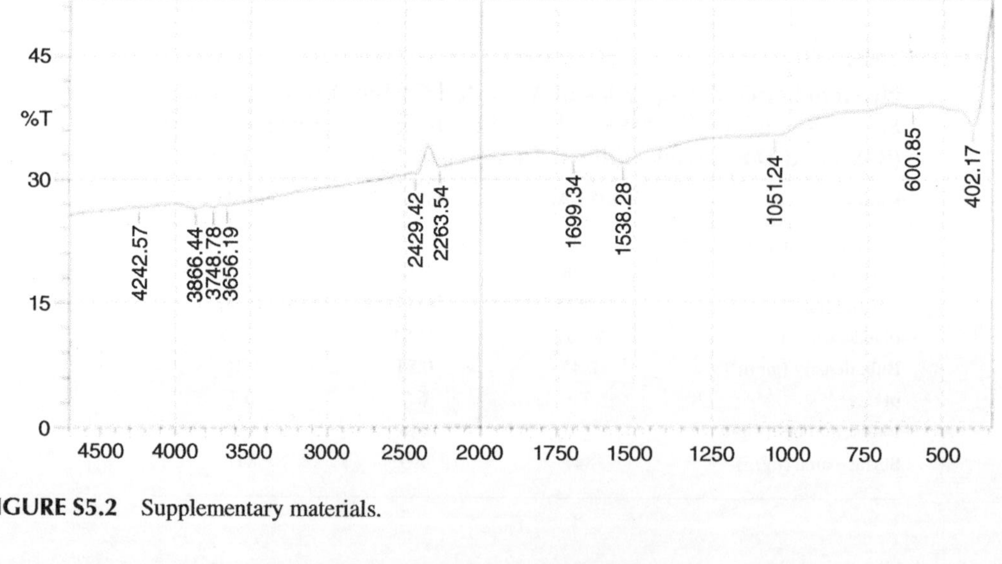

FIGURE S5.2 Supplementary materials.

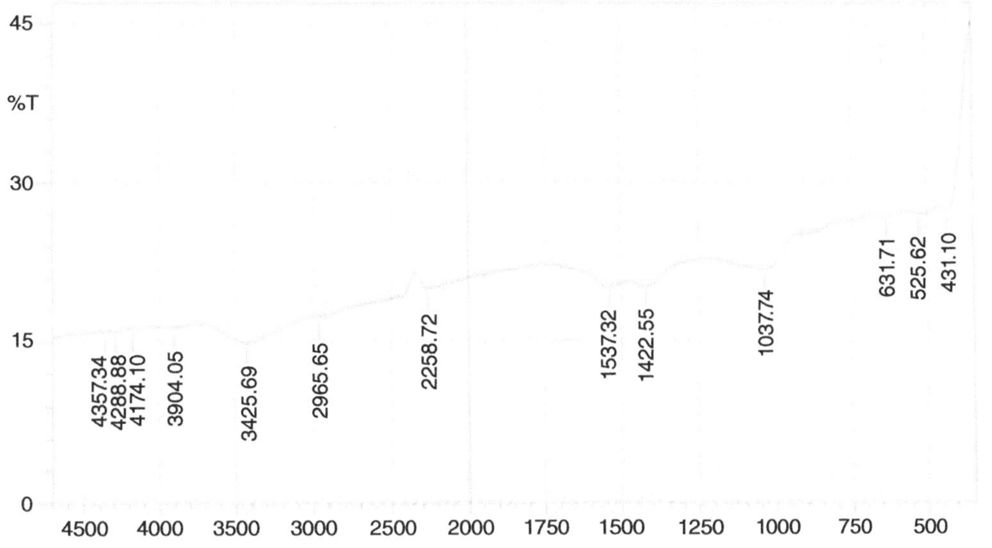

FIGURE S5.3 Supplementary materials.

TABLE 5.3
Summary of FTIR Peaks in Activated Carbon from *Mucuna Pruriens* Seed Shells (MSS), *Hevea Brasiliensis* (Rubber) Seed Shells (RSS), and *Dacryodes Edulis* (African Pear) Seed Kernels (APS)

Wave Number (cm^{-1})	Assignment	Functional Group
	MSS-AC	
1115.86	C – N stretching vibrations	Aliphatic amines
1542.14	N – O asymmetric stretching vibrations	-
3464.27	O – H stretching vibrations	Alcohols and phenols
3635.94	O – H stretching vibrations	Alcohols and phenols
3748.78	O – H stretching vibrations	Alcohols and phenols
3852.94	O – H stretching vibrations	Alcohols and phenols
	RSS-AC	
1051.24	C – N stretching vibrations	Aliphatic amines
1538.28	N – O asymmetric stretch	-
1699.34	C = O stretching vibrations	α, β – unsaturated aldehydes, ketones
3656.19	O – H stretching vibrations	Alcohols and phenols
3748.78	O – H stretching vibrations	Alcohols and phenols
3866.44	O – H stretching vibrations	Alcohols and phenols
	APS-AC	
1037.74	C – N stretching vibrations	Aliphatic amines
1537.32	N – O asymmetric stretching vibrations	Nitro compounds
2965.65	C – H stretching vibrations	Alkanes
3425.69	O – H stretching vibrations	Alcohols and phenols
3904.05	O – H stretching vibrations	Alcohols and phenols

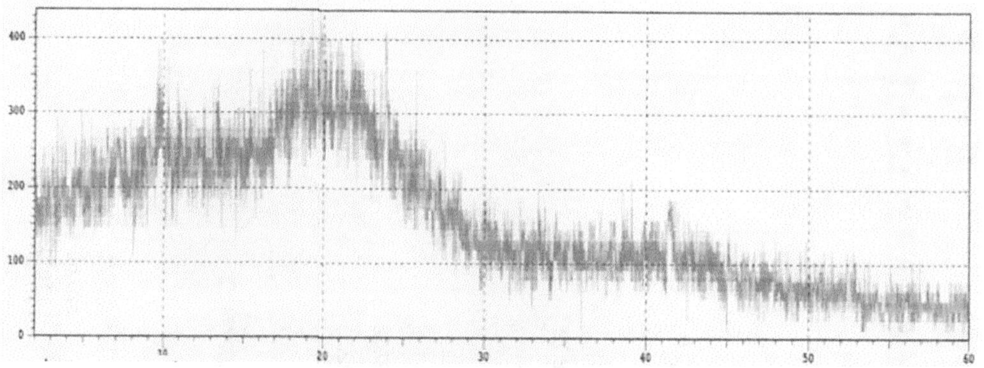

FIGURE S5.4 Supplementary materials.

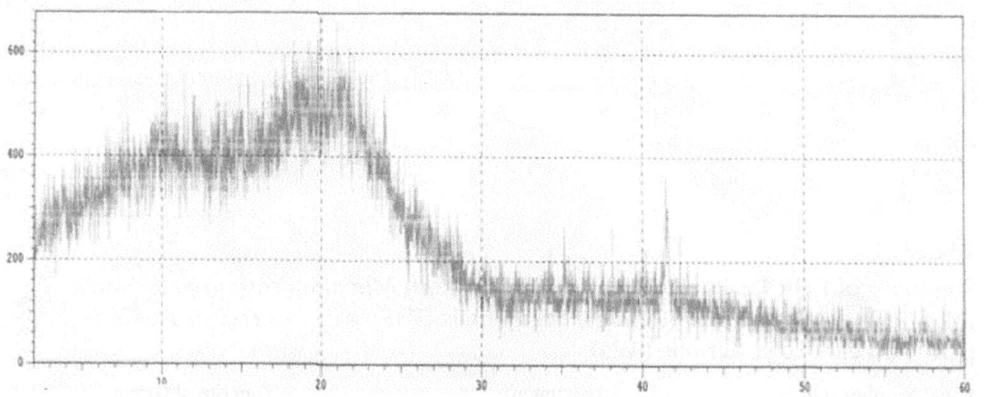

FIGURE S5.5 Supplementary materials.

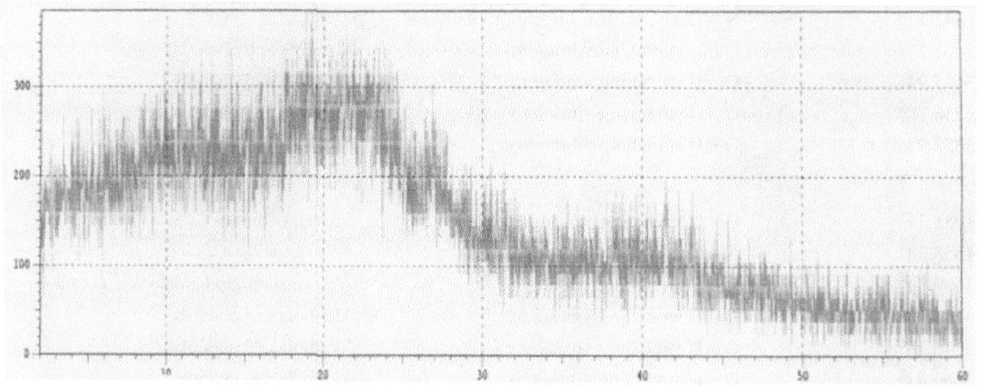

FIGURE S5.6 Supplementary materials.

Table 5.4. These help to elucidate the crystallites present in the activated carbon. It can be observed that MSS-AC and RSS-AC are majorly crystallites of sodium, aluminium and silicon. For APS-AC, there are further crystallites of chromium and magnesium. The crystalline composition is usually a representation of the ash content of the carbon. It is observed that the composition of each of these

TABLE 5.4
Summary of XRD Peaks in Activated Carbon from *Mucuna Pruriens* Seed Shells (MSS), *Hevea Brasiliensis* (Rubber) Seed Shells (RSS), and *Dacryodes Edulis* (African Pear) Seed Kernels (APS)

Mineral Name	Mineral Formula	Composition (%)
	MSS-AC	
Mordenite	$Na_2Al_2Si_{13}.3O_{29}.6+x$	0.765
Halloysite-7	$Al_2Si_2O_5(OH)_4$	0.622
Dickite-2	$Al_2Si_2O_5(OH)_4$	0.736
Sanid	$(Na, K)Si_3AlO_8$	0.506
Orthoclase	$KAlSi_3O_8$	0.697
Xonotlite	$Ca_6Si_6O_{17}(OH)_2$	0.595
Barite	$BaSO_4$	0.643
Petalite-1	$LiAlSi_4O_{10}$	0.658
	RSS-AC	
Halloysite-7	$Al_2Si_2O_5(OH)_4$	0.643
-	$Mg_3Si_4O_5(OH)_4$	0.679
Orthoclase	$KAlSi_3O_8$	0.644
Antigorite	$Mg_3Si_4O_5(OH)_4$	0.441
-	$Al_2Si_2O_5(OH)_4$	0.628
Cristobalite	SiO_2	0.545
	APS-AC	
Eskolaite	Cr_2O_3	0.602
Chrysotile	$Mg_3[Si_2\text{-}xO_5]\ (OH)_{4\text{-}4x}$	0.556
Antigorite	$Mg_3Si_4O_5(OH)_4$	0.517
Talc-2	$Mg_3Si_4O_{10}(OH)_2$	0.565
-	$(Na,Ca)_2(Fe,Mn)_3Fe_2(Si, Al)_8O_{22}(OH,F)$	0.551

species was usually <1%. The carbon itself is amorphous (El-Sayed & Nada 2017) hence it can be easily deduced that these crystals are for the ash content. They are usually non-volatile hence are retained in the solid product after much of the volatile matter has been released through the carbonisation process.

5.3.4 Morphology of the Activated Carbons

The 3D reconstruction of the surface morphology of MSS-AC, RSS-AC and APS-AC are shown in Figure 5.2(a–c). It can be observed that the activated carbons had a heterogeneous surface morphology. A lot of features can be observed on the surfaces and these range from ridges formations to short pikes and deep holes. Such surface heterogeneity is usually consistent with materials of biomass origin having undergone some form of thermochemical processing (Carrales-Alvarado et al. 2020; Marques et al. 2017). The extent of heterogeneity of the surfaces can also be compared quantitatively by computing the average surface roughness of the products in μm. The average roughness (R_a) value of MSS-AC, RSS-AC and APS-AC was computed to be 21.3×10^3 μm, 28.1×10^3 μm and 26.1×10^3 μm respectively. This suggests that RSS-AC had a more heterogeneous surface morphology in comparison to the others.

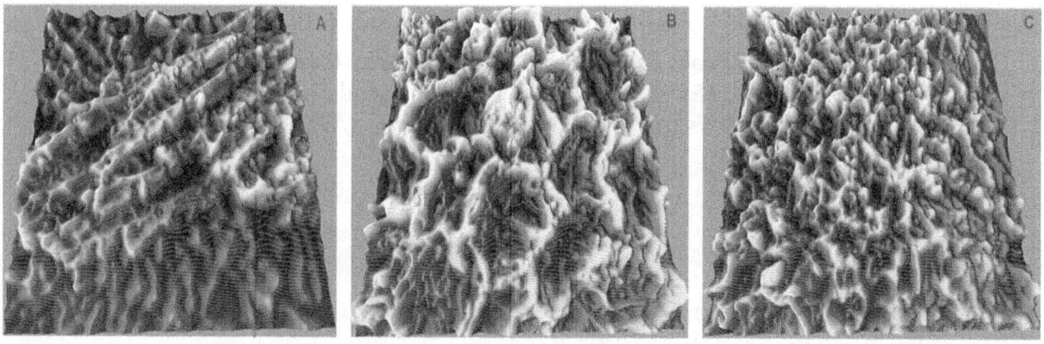

FIGURE 5.2 (a–c): 3D Reconstructed SEM micrographs of the surface morphology of activated carbon from *Mucuna pruriens* seed shells (a), *Hevea brasiliensis* (rubber) seed shells (b), and *Dacryodes edulis* (African pear) seed kernels (c). All micrographs are at a magnification of ×4,000.

5.3.5 Comparison of Product Quality with Other Activated Carbons

In this section, we compare the specific surface area obtained for MSS-AC, RSS-AC and APS-AC against those obtained from other agricultural waste. A comparison is shown in Table 5.5 and the activation agent, temperature and time were reported. The results were sorted in decreasing order of specific surface area. It can be observed that the specific surface area obtained is intermediate compared to those from other agricultural wastes. It should be noted that factors such as the type of activation agent, process temperature and time can seriously affect the specific surface area obtained (Gratuito et al. 2008; Srinivasakannan & Bakar 2004). Hence it will be difficult (and erroneous) to make generalisations from the comparison. The observations from the current study are positive because a lower temperature is required for the achievement of the results. This also suggests a lower energy requirement for the process. Furthermore, the activation agent used in this study is eco-friendly.

TABLE 5.5
Comparison of Specific Surface Area Results with Those of Other Agricultural Wastes

Feedstock	Specific Surface Area (m^2/g)	Activation Agent	Temp	Carbonisation Time	Ref.
Cocoa pod husk	1162	KOH	700°C	0.5 h	(Tsai, Bai, et al. 2019)
Hevea brasiliensis seed shells	999	H_3PO_4	300°C	3 h	This study
Soybean residue	927	K_2CO_3	600°C	2 h	(Wang et al. 2019)
Mucuna pruriens seed shells	847	H_3PO_4	300°C	3 h	This study
Dacryodes edulis seed kernels	841	H_3PO_4	300°C	3 h	This study
Sugarcane bagasse	829	CO_2	900°C	1 h	(Giusto et al. 2017)
Cocoa pod husk	716	CO_2	900°C	2 h	(Tsai, Jiang, and Lin 2019)
Cocoa pod husk	744	KOH	600°C	0.5 h	(Tsai, Bai, et al. 2019)

5.3.6 Practical Implications of the Study

In this section, we discuss the practical relevance of the observations of the current study. Firstly, the investigation is advantageous to solid waste management endeavours as it proffers a utilisation route for the specific agricultural wastes studied. Furthermore, since the feedstock is agricultural waste, we see that it helps in resource conservation and the achievement of the circular economy initiative (Umenweke et al. 2021). We have been able to identify that a good yield of high-quality activated carbon can be obtained. This yield and quality are very important in underlying the viability of the feedstock. When the yield is too low then energy cost increases. When the quality is also low, then potential applications are lesser and performance in these applications would not be up to standard. This leads us to the need to identify potential applications for the current product. Based on the properties, its use as an adsorbent is recommended (Rocha et al. 2020). The study is also able to derive a high specific surface area from the biomass converted. The high specific surface area and micropore volume ensure that the material can be used effectively for the removal of heavy metals, dyes and pharmaceuticals (Su 2017). Furthermore, the functional groups would serve as good points for physico-chemical interactions with these adsorbate species. The product will also be effective as a rheological and tribological stabiliser of lubricating oils due to its mineral content (Adeniyi et al. 2021).

5.3.7 Potential Application in Water Treatment

As a proof of concept and preliminary trial, the activated carbons prepared from agricultural wastes were tested for wastewater treatment considering a solution of Congo red dye as the pollutant. Dyes are among the most important wastewater pollutants in the world and adsorption is a key process in removing them from water (Al-Qahtani et al. 2022; Anastopoulos et al. 2021, 2023; Hevira et al. 2023). This was done to investigate the effectiveness of using MSS-AC, RSS-AC and APS-AC for dye removal from aqueous media. This is a very important property of adsorbents in the wastewater management sector. The samples used were those obtained at optimum production parameters and tests were done at similar experimental parameters. The test was conducted using a 100 mg/L congo red solution, 1.0 g/L activated carbon dosage, pH 2 (Yek et al. 2020), agitation speed of 120 rpm, contact time of 60 min and at room temperature (25 ± 2°C). The respective removal efficiencies for MSS-AC, RSS-AC and APS-AC were 97.8, 97.56 and 99.70 % at 60 minutes contact time. A low pH was optimum for the study because the pKa of Congo red is 4.1. Though Congo red is an acidic dye, it contains negatively charged sulfonated group ($–SO^{3-}$ Na^+) (Chatterjee et al. 2007). Below a pH of 4.1, the dye is protonated and its uptake is enhanced by electrostatic interactions and cation exchange with activated carbons. Other mechanisms are also involved in the uptake of Congo red by the activated carbon (summarised in Figure 5.3). Biochar usually has an amount of aromaticity based on its H/C ratio. Also, Congo red is a polycyclic aromatic compound. This suggests that Congo red can be adsorbed by π–π EDA interactions with MSS-AC, RSS-AC and APS-AC. The presence of heteroatoms like sulphur and nitrogen in the Congo red structure can also facilitate uptake by the formation of weak hydrogen bonds in the adsorptive system. Due to the porous nature of activated carbon, pore diffusion and filling is also significant in this case (Liu et al. 2022). From these values, it shows that the APS-AC performed exceptionally for Congo red reduction. This may be due to the more prominent pores observed on its surface viewed via the SEM analysis (Rangabhashiyam et al. 2022). Though RSS had the highest specific surface area, this did not transform to the highest removal efficiency. This justifies the complex nature of the physico-chemical interactions between the activated carbon and the dye (Adeyanju et al. 2022).

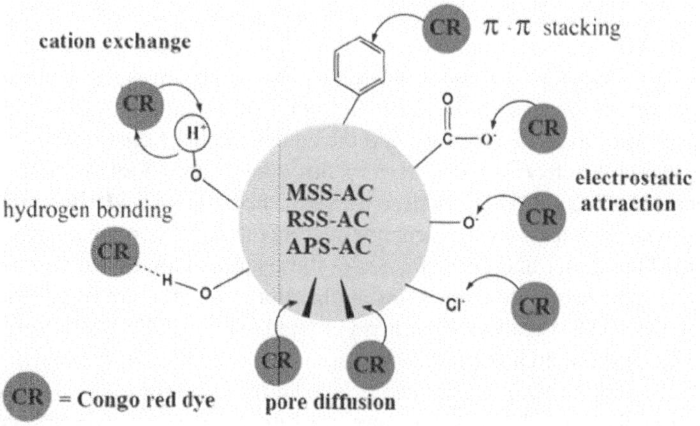

FIGURE 5.3 Summary of adsorption mechanisms for congo red adsorption onto activated carbon from *Mucuna pruriens* seed shells (MSS), *Hevea brasiliensis* (rubber) seed shells (RSS), and *Dacryodes edulis* (African pear) seed kernels (APS).

5.4 CONCLUSIONS

In this study, the optimum process time and temperature for the conversion of some West African agricultural wastes into activated carbon were studied. The highest yield of activated carbon was achieved at 300°C and 1 h. This yield was 34.2 wt% for *Mucuna pruriens* seed shells (MSS), 38.3 wt% for *Hevea brasiliensis* seed shells (RSS) and 35.0 wt% for *Dacryodes edulis* seed kernels (APS). Lower temperatures led to a higher yield of activated carbon. This was due to the nature of the carbonisation reactions which are favoured at higher temperatures. Furthermore, higher temperatures lead to a greater intensity of breakdown of the agricultural wastes. Also, higher process time led to less yield. This is because the longer residence time facilitates the oxidation of the carbon species in the agricultural wastes and their subsequent volatilisation. The specific surface area obtained was 841 m^2/g for APS-AC, 847 m^2/g for MSS-AC and 999 m^2/g for RSS-AC. Similar FTIR peaks were observed for the carbons. Furthermore, XRD revealed that the inorganic ash in the activated carbons was majorly crystallites of sodium, aluminium, silicon, chromium and magnesium. For a quantitative comparison of the surface heterogeneity, the average roughness (R_a) value of MSS-AC, RSS-AC and APS-AC was computed to be 21.3×10^3 µm, 28.1×10^3 µm and 26.1×10^3 µm respectively. Based on the findings of the study, it was discussed that the product can be used as an adsorbent for water treatment and as a rheological and tribological stabiliser of lubricating oils. Based on preliminary tests, Congo red removal efficiency for MSS-AC, RSS-AC and APS-AC were 97.8, 97.56 and 99.70 % respectively at 60 minutes' contact time. This confirms the assertion that the adsorbent can be effectively used for water treatment.

Disclosure statements

Availability of data and materials: Associated data is available as an electronic supplementary material to the article

Competing interests/Conflict of interest: We declare that there are no conflicts of interest.

Funding: Not applicable

Authors' contributions: CAI did the study conception and design. Material preparation, data collection and analysis were performed by CAI and JOI. The first draft of the manuscript was written by JOI and CAI. All authors read and approved the final manuscript.

Acknowledgements: All cited authors are acknowledged

REFERENCES

Abdul Shukor N.S., Ismail K., Deris R.R.R., Alis A.B., Mohd Ishak M.A. (2017) Production and Characterisation of Single and Mixed Activated Carbons from Coconut Shell and Rubber Seed Pericarp Using Microwave-Induced Chemical Activating Agent. In: *Materials Science Forum*. Trans Tech Publications Ltd., Vol. 889, pp 209–214. https://doi.org/10.4028/www.scientific.net/MSF.889.209

Abdulyekeen K.A., Umar A.A., Patah M.F.A., Daud W.M.A.W. (2021) Torrefaction of biomass: Production of enhanced solid biofuel from municipal solid waste and other types of biomass. *Renewable and Sustainable Energy Reviews* 150:111436. https://doi.org/10.1016/j.rser.2021.111436

Adeniyi A.G., Abdulkareem S.A., Ighalo J.O., Amosa M.K., Popoola A.O., Ogunniyi S., Abdulkareem M.T. (2021) Usage of biomass-based carbon materials as lubricant additive: Effects on rheological and tribological properties. *Letters in Applied NanoBioScience 10*:2861–2868. https://doi.org/10.33263/LIANBS104.28612868

Adeniyi A.G., Ighalo J.O., Iwuozor K.O., Amoloye M.A. (2022) A study on the thermochemical co-conversion of poultry litter and elephant grass to biochar. *Clean Technologies and Environmental Policy* 24: 2193–2202. https://doi.org/10.1007/s10098-022-02311-3

Adeniyi A.G., Iwuozor K.O., Adeleke J., Emenike E.C., Micheal K.T., Ighalo J.O. (2023) Production and characterization of neem leaves: Effect of two different retort carbonization systems. *Bioresource Technology Reports*:101597. https://doi.org/10.1016/j.biteb.2023.101597

Adeyanju C.A., Ogunniyi S., Rangabhashiyam S., Oniye M.M., Ajala O.J., Adeniyi A.G., Igwegbe C.A., Ighalo J.O. (2022) Recent advances on the aqueous phase adsorption of carbamazepine. *ChemBioEng Reviews* 9:231–247. https://doi.org/10.1002/cben.202100042

Ali J., Bakhsh E.M., Hussain N., Bilal M., Akhtar K., Fagieh T.M., Danish E.Y., Asiri A.M., Su X., Khan S.B. (2022) A new biosource for synthesis of activated carbon and its potential use for removal of methylene blue and eriochrome black T from aqueous solutions. *Industrial Crops and Products* 179:114676. https://doi.org/10.1016/j.indcrop.2022.114676

Al-Qahtani S.D., Snari R.M., Alamrani N.A., Aljuhani E., Bayazeed A., Aldawsari A.M., El-Metwaly N.M. (2022) Synthesis and adsorption properties of fibrous-like aerogel from acylhydrazone polyviologen: Efficient removal of reactive dyes from wastewater. *Journal of Materials Research and Technology* 18:1822–1833

Anastopoulos I., Ahmed M.J., Ojukwu V.E., Danish M., Stylianou M., Ighalo J.O. (2023) A comprehensive review on adsorption of reactive red 120 dye using various adsorbents. *Journal of Molecular Liquids* 394:123719. https://doi.org/10.1016/j.molliq.2023.123719

Anastopoulos I., Giannopoulos G., Islam A., Ighalo J.O., Iwuchukwu F.U., Pashalidis I., Kalderis D., Giannakoudakis D.A., Nair V., Lima E.C. (2022) Potential Environmental Applications of Hellianthus Annuum (Sunflower) Residue - Based Adsorbents for Dye Removal in (Waste) Waters. In: Anastopoulos I, Lima EC, Meili L, Giannakoudakis DA (eds) *Biomass-Derived Materials for Environmental Applications*. Elsevier, Chennai, India, pp 307–318. https://doi.org/10.1016/B978-0-323-91914-2.00008-8

Anastopoulos I., Ighalo J.O., Igwegbe C.A., Giannakoudakis D.A., Triantafyllidis K.S., Pashalidis I., Kalderis D. (2021) Sunflower-biomass derived adsorbents for toxic/heavy metals removal from wastewater. *Journal of Molecular Liquids* 342:117540. https://doi.org/10.1016/j.molliq.2021.117540

Atikah M., Harun R. (2019) Simulation and optimization of chlorella vulgaris gasification using aspen plus. *Process Integration and Optimization for Sustainability*:1–9. https://doi.org/10.1007/s41660-019-0080-7

Azeez L., Adejumo A.L., Asaolu S.S., Adeoye M.D., Adetoro R.O. (2020) Functionalization of rice husks with ortho-phosphoric acid enhanced adsorptive capacity for anionic dye removal. *Chemistry Africa* 3:457–467. https://doi.org/10.1007/s42250-020-00142-7

Bhatnagar A., Sillanpää M., Witek-Krowiak A. (2015) Agricultural waste peels as versatile biomass for water purification–a review. *Chemical Engineering Journal* 270:244–271. https://doi.org/10.1016/j.cej.2015.01.135

Carrales-Alvarado D.H., Leyva-Ramos R., Rodríguez-Ramos I., Mendoza-Mendoza E., Moral-Rodríguez A.E. (2020) Adsorption capacity of different types of carbon nanotubes towards metronidazole and dimetridazole antibiotics from aqueous solutions: Effect of morphology and surface chemistry. *Environmental Science and Pollution Research* 27:17123–17137. https://doi.org/10.1007/s11356-020-08110-x

Chatterjee S., Chatterjee S., Chatterjee B.P., Guha A.K. (2007) Adsorptive removal of congo red, a carcinogenic textile dye by chitosan hydrobeads: Binding mechanism, equilibrium and kinetics. *Colloids and Surfaces A: Physicochemical and Engineering Aspects* 299:146–152. https://doi.org/10.1016/j.colsurfa.2006.11.036

Dabagh A., Bagui A., Abali M.H., Aziam R., Chiban M., Sinan F., Zerbet M. (2021) Increasing the adsorption efficiency of methylene blue by acid treatment of the plant carpobrotus edulis. *Chemistry Africa* 4: 585–598. https://doi.org/10.1007/s42250-021-00233-z

Dai Y., Sun Q., Wang W., Lu L., Liu M., Li J., Yang S., Sun Y., Zhang K., Xu J. (2018) Utilizations of agricultural waste as adsorbent for the removal of contaminants: A review. *Chemosphere* 211:235–253. https://doi.org/10.1016/j.chemosphere.2018.06.179

Dungani R., Karina M., Subyakto A.S., Hermawan D., Hadiyane A. (2016) Agricultural waste fibers towards sustainability and advanced utilization: A review. *Asian Journal of Plant Sciences* 15:42–55. https://doi.org/10.3923/ajps.2016.42.55

El-Sayed M., Nada A.A. (2017) Polyethylenimine– functionalized amorphous carbon fabricated from oil palm leaves as a novel adsorbent for Cr (VI) and Pb (II) from aqueous solution. *Journal of Water Process Engineering* 16:296–308. https://doi.org/10.1016/j.jwpe.2017.02.012

Farhadi A., Ameri A., Tamjidi S. (2021) Application of agricultural wastes as a low-cost adsorbent for removal of heavy metals and dyes from wastewater: A review study. *Physical Chemistry Research* 9:211–226. https://doi.org/10.22036/pcr.2021.256683.1852

Geremew B. (2017) A review on elimination of heavy metals from wastewater using agricultural wastes as adsorbents. *Science Journal of Analytical Chemistry* 5:72–75. https://doi.org/10.11648/j.sjac.20170505.12

Giusto L.A.R., Pissetti F.L., Castro T.S., Magalhães F. (2017) Preparation of activated carbon from sugarcane bagasse soot and methylene blue adsorption. *Water, Air, & Soil Pollution* 228:249. https://doi.org/10.1007/s11270-017-3422-5

Gratuito M.K.B., Panyathanmaporn T., Chumnanklang R.A., Sirinuntawittaya N., Dutta A. (2008) Production of activated carbon from coconut shell: Optimization using response surface methodology. *Bioresources Technology* 999:4887–4895. https://doi.org/10.1016/j.biortech.2007.09.042

Hevira L., Ighalo J.O., Sondari D. (2023) Chitosan-based polysaccharides for effective synthetic dye adsorption. *Journal of Molecular Liquids* 393:123604. https://doi.org/10.1016/j.molliq.2023.123604

Ighalo J.O. (2024) Biowastes and derived green sorbents for water decontamination: insights on thermochemical conversion strategies. *Current Opinion in Green and Sustainable Chemistry*:100880. https://doi.org/10.1016/j.cogsc.2024.100880

Ighalo J.O., Adeniyi A.G. (2020) Biomass to biochar conversion for agricultural and environmental applications in nigeria: Challenges, peculiarities and prospects. *Materials International* 2:111–116. https://doi.org/10.33263/Materials22.111116

Ighalo J.O., Adeniyi A.G., Igwegbe C.A. (2021) 3D reconstruction and morphological analysis of electrostimulated hyperthermophile biofilms of Thermotoga neapolitana. *Biotechnology Letters* 43:1303–1309. https://doi.org/10.1007/s10529-021-03123-z

Ighalo J.O., Adeniyi A.G., Igwegbe C.A. (2022) Computer-aided modelling of solid waste conversion: case study of maize (Zea mays) residues air gasification. In: Marques G, Ighalo JO (eds) *Current Trends and Advances in Computer-aided Intelligent Environmental Data Engineering*. Elsevier, Chennai, India, pp 381–391. https://doi.org/10.1016/b978-0-323-85597-6.00017-3

Ighalo J.O., Igwegbe C.A., Adeniyi A.G. (2022) The Utilisation of Rubber (Hevea brasiliensis) Seed Shells as Adsorbent for Water Pollution Remediation. In: Anastopoulos I, Lima EC, Meili L, Giannakoudakis DA (eds) *Biomass-Derived Materials for Environmental Applications*. Elsevier, Chennai, India, pp 13–27. https://doi.org/10.1016/B978-0-323-91914-2.00005-2

Ighalo J.O., Iwuchukwu F.U., Eyankware O.E., Iwuozor K.O., Olotu K., Bright C.E., Igwegbe C.A. (2022) Flash pyrolysis of biomass: A review of recent advances. *Clean Technologies and Environmental Policy* 24:2349–2363. https://doi.org/10.1007/s10098-022-02339-5

Ighalo J.O., Ogunniyi S., Adeniyi A.G., Igwegbe C.A., Sanusi S.K., Adeyanju C.A. (2022) Competitive adsorption of heavy metals in a quaternary solution by sugarcane bagasse – LDPE hybrid biochar: Equilibrium isotherm and kinetics modelling. *Chemical Product and Process Modelling* 18:231–246. https://doi.org/10.1515/cppm-2021-0056

Ighalo J.O., Omoarukhe F.O., Ojukwu V.E., Iwuozor K.O., Igwegbe C.A. (2022) Cost of adsorbent preparation and usage in wastewater treatment: A review. *Cleaner Chemical Engineering* 3:100042. https://doi.org/10.1016/j.clce.2022.100042

Ighalo J.O., Onifade D.V., Adeniyi A.G. (2021) Retort-heating carbonisation of almond (terminalia catappa) leaves and LDPE waste for biochar production: Evaluation of product quality. *International Journal of Sustainable Engineering* 14:1059–1067. https://doi.org/10.1080/19397038.2021.1886371

Ighalo J.O., Rangabhashiyam S., Dulta K., Umeh C.T., Iwuozor K.O., Aniagor C.O., Eshiemogie S.O., Iwuchukwu F.U., Igwegbe C.A. (2022) Recent advances in hydrochar application for the adsorptive removal of wastewater pollutants. *Chemical Engineering Research and Design* 184:419–456. https://doi.org/10.1016/j.cherd.2022.06.028

Iwuozor K.O., Ighalo J.O., Ogunfowora L.A., Adeniyi A.G., Igwegbe C.A. (2021) An empirical literature analysis of adsorbent performance for methylene blue uptake from aqueous media. *Journal of Environmental Chemical Engineering* 9:105658. https://doi.org/10.1016/j.jece.2021.105658

Jabar J.M., Odusote Y.A., Ayinde Y.T., Yılmaz M. (2022) African almond (Terminalia catappa L) leaves biochar prepared through pyrolysis using H3PO4 as chemical activator for sequestration of methylene blue dye. *Results in Engineering* 14:100385

Kan T., Strezov V., Evans T.J. (2016) Lignocellulosic biomass pyrolysis: A review of product properties and effects of pyrolysis parameters. *Renewable and Sustainable Energy Reviews* 57:1126–1140. https://doi.org/10.1016/j.rser.2015.12.185

Kearns J.P., Shimabuku K.K., Knappe D.R., Summers R.S. (2019) High temperature co-pyrolysis thermal air activation enhances biochar adsorption of herbicides from surface water. *Environmental Engineering Science* 36:710–723

Kumari M., Chaudhary G.R., Chaudhary S., Umar A. (2022) Transformation of solid plastic waste to activated carbon fibres for wastewater treatment. *Chemosphere*:133692. https://doi.org/10.1016/j.chemosphere.2022.133692

Lachos-Perez D., César Torres-Mayanga P., Abaide E.R., Zabot G.L., De Castilhos F. (2022) Hydrothermal carbonization and liquefaction: differences, progress, challenges, and opportunities. *Bioresource Technology* 343:126084. https://doi.org/10.1016/j.biortech.2021.126084

Lam S.S., Tsang Y.F., Yek P.N.Y., Liew R.K., Osman M.S., Peng W., Lee W.H., Park Y.-K. (2019) Co-processing of oil palm waste and waste oil via microwave co-torrefaction: A waste reduction approach for producing solid fuel product with improved properties. *Process Safety and Environmental Protection* 128:30–35.

Liao W., Zhang X., Ke S., Shao J., Yang H., Zhang S., Chen H. (2022) Effect of different biomass species and pyrolysis temperatures on heavy metal adsorption, stability and economy of biochar. *Industrial Crops and Products* 186:115238. https://doi.org/10.1016/j.indcrop.2022.115238

Lima D.R., Hosseini-Bandegharaei A., Thue P.S., Lima E.C., de Albuquerque Y.R.T., dos Reis G.S., Umpierres C.S., Dias S.L.P., Tran H.N. (2019) Efficient acetaminophen removal from water and hospital effluents treatment by activated carbons derived from Brazil nutshells. *Colloids and Surfaces A: Physicochemical and Engineering Aspects* 583:123966. https://doi.org/10.1016/j.colsurfa.2019.123966

Liu L., Cui W., Lu C., Zain A., Zhang W., Shen G., Hu S., Qian X. (2020) Analyzing the adsorptive behavior of Amoxicillin on four Zr-MOFs nanoparticles: Functional groups dependence of adsorption performance and mechanisms. *Journal of Environmental Management* 268:110630. https://doi.org/10.1016/j.jenvman.2020.110630

Liu T., Lawluvy Y., Shi Y., Ighalo J.O., He Y., Zhang Y., Yap P.-S. (2022) Adsorption of cadmium and lead from aqueous solution using modified biochar: A review. *Journal of Environmental Chemical Engineering* 10:106502. https://doi.org/10.1016/j.jece.2021.106502

Marques S.C., Marcuzzo J.M., Baldan M.R., Mestre A.S., Carvalho A.P. (2017) Pharmaceuticals removal by activated carbons: Role of morphology on cyclic thermal regeneration. *Chemical Engineering Journal* 321:233–244. https://doi.org/10.1016/j.cej.2017.03.101

Mbarki F., Selmi T., Kesraoui A., Seffen M. (2022) Low-cost activated carbon preparation from Corn stigmata fibers chemically activated using H3PO4, ZnCl2 and KOH: Study of methylene blue adsorption, stochastic isotherm and fractal kinetic. *Industrial Crops and Products* 178:114546. https://doi.org/10.1016/j.indcrop.2022.114546

Mokrzycki J., Lorenc-Grabowska E., Kordek-Khalil K., Rutkowski P. (2020) Hydrothermal and pyrolytic biochars from waste milk thistle (Silybum marianum) extrudates as precursors for production of effective isoproturon adsorbents. *Journal of Water Process Engineering* 37:101459.

Nguyen D.T., Tran H.N., Juang R.S., Dat N.D., Tomul F., Ivanets A., Woo S.H., Hosseini-Bandegharaei A., Nguyen V.P., Chao H.P. (2020) Adsorption process and mechanism of acetaminophen onto commercial activated carbon. *Journal of Environmental Chemical Engineering* 8:104408. https://doi.org/10.1016/j.jece.2020.104408

Nwabanne J.T., Igbokwe P.K. (2012) Thermodynamic and kinetic behaviors of lead (II) adsorption on activated carbon derived from Palmyra palm nut. *International Journal of Applied Science and Technology* 2:245–254.

Nzediegwu C., Naeth M.A., Chang S.X. (2021) Lead (II) adsorption on microwave-pyrolyzed biochars and hydrochars depends on feedstock type and production temperature. *Journal of Hazardous Materials* 412:125255.

Oemar B., Chang W.-C. (2020) Taguchi method for optimizing process parameters in the production of activated carbon from rubber seed shell. *International Journal Of Advanced Manufacturing Technology* 107:4609–4620. https://doi.org/10.1007/s00170-020-05344-4

Qu J., Wang S., Jin L., Liu Y., Yin R., Jiang Z., Tao Y., Huang J., Zhang Y. (2021) Magnetic porous biochar with high specific surface area derived from microwave-assisted hydrothermal and pyrolysis treatments of water hyacinth for Cr (VI) and tetracycline adsorption from water. *Bioresource Technology* 340:125692.

Rangabhashiyam S., dos Santos Lins P. V., de Magalhães Oliveira L. M., Sepulveda P., Ighalo J. O., Rajapaksha A. U., Meili L. (2022) Sewage sludge-derived biochar for the adsorptive removal of wastewater pollutants: A critical review. *Environmental Pollution* 293:118581. https://doi.org/10.1016/j.envpol.2021.118581

Rocha L.S., Pereira D., Sousa E., Otero M., Esteves V.I., Calisto V. (2020) Recent advances on the development and application of magnetic activated carbon and char for the removal of pharmaceutical compounds from waters: A review. *Science of the Total Environment* 718:137272. https://doi.org/10.1016/j.scitotenv.2020.137272

Saliu O.D., Adeniyi A.G., Mamo M., Ndungu P., Ramontja J. (2022) Microwave exfoliation of a biochar obtained from updraft retort carbonization for supercapacitor fabrication. *Electrochemistry Communications* 139:107308.

Souza E.C., Pimenta A.S., Silva A.J.F., Nascimento P.F.P., Ighalo J.O. (2021) HNO3-treated eucalyptus charcoal: A sustainable biosorbent for removing heavy metals from aqueous solutions. *Biomass Conversion and Biorefinery* 13:4105–4119. https://doi.org/10.1007/s13399-021-01431-y

Srinivasakannan C., Bakar M.Z.A. (2004) Production of activated carbon from rubber wood sawdust. *Biomass Bioenergy* 27:89–96. https://doi.org/10.1016/j.biombioe.2003.11.002

Su C. (2017) Environmental implications and applications of engineered nanoscale magnetite and its hybrid nanocomposites: A review of recent literature. *Journal of Hazardous Materials* 322:48–84. https://doi.org/10.1016/j.jhazmat.2016.06.060

Sud D., Mahajan G., Kaur M. (2008) Agricultural waste material as potential adsorbent for sequestering heavy metal ions from aqueous solutions–A review. *Bioresource Technology* 99:6017–6027. https://doi.org/10.1016/j.biortech.2007.11.064

Tezer Ö., Karabağ N., Öngen A., Çolpan C.Ö., Ayol A. (2022) Biomass gasification for sustainable energy production: A review. *International Journal of Hydrogen Energy* 47:15419–15433. https://doi.org/10.1016/j.ijhydene.2022.02.158

Tsai W.-T., Bai Y.-C., Lin Y.-Q., Lai Y.-C., Tsai C.-H. (2019) Porous and adsorption properties of activated carbon prepared from cocoa pod husk by chemical activation. *Biomass Conversion and Biorefinery* 10:35–43. https://doi.org/10.1007/s13399-019-00403-7

Tsai W.-T., Jiang T.-J., Lin Y.-Q. (2019) Conversion of de-ashed cocoa pod husk into high-surface-area microporous carbon materials by CO 2 physical activation. *Journal of Material Cycles and Waste Management* 21:308–314. https://doi.org/10.1007/s10163-018-0791-9

Tu R., Sun Y., Wu Y., Fan X., Cheng S., Jiang E., Xu X. (2022) The fuel properties and adsorption capacities of torrefied camellia shell obtained via different steam-torrefaction reactors. *Energy* 238:121969. https://doi.org/10.1016/j.energy.2021.121969

Umenweke G., Ighalo J.O., Anusi M., Itabana B., Ekeh L. (2021) Selected thermo-chemical biorefining: Evaluation of the current trends and progressions. *European Journal of Sustainable Development Research* 5:em0154. https://doi.org/10.21601/ejosdr/10812

Vuppaladadiyam A.K., Vuppaladadiyam S.S.V., Awasthi A., Sahoo A., Rehman S., Pant K.K., Murugavelh S., Huang Q., Anthony E., Fennel P. (2022) Biomass pyrolysis: A review on recent advancements and green hydrogen production. *Bioresource Technology*:128087. https://doi.org/10.1016/j.biortech.2022.128087

Wang B., Zhai Y., Wang T., Li S., Peng C., Wang Z., Li C., Xu B. (2019) Fabrication of bean dreg-derived carbon with high adsorption for methylene blue: Effect of hydrothermal pretreatment and pyrolysis process. *Bioresource Technology* 274:525–532. https://doi.org/10.1016/j.biortech.2018.12.022

Wanga S., Daia G., Yang H., Luo Z. (2017) Lignocellulosic biomass pyrolysis mechanism: Astate-of-the-art review. *Progress in Energy and Combustion Science* 62:33–86.

Xu Y., Lu Z., Sun W., Zhang X. (2021) Influence of pore structure on biologically activated carbon performance and biofilm microbial characteristics. *Frontiers of Environmental Science & Engineering* 15:1–13. https://doi.org/10.1007/s11783-021-1419-1

Yek P.N.Y., Peng W., Wong C.C., Liew R.K., Ho Y.L., Mahari W.A.W., Azwar E., Yuan T.Q., Tabatabaei M., Aghbashlo M. (2020) Engineered biochar via microwave CO2 and steam pyrolysis to treat carcinogenic Congo red dye. *Journal of Hazardous Materials* 395:122636. https://doi.org/10.1016/j.jhazmat.2020.122636

6 Global Situation of Environmental Pollution and Its Sustainable Remedial Solutions

João Paulo Tenório da Silva Santos, Luiz Daniel da Silva Neto, Adeildo Junior de Oliveira, Thamyres Fernandes Messa Moreira, Vinicius Del Colle, Lucas Meili, and Danilo Henrique da Silva Santos

6.1 INTRODUCTION

Globalization, industrial activities, technological development, free international trade, and the economic environment are some factors responsible for the exceptional progress of the global economy in recent years (Mohammed et al., 2024). The interplay of these factors, coupled with the facilitation of free international trade, has fostered remarkable advancements in various sectors, fostering economic development on a global scale. However, although they play a crucial role in promoting the development and growth of countries, regions, and communities, the progress of economic activities has contributed negatively to the preservation of the environment, currently being considered the most serious threat to the global economy, society, and human beings (F. Wang & Taghvaee, 2023).

Drastic environmental changes, such as the degradation of ecosystems, soil desertification, and, mainly, rapid climate change – caused by large emissions of greenhouse gases, such as carbon dioxide (CO_2), nitrous oxide (N_2O), and methane (CH_4), which have directly influenced the availability and accessibility of natural resources essential to human life, such as water, increasing its demand, while limiting its availability (de Liz Arcari et al., 2023; Kurniawan et al., 2023) – are some of the main global issues arising from economic development (Ibrahim & Vo, 2021; Yu et al., 2022). According to the World Health Organization, this will have an important impact on the environment and the population's health since around 25% of diseases that affect humans are caused by environmental pollution in the form of water, soil, and air contamination (Nayak & Pathan, 2023).

In this sense, developing more sustainable economic policies that promote economic growth while protecting the environment has become increasingly urgent and has been widely discussed with global leaders, including governments, companies, nonprofit organizations, and individuals, to achieve these objectives (Yu & Zhao, 2022). The imperative to develop more sustainable economic policies that simultaneously foster economic growth and safeguard the environment has gained widespread recognition in recent years. This urgency has prompted extensive discussions among global leaders, including governments, corporations, nonprofit organizations, and individuals, as they collectively seek to address the pressing challenges posed by the intersection of economic development and environmental preservation.

One key aspect of this discourse revolves around the need for a paradigm shift in policymaking, wherein sustainability becomes a central tenet of economic strategies. Governments worldwide are being called upon to take a proactive role in formulating and implementing policies that prioritize

DOI: 10.1201/9781003668961-6

ecological integrity alongside economic advancement. This involves crafting regulations encouraging environmentally responsible practices, penalizing unsustainable activities, and fostering innovation in green technologies. As pivotal actors in the global economy, corporations are increasingly held accountable for their environmental impact. The business sector is urged to adopt sustainable practices, embracing a triple-bottom-line approach that considers not only financial gains but also social and environmental factors. This paradigm shift often involves incorporating circular economy principles, investing in renewable energy sources, and reevaluating supply chain practices to minimize carbon footprints.

Considered one of the biggest problems today, given its crucial importance for global socioeconomic development (Nayak & Pathan, 2023), the global water crisis is a multifaceted challenge threatening communities, ecosystems, and economies worldwide. Rapid population growth, urbanization, and industrialization have placed an unprecedented strain on freshwater resources. The consequences of the water crisis are far reaching, impacting agriculture, energy production, and public health. Insufficient access to clean and safe water perpetuates a cycle of poverty and exacerbates social inequalities, particularly affecting marginalized communities that often bear the brunt of inadequate water infrastructure and services. The global water crisis originates from human activities and, mainly, from the poor management of water resources by conventional treatment systems, which generates a great impact on the quality of the water distributed since these systems are the junction point between sewage and natural sources of water, such as rivers, lakes, reservoirs, and groundwater (Huang et al., 2022; Jiang et al., 2018). Addressing the water crisis requires a comprehensive and collaborative approach. Sustainable water management practices, including efficient water use in agriculture, protecting watersheds, and implementing water-saving technologies, are crucial for ensuring long-term water availability. Additionally, investing in water treatment and distribution infrastructure is essential to provide safe drinking water to all. International cooperation is vital to tackle transboundary water issues and share best practices in water management.

The presence of various hazardous materials such as trace metals, pesticides, dyes, radionuclides, organic, inorganic, microplastics, radioactive nuclides, pharmaceutical compounds, and emerging and diverse pollutants in aquatic environments can cause many health problems for humanity and other living beings (Raheem et al., 2023). It is necessary to develop alternative methodologies to remove these varied classes of pollutants resistant to conventional treatments and ensure that the treated water is healthy for reuse or consumption (Kıdak and Doğan, 2018; Paulino et al., 2022). Various processes, including ion exchange, co-precipitation, adsorption, filtration, chemical oxidation, biochemical processes, membrane separation, and so on, have been studied and improved for these purposes. In this sense, this chapter intends to outline the situation of global environmental pollution and the impacts on global socioeconomic development and public health, focusing on public policies adopted to mitigate and eliminate these problems.

In this chapter, we will discuss 1) the relationship between environmental pollution, health, and development; and 2) the main conventional remediation technologies, presenting their advantages and disadvantages; and 3) explore innovative approaches to new and existing pollutant removal technologies, focusing on a comparative analysis of advanced oxidation processes and conventional techniques. More specifically, our goal is to improve understanding of the synergistic effects of these treatment methods, as the varying chemical compositions of many substances classified as pollutants require distinct approaches to their degradation.

6.2 ENVIRONMENTAL POLLUTION, HEALTH, AND DEVELOPMENT

Environmental pollution is one of the most discussed topics by scientists worldwide. The tremendous demographic and industrial expansion observed in recent times has increased human activities over natural resources. Currently, practically all ecosystems have already suffered some influence, directly or indirectly, from human actions, resulting in the reduction and loss of biodiversity. Contamination of environmental compartments, such as water, soil, and atmosphere, poses severe risks to human health and the environment. The need to prevent and remedy environmental contamination stimulates research and laws in search of solutions in the most diverse areas of science. Understanding the complexity of treating each polluting agent makes it possible to choose the technology that will effectively remedy the environmental damage caused by introducing, without adequate treatment, high concentrations of unwanted compounds into ecological ecosystems (Steibl and Laforsch, 2019; Wiedmann et al., 2020; Wu et al., 2020).

Environmental pollution stems from various sources beyond industrialization, urbanization, population growth, exploration, and mining. It also arises from the transboundary movement of pollutants, involving the exchange of contaminants between neighboring countries and the transfer from developed to developing nations, or vice versa. In a general overview, polluting industries and hazardous waste sites are disproportionately located in poor, minority, and marginalized communities. A phenomenon that has been called "environmental injustice" is being seen at various social levels. Due to a lack of resources, economic development, political structure, or even knowledge, developing countries are used by large industries in developed countries to produce through practices not permitted in their own countries of origin. Because of this, transboundary pollution is part of the reason why pollution is a global challenge (Landrigan & Fuller, 2016; Spellman, 2017; Ukaogo et al., 2020).

The relationship between the economy and the environment is intrinsic and complex, significantly impacting human health, ecosystems, and the economic development of countries. The enormous demands on urban utility services, transport, social services, waste, and so on, generate pressure on geopolitics.

The impacts are not only on human beings but also on the pollution affecting aquatic and terrestrial biomes, which are necessary for supporting and balancing the planetary ecosystem. As the demand for raw materials such as minerals and fossil fuels increases, exploring more remote areas and using more intensive extraction methods is necessary, resulting in significant environmental damage. Thus, any economic activity begins with looting some environmental assets: land, minerals, vegetation, air, water, and animals. Furthermore, throughout the production process, part of what was withdrawn is discarded into the environment as a waste of solid, liquid, or gaseous products. The degradation of ecosystems, air and water pollution, and biodiversity loss are just some direct consequences of this rampant exploitation.

Concerned about environmental and climate issues, nations began to be on alert in the second half of the last century. During the 1960s, environmental movements emerged, motivated by water and air contamination in industrialized countries. After the Stockholm Conference on the Environment in the 1970s, nations began to structure their environmental bodies and establish legislation to control environmental pollution, and polluting began to be considered a crime in several countries. Sustainable development gained an economic bias in the 1990s, when British standards BS7750 (specification for environmental management systems) came into force, serving as the basis for developing a system of environmental standards at a global level. Thus, sustainable development's main role is to balance economic development with environmental sustainability, aiming for perfect harmony between the environment and efficient consumption (Muralikrishna & Manickam, 2017b).

Although research in environmental health has been growing since the second half of the 20th century, this subject still lacks the robustness of data for a broader understanding, especially in the social area. Generally, environmental pollution is greater in less-developed countries, possibly due to poverty, inefficient legislation, and the low level of education of the inhabitants, mainly in matters related to health and public and environmental policies (Muralikrishna & Manickam, 2017a). The most vulnerable population suffers the impact of pollution on the region's job market and economic activities. Pollution seriously affects the GDP growth of developing countries such as Argentina, Bangladesh, Bolivia, Brazil, China, Colombia, India, Laos, Morocco, Nepal, Pakistan, Peru, and Zambia at national and local levels. More than half of global premature deaths occur due to high air pollution in South Asian countries. Key contributing factors include rapid urbanization and industrialization, which drive elevated levels of air pollution—particularly particulate matter. This pollution impairs cognitive and lung development, increases mortality from respiratory infections, and negatively impacts mental health in children (Greenstone et al., 2022; Kundu et al., 2024).

There is a correlation between waterborne diseases and socio-environmental indicators, with the origin, quality of water, sewage collection and treatment, frequency of floods, and lack of water supply. In places without basic sanitation, workers' income suffers, they are forced to leave for health reasons, children perform less well at school, properties are less valued, tourism is not developed properly, and so forth. Furthermore, there are cases of racial environmental injustice, in which part of the population receives a more significant burden of pollution, highlighting an unequal health experience for the marginalized population from a political-historical perspective (Cesa et al., 2016; de Jesus, 2020; Khasawneh and Palaniandy, 2021; Moniz et al., 2017; Muralikrishna & Manickam, 2017b). Faced with environmental problems, several international organizations, authorities, and governments created and established strategic measures and guidelines to manage and control environmental pollution (Figure 6.1) (Khasawneh & Palaniandy, 2021).

Knowledge of the causes and consequences of environmental pollution is fundamental, but the cost of inaction is enormous. Some types of pollution were identified in the main matrices of environmental pollution that impact health, which are closely related to the Sustainable Development Goals proposed by the United Nations in 2015. The environmental impacts of pollution on health depend on the concentration of pollutants, response coefficients to exposure, and other related attributes (Feng et al., 2019). Furthermore, the health effects of air pollution vary significantly among groups of highly susceptible individuals. Water pollution is also another vital aspect that substantially affects human health. It is important to note that air and water pollution affect crops and animals through food and inhalation. Thus, pollutants are present in plants and animals for human consumption, indirectly harming human health (Feng et al., 2019; Ugya et al., 2019).

It is worth highlighting that noise pollution has been evaluated and recognized as an environmental pollution problem, especially in developing countries. Noise pollution has increasingly affected the global population and has intensified since the Industrial Revolution in the 18th and 19th

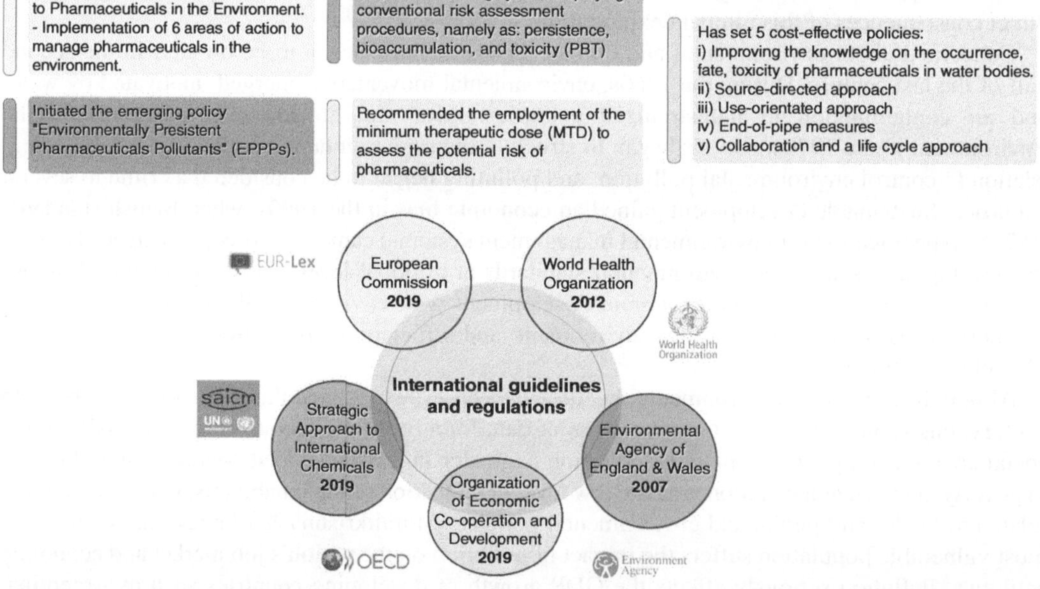

FIGURE 6.1 Main guidelines and regulations issued by international organizations and committees. **(Khasawneh & Palaniandy, 2021)**

centuries due to the increasing use of machines (de Souza et al., 2020). This matrix of environmental pollution is still little explored and valued in society, in general, by environmental managers and the population itself. Exposure to loud sounds, even for short periods, can damage sensitive structures in the inner ear and cause hearing loss, which can be immediate or delayed, temporary or permanent, in one ear or both. Furthermore, there is evidence of a relationship between this type of pollution and difficulties with attention and concentration, anxiety, headache, and hearing loss (de Souza et al., 2020; Khasawneh and Palaniandy, 2021; Muralikrishna & Manickam, 2017b; Spellman, 2017).

Different physical, chemical, and biological approaches have been applied to rid the environment of pollution, but most of them create additional environmental problems or are costly. For example, several processes and technologies are adopted for effluent treatment, such as preliminary treatment, aeration, neutralization, flocculation and coagulation, biological degradation, filtration, and recycling. Pharmaceutical pollutants are physically removed from different materials through adsorption phenomena. However, contamination by medicines in the environment becomes more complex due to the aqueous matrix created by the mixture of pharmaceuticals present and other processes used to remove this type of contaminant (Taoufik et al., 2020).

More advanced mechanisms are needed to treat different kinds of waste. The integration of several biological and more advanced treatment methods helps mitigate the presence of potentially dangerous substances (Madhura et al., 2019). Therefore, the focus should be on eliminating pollution in the bud so that the remediation of the already affected environment is quick and viable, with ecological and economic approaches that generate fewer secondary by-products. Biological methods based on the activity of microorganisms and higher plants can degrade accumulated pollutants, leading to their mineralization, immobilization, or removal.

6.3 CONVENTIONAL TECHNOLOGIES FOR MITIGATION AND ELIMINATION OF POLLUTANTS AROUND THE WORLD

The quest to minimize the generation of chemical waste and address the necessity for its management has prompted researchers and businesses to explore and create approaches that can eradicate or alleviate contaminant levels in diverse types of waste and affected environments. There is a growing emphasis on discovering effective treatment methods, with the ultimate goal of recommending to sectors responsible for contaminant generation the adoption of efficient technologies for decontaminating environments and waste materials derived from various industrial processes, urban pursuits, and agricultural activities, among other sources.

In view of the increasing contamination of water and soil, the concern that has led the scientific community to think about solutions to this problem is evident. These pollutants include medicines, contaminated packaging, veterinary products, substances present in hygiene products, cosmetics, agrochemicals, various industrial contaminants, and so on. Products like these are often present in soil and water matrices and can even undergo different types of treatment without being completely removed (Sutherland & Ralph, 2019). Faced with this problem, and since activities that produce such contaminants cannot cease, there is a need to propose methods aimed at remediating environments and waste that frequently receive products that cause contamination with xenobiotic chemical species.

Treatment and remediation methods primarily focus on addressing organic compounds and potentially toxic metals, which are the key chemical species responsible for contamination. A considerable number of organic contaminants are normally persistent in the environment. It can assume stable forms that are difficult to naturally degrade and reduced to species that are nontoxic to living beings. This makes them targets of the need to apply treatment and remediation processes to eliminate contaminated materials and environments. In terms of chemical degradation, compounds of an organic nature are normally not decomposed under mild temperature conditions and weakly oxidizing media. Strong acidic and oxidizing media are required for its complete reduction to simpler inorganic species. On the other hand, using some weaker oxidants combined with certain physical and chemical factors may potentially promote the transformation of these compounds into the chemical species of interest. In view of such properties, eliminating recalcitrant and refractory contaminants requires technologies that incorporate a combination of factors capable of favoring their decomposition and recovering contaminated matrices and environments. Furthermore, other means of removing such chemical species do not aim at their decomposition, and this is the case with processes that employ adsorbent materials. Their transformation to form the corresponding mineral components, such as H_2O and CO_2, is particularly interesting. However, the complete degradation of contaminants in situ is not essential to meet all treatment demands since, depending on the case, the simple removal of polluting species from the environment or waste, followed by their appropriate disposal, can be satisfactory.

Conventional remediation technologies are mainly applied to treating effluents, solid waste, and contaminated soil. Their execution aims to eliminate contaminating chemical species or considerably reduces their concentrations in matrices and treated environments to safe levels for the protection of humans and animals, which occurs most of the time. Remediation of a polluted environment can be carried out on-site or off-site. In the first case, the treatment is carried out at the contamination site, which is called in situ remediation. In the other situation, the contaminated material (water, soil, etc.) is extracted and transported to a location where decontamination procedures are carried out, such as ex-situ remediation (Liu et al., 2018). In-situ treatment is preferable, as the costs for its application are normally lower.

The most common processes applied in remediation are, for the most part, aimed at treating certain environmental targets determined by the type of environment or properties of the environmental matrices to be treated, such as physical state composition, among other aspects. There are methodologies developed for soil treatment, and others are appropriate for water environments, such as effluents, streams, and underground aquifers. Others are suitable for removing pollutants from other types of contaminated materials. Cases such as chemical spills in soil and water require specific treatment technologies for each case that consider the dimension of the environmental impact.

This section will present some main technologies dedicated mainly to soil and water remediation. The processes most used in remediation consist of extracting pollutants or degrading them. These technologies include soil washing, electrokinetic remediation, thermal processes, pumping and treatment, multiphase extraction, solidification and stabilization, permeable reactive barriers, and bioremediation.

6.3.1 Soil Washing

This can be an efficient way of removing pollutants from the soil. However, it is one of the most aggressive techniques applied in its recovery and can present high costs as it requires its facilities to have the capacity to treat large volumes of material. It is recommended for soils with extreme contamination levels mainly affected by toxic metals. It can also be useful for extracting pesticides and hydrocarbons, among other pollutants.

The process involves transferring soil contaminants to an aqueous liquid fraction, more precisely, the washing water. Its execution is basically divided into two basic steps, mechanical and physical-chemical, consisting of an ex-situ remediation method. It can be carried out in two ways: on-site, in which the soil is moved and treated in the same contaminated location, or it can be off-site, when the soil is transported to another location where a specialized treatment unit is located (Dermont et al., 2008).

Initially, the soil is removed by excavation and subjected to fractionation to separate the smaller particles, which keeps most of the pollutants adsorbed. Then, the fraction to be treated is subjected to washing with appropriate solvents to remove contaminants, using water as solvents, which can be at different temperatures, surfactants, acidic or basic solutions, and organic solvents. Some reagents, such as chelating agents for metal complexation, are used to increase extraction efficiency. In addition, biodegradable chemical agents can also be used in the extraction, such as some low molecular weight organic acids like citric and oxalic acids (Xiao et al., 2019). After dispersing the polluting agents in the washing fluid and separating the phase of interest, it must be subjected to a purification process to recover the solvents used.

6.3.2 Electrokinetic Remediation

Electrochemistry plays a crucial role in the remediation of contaminated environments, offering versatile applications in various remediation processes. One approach involves utilizing potential or electric current to facilitate the transportation of chemical species in soil and groundwater environments. Another way of using it is the application of electric current together with chemical components possessing oxidizing and/or catalytic properties, which are capable of promoting oxidative degradation of the chemical species of interest. In the first case, only electrochemical components (electrical potential and potential) are normally used to remove metal ions in the environment to be treated through an in-situ technique known as electrokinetic remediation. The property of electrically charged chemical species is to be influenced by an electrical potential capable of promoting their displacement in a material medium, such as soil or water, which is used to extract ionic chemical species from the environment. With its main focus of application on the treatment of soils and sediments, it is intended for removing potentially toxic metals. However, it is also feasible to mobilize certain organic compounds.

When carrying out this technique, a certain number of wells are excavated in the ground, which receives the extracted ion solution. Electrodes are introduced between these wells, and a low-intensity current or potential gradient is applied, capable of promoting the mobilization of ions, which are concentrated and removed from the wells and subjected to appropriate treatments. The process is guided by electroosmosis, electromigration, and electrophoresis. Electroosmosis consists of the movement of groundwater toward the cathode; electromigration corresponds to the transport of ions and ionic complexes, whose movements are directed to electrodes with opposite charges; electrophoresis occurs with the movement of larger systems, such as charged particles or colloids, promoted by the action of an applied electric field (Virkutyte et al., 2002).

One of the main advantages is its use in situ, which helps to avoid secondary contamination. Furthermore, the technique can also be applied ex-situ and can be used to remove inorganic and organic contaminants. With the growing trend of using solar energy, the technique offers the possibility of increasing its sustainability potential in some places with electricity shortages. An important disadvantage of this technique is the possibility of causing changes in soil pH, requiring the addition of chemicals to correct it, which can affect the treated environment (Li et al., 2012).

6.3.3 Thermal Process

Thermal remediation processes constitute another effective alternative in the recovery of polluted areas and environments, focusing on soil, mud, and the treatment of waste of industrial origin. It is based on heating the contaminated material to eliminate the contaminants present, which can undergo thermal decomposition and oxidation or be transferred to the vapor phase, followed by their removal or oxidation.

Thermal desorption is a thermal phenomenon commonly used in the remediation of solid material contaminated by pollutants of an organic nature, such as hydrocarbons and pesticides, but it can also be applied to remove mercury from soil. It consists of the detachment of pollutants from the environment on which they are adsorbed. The increase in temperature is also responsible for volatilizing them when they are solubilized in aqueous matrices. In these processes, heat transfer can be conducted in different ways. The most common are heating by electrical resistance, heat conduction, and steam injection into the material or environment to be treated (Ding et al., 2019).

A thermal desorption system is conducted in a few steps. It begins with heating the solid to separate contaminants from the matrix. These, together with any water in the solid, are desorbed onto the surface of the particles. By maintaining heating, these components are separated from the solid and transferred to a vapor phase. In the second stage, the vapors are captured by an exhaust system and transferred to a treatment system, which may include filtration, precipitation, and condensation steps or go to a combustion chamber to oxidize them. Pyrolysis and incineration are commonly employed as heat treatment steps. Pyrolysis involves heating the material without oxygen, aiming for its decomposition. During incineration, contaminants are oxidized to less hazardous products, such as H_2O and CO_2 (Wei et al., 2020). This process presents onerous operational costs, mainly due to the high energy demand. Another disadvantage of its use is the significant changes it can cause to the soil, affecting its chemical and biological properties and available organic matter since heating eliminates microorganisms, important metabolites, and nutrients (O'Brien et al., 2018).

6.3.4 Multiphase Extraction

Widely used in the remediation of soils and groundwater contaminated with numerous pollutants, including hydrocarbons from the dispersion of fossil fuels such as gasoline and diesel oil, multiphase extraction is carried out in situ using bio-ventilation techniques and pollutant removal under low pressure. Through the technology, it is possible to extract contaminants in different phases: free phase, vapor phase, and dissolved phase.

Through the excavation of wells subjected to pressure reduction produced with vacuum pumps, underground fluids (liquid and gaseous) are moved toward these wells, transporting pollutants. Low pressure can increase soil aeration and, consequently, raise oxygen levels, promoting aerobic microbiological activity in the environment and contributing to the degradation of the residual contaminant fraction. Vapors originating from the ground are collected and, together with the extracted liquid containing the free phase of contaminants and their dissolved phase, are sent to a vapor treatment system, followed by reintroduction of air to the environment. Once collected, the extracted liquid containing the free and dissolved phase pollutants is separated from the water and reused or introduced into the environment under appropriate conditions (Baldwin et al., 2009).

Simple vapor extraction is also a technique to remove volatile soil pollutants. Unlike multiphase extraction, its purpose is to remove contaminants from the matrix only in the gas phase. It can be used to remediate the surface phase of the soil (unsaturated zone) and treat the saturated zone with a greater presence of groundwater. Multiphase extraction is one of the most efficient in situ technologies for extracting hydrocarbons dispersed in the soil. The possibility of extracting pollutants in different phases gives it an appreciable advantage compared to simply removing the volatile fraction of pollutants since hydrocarbons with larger chains are less volatile and less susceptible to being entrained in the form of vapors (Qi et al., 2020).

6.3.5 Solidification and Stabilization

When specific contaminants in the soil pose a risk of spreading throughout the site, with a potential risk of increasing environmental damage, it is sometimes necessary to resort to remediation technologies capable of containing their spread in the environment. Addressing the issue of contamination, stabilization, and solidification of pollutants presents a viable alternative. These techniques are employed to disrupt or mitigate the leaching of contaminants found in soil or mud, specifically targeting substances like toxic metals, certain organic compounds, and even radioactive chemical species. In the solidification immobilization technique, binders such as cement or clay are utilized in soil or mud to create rigid structures that encapsulate and contain the pollutants. In this way, solidification accompanied by the consequent reduction in permeability contributes to their reduced ability to disperse in the environment (Liu et al., 2018).

Stabilization involves converting pollutants into chemical species with a reduced capacity to diffuse into the soil. This technique aims to modify such species chemically. To achieve this, chemical reagents can interact with pollutants, aiming to form less soluble products that are less prone to leaching into the soil. Stabilization can be accomplished with solidification in the same environmental matrix. Having the advantage of being carried out in situ, these techniques can also be applied ex-situ, depending on aspects such as soil characteristics, type of waste, and depth of contamination. On the other hand, since the contaminants remain in the treated matrices, the environment in which they are located requires frequent monitoring of these chemical species, as it can generate problems in the long term (Wang et al., 2018).

6.3.6 Permeable Reactive Barriers

Permeable reactive barriers comprise a remediation technology that uses physical barriers to contain the spread of pollutants. These barriers are constructed across the contamination plume against underground water flow and are made of material with physical-chemical properties favorable to water permeabilization and blocking the passage of contaminants of interest. As an in-situ technique, it is suitable for containing the dispersion of organic and inorganic pollutants in soils and contributing to their stabilization and degradation. Various materials can be used to construct these barriers, and their choice depends on the type of pollutant considered. In the barriers, materials with different properties are used, such as adsorbent materials like activated carbon, to retain organic compounds and others capable of promoting reactional processes on the surface of the barrier involving pollutants (Chen et al., 2016).

Different interaction mechanisms are present in the remediation process involving a permeable reactive barrier. As water passes and contaminants block the barrier, some are retained by sorption. The degradation of organic compounds is also a process that can occur when oxidizing material is used to construct barriers. It is known that the use of metallic iron in its structure conditions the decomposition of organochlorine compounds. Some barriers promote changes in pH on their surface, which can precipitate contaminating chemical species, such as potentially toxic metals. Other materials are incorporated into the barriers to induce oxidation-reduction reactions in metal ions, transforming them into less toxic species (Chen et al., 2016; Jin Suk et al., 2009).

The different mechanisms that can be used in a permeable reactive barrier for remediation constitute an advantage of the technique since there are countless possibilities for using new materials, many of which are objects of development, giving the technique a high potential. On the other hand, it is a technology that still has a high cost of execution, in addition to presenting difficulties due to the complexity of the underground matrices, which offer numerous interferences that can compromise the technique's efficiency.

6.3.7 Bioremediation

Considered a technology with high potential for sustainability in the treatment of contaminated environments and waste, bioremediation consists of using biological agents to carry out the biochemical transformation of polluting substances into other components with little or no toxicity. It is a process driven mainly by microorganisms that use contaminants in metabolic pathways as carbon sources for their development. Effective for the remediation of soils, groundwater, and surface waters, it can be carried out in situ or ex-situ. This last form offers the advantage of better control of the conditions that enable the biodegradation of pollutants or increase their efficiency. However, it presents higher operational costs (Azubuike et al., 2016).

Bacteria, filamentous fungi, and yeasts are widely used in bioremediation, which may be natural to the environment or added to it. Products such as H_2O and CO_2 are obtained by the complete biotransformation of organic compounds promoted by these microorganisms (Singh, 2006). The main ex-situ bioremediation applications are conducted in composting systems and bioreactors. Numerous wastes with high loads of natural organic compounds can be decomposed into simpler inorganic components with treatments using these systems.

Bioremediation can be used in different ways in situ. The techniques of natural attenuation, biostimulation, bioaugmentation, phytoremediation, and landfarming constitute specific ways of implementing this technology. In natural attenuation, the biodegradation process occurs in the most natural way possible, carried out by the action of microorganisms from the contaminated site without adding additives to the environment, influenced by physical and chemical factors characteristic of the environment. Through a combination of processes such as biodegradation, dilution, dispersion, and adsorption, pollutants can undergo transformation or reduction in their concentration in soil, groundwater, and surface waters, such as lakes and effluents, with biodegradation being the process responsible for their degradation. As it is a natural process without control methodology, natural attenuation is considerably slow in reducing the concentration of pollutants. It may prove inefficient when the situation demands urgency in containing environmental contamination levels.

There are ways to promote bioremediation in environments by controlling certain factors. Additional materials are required in some situations to increase biodegradation efficiency. Examples of this occur in the processes of bio-stimulation and bioaugmentation. First, the biological activity of contaminant decomposition is optimized by adding specific supplements on-site, containing macro and micronutrients, vitamins, and/or enzymes capable of stimulating the development of existing colonies in the environment. In bioaugmentation, selected microorganisms are introduced into the environment with a recognized ability to promote the biochemical decomposition of contaminants present in the environment. Both methods can be performed together. They are widely applied in effluent treatment (Hassanshahian et al., 2014).

Phytoremediation is a bioremediation method employing particular plant species that can alleviate or eliminate specific pollutants. The process involves various mechanisms through which plants can reduce contamination levels, encompassing both inorganic and organic contaminants. The absorption of certain pollutants, such as some metallic ions, carried out by roots and aerial parts of plants; the incorporation of organic compounds into lignin; and the increase in microbiological activity on the degradation of organic compounds stimulated by growing roots constitute some of these mechanisms of action of plants on pollutants (Peuke and Rennenberg, 2005).

Compared to other technologies, bioremediation offers the advantage of being a safe method for the environment, ensuring the maintenance of populations of other species in the vicinity of the treated site. Furthermore, it is considered a low-cost process compared to other methods. Despite the advantages, it can be considered slow in terms of the pollutant degradation process, and when carried out in situ, it can pose the risk of causing, in some cases, ecological imbalance when microorganisms that do not inhabit the place to be treated are used.

6.4 INNOVATION IN SUSTAINABLE APPROACHES TO ENVIRONMENTAL POLLUTION

Combining the conventional approach with the advanced oxidation process is necessary since the last one is still expensive. Different efforts must be made to face the wide challenges of adequate concentration limits of pollutants of emerging concern fixed by environmental regulations. Conventional biological and physical-chemical technologies leave a gap in reaching these values. We will show the possibilities of combining advanced oxidation processes (AOP) and conventional treatments (CT) to eliminate pollutants of emerging concern (PEC), which is a big challenge since these pollutants are not biodegradable.

The pressing need to address the challenges associated with PEC has sparked a growing recognition that a hybrid approach, combining conventional methods with AOP, is essential. While highly effective, this is particularly crucial as AOP is still considered expensive, necessitating a pragmatic strategy that optimally integrates both CT and AOP. Environmental regulations have established concentration limits for PEC, reflecting the imperative to mitigate their adverse effects on ecosystems and human health. Biological and physical-chemical conventional technologies, which have been the mainstay in wastewater treatment, often fall short when meeting these stringent concentration limits. This creates a notable gap in the efficacy of traditional methods, especially concerning pollutants that resist easy degradation and are not amenable to conventional treatment approaches.

Incorporating advanced oxidation processes into the wastewater treatment landscape offers a promising avenue for tackling this challenge. Advanced oxidation processes generate powerful reactive species, such as hydroxyl radicals, capable of breaking down recalcitrant pollutants into less-harmful by-products. Despite their effectiveness, the cost implications associated with AOP have hindered their widespread adoption as standalone solutions. The hybrid approach, therefore, becomes a strategic response to the financial constraints of employing AOP exclusively. By integrating AOP with conventional treatments, it becomes possible to optimize the overall efficiency of pollutant removal while managing costs effectively. Conventional treatments, such as activated sludge processes or chemical precipitation, can handle the bulk of pollutants, preparing the wastewater for the subsequent AOP stage, where more resistant contaminants are targeted and broken down.

This integration is particularly relevant for PEC, which often exhibit recalcitrant characteristics and resist degradation through traditional means. Pesticides, pharmaceutical residues, and personal care products are examples of pollutants that pose unique challenges to CT methods. In this comprehensive approach, the sequence and design of the treatment stages become critical. Conventional treatments are a pre-treatment step to remove readily biodegradable components and reduce the overall pollutant load. The effluent from these processes is then subjected to AOP, where the remaining refractory pollutants are effectively dismantled.

The need to address emerging pollutants within the constraints of environmental regulations and economic feasibility has underscored the importance of combining advanced oxidation processes with conventional treatments. This hybrid strategy represents a sophisticated and cost-effective approach to achieving the stringent concentration limits set by regulations. By strategically integrating these techniques, we pave the way for a more sustainable and comprehensive solution to eliminating PEC, contributing to the overall improvement of water quality and environmental health.

6.4.1 Important Variables for Pollutants of Emerging Concern Removal by AOP

The removal of PEC from diverse environmental matrices is a hot topic because, apart from being not biodegradable, they have biological effects, even if found in small quantities (Antunes et al., 2021). The so-called advanced oxidation processes represent a category of treatments capable of eliminating PEC. This treatment class generates potent oxidative species that can degrade PEC partially or completely. AOP tech share similarities, yet they are not identical; that is, the reaction conditions involved in the production of oxidative species vary. The AOP–based tech mainly produces hydroxyl radicals but not exclusively (other oxidative species can also be produced) (Dewil et al., 2017).

The hydroxyl radical stands out due to its substantial reduction potential (potential (2.8 V). Furthermore, its non-selectivity allows for a broad spectrum of action in the oxidation of PEC. Concurrently, various other reactive species can be generated through AOP, highlighting the significance of this process family in PEC elimination. Notable AOP variants include Fenton and Fenton-like reactions, heterogeneous and homogeneous photocatalysis, electrochemical processes, and wet oxidation (Dewil et al., 2017).

Integrating AOP with other techniques, including CT, has attracted significant attention from the scientific community. This interest is particularly pronounced because different arrangements of these approaches can lead to enhanced pollutant degradation performance. The synergistic effect between AOP and CT is a current hot point.

The scientific community has increasingly focused on integrating AOP with other techniques, notably CT, creating a dynamic field of study with promising implications for environmental remediation. The interest in this integration stems from the fact that it offers a pathway to address the challenges associated with refractory pollutants, where enhanced degradation can be achieved through strategic synergy. One of the key drivers behind the exploration of integrated AOP and CT is the diverse nature of pollutants present in wastewater. While effective for certain pollutants, conventional treatments may fall short when dealing with persistent and complex contaminants, such as pharmaceuticals, pesticides, and personal care products. AOP, known for their ability to generate highly reactive species, provide a complementary mechanism to target and break down these recalcitrant pollutants effectively.

The arrangements and permutations of these integrated approaches hold considerable promise. Researchers actively explore the optimal sequence and combination of AOP and CT to achieve the most efficient pollutant removal. The idea is to leverage the strengths of each technique, using conventional treatments to handle the bulk of pollutants and then applying AOP to tackle the more resistant and challenging components of wastewater (Garrido-Cardenas et al., 2019). Moreover, the integration of AOP with other treatment methods can lead to improvements in the degradation of refractory species during oxidation. The reactive radicals generated in AOP play an important role in breaking down complex organic molecules that may persist through traditional treatment methods. This synergistic effect becomes particularly pronounced when considering pollutants with intricate molecular structures that resist degradation through singular treatment approaches.

Understanding the mechanisms behind the synergistic effects is the focus in this interdisciplinary field. Researchers are delving into the intricate details of how AOP and CT interact, exploring the chemical kinetics and reaction pathways involved. This in-depth understanding is crucial for optimizing the design of treatment systems, ensuring that the integrated approach is effective and efficient in terms of resource utilization and cost-effectiveness. Exploring integrated AOP and CT extends beyond laboratory-scale studies, with a growing emphasis on real-world applications and scalability. Researchers are working toward translating promising results from controlled environments to practical and implementable solutions for wastewater treatment plants and industrial facilities (Babu Ponnusami et al., 2023).

Integrating AOP with other techniques, particularly conventional treatments, has become a focal point in environmental science and engineering. The synergistic effects arising from these combinations hold significant promise for addressing the challenges of refractory pollutants. This research contributes to the advancement of water treatment technologies and underscores the importance of interdisciplinary collaboration in developing comprehensive and sustainable solutions for environmental remediation.

The efficiency of AOPs in wastewater treatment is a complex interplay of several variables, necessitating a comprehensive understanding of the intricate factors that influence their performance. Various elements, including pollutant concentration, environmental matrix, reactor configuration, and chemical reactants, collectively shape the efficacy of AOP. Each of these factors plays a crucial role, and a meticulous examination of their interactions is imperative to design and implement effective treatment strategies.

Firstly, the concentration of pollutants in the wastewater directly influences AOP efficiency. Higher pollutant concentrations may overwhelm the oxidative capacity of AOP, limiting their effectiveness. Conversely, the radical species generated during AOP may be underutilized at lower concentrations, impacting the overall efficiency of pollutant degradation. Striking the right balance and optimizing AOP conditions concerning pollutant concentrations is vital for achieving optimal treatment outcomes (Seibert et al., 2020).

The environmental matrix, encompassing the composition of the wastewater and its physical characteristics, is another critical factor. Interfering substances, such as organic and inorganic compounds, can either enhance or hinder AOP performance (Lado Ribeiro et al., 2019). Understanding the complex interactions within the environmental matrix is essential for predicting how AOP will respond in real-world wastewater scenarios. Additionally, factors like pH, temperature, and the presence of coexisting ions can significantly influence the kinetics of AOP reactions, requiring careful consideration in process design (Boczkaj and Fernandes, 2017; Sonawane et al., 2022; Wang & Wang, 2021).

The reactor configuration represents an important aspect of the application of AOP. The design of the reactor system, including its size, shape, and arrangement of components, determines the distribution of radicals and their contact time with pollutants. Variations in reactor configuration can lead to differences in mass transfer efficiency and reaction kinetics, ultimately affecting the overall treatment efficiency. Tailoring the reactor design to the specific characteristics of the pollutants and the desired treatment outcomes is crucial for maximizing AOP effectiveness (Nava & Ponce de León, 2018).

Chemical reactants, including the choice of oxidants and catalysts, constitute another set of variables influencing AOP performance. The selection of appropriate reactants depends on the nature of the pollutants targeted and the desired reaction pathways. Factors such as the reactivity and stability of oxidants and the potential for by-product formation must be carefully evaluated. Additionally, the cost and availability of chemical reactants play a role in determining the practical feasibility of AOP applications (Huang et al., 2018).

The successful application of AOP in wastewater treatment demands a global vision of these diverse variables. Researchers and practitioners continually refine and optimize AOP by conducting detailed investigations into the interdependence of these factors. This multidimensional approach ensures that AOP can be tailored to specific wastewater characteristics and pollutant profiles, ultimately enhancing their efficiency and applicability in diverse environmental contexts.

6.4.2 Synergism in Advanced Oxidation Process

The main purpose of combining different AOP is to increase the efficiency of the overall process. AOP combined may increase the production of reactive oxygen species and the possibility of positive interactions, the so-called synergistic effect (Dewil et al., 2017). A way to measure the increase or not in efficiency can be estimated by Equation 6.1:

$$S = \frac{k_{combined} - \sum_{1}^{n} k_i}{k_{combined}} \tag{6.1}$$

In Equation 6.1, S is the synergistic effect, $k_{combined}$ is the kinetic constant for a combination of AOP and the sum of separated AOP (k_i). The S can assume a positive value when the synergism occurs, a negative value when an antagonistic synergism occurs, or a null value when the cumulative effect is simpler (Dewil et al., 2017). Correlating kinetic constants by Equation (6.1) seems a good tool for knowing the synergistic effect.

The use of percarbonate-based (PC) AOP has proven to be a good alternative to H_2O_2 since the main reactant of this approach, sodium percarbonate (SPC), has a pH range much bigger than H_2O_2. Furthermore, PC transportation does not risk spilling, splashing, or corrosion (Liu et al., 2021). Askarniya et al. (2023) studied the degradation of bisphenol *S* by a combination of ozone and SPC-based AOP. The authors found a synergistic factor of 3.84, which indicates a good synergism between O_3 and SPC techniques (Askarniya et al., 2023). The reason for better efficiency is the increase in oxidation species production when processes are combined.

Analyzing a work by Nejumal et al. (2023), we found a synergistic effect between persulfate active by ultrasound (US/PS) for bisphenol *S* degradation. In this work, authors observed increased kinetic constants in different approaches. As the concentration of persulfate increased, the kinetic constants also increased. Using Equation 6.1, the S value is positive for all conditions, confirming the synergistic effect. The US/PS approach brings additional advantages in comparison to the isolated process, and the combined AOP produces both ($^{\bullet}OH$) and $(SO_4)^{\bullet-}$ radicals, which lead to efficient degradation of bisphenol S and its intermediates.

Guo found a relationship between the increase in SPC dosage and the efficiency in sulfamethoxazole degradation in all conditions evaluated, and there was an improvement in the oxidation of this drug when O_3 and SPC were combined (Guo et al., 2021). The synergetic factor is a good way to estimate energy consumption costs since economic factors are essential for the implementation of a treatment. In this work, the S values were also positive, and the author found a better degradation for the combined process.

6.4.3 Biological Treatment Featuring Advanced Oxidation Processes

Unfortunately, the main bottleneck of AOP implementation is the economic costs since the price of equipment and reactants used are higher than those used for conventional technologies (CT). Meanwhile, it is worth highlighting that AOP is the most promising alternative for PEC elimination. AOP is a versatile approach that permits the combination with conventional technologies, since CT still shows a good cost benefit. The combination of AOP and CT has shown to be a good opportunity to innovate in PEC elimination because AOP can turn PEC into biologically treatable (Babu Ponnusami et al., 2023).

Mukherjee et al. showed in their review that the AOP and CT combination could be in three ways: 1) AOP acting as a pre-treatment; 2) AOP acting as post-treatment; 3) AOP acting as simultaneous treatment (Mukherjee et al., 2023). We can say that the versatility of AOP relates to the fact that they can generate intermediate species that are more biodegradable and, thus, more treatable by CT.

Despite the widespread use of different pharmaceutical products and the global popular growth, these products' presence in the environment has increased. When designing a drug, the primary focus is targeting a specific biological organism. However, a significant issue arises when nontarget organisms come into contact with these compounds. These substances can have severe consequences for various trophic levels, even at low concentrations. Consequently, the scientific community has joined forces to address this issue, identifying the most commonly occurring environmental drugs. Around the world, different drug classes have been identified, such as acetylsalicylic acid, antibiotics, antineoplastic, antidiabetic, psychiatric drugs, and lipid regulators (González-González et al., 2022).

Around the world, regulatory agencies have intensified legislation searching for high efficiency in PEC elimination since the conventional active sludge (CAS) is still the most used approach. A membrane bioreactor (MBR) is an alternative to the CAS process, and MBR can help achieve an effluent with a better quality for coupling with AOP. There is a variation of MBR, the hybrid moving bed biofilm reactor–membrane bioreactor (MBBR-MBR). In this approach, a filter helps the growth of microorganisms better and safeguards them from PEC. MBR and MBBR-MBR are technologies with great potential in eliminating PEC when combined with H_2O_2/UV AOP. The hybrid process MBR with H_2O_2/UV and MBBR-MBR with H_2O_2/UV showed a decrease in biological demand oxygen (BOD) (<0.5 mg O_2L^{-1}), suspended solids (<1 mgL^{-1}), turbidity (<1 NTU), and no presence of *E. coli*. MBBR-MBR with H_2O_2/UV is more efficient in PEC elimination. The PEC compounds was degraded at the end of the investigation (Monteoliva-García et al., 2020).

The best efficiency of MBBR-MBR seems to be a better approach, which appears to be linked to a greater presence and variety of microorganisms. Therefore, it was also shown to be an adequate alternative for wastewater secondary treatment and the advanced oxidation process pre-treatment. Both algae and the photocatalytic process rely on light for their operation; thus, their combination is highly promising (Li et al., 2022). Bismuth vanadate ($BiVO_4$) is an excellent photocatalytic material, and when integrated with algae biodegradation, it enhances overall process efficiency. Moreover, this material facilitates increased production of oxidation species, including the generation of a triplet state in dissolved organic matter, identified as a key contributor to antibiotic oxidation (Chen et al., 2020).

As a pre-treatment strategy, AOP such as Fenton's reaction may enhance the biodegradability of PECs. Combining Fenton's reaction with biological treatment appears to be an effective approach to eliminating antibiotics. Fenton's reaction improves antibiotic elimination by breaking down the initial PEC structure into smaller organic fragments, which are more readily degradable by microorganisms (Li et al., 2022).

Another possibility is to combine the ozonation process with biological treatment. Ozonation – the insertion of ozone (O_3), a strong oxidizing agent – can break down PEC into more easily biodegradable intermediaries, enhancing biological treatment. In addition, this combination can oxidize a wider range of PEC and biodegradable pollutants, resulting in a more effective wastewater treatment (Babu Ponnusami et al., 2023).

The versatility of AOP allows combining more than two approaches. An interesting alternative is the usage of AOP post-treatment after the effluent has been submitted for biological treatment. Photoelectrocatalytic ozonation was achieved after 240 min with 43% total organic carbon removal in synthetic wastewater, suggesting which particulates presence hinders an efficient PEC degradation and showing the necessity of approach adequation for implementation of how treatment will occur. In this scenario, the best condition is to use AOP as tertiary treatment, a type of effluent polishing (Palomares-Reyna et al., 2022).

Coupling the Fenton reaction with the electrochemical oxidation process could be feasible to enhance biological treatment. Moreover, when the solar light source is used, the so-called solarphotoelectron-Fenton process (SPEF), there is an upgrade in PEC oxidation effectiveness (Brillas, 2020). The sequential anaerobic-SPEF treatment has shown an effective approach for reducing biological oxygen demand (Vidal et al., 2016).

6.5 CONCLUSIONS

Therefore, the intricate relationship between environmental pollution, public health, and socioeconomic development underscores the urgency of adopting comprehensive and sustainable strategies. The adverse impacts of pollution on human health are undeniable, ranging from respiratory diseases to long-term chronic conditions. Concurrently, the environmental degradation resulting from pollution poses significant threats to ecosystems and biodiversity. Addressing these challenges requires a concerted effort to implement stringent environmental regulations, promote cleaner technologies, and raise public awareness. By prioritizing sustainable development practices that mitigate pollution, we can foster healthier communities, protect ecosystems, and lay the foundation for a more resilient and prosperous future. The interconnected nature of environmental health, public well-being, and development necessitates an approach recognizing the importance of balancing economic progress with ecological preservation and human health.

The widespread use of conventional technologies to mitigate and eliminate pollutants worldwide has played a pivotal role in addressing environmental challenges. These technologies, ranging from traditional wastewater treatment plants to air pollution control devices, have demonstrated efficacy in reducing the impact of pollutants on ecosystems and public health. However, as environmental issues evolve and become more complex, there is a growing need for continued research and innovation to enhance the efficiency and sustainability of these conventional approaches. Additionally, recognizing the limitations of conventional methods underscores the importance of exploring and integrating emerging technologies to create a more resilient and adaptive global framework for pollution control. By combining the strengths of conventional and cutting-edge solutions, we can better navigate the evolving landscape of environmental challenges and work toward a cleaner, healthier, and more sustainable future.

The innovation in pollutants of emerging concern by AOP coupled with CT is very important since both categories have bottlenecks that can be overcome when combined. Synergism between AOP and CT technologies represents a promising alternative for addressing the challenges of pollutants in various environmental matrices.

Different factors such as pollutant concentration, environmental matrix, reactor configuration, and chemical reactants underscore the need for a nuanced approach to optimizing the efficiency of wastewater treatment. By integrating AOP with conventional methods, we can harness the strengths of each approach to create a comprehensive and tailored solution for emerging concerns of pollutant removal.

The adaptability of AOP to diverse wastewater conditions and their capacity to target refractory pollutants make them invaluable tools in environmental remediation. However, recognizing the cost implications of AOP, the integration with conventional treatments becomes not only pragmatic but also economically viable. This hybrid approach not only maximizes the removal of contaminants but also allows for the scalability and applicability of the treatment strategy in real-world settings.

REFERENCES

Antunes, E., Vuppaladadiyam, A. K., Sarmah, A. K., Varsha, S. S. V., Pant, K. K., Tiwari, B., & Pandey, A. (2021). Application of biochar for emerging contaminant mitigation. *Advances in Chemical Pollution, Environmental Management and Protection*, *7*, 65–91.

Askarniya, Z., Rayaroth, M. P., Sun, X., Wang, Z., & Boczkaj, G. (2023). Degradation of bisphenol S – a contaminant of emerging concern - by synergistic ozone and percarbonate based AOP. *Water Resources and Industry*, *29*, 100208.

Azubuike, C. C., Chikere, C. B., & Okpokwasili, G. C. (2016). Bioremediation techniques–classification based on site of application: principles, advantages, limitations and prospects. *World Journal of Microbiology and Biotechnology*, *32*(11), 1–18.

Babu Ponnusami, A., Sinha, S., Ashokan, H., Paul, M. V., Hariharan, S. P., Arun, J., Gopinath, K. P., Hoang Le, Q., & Pugazhendhi, A. (2023). Advanced oxidation process (AOP) combined biological process for wastewater treatment: A review on advancements, feasibility and practicability of combined techniques. *Environmental Research*, *237*, 116944.

Baldwin, B. R., Nakatsu, C. H., Nebe, J., Wickham, G. S., Parks, C., & Nies, L. (2009). Enumeration of aromatic oxygenase genes to evaluate biodegradation during multi-phase extraction at a gasoline-contaminated site. *Journal of Hazardous Materials*, *163*(2–3), 524–530.

Boczkaj, G., & Fernandes, A. (2017). Wastewater treatment by means of advanced oxidation processes at basic pH conditions: A review. *Chemical Engineering Journal*, *320*, 608–633.

Brillas, E. (2020). A review on the photoelectro-Fenton process as efficient electrochemical advanced oxidation for wastewater remediation. Treatment with UV light, sunlight, and coupling with conventional and other photo-assisted advanced technologies. *Chemosphere*, *250*, 126198.

Cesa, M., Fongaro, G., & Barardi, C. R. M. (2016). Waterborne diseases classification and relationship with social-environmental factors in Florianópolis city - Southern Brazil. *Journal of Water and Health*, *14*(2), 340–348.

Chen, J. S., Hsu, S. Y., Li, M. H., & Liu, C. W. (2016). Assessing the performance of a permeable reactive barrier–aquifer system using a dual-domain solute transport model. *Journal of Hydrology*, *543*, 849–860.

Chen, S., Yuan, M., Feng, W., Liu, W., Zhang, W., Xu, H., Zheng, X., Shen, G., Guo, C., & Wang, L. (2020). Catalytic degradation mechanism of sulfamethazine via photosynergy of monoclinic BiVO4 and microalgae under visible-light irradiation. *Water Research*, *185*, 116220. https://doi.org/10.1016/J.WATRES.2020.116220

de Jesus, V. (2020). Racializing the (sociological) view on environmental health in the sanitation of the black population: A colonial continuum called environmental racism. *Saude e Sociedade*, *29*(2), 1–15.

de Liz Arcari, A., Tavora, J., van der Wal, D., & Salama, M. S. (2023). The Wastewater Contamination Index: A methodology to assess the risk of wastewater contamination from satellite-derived water quality indicators. *Frontiers in Environmental Science*, *11*, 1130655.

de Souza, T. B., Alberto, K. C., & Barbosa, S. A. (2020). Evaluation of noise pollution related to human perception in a university campus in Brazil. *Applied Acoustics*, *157*, 107023.

Dermont, G., Bergeron, M., Mercier, G., & Richer-Laflèche, M. (2008). Soil washing for metal removal: A review of physical/chemical technologies and field applications. *Journal of Hazardous Materials*, *152*(1), 1–31.

Dewil, R., Mantzavinos, D., Poulios, I., & Rodrigo, M. A. (2017). New perspectives for advanced oxidation processes. *Journal of Environmental Management*, *195*, 93–99.

Ding, D., Song, X., Wei, C., & LaChance, J. (2019). A review on the sustainability of thermal treatment for contaminated soils. *Environmental Pollution, 253*, 449–463.

Feng, Y., Cheng, J., Shen, J., & Sun, H. (2019). Spatial effects of air pollution on public health in China. *Environmental and Resource Economics, 73*(1), 229–250.

Garrido-Cardenas, J. A., Esteban-García, B., Agüera, A., Sánchez-Pérez, J. A., & Manzano-Agugliaro, F. (2019). Wastewater treatment by advanced oxidation process and their worldwide research trends. *International Journal of Environmental Research and Public Health, 17*(1), 170.

González-González, R. B., Sharma, P., Singh, S. P., Américo-Pinheiro, J. H. P., Parra-Saldívar, R., Bilal, M., & Iqbal, H. M. N. (2022). Persistence, environmental hazards, and mitigation of pharmaceutically active residual contaminants from water matrices. *Science of the Total Environment, 821*, 153329.

Greenstone, M., Hasenkopf, C., Lee, K. (2022). Annual update. Air Quality Life Index. (30 pages).

Guo, H., Li, D., Li, Z., Lin, S., Wang, Y., Pan, S., & Han, J. (2021). Promoted elimination of antibiotic sulfamethoxazole in water using sodium percarbonate activated by ozone: Mechanism, degradation pathway and toxicity assessment. *Separation and Purification Technology, 266*, 118543.

Hassanshahian, M., Emtiazi, G., Caruso, G., & Cappello, S. (2014). Bioremediation (bioaugmentation/biostimulation) trials of oil polluted seawater: A mesocosm simulation study. *Marine Environmental Research, 95*, 28–38.

Huang, H., Dong, C., Feng, W., Wang, Y., Huang, B., & Chen, Y. (2022). Biomedical engineering of two-dimensional MXenes. *Advanced Drug Delivery Reviews, 184*, 114178.

Huang, W., Bianco, A., Brigante, M., & Mailhot, G. (2018). UVA-UVB activation of hydrogen peroxide and persulfate for advanced oxidation processes: Efficiency, mechanism and effect of various water constituents. *Journal of Hazardous Materials, 347*, 279–287.

Ibrahim, M., & Vo, X. V. (2021). Exploring the relationships among innovation, financial sector development and environmental pollution in selected industrialized countries. *Journal of Environmental Management, 284*, 112057.

Jiang, M., Qi, Y., Liu, H., & Chen, Y. (2018). The role of nanomaterials and nanotechnologies in wastewater treatment: A bibliometric analysis. *Nanoscale Research Letters, 13*(1), 1–13.

Jin Suk, O., Jeen, S. W., Gillham, R. W., & Gui, L. (2009). Effects of initial iron corrosion rate on long-term performance of iron permeable reactive barriers: Column experiments and numerical simulation. *Journal of Contaminant Hydrology, 103*(3–4), 145–156.

Khasawneh, O. F. S., & Palaniandy, P. (2021). Occurrence and removal of pharmaceuticals in wastewater treatment plants. *Process Safety and Environmental Protection, 150*, 532–556.

Kıdak, R., & Doğan, Ş. (2018). Medium-high frequency ultrasound and ozone based advanced oxidation for amoxicillin removal in water. *Ultrasonics Sonochemistry, 40*, 131–139.

Kundu, S. K. et al. (2024). Urgent Call to Ensure Clean Air in South Asia—A Growing but Neglected Public Health Emergency. *International Journal of Public Health, 69*, 1–3.

Kurniawan, T. A., Lo, W., Liang, X., Goh, H. H., Othman, M. H. D., Chong, K. K., & Chew, K. W. (2023). Remediation technologies for contaminated groundwater due to arsenic (As), mercury (Hg), and/or fluoride (F): A critical review and way forward to contribute to carbon neutrality. *Separation and Purification Technology, 314*, 123474.

Lado Ribeiro, A. R., Moreira, N. F. F., Li Puma, G., & Silva, A. M. T. (2019). Impact of water matrix on the removal of micropollutants by advanced oxidation technologies. *Chemical Engineering Journal, 363*, 155–173.

Landrigan, P. J., & Fuller, R. (2016). Pollution, health and development: the need for a new paradigm. *Reviews on Environmental Health, 31*(1), 121–124.

Li, G., Guo, S., Li, S., Zhang, L., & Wang, S. (2012). Comparison of approaching and fixed anodes for avoiding the 'focusing' effect during electrokinetic remediation of chromium-contaminated soil. *Chemical Engineering Journal, 203*, 231–238.

Li, S., Show, P. L., Ngo, H. H., & Ho, S. H. (2022). Algae-mediated antibiotic wastewater treatment: A critical review. *Environmental Science and Ecotechnology, 9*, 100145.

Liu, L., Li, W., Song, W., & Guo, M. (2018). Remediation techniques for heavy metal-contaminated soils: Principles and applicability. *Science of The Total Environment, 633*, 206–219.

Liu, X., He, S., Yang, Y., Yao, B., Tang, Y., Luo, L., Zhi, D., Wan, Z., Wang, L., & Zhou, Y. (2021). A review on percarbonate-based advanced oxidation processes for remediation of organic compounds in water. *Environmental Research, 200*, 111371.

Madhura, L., Singh, S., Kanchi, S., Sabela, M., Bisetty, K., & Inamuddin. (2019). Nanotechnology-based water quality management for wastewater treatment. In Eric Lichtfouse, Jie Han, Jan Schwarzbauer, Didier Robert (Eds.), *Environmental Chemistry Letters* (Vol. 17, Issue 1). Springer International Publishing.

Mohammed, S., Gill, A. R., Ghosal, K., Al-Dalahmeh, M., Alsafadi, K., Szabó, S., Oláh, J., Alkerdi, A., Ocwa, A., & Harsanyi, E. (2024). Assessment of the environmental kuznets curve within EU-27: Steps toward environmental sustainability (1990–2019). *Environmental Science and Ecotechnology*, *18*, 100312.

Moniz, M. D. A., Sabóia, V. M., Carmo, C. N. D., & Hacon, S. D. S. (2017). Participatory environmental diagnosis and of health risks from the surrounding communities the Petrochemical Complex of Rio de Janeiro, Brazil. *Ciencia e Saude Coletiva*, *22*(11), 3793–3805.

Monteoliva-García, A., Martín-Pascual, J., Muñío, M. M., & Poyatos, J. M. (2020). Effects of carrier addition on water quality and pharmaceutical removal capacity of a membrane bioreactor – Advanced oxidation process combined treatment. *Science of the Total Environment*, *708*, 135104.

Mukherjee, J., Lodh, B. K., Sharma, R., Mahata, N., Shah, M. P., Mandal, S., Ghanta, S., & Bhunia, B. (2023). Advanced oxidation process for the treatment of industrial wastewater: A review on strategies, mechanisms, bottlenecks and prospects. *Chemosphere*, *345*, 140473.

Muralikrishna, I. V., & Manickam, V. (2017a). Analytical methods for monitoring environmental pollution. In I. V. Muralikrishna & V. Manickam (Eds.), *Environmental Management* (1st ed.). Elsevier Inc.

Muralikrishna, I. V., & Manickam, V. (2017b). Environmental policies and legislation. In I. V. Muralikrishna & V. Manickam (Eds.), *Environmental Management* (1st ed.). Elsevier Inc.

Nava, J. L., & Ponce de León, C. (2018). Reactor design for advanced oxidation processes. *Handbook of Environmental Chemistry*, *61*, 263–286.

Nayak, T., & Pathan, A. (2023). Environmental remediation and application of carbon-based nanomaterials in the treatment of heavy metal-contaminated water: A review. *Materials Today: Proceedings*, *92*, 1659–1670.

Nejumal, K. K., Satayev, M. I., Rayaroth, M. P., Arun, P., Dineep, D., Aravind, U. K., Azimov, A. M., & Aravindakumar, C. T. (2023). Degradation studies of bisphenol S by ultrasound activated persulfate in aqueous medium. *Ultrasonics Sonochemistry*, *101*, 106700.

O'Brien, P. L., DeSutter, T. M., Casey, F. X. M., Khan, E., & Wick, A. F. (2018). Thermal remediation alters soil properties: A review. *Journal of Environmental Management*, *206*, 826–835.

Palomares-Reyna, D., Sosa-Rodríguez, F. S., Goyes, R. E. P., Fuentes-Camargo, I., Elizalde, I., Ramírez-López, M., Lartundo-Rojas, L., & Vazquez-Arenas, J. (2022). Concomitant abatement of ciprofloxacin, sulfamethoxazole and cefadroxil in synthetic wastewater using hybrid photoelectrochemical ozonation driven on TiO2 containing C, N and P species. *Journal of Hazardous Materials Advances*, *8*, 100179.

Paulino, J., Silva, A., Santos, D., Nagliate, P., & Meili, L. (2022). Hotspots and trends of layered double hydroxide-based adsorbents for polluted water treatment: Insights from bibliometric analysis. *Environmental Management*, *71*(5), 1098–1109.

Peuke, A. D., & Rennenberg, H. (2005). Phytoremediation. *EMBO Reports*, *6*(6), 497–501.

Qi, S., Luo, J., O'Connor, D., Wang, Y., & Hou, D. (2020). A numerical model to optimize LNAPL remediation by multi-phase extraction. *Science of the Total Environment*, *718*, 137309.

Raheem, I., Mubarak, N. M., Karri, R. R., Solangi, N. H., Jatoi, A. S., Mazari, S. A., Khalid, M., Tan, Y. H., Koduru, J. R., & Malafaia, G. (2023). Rapid growth of MXene-based membranes for sustainable environmental pollution remediation. *Chemosphere*, *311*, 137056.

Seibert, D., Zorzo, C. F., Borba, F. H., de Souza, R. M., Quesada, H. B., Bergamasco, R., Baptista, A. T., & Inticher, J. J. (2020). Occurrence, statutory guideline values and removal of contaminants of emerging concern by Electrochemical Advanced Oxidation Processes: A review. *Science of the Total Environment*, *748*, 141527.

Singh, H. (2006). *Mycoremediation: Fungal Bioremediation* (pp. 1–592). John Wiley & Sons.

Sonawane, S., Rayaroth, M. P., Landge, V. K., Fedorov, K., & Boczkaj, G. (2022). Thermally activated persulfate-based Advanced Oxidation Processes — recent progress and challenges in mineralization of persistent organic chemicals: A review. *Current Opinion in Chemical Engineering*, *37*, 100839.

Spellman, F. R. (2017). *The science of environmental pollution* (F. R. Spellman, Ed.; 3rd ed.). Taylor & Francis.

Steibl, S., & Laforsch, C. (2019). Disentangling the environmental impact of different human disturbances: A case study on islands. *Scientific Reports*, *9*(1), 1–8.

Sutherland, D. L., & Ralph, P. J. (2019). Microalgal bioremediation of emerging contaminants - Opportunities and challenges. *Water Research*, *164*, 114921.

Taoufik, N., Boumya, W., Janani, F. Z., Elhalil, A., Mahjoubi, F. Z., & Barka, N. (2020). Removal of emerging pharmaceutical pollutants: A systematic mapping study review. *Journal of Environmental Chemical Engineering*, *8*(5), 104251.

Ugya, A. Y., Hua, X., Agamuthu, P., & Ma, J. (2019). Molecular approach to uncover the function of bacteria in petrochemical refining wastewater: A mini review. *Applied Ecology and Environmental Research*, *17*(2), 3645–3665.

Ukaogo, P. O., Ewuzie, U., & Onwuka, C. V. (2020). Environmental pollution: Causes, effects, and the remedies. In Pankaj Chowdhary, Abhay Raj, Digvijay Verma, Yusuf Akhter (Eds.). *Microorganisms for Sustainable Environment and Health*. INC.

Vidal, J., Huiliñir, C., & Salazar, R. (2016). Removal of organic matter contained in slaughterhouse wastewater using a combination of anaerobic digestion and solar photoelectro-Fenton processes. *Electrochimica Acta*, *210*, 163–170.

Virkutyte, J., Sillanpää, M., & Latostenmaa, P. (2002). Electrokinetic soil remediation — critical overview. *Science of the Total Environment*, *289*(1–3), 97–121.

Wang, F., & Taghvaee, V. M. (2023). Impact of technology and economic complexity on environmental pollution and economic growth in developing and developed countries: Using IPAT and STIRPAT models. *Environmental Science and Pollution Research*, *30*(29), 73349–73360.

Wang, J., & Wang, S. (2021). Effect of inorganic anions on the performance of advanced oxidation processes for degradation of organic contaminants. *Chemical Engineering Journal*, *411*, 128392.

Wang, L., Chen, L., Tsang, D. C. W., Li, J. S., Yeung, T. L. Y., Ding, S., & Poon, C. S. (2018). Green remediation of contaminated sediment by stabilization/solidification with industrial by-products and CO2 utilization. *Science of The Total Environment*, *631–632*, 1321–1327.

Wei, Y., Wang, F., Liu, X., Fu, P., Yao, R., Ren, T., Shi, D., & Li, Y. (2020). Thermal remediation of cyanide-contaminated soils: Process optimization and mechanistic study. *Chemosphere*, *239*, 124707.

Wiedmann, T., Lenzen, M., Keyßer, L. T., & Steinberger, J. K. (2020). Scientists' warning on affluence. *Nature Communications*, *11*(1), 1–10.

Wu, C. F., Li, F., Hsueh, H. P., Wang, C. M., Lin, M. C., & Chang, T. (2020). A dynamic relationship between environmental degradation, healthcare expenditure and economic growth in wavelet analysis: Empirical evidence from Taiwan. *International Journal of Environmental Research and Public Health*, *17*(4), 1386.

Xiao, R., Ali, A., Wang, P., Li, R., Tian, X., & Zhang, Z. (2019). Comparison of the feasibility of different washing solutions for combined soil washing and phytoremediation for the detoxification of cadmium (Cd) and zinc (Zn) in contaminated soil. *Chemosphere*, *230*, 510–518.

Yu, D., & Zhao, P. (2022). Global value chain governance of the apparel design industry under the background of global sustainable economic development. *Journal of Environmental and Public Health*, *2022*, 9848627.

Yu, Y., Duan, C., Li, S., Peng, C., Yang, J., Yan, K., Bi, X., & Zou, P. (2022). Relationship between environmental pollution and economic development in late-developing regions shows an inverted V. *Science of the Total Environment*, *838*, 156295.

7 Microbial Rejuvenation of Heavily Polluted Soil

Hira Singh Gariya, Himanshi Rawat, and Arun Bhatt

7.1 INTRODUCTION

In the face of escalating environmental challenges, soil pollution stands as a significant threat to ecosystem health and biodiversity. The increasing anthropogenic activities, ranging from industrial discharges to improper waste disposal, have left a lasting imprint on soil quality, rendering vast expanses heavily contaminated. In this context, the microbial rejuvenation of polluted soil emerges as a beacon of hope and a sustainable solution to remediate and restore these compromised environments. Microorganisms, with their inherent capacity to adapt and thrive in diverse conditions, play a crucial role in the natural cycles of nutrient cycling and organic matter decomposition. The concept of microbial remediation harnesses the power of these tiny yet resilient agents to mitigate the adverse effects of soil pollution. This chapter investigates how microbial communities employ multiple strategies to revitalise contaminated soils, seeking to reveal the mechanisms that enable ecological restoration.

In exploring the complex network of interactions within the microbial world, the goal is not only to comprehend the microbial diversity thriving in polluted soils but also to identify key microbial players capable of detoxifying pollutants and enhancing soil fertility. This investigation encompasses a spectrum of microbial activities, including biodegradation, bioaccumulation, and nutrient cycling, all of which contribute to the overall revitalisation of contaminated soil. Understanding the ecological dynamics and microbial communities involved is essential for developing effective bioremediation strategies. Through an analysis of established research and modern technological advances, this chapter illustrates the promise of microbial rejuvenation as a sustainable method for remediating contaminated soils.

7.1.1 Soil Pollution: Causes and Impacts

All the undesirable changes that occur in the environment due to the various types of human activities are environmental pollution (Al-Taai, 2021). Soil is the thin layer of organic and inorganic materials that covers the Earth's rocky surface. The organic portion is concentrated in the dark uppermost topsoil, which is derived from the decayed remains of plants and animals. The inorganic portion is made up of rock fragments formed over thousands of years by physical and chemical weathering of bedrock. To supply the world with sufficient food, productive soils are necessary for agriculture (Havugimana et al., 2017). Soil pollution can be defined as the entry of foreign bodies into the soil that leads to a change in the chemical and physical composition of soil (Al-Taai, 2021). Any factor that deteriorates the quality, texture, and mineral content of the soil or disturbs the biological balance of the organisms in the soil is called a soil pollutant. These foreign bodies can be persistent toxic compounds, organic contaminants (hydrocarbons, polynuclear aromatic hydrocarbons, polychlorinated biphenyls, chlorinated aromatic compounds), chemicals, salts, radioactive materials, or disease-causing agents, which have adverse effects on plant growth and animal health

DOI: 10.1201/9781003668961-7

(Aqeel et al., 2014). Soil is a vital part of the natural environment. It influences the distribution of plant species and provides a habitat for a wide range of organisms. It controls the flow of water and chemical substances between the atmosphere and the earth and acts as both a source and store for gases (like oxygen and carbon dioxide) in the atmosphere. Soils not only reflect natural processes but also record human activities both at present and in the past (Mishra et al., 2016). There are many ways by which soil can become polluted, such as seepage from a landfill; discharge of industrial waste and percolation of contaminated water into the soil; rupture of underground storage tanks; excess application of pesticides, herbicides, or fertilizer; solid waste seepage; oil and fuel dumping; deforestation and soil erosion, and so on (Aqeel et al., 2014). Soil pollution can also lead to water and air pollution. When toxic chemicals leach into groundwater or if contaminated runoff reaches streams, lakes, or oceans through soil, it leads to water pollution. The release of volatile compounds into the atmosphere such as nitrogen lost through ammonia volatilization and denitrification, or sulphur dioxide and other sulphur compounds generated during the decomposition of organic matter in soil contributes to air pollution and can lead to acid rain (Mishra et al., 2016).

7.2 POLLUTANTS

A pollutant is a substance or energy introduced into the environment that has undesired effects or which adversely affects the usefulness of a resource. A pollutant may cause long- or short-term damage by changing the growth rate of plant or animal species, by interfering with environmental good health, or by interfering with human comfort, health, or property values. Some pollutants are biodegradable and therefore will not persist for the long term in the environment, while the degradation products of some pollutants are themselves polluting. One such example is the products dichloro-diphenyl-dichloroethylene (DDE) and dichloro-diphenyl-dichloro-ethane (DDD) produced from the degradation of dichlorodiphenyltrichloroethane (DDT). The most common chemicals known to be involved in causing soil pollution are petroleum hydrocarbons, heavy metals, pesticides, and solvents (Aqeel et al., 2014). Human activities directly or indirectly affect the environment negatively in different ways (Havugimana et al., 2017). A soil pollutant is any factor that deteriorates the quality, texture, and mineral content of the soil or that disturbs the biological balance of the organisms in the soil (Havugimana et al., 2017). Soil pollution may be classified into natural causes and man-made causes (Anthropogenic sources).

7.2.1 Sources of Soil Pollutants

7.2.1.1 Natural, Geogenic Sources

Soil is formed from the parent material (bedrock) by mechanical weathering of rocks, temperature changes, abrasion, wind, moving water, glaciers, chemical weathering activities, and lichens (Havugimana et al., 2017). Several soil parent materials are natural sources of certain heavy metals and other elements, such as radionuclides, which can pose a risk to the environment and human health when present in elevated concentrations (Rodríguez-Eugenio et al., 2018). Arsenic (As) contamination is one of the major environmental problems around the world. Natural sources of As include volcanic releases, weathering of As–containing minerals and ores, and naturally occurring mineralized zones of arsenopyrite (gossans) formed by the weathering of sulphide-bearing rock. Many of these minerals present a high spatial variability, and most of them can be found in higher concentrations in deeper layers (Rodríguez-Eugenio et al., 2018). Arsenic is considered nonessential for plants and other organisms. It can affect growth and productivity in plants due to various morphological, physiological, biochemical, and molecular alternations. The most dangerous biochemical effect of As at the subcellular level is the production of the reactive oxygen species (ROS) such as superoxide radical (O_2^-), hydroxyl radical (OH), and hydrogen peroxide (H_2O_2). These ROSs are quite dangerous for plant metabolism and can cause irreparable damage to important macromolecules, including lipids, proteins, carbohydrates, and DNA. It has been noticed that the generation of ROS in plants is linked to the conversion of As (V) to As (III). Researchers have found that amongst its inorganic forms, As (III) is more soluble and 60 times more toxic and mobile than As (V). As (III) has been shown to react with sulfhydryl (–SH) groups of proteins and enzymes, culminating in inhibition of cellular function and, eventually, cell death. As also has been reported to inhibit the rate of photosynthesis in plants and reduce the ability of cells to produce ATP and carry out normal metabolism. Chemically, As (V) is quite like Pi molecules which can disrupt the phosphate-dependent metabolic aspects during ATP synthesis when present in high concentration as it can compete with Pi to bind with transport proteins available on the root surface. This can result in the formation of As (V)-ADP, which is short-lived and unstable; thus, rapid autohydrolysis of As (V)-ADP establishes a futile cycle that uncouples oxidative phosphorylation and photophosphorylation, reducing the ability of cells to produce ATP and carry out normal metabolism (Abbas et al., 2018). Human exposure to arsenic in drinking water has been associated with cancers (lung, bladder, and skin), chronic diseases of the skin, heart, lungs, and nervous system, and diabetic effects. Goats experimentally induced with As were characterized by increased heart rate, respiration rate, absence of ruminal motility, diarrhoea, drooling of saliva, stiff gait, paresis of hindquarter, lameness, tremor, convulsion, coffee-coloured urine, congested mucous membrane, paresis of limb, and death, whereas chronic arsenic exposure causes reduced body weight, coffee-coloured urine, congested mucous membrane, and polyuria. It is also able to suppress the replication of DNA with altered repair enzymes. The dangerous effect of arsenic is alteration of mitochondrial membrane potential and ATP formation during glycolysis, and induction of apoptosis in various cells (Mandal, 2017).

Some other natural causes of soil pollution are earthquakes, volcanic eruptions, landslides, forest fires, hurricanes, and floods. Such natural disasters also cause severe damage to the composition of soil and can add many toxic elements, including dioxin-like compounds and polycyclic aromatic hydrocarbons (PAHs), into the environment (Havugimana et al., 2017, Rodríguez-Eugenio et al., 2018). High levels of heavy metals including mercury (Hg), or with the weathering of the parent material, where high levels of chromium (Cr), copper (Cu), nickel (Ni), and zinc (Zn), have been identified in volcanic soils in Réunion (part of the Mascarene Islands) that can be associated with the active volcanic activity. High levels of Cr and Ni have also been reported in volcanic Indonesian soils. These are associated with pedo-geochemical origins. However, this natural pollution does not normally cause environmental problems due to the regenerative ability and the adaptation capacity of plants, but the problems arise when their resilience and response ability are altered when the ecosystems are subject to external pressures. Polycyclic aromatic hydrocarbons can also occur naturally in soils which are usually of cosmogenic origin, being relatively common in cosmic dust samples and meteorites, or derive from the diagenetic alteration processes of waxes contained in soil organic matter. Biogenic production of PAHs is favoured under reducing conditions (Rodríguez-Eugenio et al., 2018).

7.2.2 Anthropogenic Sources

7.2.2.1 Industrial Activity

Industrial activity is the biggest contributor to soil pollution in the last century, especially due to the increased amount of mining and manufacturing. Most industries are dependent on extracting minerals from the earth, whether it is iron ore or coal, the by-products are contaminated and are not disposed of in a manner that can be considered safe. As a result, the industrial waste lingers on the soil surface for a long time and makes it unsuitable for use (Havugimana et al., 2017). Gaseous pollutants and radionuclides are released to the atmosphere due to industrial activities that enter the soil directly through acid rain or atmospheric deposition. The industrial land can also be polluted by incorrect chemical storage or direct discharge of waste into the soil. Heavy metals from anthropogenic activities are also frequent in industrial sites. These can arise from dust and spillages of raw materials, wastes, final products, fuel ash, and fires (Rodríguez-Eugenio et al., 2018). Salinization (an excessive accumulation of water-soluble salts), which is another major threat to global soils, affects many soils which are close to certain industrial activities, mainly those associated with chlor-alkali, textiles, glass, rubber production, animal hide processing and leather tanning, metal processing, pharmaceuticals, oil and gas drilling, pigment manufacture, ceramic manufacture, and soap and detergent production (Rodríguez-Eugenio et al., 2018). The sustainable development of regional economies and agriculture are restricted due to soil salinization, mainly by accelerating desertification, inhibiting the growth of crops, hindering agricultural development, and reducing soil quality (Yang et al., 2020).

7.2.2.1.1 Case Study: Heavy Metal Pollution in the Wetland of Thane Creek

Heavy metals like Fe, Cu, Zn, Ni, and other trace elements are important for the proper functioning of biological systems, and their deficiency or excess could lead to various disorders. These heavy metals not only have a marked effect on the aquatic flora and fauna but also on humans, which through biomagnification enter the food chain and ultimately affect human beings as well. Food chain contamination by heavy metals has become a burning issue in recent years because of their potential accumulation in biosystems through contaminated water, soil, and air. The main sources of heavy metals in vegetable crops are their growth media (soil, air, nutrient solutions) from where these are taken up by the roots or foliage. It is found that the presence of toxic heavy metals like Fe, Pb, and Hg also reduce soil fertility and agricultural output. Heavy metals are present in both natural and contaminated environments, although their concentration is low in natural environments. However, in contaminated environments the concentrations that impact public health are high. The elements of concern include iron, mercury, cadmium, arsenic, zinc, nickel, and copper.

In an investigation conducted by Pravin et al. subjected to the physico-chemical characteristics and heavy metal pollution in the soil from one of Asia's biggest industrial districts, the Thane–Belapur Industrial Complex, located at the south of Mumbai harbour along the west coast of India. Thane Creek lies in the southern part of the Deccan belt of India between latitude 18° 53′ to 19° 04′ N, longitude 72° 48′ to72° 53′E. The soil samples were analyzed for the presence of various heavy metal contents like nickel, zinc, cadmium, copper, iron, arsenic, and mercury and for physico-chemical characteristics like bulk density, moisture content, pH, electrical conductivity, alkalinity, and chlorinity. It was observed that the concentration of these heavy metals increased gradually in dry seasons, followed by a sharp decrease during the rainy season. The pH of the soil samples collected for different sessions ranged from 8.10 to 8.94. The higher chlorinity is an indicator of higher pollution due to higher organic waste of animal origin, which was found to be between averages of 23–31 mg/L in the collected samples. An average of various heavy metals in dry and rainy seasons was found as Ni 78 and 62 μg/g; Zn 196 and 82 μg/g; Cd 21 and 8 μg/g; Cu 121 and 107 μg/g; As 253 and 228 μg/g; Hg 28 and 9 μg/g while Fe 7.67% and 7.41% respectively. This experimental data on the soil pollution status (2008) along the Kalwa bridge of Thane creek suggest a need for regular monitoring of water resources and further improvement in the industrial waste water treatment methods. The Thane Creek area may soon become ecologically inactive if the conditions continue for a long period (Singare et al., 2010).

7.2.3 Agricultural Activities

Chemical utilization has gone up tremendously since technology provided us with modern pesticides and fertilizers. They are full of chemicals that are not produced in nature and cannot be degraded naturally. As a result, they seep into the ground after they mix with water and slowly reduce the soil fertility. Other chemicals are found to damage the composition of the soil and make it easier to erode by water and air (Havugimana et al., 2017). The improper and continuous use of herbicides, pesticides, and fungicides to protect the crops from pests, fungi, and so on alter the basic composition of the soils and make the soil toxic for plant growth. Similarly, organic insecticides like DDT, aldrin, benzene hexachloride, and others used against soil-borne pests accumulate in the soil and are not easily degraded by the soil and water bacteria. Consequently, they have a very deleterious effect on plant growth, causing stunted growth and reducing the yield and size of fruit. Their degradation products may also be absorbed by the plants from where they reach the animals and man through the food chains (Mishra et al., 2016) and when they decompose, they cause soil pollution since they become a part of the land (Havugimana et al., 2017).

Mixed fertilizers often contain ammonium nitrate (NH_4NO_3), phosphorus as P_2O_5, and potassium as K_2O. Since the metals are not degradable, their accumulation in the soil above toxic levels due to excessive use of phosphate fertilizers becomes an indestructible poison for crops. The overuse of NPK fertilizers reduces the yield of vegetables and crops grown on the soil over the years. It also reduces the protein content of wheat, maize, grains, and so on grown on that soil and the carbohydrate quality of such crops. Excess potassium content in soil decreases vitamin C and carotene content in vegetables and fruits. Vegetables and fruits grown on over-fertilized soil are more prone to attacks by insects and disease (Havugimana et al., 2017). The first widespread use of the insecticides dichlorodiphenyltrichloroethane (DDT) and gammaxene began at the end of World War II. Insects soon became resistant to DDT, and due to its nonbiodegradable nature, it persisted in the environment. Since it was soluble in fat rather than water, it biomagnified up the food chain and was found to disrupt calcium metabolism in birds, causing eggshells to be thin and fragile. DDT has now been banned in most developed countries but is still produced and exported to other developing nations whose needs outweigh the problems caused by it (Havugimana et al., 2017).

7.2.4 Waste Disposal

Every human produces a certain amount of personal waste products by way of urine and faeces. The sewer system ends at the landfill, where the biological waste pollutes the soil and water. Our bodies are full of toxins and chemicals that are now seeping into the land and causing soil pollution (Havugimana et al., 2017). Human and animal excreta, which are used as organic manure to promote crop yield, also pollute the soil by contaminating the soil and vegetable crops with the pathogens that may be present in the excreta (Mishra et al., 2016). Some pathogens may originate from animal faeces entering soil, thus making soil the main pathway of contamination through dermal contact or contact with contaminated water and food (Rodríguez-Eugenio et al., 2018). In general, solid waste, including garbage, domestic refuse, and discarded solid materials such as those from commercial, industrial, and agricultural operations is also one of the causes of soil pollution. They contain increasing amounts of paper, cardboard, plastics, glass, old construction material, packaging material, and toxic or otherwise hazardous substances. A significant amount of urban solid waste tends to be paper and food waste, and agricultural waste is recyclable or biodegradable in landfills. The portion of solid waste that is hazardous such as oils, battery metals, heavy metals from smelting industries, and organic solvents is the one we must pay particular attention to. These get deposited into the soils of the surrounding area and pollute them by altering the chemical and biological properties of the soil (Havugimana et al., 2017). Lead-based paint is a major legacy source of lead (Pb) contamination in urban areas. Soils become contaminated during renovations or demolition, when lead-based paint is pulverized into dust or small particles and then enters the environment (Rodríguez-Eugenio et al., 2018). Plastics are also a major source of pollution. They are widely used in various common items used in daily life. Plastics have a strong presence in the environment globally and are extremely persistent in the environment. They widely accumulate in oceans and landfills, but also in soils where producing factories are located (Rodríguez-Eugenio et al., 2018). Improper handling of the plastic waste has created numerous environmental problems. The waste plastic residues from plastic mulching practices in agriculture are left over after use, which, by the action of physical chemical and biological processes such as ultraviolet radiation, by water or air erosion, and by earthworms in the soil environment are broken into small particles, which eventually reduce in size to <5 mm, known as microplastics (MPs). Also, the residual plastics are further converted to even tiny particles by photo-oxidative degradation. Due to plastic ignorance, pollution by MPs has become an environmental concern worldwide. Numerous MPs sources have been reported in the literature, such as the aerial deposition, sewage sludge practices, compost, plastic film mulching, and abrasion of tyres. Mulching and sludge operations are the main sources of MPs in the soil. MPs are composed of synthetic polymers generally found in the form of solid particles, in a size range of 1–5,000μm which are insoluble in water with thermoplastic characteristics. These residues may integrate within soils and further break down into pieces to generate particles down to a nanoscale or smaller.

The accumulation of residual MPs in field soil can cause various negative impacts on crop production as they:

1. Destroy the soil structure, alter the physico-chemical properties of soil, like soil aggregation, water-holding capacity, nutrient mobility, and moisture diffusion in soil
2. Reduce the availability of nutrients, alter microbial activities, reduce seed germination, and affect crop development and growth
3. Slow down development of roots during germination
4. Influence greenhouse gas emissions
5. Hinder the sowing and growth of plants in the soil environment
6. Affect soil flora and fauna and terrestrial organisms.

(Sajjad et al., 2022)

The use of sewage sludge to irrigate soils may be beneficial, as it adds organic matter and nutrients to soils (Rodríguez-Eugenio et al., 2018). However, irrigating land with sewage water without pretreatment of sewage causes profound changes in the irrigated soils. Various changes that are brought about in the soil as a result of sewage irrigation include physical changes like leaching, changes in humus content, and porosity and chemical changes like soil reaction, base exchange status, salinity, quantity, and availability of nutrients like nitrogen, potash, phosphorus, and so on. Sewage sludges also pollute the soil by accumulating metals like lead, nickel, zinc, cadmium, and others. This may lead to the phyto-toxicity of plants (Aqeel et al., 2014).

7.2.4.1 Case Study: Contaminated Vegetables from Wastewater-Irrigated Area of Varanasi, India

An investigation conducted by Anita et al. in the suburban area of Varanasi, India, where wastewater from Dinapur sewage treatment plant is used for irrigating vegetable plots, found that the concentrations of certain heavy metals in all the vegetables grown have crossed the safe limits for human consumption.

The following was the observation:

1. Among fruits and vegetables, lady's finger had the highest Zn concentration (122.3–132.7μg g^{-1}).
2. The maximum concentration of Cu (17.94 μg g^{-1}) was found in tomato.
3. Cd concentration in brinjal was recorded 1.55–13.80 μg g^{-1}.
4. The ranges of Zn and Cu concentrations were 9.00–30.05 μg g^{-1} and 4.55–17.00 μg g^{-1} respectively in bottle gourd and 1.40–1.70 μg g^{-1} and 4.60–5.55 μg g^{-1} respectively in bitter gourd.
5. Among leafy vegetables (palak, cabbage, and amaranthus), palak had the highest Ni concentration (10.45–39.25 μg g^{-1}).
6. Radish was observed to have concentration in μg g^{-1} of 2.19 for Cd, 12.20 for Pb, 3.69 for Cr, and 13.75 for Cu.

7. When the concentrations of metals were compared with permissible Indian Standard limits and safe limits given by WHO/FAO (2007), it was found that the concentrations of Cd, Pb, and Ni was higher in all the vegetables, whereas Zn concentration was higher in lady's finger and cabbage at wastewater irrigated sites. The concentration of Cd and Pb was also higher in all the vegetables as compared to the safe limits given by the commission regulation.
8. Among different metals, Cd showed the maximum transfer factor value, which ranged from 0.61 (cauliflower) to 2.96 (cabbage) at wastewater irrigated sites and a minimum for Cr, ranging from 0.073 (bitter gourd) to 0.356 (bottle gourd). Transfer factor for Cu, Pb, Zn, and Ni was highest for lady's finger (2.08) followed by cauliflower (1.12), cabbage (1.36), and brinjal (0.890).

(Singh et al., 2010)

This study showed that irrigating an area with wastewater led to the accumulation of heavy metals in soil and consequently in the vegetables. The concentrations of Cd, Pb, and Ni were above the national and international permissible limits in all the vegetables, posing a potential threat to human health. The percentage of daily intake rate of Cu, Ni, Pb, and Cr was found to be higher in fruit and vegetables, whereas that of Cd and Zn was higher in leafy vegetables. Consumption of these vegetables with elevated levels of heavy metals may lead to a high level of accumulation of these metals in the body, causing related health disorders. Thus, according to the study, regular monitoring of heavy metal contamination in the vegetables grown in wastewater irrigated areas is necessary, and consumption of contaminated vegetables should be avoided. Also, the wastewater treatment technology should involve methods to remove heavy metals from the effluent (Singh et al., 2010).

7.2.5 Oil Pollution

Petroleum is one of the most important raw materials in the world and is essential to produce various compounds, including fuels, drugs, solvents, fertilizers, and plastics. Crude oil is a complex mixture that contains thousands of different hydrocarbon compounds like alkanes (paraffins), cycloalkanes (naphthenes), and various aromatic hydrocarbons and inorganic elements including metals (such as Fe, Ni, Cu, V), sulphur, and nitrogen, which make up about 80% of crude oil's composition (da Silva Correa et al., 2022; Saha and Kumar, 2023). Oil pollution is an inevitable outcome because of rapid population growth and the industrialization process. Oil hydrocarbon materials are vastly produced around exploration and refining installations; transfer of these materials and accidental oil spills are the major causes of soil pollution (Havugimana et al., 2017). Oil contamination reduces the strength, permeability, fertility, maximum dryness, nutrient retention, water availability, soil compaction, biomass production, optimum moisture content, and changes to soil porosity, pH, and salinity. Oil deposited on the leaves and stems of plants blocks transpiration pathways, affecting the entire process of photosynthesis by covering the stomata and preventing oxygen diffusion and CO_2 entry. The blocking of stomata reduces stomatal conductance, increasing leaf temperature and resulting in thermal stress, which leads to oxidative stress. It may penetrate the inner layer of soil, where it directly affects the plant root system and microbial population and thus depletes soil oxygen content. Crude oil may cause a reduction in the level of available plant nutrients, developing a pseudo nutrient deficiency or may elevate the concentration of certain elements to toxic levels. Oil usually limits soil particles and blocks spaces, which reduces aeration and water infiltration and increases bulk density, developing an anaerobic environment in soil. Such anoxic situations affect basic soil chemical and physical indicators and thus, the soil microbial communities. Heavy crude oil contamination can also result in complete mortality of wetland vegetation. The presence of oil in the soil also affects water absorption capacity and produces stunted roots, preventing their expansion. This is because oil inhibits the action of auxins and the metabolism of cellular organelles, preventing cellular expansion in the roots (da Silva Correa et al., 2022, Saikia et al., 2023). In addition to the direct emission of these pollutants, dust from burning gases along with oil have managed to add toxic and harmful substances to the soils of the region for many years. The chemicals present in the fuel deteriorate the quality of soil and make it unsuitable for cultivation. Additionally, these chemicals can enter the groundwater through soil and make the water undrinkable (Havugimana et al., 2017).

The biodegradation of hydrocarbons is naturally very slow and thus oil persists in the environment for a long time and thus causes oil toxicity. For this reason, oil is considered a persistent pollutant not only in soil but water also. It not only impacts the environment, but oil also affects human health and produces cytotoxic, genotoxic, carcinogenic, and mutagenic effects. Hydrocarbons can penetrate seeds and alter their metabolic reactions, such as inhibiting the activity of amylase and starch phosphorylase (da Silva Correa et al., 2022). The more oil materials penetrate the greater depth of soil, then its removal will be a very difficult task and will cost dearly (Havugimana et al., 2017).

7.2.5.1 Case Study: Impact of Oil Contamination on Physico-Chemical Properties of Soil Around Sildubi Oil Spill, Assam

In a study conducted by Tanuj et al., the effects of crude oil pollution on the physico-chemical properties of soil were studied in the Sildubi oil field area in the Jorhat district of upper Assam, where oil exploration activities are conducted by Oil & Natural Gas Cooperation Limited (ONGCL). Soil samples were collected at 0–15 cm depth from the top surface of the drilling site. Four samples were also collected from North (N), South (S), East (E), and West (W) zones at a radius of 500 and 1000 m from the drilling site in circular mode for comparison. After analysis of the soil collected from oil-impacted sites and comparison of soil samples with samples collected from the radius of 500 and 1000 m from zone of drilling, the following results were obtained:

1. The soil was severely contaminated by hydrocarbons. Hydrocarbon contamination can increase soil total organic carbon, change soil chemical properties including pH, and alter soil mineral nutrients such as sodium, potassium, sulphate, phosphate, and nitrate, thus indirectly affecting the growth and development of plants and microorganisms.
2. The alteration in soil's physical and chemical properties had lowered the crop productivity of neighbouring agricultural lands.
3. Soil pH, organic carbon, and organic matter were found to be significantly higher at the drilling site.
4. A lower bulk density in crude oil–contaminated soil was recorded.
5. Soil moisture contents at the contaminated site were appreciably lower than the samples collected at 500 and 1000 m radii.
6. Lower water-holding capacity with reduced porosity was recorded at the highly polluted site.
7. The soil pH value at the drilling site was 7.85, which was significantly higher than the samples collected from 500 and 1000 m radius. The higher pH values in the soil might be due to the hydrophobic nature of crude oil, which might induce a potential drought in the surface layers of polluted soil.
8. Organic carbon content was found to be higher in the oil-contaminated soils, while the soil samples collected at a 1000 m radius was found to be the lowest. It can be concluded that high organic carbon in crude oil–contaminated soil might be due to the slow decomposition rate by soil organisms due to poor soil aeration.
9. Total nitrogen, phosphorus, and calcium content in the soils of the drilling site was found to be higher than the samples collected from 500 and 1000 m radii, while the concentration of potassium was found to be the lowest at the drilling site.
10. The alkaline pH of the oil-contaminated soil might be due to the accumulation of high sodium that during heavy rains, entered the nearby rice fields.
11. A sink in crop production in neighbouring agricultural lands can be due to developing Fe stress situations either by Fe deficiency in alkaline pH or by toxicity in acidic pH.

(Saikia et al., 2023)

7.2.6 Radioactive Pollutants

Living organisms are continuously exposed to a variety of radiation called background radiation. If the level of radioactive radiation increases above a certain limit (called radioactive pollution) it causes harmful effects to living beings (Havugimana et al., 2017). Radioactive waste from mining and nuclear processes may reach the soil via water or as "fall-out". From the soil they reach the plants and then grazing animals (livestock), ultimately reaching man through milk and meat, resulting in retarded and abnormal growth (Mishra et al., 2016).

7.2.7 Mining

Mining has greatly affected soil, water, and living organisms since ancient times. The main source of metal pollutants in soils is mining and smelting activities (Al-Taai, 2021, Havugimana et al., 2017). Metal smelting to separate minerals has introduced many pollutants into the soil. Mining and smelting facilities release huge quantities of heavy metals and other toxic elements to the environment that persist for long periods, even long after the end of these activities (Rodríguez-Eugenio et al., 2018). Mining operations routinely modify the surrounding landscape by exposing previously undisturbed earthen materials, erosion of exposed soils, extracted mineral ores, tailings, and fine material in waste rock piles that can result in substantial sediment loading to surface waters and drainage ways. In addition, spills, and leaks of hazardous materials and the deposition of contaminated windblown dust can lead to soil contamination (Havugimana et al., 2017).

Gold mining is generally carried out by using traditional methods for separating the gold from the raw material in developing countries. The main waste product from this process is mud and rubble containing a high concentration of mercury. These wastes are directly disposed of in the river, which is also used as a source of irrigation water or in domestic activities. It was found in a soil survey conducted in some areas near a mining site that the soil surrounding the traditional mining was polluted by mercury; the concentration of mercury in soil near the mining areas was three times more than in distant soil (Havugimana et al., 2017). Toxic mining wastes are stocked up in tailings, mainly formed by fine particles that can have different concentrations of heavy metals. These polluted particles can be dispersed by wind and water erosion, sometimes reaching agricultural soils. A study conducted in 2014 by Mileusnić et al. has found high levels of lead and copper in agricultural fields located near a tailings dam in Namibia. Toxic concentrations of chromium and nickel were found in agricultural soils near an abandoned chromite-asbestos mine waste in India and in the crops grown in those soils, resulting in a high risk to human and livestock health (Rodríguez-Eugenio et al., 2018). In the production of fertilizers, phosphate rocks are used, which are naturally rich in radioactivity and generate a by-product called phosphogypsum, which maintains nearly 80% of its original radioactivity due to 238U decay products such as radon, 226Ra, and polonium, 210Po. These industries are not only polluting the environment but also generating a radioactive source of pollution, which constitutes a threat to the surrounding ecosystems and organisms (Rodríguez-Eugenio et al., 2018).

7.2.8 Impacts

Many of the compounds at high concentrations or following prolonged exposure have adverse effects in humans and other organisms which include the danger of acute toxicity, mutagenesis (genetic changes), carcinogenesis, and teratogenesis (birth defects) for humans and other organisms. Some of these man-made toxic compounds are also resistant to physical, chemical, or biological degradation and thus are an environmental burden of considerable magnitude (Mishra et al., 2016). Some carbon-based organic contaminants can undergo chemical changes or degrade into products that may be more toxic than the original compound. Chemical elements such as heavy metals cannot break down, but changes in their characteristics can be easily taken up by plants or animals (Havugimana et al., 2017).

Different contaminants vary in their tendency to:

1. End up in water held in the soil or in the underlying groundwater by leaching through the soil.
2. Volatilize (evaporate) into the air.
3. Bind tightly to the soil.

(Havugimana et al., 2017)

Although some of the heavy metals such as Zn, Mn, Ni, and Cu act as micronutrients at lower concentrations, they become toxic when present at higher concentrations. Studies have shown that crops and vegetables grown in soils contaminated with heavy metals have greater accumulation of heavy metals than those grown in uncontaminated soil. Dietary intake of heavy metals through contaminated crops may lead to various chronic diseases. The bio-toxic effects of heavy metals depend upon various factors, including the concentrations and oxidation states of heavy metals, the kinds of sources, and the modes of deposition. Severe exposure to Cd may result in pulmonary effects such as emphysema, bronchiolitis, and alveolitis and renal effects from subchronic inhalation of Cd. Studies have found that Pb toxicity causes a reduction in haemoglobin synthesis; disturbance in the functioning of the kidneys, joints, reproductive and cardiovascular systems; and chronic damage to the central and peripheral nervous systems. Higher concentrations of Zn are related to impairment of growth and reproduction (Wang et al., 2015, Table 7.1).

Some of the effects of soil pollution are listed as follows:

7.2.9 Effects on Soil Environment

- Contamination of the soil affects the balance of the ecological system.
- Imbalance of flora and fauna inhabiting the soil.
- Contaminated soil decreases soil fertility and hence there is a decrease in the soil yield.
- Loss of natural nutrients in soil.

TABLE 7.1
Guideline for Safe Limit of Heavy Metal (Singh et al., 2010; Wang et al., 2015)

Sample	Standard	Cd	Cu	Pb	Zn	Mn	Ni	Cr	As	Hg
Soil (µg g^{-1})	Indian Standard	3–6	35–270	250–500	300–600	–	75–150	–		
	WHO (1996)	0.8	36	85	50	–	35	100	–	–
	European Union Standards (EU 2002)	3.0	140	300	300	–	75	150	–	–
Water (µg ml^{-1})	Indian Standard	0.01	0.05	0.10	5.0	0.10	–	0.05		
	FAO (1985)	0.01	0.20	5.0	2.0	0.20	0.20	0.10		
	European Union Standards (EU 2002)									
Plant (µg g^{-1})	Indian Standard	1.5	30.0	2.5	50.0	–	1.5	20.0		
	WHO	0.02	10	2	0.60	–	10	1.30	0.15	0.02
	European Union Standards (EU 2002)	0.2	–	0.2	–	–	–	–	0.1	0.02

Safety limit guidelines for heavy metals in environmental and health contexts, including recommended thresholds for Cd (Cadmium), Cu (Copper), Pb (Lead), Zn (Zinc), Mn (Manganese), Ni (Nickel), Cr (Chromium), As (Arsenic), and Hg (Mercury).

- Reduced nitrogen fixation.
- Increased soil erosion.
- Affects ionic exchange capacity.
- Increase in soil salinity, which makes it unfit for cultivation.
- Creation of toxic dust.
- Foul odor due to industrial chemicals and gases.
- Alteration in soil structure can lead to the death of organisms in it. For example, Cd causes inhibitory effects on soil microbial activities, microbial growth, and microbial metabolic processes while Cu toxicity leads to the destruction of cell membranes and causes protein denaturation in microbes.
- Reduction in soil fertility. For example, Pb affects soil sorption capacity and humic acid content in the soil.
- Agriculture on contaminated soil results in fruits and vegetables that lack quality nutrients.
- Affects the food chain by introducing toxic components.
- Crops grown on polluted soil cause health problems when consumed.
- Reduces aeration and water infiltration, and increases bulk density, developing an anaerobic environment in the soil.

(Alengebawy et al., 2021, Mishra et al., 2016, Yang et al., 2020)

7.2.10 Effects on Humans

- Consumption of crops and plants grown on polluted soil causes health hazards
- Chronic exposure to heavy metals, petroleum, solvents, and agricultural chemicals can be carcinogenic.
- Higher incidence of leukaemia is associated with long-term exposure to benzene.
- Mercury causes higher incidences of kidney damage.
- Cyclodienes are linked to liver toxicity.
- Organophosphates can lead to neuromuscular blockage.
- Chlorinated solvents induce damage to the liver and kidneys and depression of the central nervous system.
- Exposure to arsenic is associated with cancers (lung, bladder, and skin), chronic diseases of the skin, heart, lungs, and nervous system, and diabetic effects. It is also able to suppress the replication of DNA with altered repair enzymes. The most dangerous effect of arsenic is alteration of mitochondrial membrane potential and ATP formation during glycolysis, and induction of apoptosis in various cells.
- Even low concentrations of some heavy metals like lead, cadmium, and mercury are toxic to living organisms, contributing to metabolic abnormalities in organisms, especially consumers of food from plants and other crops grown from contaminated soil.
- Bioaccumulation of Pb in the human body interferes with the functioning of mitochondria, thereby impairing respiration. It also causes constipation, swelling of the brain, paralysis, and eventual death.
- Cd alters the protease, urease, and alkaline phosphatase activity. Cd toxicity can cause renal damage, mitochondrial dysfunction, osteoporosis, paediatric cancer, and stunted development in children.
- Pb toxicity in infants leads to permanent damage to multiple organ systems.
- Heavy metals are associated with human systems dysfunction, neurodegenerative disorders, musculoskeletal diseases, reproductive hormonal imbalance, infertility, oxidative stress, DNA damage, and various other diseases in adults.
- Pesticides are highly carcinogenic and related to numerous malignancies, such as neuroblastoma, leukaemia, soft tissue sarcoma, Burkitt lymphoma, non-Hodgkin lymphoma, Wilm's tumour, lung cancer, ovarian cancer, and rectal cancer.

- Genetic disruption by pesticides can be generally divided into three main classes: (i) pre-mutagenic damages (like DNA strand breaks, DNA adducts); (ii) gene mutations (insertions, deletions, inversions, and translocations); (iii) chromosomal aberrations [loss or gain of the entire chromosome (aneuploidy), deletion or breaks (clastogenicity), and chromosomal rearrangements].

(Alengebawy et al., 2021; Mandal 2017; Mishra et al., 2016; Nyiramigisha and Komariah, 2021)

7.2.11 Effects on Plants

- Heavy metal accumulation in the soil and plants affects the physiological activities of plants including photosynthesis, gaseous exchange, and nutrient absorption which result in reduced plant growth.
- Inhibition of seed germination.
- Chlorophyll content reduction.
- Damage such as leaf rolls, chlorosis, growth inhibition, root tips browning, and even death of the plant.
- Heavy metal phytotoxicity in plants leads to chlorosis, poor plant growth, plant depression, reduced nutrient uptake, plant metabolism disturbance, negative effect on seed germination, reduced possibility of harvest production, and reduced ability to repair molecular nitrogen in leguminous plants.
- The heavy metals are also associated with the production of free radicals, resulting in elevation of intracellular levels of reactive oxygen species (ROS), causing oxidative stress, which causes damage to biological molecules (e.g., proteins, nucleic acids, lipids, and enzymes).
- Ni can result in height and roots-length reduction, decrease of chlorophyll content, reduction of photosynthetic pigments, and the accumulation of Na^+, K^+, and Ca^{2+} in the plant.
- Cd disturbs the guard cell regulation, thus affecting the water status of the plant; inhibits seed germination, reduces stem conductivity, and has a toxic influence on plant phenotype (e.g., reducing plant weight and length of roots and shoots), cytotoxicity (e.g., reducing chlorophyll content and inhibiting photosynthetic performance), and metabolic processes (e.g., chlorosis and cell damage).
- Cu toxicity results in root deformation, biosynthesis of chlorophyll, decreases shoot length, reduces polypeptides, and causes changes in lipid content and plant productivity.
- Pb gradually slows down seed germination and causes abnormality of plant metabolism, morpho-physiological features, lowers chlorophyll biosynthesis, imbalances hormones, decreases protein content, induces over-production of reactive oxygen species (ROS), and affects plant growth and productivity.
- Al results in retardation of cell division, loosening of the cell wall, destruction of the plasma membrane, and alteration of calcium homeostasis.
- The pesticide residues and degradation by-products taken up through the roots via xylem to the entire plant parts cause various effects on plants including overproduction of ROS, oxidative stress, DNA damage, photosynthetic blockage, necrosis, chlorosis, and leaves twisting, and ultimately end with plant death.
- Excessive use of insecticides causes a reduction in grain protein content, blockage of stomatal conductance, and alterations in the photosynthetic process.
- Herbicides cause alteration of the physiological and biochemical plant efficiency, increasing the susceptibility of plants toward diseases, while fungicides result in reduction of chlorophyll and carotenoid concentrations, destruction of chloroplasts, stomatal closure, and electron transfer suppression.

- Pesticides can cause chlorosis, necrosis, leaves twisting, and photosynthesis malfunction due to oxidative stress.
- Oil on leaves and stems blocks transpiration pathways, affecting the entire process of photosynthesis by covering the stomata and preventing oxygen diffusion and CO_2 entry.

(Alengebawy et al., 2021, Alori et al., 2022; Ayilara & Babalola, 2023; da Silva Correa et al., 2022; Nyiramigisha and Komariah, 2021)

7.3 MECHANISM OF MICROBIAL DEGRADATION

Numerous in situ (on-site, in the soil) and off-site (removal of contaminated soil for treatment) techniques are being developed to decontaminate polluted soils. None is ideal for remediating contaminated soils, and often, more than one of the techniques may be necessary to optimize the cleanup effort (Mishra et al., 2016). The most common decontamination method used is to remove the soil and deposit it in landfills or to incinerate it. These methods, however, often cause various other problems; landfilling merely confines the polluted soil while doing little to decontaminate it, while incineration removes toxic organic chemicals from the soil but subsequently releases them into the air, causing air pollution (Mishra et al., 2016).

Various soil washing techniques have been developed for the removal and recovery of heavy metals, including physical methods such as attrition scrubbing and wet-screening, and chemical methods consisting of treatments with organic and inorganic acids, bases, salts, and chelating agents. For example, chemicals like hydrochloric, nitric, phosphoric, and citric acids, sodium carbonate, and sodium hydroxide, and the chelating agents EDTA and DTPA are used to extract radionuclides and toxic metals. The problem with these methods again is that they generate secondary waste products that may require additional hazardous waste treatments (Mishra et al., 2016). Bioremediation is a biological method of reclaiming a polluted environment by decontaminating and degrading heavy metals and xenobiotics. It is less expensive, more sustainable, and safer than other conventional methods. This biological method of remediation is a productive alternative for cleaning, debugging, managing, and rehabilitating and consequently improves contaminated environments via judicious utilization of microbial activities. The rate of degradation of waste substances is usually dictated by competitiveness among biological agents, suboptimal supply of essential nutrients, abiotic conditions (including temperature, aeration, pH, the presence of oxygen or different electron acceptors, and supplements and moisture), and pollutant type (includes pesticides, agrochemicals, chlorinated compounds, heavy metals, xenobiotic compounds, organic halogens, greenhouse gases, hydrocarbons, nuclear waste, dyes, plastics, and sludge). Bioremediation is often effective only under conducive environmental conditions favourable for microbial growth and development. It has been successfully used in various parts of the world (Alori et al., 2022; Meena et al., 2021; Sharma, 2021).

Bioremediation can be carried out outside the pollution site, ex situ, or at the site of pollution, in situ (Alori et al., 2022; Figure 7.1).

7.3.1 Ex-situ Methods of Bioremediation

In the case of ex-situ methods, polluted soil is removed from the site of pollution and taken to another site for treatment. This method is considered when favoured by the depth of pollution, concentration of pollutant, cost of treatment, pollutant type, and geographical location of the polluted site. The ex-situ technique can be carried out using solids containing bioremediators or slurry containing bioremediators (Alori et al., 2022).

7.3.1.1 Treatment with Solids Containing Bioremediators

In this method, polluted soil is excavated, placed into piles, and mixed with solids containing bioremediators, including bacteria, fungi, organic waste such as animal manures, leaves, domestic and industrial wastes, and so on. Air or ventilation through the piles is allowed to enhance microbial

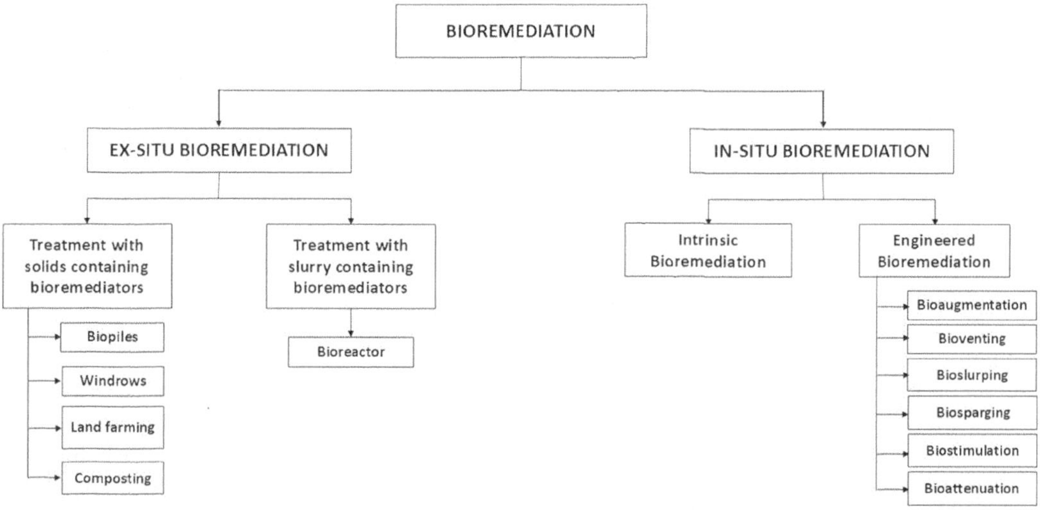

FIGURE 7.1 Classification of Bioremediation Based on Site or Location.

respiration. This system of remediation requires a large space and clean-up takes longer to complete as compared to the slurry-containing bioremediator. Examples of the use of solids containing bioremediators include biopiles, windrows, land farming, composting, and others (Alori et al., 2022).

7.3.1.1.1 Biopile

Polluted soil is excavated and piled above the ground at another site. At the site of treatment, the growth and activities of microbes present in the soil are then stimulated by the addition of air and nutrients. Biopile treatment is employed for treating volatile and low molecular weight pollutants. It can also be used to remediate polluted extreme environments such as in very cold regions. This system is preferred as it is cost-effective and the rate of biodegradation is pH, nutrient, temperature, and aeration dependent. A basic biopile system includes a treatment bed, an aeration system, an irrigation or nutrient system, and a leachate collection system. A heating system can be integrated into the biopile to increase microbial activities that further increase the rate of biodegradation. The addition of organic materials such as straw, saw dust, bark, or wood chips to improve permeability and moisture retention may enhance the remediation process. Excessive heating of air may lead to excessive drying of the soil which will inhibit microbial activities; therefore, the process is known to stimulate volatilization rather than biodegradation (Alori et al., 2022; Hazen et al., 2003; Lin et al., 2022).

7.3.1.1.2 Windrows

Piled polluted soil is rotated periodically to improve the degradation of pollutants in the soil by microorganisms. The periodic turning of polluted soil, along with the addition of water, results in an increase in aeration, uniform distribution of pollutants, nutrients, and microbial degradative activities, thus speeding up the bioremediation rate, which can be accomplished through assimilation, biotransformation, and mineralization. Windrow treatment is more effective in the removal of hydrocarbons from the soil when compared to biopile treatment. The windrows should not be considered for the remediation of toxic volatile compounds because of the periodic turning associated with windrow treatment. The use of windrow treatment has been implicated in the release of greenhouse gas, CH_4, due to the development of an anaerobic zone within piled polluted soil, which usually occurs as result of reduced aeration (Alori et al., 2022; Lin et al., 2022).

7.3.1.1.3 Land Farming

In land farming, aerobic biodegradation of polluted soil is initiated by placing the polluted soil on a fixed layer above the ground surface to allow aerobic biodegradation of pollutants by autochthonous microorganisms. The major operations that stimulate activities of autochthonous microorganisms include aeration, addition of nutrients (nitrogen, phosphorus, and potassium), and irrigation to enhance bioremediation during land farming. It is regarded as ex-situ bioremediation in most cases but sometimes it can be considered an in situ bioremediation technique. When the excavated polluted soil is treated on-site, it is regarded as in situ bioremediation whereas if the polluted soil is treated in another location, it is regarded as ex situ. It has been reported that when a pollutant lies <1 m below the ground surface, bioremediation might proceed without excavation whereas the pollutant lying at >1.7 m needs to be transported to the ground surface for effective bioremediation. The land farming technique is very simple to design and implement, it requires low energy and capital input and a large volume of polluted soil can be treated with minimal environmental impact. The construction of a suitable land farming design with an impermeable liner minimizes leaching of pollutants into neighbouring areas during the bioremediation operation. Land farming is usually used for the remediation of sites polluted with hydrocarbon including polyaromatic hydrocarbons (Alori et al., 2022; Lin et al., 2022).

7.3.1.1.4 Composting

Composting is a self-heating biological process that has been used for centuries as a solution for management of organic waste. Apart from managing organic waste, the product from composting can be used as a soil amendment and organic fertilizer. Studies have shown that the diverse microbial communities present in composting materials are also responsible for the biodegradation of recalcitrant organic contaminants. For instance, some bacterial species including *Acinetobacter lwoffii*, *Bacillus subtilis*, and *Raoultella ornithinolytica* are found to degrade crude oil. Temperature rise during composting (thermophilic phase), due to energy release during the degradation of organic material, also enhances degradation by stimulating the activities of both mesophilic and thermophilic microorganisms and making the contaminants less viscous and more bioavailable. Moreover, biosurfactants like rhamnolipids produced by certain microbial species during composting, also enhance biodegradation by solubilizing the organic contaminants (Alori et al., 2022; Lin et al., 2022).

7.3.1.2 Treatment with Slurry Containing Bioremediators

In this mode of treatment, first stones and rubbles in the excavated polluted soil are removed, and then nutrients and water are mixed with the polluted soil in an aerated bioreactor. This provides a favourable environmental condition for the soil microorganisms to grow and break down the pollutants contained in the soil. Excellent control of bioprocess parameters including temperature, pH, agitation and aeration rates, substrate, and inoculum concentrations is one of the major advantages of bioreactor-based bioremediation. This process can be used to treat soil or water polluted with volatile organic compounds including benzene, toluene, ethylbenzene, and xylenes (BTEX); after the degradation of the pollutant, soil is removed and dried via vacuum filters, pressure filters, and centrifuges. After that the cleaned soil is deposited and the resultant fluid is further treated. Genetically modified microorganisms can be used for bioaugmentation in a bioreactor after which the organism can be destroyed before treated soils are returned to the field for landfilling. The bioreactor is an enclosed system, and destruction of microorganisms before returning to field will ensure that no foreign gene escapes into an environment after bioremediation.

This mode of treatment has a faster rate of remediation compared to the other treatment processes. The amount of water to be added to the polluted soil to initiate the degradation process depends on the property of the soil and pollutant concentration. Biodegradation of contaminants carried out in a bioreactor is usually more manageable, controllable, and predictable. Moreover, some of the limiting factors of the bioremediation process including controlled bioaugmentation, nutrient addition, increased pollutant bioavailability, and mass transfer (contact between pollutant and microbes) can effectively be established in a bioreactor, thus making bioreactor-based bioremediation more efficient (Alori et al., 2022; Azubuike et al., 2016).

7.3.2 In Situ Methods of Bioremediation

In case of in situ methods, soil does not need to be excavated, and therefore the chance of causing further environmental harm is minimized. In situ biodegradation involves the enhancement of naturally occurring microorganisms by artificially stimulating their numbers and activity which assist in degrading the soil contaminants. Several environmental, chemical, and management factors, including moisture content, pH, temperature, the microbial community that is present, and the availability of nutrients, affect the biodegradation of soil pollutants. Of all the decontamination methods, bioremediation appears to be the least damaging and most environmentally acceptable technique (Mishra et al., 2016). In situ bioremediation techniques have been effectively used to treat sites polluted with chlorinated solvents, heavy metals, dyes, and hydrocarbons (Azubuike et al., 2016). Based on whether the bioremediator is within the polluted soil or is introduced from an external source, in situ bioremediation could be categorized into two types – intrinsic and engineered bioremediation (Alori et al., 2022).

7.3.2.1 Intrinsic Bioremediation

Intrinsic bioremediation, also known as natural reduction, is an in situ bioremediation technique, which is carried out by the stimulation of indigenous microbes of the polluted sites without any external force (human intervention). This remediation could be done by both aerobic and anaerobic processes. The technique is less expensive compared to engineered techniques as no external force of any sort is applied. It has the capacity to remove recalcitrant pollutants because of the anaerobic process that could also be involved (Alori et al., 2022; Sharma, 2021). One of the major limitations of intrinsic bioremediation is that because no external force is incorporated to expedite the remediation process, it might take a longer time to achieve the target level of pollutant concentration (Azubuike et al., 2016).

7.3.2.1.1 Engineered Bioremediation

Some engineered in situ bioremediation techniques in use are bioaugmentation, bioventing, biosparging, bioslurping, biostimulation, and bioattenuation (Azubuike et al., 2016; Sharma, 2021).

7.3.2.1.2 Bioventing

Bioventing techniques involve controlled stimulation of airflow by delivering oxygen to the unsaturated (vadose) zone to increase indigenous microbes' activity for bioremediation. Nutrients and moisture are added to increase the rate of bioremediation, which will achieve microbial transformation of pollutants to a harmless state (Azubuike et al., 2016; Sharma, 2021).

7.3.2.1.3 Bioslurping

This technique combines vacuum-enhanced pumping, soil vapor extraction, and bioventing to achieve soil and ground water remediation by indirect provision of oxygen. This technique is used for recovering products from remediating capillary, light non-aqueous phase liquids (LNAPLs), and both unsaturated and saturated zones. This technique is used to remediate soils contaminated with volatile and semi-volatile organic compounds. The system uses a "slurp" that extends into the free product layer and draws up liquids (free products and soil gas) from this layer. The pumping machine transports LNAPLs to the surface by upward movement, where it gets separated from air and water. Excessive soil moisture limits air permeability and decreases oxygen transfer rate, which results in decrease in microbial activities. This technique is not suitable for the remediation of soil with low permeability but it is cost effective as the operation procedure results in less use of ground water and thus minimizes storage, treatment, and disposal costs. The major concerns of this particular in situ technique are saturated soil lenses which are created by establishing a vacuum on a deep high-permeable site and fluctuating water table, which are difficult to aerate (Azubuike et al., 2016; Sharma, 2021).

7.3.2.1.4 Biosparging

In the biosparging technique, microbial activities in polluted sites are improved by injecting air into the saturated zone, which can cause upward movement of volatile organic compounds to the unsaturated zone to promote biodegradation. The effectiveness of biosparging is dependent on two major factors – soil permeability, which determines pollutant bioavailability to microorganisms, and pollutant biodegradability. Biosparging is similar in operation with a closely related technique known as in situ air sparging where high airflow is required for volatilization of pollutants, whereas injected air in biosparging stimulates biodegradation. Biosparging can be widely used for the treatment of aquifers contaminated with petroleum products, especially diesel and kerosene. In a study conducted by Kao et al. (2005), it was reported that biosparging resulted in reduction (>70%) of BTEX in a contaminated aquifer. The major limitation of the technique is predicting the direction of airflow (Alori et al., 2022;, Azubuike et al., 2016).

7.3.2.1.5 Bioaugmentation

This technique is characterized by the increase of native microbiota by the inoculation of exogenous microorganisms. Generally, the microorganisms used are bacteria, filamentous fungi, and yeasts (Alori et al., 2022). In this method of remediation microorganisms are collected from the remediation site, autonomously refined, genetically changed, and returned to the site. Firstly, this is done to ensure complete wipe out and change the impurities into less toxic compounds by in situ microorganisms. For instance, all fundamental microorganisms found in soil and groundwater contaminated with chlorinated ethenes, like tetrachloroethylene and trichloroethylene are assembled and used to ensure that the in situ microorganisms can completely wipe out and change these impurities to ethylene and chloride, which are non-destructive. Secondly, genetically engineered microorganisms can extend the degradative capability of a wide extent of characteristic toxic compounds (Meena et al., 2021). In addition, there are various cases where recalcitrant compounds may be new, and as a result, microorganisms may not have yet adapted to use them as a substrate. Bioaugmentation can overcome these challenges; one of its main advantages is that treatment can be tailored to a specific pollutant that is dominant in the environment. Thus, this approach is attractive for remediating both the increasing number of emerging pollutants as well as pollutants that are present at high concentrations (Nzila et al., 2016).

Bioaugmentation can be classified into two different approaches: (i) cell bioaugmentation, which relies on the survival and catabolic activity of the inoculated strains that perform the degradation of the target contaminants, and (ii) genetic bioaugmentation, which is based on the spread of catabolic genes such as plasmids, integrons, and transposons, located in mobile genetic elements, into native microbial populations through mechanisms of horizontal gene transfer. Wang and Mulligan, in one of their studies conducted in 2004, reported an accelerated removal of quinoline after the inoculation of *Burkholderia pickettii*, and Mrozik et al. in 2011 showed that cell bioaugmentation with *Pseudomona* sp. JS150 significantly enhanced phenol degradation in soil, thereby reducing the possibility of formation of phenoxyl radicals (Garbisu et al., 2017). Effective results were found with genetic bioaugmentation too. A microcosm study conducted by Halden et al. (1999) detected that transfer of plasmids pPOB and pD30.9 from *Pseudomonas pseudoalcaligenes* POB310 (pPOB), *Pseudomonas* sp. B13-D5 (pD30.9), and B13ST1 (pPOB) to recipient soil bacteria showed an enhanced degradation of 3-phenoxybenzoicacid (3-POB). Similarly, a study conducted by Dejonghe et al. (2000) reported degradation of 2,4-D in soil under microcosm conditions, using *P. putida* as a donor strain of two catabolic plasmids (pEMT1 and pJP4). There are various such examples of plasmid-mediated bioaugmentation that showed successful degradation of different contaminants (Garbisu et al., 2017).

7.3.2.1.6 Biostimulation

Biostimulation includes the alteration of the bioenvironment to enhance the activity of present microorganisms and accomplish the target bioprocesses efficiently. It focuses on the induction of indigenous or already existing microorganisms and life form organization. First manures and minor components and electron acceptors are provided for improvement upgrades. Generally, inorganic additions including macro- (such as N, P, K) or micronutrients (such as Mg, S, Fe, Cl, Zn, Mn, Cu, Na) are done to recover depleted soils. Then other natural requirements like pH, temperature, and oxygen are provided to enhance their assimilation rate and pathways. Additionally, the presence of a restricted amount of poison can act as an energizer by turning on the operons for bioremediation of synthetic substances. These enhancements are the basic structural requirements of life that license microorganisms to make the key needed including imperativeness and cell biomass and stimulate them to degrade the toxin (Meena et al., 2021, Mohamed & Samer, 2023, Vásquez-Murrieta et al., 2016).

7.3.2.1.7 Bioattenuation

Bioattenuation relies on natural processes in which the indigenous microbial populations degrade recalcitrant or xenobiotic compounds based on their metabolic processes. This method of bioremediation includes a variety of chemical, physical, and biological processes that reduce the mass, toxicity, volume, or concentration of contaminants. These processes include aerobic and anaerobic biodegradation, sorption, volatilization, and chemical or biological stabilization or transformation of contaminants. Time is not a limiting factor and is generally applied on sites with low concentrations of contaminants, where no other remedial techniques are applicable (Vásquez-Murrieta et al., 2016).

7.3.3 Microorganisms Used in Bioremediation

Microorganisms are ubiquitous in nature, and the nutritional versatility of microorganisms is a property that can be exploited to biodegrade pollutants, making them a great choice for bioremediation. Here certain microorganisms convert, modify, and utilize toxic pollutants to obtain energy and biomass. Bioremediation is a process in which microbes specifically break down the pollutants, transforming them into less toxic or non-toxic elemental and compound forms (Sarao & Kaur, 2022). Adaptability is another characteristic of microorganisms that makes them suitable for bioremediation. Microbes can grow at temperatures below zero as well as in extreme heat in the presence of hazardous compounds or any waste stream (Sharma, 2021). Microorganisms also restore the original natural surroundings by preventing further pollution (Mrozik & Piotrowska-Seget, 2010). The bacterial strain *Mycobacterium vanbaalenii* PYR-1 was reported as the first isolated bacteria that was capable of totally mineralizing pyrene by mono- and dioxygenase reactions (Elyamine et al., 2021).

There are groups of microbes which are used in bioremediation:

7.3.3.1 Aerobic Bacteria

These have degradative capacities to degrade the complex compounds (like pesticides, hydrocarbons, alkanes, and polyaromatic compounds) including *Pseudomonas*, *Acinetobacter*, *Sphingomonas*, *Nocardia*, *Flavobacterium*, *Rhodococcus*, and *Mycobacterium*. Many of these bacteria use the contaminants as carbon and energy sources (Sharma, 2021). It has been reported that different aerobic bacteria degrade PAHs through three main steps, including the ring cleavage process, side chain metabolism process, and central metabolism process. In each step, different cascade enzymatic reactions succeed each other including various enzymes, mainly dioxygenase, dihydrodiol dehydrogenase, ring-breaking dioxygenase, epoxide hydrolase, alcohol dehydrogenase, and acetaldehyde dehydrogenation enzymes and decarboxylase. Oxygen is the final electron acceptor in the aerobic degradation process (Elyamine et al., 2021).

7.3.3.2 Anaerobic Bacteria

These are not as regularly used as aerobic bacteria. This microbial life in the absence of oxygen has significant potential for solving one of the important present-day problems of environmental pollution and degradation. In anaerobic degradation, nitrogen, metal ions such as ferric, and manganese, sulphide, chlorate, trimethylamine oxide, perchlorate, carbon dioxide, and fumarate instead of oxygen constitute the final electron acceptors to oxidize organic compounds to carbon dioxide. If the natural rates of degradation of contaminants are too slow then increased levels of alternate electron acceptors may accelerate the rate of anaerobic contaminant degradation. This is because electron acceptors like sulphate and nitrate are highly soluble and are not consumed by non-biological processes. Normally the hydrocarbon-contaminated soils lack oxygen, and sometimes for biostimulation urea and ammonia-based fertilizers are added, which potentially exert an oxygen demand due to biological ammonia oxidation. For such situations anaerobic bioremediation will be a great choice. Several denitrifying, manganese-iron and sulfate-reducing bacteria have been isolated that could degrade simple aromatic or aliphatic hydrocarbons under anoxic conditions (Balagurusamy, 2005; Elyamine et al., 2021). In a study the strain *Clostridium* sp. ER9, a strict rod-shaped Gram-positive bacterium, was reported to metabolize pyrene up to 93% within 49 days of incubation anaerobically. In another study, a facultative anaerobic bacterium, *Paracoccus denitrificans* M^{-1}, showed a capacity to degrade pyrene under anoxic conditions with different nitrate/nitrite (NO_3–/NO_2–) ratios within 5–6 days of incubation, and a study conducted by Yan et al. reported that the strain *Hydrogenophaga* sp. PYR1 is capable of anaerobically removing pyrene under iron reduction conditions with an efficiency of 22% (Elyamine et al., 2021).

7.3.4 Genetically Engineered Microbes for Remediation

The term genetically modified organism (GMO) represents any living organism with genetic material that has been altered using genetic engineering techniques. At first, genetically modified bacteria were used in medicine, pharmaceuticals, and agriculture, but they have become a powerful tool in control of toxic and dangerous waste. The aim of bioremediation is to encourage microbes to destroy the contaminants present in the environment by providing them with nutrients and other chemicals. Scientists are investigating procedures to augment contaminated sites with non-native microorganisms, for which genetically modified microorganisms (GMMs) can be a suitable choice. Construction of GMMs used in bioremediation can be designed to make them capable of degrading pollutants at the site of contamination. Due to regulatory risk assessment concerns and insecurity of their practical impact and delivery under field conditions, there are not that many field tests of the application of the GMMs in bioremediation.

There are at least four principal approaches to developing genetically modified microbes for bioremediation applications, including

1. Modification of enzyme specificity and affinity
2. Pathway construction and regulation
3. Bioprocess development, monitoring, and control
4. Bio affinity bioreporter sensor applications for chemical sensing, toxicity reduction, and end-point analysis.

(Azad et al., 2014)

For example, mercury is the most toxic heavy metal to be released into the environment. It has been observed that genetically engineered (GE) *Escherichia coli* strain JM109 can remove mercury from contaminated water, soil, or sediment. GE bacteria containing the MerA gene can remove mercury from a contaminated site and the transgenic bacteria expressing metallothioneins and polyphosphate kinase can promote effective mercury bioremediation. GE bacteria also are capable of metabolizing chlorinated organic compounds such as lindane and trichloroethylene. Lindane (c-hexachlorocyclohexane) is extremely toxic to humans as well as deleterious for the environment. A study conducted by Chaurasia et al. reported that more than 98% of the lindane was degraded from paddy fields within 6–10 days by the recombinant *Anabaena*. GE *E. coli* and *P. putida* have also been reported to degrade trichloroethylene, via their transformation through the insertion of a range of phenol catabolic genes (pheA, pheB, pheC, pheD and pheR). Another study conducted by Marconi et al. reported that GE *P. putida* S12 can degrade naphthalene, toluene, and biphenyl following insertion of the plasmids, encoding genes for the catabolism of these contaminants (Azad et al., 2014). Genetically modified *P. putida* KT2440 has also been tested for bioremediation of organophosphate and pyrethroid (Rafeeq et al., 2023). Recombinant DNA technology is a promising method for the preparation of GE organisms that can combat environmental pollutants for successful bioremediation. DNA shuffling, which is a powerful mutagenesis technique, can create new enzymatic activity and biocatalysts with a higher rate of degradation for chlorinated ethane and polyaromatic hydrocarbons. The DNA techniques have been applied to evolve enzymes such as dioxygenases and monooxygenases for bioremediation. This technology can improve catabolic pathways that can overcome the toxic effects of pollutants (Azad et al., 2014). The potent technology CRISPR has made genome editing possible. Scientists can readily change DNA sequences and alter gene activity in microorganisms that can be used for the removal of organic pollutants by bacteria. For biodegradation of the sulfonated azo dye Green HE4B, Joshi and Thagard (2013) created a UV–irradiated mutant of *Pseudomonas* sp. 1 F. This mutant strain took 25% less time to fully break down the resistant dye GHE4B than the wild-type strain. UV irradiation causes a rise in the enzymatic activity of lignin peroxidase, laccase, veratryl alcohol, NADH- (dichlorophenol-indophenol) reductase, and oxidase. Site-directed mutagenesis can be used to modify regions of DNA to develop better strains suited to certain purposes. Genetically modified organisms are also needed for bioremediation of radioactive chemicals because they are more resilient and have a greater chance of surviving in highly radioactive waste sites than wild microorganisms. Till now there are no reports of wild type bacteria being able to break down or transform radioactive materials into safer forms. The bacteria *Deinococcus radiodurans* has been observed to consume and digest radioactive iodine (>99%) and extremely radioactive mercury and toluene, which scientists believe is the most radio-resistant organism ever discovered. Thus, genetically modified bacteria be considered the safest and fastest alternative for converting or degrading radioactive materials (Rafeeq et al., 2023; Table 7.2).

7.3.5 Mechanisms of Microbial Remediation

Microbes can remove pollutants from the environment using two different mechanisms, namely mobilization and immobilization. Mobilization is a process that involves enzymatic oxidation, bioleaching, biostimulation, bioaugmentation, and enzymatic reduction procedure whereas immobilization includes bioaccumulation, complexation, biosorption, and precipitation (solidification). In mineralization, microbes transform pollutants into end products such as carbon dioxide and water or other intermediate metabolic substances. Similarly, immobilization is the conversion of compounds into a form where it will be unavailable in the environment like the conversion of nitrate into organic nitrogen (Ayilara & Babalola, 2023). Immobilization can be carried out using the in situ and the ex-situ methods. The ex-situ process involves the removal of contaminants from polluted soils in another location while in situ procedure takes place on site (Ayilara and Babalola, 2023).

TABLE 7.2
Some Genetically Engineered Bacterial Strains and Contaminants Degraded by Them (Rebello et al., 2021, Azad et al., 2014, Rafeeq et al., 2023)

Genetically Engineered Bacterial Species	Contaminant Degraded
Ralstonia eutropha CH34, *Mesorhizobium huakuii* B3	Cd^{2+}
Deinococcus radiodurans strains, *E. coli* JM109, *Achromobacter* sp. AO22, *Pseudomonas* K^{-62}	Hg
Escherichia coli and *Moraxella* sp.	Cd and Hg
P. fluorescens 4F39, *E. coli* SE5000	Ni
P. putida 06909, *Caulobacter crescentus* JS4022/p7236H, *B. subtilis* BR151 (pTOO24)	Cd
Sphingomonas desiccabilis and *Bacillus idriensis* strains	As
Recombinant *Anabaena*	Lindane
GE *E. coli* and *P. putida*	Trichloroethylene
P. putida S12	Naphthalene, toluene, and biphenyl
Recombinant *Rhodococcus erythropolis*	Nitrogen and organic matter in landfill leachate
Recombinant *Deinococcus radiodurans*	Cadmium and uranium bioaccumulated in the cytoplasm
Pseudomonas sp. WBC-3 and *Sphingobium japonicum*	Methyl parathion and -hexachlorocyclohexane

Genetically engineered bacterial strains designed to degrade specific environmental contaminants, showcasing the application of biotechnology in the bioremediation of pollutants like hydrocarbons, heavy metals, and toxic chemicals.

7.3.6 Enzymatic Oxidation

Enzymatic oxidation is the process in which the pollutant compounds are oxidized from a higher oxidation state to a lower one, during which heavy metals lose an electron and become less toxic. This process utilizes an enzyme called oxidoreductase released by the microbes. This method is highly effective in remediating dyes, phenols, and other pollutants that are not easily degraded by bacteria. Laccase is an example of an oxidoreductase enzyme that catalyzes the oxidation of aromatic amines. Laccase production has been reported in some species including *Pycnoporus* sp. and *Leptosphaerulina* sp., where it was outlined to degrade heavy metals (Ayilara & Babalola, 2023).

7.3.7 Enzymatic Reduction

In this process the pollutants are converted to an insoluble reduced oxidized state, which is carried out by obligate and facultative anaerobes. This method is effective in the bioremediation of compounds such as polychlorinated dibenzo-p-dioxins and dibenzofurans. Some examples include the reduction of hexavalent chromium to trivalent chromium catalyzed by chrome reductase and the reduction of the azo compounds by cleaving to azo bonds by azoreductase (Ayilara & Babalola, 2023). The most representative enzymes from microorganisms and genetically engineered microorganisms involved in the degradation of the different classes of pollutants include cytochrome P450, laccases, hydrolases, dehalogenases, dehydrogenases, proteases, and lipases (Bhandari et al., 2021).

Microbial enzymes involved in bioremediation and their functions (Bhandari et al., 2021; Karigar & Rao, 2011).

7.3.7.1 Cytochrome P450

Cytochrome P450 is responsible for electron transfer reactions and catalysis by reduction or oxidation of heme iron. It utilizes pyridine nucleotides as electron donors and molecular oxygen and NADH or NADPH as a cofactor to produce carbon substrates and oxidized products:

$$\text{NAD}(\text{P})\text{H} + \text{O2} + \text{R} \rightarrow \text{NAD}(\text{P}) + {+}\text{RO} + \text{H}_2\text{O}$$

They can also use ferredoxin and ferredoxin reductase as a source of electrons for catalytic function. P450s have an intrinsic capacity to degrade xenobiotics through chemical transformations like aliphatic hydroxylation and epoxidations, dealkylations, dehalogenation, and various other inactivation-based mechanisms. CYP101, CYP102, CYP1A1, CYP1A2, and CYP1B1 are known to metabolize PAHs. Out of these CYP1A1 shows high activity towards dibenzo-p-dioxin (DD) and mono-, di-, and trichloro-DDs whereas CYP1A1 mutant, F240A, shows activity towards 2,3,7,8-tetrachloro-DD.

7.3.7.2 Laccases (benzenediol oxygen oxidoreductases)

These are multi-copper-containing extracellular enzymes that consist of monomeric, dimeric, and tetrameric glycoproteins found in plants, bacteria, and fungi. Laccase has the catalytic capacity to oxidize phenolic compounds (ortho- and para-diphenol), aromatic amine, and their substituted compounds, having various functional groups by the formation of two water molecules with concomitant electron loss of a single oxygen molecule. It can also result in oxidation of nonphenolic compounds that are less soluble and more stable, which helps to control groundwater and underwater pollution. It not only oxidizes phenolic and methoxy phenolic acids but also decarboxylates them and attacks their methoxy groups that causing demethylation.

The substrate specificity and affinity of laccase can vary with changes in pH and is inhibited by various reagents including halides (excluding iodide), azide, cyanide, and hydroxide.

7.3.7.3 Dehalogenase

A wide range of halogenated compounds is degraded using the dehalogenase enzyme. It cleaves C-X bonds through three mechanisms.

- **Hydrolytic mechanism**: Halogen substituent is replaced in a nucleophilic substitution reaction by a hydroxyl group, where a water molecule serves as a cofactor.
- **Oxygen-lytic mechanism**: Catalyzed by mono- or di-oxygenase enzymes that incorporate one or two atoms of molecular oxygen into the substrate.
- **Reductive mechanism**: Halogen is substituted by hydrogen under aerobic conditions, where organohalides are used as terminal electron acceptors.

7.3.7.4 Dehydrogenase

This enzyme belongs to the family of oxidoreductases and is found in organisms ranging from bacteria, yeast, and plants to animals, including humans. It utilizes coenzymes such as NAD^+/$NADP^+$ or flavin such as FAD and FMN as an electron acceptor and transfers two hydrogen atoms from organic compounds to electron acceptors resulting in oxidation of organic compounds and generating energy.

7.3.7.5 Hydrolase

Hydrolases facilitate the cleavage of chemical bonds including C–C, C–O, C–N, S-S, S-N, S-P, C-P, and bond utilizing water and thus convert larger molecules to smaller ones to reduce pollutants' toxicity. Microbial hydrolytic enzymes including lipases, esterases, amylases, proteases, cellulases, nitrilases, peroxidases, cutinase, and others can be used for the management of waste produced during food processing, degradation of plastics and insecticides, treatment of biofilm deposits, and oil-contaminated soils. The breakdown of ester, amide, and peptide bonds by esterases, amidases, and proteases may lead to products with little or no toxicity and thus can help in bioremediating a contaminated site. Organophosphate compounds, which are highly lethal neurotoxins, are widely used as pesticides in agriculture. These represent a great threat in the biotic environment, where pyrethroids and malathion are unsafe for living environments. Organophosphate pesticides can be detoxified by the hydrolysis of phosphodiester bonds, whereas pyrethroids and malathion can be degraded by the hydrolysis of carboxyl ester bonds. These pesticides usually consist of phosphorus either as a phosphate ester or a phosphonate, and being ester, the principal reactions for detoxification involved are hydrolysis, oxidation, alkylation, and dealkylation. Thus, microbial degradation through hydrolysis of P-O-alkyl and P-O-aryl bonds is considered as the most significant step in detoxification.

A study concluded that the gene for organophosphorus hydrolase from *Pseudomonas diminuta* expressed in *E. coli* can result in detoxification of methyl parathion ranging between 10–80% and 3.6–45%. Lipase that can cause hydrolysis of triacylglycerols to glycerols and free-fatty acids can be used in bioremediation oil spills whereas protease is capable of degrading α-ester bonds, poly (hydroxybutyrate) (PHB) depolymerase ß-ester bonds, and lipase γ-ω bonds and thus can be used in bioremediation for the degradation of polymers (Table 7.3).

7.3.8 Factors Affecting Microbial Bioremediation

One of the most important points for bioremediation efficacy is maintaining the proper number and biomass of the introduced strain. A study concluded that using activated carbon and zeolite in crude oil-contaminated soil stimulated microbial colonization and enhanced hydrocarbon degradation. It

TABLE 7.3
Enzymes Produced by Microorganisms to Degrade Contaminants (Ayilara & Babalola, 2023, Bhandari et al., 2021)

Microorganisms	Enzymes	Contaminant Remediated
T. fusca, *Pseudomonas* sp., *Burkholderia* sp., *Ralstonia* sp., *Achromobacter* sp., *Sphingomonas* sp., *Comamonas* sp.	Hydrolases	Polyester plastics and Hydrocarbon
Bacillus safenis	Oxidoreductase	Xenobiotics
Brevundimonas diminuta	Phosphodiesterase	Pesticides
Baciluus pumilus	Lipase	Oil containing industrial wastewater
Pseudomonas putida	Laccase	Synthetic dyes
E. coli and *Bacillus* sp. F31	Lignin peroxidase	Synthetic dyes
E. coli and *S. rhizophila*	Dehydrogenase	Steroids and Polyvinyl alcohol
Bacillus subtilis	Protease	Casein and feather
Bacillus cereus	Amylase	Waste water pollutants
Pseudomonas sp.	Oxygenase	Pesticides
Bacillus pumilus	Lipase	Palm oil
Bacillus megaterium	CYP102A1 (Cytochrome P450)	PAHs
Achromobacter, *Pseudomonas*, *Flavobacterium*, *Nocardia*, and *Bacillus cereus*	Hydrolase (such as carbamate or parathion hydrolase)	Carbofuran, carbaryl or parathion, diazinon, and coumaphos

Enzymes produced by microorganisms specifically designed to break down and degrade environmental contaminants, facilitating the bioremediation process.

was revealed that bacterial biomass and dehydrogenase activity in activated carbon and zeolite were 12 and 3 times higher, respectively. Introduction of immobilized cells may induce more rapid degradation of aromatic compounds. Such treatment offers the protection of inoculated microorganisms from suboptimal environmental conditions (such as improper pH, presence of toxic substances) and reduces competition with indigenous microflora. Moreover, immobilization increases the biological stability of cells, including plasmids. To immobilize both natural (like dextran, agar, agarose, alginate, chitosan polyacrylamides, κ-carrageenan) and synthetic materials (like poly (carbamoyl) sulphonate, polyacrylamide, and polyvinyl alcohol) can be used (Mrozik & Piotrowska-Seget, 2010). When microorganisms have access to a variety of materials to help them generate energy and nutrients to build more cells, they efficiently act against the pollutants. Several factors including the chemical nature and concentration of pollutants, the physicochemical characteristics of the environment, and their availability to microorganisms, determine the efficiency of the bioremediation (Sarao & Kaur, 2022).

Various factors affecting the rate of microbial remediation are summarized in the following subsections.

7.3.8.1 Biological Factors

Competition between microorganisms for limited carbon sources, antagonistic interactions between microorganisms, or the predation of microorganisms by protozoa and bacteriophages are some factors that affect the degradation of organic compounds. The rate of contaminant degradation often depends on the concentration of the contaminant and the amount of "catalyst" present. Here, the amount of "catalyst" represents the number of effective organisms metabolizing the contaminant as well as the number of enzymes (s) produced by each cell. The degradation rate of the contaminant is determined by the expression of specific enzymes by the cells. Other important biological factors responsible for affecting bioremediation rate included mutation, horizontal gene transfer, enzyme activity, interaction (competition, succession, and predation), its own growth until critical biomass is reached, population size, and composition (Sarao & Kaur, 2022).

7.3.8.2 Environmental Factors

The rate at which a contaminant is taken, metabolized, and transferred to the cell (i.e., mass transfer) depends on the rate at which microbial cells can readily convert contaminants during the bioremediation process (Alori et al., 2022). For the pollutants to undergo biological degradation, they must be bioavailable. Bioavailability is associated with the physical state of the contaminant and the possibility of efficient interaction between the microorganism and the contaminant. This contact is best when the microorganism–contaminant interface is maximised (Ali et al., 2015). The possible interactions during the bioremediation process are determined by the metabolic characteristics of the microorganisms and physicochemical properties of the targeted contaminants. The environmental condition at the site of contamination determines the success of interaction (Sarao & Kaur, 2022). The bioavailability of a contaminant is also regulated by some physical and chemical processes such as sorption and desorption, diffusion, and dissolution (Alori et al., 2022).

Properties of contaminants are critical for contaminant-soil interactions, contaminant mobility, and the ability of treatment technologies to remove, destroy, or immobilize contaminants. Some of the important contaminant properties include solubility in water, dielectric constant, diffusion coefficient, molecular weight, vapor pressure, density, and aqueous solution chemistry (Ali et al., 2015).

For any bioremediation technique to be successful, the methods of bioremediation to be employed for a judicious degradation to occur depend on the right microbes in the right place, with the right environmental factors. Here, the right microorganisms are bacteria and/or fungi, which have the physiological and metabolic capabilities to degrade pollutants at the contaminated site (Ali et al., 2015). Soil pH, temperature, moisture, soil structure, solubility in water, nutrients, site characteristics, redox potential, and oxygen content are some abiotic factors that directly affect the growth and activity of microorganisms, thus affecting the rate of bioremediation (Sarao & Kaur, 2022).

7.3.8.3 pH

Bioremediation can occur under a wide range of pH, but the ideal pH for biodegradation is 6.5 to 8.5 in most aquatic and terrestrial systems (Sarao & Kaur, 2022).

7.3.8.4 Temperature

Microbial metabolism is substantially affected by temperature. Most microorganisms grow well in the range of 10°C–38°C. Technically it is extremely difficult to control the temperature of in situ processes, while the temperature of ex-situ processes can only be moderately influenced. Temperatures of the subsurface (within 10 m) may fluctuate seasonally, while subsurface temperatures down to 100 m typically remain within 1°C–2°C of the mean annual surface temperature. This suggests that bioremediation within the subsurface would occur more quickly in temperate climates (Ali et al., 2015). Temperature also influences the bio-chemical reaction rates, and the rates of many of them double for each 10°C rise in temperature. The biological enzymes also have an optimum temperature at which they participate in the degradation pathway and can get inactivated with a change in temperature. Cells also die above a certain temperature. For instance, in a very cold environment, because of the sub-zero temperature of water the transport channels within the microbial cells shut down. This may even freeze the entire cytoplasm, leading those results in metabolically inactivation of oleophilic microbes. Hence, temperature can either speed up or slow down the bioremediation rate by affecting the physiological properties of microbes (Sarao & Kaur, 2022).

7.3.8.5 Nutrients

Although microorganisms are ubiquitous in nature and present in contaminated soil, they cannot necessarily be there in the numbers required for bioremediation of the site. Therefore, the growth and activity of the microorganisms are stimulated by providing the essential nutrient balance for microbial growth and reproduction. This however determines the rate of biodegradation and its effectiveness. Several nutrients, including carbon, nitrogen, and phosphorous are required by the microbes to survive and continue their activities. The supply of essential nutrients such as N and P can bio-optimize the bacterial C: N: P ratio. Moreover, in cold environments, the addition of an appropriate quantity of nutrients can come up as a useful strategy for improving the metabolic activity of microorganisms and hence the biodegradation rate (Sarao & Kaur, 2022).

7.3.8.6 Oxygen

Biodegradation can be carried out in both aerobic and anaerobic conditions and the amount of oxygen available will determine whether the system is aerobic or anaerobic. In most applications, bioremediation is an oxidation process and during the oxidation of contaminants, microorganisms extract energy via electron transfer. Electrons are removed from the contaminant and transferred to a terminal electron acceptor, which is oxygen in the case of aerobic biodegradation. Oxygen concentrations during decomposition of the organic substrate in the subsurface may become reduced, which can make it a major kinetic limitation on the degradation process. This is more intense in the cases of organic molecules with high oxygen demand such as petroleum hydrocarbons (Ali et al., 2015). To increase the amount of oxygen in the soil it is possible to till or sparge air. In some cases, hydrogen peroxide or magnesium peroxide is introduced into the environment to increase oxygen demand (Sarao & Kaur, 2022). Application of hydrogen peroxide alone can provide more than 100 mg/L oxygen (Ali et al., 2015).

However, there are some contaminants like chlorate compounds that are degraded only in an anaerobic environment (Sarao & Kaur, 2022). Therefore, in the absence of molecular oxygen, anaerobic microorganisms use other forms of combined oxygen. For example, denitrifying bacteria use nitrate (NO_3^-), nitrite (NO_2^-), or nitrous oxide (N_2O); dissimilatory metal-reducing bacteria use manganese or ferric iron oxides (e.g., MnO_2, Fe $(OH)_3$, orFeO_2^-); sulfate-reducing bacteria use sulfate ($(SO_4)_2^-$); and methanogens use carbon dioxide (CO_2) or bicarbonate (HCO_3) as electron acceptors (Ali et al., 2015).

7.3.8.7 Moisture Content

Adequate water is essential for the growth of microorganisms. Therefore, the soil moisture content has an adverse effect on the biodegradation agent (Sarao & Kaur, 2022). Along with this, the moisture content of soil also alters the bioavailability of contaminants, the transfer of gases, the effective toxicity level of contaminants, and species distribution. It has been reported that in cases where no extra source of oxygen is being provided (for example, in bioremediation of surface contamination), 20% moisture may be adequate, whereas if a continuous recirculation system (pipe networks) is being used for deeper contamination, 80% water content would be more appropriate. Also, if the water content is too high during bioremediation, this will make it difficult for atmospheric oxygen to penetrate the soil and will affect the growth efficiency of organisms (Ali et al., 2015).

7.3.8.8 Metal Ions

Bacteria and fungi require small amounts of metals for their metabolic processes but the presence of higher amounts can inhibit the same, which can directly or indirectly affect the rate of degradation (Sarao & Kaur, 2022).

7.3.8.9 Soil Parameters

Different soil parameters, including the soil region, moisture-holding capacity texture, and particle size, affect the rate of microbial biodegradation. There is a higher population and diversity of microbes at the top layer of the soil (0–10 cm) due to the increased availability of oxygen and organic matter, which is the opposite of what happens in sediment soils. In soils with fine particles, such as clayey soils, hydrocarbon retention takes place more at the surface, due to which the nutrients of the soil and oxygen becomes unavailable. Therefore, the well-drained soil is the best soil texture to promote increased microbial biodegradation, which supports oxygen availability and inhabits more soil microbes (Ayilara & Babalola, 2023).

7.3.8.10 Salinity

The presence of salinity influences the hydrocarbon clastic activity of the halotolerant and halophilic microbes. It also exposes the soil microbes to stress conditions. The osmotic pressure of microorganisms increases as the saline concentration of an environment increases, which has a direct negative impact on the metabolic activities of the microbes as well as the transportation system and solubility of the heavy metals (Ayilara & Babalola, 2023; Table 7.4).

TABLE 7.4
Microorganisms and Compounds Degraded by Them (Meena et al., 2021; Mrozik & Piotrowska-Seget, 2010; Elyamine et al., 2021; Ayilara & Babalola, 2023; Ayilara & Babalola, 2023)

Microorganisms	Contaminants
Penicillium chrysogenum, *Pseudomonas putida*	Monocyclic aromatic hydro carbons, benzene, toluene, ethylbenzene and xylene, phenol compounds
P. alcaligenes, *P. mendocina*, and *P. putida*, *P. veronii*, *Achromobacter*, *Flavobacterium*, *Acinetobacter*	Petrol and diesel polycyclic aromatic hydrocarbons toluene
Phanerochaete chrysosporium, *Pseudomonas fluorescens*	Biphenyl, triphenylmethane and polychlorinated biphenyl
A. niger, *A. fumigatus*, *F. solani*, and *P. Funiculosum*, *Tyromyces palustris*, *Gloeophyllum trabeum*, *Trametes versicolor*	Hydrocarbon
Rhodococcus, *Pseudomonas*, *Sphingomonas*, *Bacillus*, *Flavobacterium*, *Mycobacterium*, *Nocardia*, *Vibrio*, *Micrococcus*, and *Acinetobacter*	PAHs

(*Continued*)

TABLE 7.4 *(Continued)*

Microorganisms and Compounds Degraded by Them (Meena et al., 2021; Mrozik & Piotrowska-Seget, 2010; Elyamine et al., 2021; Ayilara & Babalola, 2023; Ayilara & Babalola, 2023)

Microorganisms	Contaminants
Coprinellus radians	Methylnaphthalenes, and dibenzofurans
Alcaligenes odorans, *Bacillus subtilis*, *Corynebacterium propinquum*, *Pseudomonas aeruginosa*	Phenol
Candida viswanathii	Phenanthrene, benzopyrene
Cyanobacteria, green algae and diatoms and *Bacillus licheniformis*, *Bacillussonorensis*, *Pseudomonas putida* KT2442	Naphthalene
Gleophyllum striatum	Lignin peroxidase
Aspergillus niger, *Candida glabrata*, *Candida krusei and Saccharomycescerevisiae*, *B. brevis*, *P. aeruginosa* KH6, *B. licheniformis*, and *B. Sphaericus*, *Comamonas testosteroni BR60*	Crude oil
Bacillus cereus A, *B. coagulans Pseudomonas aeruginosa*, *P. putida*, *P. cepacia*, *Arthobacter* sp. and *Bacillus* sp.	Diesel oil
Pseudomonas cepacia, *Bacillus cereus*, *Bacillus coagulans*, *Citrobacter koseriand*, *Serratia ficaria*	Crude oil, diesel oil
B. subtilis strain NAP1, NAP2, NAP4	Oil-based based paints
Myrothecium roridum IM 6482, *Pycnoporus sanguineous*, *Phanerochaete chrysosporium* and *Trametes trogii*, *Penicillium ochrochloron*	Industrial dyes
Micrococcusluteus, *Listeriadenitrificans* and *Nocardiaatlantica*, *Bacillus* spp. ETL-2012, *Pseudomonas aeruginosa*, *Bacillus pumilus* HKG212, *Exiguobacterium indicum*, *Exiguobacterium aurantiacums*, *Bacillus cereus*, and *Acinetobacter baumanii*	Textile dyes
Saccharomyces cerevisiae, *Cunninghamella elegans*, *Pseudomonas fluorescens*, and *Pseudomonas aeruginosa*, *Microbacterium profundi strain* Shh49T, *Aspergillus versicolor*, *A. fumigatus*, *Paecilomyces* sp., *Paecilomyces* sp., *Terichoderma* sp., *Microsporum* sp., *Cladosporium* sp., *Geobacter* spp., *Aerococcus* sp., *Rhodopseudomonas palustris*	Heavy metals
Bacillus, *Staphylococcus*	Endosulfans
Enterobacter, *Acenetobactor* sp., *Pseudomonas* sp., *Enterobacter* sp. and *Photobacterium* sp.	Chlorpyrifos and methyl parathion
Escherichia coli AtzA	Atrazine
Pseudomonas fluorescens MP, *Pseudomonas fluorescens* RE	2,4-Dinitrotoluene
Arthrobacter chlorophenolicus A6L	4-Chlorophenol
Absidia cylindrospora	Fluorene
Burkholderia sp. FDS-1	Fenitrothion
Aspergillus sp. LEBM1 and LEBM3	Chlorobenzene
Rhodococcus HCCS, *Sphingomonas* MWFG, and *Paracoccus SPNT*, *Arthrobacter* sp. MAL3, *Microbacterium* sp. MAL2, *Stenotrophomonas* sp. MAL1, *Bacillus thurigiensis*, *Bacillus pumilus*, *Rhodococcus hoagie*	Phenanthrene, fluoranthene, and pyrene
E. coli	Hexavalent chromium, Atrazine pesticide, Cd, As, Hg, and Zn
Escherichia coli DH5α	Uranium and chromium

(Continued)

TABLE 7.4 *(Continued)*

Microorganisms and Compounds Degraded by Them (Meena et al., 2021; Mrozik & Piotrowska-Seget, 2010; Elyamine et al., 2021; Ayilara & Babalola, 2023; Ayilara & Babalola, 2023)

Microorganisms	Contaminants
Acinetobacter baumannii S30 pJES	Hydrocarbon
Streptomyces coelicolor M145	n-Hexadecane
Acinetobacter sp. BS3	Aromatic hydrocarbons
Sphingomonas sp. BHC-A	Hexachlorocyclohexane (HCH) and methyl parathion

Microorganisms and the specific compounds they degrade, illustrating their role in breaking down environmental pollutants.

7.3.8.11 Case Study Related to Successful Microbial Rejuvenation of Polluted Soil

CASE 1 Bioremediation of Soils Contaminated with Heavy Metals (Atuchin et al., 2023)

In the study, the promising strains that can remove heavy metals from experimental media were selected included *Pantoea* sp., *Achromobacter denitrificans*, *Klebsiella oxytoca*, *Rhizobium radiobacter*, and *Pseudomonas fluorescens*. Consortia were compiled and investigated for the ability to remove heavy metals from nutrient media, as well as to produce phytohormones. For this, samples of technogenically disturbed soils were taken on the territory of the Osinnikovskoye deposit (53°26′ N; 87°25′ E) of the coal company OJSC Kuzbassrazrezugol, a branch of the Kaltan coal mine. The dump was laid in the village of Malinovka, Novokuznetsk district, Kemerovo region (Russia). The microorganisms were isolated on a liquid nutrient medium consisting of 0.4 g KH_2PO_4, 3.0 g $(NH_4)_2SO_4$, 3.0 g $Na_3C_6H_5O_7$, 0.01 g $CuSO_4$, 0.005 g $ZnSO_4$, 0.001 g $FeSO_4$, 0.2 g $CdCl_2$, 0.5 g $MgSO_4$, 0.1 g $MnSO_4$, 8.0 g peptone, 5.0 g maltose, 20.0 g agar-agar, and ≤1 L water with heavy metals. 1 g of soil was added to 100 mL of sterile medium and cultivated in an LSI-3016A shaker-incubator at 100 rpm and 25°C for 24 h. To obtain pure isolates, 1 mL of the liquid culture was sown by the surface method on a solid medium with the abovementioned composition in a TSO–1/80 SPU thermostat at 25°C for 24 h. To determine the morphological features, single colonies were obtained by the streak plate method on plain agar. The biochemical profile of the cell wall was described by the Gram method, and microscopy of the samples involved an Axio Scope A1 microscope.

To select strains capable of degrading heavy metals, the samples were examined for the ability to absorb heavy metals from the nutrient medium while maintaining the rate of biomass growth in comparison with the control sample (containing no heavy metals in the medium). The heavy metal degrading microorganisms were identified and tested for biocompatibility. In the samples of polluted soils, 10 pure microbial cultures were isolated, which turned out to be resistant to heavy metals represented mainly as Gram-negative bacilli. Inoculums of isolated microorganisms in beef extract broth were tested for the uptake potential of Pb, Ni, As, Cd, and Hg. 10 mL of the working solution of metal salts Na_3AsO_4, Pb $(NO_3)_2$, Ni (NO_3), Cd $(NO_3)_2$, and Hg $(NO_3)_2$ had a concentration of 1 mg/mL which was added to 90 mL of the beef extract broth. To study the biomass accumulation, the culture liquid was centrifuged at 8000 rpm for 15 min and the resulting biomass was kept in a VD 23 BINDER oven at 60°C until constant weight. The biomass accumulation was determined by the following equation:

$$\text{Biomass accumulation} = \text{dry biomass in g/medium volume in L}$$

The amount of indole-3-acetic acid synthesized by the consortium was determined according to the method developed by Sarmiento-Lopez et al. in which 0.1% L-tryptophan and 5% of the bacterial consortium were added to the beef extract broth and the mix was thermostated for 48 h. After that, the biomass was separated by centrifugation at 10,000 rpm for 5 min. After that 1 mL of Salkowski reagent (dissolved 0.1 g of $FeCl_3$ in 100 mL of 50% H_2SO_4) was mixed with 1 mL of the supernatant and kept at room temperature for 30 min until the solution turned pink. The optical density of the resulting solution was measured on a spectrophotometer at 535 nm and the amount of indole-3-acetic acid was determined from the calibration curve of the standard solution. Indole-3-butyric acid was registered by HPLC using a Shimadzu LC-20AD chromatograph according to the Martinez-Morales method. During formation of consortia, biocompatibility of the microorganisms was checked, which had the following observations:

1. *Pantoea* sp. and *Pseudomonas fluorescens* could not be used together in one consortium.
2. *Pantoea* sp. inhibited *Achromobacter denitrificans* and *Rhizobium radiobacter*, while exhibiting active co-growth with *Klebsiella oxytoca* and *Pseudomonas fluorescens*.
3. *Pseudomonas fluorescens* did not enter the prospective consortium as it showed no signs of growth except when accompanied by *Pantoea* sp.
4. *Achromobacter denitrificans*, *Klebsiella oxytoca*, and *Rhizobium radiobacter* did not inhibit each other and were found to promote each other's biomass accumulation.
5. *Achromobacter denitrificans*, *Klebsiella oxytoca*, and *Rhizobium radiobacter* were proven to be the most suitable for a consortium.

After biodegradation tests the following observations were made:

1. *Achromobacter denitrificans* utilized 40.03 mg/L As from the nutrient medium and was also effective in removing Hg (73.08 mg/L), Pb (71.10 mg/L), and Cd (62.41 mg/L).
2. *Klebsiella oxytoca* utilized 51.14 mg/L Pb, 51.33 mg/L Hg, and 78.02 mg/L Cd.
3. *Rhizobium radiobacter* had the best results for Ni (56.94 mg/L), Hg (61.57 mg/L), and Pb (74.26 mg/L).
4. These microorganisms proved to be biocompatible and did not inhibit each other's growth.
5. Consortium constituting *Achromobacter denitrificans*, *Klebsiella oxytoca*, and *Rhizobium radiobacter* with a ratio 1:1:2 removed 56.39 mg/L Cd, 58.03 mg/L Hg, 61.17 mg/L As, 91.13 mg/L Pb, and 98.22 mg/L Ni from the medium with individual heavy metals. The same consortium also utilized 47.33–83.26 mg/L of composite pollutants which consist of Pb, As, Hg, Ni, and Cd. No significant decrease in biomass accumulation was spotted in this consortium during cultivation in environments with separate and composite heavy-metal–contaminated media. The optimal pH for the consortium was pH = 7, while the greatest biomass accumulation was observed at 25–30°C. The ability of this consortium to produce indole-3-acetic acid and indole-3-butyric acid was 18.03 and 2.02 μg/L, respectively which concludes that it has good prospects for the phytoremediation method, but this fact needs additional investigation.

The study concluded that temperatures ≥ 35°C had a negative impact on biomass production and a temperature of 20°C also inhibited microbial growth. The optimal cultivation temperature was concluded to be 25–30°C.

CASE 2 Bioremediation of Soil Polluted with Pesticide (Kumar et al., 2015)

The rapid industrialization in the Tarai Bhavan region of Udham Singh Nagar, Uttarakhand, has exposed the soil and water to industrial effluent rich in pesticides like chlorophenols that are adversely affecting the ecosystem and disturbing the food chain. Pentachlorophenol (PCP) is a polyhalogenated aromatic hydrocarbon of the chlorophenol family (chlorophenols are phenols carrying one chlorine atom attached to the benzene ring) having chemical formula C6HCl5O and it has no isomers. PCP has been used extensively as a wood preservative, fungicide, bactericide, herbicide, molluscicide, algaecide, and insecticide. This extensive use of PCP has led to the contamination of many terrestrial and aquatic ecosystems worldwide. Although numerous reports have shown that PCP undergoes biodegradation, its rate of biodegradation in the environment is often slow.

The pure culture of *Pseudomonas fluorescence* was obtained from gene pool, G. B. Pant University of Agriculture and Technology, Department of Microbiology, Pantnagar and the pure culture of *Phanerochete chrysosporium* was obtained from Institute of Microbial Technology (CSIR Laboratory), Chandigarh. Nutrient agar medium was used to culture bacterial strains and fungal medium was used for culturing fungi. Continuous enrichment of bacterial strains was facilitated by minimal salt medium constituting the given components in g/l: $Na_2HPO_4.2H_2O$- 7.8; K_2HPO_4 - 6.8; $MgSO_4$-0.2; $NaNO_3$-0.085; Ca $(NO_3)_2$ $4H_2O$- 0.050; Ferrous ammonium citrate - 0.01; Trace elements- 1ml. Trace element (1 ml) was added and the final volume was made 1 liter by adding distilled water and the pH was adjusted to 7.0. For continuous enrichment of the fungal strain, the minimal salt medium constituting D-glucose - 10; NH_4NO_3 - 0.079; KH_2PO_4 - 2; $MgSO_4$ - 0.05; $CaCl_2.2H_2O$ 0.1; Mineral solution- 1ml; Vitamin solution- 0.5 ml, pH- 5.6 was used. The bacterial cells with maximum CFU were characterized based upon morphology of colonies, diameter, colour, opacity, form elevation, margin, smoothness, texture, and spreading nature. After that isolates were stored on nutrient agar slant as well as in 25% glycerol at 20°C. Indirect staining with nigrosine and Gram staining was performed for characterization of bacterial strain. Fungal strain having maximum growth in basal mineral medium containing PCP was taken and identified by visual observation of colony growth on potato dextrose agar plate and then microscopic examination was performed. Different concentrations of PCP were inoculated with microorganisms and were analyzed for chloride release and Ring Cleavage at different time periods. Degradation of PCP was studied in laboratory and field conditions through the following experiments:

1. **Degradation of pentachlorophenol**: After enrichment the bacterial and fungal were inoculated in fresh mineral salt medium with PCP (100 ppm) as a sole carbon source for a few days in a gyratory shaker. The samples were taken out at 4, 8, 16, 32, hours and growth of microbes was measured.
2. **Estimation of chloride ions release (Argentometric method)**: 24 ml of sample was taken in a beaker and 5 drops of potassium chromate was added as an indicator solution. With the help of burette, $AgNO_3$ (0.0141 N) solution was added to the beaker. As a result, a red colour formed that disappeared soon. At a point where all the chloride ions were precipitated, a stable red colour appeared that refers to the end point of the reaction. The amount of chloride ions in the sample was calculated:

$$C1V1(AgNO_3) = C2V2(\text{Sample})$$

With the help of this formula C2 concentration was calculated. The amount of chloride ions present in the sample was calculated by multiplying 35.5 (atomic weight of chloride) with the concentration of sample (C2).

3. **Estimation of ring cleavage**: 4 ml of cell suspension was dissolved in 0.02 M Tris buffer (0.1) and EDTA (0.1 ml) for lyses of bacterial and fungal cells with pH adjusted to 7.8. The mixture was treated with toluene and 0.1 M catechol (4.0 ml) and the colour change was observed. The appearance of yellow colour suggests presence of meta cleavage. If mixture is shaken for 1 hour at 170 rpm and result in formation of ß-ketoadipic acid (Rothera Reaction) indicates the presence of ortho fission. In this procedure, the above mixture was acidified with 2 mL of 10.2M HCl followed by addition of 1ml $NaNO_3$ (1%). After 2 min, concentrated ammonia (15 ml) and 10% ferrous sulfate solution were added. The development of reddish-brown colour indicated typical Roth era reaction and the presence of ortho cleavage.

Mode of ring fission of pentachlorophenol by both the organisms was assessed by observing meta and ortho ring cleavage. No yellow colour appeared after mixing catechol in cell suspension which indicates that meta cleavage is absent in both the organisms. However, positive Rothera test (as violet colour appeared due to β ketoadipic acid formation) for both the organisms, which indicated that ring cleavage occurred at ortho position and confirmed dichlorination of PCP. It was concluded from the study that the flask inoculated with bacteria showed lower efficiency of ring cleavage than fungi. The results of ring cleavage and chloride release in liquid batch culture concluded that both microorganisms could utilize PCP.

CASE 3 Pentachlorophenol Remediation by *Enterobacter* sp. (Karn & Geetanjali, 2014)

A similar experiment was run by Santosh et al. for bioremediating soil contaminated with Pentachlorophenol (PCP) by *Enterobacter* sp. Sludge samples were collected from industrial dumpsites of distilleries industries at Noida (near Delhi), India. For screening of PCP degrading bacteria nutrient enrichment technique in mineral salt medium (MSM) was used that contained PCP as sole source of carbon and energy. Colonies that appeared after 48 h were selected. The PCP degrading bacterial strain was isolated by the serial dilution method and purified by repeated streaking on MSM plates through which a total of seven bacteria were isolated on basis of their morphology from sludge sample and were represented as SG1 to SG7. Only SG1 strain was selected for further degradation studies because other strains were not able to grow in media containing PCP.

The identification of PCP degrading bacterium was carried out according to Bergey's Manual on Systematic Bacteriology by Holt et al. (1994). The following tests were done:

1. Catalase activity was determined by detecting bubble formation with 3% H_2O_2 solution.
2. Oxidase was determined by using paper disc with tetramethyl-p-phenylenediamine.
3. Nitrate reduction was determined using nitrate agar plate.
4. DNAase test was done using DNAase agar medium with methyl green as an indicator.
5. Starch hydrolysis was determined on starch hydrolyzing agar by detecting cleared zones formed around the colonies.
6. Antibiotic profiling was also done using Hexa disc G plus 1 and minus 9 (Hi-Media Laboratories, Mumbai, India) having 12 different antibiotics of various concentrations.
7. Genomic DNA extraction, amplification of 16S rRNA gene and sequence analysis was performed where the sequences were compared against the available DNA sequences in GeneBank (http://www.ncbi.nlm.nih.gov/) using the BLASTN tool.

Degradation of PCP was studied through following experiments:

1. **Chloride release assay**: Titration method was used to measure the chloride ions released during degradation of PCP in MSM. In this method, to 1.0 mL of culture filtrate in titration dish, 2–3 drops of phenolphthalein indicator were added. In case a pink colour appeared, it was titrated with N/50 sulfuric acid until the colour changed from pink to that of the original sample. After that, 10–15 drops of potassium chromate indicator were added to give the filtrate a bright yellow colour. Silver nitrate (0.028 N) was added dropwise with the help of a pipette, stirring continuously with a stirring rod, until the sample just turned from yellow to an orange-red. A calculation was done by part per million (ppm) = mL of silver nitrate used × 1000.
2. **Bromothymol blue test and determination of ring cleavage**: Presumptive dissimilation tests were performed in agar plates containing MSM with 20 ppm of PCP and 20 mg/L^1 of bromothymol blue as an indicator. This test is based on the fact that when dichlorination takes place HCl is released, which results in acidification of the medium. Overnight grown cells were taken to determine ring cleavage activity. 10 mL cell suspension was centrifuged at 7000 rpm for 10 min to separate the biomass and the supernatant and biomass were collected for extracellular and intracellular activity respectively. To check intracellular activity, bacterial cells were lysed with lysozyme (2 mg mL^{-1}) in 1ml TES buffer and kept for 1 h for proper lysis of bacterial cell. The mixture was treated with toluene (1 mL) and 0.01 M catechol (1 mL). If yellow colour is observed it means the meta cleavage has occurred. In absence of colour, the mixture is shaken for 1 hour at 170 rpm for the formation of b-ketoadipic acid (Rothera reaction) which indicates the presence of ortho fission. In this procedure, the above mixture was acidified with 2 mL of 10.2M HCl. 1 mL $NaNO_3$ (1%) was added to the above followed by 15 mL conc. ammonia and 1 mL ferrous sulfate solution (10%).

The same process was followed for supernatant for identification of meta and ortho cleavage. The development of a reddish-brown colour indicated a typical Rothera reaction and the presence of ortho cleavage. MSM containing PCP with no bacterial culture was used as a control for the ring cleavage assay. The degradation ability of the isolated strain was analyzed in 250 mL of Erlenmeyer flask containing 50 mL of MSM supplemented with 0.4, 0.8,1.5 and 2.0 mM of PCP separately and 5% inoculum which were incubated on a shaker at 37 °C and 120 rpm/min. Degradation efficiency was determined by consumption of PCP by the bacterial strain from the culture medium. In the extraction of PCP, the cell suspension was centrifuged at 8,000 rpm for 5 min. The cell free supernatant fractions were extracted three times with an equal volume of n-hexane which were shaken vigorously for 15 min in a standard separating funnel. Further, the organic layer was dried with anhydrous sodium sulfate and the residue was finally dissolved in 50 μL mixture of n-hexane: ethyl acetate (10:1), which was then immediately analyzed on a GC. The GC analysis was performed in electron capture detection mode with a gas chromatograph Nucon GC-5765 (Centurion Scientific, New Delhi, India) capillary column DB-5 (30 m length 0.025 mm i.d. 0.25 μm film thickness) at a temperature programme of 50°C (2 min), then raised to 10°C/min^1 to 280°C for 10 min. Helium with a constant flow of 1.2 mL/min was used as the carrier gas. The samples (injected volume 0.1 μL) were analyzed in split less mode at an injection temperature of 250°C and detector temperature 280°C.

After that the data were statistically analyzed by analysis of variance (ANOVA) and the mean differences were compared by Tukey-Kramer Multiple Comparison Test at $p < 0.05$. All experiments were performed in three replicates and analyses were performed using GraphPad Prism (v 4.03) software. It was found that the isolated SG1 was Gram-negative rod shaped that gave positive catalase and nitrate reduction test. This isolate was sensitive to three antibiotics including Gentamicin (10), Chloramphenicol (30), and Co-Trimoxazole (25 mcg/disc). Partial sequence data of the isolate was analyzed by BLAST search and the BLAST analysis of this sequence showed 99% of sequence similarity with *Enterobacter* sp. This strain could tolerate up to 2 mM of PCP and degraded 85% of PCP in the culture medium. An increasing concentration had negative effect on the growth of bacteria and degradation ability. Furthermore, *Enterobacter* sp. was able to degrade 70% of PCP at 100 mg L^{-1}. In this study, ring fission of PCP was determined where yellow colour did not appear after mixing catechol in cell suspension indicating that meta cleavage is absent. However, a positive Rothera test was observed after 144 h. The exact mechanism for ring cleavage is not clear but the tests confirmed degradation of PCP by strain SG1. Additionally, the degradation was also confirmed by the GC analysis. This study concluded that *Enterobacter* sp. is an efficient, stable bacterial strain capable of degrading PCP but scale up studies need to be tested for bioremediation of PCP at an industrial scale and the genes responsible for biodegradation of these compounds and their expression need to be found.

Various studies were conducted earlier that had the following results:

Earlier *Nandish* (2005) observed *Enterobacter* sp. was able to degrade 70% of PCP at 100 mg L^{-1}; Saber and Crawford (1985) isolated *Flavobacterium* by enrichment method from PCP contaminated soils which were able to mineralize 100–200 ppm PCP as a sole source of carbon and energy and liberated chlorine as chloride into the medium; Radehaus and Schmidt (1992) reported complete mineralization of PCP (160 mg L^{-1}) in a week by *Pseudomonas* sp. strain RA2; Singh et al. (2007) reported that *Serratia marcescens*, the bacterial isolated from pulp and paper mill waste was able to degrade 90.3% of PCP within 168 h when grown with 300 mg L of PCP; Shah and Thakur (2002) observed 72% PCP removal by *Pseudomonas fluorescence* (TE3) in 96 h when grown with 100 mg L^{-1} PCP; Sharma and Thakur (2008) observed that *Pseudomonas aeruginosa* (PCP2) utilized 60% of PCP within 96 h of incubation; Premlatha and Rajkumar (1994) recorded the complete degradation of PCP (800 mg L^{-1}) by *Pseudomonas aeruginosa* in 6 days with glucose as a co-substrate.

CASE 4 Borhola Oil Fields Contamination Case (Gogoi et al., 2003)

Borhola oil fields, under the Damodar Valley project (DVP) of Eastern (India) Regional Business Centre (ERBC), Oil and Natural Gas Corporation Limited (ONGCL), India, with an operational area of approximately 300 acres, have been producing crude oil since 1972 at an estimated capacity of two million tonnes per annum. During normal operation, leakage, and spillage of crude oil result in soil contamination at various locations including oil wells, sumps and pits, tank batteries, gathering lines, and pump stations. Furthermore, the waste crude oil generated during field operations is collected in a waste pit constructed near the drilling site. Even after removal of waste there is some contaminated soil remaining in the waste pit which is left to be degraded by natural processes. During the rainy season crude oil from the waste pit may spread to the surrounding fields causing pollution. In this study a systematic laboratory treatability and field pilot scale study was conducted towards the development of a bioremediation protocol for the oil contaminated soil at Borhola oil fields.

Procedure: Here, a site of 0.25 acres was selected, eight non-overlapping square grids were taken of equal 4 × 4 ft. surface area and samples were collected from the centre of each grid. Accordingly, three samples, weighing 3 kg each, were collected from the centre of each grid, from different depths going up to a maximum of 2 ft. and were stored in plastic buckets and homogenized in a mixer before analysis. The soil was analyzed for compounds like BTEX (benzene, toluene, ethylbenzene, and xylene) or PNAS (polynuclear aromatic hydrocarbons), oil and grease (O&G) and total petroleum hydrocarbons (TPH) concentrations. Oil and moisture content in soil were determined.

For the component analysis, using n-pentane as the antisolvent, the ashphaltenes were precipitated from the extracted oil samples and Borhola crude oil. After the separation of ashphaltenes, the extracted oil and Borhola oil were further fractionated into saturated and aromatic hydrocarbons. This was done using column chromatography on an activated alumina column by elution with petroleum ether (40–60°C boiling range)–benzene mixture (2:1 v/v) and methanol in sequence. The lower percentage of saturates with a relative increase in their aromatic content in the samples may be attributed to natural biodegradation of the crude. It is known that saturates are degraded faster than aromatics and others. To draw further inferences regarding natural biodegradation, chromatographic analysis of the saturate fraction was conducted with a Shimadzu GC RIA gas chromatograph, with a metallic capillary column (30 m = 0.32 mm i.d.) coated with OV-101 and temperature programmed from 100 to 280°C at 4°C/min. Air permeability of the contaminated soil was determined by a method based on Darcy's law for steady flow in a packed bed. The levels of N, P, K, Fe, Mg, and C, etc., along with the pH were estimated by standard procedures suggested in the textbook (Partha & Bossert, 1984). It was observed that the concentrations of inorganic nutrients like N, P, Fe, and Mg, were low, which could limit the metabolism of the existing microorganisms capable of degrading hydrocarbons in the soil environment. Therefore, application of extraneous nutrients may be required for developing a feasible bioremediation method. The porosity and particle density of soil samples was determined by the usual gravimetric methods and were found to be 38% and 2.6 g/cm^3, respectively. The average soil particle size determined by sieve analysis was found to be 0.5 mm and the sand and clay-silt content of the sand were in the ranges of 68%–79% and 25%–31%, respectively. Total microbial content was determined by the Agar Plate.

Method: The preliminary screening of the contaminant-specific degraders was performed in a medium that provides only the necessary inorganic nutrients required for microbial growth without an intrinsic carbon source, so that the contaminant supplement serves as the sole source of carbon. Following incubation, the cultures were evaluated for the presence and relative abundance of microbial growth. Isolates demonstrating little or no growth were judged to be poor or non-degraders of the contaminants and eliminated from further studies while the remaining isolates were then evaluated for the effects of environmental parameters and for performance in a bench scale bioremediation test.

The basal medium used for isolation of hydrocarbon degrading microbes was having following composition (g/L): 4.74 K_2HPO_4; 0.56 K_2HPO_4; 0.50 $MgSO_4.7H_2O$; 1 NaCl; 0.01 $FeCl_3$; 0.01 $CaCl_2$. $2H_2O$; 2.5 NH_4NO_3; 1 mL of a trace element solution consisting of (g/L) 10 $CuSO_4.5H_2O$; 10 H_3BO_3; 10 $MnSO_4$. $5H_2O$; 70 $ZnSO_4$. $7H_2O$; 10 MoO°C with a medium pH of 7.0. Fifty mL of the medium taken in a 250-mL flask was sterilized by autoclaving at 120°C for 15 min and the degradation capacity of the indigenous microorganisms was assessed qualitatively by observing the growth and colour change of the enriched culture media. One gram each of the oil-contaminated soil samples was then added to 50 mL of medium and 1% v/v crude oil in a flask. The flask was then incubated in a shaker at 30°C and 280 rev/min for 12 days. After that 1 mL of culture broth from each flask was transferred to a fresh medium of the same composition and incubated under the same conditions for another 12 days. The growth of bacterial population (consortia) and the extent of degradation of crude oil in each flask was assessed visually. Thereafter, a comparative assessment was done to find the potential of bacterial consortia to utilize individual classes of hydrocarbon compounds using n-hexadecane, BTEX, and phenanthrene as representatives for the groups of aliphatic, aromatic, and PNA hydrocarbons, respectively. It was concluded from the above study that alkane degrading microorganisms were more prevalent than aromatic/polyaromatic hydrocarbon degraders in the contaminated soil samples. Pure cultures were isolated from the consortia of soil having maximum degradation of the pollutant like. In this case it was sample nos. 4 and 5 that exhibited biodegradation capability for hexadecane and BTEX. The relative potentials for bioremediation of the major types of petroleum compounds in decreasing order:

Monoaromatics > straight chain alkanes > branched alkanes > naphthenes > polynuclear aromatics > polars

In general, laboratory pan microcosm and/or column tests are conducted to determine the kinetics of biodegradation, concluding whether an ex- or in situ approach is desired. The tests were conducted using fresh sand contaminated with crude oil in an open batch reactor. It was observed that added bacteria degraded the hydrocarbon content of the crude oil, but the extent of degradation of individual hydrocarbons needs to be quantified. After that the field pilot test was conducted for the duration of 1 year. For the field pilot study six test cells with an air sparger were used, designated as A, B, C, D, E, and F, which were filled with approximately 500 kg each of crude oil–contaminated soil prepared by mixing in a Banbury mixer. The soil pH was maintained between 6 and 8 by adding approximately 300 g of calcium carbonate during the mixing operation. The experimental conditions in each cell were kept different to assess the effects of aeration, nutrient supply and added bacterial consortia. In test cells E and F, as a source of naturally adapted hydrocarbon degrading microbial consortia, 25 kg each of oil-contaminated soil collected from the top surface of the drainage channel at the waste pit was added. Aeration was provided every day for 1 h at a rate of 100 m^3/h and the moisture content in each cell was monitored and maintained between 50% and 65% during the entire period by adding water time to time.

This study concluded that nutrient and extraneous consortia addition have an advantageous effect on the bioremediation process. The field pilot study also showed that by applying a nutrient and microbe enriched solution, the crude oil components were reduced by 75% under aerated conditions within a time span of 1 year. The experimental results of this study can be utilized to assess the role of various factors that control the success of bioremediation including the availability of microorganisms that can metabolize the contaminant and utilize it as a carbon source through solubilization or desorption under favourable conditions of temperature, pH, sufficient metabolic nutrients, and moisture.

7.4 CHALLENGES AND LIMITATIONS IN MICROBIAL REJUVENATION OF HEAVILY POLLUTED SOIL

Challenges and limitations in microbial rejuvenation of heavily polluted soil, addressing obstacles in effectively restoring contaminated environments using microbial techniques.

1. **Complexity of Pollutants**: Challenge: Heavily polluted soils often contain a complex mixture of pollutants, making it challenging for microbial communities to effectively target and degrade all contaminants.
   Limitation: Microorganisms may have specific capabilities, and some pollutants might remain recalcitrant, requiring additional strategies for complete remediation.
2. **Microbial Adaptation**: Microbial communities may take time to adapt to the pollutants present, slowing down the remediation process.
   Limitation: Rapid adaptation may not always occur, hindering the efficiency of microbial rejuvenation, especially in the presence of persistent pollutants.

3. **Nutrient Limitations**: The availability of essential nutrients, such as nitrogen and phosphorus, is crucial for microbial activity. Polluted soils may lack these nutrients, affecting microbial growth and remediation efficiency.
   Limitation: Nutrient deficiencies can limit the microbial population and their ability to carry out effective pollutant degradation.
4. **Unforeseen Ecological Consequences**: Introducing microbial agents for rejuvenation may have unintended consequences on the existing soil ecosystem, potentially disrupting native microbial communities or interactions. Limitation: Balancing the introduction of beneficial microorganisms with potential ecological impacts requires careful consideration and monitoring.
5. **Scale-Up Challenges**: Scaling up microbial rejuvenation strategies from laboratory experiments to field applications presents logistical and technical challenges.
   Limitation: Practical implementation on a large scale may be resource-intensive, and the efficacy observed in controlled settings may not always translate seamlessly to real-world scenarios.
6. **Regulatory Hurdles**: Compliance with environmental regulations and obtaining approvals for the use of microbial agents in soil remediation can be a lengthy and bureaucratic process. Limitation: Regulatory constraints may hinder the timely deployment of microbial rejuvenation strategies in polluted areas. Understanding and addressing these challenges and limitations are essential for the successful implementation of microbial rejuvenation strategies in heavily polluted soils, ensuring a more sustainable and effective approach to soil remediation.

7.5 ADVANCEMENTS IN BIOREMEDIATION TECHNIQUES

Explore and develop more sophisticated bioremediation strategies, including the engineering of microbes with enhanced pollutant degradation capabilities, to elevate the effectiveness of soil cleanup.

1. **Studies on Microbial Consortia**: Investigate the collaborative effects of microbial consortia in pollutant degradation, aiming to understand the interactions among different microorganisms for more efficient and sustainable soil rejuvenation.
2. **Application of Omics Technologies**: Employ advanced omics technologies (genomics, metagenomics, meta-transcriptomics) to analyze the genetic and functional potential of soil microbial communities. This can aid in pinpointing key microbial contributors and their metabolic pathways for targeted interventions.
3. **Strategies for Biostimulation and Bioaugmentation**: Develop customized biostimulation and bioaugmentation approaches to boost the resilience and metabolic activity of native microbial communities, fostering natural soil rejuvenation processes.
4. **Integration of Ecological Considerations**: Merge ecological principles into microbial rejuvenation strategies to ensure the restoration of a balanced and sustainable soil ecosystem. Consider the impact on non-target organisms and the lasting ecological consequences of interventions.
5. **Implementation of Monitoring and Assessment Tools**: Design and deploy advanced monitoring tools, such as remote sensing and sensor technologies, for real-time assessment of changes in soil quality and microbial activity. This enables adaptive management and timely interventions.
6. **Adaptation to Climate Change**: Consider the implications of climate change on soil microbial communities and tailor rejuvenation strategies to withstand evolving environmental conditions.
7. **Public Engagement and Policy Integration**: Engage local communities and stakeholders in the soil rejuvenation process, creating awareness about the significance of microbial

interventions. Collaborate with policymakers to integrate microbial solutions into broader environmental management frameworks.

8. **Educational Initiatives**: Promote educational and awareness programs to train professionals and the public on the benefits and risks associated with microbial rejuvenation techniques. Foster a deeper understanding of sustainable soil management practices.
9. **Interdisciplinary Collaborations**: Encourage collaborations among microbiologists, ecologists, engineers, and other relevant disciplines to bring diverse expertise to the development of innovative and holistic approaches for soil rejuvenation.

7.6 FUTURE DIRECTIONS

The microbial rejuvenation of heavily polluted soil represents a promising approach in environmental science and sustainable practices. In recent years, researchers have made significant strides in understanding the pivotal role of microorganisms in remediating soil contaminated with diverse pollutants. The utilization of bacteria, fungi, and other microbial agents has shown promise in breaking down pollutants, metabolizing toxic compounds, and enhancing soil health. Looking ahead, several key aspects can be identified that will shape the development of this field.

1. **Advanced Microbial Consortia**: Future research in soil rejuvenation may emphasize developing specialized microbial consortia designed for specific pollutants. Combining microbial communities with complementary metabolic functions could improve the effectiveness and scope of soil remediation.
2. **Genetically Engineered Microbes**: Harnessing the power of genetic engineering opens new frontiers in microbial-assisted soil rejuvenation. Engineered microbes with enhanced pollutant-degrading capabilities or stress tolerance could be designed to thrive in harsh soil conditions, offering a targeted and sustainable approach to remediation.
3. **Integration of Emerging Technologies**: The synergy between microbial rejuvenation and emerging technologies, such as artificial intelligence, sensors, and precision agriculture, presents exciting possibilities. Smart monitoring systems could provide real-time data on microbial activities, aiding in adaptive management strategies for soil remediation.
4. **Eco-friendly Soil Amendments**: Investigating and developing eco-friendly soil amendments that promote the growth and activity of beneficial microorganisms could be a game-changer. These amendments, derived from natural sources, have the potential to create a conducive environment for microbial communities to flourish.
5. **Long-Term Ecosystem Resilience**: It is essential to focus on the long-term resilience of ecosystems. Future research should investigate how microbial rejuvenation influences soil structure, biodiversity, and ecosystem functions over extended periods to ensure that its benefits are sustained.

7.7 CONCLUSION

In conclusion, the microbial rejuvenation of heavily polluted soil not only addresses the urgent need for environmental restoration but also contributing to the advancement of sustainable practices. As we navigate the future of soil remediation, a holistic approach that integrates advanced technologies, genetic innovation, and ecological principles promises a brighter and more resilient environmental landscape. This journey holds the potential to transform polluted landscapes into thriving ecosystems, showcasing the power of microbial communities in shaping the sustainable future of our planet.

REFERENCES

Abbas, Ghulam, Behzad Murtaza, Irshad Bibi, Muhammad Shahid, Nabeel Khan Niazi, Muhammad Imran Khan, Muhammad Amjad, Munawar Hussain, and Natasha Khan (2018). Arsenic Uptake, Toxicity, Detoxification, and Speciation in Plants: Physiological, Biochemical, and Molecular Aspects. *International Journal of Environmental Research and Public Health*. January 2;15(1), 59.

Alengebawy, Ahmed, Sara Taha Abdelkhalek, Sundas Rana Qureshi, and Man-Qun Wang (2021). Heavy Metals and Pesticides Toxicity in Agricultural Soil and Plants: Ecological Risks and Human Health Implications. *Toxics*, 9(3), 42.

Ali, Uqab, Syeed Mudasir, Saleemm Farooq, and Ruqeya Nazir (2015). Factors Affecting Bioremediation. *Journal of Research & Development*, 15, 102–109.

Alori, Elizabeth Temitope, Alhasan Idris Gabasawa, Chinyere Edna Elenwo, and Oluwadolapo Ololade Agbeyegbe; (2022). Bioremediation Techniques as Affected by Limiting Factors in Soil Environment. *Frontiers in Soil Science, Soil Pollution & Remediation*, 2, 937186.

Al-Taai, Suaad Hadi Hassan (2021). Water pollution its causes and effects, *IOP Conference, Iraq, Series: Earth and Environmental Science*, Vol. 790, 012009.

Aqeel, Ashraf Muhammad, Mohd Jamil Maah, and Ismail Yusoff (2014). Soil Contamination, Risk Assessment and Remediation. *Environmental Risk Assessment of Soil Contamination*. InTech, 57287.

Atuchin, Victor V., Lyudmila K. Asyakina, Yulia R. Serazetdinova, Anna S. Frolova, Natalia S. Velichkovich, and Alexander Yu Prosekov, (2023). Microorganisms for Bioremediation of Soils Contaminated with Heavy Metals. *Microorganisms*, 11, 864.

Ayilara Modupe S., Olubukola O. Babalola (2023). Bioremediation of Environmental Wastes: The Role Of Microorganisms. *Frontiers in Agronomy*, 5, 1183691.

Azad, Abul Kalam Md., Latifah Amin, and Nik Marzuki Sidik. (2014). Genetically Engineered Organisms for Bioremediation of Pollutants in Contaminated Sites. *Chinese Science Bulletin*, *59*(8), 703–714.

Azubuike, Christopher Chibueze, Chioma Blaise Chikere, and Gideon Chijioke Okpokwasili (2016). Bioremediation Techniques-Classification Based on Site of Application: Principles, Advantages, Limitations, and Prospects. *World Journal of Microbiology and Biotechnology*, *32*(11), 180.

Balagurusamy, Nagamani (2005). Anaerobic Bioremediation–An Untapped Potential Biorremediacion. *Revista Mexicana De Ingenieria Quimica*, 4, 273–287 anaerobica-un potencial sin explotar.

Bhandari, Sobika, Darbin Kumar Poudel, Rishab Marahatha, Sonika Dawadi, Karan Khadayat, Sitaram Phuyal, Shreesti Shrestha, Santosh Gaire, Kusum Basnet, Uddhav Khadka, and Niranjan Parajuli (2021). Microbial Enzymes Used in Bioremediation. *Journal of Chemistry*, 2021, 8849512.

da Silva Correa, Hauane, Christopher Thomas Blum, Franklin Galvão, and Leila Teresinha Maranho (2022). Effects of Oil Contamination on Plant Growth and Development: A Review. *Environmental Science and Pollution Research*, 29, 43501–43515.

Dejonghe, W., J. Goris, S. El Fantroussi, M. Hofte, P. De Vos, W. Verstraete, et al.; (2000). Effect of dissemination of 2,4-dichlorophenoxyacetic acid (2,4-D) degradation plasmids on 2,4-D degradation and on bacterial community structure in two different soil horizons.; Applied and Environmental Microbiology, Vol. 66, pp. 3297–3304.

Elyamine, Ali Mohamed, Jie Kan, Shanshan Meng, Peng Tao, Hui Wang, and Zhong Hu (2021). Aerobic and Anaerobic Bacterial and Fungal Degradation of Pyrene: Mechanism Pathway Including Biochemical Reaction and Catabolic Genes. *International Journal of Molecular Sciences*, *22*(15), 8202.

European Union (2002). Heavy metals in waste.; European Commission Directorate General Environment, EU Standard 2000/532/EC, Brussels.

FAO (1985). Water quality for agriculture.; Food and Agriculture Organization of the United Nations, Irrigation and Drainage Paper No. 29, Rev. 1, Rome.

Garbisu, Carlos, Garaiyurrebaso Olatz, Epelde Lur, Grohmann Elisabeth, and Alkorta Itziar (2017). Plasmid-Mediated Bioaugmentation for the Bioremediation of Contaminated Soils. *Frontiers in Microbiology*, 8, 8202.

Gogoi, Gogoi, Binod Kumar, Narendra Nath Dutta, Pranab Goswami, and T.R. Krishna Mohan. (2003). A Case Study of Bioremediation of Petroleum-Hydrocarbon Contaminated Soil at a Crude Oil Spill Site. *Advances in Environmental Research*, 7, 767–782.

Halden, R.U., S. Tepp, B.G. Halden, and D.F. Dwyer; (1999). Degradation of 3-Phenoxybenzoic Acid in Soil by *Pseudomonas pseudoalcaligenes* POB310(pPOB) and Two Modified *Pseudomonas* Strains.; *Applied and Environmental Microbiology*, Vol. 65, pp. 3354–3359.

Havugimana, Erneste, Balkrishna Sopan Bhople, Anil Kumar, Emmanuel Byiringiro, Jean Pierremugabo, and Arun Kumar (2017). Soil Pollution: Major Sources and Types of Soil Pollutants, Chapter. *Environmental Science and Engineering*, 11, 53–86.

Hazen, T. C., A. J. Tien, A. Worsztynowicz, D. J. Altman, K. Ulfig, and T. Manko, (2003). Biopiles for remediation of petroleum-contaminated soils: a polish case study. *Conference: Proceedings of the NATO Advanced Research Workshop on The Utilization of Bioremediation to Reduce soil Contamination: Problems and Solutions Proceedings of the NATO Advanced Research Workshop on The Utilization of Bioremediation to Reduce soil Contamination: Problems and Solutions.*

Holt, J.G., N.R. Krieg, P.H.A. Sneath, J.T. Staley, and S.T. Williams; (1994). *Bergey's Manual of Determinative Bacteriology*, 9th Edn.; Williams and Wilkins, Baltimore, pp. 626–640.

Joshi, R.P., and S.M. Thagard; (2013). Streamer-like Electrical Discharges in Water: Part II. Environmental Applications. *Plasma Chemistry and Plasma Processing*, Vol. 33, Issue (01), pp. 17–49.

Karigar, Chandrakant S., and Shwetha S. Rao (2011). Role of Microbial Enzymes in the Bioremediation of Pollutants: A Review. *Enzyme Research 7*, 805187.

Karn, Santosh Kumar, Geetanjali. (2014). Pentachlorophenol Remediation by Enterobacter sp. SG1 Isolated from Industrial Dump Site. *Pakistan Journal of Biological Sciences*, *17*(3), 388–394.

Kao, C.M., J.K. Liu, Y.L. Chen, C.T. Chai, and S.C. Chen; (2005). Factors Affecting the Biodegradation of PCP by *Pseudomonas mendocina* NSYSU. *Journal of Hazardous Materials*, Vol. 124, pp. 68–73.

Kumar, Rajdeo, Rajdeo Kumar, Ashish Chauhan Ashish Chauhan, Nisha Yadav Nisha Yadav, Laxmi Rawat Laxmi Rawat, and Manish Kumar Goyal (2015). Bioremediation of Polluted Soil obtained from Tarai Bhavan Region of Uttarakhand, India. *Journal of Bioremediation and Biodegradation*, 6, 276.

Lin, Chitsan, Nicholas Kiprotich Cheruiyot, Xuan-Thanh Bui, and Huu Hao Ngo (2022). Composting and Its Application in Bioremediation of Organic Contaminants. *Bioengineered*, 13(1), 1073–1089.

Mandal, Paramita (2017). An Insight of Environmental Contamination of Arsenic on Animal Health. *Emerging Contaminants*, *3*(1), 17–22.

Meena, Tanmay, Deepesh Kumar Neelam, Varsha Gupta, Devki, and Ravikant Rahi (2021). Bioremediation-An Overview. *International Journal of Recent Scientific Research*, *12*(03) (A), 41125–41133.

Mileusnić, Marta, Benjamin Siyowi Mapani, Akalemwa Fred Kamona, Stanko Ružičić, Isaac Mapaure, and Percy Maruwa Chimwamurombe, (2014). Assessment of Agricultural Soil Contamination by Potentially Toxic Metals Dispersed from Improperly Disposed Tailings, Kombat Mine, Namibia., *Journal of Geochemical Exploration*, *144*, Part C, 409–420.

Mishra, Rajesh Kumar, Naseer Mohammad, and Nilanjan Roychoudhury (2016). Soil pollution: Causes, effects and control. *Van Sangyan*, *3*, 1–14.

Mohamed, Badr Ali, Mohamed Samer (2023). Biobutanol production from agricultural wastes, Editor (s): Rangabhashiyam Selvasembian, Nur Izyan Wan Azelee, Saravanan Ramiah Shanmugam, Ponnusami Venkatachalam, Ajay Kumar Mishra, *Valorization of Wastes for Sustainable Development*, Elsevier, pp. 181–200.

Mrozik, Agnieszka, Zofia Piotrowska-Seget (2010). Bioaugmentation as a Strategy for Cleaning Up of Soils Contaminated with Aromatic Compounds. *Microbiological Research*, *165*(5), 363–375.

Mrozik, A., S. Miga, and Z. Piotrowska-Seget; (2011). Enhancement of phenol degradation by soil bioaugmentation with *Pseudomonas* sp. JS150.; *Journal of Applied Microbiology*, Vol. 111, pp. 1357–1370.

Nandish, M.S.; (2005). Microbial degradation of phenol and pentachlorophenol. Master's Thesis, University of Agricultural Sciences, Dharwad, India, pp. 1–80.

Nyiramigisha, Philomene, Komariah, Sajidan (2021). Harmful Impacts of Heavy Metal Contamination in the Soil and Crops Grown Around Dumpsites, *Reviews in Agricultural Science*, 9, 271–282.

Nzila, Alexis, Shaikh Abdur Razzak, and Jesse Zhu (2016). Bioaugmentation: An Emerging Strategy of Industrial Wastewater Treatment for Reuse and Discharge. *International Journal of Environmental Research and Public Health*, *13*(9), 846.

Partha, R., and I. Bossert; (1984). Treatment and disposal of petroleum refinery waste. In: Atlas, R.M. (Ed.), *Petroleum Microbiology*, McMillan Publishing Co, New York, pp. 553–577.

Premlatha, A., and G.S. Rajkumar; (1994). Pentachlorophenol degradation by *Pseudomonas aeruginosa. World Journal of Microbiology and Biotechnology*, Vol. 10, pp. 334–337.

Radehaus, P.M., and S.K. Schmidt (1992). Characterization of a novel *Pseudomonas* sp. that mineralizes high concentrations of pentachlorophenol. Applied and Environmental Microbiology, *58*, pp. 2879–2885.

Rafeeq, Hamza, Nadia Afsheen, Sadia Rafique, Arooj Arshad, Maham Intisar, Asim Hussain, Muhammad Bilal, and Hafiz M.N. Iqbal. (2023). Genetically Engineered Microorganisms for Environmental Remediation. *Chemosphere*, *310*, 136751.

Rebello, Sharrel, Vinod Kumar Nathan, Raveendran Sindhu, Parameswaran Binod, Mukesh Kumar Awasthi, and Ashok Pandey. (2021). Bioengineered Microbes for Soil Health Restoration: Present Status and Future. *Bioengineered*, *12*(2), 12839–12853.

Rodríguez-Eugenio, N., Michael McLaughlin, and Daniel Pennock (2018). *Soil Pollution: A Hidden Reality*. Rome, FAO. 142 p.

Saber, D.L., and R.L. Crawford; (1985). Isolation and Characterization of *Flavobacterium* Strain that Degrades Pentachlorophenol. *Applied and Environmental Microbiology*, Vol. 50, pp. 1512–1518.

Saha, Anuradha, Rupak Kumar. (2023). Environmental Mishaps: A Case Study of Oil Spill. https://www.researchgate.net/publication/369146456_Environmental_Mishaps_A_Case_Study_of_Oil_Spill

Saikia, Tarun, Jhorna Bora, and Cinu Gogoi. (2023). Impact of Crude Oil Contamination on Soil Physicochemical Properties around the Sildubi Oil Spill, Borhola, Assam. *International Journal of Plant & Soil Science*, *35*(22), 241–253.

Sajjad, Muhammad, Qing Huang, Sardar Khan, Muhammad Amjad Khan, Yin Liu, Junfeng Wang, Faqin Lian, Qingqing Wang, and Genmao Guo (2022). Microplastics in the Soil Environment: A Critical Review. *Environmental Technology & Innovation*, 27, 102408.

Sarao, Loveleen Kaur, Sandeep Kaur (2022). Microbial Bioremediation. In: Maulin P. Shah (Ed.), *New Vistas in Microbial Sciences-1*, Integrated Publications, 47–66.

Shah, S., and I.S. Thakur; (2002). Enrichment and characterization of microbial community of tannery effluent for the degradation of pentachlorophenol. *World Journal of Microbiology and Biotechnology*, *18*, pp. 693–698.

Sharma, Indu (2021). Bioremediation Techniques for Polluted Environment: Concept, Advantages, Limitations, and Prospects. In: Mario Alfonso Murillo-Tovar, Hugo Albeiro Saldarriaga Noreña, and Agnieszka Saeid (Eds.), *Trace Metals in the Environment - New Approaches and Recent Advances*, Chapter 12. Rijeka: IntechOpen.

Sharma, A., and I.S. Thakur; (2008). Characterization of pentachlorophenol degrading bacterial consortium from chemostat. *Bulletin of Environmental Contamination and Toxicology*, *81*, pp. 182–188.

Singare, Pravin U., Ram S. Lokhande, and Pragati P. Pathak.(2010) Soil Pollution along Kalwa Bridge at Thane Creek of Maharashtra, India. *Journal of Environmental Protection*, *1*(2).

Singh, S., R. Chandra, D.K. Patel, and V. Rai; (2007). Isolation and characterization of novel *Serratia marcescens* (AY927692) for pentachlorophenol degradation from pulp and paper mill waste. *World Journal of Microbiology and Biotechnology*, *23*, pp. 1747–1754.

Singh, Anita, Rajesh Kumar Sharma, Madhoolika Agrawal, and Fiona M. Marshall. (2010) Risk Assessment of Heavy Metal Toxicity Through Contaminated Vegetables From Waste Water Irrigated Area of Varanasi, India. *Tropical Ecology*, *51*(2S), 375–387.

Vásquez-Murrieta, María S., Oscar J. Hernández-Hernández, Juan A. Cruz-Maya, Juan C. Cancino-Díaz, and Janet Jan-Roblero (2016). Approaches for Removal of PAHs in Soils: Bioaugmentation, Biostimulation and Bioattenuation [Internet]. In: Marcelo L. Larramendy and Sonia Soloneski (Eds.), *Soil Contamination*, Chapter 17. Rijeka: IntechOpen.

Wang, S., and C.N. Mulligan; (2004). An Evaluation of Surfactant foam Technology in Remediation of Contaminated Soil. *Chemosphere*, 57, pp. 1079–1089.

Wang, Zuwei, Xiangfeng Zeng, Mingshuo Geng, Chunyi Chen, Jianchao Cai,Xiaoman Yu, Yingying Hou, and Hui Zhang (2015). Health Risks of Heavy Metals Uptake by CropsGrown in a Sewage Irrigation Area in China. *Polish Journal of Environmental Studies*, *24*(3), 1379–1386.

World Health Organization (WHO) (1996). Guidelines for Drinking-Water Quality, 2nd ed.; Vol. 2: *Health Criteria and Other Supporting Information*, WHO, Geneva.

World Health Organization (WHO) and Food and Agriculture Organization (FAO); (2007). *CodexAlimentarius: General Standard for Contaminants and Toxins in Food and Feed.; Codex Stan 193-1995, Rev. 4-2007*, FAO/WHO, Rome.

Yang, Jianxia, Jun Zhao, Guofeng Zhu, Yuchun Wang, Xinggang Ma, Jianbang Wang, Huiwen Guo, and Yu Zhang, (2020). Soil Salinization in the Oasis Areas of Downstream Inland Rivers—Case Study: Minqin Oasis. *Quaternary International*, 537, 69–78.

8 Sustainable Energy Production through Microbial Fuel Cells and Waste Management

Namrata Bist, Anirbid Sircar, and Tithee Bhavsar

8.1 INTRODUCTION

The consumption of energy around the world has been on a strong trend in recent decades (Rahimnejad et al., 2011). There are three types of energy available, fossil fuels, renewable sources and nuclear resources, which could be classified into two major categories in the case of nonrenewable energies that account for an enormous amount of energy consumption (Akdeniz et al., 2002). Due to the emission of carbon dioxide, fossil fuels have a negative impact on nature. In fact it follows logically from what has been claimed that the consumption of fossil fuels is seriously undermining human life, in particular through its catastrophic consequences such as global warming and atmospheric pollution (Rahimnejad et al., 2012).

In actual fact, fuel cells (FCs) have a plethora of advantages over other kinds of energy generators, for example no emissions of environmental polluting gases (such as SOx, NOx, CO2 and CO), higher efficiency, no existence of mobile parts, lack of sonic pollution, and so forth (Peighambardoust et al., 2010). Microfuel cells that use active microorganisms as biocatalysts in anaerobic anode compartments for the production of bioelectrical energy are one type of FC (Rahimnejad et al., 2012).

The promise for treating trash, producing power, and producing biofuels all at the same time is presented by microbial fuel cells (MFCs) (Logan & Rabaey, 2012). Fuels in MFCs, primarily organic substrates, oxidize, produce electrons, and deposit those electrons on the anode electrode (Jančula & Maršálek, 2012). In bio-electrochemical systems, the kind and concentration of a substrate are crucial since they influence the power output and the makeup of the microbial population. For simplicity of handling and power output, simple organic molecules like acetate are utilized (Demirbas et al., 2011). However, more affordable and accessible feedstock needs to be investigated in order to generate a significant amount of electricity. One such choice is microalgae biomass (Logan, 2005). The biocatalyst used in MFC is made up of bacteria that are electrochemically active and exist as biofilms on the anode electrode surface. According to Rashid et al. (2009), these microorganisms are known as exoelectrogens and are able to transmit and deposit electrons straight onto the anode.

According to Kim et al. (2002), the microbial fuel cell (MFC) is a device that converts the chemical energy of a fuel, usually organic substrates, into electrical energy through biocatalytic reactions carried out by microorganisms. This process provides new opportunities for the sustainable production of energy, in the form of electricity produced directly from biodegradable compounds. MFCs offer a promising sustainable energy alternative to combat issues about non-renewable energy consumption, climate change, and environmental pollution (del Campo et al., 2013). The anode and cathode compartments of microbial fuel cells (MFCs) are divided by an ion conducting membrane. In the anode compartment, protons and electrons are produced as anaerobic bacteria break down organic matter into carbon dioxide. An external circuit conducts electron flux, producing an electrical current. To reach the cathode, protons travel across the membrane. Protons and electrons mix

 DOI: 10.1201/9781003668961-8

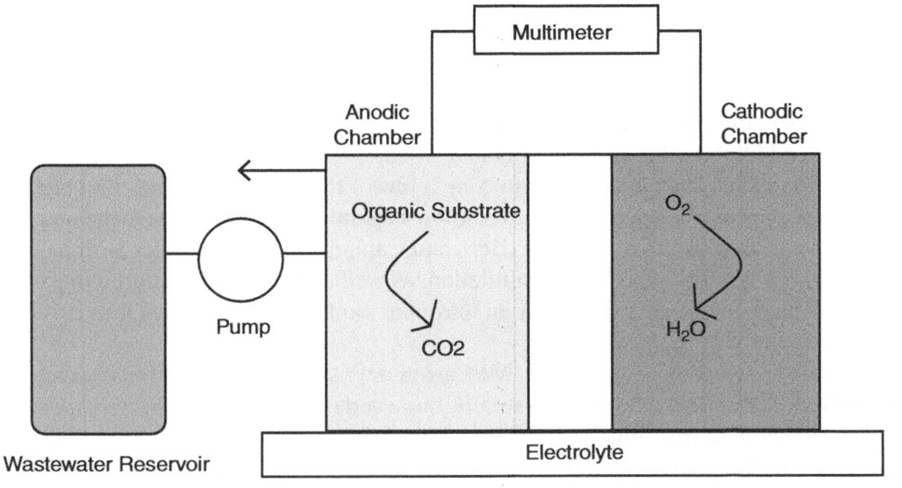

FIGURE 8.1 Schematic setup of MFC.

with an electron acceptor, often oxygen, in the cathode compartment to form water (del Campo et al., 2013) (Figure 8.1).

8.2 LITERATURE STUDY

The United States continues to use aerobic treatment technologies for wastewater treatment despite the rising energy costs. However, conventional anaerobic treatment technologies are becoming more popular. Unfortunately, many wastewaters are too dilute to make current anaerobic treatment technologies, based on methane production, economical. The cost of wastewater treatment in the United States is substantial, with an annual expenditure of $25 billion for all types of wastewater treatment (Logan, 2005).

Recently, it was discovered that bacteria in an MFC can generate electricity (Kim et al., 1999). Organic matter in MFCs is oxidized by bacteria, which then transfer the electrons to an electrode (anode). Only when electrons are transferred to the electrode can electricity be produced. Additionally, protons are produced; these protons need to be allowed to move to the counter electrode in order to mix with oxygen and electrons to generate water. The MFC's foundation is created by current generation and the potential difference between the anode and cathode chambers. The greatest potential is between 0.5 and 0.8 V overall, which is comparable to the potential produced in a hydrogen fuel cell. By connecting the reactors in series, this low voltage can be changed to create a greater voltage (Logan, 2005).

8.2.1 Various Configurations of MFC

8.2.1.1 Bacterioalgal MFC

An approach for improving the environmental and carbon capture characteristics of this technology is called bacterioflagal MFC. Using anode-reducing exodenogenic bacteria as an anolyte and autotrophic and phototrophic organisms as catalysts, the process works. It is possible to add algae to the MFC's anodic or cathodic compartments in order to produce oxygen or act as a source of carbon (substrate) (Saba et al., 2017). Passive CO2 capture from the atmosphere, or an artificial compression of more diffused CO2 streams into a cathode chamber may be used to cultivate it (Cui et al., 2014). Under laboratory conditions, the necessary sunlight can be obtained from natural sources or through grow lights. Various media, for example glucose, formate, acetate, and dry algae biomass can obtain a carbon substrate (Saba et al., 2017). It is problematic to use dry algae as a substrate due to bacterial difficulties in hydrolysing algal cell walls (Cui et al., 2014).

Simple sources of carbon available to bacteria are glucose, gluteic acid and acetate substrates but they are more costly than other resource materials like sewage sludges, landfill leachates, food processing waste in addition to drinking water (Juang et al., 2012).

Algal ponds, lagoons, and photobioreactors are a few of the artificial settings in which algae can develop. Because they may consume different carbon substrates, microalgae have the unusual capacity to develop heterotrophically in the dark (Berk and Canfield, 1964). Together with its other capabilities, like oxygen synthesis and CO2 assimilation, as well as its logarithmic growth pattern, algae cultivation in photobioreactors is done in a mixotrophic mode, which qualifies it as a viable choice for an MFC biocatholyte.

Blue green algae, which is being used in MFCs, not only creates pigments but also releases powerful toxins (Berk & Canfield, 1964). If algae cells burst and die, they release micocystins, the toxins in lake waters which arise from blue green algae. Conventional methods of treatment, such as coagulation, flocculation, deposition and filtration, cannot be used for the treatment of micocystin.

Based on the oxidation of substrate by bacteria in anolyte and the extraction of oxygen from algae in catholyte, the bacterioalgal MFC functions. Through photosynthesis, the oxygen acts as an electron absorber, lowering CO2 and producing more water and algal biomass (Saba et al., 2017). Among the chemical reactions at play are:

Light reaction:

$$6CO_2 + 12H + +12e- \rightarrow C_6H_{12}O_6\left(\text{biomass}\right) + 3O_2$$

Dark reaction:

$$C_6H_{12}O_6\left(\text{biomass}\right) + 6O_2 \rightarrow 6CO_2 + 6H_2O$$

Cathodic reaction:

$$O_2 + 4H + +4e- \rightarrow 2H_2O$$

Although current wastewater treatment methods consume a lot of energy, MFCs can be used to generate energy for wastewater prior to treatment. Wastewater can be integrated into the current wastewater treatment infrastructure, support significant biofuel production capacity, and help ensure a sufficient supply of nutrients. Reduction of the chemical oxygen demand (COD) is one metric used to assess the effectiveness of wastewater treatment. It was found that wastewater microbial consortia eliminated over 80% of COD in MFC (del Campo et al., 2013).

Though the removal of bacteriogal CODs has been the subject of only a few research studies, a promising component of the technology may make bacteriogal COD MFCs more significant and useful in their commercial application.

8.2.1.2 Microalgae-Based MFC

As a potential feedstock to produce biofuels, microalgae have been considered. Currently, it is not economically feasible to scale up the use of algae-derived biofuels. Integration of biofuels with other technologies, for example waste management and electricity production, is a clear way to cut biofuel costs. Such potential for waste treatment, generation of electricity and production of biofuels simultaneously is offered by microbial fuel cells (ElMekawy et al., 2013).

Discharging microalgae directly into the sewer system can pose a threat to human health. Therefore, it is necessary to remove microalgae from water bodies and dispose of them properly. The process of producing biofuels from microalgae biomass is not yet commercially viable, despite its use. As an alternative, microalgae can be employed as a substrate in MFC (Figure 8.2), which has the ability to both treat waste and produce power (Moqsud et al., 2013). As a result, using

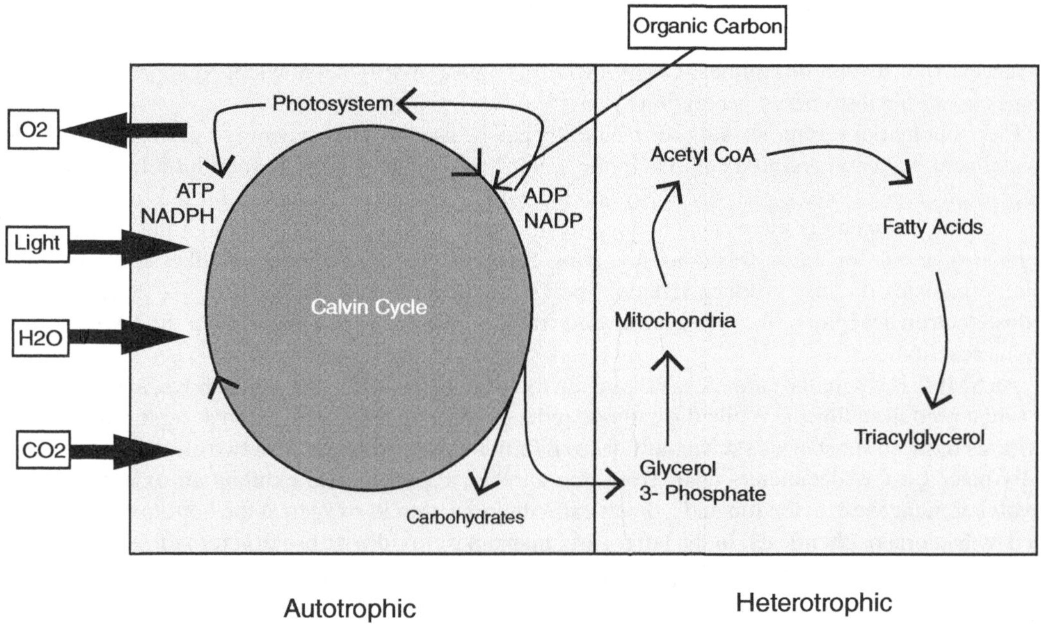

FIGURE 8.2 Autotrophic and heterotrophic MFC.

microalgae biomass has two advantages: it reduces pollution and serves as a fuel for MFC. Proteins (32%) and carbohydrates (51%), which are abundant in microalgae biomass, are easily broken down by electrogens to create energy (Rashid et al., 2013).

Despite the fact that microalgae are the primary cause of eutrophication and stream pollution, a significant amount of microalgae biomass is produced globally during wastewater treatment.

Algae contribute more than half of Earth's primary photosynthetic throughput by converting solar energy into a variety of metabolic energy sources (Lal, 2008). Microalgae are becoming more and more popular around the world because of their rapid growth and up to 6.24 kg/m3/day CO2 fixation rates (Rosenbaum et al., 2005). This demonstrates the enormous potential for solar energy conversion by photosynthetic microbes. Algae and plants naturally do photosynthesis, which is a complicated biological redox reaction. They may use solar energy to create various redox reactions that yield oxygen and carbohydrates, as well as other compounds that can be used as building blocks for other molecules or as a source of energy (ElMekawy et al., 2013).

The autotrophic growth approach, which involves growing microalgae in naturally or artificially lighted conditions, is widely used to cultivate microalgae. In this mode, solar energy is captured by the cells and CO2 is used as a carbon source. As an alternative, mixotrophic growth mode, which mixes autotrophic and heterotrophic modes, can be used to develop microalgae (ElMekawy et al., 2014). The photosynthetic and respiratory metabolisms run concurrently with the simultaneous uptake of CO2 and organic carbon, respectively, in the mixotrophic growth mode. The primary lack of illuminated autotrophic photobioreactors is addressed by the heterotrophic growth mode, which enables any bioreactor to be used without the requirement for specialized design or equipment (ElMekawy et al., 2014).

There are a number of drawbacks to heterotrophic microalgal culture, such as the restricted range of species that can be employed, higher energy costs because an organic substrate is required, and the potential for growth inhibition because of an excess of organic substrate and an inability to produce light-induced metabolites. This cultivation method is also vulnerable to contamination and competition from other microbes (Chen, 1996).

8.2.1.3 Sediment Microbial Fuel Cell (SMFC)

Sediment microbial fuel cell technology is a relatively recent field of study that incorporates advances from several disciplines. Using SMFCs is a customarily acknowledged approach when it comes to sustainable energy generation.

Electropotential variations that occur naturally can be used to harvest energy by burying an anode in sediment and submerging a cathode in the water body above (Figure 8.3) (Whitfield, 1972). We refer to this type of setup as an SMFC or benthic MFC. In the event that the SMFC is not receiving any current, an open circuit voltage of roughly 750 mV is often reached. Through the oxidation of organic materials or redox reactions involving inorganic reduced chemicals, like sulfur species, microorganisms directly produce reducing power in the sediment (Reimers et al., 2001). Lower redox electron acceptors, like oxygen released from the water, are reduced during the SMFC's oxidation reaction.

An SMFC is where the cathode is floating on top of or below the water layer above, and the anode is submerged in sediment. While the cathode reduces oxygen, the last electron acceptor, the anode, oxidizes reduced substances such as sulfides and fermentation products. The two electrodes are typically placed a few decimeters apart from one another. Each cathode exhibits an oxidation cycle involving manganese at the top and a direct cathodic reduction in oxygen at the bottom that is mediated by bacteria or chemicals. In the latter case, manganese-oxidizing bacteria reoxidize and deposit the electrons that were previously provided from the cathode to reduce manganese oxides. Such reactions seem to take place in a biofilm on the surface of the cathode. Small organisms are represented by ovals (De Schamphelaire et al., 2008).

Producing electrical energy in the form of easily accessible electricity is the primary use of SMFCs. Long-term electricity generation is feasible in rural locations when done sustainably. They enable small electric devices to produce electricity on-site. This implies that SMFCs can power sensors and data transmitters, for instance, at the ocean's bottom. Weak currents or voltages can be

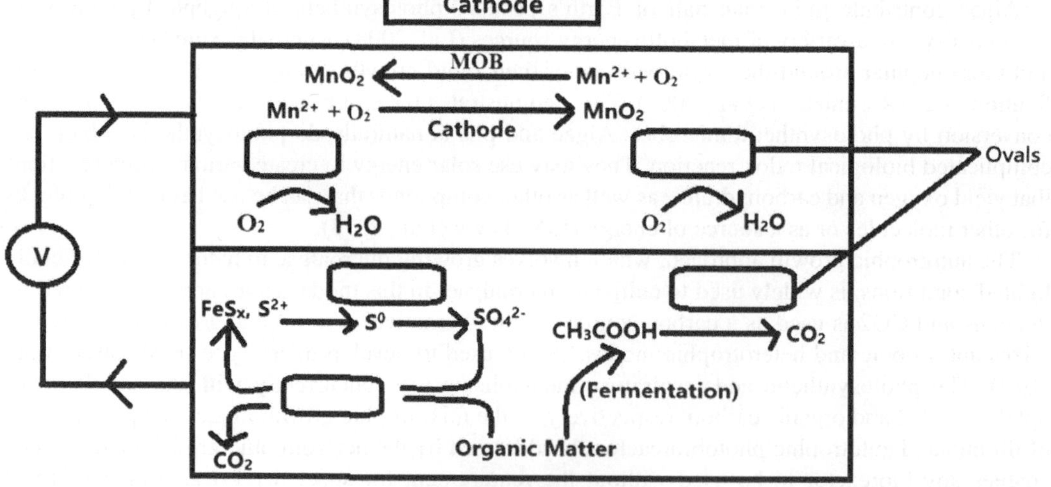

FIGURE 8.3 Schematic setup of SMFC.

handled by adding converters and capacitors (Shantaram et al., 2005). A meteorological buoy and data transmitter with an average power usage of 18 mW were powered by an SMFC with a bio-anode, demonstrating this principle recently through the application of a power conditioner and a capacitor (Tender et al., 2008).

Sediment fuel cell research has focused on the marine environment, but freshwater habitats may potentially contribute to the continued generation of power. There have been reports of lower production values, which are probably caused by a drop in conductivity and the fact that saltwater performs better at the cathodes than freshwater because it improves virtual corrosion at the cathodes (Bergel et al., 2005).

While sulfur compounds in sulfide-rich sediments are linked to the oxidation of organic materials, they can also serve as a significant supply of electrons for the anodes. These sediments contain sulfide because of the oxidation of organic materials and the reduction of microbial sulfate. A fraction of the generated sulfide will react with Fe_2^+ to produce minerals that are reduced in sulfide, like FeS or FeS_2. Both the oxidation of sulphites contained in iron minerals and the oxidation of dissolved sulphide to S^0 at an anode are feasible (Ryckelynck et al., 2005). Thus, it is quite probable that abiotic or microbially driven additional oxidation of S0 to SO42 is one of the ongoing production processes in this kind of material. Sulfate regenerates overall as an electron acceptor of organic matter oxidation as a result of these processes. Furthermore, elemental sulfur can be employed as an electron acceptor in the microbial breakdown of organic materials (Ho et al., 2004). As a result, in anodic half cells, sulfur compounds can act as electron shuttles for the oxidation of organic matter, and the increased activity of the sulfur cycle at anodes may promote the oxidation of organic matter in sediment (see also Figure 8.3).

8.2.1.4 Laminaria-Based Microbial Fuel Cell (LbMs)

Marine algae are an environmentally friendly feedstock for microbial fuel cells because of their high carbohydrate content, increased potential for adsorption of carbon dioxide, and low water and energy requirements during cultivation. Two MFCs have been tested in batch fed mode using *Laminaria saccharina* as a model for algal electron donors and mixed based microbial consortia as biocatalysts in the anode compartment (Gadhamshetty et al., 2013).

The bacteria oxidize electron donors and respire using solid anodes to accept terminal electrons in the MFC's anode. In addition to offering potential benefits for waste stabilization, renewable energy generation and costs saving in operational activities, the use of MSWs as electron donors offers an opportunity to mitigate the demand for pure organic compounds from MFCs.

Given its high carbohydrate content, *Laminaria* is a form of brown algae that has been identified as potential feedstock to produce biofuels. For instance, in *Laminaria japonica* the carbohydrate content was almost 66.7% (Zhang et al., 2010). This is significant because of its increased lignin content in common SOWs, it is less able to interact with metabolic pathways and reduce the potential for application of SOW–based bioenergy. In the MFC there are no reports of marine algae yet.

8.2.1.5 Mediatorless Microbial Fuel Cell

The performance of a mediatorless microbial fuel cell (Figure 8.4) has been optimized considering the different operating circumstances. Current generation has been caused by a number of variables, including pH, resistance, electrolyte consumption, and the concentration of dissolved oxygen in the battery compartments. The strongest currents were produced at pH 7 (Gil et al., 2003).

Protons and electrons are transferred into the cathode compartment via the circuit and through the membrane. By consuming protons and electrons in the cathode chamber, oxygen is reduced to water (Lithgow et al., 1986). Since the majority of bacteria lack electrochemical activity, mediators including thionine, methyl viologen, hydrogen acid, and others are useful in facilitating the

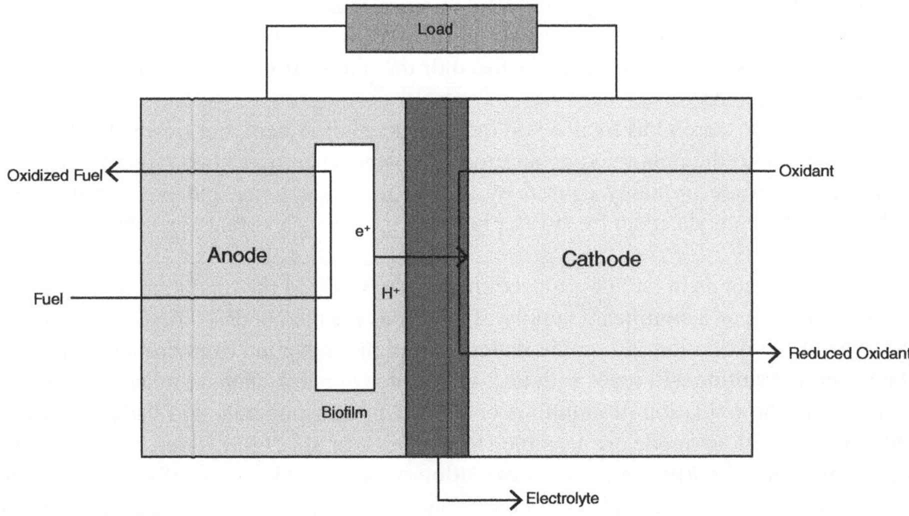

FIGURE 8.4 Schematic setup of mediatorless MFC.

transport of electrons from bacterial cells to an electrode. Due to the high cost and toxicity of various mediators, a commercially viable microbial fuel cell has not been produced (Delaney et al., 1984).

It has also been demonstrated that the Fe(III)-reducer *Shewanella putrefaciens* is an electrochemically active catalyst that may be employed in a mediatorless microbial fuel cell (Kim et al., 1999). Additionally, fuel cells were employed to enrich electrochemically active bacteria by employing activated sludge as the source of bacteria and wastewater from a starch processing factory as an electron donor. The cells have isolated an electrochemically active bacteria from the mediatorless microbial fuel (Park et al., 2001).

8.2.1.6 Photosynthetic Microbial Fuel Cell (PMFC)

A novel architecture has been designed for photosynthetic microbial fuel cells, or PMFCs. With graphite carbon material acting as the anode and polishing gold mesh serving as the cathode, this device lacks a mediator and membrane. A thin biofilm of *Spirulina platensis* is formed on the anode by aggregating and depositing it there.

The basic configuration of the MFC system consists of three key parts: the separator, cathode, and anode (Osman et al., 2010). Proton exchange membranes (PEMs) have historically been employed as insulators to preserve the redox potential and permit the exchange of particular ions. Even though PEM has been used in numerous studies to lower MFC internal resistance, its use has been demonstrated to have disadvantages, such as high prices, scalability issues, and instability, which restrict the use of PEM in MFA applications (Liu and Logan, 2004). Furthermore, it has been proposed that chemical mediators such as red unsaturated nitric oxide or ferric oxide may be employed to hasten oxidation, ion penetration into the cell membranes, and chemical stability of MFCs. It has been demonstrated that these substances are not appropriate for long-term use. In wastewater treatment procedures connected to electricity production, a mediator-free and membrane-less MFC has been presented as a solution to the aforementioned issues.

The photosynthetic microbial fuel cell, or PMFC, is a type of bioelectrical device that uses the biological organisms' active catalytic activities to turn sunlight into electricity (Madiraju et al., 2012). It is typically applied to photosynthetic and cyanobacteria strains for this reason. The photosynthetic microalga *Spirulina platensis* uses chlorophyll (Chl) as a pigment for photosynthetic autotrophic growth. In general, chlorophyll absorbs light to release energy in the dark and convert it to glucose when there is light (Xie et al., 2011).

A range of electrolytes, including carbon fibers, carbon brushes, graphite from different species, Pt, Cu, and CuAu (Bullen et al., 2006), were employed to investigate the electromagnetic field performance in MFCs. Graphite felt electrodes were utilized by (Ringeisen et al., 2007), to generate electricity in the presence of a mediator and PEM. With S. platensis linked to an anodic electrode for direct electrical generation in the dark using a Pt–Pt electrode, Fu et al. (2010) have developed a mediator-less and membrane-less PMFC.

Although there have been numerous reports on the mechanisms of MFC for energy generation, little focus has been placed on biomass behaviour including growth degradation and in relation to PMFC's longer term operation.

Few studies have examined biomass behavior, including growth degradation and its relationship to PMFC's longer-term operation, despite the abundance of papers on the processes of MFC for energy generation. *S. platensis* was directly cultivated to reestablish PMFC under light conditions. Six sets of PMFCs were employed for repetitive energy generation in dark and then light circumstances for long-term operational reasons. In the absence of PEM and mediator, PMFC was utilized in a continuous process of producing electricity (Lin et al., 2013).

8.2.2 Anolyte and Catholyte Electrodes

It was discovered only a few years ago that electricity could be generated using bacteria in an MFC. The MFC is essentially anaerobic treatment because the bacteria do not have access to oxygen in an electrode chamber. The bacteria oxidize organic materials and transfer the electrons to an electrode (anode) in order to produce energy. Even if fermentation is occurring in this chamber, the electrons' transmission to an electrode merely produces electricity. In order for protons to sneak onto the counter electrodes and produce water, they must be able to combine with oxygen or electrons in order to maintain a power balance (Sharma and Li, 2010).

The base of MFC is created by the generation of current, which may be different between cathode and anode chambers. In general, the maximum potential is available. It is the equivalent of a 0.5 to 0.8 volts, similar to what's produced in hydrogen fuel cells.

It must be noted that the MFC is a kind of combination system, which differs in many respects from traditional aerobic or anoxic wastewater treatment systems. This is an anaerobic treatment technology at the microbial level. In order to generate power, the bacteria shall be produced in an anaerobic environment (Kim et al., 1999). This is because it's an anaerobic system when the oxygen gets into the cathode, but its use of oxygen isn't coupled to respiration by bacteria. No aeration is needed for wastewater when using an air cathode and this saves the cost of running aerobic processes. The system needs to be designed to minimize oxygen leakage into the system, particularly across the cathode (air-cathode system) and the proton exchange membrane (aqueous cathode system), so that there is a minimal amount of aerobic treatment occurring unintentionally. However, the bacteria in the anode chamber are rapidly absorbing oxygen and thus maintain low redox potential even when it leaks into the anode chamber (Bergel et al., 2005).

8.2.2.1 Two-Chambered MFC

The most basic kind of MFC has two chambers divided by a substance that transfers protons between the chambers. An impermeable ceramic plate or perhaps a salt bridge could serve as the substance that conducts protons. The material should, in theory, allow protons to diffuse across chambers but prevent oxygen, bacteria, or substrates from passing through them. PEMs developed for hydrogen fuel cells are the best suited to such a task, although they allow oxygen to be dispersed in an anode chamber. Any conducting, noncorrosive materials are capable of being made into the electrodes (Oh et al., 2004).

The anode is made up of ordinary carbon paper, charcoal cloth or graphite according to the system. A catalyst for the formation of water from protons, electrons and oxygen shall be present in the cathode which is usually Pt held on a carbon surface using an adsorbent. If the wire does not come into contact with bacteria, the electrodes may be connected by any type of wire (Logan and Rabaey, 2012). The Pt wire is the most suitable choice, but it's expensive, so many copper wires are used on all surfaces, which have been coated with a nonconductive epoxy. Copper wire can be expected to fail in the system even if it is coated in this way. The carbon electrodes may be used in the outer part of the chamber to prevent any wire from being inserted inside, so that a separate cable and clip can be installed on an electrode. Each electrode shall be placed in its own but adjacent chamber. The anode chamber contains the bacteria and is enclosed in a tight seal to avoid diffusion of oxygen. To prevent air from entering the chamber, headspace shall be flushed with nitrogen gas. Air bubbles appear in the water when the cathode is submerged in it (Min, Kim, et al., 2005).

8.2.2.2 Single Chambered MFC

Putting an electrode right into PEM instead of a cathode chamber could simplify and improve the efficiency of MFCs. In this approach, the necessity for aerated water is eliminated because the oxygen in the air may be transported straight to the cathode. There are several possible designs for this system. The cathode, which can be a large SCMFC, was positioned in the center of the cylinder such that the anode chamber formed a circular cylinder around the cathode. The cathode was connected to an external circuit that included a resistor and extended the graphite rods that were inserted in the anode chamber outside of it. Air has to be able to pass freely through the central tube in order to react at a cathode. To give support and interact with the solution inside the anode chamber, the Nafion membrane was heated to a cathode and then wrapped around one of the perforated plastic tubes (Liu and Logan, 2004).

The second version of SCMFC was a single tube that had two circular electrodes placed on it at the tube's opposing ends. One end of the anode must be secured to prevent oxygen from diffusing into the chamber, and the other end must be open to face the solution in the anode chamber and the air when the cathode is linked to PEM (Liu et al., 2004).

8.2.2.3 Effects of Anode in MFC

Bacteria produce electrons and have a significant role in the anode chamber. These produced electrons are used to reduce the electron acceptors in the cathode as they move through the circuit that is external. Similar to this, in order to complete the circuit created, protons must be pulled into the PEM from the anode to the cathode (Park et al., 2014).

One of the main parts of MFCs is an anaerobic anode compartment. The anode chamber shall provide all the necessary conditions for the degradation of biomass. Substrate, mediator, microbes, and the anode electrode—which serves as an electron acceptor—are all present in this compartment (Park et al., 2014).

In particular, it is important to note that a commensurate catalyst must reduce the activation energy needed for anodic reactions. The majority of the time, the anode chamber's resident bacteria are employed as catalysts. One of the most effective variables influencing the MFC's efficiency is anodic microbial electron transfer, which may be increased through a variety of applied techniques, including the addition of electron mediators and the modification of the electrode and cell design (Rahimnejad et al., 2012).

The optimal electrode material should possess the following qualities: adequate electrical conductivity, chemical stability, low resistance, anti-corrosion, high surface area, and mechanical toughness and stress. Graphite fiber brush, carbon cloth, graphite rod, carbon paper, reticulated vitreous carbon, and carbon felt are among the most commonly used carbon-based substances in anodes because of their large surface area, high electric conductivity, and stability in microbial cultures.

Moreover, activated granular carbons, also known as graphite granulates (GGs), have significant levels of microporosity and catalytic activity. With increased conductivity, GGs are also less expensive (Hassan et al., 2014).

In order to promote the performance of MFCs, it would be appropriate to alter the anode electrode. Changing the carbon and metal base anodes to conductor polymers is one efficient way of improving MFC output power. Attention needs to be paid to ensuring that the modified electrode stability is certified when using such organic polymers with microorganism of substrate (Logan et al., 2006).

8.2.2.4 Effects of Cathode in MFC

To complete the electrical circuit, a proton exchange membrane transports protons generated in the anode to the cathode. After being taken up into a cathode chamber, the electrons are transported to oxygen. This radical oxygen is supplied by positive ions into the anode through the following process, which forms water and is disseminated by the ion-permeable membrane on the cathode with the help of catalysts:

$$H_2 \rightarrow 2H^+ + 2e^-$$

$$O_2 + 4H^+ + 4e^- \rightarrow 2H_2O$$

The cathode's reaction yield is influenced by the oxidant's concentration and species, proton availability, catalyst performance, electrode structure, and catalytic ability. Catalysis is required in anodic and catalytic processes (Bettin, 2006). Furthermore, the presence of a suitable catalyst must be ensured because it is capable of reducing activation energy and improving reaction speed (Zhou et al., 2014).

Oxygen's high oxidation potential, accessibility, and lack of chemical waste product formation (water is the only byproduct) (Logan et al., 2006) have made it a common final electron acceptor in the cathode. One of the constraining agents in MFCs is the slow reduction kinetic of oxygen on plain graphite, which typically requires the use of a comparable catalyst and results in high potential. To lessen this issue, potassium ferricyanide K3[Fe(CN)6)] has been suggested. However, because oxygen is insufficient to fully oxidize K3[Fe(CN)6], regeneration may be problematic; as a result, it needs to be replenished regularly (Franks and Nevin, 2010).

Platinum, due to its cathodic action, is also used on a frequent basis as an antibacterial catalyst. The fact that platinum is not an adequate catalyst for MFCs because of its toxicity toward certain substances in the substrate solution should be considered (Park and Zeikus, 2003).

The objective of studying biocatalysts is to use oxygen oxidation on the cathode to avoid using a catalytic converter. Rather than using artificial mediators or catalysts like nitrate, sulfate, tetra chloroethene, fumarate, or perchlorate, a biocathode in which cathodic reactions are catalyzed using microorganisms can be adopted into MFCs to improve the cathode performance. This is because biocathodes have been used to improve electricity production in MFCs (Chang et al., 2005).

Compared to abiotic cathodes, biocathodes are more advantageous because they can be used to reduce the configuration and costs, thus eliminating the need for expensive catalysts, for example Pt. and mediators. Utilizing microbial metabolism to produce useful products or eliminate byproducts is another benefit of biocathodes (Lefebvre et al., 2009).

There are two primary categories of biocathodes: anaerobic and aerobic. The biocathode in aerobic biocathodes has the ability to decrease oxygen. Additionally, the biofilm on the cathode surface can catalyze the oxidation of transition metal compounds such as Fe(II) and Mn(II). In addition, a higher density of power can be achieved by the MFC using aerobic biocathodes as opposed to anaerobic biocathodes (Watanabe, 2008). Likewise, the excess effluent stream in a cathodic chamber is treated potentically with an MFC made up of biocathode. In addition, the accumulation of metabolites by microbes may prevent bacteria from performing their activities in the cathode chamber.

8.2.3 Significant Parameters Influencing Performance of MFC

The efficiency of fuel cells produced by bacteria can be influenced by several factors. These include the pace at which fuel oxidizes and the microorganisms' transmission of electrons to the electrode, the circuit's resistance, the proton transfer across the membrane to the cathode, the oxygen supply, and the cathode's reduction.

8.2.3.1 pH of Anode Compartment

Prior to recording potential, all of the anode content was replaced with wastewater that had been diluted with distilled water and had a COD value of 400 mg l^{-1} and a pH range of 5 to 9. The values were lower at pH 9 and below 6, and the maximum current was seen between pH 7 and 8. This demonstrates that microbial activity is slower in the absence of ideal pH than it is at optimal pH. The low results at pH 9.0 could be the consequence of a weak proton transfer in the lower concentration of protons throughout the membranes (Gil et al., 2003).

8.2.3.2 Resistance

Gil et al. (2003) reported that fuel cells functioned based on their own resistance between the anode and cathode in water. The largest current was determined to have the lowest resistance. He carried out laboratory tests that demonstrate how the lowest current was consistently maintained against resistance of more than 500 V for more than 30 hours. On the other hand, the current peaked right after the fuel delivery and dropped off quickly in two shoulders at low resistance values of 10 and 100 V.

The findings indicate that at more than 500 V, resistance acts as a rate-limiting step in the generation of current, while at lower resistance levels than 200 V, the cathode's rate of electron consumption is less than the transfer rate. This could be the result of a limited supply of oxygen or protons. His investigations further show that the coulomb yield increases with decreasing resistance. There is another mechanism consuming some electrons besides the cathode reaction, as indicated by the difference in coulomb yield.

The microbial fuel cell produced lower coulombs when it was run without any nitrogen gas entering the anode compartment as opposed to nitrogen gassing.

8.2.3.3 Effect of an Electrolyte

Microbial fuel cells were operated using water that had been distilled into various salt solutions to a COD of 100 parts per million. A 50 M pH 7.0 buffer containing 100 mN NaCl has been used to test the effects of phosphate buffers and NaCl (Gil et al., 2003). The same salt solution has been utilized as with the cathode electrolyte. A peak current resulted by using phosphate buffer from the experiment. NaCl is used as an electrolyte, and then phosphate buffer and purified water are added. Low current and coulomb were found in the fuel cells containing the NaCl solution.

According to the studies of Gil et al. (2003), To determine the changes of pH, a second set of experiments have been carried out which used 400 ppm municipal wastewater diluted with potable water as opposed to controls. Test was with 50 mM phosphate buffer, which contains 100 mM NaCl in a pH of 7.0. While the pH in the fuel cell without the buffer was noticeably higher than in the control experiment, the pH in the cathode and anode of his experiment was less than 0.5 after the fuel was introduced.

The pH in the cathode and anode compartments progressively rose to about 9.5 after the fuel delivery before falling to 5.4. These findings demonstrate that proton transport across the membrane proceeds more slowly than both the anode's production rate and the cathode compartments' consumption rate, and that the buffer can accelerate the slow proton transport rate.

Since it has been shown that microbial fuel cells function best at a pH of neutral, a buffer is required for both optimal microbial activity and to offset the slow passage of protons through membranes.

8.2.3.4 Aeration Rate of Cathode Compartment

The cathode compartment aeration rate in microbial fuel cells ranged from 10 to 200 ml min^{-1}. Nitrogen was pumped into the anode compartment at a rate equal to that of the aeration rate to the cathode compartment (Gil et al., 2003). The poor performance of the microbial fuel cell at high gassing rates may be caused by the shear stress that disrupts the immobilized bacteria on the anode, as suggested by Gil et al.

8.2.3.5 Dissolved Oxygen (DO)

Rather than aerating, air-saturated electrolyte was continuously fed into the cathode compartment to power the microbial cells and track the amount of dissolved oxygen. Fuel was produced from wastewater treated with starch that had 400 mg 1^{-1} COD levels. Current and DO concentrations were tested at different electrolyte flow rates (Gil et al., 2003).

The current dropped back to its original levels as soon as aeration ceased, while the decrease in DO was less pronounced. As soon as the aeration was terminated, high DO readings were caused by air bubbles in the oxygen electrode area. These results show that DO is a major limiting factor in the case of a microbial fuel cell, where the electrode material is graphite. Microbial fuel cells were operated in various wastewater concentrations to ascertain the correlation between fuel concentration and maximum currents. The greatest current increased with the wastewater concentration (Gil et al., 2003).

8.2.3.6 Effect of Conecntration on MFC Performance

Variations in substrate concentrations will affect the amount of organic matter accessible to bacteria for degradation and consequent electrical energy production. To maximize power production, it is consequently required to optimize the substrate concentration. The performance of the MFC at different substrate concentrations of microalgae powder will be investigated here (Gil et al., 2003).

Microbial fuel cells were operated in various wastewater concentrations to ascertain the correlation between fuel concentration and maximum currents. The greatest current increased with the wastewater concentration.

8.2.4 Various Membranes Used for MFC Applications

Bioelectricity in MFCs is produced by a bio-potential that develops between the electron-generating reduction reaction of the bacterial metabolic activity and the conditions of the electron acceptor, which are divided by a membrane. The cathode of a single-chamber MFC and the anodic biological processes are physically separated by a separator, one of the key components of MFCs (Kim et al., 2007).

Apart from the benefits of employing separators, there may be some adverse effects associated with their use. Firstly, a separator has been used for transfer of protons from an atom chamber and the permeation of substrates and oxygen is impeded by it. In the last decade there has been ongoing development of different separators to overcome these problems. To that end, it is vital to mention that substances from various separators fall into three categories based on the features that allow for filtration.

Ion exchange membranes can be broadly classified into two types: anion exchange membranes and cation exchange membranes, depending on the type of ionic groups bonded to the membrane substrate. Glass fibers, microfiltration membrane (MFM), salt bridge, bipolar membrane (BP2M), cation exchange membrane (CEM), anion exchange membrane (AEM), ultrafiltration membrane (UFM), porous textiles, and other coarse pore filter materials are typically found in these separators (Rahimnejad et al., 2012).

Microfiltration membranes are filters that are applied in a low pressure-driven membrane microfiltration technique which removes suspended particles in the size range, MFM ≥0.01 μm inpore size. MFM are becoming more popular for separating microorganisms, colloids, suspended solids, and macromolecule compounds from water. They can also be used as a substitute for the proton exchange membrane (PEM) in MFCs (Sun et al., 2009).

8.2.4.1 Cation Exchange Membrane (CEM)

MFCs' function as CEMs is one of the key elements affecting how well they perform. Put differently, they have to work as a whole in order for produced protons to be transmitted to a cathode chamber at MFCs. CEMs must also have the ability to stop the transfer of extra materials, such as substrates or oxygen, between the anode and cathode compartments (Min, Cheng, and Logan, 2005).

A common ion-penetrable separator, CEM requires the transfer of positive charges in order to be rendered available. Cation exchange membranes (e.g., $-PO_3^-$, $-COO^-$, $-C_6H_4O^-$) are essentially the groups with negative charges. These membranes bind to the membrane's backbone and permit ions to pass through them without being rejected. Consequently, they are commonly known as proton exchange membranes, or PEMs, and are widely employed as membrane separators in MFCs (Li et al., 2011). Ultrex, nafion, bipolar membranes, dialysing membranes, polystyrene divinylbenzene with hydrochloric acid group, glass wool nanoporous filters, and microfiltration membranes are just a few examples of the many materials that have been used to create CEMs in MFCs. It is widely acknowledged that Nafion, one of the most widely used CEMs in MFCs, was initially created by Dupont in 1970 (Grot, 1974). As a result, the fuel system and the chlor-alkali production business both made use of this membrane.

Nafion is one of the most widely used CEMs. It is a perfluorosulfonic acid membrane with a hydrophobic fluorocarbon backbone (–CF2-CF2–) and hydrophilic sulfonate groups ($SO3^-$) attached. Because of these negatively charged sulfonate groups, Nafion exhibits high conductivity to various cations in charged sulfonate groups. In addition, nafion does not react well with neutral pH values or with cation species like Na^+, K^+, or $NH4^+$, which raises their concentrations by 10^5 times that of H^+. The cathode chamber's pH rises as a result of this process because these species are more likely than protons to be transmitted over a membrane (Rahimnejad et al., 2012).

Other substitute cut-rate CEMs are Zirfon and Hyflon (Solvay-Solexis, Italy). Zirfon is an organ mineral substance that is macro porous and is composed of 15% polysulfone and 85% hydrophilic ZrO2 powder (Peighambardoust et al., 2010). Zirfon is less harmful to anodic processes than Nafion because of its higher oxygen permeability, but it has a much lower specific resistance.

8.2.4.2 Anion Exchange Membrane (CEM)

AEM which uses carbonate and phosphorus as a pH buffer (Zuo et al., 2008) with the aim of improving proton transfer. To use a buffer anion for AEM separator, the substrates' permeability is usually higher compared to CEM. AEM includes ions with positive charges (e.g. $-PR_3^+$, $-SR_2^+$, $-NH_3^+$, COO^-) that join the membrane and transmit anions through it (Peighambardoust et al., 2010). Another investigation demonstrated that when CEM was used as cathodes for membranes, the AEM membrane-based assembly had a much higher power.

8.2.4.3 Bipolar Membrane (BPM)

BPM, which consists of two monopolar membranes, CEM and AEM, is another alternative separator that has been employed in MFCs with proton and hydroxide ions. It was applied as a follow-up to treat water with a high salinity level. Principal anxiety generally lowers weight, particularly cost, and increases electrical conductivity in bipolar membrane investigations. Therefore, bipolar plate membranes have greatly benefited from the use of metallic materials like graphite and stainless steel (Zhuang et al., 2012).

Graphite is a widely used bipolar membrane plate because of its low density, high conductivity, and manufacturing. Stainless steel has the electrical conductivity and chemical stability of graphite, which makes it suitable for use as a secondary bipolar plate (Dihrab et al., 2009).

According to M. Rahimnejad et al., every bipolar plate supporting two adjacent cells in the stack version of MFC is thought to perform five typical functions: (i) separating individual cells in the stack; (ii) facilitating water management inside the cell; (iii) transporting current away from the cell; (iv) distributing fuel and fuel oxidants inside the cell; and (v) facilitating heat management.

8.3 CHALLENGES IN MFC BIOELECTRICITY GENERATION

Regardless the fact that a variety of electrochemical parameters, including power density, cell voltage, and biological parameters, like substrate loading rate in continuous systems, can be used to characterize MFCs, the following factors primarily affect MFC performance: (1) oxygen supply and consumption in the cathode chamber; (2) substrate oxidation in the anode chamber; (3) electron shuttle from the anode compartment to the anode surface; and (4) permeability of the proton exchange membrane (Rahimnejad et al., 2012).

MFC technology has advanced significantly in the last few decades. Nevertheless, there have been a number of obstacles to its practical application and scaling up, such as membrane resistance during the proton transport process and turbulence at every compartment. In addition to these, MFCs have had to contend with two issues related to bottlenecks in the electricity generation process. Although there is a great deal of overlap, there are direct correlations in each system between the substrate concentrations and the power generated by the MFCs. Additionally, power generation will not occur above and beyond the designated substrate concentration (Berk and Canfield, 1964).

MFCs have a restricted output, but a sizable amount of the energy produced in this kind of device is consumed by internal resistance, which is high. Proton exchange membrane, which divides the anode and cathode chambers, is found to be the primary cause of excessive internal resistance (R_{in}) in two MFCs. It has been proposed that new MFC designs, such as single chamber MFC (SCMFC), stacked MFC, and up flow MFC, can boost energy production while minimizing internal resistance by removing PEM.

The cathode's performance is thought to be the primary constraint. To make MFC scalable, designing the cathode is an extremely difficult task. However, the surface area of the cathode does not significantly affect the power production and can be made more efficient by employing high surface area materials or granular materials like graphite. However, a major problem for MFC setup is finding materials that maximize power output and columbic efficiency while also lowering expenditure.

Carbon paper, brush, fiber, and various forms of graphite are among the materials used in the cathode. Other materials include Pt, a commonly used catalyst, and alternative polymer binders like naphiron, which has also been tested; other materials include Cu, Cu–Au, tungsten carbide, granular graphite (which is said to be an excellent material), and reticulated vitreous carbon.

Different factors influence power generation's response to cathodic biofilm thickness; power generation was shown to decrease gradually as cathode biofilm thickness increased on both the electrode and the current collector. The reduction in biocathode start-up time has limited the usefulness of this technology. In relation to biocathodic microorganism, alterations in pH may alter the surface properties of cells with a biofilm which can include nucleation charge, hydrophobicity, cell morphology and size, as well as bacterial structures that could interrupt biological activity. However, it has proved to be one of the difficulties for sustainable microplastics with biocathode because they must operate at high rates and have little dependence on chemical use despite the fact that using a buffer system is required in order to control pH levels in bioanode or biocathode. Another important concern is the role played by carbon sources, as limited carbon availability has an impact on biocathode performance.

Furthermore, high biocathode MFC performance is hampered by molecular diffusion and electroosmosis because organic molecules converge with anodes and cathodes through membranes separating them.

Conversely, the cultivation of microalgae in open ponds or bioreactors, where they rely on photosynthesis to develop and proliferate, underscores the pressing necessity for novel MFC designs that can customize the microalgae's access to light, CO2, and nutrients. Since illumination directly affects the biomass yield of many algae—which is limited by thermodynamics—it is primarily credited with the foundation of these new structures. This is because illumination can achieve an optimal photodynamic efficiency by converting one CO2 molecule to carbohydrates with eight photons of photosynthesise active radiation, or PAR.

In terms of power generation, the biological catalyst cathodes have a slightly lower output than those made from chemical catalysts, which is considered to be an important challenge for microalgal cells in the short term.

8.4 UTILIZATION OF MFC IN POWER GENERATION AND WASTEWATER TREATMENT

In wastewater, which includes a wide variety of organic substrates, there is considerable potential as an energy source. Renewable energy forms from biodegradable organic matter are created in the different types of wastewaters, for example sanitation water, food waste, pigs' sewage and corn stover.

The MFC technology, which has been identified as a completely different method for wastewater treatment at the beginning of 1991, is positive because it can capture energy via electricity or hydrogen gas. Scientists have indicated that bioassay is widespread in use as a reliable and very cost-effective way of removing nitrogen and organic matter from leachate. Anaerobic digestion with a long time of detention, suitable for high strength methane wastewaters, is simultaneous anaerobic digestion and the production of electricity from waste materials.

The ability of MFC technology to produce power simultaneously as well as it can remove salt from wastewater has been demonstrated. The findings showed that both energy production and CE were reduced because of greater serenity concentration. In addition, it was demonstrated that MFC–based technology has been shown to increase the rate of removal of odor when used as an efficient method for electricity generation and odour elimination. Maximum power generation capacity of 228 mW/mm^2 is reached (Kim et al., 1999).

8.4.1 Electricity Generating from Wastewater

In practical terms, MFC application in wastewater treatment offers the most promise. The discovery that mediators—chemicals that transfer electrons—did not need to be incorporated into MFCs prompted the creation of MFCs for wastewater treatment, proving for the first time that regular home wastewater could be used to generate energy (Liu and Logan, 2004).

According to B.E. Logan et al., continuous treatment of household wastewater in the big SCMFC produced up to 26 mW/m^2. Not only did it generate electricity, but it also removed up to 80% of the Biochemical Oxygen Demand (BOD). Throughout the studied range, power generation was linear and proportionate to wastewater strength (COD). As a result, it would have been expected that this system's energy density would have grown with a more concentrated wastewater. It's been determined that for the best possible system performance, air can be passively transferred out of the cathode and does not need to be injected into it (Logan et al., 2006).

A novel approach to wastewater treatment is predicated on the production of electricity. To set this type of system apart from traditional water treatment methods, it is now called the bacterial POWER reactor producing wastewater energy. By injecting bacteria into wastewater systems, this moniker suggests that energy is produced. An analysis of the power generated in a city of 100,000 people might be conducted to determine the economic possibilities of this kind of wastewater treatment system for residential use.

In order to process wastewater, current wastewater treatment technologies require a lot of energy. MFCs can supply an energy source in the form of sewage pretreatment systems. Furthermore, the integration of wastewater with existing wastewater treatment systems can provide a significant amount of nutrients and increase the generation of biofuels.

One technique to gauge the effectiveness of wastewater treatment is to look at the decrease in COD. Anaerobic consortiums find it simple to reduce COD in the MFC's anionic chamber and break down organic substrates. A few MFC investigations have demonstrated removals of up to 90% of COD.

8.5 CONCLUSION

The energy crisis has led to a growing interest in alternative energies among researchers around the world as oil reserves are running out. In addition, emissions from the use of fossil fuels can have an impact on the environment. With new sources of energy that are free from pollution, clean fuels such as hydrogen fuel cells and biofuels would have been appropriate replacements for conventional fossil fuels. Different types of fuel cells are MFCs, which generate energy through the use of active biocatalysts like bacteria or enzymes. Among the most recent methods to produce power from a range of substrates are MFCs. The prospect of producing sustainable energy from a wide range of interchangeable materials, such as agricultural waste, has spurred research interest in this field in recent years. The energy produced is utilized to power the MFCs in a variety of ways. For MFCs to be commercially feasible, there is still not enough generating power, hence researchers are trying to increase this capacity.

Although still in its early stages of research, the microbial POWER technology holds significant potential for treating wastewater and producing power. The price of building and maintaining the system will be a key consideration for its utilization in the future. We are aware that a power plant's capital expenses will be roughly $1000.00 per kW. As we work on the cost to produce electricity, the costs of a combined wastewater treatment and energy production facility can therefore be compared with those required for separate facilities. It is obvious that the integration of energy generating and wastewater treatment technologies will provide an interesting and demanding engineering challenge.

REFERENCES

Akdeniz, F., Çağlar, A., & Güllü, D. (2002). Recent energy investigations on fossil and alternative nonfossil resources in Turkey. *Energy Conversion and Management*, 43(4), 575–589.

Bergel, A., Féron, D., & Mollica, A. (2005). Catalysis of oxygen reduction in PEM fuel cell by seawater biofilm. *Electrochemistry Communications*, 7(9), 900–904.

Berk, R. S., & Canfield, J. H. (1964). Bioelectrochemical energy conversion. *Applied Microbiology*, 12(1), 10–12.

Bettin, C. 2006. Applicability and feasibility of incorporating microbial fuel cell technology into implantable biomedical devices (Doctoral dissertation, The Ohio State University).

Bullen, R. A., Arnot, T. C., Lakeman, J. B., & Walsh, F. C. (2006). Biofuel cells and their development. *Biosensors and Bioelectronics*, 21(11), 2015–2045.

Chang, I. S., Moon, H., Jang, J. K., & Kim, B. H. (2005). Improvement of a microbial fuel cell performance as a BOD sensor using respiratory inhibitors. *Biosensors and Bioelectronics*, 20(9), 1856–1859.

Chen, F. (1996). High cell density culture of microalgae in heterotrophic growth. *Trends in Biotechnology*, 14(11), 421–426.

Cui, Y., Rashid, N., Hu, N., Rehman, M. S. U., & Han, J. I. (2014). Electricity generation and microalgae cultivation in microbial fuel cell using microalgae-enriched anode and bio-cathode. *Energy Conversion and Management*, 79, 674–680.

De Schamphelaire, L., Rabaey, K., Boeckx, P., Boon, N., & Verstraete, W. (2008). Outlook for benefits of sediment microbial fuel cells with two bio-electrodes. *Microbial Biotechnology*, 1(6), 446–462.

del Campo, A. G., Cañizares, P., Rodrigo, M. A., Fernández, F. J., & Lobato, J. (2013). Microbial fuel cell with an algae-assisted cathode: A preliminary assessment. *Journal of Power Sources*, 242, 638–645.

Delaney, G. M., Bennetto, H. P., Mason, J. R., Roller, S. D., Stirling, J. L., & Thurston, C. F. (1984). Electron-transfer coupling in microbial fuel cells. 2. performance of fuel cells containing selected microorganism—mediator—substrate combinations. *Journal of Chemical Technology and Biotechnology*, 34(1), 13–27.

Demirbas, M. F., Balat, M., & Balat, H. (2011). Biowastes-to-biofuels. *Energy Conversion and Management*, 52(4), 1815–1828.

Dihrab, S. S., Sopian, K., Alghoul, M. A., & Sulaiman, M. Y. (2009). Review of the membrane and bipolar plates materials for conventional and unitized regenerative fuel cells. *Renewable and Sustainable Energy Reviews*, 13(6–7), 1663–1668.

ElMekawy, A., Diels, L., De Wever, H., & Pant, D. (2013). Valorization of cereal based biorefinery byproducts: Reality and expectations. *Environmental Science & Technology*, 47(16), 9014–9027.

ElMekawy, A., Hegab, H. M., Vanbroekhoven, K., & Pant, D. (2014). Techno-productive potential of photosynthetic microbial fuel cells through different configurations. *Renewable and Sustainable Energy Reviews*, 39, 617–627.

Franks, A. E., & Nevin, K. P. (2010). Microbial fuel cells, a current review. *Energies*, 3(5), 899–919.

Fu, C. C., Hung, T. C., Wu, W. T., Wen, T. C., & Su, C. H. (2010). Current and voltage responses in instant photosynthetic microbial cells with Spirulina platensis. *Biochemical Engineering Journal*, 52(2–3), 175–180.

Gadhamshetty, V., Belanger, D., Gardiner, C. J., Cummings, A., & Hynes, A. (2013). Evaluation of Laminaria-based microbial fuel cells (LbMs) for electricity production. *Bioresource Technology*, 127, 378–385.

Gil, G. C., Chang, I. S., Kim, B. H., Kim, M., Jang, J. K., Park, H. S., & Kim, H. J. (2003). Operational parameters affecting the performannce of a mediator-less microbial fuel cell. *Biosensors and Bioelectronics*, 18(4), 327–334.

Grot, W. 1974. *Laminates of Support Material and Fluorinated Polymer Containing Pendant Side Chains Containing Sulfonyl Groups*. Google Patents. Washington, DC: U.S. Patent and Trademark Office.

Hassan, S. H., El-Rab, S. M. G., Rahimnejad, M., Ghasemi, M., Joo, J. H., Sik-Ok, Y., Kim, I. S. & Oh, S. E. (2014). Electricity generation from rice straw using a microbial fuel cell. *International Journal of Hydrogen Energy*, 39(17), 9490–9496.

Ho, T. Y., Taylor, G. T., Astor, Y., Varela, R., Müller-Karger, F., & Scranton, M. I. (2004). Vertical and temporal variability of redox zonation in the water column of the Cariaco Basin: implications for organic carbon oxidation pathways. *Marine Chemistry*, 86(1–2), 89–104.

Jančula, D., & Maršálek, B. (2012). The toxicity of phthalocyanines to the aquatic plant Lemna minor (duckweed)–Testing of 31 compounds. *Chemosphere*, 88(8), 962–965.

Juang, D. F., Lee, C. H., & Hsueh, S. C. (2012). Comparison of electrogenic capabilities of microbial fuel cell with different light power on algae grown cathode. *Bioresource Technology*, 123, 23–29.

Kim, B. H., Kim, H. J., Hyun, M. S., & Park, D. H. (1999). Direct electrode reaction of Fe (III)-reducing bacterium, Shewanella putrefaciens. *Journal of Microbiology and Biotechnology*, 9(2), 127–131.

Kim, H. J., Park, H. S., Hyun, M. S., Chang, I. S., Kim, M., & Kim, B. H. (2002). A mediator-less microbial fuel cell using a metal reducing bacterium, Shewanella putrefaciens. *Enzyme and Microbial Technology*, 30(2), 145–152.

Kim, J. R., Cheng, S., Oh, S. E., & Logan, B. E. (2007). Power generation using different cation, anion, and ultrafiltration membranes in microbial fuel cells. *Environmental Science & Technology*, 41(3), 1004–1009.

Lal, R. (2008). Sequestration of atmospheric CO2 in global carbon pools. *Energy & Environmental Science*, 1(1), 86–100.

Lefebvre, O., Ooi, W. K., Tang, Z., Abdullah-Al-Mamun, M., Chua, D. H., & Ng, H. Y. (2009). Optimization of a Pt-free cathode suitable for practical applications of microbial fuel cells. *Bioresource Technology*, 100(20), 4907–4910.

Li, W. W., Sheng, G. P., Liu, X. W., & Yu, H. Q. (2011). Recent advances in the separators for microbial fuel cells. *Bioresource Technology*, 102(1), 244–252.

Lin, C. C., Wei, C. H., Chen, C. I., Shieh, C. J., & Liu, Y. C. (2013). Characteristics of the photosynthesis microbial fuel cell with a Spirulina platensis biofilm. *Bioresource Technology*, 135, 640–643.

Lithgow, A. M., Romero, L., Sanchez, I. C., Souto, F. A., & Vega, C. A. (1986). Interception of the electron-transport chain in bacteria with hydrophilic redox mediators. I: Selective improvement of the performance of biofuel cells with 2, 6-disulphonated thionine as mediator. *Journal of Chemical Research*, (5), 178–179.

Liu, H., & Logan, B. E. (2004). Electricity generation using an air-cathode single chamber microbial fuel cell in the presence and absence of a proton exchange membrane. *Environmental Science & Technology*, 38(14), 4040–4046.

Liu, H., Ramnarayanan, R., & Logan, B. E. (2004). Production of electricity during wastewater treatment using a single chamber microbial fuel cell. *Environmental Science & Technology*, 38(7), 2281–2285.

Logan, B. E. (2005). Simultaneous wastewater treatment and biological electricity generation. *Water Science and Technology*, 52(1–2), 31–37.

Logan, B. E., Hamelers, B., Rozendal, R., Schröder, U., Keller, J., Freguia, S., Aelterman, P., Verstraete, W., & Rabaey, K. (2006). Microbial fuel cells: methodology and technology. *Environmental Science & Technology*, 40(17), 5181–5192.

Logan, B. E., & Rabaey, K. (2012). Conversion of wastes into bioelectricity and chemicals by using microbial electrochemical technologies. *Science*, 337(6095), 686–690.

Madiraju, K. S., Lyew, D., Kok, R., & Raghavan, V. (2012). Carbon neutral electricity production by Synechocystis sp. PCC6803 in a microbial fuel cell. *Bioresource Technology*, 110, 214–218.

Min, B., Cheng, S., & Logan, B. E. (2005). Electricity generation using membrane and salt bridge microbial fuel cells. *Water Research*, 39(9), 1675–1686.

Min, B., Kim, J., Oh, S., Regan, J. M., & Logan, B. E. (2005). Electricity generation from swine wastewater using microbial fuel cells. *Water Research*, 39(20), 4961–4968.

Moqsud, M. A., Omine, K., Yasufuku, N., Hyodo, M., & Nakata, Y. (2013). Microbial fuel cell (MFC) for bioelectricity generation from organic wastes. *Waste Management*, 33(11), 2465–2469.

Oh, S., Min, B., & Logan, B. E. (2004). Cathode performance as a factor in electricity generation in microbial fuel cells. *Environmental Science & Technology*, 38(18), 4900–4904.

Osman, M. H., Shah, A. A., & Walsh, F. C. (2010). Recent progress and continuing challenges in bio-fuel cells. Part II: Microbial. *Biosensors and Bioelectronics*, 26(3), 953–963.

Park, D. H., & Zeikus, J. G. (2003). Improved fuel cell and electrode designs for producing electricity from microbial degradation. *Biotechnology and Bioengineering*, 81(3), 348–355.

Park, H. S., Kim, B. H., Kim, H. S., Kim, H. J., Kim, G. T., Kim, M., Chang, I. S., Park, Y. K., & Chang, H. I. (2001). A novel electrochemically active and Fe (III)-reducing bacterium phylogenetically related to Clostridium butyricum isolated from a microbial fuel cell. *Anaerobe*, 7(6), 297–306.

Park, I. H., Christy, M., Kim, P., & Nahm, K. S. (2014). Enhanced electrical contact of microbes using Fe3O4/CNT nanocomposite anode in mediator-less microbial fuel cell. *Biosensors and Bioelectronics*, 58, 75–80.

Peighambardoust, S. J., Rowshanzamir, S., & Amjadi, M. (2010). Review of the proton exchange membranes for fuel cell applications. *International Journal of Hydrogen Energy*, 35(17), 9349–9384.

Rahimnejad, M., Ghoreyshi, A. A., Najafpour, G., & Jafary, T. (2011). Power generation from organic substrate in batch and continuous flow microbial fuel cell operations. *Applied Energy*, 88(11), 3999–4004.

Rahimnejad, M., Ghoreyshi, A. A., Najafpour, G. D., Younesi, H., & Shakeri, M. (2012). A novel microbial fuel cell stack for continuous production of clean energy. *International Journal of Hydrogen Energy*, 37(7), 5992–6000.

Rashid, N., Rehman, M. S. U., & Han, J. I. (2013). Recycling and reuse of spent microalgal biomass for sustainable biofuels. *Biochemical Engineering Journal*, 75, 101–107.

Rashid, N., Song, W., Park, J., Jin, H. F., & Lee, K. (2009). Characteristics of hydrogen production by immobilized cyanobacterium Microcystis aeruginosa through cycles of photosynthesis and anaerobic incubation. *Journal of Industrial and Engineering Chemistry*, 15(4), 498–503.

Reimers, C. E., Tender, L. M., Fertig, S., & Wang, W. (2001). Harvesting energy from the marine sediment–water interface. *Environmental Science & Technology*, 35(1), 192–195.

Ringeisen, B. R., Ray, R., & Little, B. (2007). A miniature microbial fuel cell operating with an aerobic anode chamber. *Journal of Power Sources*, 165(2), 591–597.

Rosenbaum, M., Schröder, U., & Scholz, F. (2005). In situ electrooxidation of photobiological hydrogen in a photobioelectrochemical fuel cell based on Rhodobacter sphaeroides. *Environmental Science & Technology*, 39(16), 6328–6333.

Ryckelynck, N., Stecher, H. A., & Reimers, C. E. (2005). Understanding the anodic mechanism of a seafloor fuel cell: interactions between geochemistry and microbial activity. *Biogeochemistry*, 76, 113–139.

Saba, B., Christy, A. D., Yu, Z., & Co, A. C. (2017). Sustainable power generation from bacterio-algal microbial fuel cells (MFCs): An overview. *Renewable and Sustainable Energy Reviews*, 73, 75–84.

Shantaram, A., Beyenal, H., Veluchamy, R. R. A., & Lewandowski, Z. (2005). Wireless sensors powered by microbial fuel cells. *Environmental Science & Technology*, 39(13), 5037–5042.

Sharma, Y., & Li, B. (2010). The variation of power generation with organic substrates in single-chamber microbial fuel cells (SCMFCs). *Bioresource Technology*, 101(6), 1844–1850.

Sun, J., Hu, Y., Bi, Z., & Cao, Y. (2009). Improved performance of air-cathode single-chamber microbial fuel cell for wastewater treatment using microfiltration membranes and multiple sludge inoculation. *Journal of Power Sources*, 187(2), 471–479.

Tender, L. M., Gray, S. A., Groveman, E., Lowy, D. A., Kauffman, P., Melhado, J., Tyce, R. C., Flynn, D., Petrecca, R., & Dobarro, J. (2008). The first demonstration of a microbial fuel cell as a viable power supply: powering a meteorological buoy. *Journal of Power Sources*, 179(2), 571–575.

Watanabe, K. (2008). Recent developments in microbial fuel cell technologies for sustainable bioenergy. *Journal of Bioscience and Bioengineering*, 106(6), 528–536.

Whitfield, M. (1972). The electrochemical characteristics of natural redox cells. *Limnology and Oceanography*, 17(3), 383–393.

Xie, X. H., Li, E. L., & Kang Tang, Z. (2011). Mediator toxicity and dual effect of glucose on the lifespan for current generation by Cyanobacterium Synechocystis PCC 6714 based photoelectrochemical cells. *Journal of Chemical Technology & Biotechnology*, 86(1), 109–114.

Zhang, Z., Wang, F., Wang, X., Liu, X., Hou, Y., & Zhang, Q. (2010). Extraction of the polysaccharides from five algae and their potential antioxidant activity in vitro. *Carbohydrate Polymers*, 82(1), 118–121.

Zhou, M., Yang, J., Wang, H., Jin, T., Hassett, D. J., & Gu, T. (2014). Bioelectrochemistry of microbial fuel cells and their potential applications in bioenergy. *Bioenergy Research: Advances and Applications*, 131–152.

Zhuang, L., Zheng, Y., Zhou, S., Yuan, Y., Yuan, H., & Chen, Y. (2012). Scalable microbial fuel cell (MFC) stack for continuous real wastewater treatment. *Bioresource Technology*, 106, 82–88.

Zuo, Y., Cheng, S., & Logan, B. E. (2008). Ion exchange membrane cathodes for scalable microbial fuel cells. *Environmental Science & Technology*, 42(18), 6967–6972.

9 Insights into Omics Approaches in Plant Phytoremediation of Heavy Metals

Yuh Nien Chow and Keng Yuen Foo

9.1 INTRODUCTION

Today, the biological, physical, and molecular interactions between plants and ecosystems have attracted an aesthetic concern among the scientific community. Amongst the abiotic stressors, heavy metals (HMs), mainly in the form of transition metals, lanthanides, actinides, and metalloids, have persisted as severe environmental problems at a global scale. With an atomic mass of more than 20 (except the alkali metals), and the atomic density above 5g/cm^3, HM ions well recognized as having a long half-life exceeding two decades, high persistency in nature, non-biodegradable, toxic, have prolonged tenacity in the environment, high water solubility, and bioaccumulative property in the aquatic life. In the soil ecosystem, HMs will be accumulated in the living tissues or cells via the contaminated underground water, either from the leaching, physical processes of weathering, or chemical redox reactions.

Naturally, HMs are found in rocks, the weathering of which will transform them into part of the soil media. Anthropogenic activities, particularly combustion or disposal of coal or waste tyres, agrochemicals application, petrochemical spillage, mining, industrial emissions, atmospheric deposition, and emission of sewage sludge and animal manure are the major sources of these metallic elements in the environment. This is further exacerbated by excessive industrialization, rapid urbanization, growing population, and the practice of unregulated wastewater irrigation, which have deteriorated the water streams or watercourses, in which these HMs would eventually find their way to the food chain, turning them into intractable pollutants and jeopardizing human health.

The remediation technologies of the HMs–polluted environment encompass biological, chemical, and physical strategies. Physico-chemical techniques, particularly encapsulation, electro kinetics, solidification, soil washing, excavation, soil flushing, and reburial of contaminated matrices, are generally costly, inefficient for low concentration sites, and destructive to the arable land properties. The biological remediation approach promotes plant growth and restoration of the contaminated areas. Likewise, bioremediation is a highly feasible remediation method as compared to other treatment processes.

Phytoremediation, or "eco-agriculture", is renowned as a green viable solution for environmental clean-up. It is referred to the remediation of hazardous pollutants in the environment via plants and microorganisms in soils, mainly through absorption, detoxification, degradation, bioaccumulation, and volatilization of HMs. The unique ability of aquatic hydrophytes, specifically *E. crassipes*, *P. australis*, *L. stolonifera*, and *E. stagnina* for lead (Pb), nickel (Ni), and cadmium (Cd) removal from polluted wetlands have been reported (Eid et al., 2020). Meanwhile, 721 plant species from 52 plant families have been identified as hyperaccumulators, which can accumulate exceptionally high concentrations of HMs in the plant tissues. As compared to conventional techniques,

DOI: 10.1201/9781003668961-9

phytoremediation is cost-effective and non-invasive to immediate matrices as the ultimate goal for restoration of the natural environment. However, the plants selection, growing conditions, and in-depth understanding of the hereditary and molecular mechanism of phytoremediation hold the utmost significance in a sustainable manner.

With the new emergence of omics and bio-informatic approaches, the plant-soil microbial communities-environment interconnections have been comprehensively recorded. Most of the available literature adopted the strategies of genomics, metabolomics, transcriptomics, ionomics, and proteomics of plants under the exposure of HMs–induced stress (Jamla et al., 2021). Transcriptomic alterations have been discovered in the roots of the zinc (Zn), Cu, Pb, and copper (Cu)–treated maize plant and barley, whilst metabolome alterations were reported in paddy driven by Cd- and Cu-induced toxicity (Navarro-Reig et al., 2015). A wide range of epigenetic regulators have been detected in plants exposed to HMs. HMs–responsive microRNAs (miRNAs) were discovered in paddy (Huang et al., 2009), wheat (Qiu et al., 2016), maize (Gao et al., 2019), and grapevine (Jiu et al., 2019). Interestingly, the discoveries in omics have further expanded the exploration of other HMs–phytoremediation strategies and development of HMs–tolerant plants. In this context, the fundamental understanding and mechanisms of phytoremediation are comprehensively documented. This chapter also presents the current trends in the state-of-the-art of omics with respect to the genomics, proteomics, transcriptomics, metabolomics, ionomics, miRNAomics, phenomics, and metallomics for phytoremediation research. The challenges, current research gaps, and future prospects are outlined to enable a sustainable plant-based solution for the mitigation of HMs contamination.

9.2 PHYTOREMEDIATION

9.2.1 The Fundamental Understanding of Phytoremediation

Phytoremediation, also known as "botanical bioremediation" or green remediation, is an environmentally friendly technique that adopts the approaches of phytoextraction, phytostabilization, rhizofiltration, phytodegradation, and phytovolatilization to remove, transfer, degrade, reduce, stabilize, and immobilize the pollutants at the polluted areas, rendering them harmless, for the restoration of the natural conditions. This scientifically proven strategy is cost-effective, species-specific, and environment friendly. Within the systems, plants own wide arrays of defence machinery to overcome the contamination-induced stress, particularly inorganic HMs, organic pollutants, and anthropogenic chemicals, notably presented as pesticides, fungicides, and herbicides. HMs are needed in a small fraction to support the biological processes in plants. At the levels exceeding the threshold limit, the impairment of plants' physiological processes, particularly reactive oxygen species (ROS) production, cellular development, reduction of photosynthetic pigments, respiration disturbance, and withering would take place. The core determinant factor for the successful phytoremediation process lies in the equilibrium of the plant's capability to accumulate and transform the toxic HMs into their non-toxic forms, and maintenance of the plant's defence system against external stresses.

Within the inorganic or organic pollutants-contaminated sites, plant roots would absorb the toxic elements, and uptake of essential nutrients, water, and pollutants could be maximized by a larger roots surface area. This triggers the plant roots to develop a defence mechanism against the HM–induced phytotoxicity, resulting in the release of organic and inorganic exudates, with a series of alterations over the soil–root interface. The changing composition and root exudates may significantly affect the availability, distribution, and soil concentrations of nutrients for plant growth, microbial population in soil, and the restoration of nutrients in nature. Specifically, the root exudates of alfalfa consisted of luteolin and flavonoids that could improve the rhizosphere's iron (Fe) bioavailability. Likewise, organic acids, namely oxalate, malate, and citrate release have been shown to enhance the bioavailability and solubility of plants' nutrients, notably aluminium (Al), phosphorus (P), and Fe. Furthermore, roots exudates may interrupt the movement of nutrients at the soil-root interface.

In rhizospheres (i) micro-organisms, with respect to the number and level of activity, (ii) soil, in relation to the particle aggregation, soil composition, and organic compounds, and (iii) availability of contaminants, are influenced by root exudates, to react as the major governing factor for plant availability. The specific soil microbial presence plays a vital role for the phytoremediation of HMs via stress regulation and tolerance, improvement of nutritional balance, and biomass management via the enhanced phytohormone synthesis, greater uptake of Fe, better minerals availability, exposure to a nitrogen (N)-stable atmospheric environment, and reduction of pathogenic attacks.

9.2.2 Unique Strategies of Phytoremediation

9.2.2.1 Phytoextraction

Phytoextraction of HMs refers to plants' HMs uptake action from the contaminated matrices, mobilization of HMs through the aerial parts, and compartmentalization of these ions in the harvestable above-ground parts. These plants will be eventually cut after the desired height is reached, followed by burning or compost addition, of which this cycle is repeatable to complete the remediation process. By absorbing, translocating, and accumulating HMs in stem and other plant compartments, this remediation, alternatively named as phytosequestration, phytoabsorption, or phytoaccumulation, is the most effective removal strategy of HMs from the growing medium as it enables a permanent evacuation of HMs from the environment. However, the efficacy of this mechanism is highly driven by the structural characteristics of the root, biomass of the stem, and type or level of HMs.

A hyperaccumulator exhibits the capability of metals accumulation in above-ground plant parts at exceptionally high concentrations with no sign of phytotoxicity. An example of hyperaccumulator is *Thlaspi caerulescens*, which can accumulate Cd to a concentration of more than one thousand times the threshold limit in stems. Meanwhile, the uptake of Cd by the Brassicae family has been mainly governed by the extensively expressed metal transporters. However, most of the hyperaccumulator species are biennial, short-lived perennial shrubs, herbs, or small-sized trees, which may be restricted by the low growth rate and biomass yield in limiting the phytoextraction efficacy.

Salix spp. and *Populus* spp. have been reported to show high accumulation of Cd and Zn in leaves. *Quercus robur*, *Populus 'Robusta'*, *Acer pseudoplatanus*, and *Fraxinus excelsior* are capable of redistributing HMs in soil, affected by two major processes: (i) the foliar accumulation of HMs, which would contribute to higher HMs level in the upper layer of soil; and (ii) soil acidification (species-specific), which could enhance metal leaching from the upper layer of soil. The efficacy of the phytoextraction process is based on the changing time constraint, and monitoring of removal rate, and any incidental inputs of external pollutants into the systems.

9.2.2.2 Phytodegradation

Phytodegradation, or phytotransformation, is a remediation technique in which the pollutants from the environment are directly degraded or metabolically transformed into inactive and less toxic forms by the enzymes either with or without the assistance of rhizosphere bacteria, metabolic processes, or roots exudates. The pollutants would be broken down internally or externally into plant nutrients. In comparison to other phytoremediation methods, phytodegradation is labor-intensive with high dependency on the soil amendments and exhibits low reliability and efficiency against inorganic contaminants.

9.2.2.3 Rhizofiltration

Rhizofiltration is a metabolic process concerned with the rhizome (root) in comparison to the seedlings (blastofilteration) or stem (caulifilteration), in which HMs are uptaken into the root cells via attachment over the roots surface through absorption process by adopting macrophytes, macroalgae, and microalgae. This technique is highly suitable for the treatment of HMs and chemical-contaminated aqueous streams. Plants with hairy, long roots and great HMs uptake capability are chosen to support the rhizofiltration process. They are initially grown in clean water with hydroponic cultivation system to maximize the expansion of roots before substitution of clean water with polluted soil or water for acclimatization, subsequent to be transferred to the contaminated on-site remediation of water pollutants. These plants will be eventually discarded after a predetermined treatment duration.

This remediation process is mediated by three governing mechanisms: (i) sorption onto plant roots' surface to induce a fast HMs removal via adsorption, ion exchange, and chelation processes, without the involvement of biological activity; (ii) plant metabolism via intracellular uptake, deposition in vacuole, and translocation to the stems, which involves low HMs removal (Salt et al., 1995); and (iii) the slowest removal process that occurs via root exudates to control HMs precipitation into insoluble compounds. Phenolics and organic acids secreted from root exudates may induce the precipitation of pollutants, resulting in the alterations of absorption, speciation, transport, and bioavailability of HMs and medium acidification.

The assessment of *Kyllinga nemoralis* and *Phragmites australis* in a rhizofiltration device at a sewage treatment plant recorded an encouraging efficacy for HMs removal from the municipal wastewater stream (Odinga et al., 2019). In a similar manner, *Hydrilla verticillata* (L.f.) *Royle* and *Lemna gibba* L. have been identified to be excellent bioindicators, while macrophages, notably *H. verticillate*, *L. gibba*, and *C. demersum*, and microphages, specifically *Wolffia globosa* (Roxb.) Hartog and Plas, *Eichhornia crassipes* (Mart.) Solms. and *Lemna minor* (L.), were scientifically reported to be a green remediation technique for HMs–polluted sites (Upadhyay et al., 2017).

9.2.2.4 Phytostabilization

Phytostabilization is an environment-friendly process that prevents contaminants leaching by root zone immobilization to reduce the bioavailability and bioaccumulation of toxic pollutants in the food chain. The plants may either induce HMs precipitation, valence reduction, sequestration in root tissue or adsorption into the root cell wall. Concurrently, the plant systems could preserve soil health over the HMs–polluted areas, and the establishment of vegetative coverage would contribute further to the minimization of leaching or dispersion of HMs into waterways.

Phytostabilizer plants, notably *Gentiana pennelliana* Fernald, *Vossia cuspidata* Roxb., *Zygophyllum fabago* L., and *Festuca rubra* L. (Griff.) were found to stabilize Cu, chromium (Cr), Zn, and Pb in the contaminated soils. Specifically, *H. petiolaris* has been reported to accumulate more than triple the quantity of Cd in the above-ground parts, and translocate high concentrations of Pb to the above-ground tissues. Unlike rhizofilteration or phytoextraction, disposal of plants is not mandatory under this technique, but selection of highly tolerant plants with good root systems for immobilization, soil structure stabilization, and soil erosion prevention is largely necessary. However, this remediation strategy has been limited for the temporary metal immobilization in the rhizosphere, as the outcome implies short-term immobilization of toxic HMs.

9.2.2.5 Phytovolatilization

During phytovolatilization, plants would absorb the pollutants and convert them into less hazardous forms by complex metabolic pathways. These compounds could be released into the atmosphere through foliage during transpiration. This strategy is less popular as compared to phytoextraction and phytofiltration. Some plants of the Brassicaceae family have been proven to be excellent candidates for volatilization of HMs, specifically silver (Hg), arsenic (As), and selenium (Se). During the process, HMs would undergo assimilation and complexation. For an example, Se is assimilated into selenoamino acids (selenocysteine and selemethionine) before further conversion to dimethylselenide (DMSe), a less toxic volatile compound, by *Brassica juncea* via biomethylation to be dispersed into the atmosphere. Hg, a highly reactive and volatile liquid at room temperature, exists mainly as HG_2C. Phytovolatilizer plants may convert Hg into ionic Hg after the absorption from the root system, before it could be released to the atmosphere as less toxic volatile compound, methyl-Hg. Comparatively, phytovolatilization is highly suitable for metalloid pollutants due to absence of bioaccumulative effect and plant disposal stage. However, HMs may be transferred to another system in this process rather than to be fully removed. In this sense, risk clarification is mandatory before the *in situ* application of the strategy. Figure 9.1 illustrates the unique strategies of phytoremediation, together with its respective features, advantages, and limitations as summarized in Table 9.1.

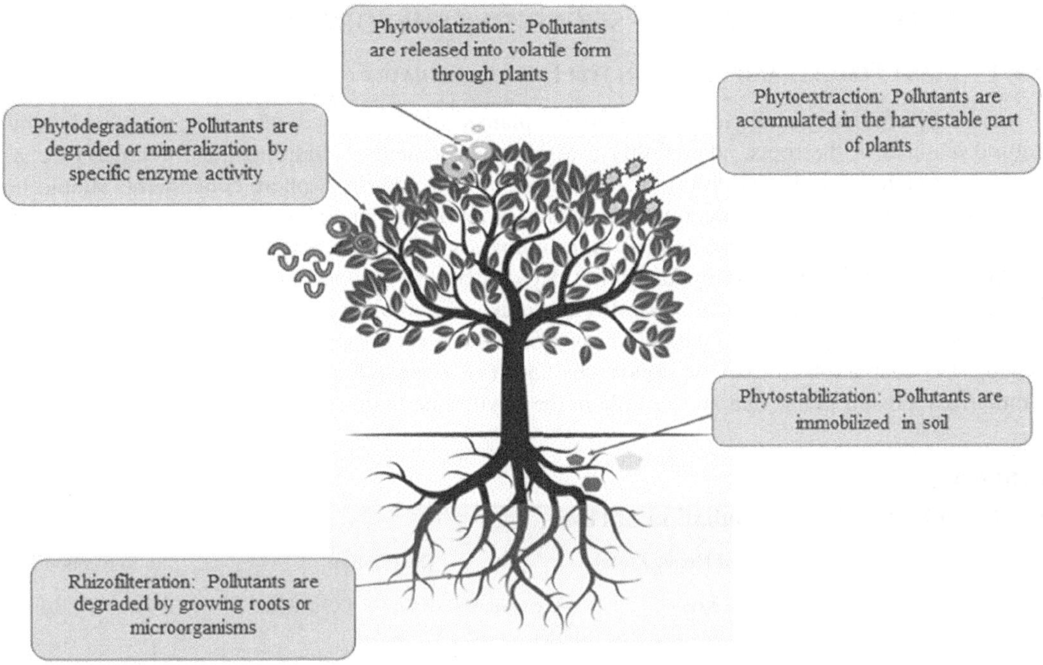

FIGURE 9.1 Unique strategies of phytoremediation.

(Adopted from Yadav et al., 2023)

TABLE 9.1
Features, Advantages, and Limitations of Different Phytoremediation strategies

Phytoremediation Strategies	Features	Application	Advantages	Limitations
Phytoextraction	Accumulation of HMs in aerial parts	Contaminated sites	Enables a permanent evacuation of HMs from the environment	Proper disposal method is required, efficacy depends on the structural features of root, biomass productivity, growth rate, and concentration or type of HMs
Phytodegradation	Degradation or metabolization of HMs into less toxic forms	Contaminated sites	Transform HMs into less toxic forms	Labor-intensive, dependent on soil amendments, and low reliability and efficiency for inorganic contaminants
Rhizofiltration	Rhizosphere-assisted accumulation of HMs	Wastewater, groundwater, wetland	Highly suitable for the remediation of aqueous streams	pH, chemical speciation, and their interactions have to be understood and well monitored
Phytostabilization	Immobilization of HMs onto the roots cell wall	Mining sites	Low cost, less disruptive, ecological restoration	Requires extensive fertilization or soil amendments and long-term maintenance to prevent leaching effect
Phytovolatilization	Volatilization of HMs via leaves	Volatile HMs	Highly suitable for metalloid contaminants, transform HMs into less toxic forms	HMs cannot be removed completely, transfer of HMs into another system might accumulate in vegetation or fruits

9.3 GOVERNING FACTORS OF SUCCESSIVE PHYTOREMEDIATION

9.3.1 Plant Features and Selection for Phytoremediation

To ensure the efficiency of phytoremediation, the plant selection, growth rate, anatomical and architectural features of the roots, adaptability towards environmental conditions, and defence mechanism, the toxicity and bioavailability of HMs, type of soil, and rhizosphere constituents should be taken into consideration. The ideal characteristics of a plant for the phytoremediation process are: (i) high HMs accretion capacity, and (ii) high biomass yield with greater accumulative capability. The plants selection for phytoremediation is decided by the (i) exceptional tolerance against the degree of contamination; (ii) great capability of uptake, accumulation, and translocation of the contaminants; (iii) rapid growth rate and biomass production; (iv) resistance against drought and water logging; (v) adaptation to extreme conditions, and (vi) compositions of the root zone. Table 9.2 summarizes a list of plants that are feasible in the phytoremediation process.

TABLE 9.2
Plants Applied in Phytoremediation of HMs

Plants	Targeted Heavy Metals	Description	Reference
Phyllanthus amarus, Althernanthera ficoides, Cyanodon dactylon	Arsenic (As)	Accumulation of Cr, Fe, Cu, and Zn.	Singh et al. (2022)
Thysanolaena maxima	Antimony (Sb)	Excellent candidate for the phytoextraction of As and Sb from polluted land.	Zhang et al. (2020)
Conyza canadensis, Nicotiana tabacum, Ricinus communis, Pontederia cordata	Cadmium (Cd)	Under the high concentration of Cd, protein and amino acids were produced by *C. canadensis* to ameliorate the phytotoxicity effect. Transgenic *Nicotiana tabacum* over-expresses gene OsMTP1 for HMs tolerance or transport protein (MTP) family, increased sequestration, induced higher yield of biomass, and tolerance against stress.	Das et al. (2016)
Ricinus communis L., *Eichhornia crassipes,*	Chromium (Cr)	Feasible for the remediation of textile wastewater.	Panneerselvam and Priya (2023)
Typha latifolia, Croton bonplandianumBaill	Cobalt (Co)	Applicable for the remediation of agricultural and industrial wastewater.	Ahmad (2021); Ray and Neogy (2022)
Artemisia absinthium L.	Copper (Cu)	Hyperaccumulator of Cu in the roots system.	Ghazaryan et al. (2021)
Pteris vittate, Cosmos bipinnatus	Lead (Pb)	Ornamental landscape plant	Majumdar et al., (2022)
Miscanthus sinensis	Mercury (Hg)	Energy crop for the phytoremediation of Hg	Zhao et al. (2019)
Pistia stratiotes (water lettuce), *Azolla filiculoides*	Nickel (Ni)	Feasible for phytoremediation.	Şentürk et al. (2023)
Hemp (*Cannabis sativa*)	Selenium (Se)	Can produce Se-fortified food products.	Stonehouse et al. (2020)
Nerium oleander, Ricinus communis	Zinc (Zn)	Good candidate for Zn remediation in soil.	He et al. (2020a)

Adopted from Yadav et al., 2023.

The contributing factors to the ability of plants in phytoremediation are listed as: (i) antioxidative activity, (ii) quantity of HMs transporters for absorption, (iii) yield of biomass, and (iv) protoplast fusion. Plant tolerance against HMs is a primary determinant that may improve the plant survival rate in contaminated areas. The up-regulation of antioxidant activity has been recognized to be a key to phytoremediation. Improvement on the uptake, sequestration, and translocation of HMs may over express the related genes affecting the antioxidant activities.

HMs penetrate through the membrane of the cell plasma, the mediation of intra-cellular surface binding sites and affinity of the protein molecules. An increased metal-transport system in plants could facilitate a higher efficacy of the genetically engineered hyperaccumulator plants. Identification of the respective genes of HMs transporters and the integration into plant would enhance uptake rates.

MTP, MATE, ZIP, and HMA families are among the recognized transporter families for metal chelation. Amongst them, the MATE family turns out to be the key player. Genomic identification of this family in paddy and *Arabidopsis thaliana* are 45 and 56 MATE genes, respectively (Wang et al., 2016). In lotus, 81 MATE genes were discovered (Ming et al., 2013). For the MTP family, 22 MTP genes have been identified in apple (Qi et al., 2022). Eight (8) HMA genes from the HMA family were detected in *A. thaliana* (Hanikenne et al., 2008) while 48 HMA genes were found in tomatoes (Hasan et al., 2022). On the other hand, 32 ZIP and 12 ZIP genes from the ZIP family were reported in apples (Qi et al., 2022) and maize (Mondal et al., 2014), respectively. In rice plants, HMs–induced stress could result in the alteration of cysteine, the major amino acid that governs the detoxification reaction and metal binding affinity.

A better plant growth rate driven by hyperaccumulator plants can be promoted via genetic engineering or time-honoured propagation. Specifically, the use of plant growth regulators in *Brassica juncea* induces a greater biomass yield and translocation of HMs for Cd and uranium removal. Protoplast fusion is among the vital techniques which may strengthen the desirable features of plants for the phytoremediation of the HMs. Scientists have adopted electrofusion to different protoplasts from hyperaccumulator species, namely *Brassica napus* and *Thlaspi caerulencens* for the phytoremediation of Zn (Brewer et al., 1999). *T. caerulescens* has been reported to produce a greater biomass with enhanced HMs accumulation and improved tolerance against HMs–induced phytotoxicity as compared to the wild species. Similarly, the development of Zn–resistant strains in *Trichoderma* has been initiated, wherein Zn removal of the fusion strains was profoundly improved.

9.3.2 Driving Factors of Phytoremediation Process

Plant species, elemental species, rhizosphere, physio-chemical and biological properties of soil, and environmental factors are among the primary determining aspects for HMs uptake and translocation in plants. Agronomic practices, including application of fertilizers and chelators, as well as pH adjustments may improve the efficiency of the phytoremediation process. Manipulation of soil pH between pH 6.5 to 7.0 could be one of the methods to minimize Pb absorption. In this section, the driving factors affecting plant ability for HMs absorption will be discussed.

Phytoremediation is a species-specific remediation method for the absorption of HMs. In a study examining the phytoremediation potential of plant species, specifically corn, field mustard, pea, and sunflower, Tariq and Ashraf (2016) concluded that hyperaccumulators exhibited different capacities for HMs accumulation. With the addition of ethylene diamine tetra acetic acid (EDTA), corn appeared to be a hyperaccumulator of cobalt (Co) and chrome, while sunflower showed a higher affinity for Cd, while *Brassica campestris* was detected to be a chrome hyperaccumulator. Pea was found to be the best Pb accumulator without EDTA treatment. These findings highlighted the necessity to conduct in-depth investigation into the region-specific plantation techniques with the greatest HMs accumulation capacity, according to the targeted pollutants.

Metallophytes, consisting of metal indicators, metal excluders, and hyperaccumulators, are the most common plants species for phytoremediation process. A metal hyperaccumulator is a group of plants growing on metalliferous land that accumulate exceptionally high concentrations of HMs in their above-ground parts, without any phytotoxicity implication (Salt et al., 1998). In comparison to non-hyperaccumulator taxa, hyperaccumulators could be distinguished by three (3) unique characteristics: (i) a profoundly higher uptake potential of HMs, (ii) a higher translocation rate, and (iii) a greater detoxification and sequestration ability. Further, the plant shall be characterized by: (i) tap or fibrous root, (ii) degree of phytotoxicity and survival, (iii) adaptability, (iv) growth rate, and (v) resistance against diseases and pests. Some hyperaccumulators can actively absorb high concentrations of HMs, as much as 100- to 1000-fold from the natural environment. For the selection of a suitable candidate for phytoextraction, bioconcentration factor and translocation factor (TF) must be taken into account.

Energy crops, notably *Miscanthus*, *Salix*, *Arundo*, and *Populus* have demonstrated excellent phytoremediation potential via phytoextraction, rhizofiltration, and phytostabilization. Accordingly, approximately 400 plant species from the class *Euphorbiaceae, Fabaceae, Scrophulariaceae, Bassicaceae, Lamiaceae*, and *Asterraceae* (Salt et al., 1998) as well as grasses, namely *Panicum maximus* (Asemoloye et al., 2020), have been rated as hyperaccumulators. *Thlaspi caerulescens* was the first identified hyperaccumulator with high tolerance for Zn (Sachs, 1865). *Sedum alfredii* was found to be the co-hyperaccumulator for Cd and Zn (Yang et al., 2004), while *Pteris vittata* can accumulate more than 5gAs/kg in the leaf (Chen et al., 2002), with the accumulation coefficient recorded at 80. Biomass-producing crops and woody plants in particular, such as *Nicotiana tabacum*, *Cannabis sativa*, and *Zea mays* have recorded great phytoextraction potential (Suman et al., 2018), mainly attributed to their high biomass yield non-edibility, unique roots for deeper penetration, and erosion resistance. *A. marina* is highly resistant to HMs, owing to a wide array of adaptation mechanisms, in particular ion exclusion. Atriplex, another metal excluder, accumulated Ni, Cu, Pb, and Zn in stems and roots at much higher concentrations than the plants grown on uncontaminated land, as these HMs are largely dispersed in root tissues, to imply its HMs exclusion mechanism.

Similar to the accumulators, metal indicators accumulate HMs in the aerial parts of plants. They play ecologically and biologically crucial roles as pollution indicators and pollutants remediators. However, these plants will wither after the continuous absorption of HMs. Bonanno and Giudice (2010) have investigated the accumulation rate of different parts of *Phragmites australis*, and evaluated their suitability to be applied for the bio-monitoring of HMs in terrestrial and aquatic environment. In the urbanized Swartkops Estuary, the HMs–accumulating capacity of *Phragmites australis*, *Spartina maritima*, and *Typha capensis* was highlighted. The potential of four aquatic macrophytes: *Eichhornia crassipes*, *Ludwigia stolonifera*, *Echinochloa stagnina*, and *Phragmites australis* for the accumulative uptake of Ni, Cd, and Pb, and the HMs signaling and phytoremediation of polluted wetlands have been verified (Eid et al., 2020).

Aquatic macrophytes, notably *Eichornnia*, *Lemna*, *Hydrilla*, *Wolfia*, *Pteris*, *Spirodela*, *Pistia*, *Crysopogon*, and *Typha*, may grow well under a polluted aquatic system for the remediation of water pollutants. Species of *Thlaspi* may survive under Ni-polluted conditions, while *T. Caerulescens* is suitable for the accumulative uptake of Cd, Ni, Pb, and Zn (Rascio and Navari-Izzo, 2011). *Spirodela polyrhiza*, a duckweed species, is recognized for As removal via phosphate uptake and sorption mechanism (Rahman et al., 2007). *Eicchornia crassipes* is a water hyacinth that remediates palladium (Pd) using rhizofiltration. Researchers have also reported that the adoption of *Spirulina* sp. And *Oenothera picensis* for the effective removal of Cu and Cr from the contaminated aquatic environment through chemisorption and phytoextraction, respectively (Jasrotia et al., 2017). Islam et al. (2015) reported the phytofiltration of Cd and As from contaminated water using *Micranthemum umbrosum*. Brake ferns have been found to grow well in As–enriched water for the phytoextraction of As. Specifically, As (V) may be reduced to As (III) in ferns roots glutathione-dependent arsenate reductase, to be translocated to stems via xylem sap (Rascio and Navari-Izzo, 2011). Table 9.3 presents the classification of plants for the phytoremediation of different targeted HMs.

TABLE 9.3
Classification of Plants for the Phytoremediation of Targeted HMs

Classification	Plant Species	Targeted HMs	References
Metal hyperaccumulator	*Salicornia maritimia*	Zn, Cu	Milić et al. (2012)
	Rhizophora mangle	As, Hg, Cd, Al	
	Laguncularia racemosa	Mg, Cr, Fe, Zn, Cu, Pb	Mejias et al. (2013)
	Eichhornia crassipes Ludwigia stolonifer	Cd, Pb, Ni	Eid et al. (2020)
	Echinochloa stagnina, Phragmites australis		
Metal excluder	*Bruguiera gymnorhiza*	Cd, Cu	Wang et al. (2013)
Metal indicator	*Phragmites austaralis*	Cu, Zn, Hg, Cd, Ni, Cr, Mn, Pb	Bonanno and Giudice (2010)
	Phragmites australis, Typha capensis, Spartina maritima	Cu, Zn, Pb	Phillips et al. (2015)
	Eichhornia crassipes	Pb, Ni, Cd	Eid et al. (2020)
	Ludwigia stolonifera		
	Echinochloa stagnina, Phragmites australis		
Aquatic macrophytes	*Spirodela polyrhiza*	As	Rahman et al. (2007)
	Utricularia gibba	Cr (VI)	Augustynowicz et al. (2015)
	Oenothera picensis, Spirulina	Cr, Cu	Jasrotia et al. (2017)
	E. crassipes	Palladium (Pd)	Laet et al. (2019)
	Zinnia elegans	Cr (VI)	Panda et al. (2020)

Adapted from Agarwal and Rani, 2022.

The governing factors of the bioavailability of HMs are cation exchange capacity (CEC), texture, soil pH, temperature, and plant nutrient. The acidic pH (higher H^+ ions) is parallel with the rising CEC reading. Under the basic condition, HMs would penetrate from the colloids fractions of natural minerals to be desorbed. For instance, lowering pH of soil has been reported to improve the effectiveness of HMs uptake by *Thlaspi caerulescens* (Rosenfeld et al., 2018). Generally, HMs could be classified into high mobility HMs (Cd, Zn, and Ni), and low mobility HMs, in particular Cu, Pb, and Cr. The mobilization of HM increases under the high soil organic matter (OM) content, which induces chelation, greater nutrient availability, and CEC reading, to increase the HMs availability for the plant uptake. The basic condition would result in the dissociation of the carbonyl, alcoholic, phenolic, and carboxyl groups to increase the affinity of ligand ions to the metal cations. Organic amendments using compost, biochar, and peat to HMs–polluted soils may enhance the soil CEC, offer extra binding sites and great affinity, and reduce the mobility of HMs. The root absorption factor could be affected by the pH condition, rhizosphere activity, moisture content, root exudation, HMs, competing ions, and soil temperature.

Plant root is of greatest importance for phytoremediation as it is the primary entry point for the pollutant uptake, followed by the metabolism and storage in plant tissues. Acidification or alkalinization, and the changing organic ligand, metal chelation, and redox potential or exudation in plant roots could affect the HMs speciation and the solubility in the plant roots. The root features, however, are governed by the temperature, precipitation, soil moisture, drought conditions, and growth rate on its own. Therefore, the evaluation of rhizosphere features and composition of exudate are the most efficient screening approaches in searching for the most appropriate accumulator. Other factors, in particular phosphate-solubilizing bacteria that promote the growth rate and osmolytes, could improve the accumulation of HMs in boosting phytoremediation efficacy.

The primary advantage of phytoremediation is its cost-effective feature, as the cost is ten times lower than other chemical and physical remediation processes. Phytoremediation is a safe technique to minimize erosion and reduce the amount of contaminated dust in stabilizing contaminated areas. Compared to bacteria- and fungi-assisted bioremediation, phytoremediation is visible and enables better monitoring and process control. The plants could react as the source of nutrients for the rhizosphere bacteria for soil quality improvement and restore the available habitat for the wildlife. In comparison to physical methods, which involve the removal of a fertile layer of topsoil, this strategy does not deteriorate the land fertility.

Additionally, biomass from plants grown within a polluted area may be harvested to be applied as biofuel, while the development of the polluted areas could be transformed into a new ecosystem for the preservation of biodiversity, atmospheric improvement, carbon dioxide fixation, and restructure of the land area. The efficiency of phytoremediation is the result of synergistic interactions between the surrounding environment and plants. Nevertheless, the implementation could be limited by the long duration requirement for significant outcomes.

9.4 OMICS APPROACHES FOR THE PHYTOREMEDIATION OF HEAVY METALS

The successive implementation of omics technology requires extensive understanding of the underlying processes, along with interaction of the molecular level heterogeneity between the individual components. In addition to the development of bioinformatics and in silico tools, omics technologies, in particular genomics, proteomics, transcriptomics, miRNAomics, and metabolomics, have produced a plethora of information that enables deeper exploration of the genes, molecules, proteins, and metabolic regulators associated with phytoremediation and HMs tolerance in plants. Generally, omics approaches contribute to the identification of resistant signalling at the genome level. Proteomics and transcriptomics are commonly adopted for elucidating the role of proteins against stress, metabolites, and controlling factors of the HMs–tolerant plants. Proteomics provides detailed information of cell proteins. The micro-array technique has been adopted for the detection of genome alterations with respect to HMs–induced stress. Metabolomics is another omics approach that highlights the phenotypic findings as metabolites are fundamental for the interrelationships between gene expression, proteins, and biological reactions. A combination of the omics approaches may lead to a competent way for pollutants removal. A summary of the overall omics approaches is depicted in Figure 9.2.

9.4.1 Genomics

The genomics approach refers to the analysis of the genome of an organism. It is a multi-disciplinary area of study which centers around the function of genomes, genome editing tools, deoxyribonucleic acid (DNA) sequencing methods, recombinant DNA, and bio-informatics tools for collecting, analyzing, and sequencing the function and structure of the genome. The integration of a genomics approach into environmental sciences has resulted in the genes detection for phytoremediation and tolerance against abiotic stressors (Das et al., 2016). Comparative genomics studies on the genes identification within the hyperaccumulator plant have discovered different classes of genes, namely YSL, HM ATPase (HMA), MATE, HM–associated proteins (HMP), ZIP families, and metal tolerance protein (MTP). In another study, 46 and 55 HMPs were identified in paddy and *Arabidopsis*, respectively (Li et al., 2020a). These HMPs were attached to HMs via cysteine residues for the transportation and detoxification reactions. Meanwhile, 36 genes from 6 sub-families were found in *Solanum tuberosum* for the uptake, translocation, and sequestration of HMs (He et al., 2020b).

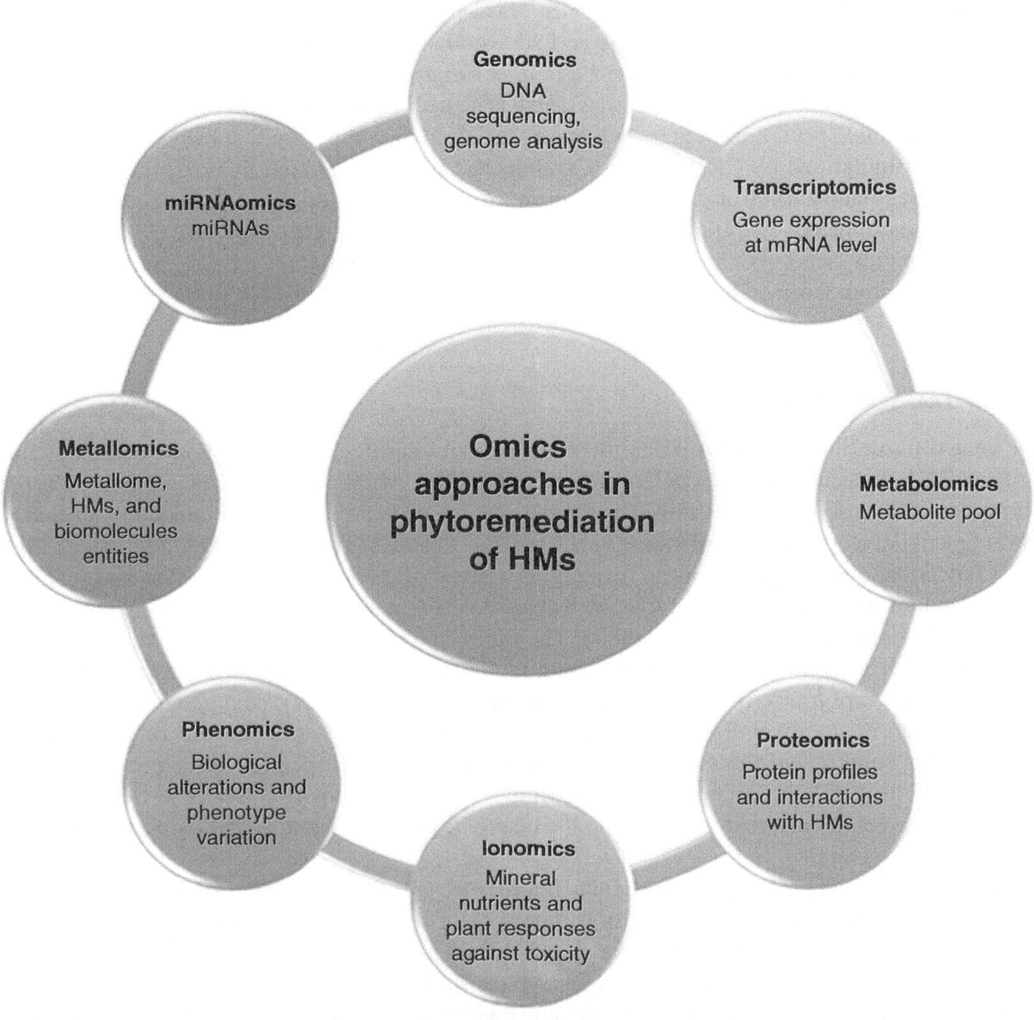

FIGURE 9.2 Recent development of omics approaches for the phytoremediation of HMs.

Quantitative trait loci (QTL) analysis is a statistical tool connecting phenotypic (selected trait measurements) and genotypic information (selected molecular markers) to elucidate the genetic variation of the complex traits. The mapping of QTL between Zn–tolerant *Arabidopsis helleri* and *Arabidopsis lyrate* (Zn–intolerant), for instance, has identified a few genomic regions contributing to Zn–tolerance (Nahar et al., 2017). Likewise, QTLs for As accumulation were detected in maize stems and leaves. Only one of the QTL was found in both tissues, indicating that different genes are responsible for As accumulation in different tissues of maize. Several trait mapping approaches, specifically RNA–sequencing, QTL–sequencing, and QTL–mapping have been established to refine the mapping sequence and separate the identical genes within the QTLs. The genetic expression and structure of Cd resistance in *Hordeum vulgare* detected a major and a minor QTL, with 47.24% and 38.59% of phenotypical variance associated with Cd resistance, respectively (Derakhshani et al., 2020).

Meanwhile, Murugaiyan et al. (2019) have identified 25 genes for the transcriptional regulation of As–induced stress using the QTL intervals. Specifically, four QTLs associated with the resistance against As–induced cytotoxicity were identified at chromosomes 8 and 12. Njobvu et al. (2020)

detected 8 QTLs related to Al tolerance in *Phaseolus vulgaris*, with up to 14.7 % phenotypic variation in the physical growth traits of roots and stems. Gene Phvul. 007G025900 was found to be related to the Al–induced phytotoxicity in *Phaseolus vulgaris*.

CRISPR–interlinked protein 9 (Cas9) systems is a specific method, with high proficiency and wide applications. Speciation in gene detection, knock-in and knock-out, substitution, insertion, deletion, and mutations by CRISPR/Cas9 system has been recognized as a reliable method to explore the regulatory roles of genes in the metabolic processes and detoxification capability of plants. One example of the application is the segmental knock-out of OsABCG3 gene in the Cd–treated paddy seedlings, which recorded profound accumulation of Cd, and the subsequent translocation from cells xylem sap as a defence strategy against Cd–induced toxicity (Fu et al., 2019).

9.4.2 Transcriptomics

The gene is transcribed to mRNA before the translation into proteins. Transcriptome refers to a full range of RNA molecules, namely rRNA, tRNA, and mRNA, while transcriptomics is defined as the expression pattern at the mRNA level. The regulation of gene expression at the molecular level may control plant development and tolerance to HMs–induced stress. These genes could be categorized as (i) functional genes, coding for the formation of metabolic compounds including sugars, alcohols, or amines for the regulation of tolerance against HMs–induced stress; or (ii) regulatory genes for the coding of transcription factors (TFs). The transcriptomics approach enables the concurrent investigation of a large quantity of genes by adopting a series of technologies, specifically the microarrays, reverse transcription polymerase chain reaction (RT-PCR), next generation sequencing (NGS), representational difference analysis, RNA sequencing, serial analysis of gene expression (SAGE), suppressive subtractive hybridization, and massively parallel signature sequencing (MPSS), to provide an in-depth understanding of plants' response towards the stressors. Transcriptome could be varied according to the changing tissue, cultivars, developmental stage, or environmental factors.

Transcriptomics approaches, in particular microarray and RT-PCR, have been applied for the identification of gene expression trends of different plant species, while NGS and transcriptomics techniques are beneficial in elucidating the epigenetic alterations contributing to phytoremediation- and stress-driven response in plants. Microarrays facilitate the hybridization of mRNA from the desired plant subjected to a specific treatment, and enable gene expression comparison under different conditions, corresponding to the treatment conditions.

An investigation examining the transcriptome of rhizosphere microorganisms with respect to the phytoremediation reaction revealed that rhizosphere microbiome was significantly altered, with higher tolerance against HMs. Comparative transcriptomics analysis has been applied for comprehensive understanding of plant toleration responses and accumulative trend of Cd. Transcriptomic analysis on KTRG-B14 (good Cd accumulator) and KTRG-B36 (weak Cd accumulator) turnip landraces detected dramatic improvement of the antioxidant system and glutathione, with up-regulation of methionine sulfoxide reductase and thioredoxins in KTRG-B14. Zn–induced facilitators, 1-like gene expression was increased in both landraces, signifying a new aspect of molecular mechanism to explain the response of plant roots against Cd–induced stress. The available literature have documented that the accumulation and movement of Cd in plants may be governed by a wide range of transporter genes. Specifically, B14 contains detoxification 1 gene while B36 is related to the ABC transporter C family member 3 genes, Cd resistance 2-like gene, and Zn transporter 10 gene, wherein the genes play a crucial role in the enhanced adsorption of Cd in the cell wall, for greater accumulation.

Similarly, comparative transcriptomics analysis on two genotypes of sweet sorghum (good Cd–accumulator [H18] and weak Cd–accumulator [L69]) in relation to sRNAs and the degradome in Cd–polluted soils indicated differential expression against Cd in both conserved and novel miRNAs. A large quantity of miRNAs and mRNAs pairs showed negatively correlated expression. Specifically, miR 408 and miR169p/q-nov 23, which are responsible for the regulation of metal tolerance and Cd uptake, transportation across transmembrane, stabilization of ROS homeostasis, construction of cell wall, and cytoskeleton activity were examined using co-expression analysis. Different cell wall components were identified in H18 and L69, which were mediated by miR169p/q, miR5565e, miR171 g/j, miR319a/b, miR 5384, and miR396a/b, via their targets, respectively. As a whole, these transcriptomics analyses have concluded that miRNAs are applicable in the production of good Cd–accumulating plants. In another comparison study of *Solanum torvum* (weak Cd–accumulator) and *Solanum nigrum* (good Cd–accumulator), detected a higher expression of genes mediating the transportation of HMs was deleted in *S. nigrum* (Wang et al., 2015).

By adopting RNA–seq analysis, transcriptional reprogramming or alterations were discovered in the roots system of Cd–accumulating wheat. Seventy-two (72) metabolism pathways, specifically phenylalanine metabolism, photosynthesis, phenylpropanoid biosynthesis, ribosome, carbon fixation, and flavonoid biosynthesis were identified in *Phytolacca americana*. The expression of TFs, in particular Zn–finger motif-containing TFs and bZIP were up-regulated in *Arabidopsis thaliana* and *Brassica juncea* (Ramos et al., 2007). In Arabidopsis, the up-regulation of ETS2 repressor factor (ERF)1 and ERF5 (AP2/ERF superfamily) expression was reported. FIT expression coupling with AtbHLH 38 or 39 have led to the activation of different transporter genes, in particular HMA3, IRT2 (Fe–regulated transporter 2), and MTP3, which have a vital role in the regulation of Fe (Wu et al., 2012). A number of research studies have verified that a specific TF could regulate the expression of various genes (Verma et al., 2022). Dehydration-responsive element-binding protein (DREB) TFs demonstrated an important role in the defense system of *Arabidopsis* root against HMs–induced phytotoxicity to reduce the flow of HMs–polluted water.

There are multiple examples that adopted transcriptomics in gene identification responsible for plant tolerance against stress. The stress response of Cr–polluted *Brassica chinensis* has been reported (Zhou et al., 2016). Yıldırım and Uylaş (2016) have adopted genome-wide transcriptome profiling on boron-treated *Populus nigra*, and a number of genes controlling the uptake, transport, and detoxification of boron were identified. Specifically, the highest induction was detected for genes encoding ATP–binding cassette transporters, metallochaperones, tyrosine aminotransferase, and glutathione S transferases, responsive for the transportation of HMs (to vacuoles), HMs binding, antioxidative defence machinery, and signalling. The assessment investigating the transcriptomic response of tobacco roots under the exposure to thorium (Th) recorded the up-regulation of 152 genes and down-regulation of 100 genes. These induced genes were associated with jasmonic production, salicylic acid signalling pathways, in connection with a number of abiotic and biotic stress-responsive genes. The up-regulation in phosphate-starvation genes and suppression in genes encoding the synthesis of phytic acid were reported, indicating obstructed phosphate uptake or its signalling under Th–induced stress.

Under As (V) and As (III) stress, stress-responsive gene expression in paddy was altered in mediating the phytotoxicity effect. The antioxidative enzymes were continuously expressed, while a large number of carriers and the enzyme-dependent pathways were identified. Shukla et al. (2018) reported that As toxicity induced differential transcriptome changes in both sensitive "Slavi-1" and tolerant "col-0" species of Arabidopsis, with pronounced over-expression of As–induced genes to minimize the phytotoxicity implications. Similarly, the role of SLIM1-TF in the regulation of cellular redox and sulfur metabolism under As–sensitivity was reported in Arabidopsis (Jobe et al., 2021). According to the transcriptome library, SLIM1 could function as a repressor or transcriptional enhancer under stress conditions.

Expression studies conducted on maize treated with Pb revealed 262 differently expressed TFs, of which bZIP TFs were identified as primarily correlated with tolerance mechanism against Pb. A hybrid transcriptomics and computation approach in Arabidopsis, rice, Glycyrrhiza, and cucumber genus recorded that a large number of WRKY-TFs were regulated under Pb and Cd stress. Molecular assessment on transcriptome mediating the unique profile of chickpea was conducted in relation to

As, Cd, and Cr phytotoxicities. The stimulation of metabolic pathways under HMs–induced stress was found, while nine genes were ascertained to be the metabolic controlling reactions via multiple expressions. In addition to the regulatory genes, the transcriptomic profiles of a wide range of functional genes, notably heat shock proteins, metallothioneins, and metal transporters (sulfate transporter, NRAMP1, glutathione-conjugated transporters, and MATE transporters) have been formed with respect to Cd, As, Hg, Al, and Pb–induced phytotoxicity (Li et al., 2017; Liu et al., 2017a).

9.4.3 Metabolomics

Metabolomics, the analysis of metabolite constituents, encompasses mainly detection, identification, characterization, quantification, and regulation of a chemical footprint of biological reactions at the cellular level. Within this framework, metabolome refers to the entire pool of active metabolites at a specific moment and condition. The plant responses under normal and stressed conditions can be restructured to be adopted in characterizing the plant's genetics (Jamla et al., 2021). By adopting the metabolomics approach, researchers are able to detect and quantify the metabolites required for plant research, phytotoxicity assessments, improvement of phytoremediation efficacy, stress responses, and detoxification machinery, owing to the availability of modern, sophisticated equipment, in particular gas chromatography mass spectrometry (GCMS), liquid chromatography-MS (LCMS), direct flow injection MS (DFIMS), high-performance thin-layer chromatography (HPTLC), capillary electrophoresis MS (CEMS), and ultra-performance liquid chromatography (UPLC) (Agarwal and Rani, 2022; Jamla et al., 2021).

Plants have numerous suitable solutes, including betaine, glycine, histidine, proline, sucrose, and trehalose, that protect the plants under distinct abiotic stressors. The possible detoxification mechanism includes delimination of ROS, cellular osmosis regulation, and stabilization of protein, enzymes, and membrane. GCMS analysis conducted on Arabidopsis exposed to Cd–induced phytotoxicity indicated an up-regulation of glycine, ascorbate, proline, and glutathione to reduce oxidative damage. Under the Cr– and Cd–induced stress, profound increase in the proline and ornithine contents were detected in *O. sativa* and *Triticum aestivum* (Meng et al., 2019). Howladar and his team (2018) found that silicon triggered the accumulation of proline to reduce cellular osmotic stress, prevent denaturation of protein, stabilize antioxidative enzymes, and regulate excess generation of ROS in cadmium-treated wheat plant. A synergetic relationship was found between the Cd concentration and proline level, with or without jasmonic acid and gibberellic acids addition in stems and roots of *Cicer arietinum*, to verify the protective role of proline during Cd–induced phytotoxicity (Ahmad et al., 2021).

Putrescine, the precursor of spermidine and spermine, is an important component in plants' response to stress. An increase in the histidine content was detected in *Alyssum lesbiacum* as a response to Ni–induced phytotoxicity (Singh et al., 2016). Purine metabolism is one first-line regulator to increase Cd absorption in K472 for land remediation, making it a crucial metabolic biomarker. Meanwhile, *L*-histidine was detected in nickel immobilization and compartmentalization within the cell sap of *N. caerulescens* root, signifying the histidine-assisted HMs detoxification pathway (Kozhevnikova et al., 2020). Dalir and Khoshgoftarmanesh (2014) reported the histidine-nickel complex in nickel-polluted nutrient medium to minimize the mobility and bioavailability of Ni and to protect wheat plants against oxidative damage.

Phenols, in particular flavonoids and tannins, are abundantly available as plants' secondary metabolites. Phenol generation during HMs exposure is correlated with the formation of phenol-HMs ions complex through the hydroxyl and carboxyl groups. The antagonistic correlation between As toxicity and phenol, and the synergistic relationship between As concentration and phenol formation, were reported in As–treated *Zea mays*, to provide structural supports to plants. The accumulation of phenolic compounds, in particular Chalcone synthase and phenylalanine ammonialyase, have been reported in *Malus pumila* (Gacnik et al., 2021), *Carthamus tinctorius* (Dehghan et al., 2014), *Solanum lycopersicum* (Pappi et al., 2021), *Viola tricolor* L. (Hassanvand et al., 2021), and

Vitis vinifera (Zhao et al., 2021) under HMs–induced phytotoxicity. Tannin has been found in its maximum quantity to confer resistance against Mn–induced phytotoxicity in *Camellia sinensis* and protect the tea plants through compartmentation and complexation in different plant compartments (Farzadfar et al., 2017). This is further supported by the flavonoids-induced improvement in the antioxidative activities of transgenic potato. Likewise, HMs, in particular Al, Co, Cu, and Mn, have resulted in an increased production of phenolic compounds in maize, *Phaseolus vulgaris*, lentil roots, *Trigonella foenum*, wheat, and tea plant (Copaciu et al., 2016).

Glutathione (GSH) is a non-protein metabolite for the regulation of free radicals scavenging conditions. Glutathione reductase (GR) may reduce the oxidized glutathione (GSSG) into the GSH form, to alleviate HMs toxicity by minimizing the HMs–cysteine-rich proteins (CRPs) complexation (Liu et al., 2017b). The conversion of GSSG into GSH, driven by NADPH, is necessary to maintain the appropriate GSH:GSSG ratio in plants under HMs–induced phytotoxicity (Singh et al., 2016). Further, the reduction of GSH into phytochelatins to chelate HMs ions may further reduce the mobility as well as the bioavailability of HMs (Singh et al., 2021). A synergistic correlation was established between GR expression and As–induced phytotoxicity in the stems of *Isatis cappadocica* desv, indicating the protective role of GR against HMs–induced phytotoxic implications (Souri et al., 2021). Similar findings have been detected in *Helianthus annuus* (Belhaj et al., 2016).

Organic acids, specifically malate, oxalate, and citrate, are metal transporters in xylem sap, which justify the metal sequestration in vacuole. Citrate is a derivative of citrate synthase, which owns a high binding affinity towards Fe, Ni, and Cd within the organelles, while malate is important for the solubilizing cytosolic ions, with high binding capacity for Zn. Ascorbate is a powerful antioxidant in the defence system against ROS. It has been found that the enhanced ascorbate content in tomato plants and Arabidopsis were associated with the alleviated oxidative stress by the synthesis and accumulation of ascorbate and related antioxidant compounds. α-tocopherol, on the other hand, is the reactive form of vitamin E for the alleviation of ROS and lipid peroxidation, and cellular damage protection against HMs–induced phytotoxicity. A number of research findings have reported the changes of α-tocopherol expression due to HMs–induced phytotoxicity. Specifically, Mishra et al. (2019) found an increased α-tocopherol level under cadmium toxicity and enhanced gene expression. Similarly, the increase of the vitamin E–regulated enzymes, specifically VTE5- and VTE2-mRNA in Arabidopsis treated with Cd and Cu was documented. They reacted by stimulating hydroxylpyruvate dioxygenase to limit the mobility and bioavailability of HMs in the root cell sap. Eventually, metabolomics serves as an implicit tool and holds a remarkable prospect in the determination of novel and distinctive information of metabolites, data analysis, and metabolic profiles, which contribute to the identification of the associated stress-response pathways. The strategy is particularly useful for the selection of plant candidates, with greater phytoremediation ability and stress tolerance against external stress.

9.4.4 Proteomics

While omics approaches using genome, transcriptome, and metabolome information have contributed to the fundamental knowledge of plant response and HMs–related stress, some of these changes could be suppressed at the protein level, not accessible by the transcriptomics analyses (Hossain and Komatsu, 2013). Proteomics, a viable approach established in the post-genomic era, is one of the major breakthroughs in omics technology that explores the understanding of the (i) complex proteins and the interaction between plants and HMs at the molecular level; and (ii) variation of proteins at different growth stages and conditions. Detailed proteomics study facilitates the (i) identification of targeted proteins for the detoxification of HMs (Chauhan et al., 2020), plant response at contaminated sites, and phytoremediation; (ii) characteristics of proteins with respect to interactions, location, and modifications at a specific duration; and (iii) generation of a new platform for the evaluation of the pollutants-induced stress, the impact on a molecular level, and the possible mechanisms. As a whole, a proteomics approach provides remediation-related proteins, genes, and the associated mechanisms to trace physiological alterations during the phytoremediation process.

The proteomics approach applied in the Populus plant driven by Cd–induced stress indicated both direct and indirect implications on the plant proteome. In that perspective, plant root is the first site exposed to external stressors, hence it is where the changes in proteome would be detected. According to the literature, the content of root proteins that regulate HMs uptake and glycosyl hydrolase family 18 varied among different accessions. In a study examining the proteomes of *Thlaspi caerulescens*, a positive interrelation between the accumulation rate of Zn and Cd with the protein photosystem II was found (Paunov et al., 2018).

Meanwhile, proteomic analysis on Cd–treated *Sorghum bichor* revealed 33 well-organized differentially expressed proteins (DEPs), namely glyceraldehyde-3-phosphate dehydrogenase, ribulose bisphosphate carboxylase small chain, cytochrome P450, glutathione S-transferase, and carbonic anhydrase. In contrast, Zn finger CCCH domain-bearing aspartate aminotransferase 3, flavonoid 3′-5′ hydroxylase, protein 14, probable F-actin-capping protein subunit beta proteins, pentatricopeptide repeat-containing protein, bergaptol O-methyltransferase, and Brevis radix-like protein 1 are some of the less-characterized proteins that exhibited up-regulation trends after the exposure to Cd. Further, a suppression of the carbon fixation, protein synthesis, and ATP production was observed.

Zhao et al. (2011) reported a down-regulation in 11 genes that are essential for glutathione metabolism and photosynthesis pathway in the Cd–polluted *Phytolacca americana*. Yannarelli et al. (2007) conducted a 2-D electrophoresis-assisted proteome analysis on Cd-–treated wheat roots and discovered the up-regulation of antioxidant proteins, whereas proteins governing the photosynthetic and ETS processes were down-regulated. The findings outlined the identification of associated protein markers between susceptible and tolerant genotypes (Hossain and Komatsu, 2013). Luo and Zhang (2019) investigated the xylem sap of brassica seedlings exposed to Cd and reported that 672 DEPs were involved in the metabolic pathways via the interaction with lipid and protein metabolism within the cell walls. A protein-like defending component, known as BnPDFL, has mediated cadmium chelation to stimulate the resistance against Cd. By adopting the iTRAQ technique, Wen et al. (2019) detected the alterations of the root proteome of maize. Likewise, several protein factors, especially tubulin, ligases, HSPs, synthetases, hydrolases, isomerases, lyases, transferases, and oxidoreductases were found to mediate Cd–induced stress in paddy (Ma et al., 2016). Under the treatment of $CdCl_2$, a number of differently abundant proteins (DAPs) including DAPs 307, 576, and 499 were extracted from roots and reported to be involved in related biological processes, particularly carbohydrate metabolism, biosynthesis of ribosome, ROS homeostasis, and cell wall organization for the regulation of Cd–induced phytotoxicity in maize. Label-free proteome analysis showed the presence of 163 and 196 DEPs in the stems and roots of Cd–exposed *Iris lactea*, respectively, which governed a wide array of functions, specifically redox reactions, TFs, GSH metabolism, ion homeostasis, and synthesis of amino acid (Wen et al., 2019).

Jian et al. (2020) have analyzed the Cd–polluted wheat roots using mass spectrometry quantitative-assisted proteomics approach, and discovered that 10,532 of the 11,651 identified proteins showed quantitative implications under stress. This finding further highlighted that DEPs in the Cd–treated wheat roots were engaged in DNA replication or repair in addition to GSH and protein metabolism. Zheng et al. (2008) have identified 40 differentially regulated spots in Al-treated *Glycine max*, and further characterized using MALDI-TOF-MS. Up-regulation of 21 proteins in antioxidant defense system, and down-regulation of 14 new proteins, which were concomitant with signal transduction, sulphur metabolism, and bio-synthesis of cysteine synthase within the plant cells were recorded. Al treatment on tomato plants indicated the stimulation of methylglyoxal (MG) in root tips, while the down-regulation of protein for gene transcription and protein translation was detected.

HMs treatment could activate the signaling cascades and defense-related genes in plants. In another study, the molecular interaction between Selenium (Se) –induced resistance against As-toxicity was revealed in theAs-exposed rice seedlings. This justified that Se could ameliorate

As–induced cell wall and membrane distortions, by down-regulating the As carrier genes (NIP2; 1, NIP1; 1, NRAMP5, ABCG5, and TIP2; 2) as well as the sulfate carrier genes (SULTR3; 6, SULTR3; 1). The up-regulation of PRX, GRX, and GST verified the protective role of Se against As–induced oxidative stress (Chauhan et al., 2020). Likewise, Se–mediated up-regulation of 172 proteins extracted by LC-MS/MS-assisted proteomics analysis was detected in the As—treated pepper plants. DEPs were identified to be interconnected with expression and processing of proteins, chaperons, metabolic processes, and changes in the post-translation.

Similarly, Cu–stressed *Eucalyptus camaldulensis* seedlings exhibited differential expressions for 26 portions, wherein 11 and 15 proteins of metabolism, translation, antioxidant enzymes, transcription, and photosynthesis were up-regulated and down-regulated, respectively, in response to the rising Cu content. According to Terzi and Yıldız (2021), Cys addition could minimize Cr–induced phytotoxicity by reducing the translocation, and the comparative proteomics of maize roots showed the involvement of 46 proteins in the regulation of ROS homeostasis. Meanwhile, sunflower seeds, leaves, and stems that are grown under different treatment conditions of Cr indicated the proteome role in stress regulation.

By adopting the 2-DE and shotgun proteomics approach, Ceballos-Laita et al. (2018) have examined the Mn–induced phytotoxicity implications on the root proteome of tomato plants. The detection of 118 and 54 proteins, with 7% of common DAPs was reported in both techniques. Notably, signaling, oxidoreductases, and protein metabolism were among the most affected processes. These findings highlighted that while Mn–induced stress could suppress a number of metabolic processes, it up-regulated oxidative phosphorylation, pyruvate metabolism, and glycolysis.

9.4.5 Ionomics

Mineral nutrients in plants, such as (i) macronutrients specifically calcium (Ca), sulfur (S), phosphorus (P), magnesium (Mg), potassium (K), and N, and (ii) micronutrients in particular Cu, Fe, Zn, molybdenum (Mo), and cobalt (Co), are the important supports for different biological processes. The regulation of these elements within the required range is essential to avoid the phytotoxicity effects yet to preserve the important defensive roles against HMs–induced toxicity. An ionomics approach focuses on the interplay between these elements, for the management of HMs–generated phytotoxicity.

In plants, K^+ion regulates the cell's anion-cation equilibrium and plays a viable role in protein and enzyme activation to facilitate a wide range of fundamental biochemical processes, controls stomatal function, and reduces the impact of oxidative damage. The addition of K could result in the rising proline and antioxidant enzymes activities, notably ascorbate peroxidase, catalase, superoxide dismutase, and ascorbate; and reduced lipid peroxidation and hydrogen peroxide in *Gladiolus grandiflora* under exposure to Cd. The synergistic phytotoxicities of Pb and Cd in spring wheat were eliminated by the application of potassium oxide, as evidenced by the reduction of the adsorbed Pb, and rising in the complexation of Mn and Fe oxides, which led to a reduction of HMs in the soil content.

Ca is an essential nutrient for plant development to stimulate enzyme activities and modulate metabolic processes. The presence of Cd may affect plant development by substituting Ca in plant cells, owing to the chemical similarity, and sharing the key presence of the same pathways and binding sites for Cd and Ca ions. Ca–mediated alleviation of HMs stress have been documented in a variety of plant species, including *Nicotiana tabacum* (Wang et al., 2021), *Solanum lycopersicum* (Guo et al., 2018), *Cicer arietinum* (Sakouhi et al., 2021), *Brassica juncea* L. (Ahmad et al., 2015), *Glycine max* roots (Chen et al., 2017), *Macleaya cordata* (Cai et al., 2021), *Sporosarcina pasteurii* (Fang et al., 2021) and pak choi seedlings (Huang et al., 2022). Calcium sulfate may alleviate the Cd–induced phytotoxicity in pak choi via the (i) regulation of sulfate absorption, (ii) enhanced bioaccumulation and synthesis of phytochelatins, and (iii) up-regulation of vacuolar sequestrating transporters (Huang et al., 2022).

Mg is a key component for the synthesis of chlorophyll and plant immune systems. The ionomics approach has recorded that Cd–induced cytotoxicity in Japanese mustard spinach could be mediated by Mg, as evidenced by the reduction of Cd deposition in roots and stems. The mechanism that lies behind this effect is mainly attributed to the Mg—modulated up-regulation of antioxidant enzymes to suppress the excessive-generation of ROS. Additionally, the amelioration of HMs–induced phytotoxicity is correlated with gene transcription in rice, namely OsZIP1, OsZIP3, and OsIRT1 (Singh et al., 2016).

N is a primary component of plant hormones, nucleic acids, vitamins, and proteins. In addition to its primary role in the synthesis of plant molecules, N is capable to minimize HMs–induced phytotoxicity by (i) enhancing the photosynthetic capacity, (ii) strengthening N–bearing molecules, and (iii) up-regulating the antioxidative enzymes activity. These are evident in the reduction of Cd–induced phytotoxicity in *Atriplex lentiformis* (Eissa and Roshdy, 2018). In addition to its role as the major component of nucleic acids and cell membranes, and involvement in phosphorylation process, P regulates HMs–induced phytotoxicity via the neutralization of metals, generation of a metal-phosphate complex to limit accessibility, and up-regulation of the non-enzymatic and enzymatic antioxidative defense machinery to suppress the over-generation of ROS.

S is available in antioxidant ferredoxin, metalloenzymes, and vitamins. Specifically, the over-expression of the S–assimilation process may reduce the Cd–induced phytotoxicity. *Brassica chinensis* L. (Li et al., 2020b), *Sorghum bicolor* (Akbudak and Filiz, 2019), and *Sedum alfredii* Hance (Guo et al., 2009) have showed greater serine acetyltransferase (SAT) and ATP–sulfurylase (ATPS) capabilities after Cd exposure, leading to an enhanced adaptation against Cd. Similarly, S could maintain the equilibrium of ascorbate-glutathione cycle and minimize the Cd–induced oxidative damage in *B. nigra*. External S application could significantly stimulate the up-regulation of OsPCS (phytochelatin synthase) and OsHMA3 (HMA) in paddy roots to increase the phytochelatin of Cd in the root cells and prevent its accumulative entry into the stem's system.

9.4.6 Phenomics, Metallomics, and miRNAomics

The phenomics approach enables the study on the interconnectedness between the alteration of an organism throughout its lifetime and the variations in its phenotypes. Specifically, phenomics represents the relation between the variants of a trait, alleles, and genes, and allows scientists to decode the plant's genetic information into a particular phenotype, including the physical growth rate and production of biomass. In other words, phenome and genome could be connected by linking the genetic information and the phenotypic traits. This could be achieved by using the high-throughput three-dimensional scan of a plant exposed to external stressors. However, there is a slow development in characterizing the plant phenotypes for identification of plant tolerance and recovery response from stress, mainly attributed to the limited effort in phenotyping, requirements of data analysis tools and equipment, and criterion-based experimental approaches.

Some of the HMs are fundamental entities for specific biological processes, including regulation, structural arrangements, and stability of protein. Proteins that consist of transition metals to ensure their optimum stability are recognized as metalloproteins, wherein the alterations of HMs ion flux would alter the structural and functional properties of metalloproteins and consequently the concentration of metallome, specifically metallometabolites, metalloenzymes, and metalloproteins.

Metallomics is a comparatively newer approach in omics that generates a metallome of a particular biological system to elucidate the functional and physiological interconnection between metallome, HMs, and biomolecule entities by adopting X-ray fluorescence (XRF), mass spectrometry (MS), synchrotron radiation X-ray absorption spectroscopy (XAS), computation, and metalloproteomic approaches. This approach has been applied for a series of environmental metal toxicity assessments, structural characterization and pharmacological assessment, HMs tolerance and homeostasis, metal transport, and biogeochemical studies, mainly attributed to its role in deciphering the accumulation and translocation of HMs in hyperaccumulating plants according to the comparison among different metallomes.

With the application of high-end technologies, some prominent results have been achieved in elucidating the HMs–plants interactions via metallomics approaches. For instance, the visualization of As in the roots of *Pteris vittate* with synchrotron radiation micro (SRM) XRF imaging indicated that As could be transported via symplast pathway, according to the direction of root growth (Kashiwabara et al., 2021). Quantification of Hg has been successful in roots and stems of rice by adopting ICP-MS coupled with high-performance liquid chromatography, while determination of Cd content in young leaves, stems, and roots, and the detection of localization pattern of Cd were conducted by ICP-MS coupling with SRM XRF mapping. The application of XAS enables the exploration of O– or S–bearing cadmium ligands in Cd–hyperaccumulating Brassicaceae members (leaf).

Under abiotic stress, plants will regulate gene expression at the transcriptional level or through the activation of miRNAs as a response to regulate stress. MiRNAs, tiny (19–24 nt) non–protein-coding RNA molecules, have been listed as a key regulator of stress tolerance, cell proliferation, transcription factors modulators, and metabolic activities. The mode of function of miRNAs is complicated and mainly via the (i) activation of repression of transcription; (ii) manipulation of the expressions of miRNAs or their targets; (iii) post-transcriptional regulation; (iv) mRNA stability, (v) epigenetic modification, and (vi) gene silencing. Mature miRNAs that are not involved in proteins coding could function as the influential genetic factors required for stress management in plants.

The roles of miRNAomics in elucidating the HMs–induced stress tolerance have been evident in a wide range of plants. Under Cd–induced stress, miR396, miR171, and miR156 were down-regulated in *Brassica napus*; miR166 and miR393 were reported as the stress regulator in *Medicago* plants; and changes have been identified in miR160, miR167, and miR397 in *Arabidopsis* for auxin synthesis (Pegler *et al*., 2021). For Al–generated toxicity in *Medicago*, an miRNAomics approach has recorded the reaction of the stress responsive genes, miR390, miR319, miR396, and miR160 in plants (Noman and Aqeel, 2017). Additionally, the interactions of miR268, miR167, and miR159 with ABC transporters and NRAMP have signified the influence of miRNAs in the uptake, transport, and protein folding in *Brassica napus* (Ding et al., 2020). Similarly, As-induced stress could trigger the interaction of miR159, miR397, and miR408 with the transcription factors including MYB, TCP, as well as ABC transporter, protein for ion binding, laccase, and scaffold proteins (Noman and Aqeel, 2017). Meanwhile, miR833, miR212, miR395, miR5658 were deactivated in cotton while miR172, miR396, and miR162 may vary in Medicago leaves according to the changing stress conditions.

9.5 CONCLUSION AND FUTURE PERSPECTIVES

Heavy metals pollution is a critical global concern that calls for an intrinsic part of major managerial solutions for sustainable ecological and environmental management. Phytoremediation is the most effective plant-based approach that allows the remediation of toxic elements with minimal damages to the terrestrial and aquatic ecosystem, especially with the integration of omics approaches. The need to intensify and accelerate phytoremediation lies in the advantages of ecological restoration and the limitations with respect to the long duration of the process, and the subsequent implications on plants and human health. Although there is a wide availability of potential plant species supporting the phytoremediation process, the shortcomings of these plants prevent them from further application in accumulative uptakes of HMs. An in-depth understanding of the complex connections of HMs tolerance and phytotoxicity in plants via omics studies, notably transcriptomics, proteomics, genomics, phenomics, metallomics, ionomics, metabolomics, and miRNAomics would enable researchers to circumvent a plethora of solutions by adopting an integrated omics approach. The novel discoveries and information deduced from the analyses of genes, transcription factors, metabolomes, proteins, microRNAs, ions, and phenomes associated with metals-induced phytotoxicity could be adopted for the development of hyperaccumulating traits in plants with greater biomass production and enhanced resistance against HMs–induced phytotoxicity. While most of the literature findings have centered around the identification of plant species for phytoremediation and the related mechanisms, less is known regarding the improvement of plant phenotype features or engineering approaches for the best remediation outcomes. Technical and innovative modification of the available phytoremediation approaches, coupled with the improvement of the inherent phytoremediator, will promote a more sustainable remediation strategy. Future research should be geared towards the investigation of the implications of various catalysts on the phytoremediation process to ensure the practicability for the on-site applications. The phytoremediation process should be optimized using plants' genetic manipulation for greater accumulation and detoxification abilities. Genetically modified plants could be an ideal solution to confer preferred characteristics required for the phytoremediation process, with adherence to public approval and field-testing assessment. Therefore, a pronounced leap in plant science with a plant ideotype portraying desirable features for phytoremediation can be anticipated.

REFERENCES

Agarwal, P., & Rani, R. (2022). Strategic management of contaminated water bodies: Omics, genome-editing and other recent advances in phytoremediation. *Environmental Technology & Innovation*, *27*, 102463.

Ahmad, P., Raja, V., Ashraf, M., Wijaya, L., Bajguz, A., & Alyemeni, M. N. (2021). Jasmonic acid (JA) and gibberellic acid (GA3) mitigated Cd-toxicity in chickpea plants through restricted cd uptake and oxidative stress management. *Scientific Reports*, *11*, 19768.

Ahmad, P., Sarwat, M., Bhat, N. A., Wani, M. R., Kazi, A. G., & Tran, L. S. P. (2015). Alleviation of cadmium toxicity in Brassica juncea L. (Czern. & Coss.) by calcium application involves various physiological and biochemical strategies. *PloS One*, *10*, e0114571.

Akbudak, M. A., & Filiz, E. (2019). Genome-wide analyses of ATP sulfurylase (ATPS) genes in higher plants and expression profiles in sorghum (Sorghum bicolor) under cadmium and salinity stresses. *Genomics*, *111*, 579–589.

Asemoloye, M. D., Chukwuka, K. S., & Jonathan, S. G. (2020). Spent mushroom compost enhances plant response and phytoremediation of heavy metal polluted soil. *Journal of Plant Nutrition and Soil Science*, *183*, 492–499.

Augustynowicz, J., Łukowicz, K., Tokarz, K., & Płachno, B. J. (2015). Potential for chromium (VI) bioremediation by the aquatic carnivorous plant Utricularia gibba L. (Lentibulariaceae). *Environmental Science and Pollution Research*, *22*, 9742–9748.

Belhaj, D., Elloumi, N., Jerbi, B., Zouari, M., Abdallah, F. B., Ayadi, H., Kallel, M. (2016). Effects of sewage sludge fertilizer on heavy metal accumulation and consequent responses of sunflower (Helianthus annuus). *Environmental Science and Pollution Research*, *23*, 20168–20177.

Bonanno, G., & Giudice, R. L. (2010). Heavy metal bioaccumulation by the organs of Phragmites australis (common reed) and their potential use as contamination indicators. *Ecological Indicators*, *10*, 639–645.

Brewer, E. P., Saunders, J. A., Angle, J. S., Chaney, R. L., McIntosh, M. S. (1999). Somatic hybridization between the zinc accumulator Thlaspi caerulescens and Brassica napus. *Theoretical and Applied Genetics*, *99*, 761–771.

Cai, B., Chen, Y., Du, L., Liu, Z., & He, L. (2021). Spent mushroom compost and calcium carbonate modification enhances phytoremediation potential of Macleaya cordata to lead-zinc mine tailings. *Journal of Environmental Management*, *294*, 113029.

Ceballos-Laita, L., Gutierrez-Carbonell, E., Imai, H., Abadía, A., Uemura, M., Abadía, J., & López-Millán, A. F. (2018). Effects of manganese toxicity on the protein profile of tomato (Solanum lycopersicum) roots as revealed by two complementary proteomic approaches, two-dimensional electrophoresis and shotgun analysis. *Journal of Proteomics*, *185*, 51–63.

Chauhan, R., Awasthi, S., Indoliya, Y., Chauhan, A. S., Mishra, S., Agrawal, L., Srivastava, S., Dwivedi, S., Singh, P.C., Mallick, S., Tripathi, R. D. (2020). Transcriptome and proteome analyses reveal selenium mediated amelioration of arsenic toxicity in rice (Oryza sativa L.). *Journal of Hazardous Materials*, *390*, 122122.

Chen, B. C., Wang, P. J., Ho, P. C., & Juang, K. W. (2017). Nonlinear biotic ligand model for assessing alleviation effects of Ca, Mg, and K on Cd toxicity to soybean roots. *Ecotoxicology*, *26*, 942–955.

Chen, T., Wei, C., Huang, Z., Huang, Q., Lu, Q., & Fan, Z. (2002). Arsenic hyperaccumulator Pteris vittata L. and its arsenic accumulation. *Chinese Science Bulletin*, *47*, 902–905.

Copaciu, F., Opris, O., Niinemets, Ü., Copolovici, L. (2016). Toxic influence of key organic soil pollutants on the total flavonoid content in wheat leaves. *Water, Air, & Soil Pollution*, *227*, 196.

Dalir, N., & Khoshgoftarmanesh, A. H. (2014). Symplastic and apoplastic uptake and root to shoot translocation of nickel in wheat as affected by exogenous amino acids. *Journal of Plant Physiology*, *171*, 531–536.

Das, N., Bhattacharya, S., & Maiti, M. K. (2016). Enhanced cadmium accumulation and tolerance in transgenic tobacco overexpressing rice metal tolerance protein gene OsMTP1 is promising for phytoremediation. *Plant Physiology and Biochemistry*, *105*, 297–309.

Dehghan, S., Sadeghi, M., Pöppel, A., Fischer, R., Lakes-Harlan, R., Kavousi, H. R., Vilcinskas, A., & Rahnamaeian, M. (2014). Differential inductions of phenylalanine ammonia-lyase and chalcone synthase during wounding, salicylic acid treatment, and salinity stress in safflower, Carthamus tinctorius. *Bioscience Reports*, *34*, e00114.

Derakhshani, B., Jafary, H., Maleki Zanjani, B., Hasanpur, K., Mishina, K., Tanaka, T., Kawahara, Y., & Oono, Y. (2020). Combined QTL mapping and RNA-Seq profiling reveals candidate genes associated with cadmium tolerance in barley. *PLoS One*, *15*, e0230820.

Ding, Y., Ding, L., Xia, Y., Wang, F., & Zhu, C. (2020). Emerging roles of microRNAs in plant heavy metal tolerance and homeostasis. *Journal of Agricultural and Food Chemistry*, *68*, 1958–1965.

Eid, E. M., Galal, T. M., Sewelam, N. A., Talha, N. I., & Abdallah, S. M. (2020). Phytoremediation of heavy metals by four aquatic macrophytes and their potential use as contamination indicators: A comparative assessment. *Environmental Science and Pollution Research*, *27*, 12138–12151.

Eissa, M. A., & Roshdy, N. M. K. (2018). Nitrogen fertilization: Effect on Cd-phytoextraction by the halophytic plant quail bush [Atriplex lentiformis (Torr.) S. Wats]. *South African journal of Botany*, *115*, 126–131.

Fang, L., Niu, Q., Cheng, L., Jiang, J., Yu, Y. Y., Chu, J., Achal, V., & You, T. (2021). Ca-mediated alleviation of Cd2+ induced toxicity and improved Cd2+ biomineralization by Sporosarcina pasteurii. *Science of the Total Environment*, *787*, 147627.

Farzadfar, S., Zarinkamar, F., & Hojati, M. (2017). Magnesium and manganese affect photosynthesis, essential oil composition and phenolic compounds of Tanacetum parthenium. *Plant Physiology and Biochemistry*, *112*, 207–217.

Fu, S., Lu, Y., Zhang, X., Yang, G., Chao, D., Wang, Z., Shi, M., Chen, J., Chao, D. Y., Li, R., & Ma, J. F. (2019). The ABC transporter ABCG36 is required for cadmium tolerance in rice. *Journal of Experimental Botany*, *70*, 5909–5918.

Gacnik, S., Veberic, R., Marinovic, S., Halbwirth, H., & Mikulic-Petkovsek, M. (2021). Effect of pre-harvest treatments with salicylic and methyl salicylic acid on the chemical profile and activity of some phenylpropanoid pathway related enzymes in apple leaves. *Scientia Horticulturae*, *277*, 109794.

Gao, J., Luo, M., Peng, H., Chen, F., & Li, W. (2019). Characterization of cadmium-responsive MicroRNAs and their target genes in maize (Zea mays) roots. *BMC Molecular Biology*, *20*, 1–9.

Ghazaryan, K. A., Movsesyan, H. S., Minkina, T. M., Nevidomskaya, D. G., & Rajput, V. D. (2021). Phytoremediation of copper-contaminated soil by Artemisia absinthium: Comparative effect of chelating agents. *Environmental Geochemistry and Health*, *44*, 1203–1215.

Guo, J., Zhou, R., Ren, X., Jia, H., Hua, L., Xu, H., Lv, X., Zhao, J., & Wei, T. (2018). Effects of salicylic acid, Epi-brassinolide and calcium on stress alleviation and Cd accumulation in tomato plants. *Ecotoxicology and Environmental Safety*, *157*, 491–496.

Guo, W. D., Liang, J., Yang, X. E., Chao, Y. E., & Feng, Y. (2009). Response of ATP sulfurylase and serine acetyltransferase towards cadmium in hyperaccumulator Sedum alfredii Hance. *Journal of Zhejiang University SCIENCE B*, *10*, 251–257.

Hanikenne, M., Talke, I. N., Haydon, M. J., Lanz, C., Nolte, A., Motte, P., Kroymann, J., Weigel, D., & Krämer, U. (2008). Evolution of metal hyperaccumulation required cis-regulatory changes and triplication of HMA4. *Nature*, *453*, 391–395.

Hasan, M. N., Islam, S., Bhuiyan, F. H., Arefin, S., Hoque, H., Jewel, N. A., Ghosh, A., & Prodhan, S. H. (2022). Genome wide analysis of the heavy-metal-associated (HMA) gene family in tomato and expression profiles under different stresses. *Gene*, 835, 146664.

Hassanvand, A., Saadatmand, S., Lari Yazdi, H., & Iranbakhsh, A. (2021). Biosynthesis of nanosilver and its effect on key genes of flavonoids and physicochemical properties of viola tricolor L. *Iranian Journal of Science and Technology, Transaction A, Science*, *45*, 805–819.

He, C., Zhao, Y., Wang, F., Oh, K., Zhao, Z., Wu, C., Zhang, X., Che, X., & Liu, X. (2020a). Phytoremediation of soil heavy metals (Cd and Zn) by castor seedlings: tolerance, accumulation and subcellular distribution. *Chemosphere*, *252*, 126471.

He, G., Qin, L., Tian, W., Meng, L., He, T., & Zhao, D. (2020b). Heavy metal transporters-associated proteins in Solanum tuberosum: genome-wide identification, comprehensive gene feature, evolution and expression analysis. *Genes*, *11*, 1269.

Hossain, Z., & Komatsu, S. (2013). Contribution of proteomic studies towards understanding plant heavy metal stress response. *Frontiers in Plant Science*, *3*, 310.

Howladar, S. M., Al-Robai, S. A., Al-Zahrani, F. S., Howladar, M. M., & Aldhebiani, A. Y. (2018). Silicon and its application method effects on modulation of cadmium stress responses in Triticum aestivum (L.) through improving the antioxidative defense system and polyamine gene expression. *Ecotoxicology and Environmental Safety*, *159*, 143–152.

Huang, S. Q., Peng, J., Qiu, C. X., & Yang, Z. M. (2009). Heavy metal-regulated new microRNAs from rice. *Journal of Inorganic Biochemistry*, *103*, 282–287.

Huang, Y., Chen, J., Sun, Y., Wang, H., Zhan, J., Huang, Y., Zou, J., Wang, L., Su, N., & Cui, J. (2022). Mechanisms of calcium sulfate in alleviating cadmium toxicity and accumulation in pak choi seedlings. *Science of the Total Environment*, *805*, 150115.

Islam, M. S., Saito, T., & Kurasaki, M. (2015). Phytofiltration of arsenic and cadmium by using an aquatic plant, Micranthemum umbrosum: Phytotoxicity, uptake kinetics, and mechanism. *Ecotoxicology and Environmental Safety*, *112*, 193–200.

Jamla, M., Khare, T., Joshi, S., Patil, S., Penna, S., & Kumar, V. (2021). Omics approaches for understanding heavy metal responses and tolerance in plants. *Current Plant Biology*, *27*, 100213. https://doi.org/10.1016/j.cpb.2021.100213

Jasrotia, S., Kansal, A., & Mehra, A. (2017). Performance of aquatic plant species for phytoremediation of arsenic-contaminated water. *Applied Water Science*, *7*, 889–896.

Jian, M., Zhang, D., Wang, X., Wei, S., Zhao, Y., Ding, Q., Han, Y., Ma, L. (2020). Differential expression pattern of the proteome in response to cadmium stress based on proteomics analysis of wheat roots. *BMC Genomics*, *21*, 1–13.

Jiu, S., Leng, X., Haider, M. S., Dong, T., Guan, L., Xie, Z., Li, X., Shangguan, L., & Fang, J. (2019). Identification of copper (Cu) stress-responsive grapevine microRNAs and their target genes by high-throughput sequencing. *Royal Society Open Science*, *6*, 180735.

Jobe, T. O., Yu, Q., Hauser, F., Xie, Q., Meng, Y., Maassen, T., Kopriva, S., & Schroeder, J. I. (2021). The SLIM1 transcription factor regulates arsenic sensitivity in Arabidopsis thaliana. *BioRxiv*, 2021-01.

Kashiwabara, T., Kitajima, N., Onuma, R., Fukuda, N., Endo, S., Terada, Y., Abe, T., Hokura, A., & Nakai, I. (2021). Synchrotron micro-X-ray fluorescence imaging of arsenic in frozen-hydrated sections of a root of Pteris vittata. *Metallomics*, *13*, mfab009.

Kozhevnikova, A. D., Seregin, I. V., & Schat, H. (2020). Accumulation of nickel by excluder Thlaspi arvense and hyperaccumulator Noccaea caerulescens upon short-term and long-term exposure. *Russian Journal of Plant Physiology*, *67*, 303–311.

Laet, C. D., Matringe, T., Petit, E., & Grison, C. (2019). Eichhornia crassipes: A powerful bio-indicator for water pollution by emerging pollutants. *Scientific Reports*, *9*, 7326.

Li, H., Pu, P., Li, X., Gong, Y., An, D., Zhang, L., & Lv, J. (2020b). Sulfur application reduces cadmium uptake in edible parts of pakchoi (Brassica chinensis L.) by cadmium chelation and vacuolar sequestration. *Ecotoxicology and Environmental Safety*, *194*, 110402.

Li, J., Zhang, M., Sun, J., Mao, X., Wang, J., Liu, H., Zheng, H., Li, X., Zhao, H., & Zou, D. (2020a). Heavy metal stress-associated proteins in rice and Arabidopsis: Genome-wide identification, phylogenetics, duplication, and expression profiles analysis. *Frontiers in Genetics*, *11*, 477.

Li, Y., Huang, J., Song, X., Zhang, Z., Jiang, Y., Zhu, Y., Zhao, H., & Ni, D. (2017). An RNA-Seq transcriptome analysis revealing novel insights into aluminum tolerance and accumulation in tea plant. *Planta*, *246*, 91–103.

Liu, W., Xiong, C., Yan, L., Zhang, Z., Ma, L., Wang, Y., Liu, Y., & Liu, Z. (2017a). Transcriptome analyses reveal candidate genes potentially involved in Al stress response in alfalfa. *Frontiers in Plant Science*, *8*, 26.

Liu, X., Zhang, H., Jiao, H., Li, L., Qiao, X., Fabrice, M. R., Wu, J., & Zhang, S. (2017b). Expansion and evolutionary patterns of cysteine-rich peptides in plants. *BMC Genom*, *18*, 1–14.

Luo, J. S., & Zhang, Z. (2019). Proteomic changes in the xylem sap of Brassica napus under cadmium stress and functional validation. *BMC Plant Biology*, *19*, 1–14.

Ma, J., Sheng, H., Li, X., & Wang, L. (2016). iTRAQ-based proteomic analysis reveals the mechanisms of silicon-mediated cadmium tolerance in rice (Oryza sativa) cells. *Plant Physiology and Biochemistry*, 104, 71–80.

Majumdar, A., Upadhyay, M. K., Ojha, M., Afsal, F., Giri, B., Srivastava, S., & Bose, S. (2022). Enhanced phytoremediation of Metal(loid)s via spiked ZVI nanoparticles: An urban clean-up strategy with ornamental plants. *Chemosphere*, *288*, 132588.

Mejias, C. L., Musa, J. C., & Otero, J. (2013). Exploratory evaluation of retranslocation and bioconcentration of heavy metals in three species of mangrove at Las Cucharillas Marsh Puerto Rico. *The Journal of Tropical Life Science*, *3*, 14–22.

Meng, Y., Zhang, L., Wang, L., Zhou, C., Shangguan, Y., & Yang, Y. (2019). Antioxidative enzymes activity and thiol metabolism in three leafy vegetables under Cd stress. *Ecotoxicology and Environmental Safety*, *173*, 214–224.

Milić, D., Luković, J., Ninkov, J., Zeremski-Škorić, T., Zorić, L., Vasin, J., & Milić, S. (2012). Heavy metal content in halophytic plants from inland and maritime saline areas. *Central European Journal of Biology*, *7*, 307–317.

Ming, R., VanBuren, R., Liu, Y., Yang, M., Han, Y., Li, L. T., Zhang, Q., Kim, M. J., Schatz, M. C., Campbell, M., & Li, J. (2013). Genome of the long-living sacred lotus (Nelumbo nucifera Gaertn.). *Genome Biology*, *14*, 1–11.

Mishra, V. K., Gupta, S., Chand, R., Yadav, P. S., Singh, S. K., Joshi, A. K., & Varadwaj, P. K. (2019). Comparative transcriptomic profiling of High-and Low-grain Zinc and Iron containing Indian wheat genotypes. *Current Plant Biology*, *18*, 100105.

Mondal, T. K., Ganie, S. A., Rana, M. K., & Sharma, T. R. (2014). Genome-wide analysis of zinc transporter genes of maize (Zea mays). *Plant Molecular Biology Reporter*, *32*, 605–616.

Murugaiyan, V., Ali, J., Mahender, A., Aslam, U. M., Jewel, Z. A., Pang, Y., Marfori-Nazarea, C. M., Wu, L. B., Frei, M., Li, Z. (2019). Mapping of genomic regions associated with arsenic toxicity stress in a backcross breeding populations of rice (*Oryza sativa* L.). *Rice*, *12*, 1–14.

Nahar, N., Rahman, A., Nawani, N. N., Ghosh, S., & Mandal, A. (2017). Phytoremediation of arsenic from the contaminated soil using transgenic tobacco plants expressing ACR2 gene of Arabidopsis thaliana. *Journal of Plant Physiology*, *218*, 121–126.

Navarro-Reig, M., Jaumot, J., García-Reiriz, A., & Tauler, R. (2015). Evaluation of changes induced in rice metabolome by Cd and Cu exposure using LC-MS with XCMS and MCR-ALS data analysis strategies. *Analytical and Bioanalytical Chemistry*, *407*, 8835–8847.

Njobvu, J., Hamabwe, S. M., Munyinda, K., Kelly, J. D., & Kamfwa, K. (2020). Quantitative trait loci mapping of resistance to aluminum toxicity in common bean. *Crop Science*, *60*, 1294–1302.

Noman, A., & Aqeel, M. (2017). miRNA-based heavy metal homeostasis and plant growth. *Environmental Science and Pollution Research*, *24*, 10068–10082.

Odinga, C. A., Kumar, A., Mthembu, M. S., Bux, F., & Swalaha, F. M. (2019). Rhizofiltration system consisting of Phragmites australis and Kyllinga nemoralis: evaluation of efficient removal of metals and pathogenic microorganisms. *Desalination and Water Treatment*, *169*, 120–132.

Panda, A., Patra, D. K., Acharya, S., Pradhan, C., & Patra, H. K. (2020). Assessment of the phytoremediation potential of Zinnia elegans L. plant species for hexavalent chromium through pot experiment. *Environmental Technology & Innovation*, *20*, 101042.

Panneerselvam, B., & Priya, K. S. (2023). Phytoremediation potential of water hyacinth in heavy metal removal in chromium and lead contaminated water. International Journal of Environmental Analytical Chemistry, 103(13), 3081–3096.

Pappi, P., Nikoloudakis, N., Fanourakis, D., Zambounis, A., Delis, C., & Tsaniklidis, G. (2021). Differential triggering of the phenylpropanoid biosynthetic pathway key genes transcription upon cold stress and viral infection in tomato leaves. *Horticulturae*, *7*, 448.

Paunov, M., Koleva, L., Vassilev, A., Vangronsveld, J., & Goltsev, V. (2018). Effects of different metals on photosynthesis: Cadmium and zinc affect chlorophyll fluorescence in durum wheat. *International Journal of Molecular Sciences*, *19*, 787.

Pegler, J. L., Oultram, J. M., Nguyen, D. Q., Grof, C. P., & Eamens, A. L. (2021). MicroRNA-mediated responses to cadmium stress in Arabidopsis thaliana. *Plants*, *10*, 130.

Phillips, D. P., Human, L. R. D., & Adams, J. B. (2015). Wetland plants as indicators of heavy metal contamination. *Marine Pollution Bulletin*, *92*, 227–232.

Qi, D., Wang, L., Liang, M., Zhang, Q., Tang, X., Geng, B., Qiao, X., & Zhang, H. (2022). Genome-wide analyses of metal tolerance protein genes in apple (Malus domestica): identification, characterization, expression and response to various metal ion stresses. *Environmental and Experimental Botany*, *201*, 104948.

Qiu, Z., Hai, B., Guo, J., Li, Y., & Zhang, L. (2016). Characterization of wheat miRNAs and their target genes responsive to cadmium stress. *Plant Physiology and Biochemistry*, *101*, 60–67.

Rahman, M. A., Hasegawa, H., Ueda, K., Maki, T., Okumura, C., & Rahman, M. M. (2007). Arsenic accumulation in duckweed (Spirodela polyrhiza L.): A good option for phytoremediation. *Chemosphere*, *69*, 493–499.

Ramos, J., Clemente, M. R., Naya, L., Loscos, J., Pérez-Rontomé, C., Sato, S., Tabata, S., & Becana, M. (2007). Phytochelatin synthases of the model legume Lotus japonicus. A small multigene family with differential response to cadmium and alternatively spliced variants. *Plant Physiology*, *143*, 1110–1118.

Rascio, N. & Navari-Izzo, F. (2011). Heavy metal hyperaccumulating plants: how and why do they do it? And what makes them so interesting? *Plant Science*, 180, 169–181.

Ray, S. & Neogy, M. (2022). Identification of vehicular emission hyper accumulator and its subsequent use as phytoremediation. *Bangladesh Journal of Botany*, *51*, 207–216.

Rosenfeld, C. E., Chaney, R. L., & Martinez, C. E. (2018). Soil geochemical factors regulate Cd accumulation by metal hyperaccumulating Noccaea caerulescens (J. Presl & C. Presl) FK Mey in field-contaminated soils. *Science of the Total Environment*, *616*, 279–287.

Sachs, J. (1865). *Handbuch der Experimental-Physiologie der Pflanzen*. Wilhelm Engelmann Verlag, Leipzig, Germany, pp. 153–154.

Sakouhi, L., Kharbech, O., Massoud, M. B., Gharsallah, C., Hassine, S. B., Munemasa, S., Murata, Y., & Chaoui, A. (2021). Calcium and ethylene glycol tetraacetic acid mitigate toxicity and alteration of gene expression associated with cadmium stress in chickpea (Cicer arietinum L.) shoots. *Protoplasma*, *258*, 849–861.

Salt, D. E., Blaylock, M., Kumar, N. P., Dushenkov, V., Ensley, B. D., Chet, I., & Raskin, I. (1995). Phytoremediation: A novel strategy for the removal of toxic metals from the environment using plants. *Bio/technology*, *13*, 468–474.

Salt, D. E., Smith, R. D., & Raskin, I. (1998). Phytoremediation. *Annual Review of Plant Biology*, *49*, 643–668.

Şentürk, İ., Divarcı, N. S. E., & Öztürk, M. (2023). Phytoremediation of nickel and chromium-containing industrial wastewaters by water lettuce. *International Journal of Phytoremediation*, *25*, 550–561.

Shukla, T., Khare, R., Kumar, S., Lakhwani, D., Sharma, D., Asif, M. H., & Trivedi, P. K. (2018). Differential transcriptome modulation leads to variation in arsenic stress response in Arabidopsis thaliana accessions. *Journal of Hazardous Materials*, *351*, 1–10.

Singh, R., Misra, A. N., & Sharma, P. (2021). Differential responses of thiol metabolism and genes involved in arsenic detoxification in tolerant and sensitive genotypes of bioenergy crop Ricinus communis. *Protoplasma*, *258*, 391–401.

Singh, S., Karwadiya, J., Srivastava, S., Patra, P. K., & Venugopalan, V. P. (2022). Potential of indigenous plant species for phytoremediation of arsenic contaminated water and soil. *Ecological Engineering*, *175*, 106476.

Singh, S., Parihar, P., Singh, R., Singh, V. P., & Prasad, S. M. (2016). Heavy metal tolerance in plants: Role of transcriptomics, proteomics, metabolomics, and ionomics. *Frontiers in Plant Science*, *6*, 1143.

Souri, Z., Karimi, N., & Ahmad, P. (2021). The effect of NADPH oxidase inhibitor diphenyleneiodonium (DPI) and glutathione (GSH) on Isatis cappadocica, under Arsenic (As) toxicity. *International Journal of Phytoremediation*, *23*, 945–957.

Stonehouse, G. C., McCarron, B. J., Guignardi, Z. S., El Mehdawi, A. F., Lima, L. W., Fakra, S. C., & Pilon-Smits, E. A. (2020). Selenium metabolism in hemp (Cannabis sativa L.)—potential for phytoremediation and biofortification. *Environmental Science & Technology*, *54*, 4221–4230.

Suman, J., Uhlik, O., Viktorova, J., & Macek, T. (2018). Phytoextraction of heavy metals: A promising tool for clean-up of polluted environment? *Frontiers in Plant Science*, *9*, 1476.

Tariq, S. R., & Ashraf, A. (2016). Comparative evaluation of phytoremediation of metal contaminated soil of firing range by four different plant species. *Arabian Journal of Chemistry*, *9*, 806–814.

Terzi, H., & Yıldız, M. (2021). Proteomic analysis reveals the role of exogenous cysteine in alleviating chromium stress in maize seedlings. *Ecotoxicology and Environmental Safety*, *209*, 111784.

Upadhyay, M. K., Gautam, A., Mallick, S., & Srivastava, S. (2017). A successive application approach for effective utilization of three aquatic plants in arsenic removal. *Water, Air and Soil Pollution*, *228*, 1.

Verma, N., Narayan, O. P., Prasad, D., Jogawat, A., Panwar, S. L., Dua, M., & Johri, A. K. (2022). Functional characterization of a high-affinity iron transporter (PiFTR) from the endophytic fungus Piriformospora indica and its role in plant growth and development. *Environmental Microbiology*, *24*, 689–706.

Wang, L., Bei, X., Gao, J., Li, Y., Yan, Y., & Hu, Y. (2016). The similar and different evolutionary trends of MATE family occurred between rice and Arabidopsis thaliana. *BMC Plant Biology*, *16*, 1–19.

Wang, S., He, T., Xu, F., Li, X., Yuan, L., Wang, Q., & Liu, H. (2021). Analysis of physiological and metabolite response of Celosia argentea to copper stress. *Plant Biology*, *23*, 391–399.

Wang, S., Wei, S., Ji, D., & Bai, J. (2015). Co-planting Cd contaminated field using hyperaccumulator Solanum nigrum L. through interplant with low accumulation welsh onion. *International Journal of Phytoremediation*, *17*, 879–884.

Wang, Y., Qiu, Q., Xin, G., Yang, Z., Zheng, J., Ye, Z., & Li, S. (2013). Heavy metal contamination in a vulnerable mangrove swamp in South China. *Environmental Monitoring and Assessment*, *185*, 5775–5787.

Wen, R. E. N., Ya, L. I. U., Zhou, M. Y., Zi, S. H. I., Wang, T. Y., Zhao, J. R., & Yu, L. I. (2019). Dynamic changes of root proteome reveal diverse responsive proteins in maize subjected to cadmium stress. *Journal of Integrative Agriculture*, *18*, 2193–2204.

Wu, H., Chen, C., Du, J., Liu, H., Cui, Y., Zhang, Y., He, Y., Wang, Y., Chu, C., Feng, Z., & Li, J. (2012). Co-overexpression FIT with AtbHLH38 or AtbHLH39 in Arabidopsis-enhanced cadmium tolerance via increased cadmium sequestration in roots and improved iron homeostasis of shoots. *Plant Physiology*, *158*, 790–800.

Yadav, R., Singh, G., Santal, A. R., & Singh, N. P. (2023). Omics approaches in effective selection and generation of potential plants for phytoremediation of heavy metal from contaminated resources. *Journal of Environmental Management*, *336*, 117730.

Yang, X. E., Long, X. X., Ye, H. B., He, Z. L., Calvert, D. V., & Stoffella, P. J. (2004). Cadmium tolerance and hyperaccumulation in a new Zn-hyperaccumulating plant species (Sedum alfredii Hance). *Plant and Soil*, *259*, 181–189.

Yannarelli, G. G., Fernández-Alvarez, A. J., Santa-Cruz, D. M., & Tomaro, M. L. (2007). Glutathione reductase activity and isoforms in leaves and roots of wheat plants subjected to cadmium stress. *Phytochemistry*, *68*, 505–512.

Yıldırım, K., & Uylaş, S. (2016). Genome-wide transcriptome profiling of black poplar (Populus nigra L.) under boron toxicity revealed candidate genes responsible in boron uptake, transport and detoxification. *Plant Physiology and Biochemistry*, *109*, 146–155.

Zhang, Y., Song, B., Zhu, L., & Zhou, Z. (2020). Evaluation of the metal(loid)s phytoextraction potential of wild plants grown in three antimony mines in southern China. *International Journal of Phytoremediation*, *23*, 781–790.

Zhao, A., Gao, L., Chen, B., & Feng, L. (2019). Phytoremediation potential of Miscanthus sinensis for mercury-polluted sites and its impacts on soil microbial community. *Environmental Science and Pollution Research*, 26, 34818–34829.

Zhao, T., Li, R., Yao, W., Wang, Y., Zhang, C., & Li, Y. (2021). Genome-wide identification and characterisation of phenylalanine ammonia-lyase gene family in grapevine. *The Journal of Horticultural Science and Biotechnology*, *96*, 456–468.

Zhao, L., Sun, Y. L., Cui, S. X., Chen, M., Yang, H. M., Liu, H. M., Chao, T. Y., & Huang, F. (2011). Cd-induced changes in leaf proteome of the hyperaccumulator plant Phytolacca americana. *Chemosphere*, *85*, 56–66.

Zheng, R., Xu, X. Y., Li, C. M., Gai, J. Y., & Yu, D. Y. (2008). Proteomic analysis of differentially expressed proteins in developmental soybean [Glycine max (L.) Merr.] seed. *Soybean Crop Intensification*, *27*, 556–563.

Zhou, Q., Guo, J. J., He, C. T., Shen, C., Huang, Y. Y., Chen, J. X., Guo, J. H., Yuan, J. G., & Yang, Z. Y. (2016). Comparative transcriptome analysis between low-and high-cadmium-accumulating genotypes of pakchoi (Brassica chinensis L.) in response to cadmium stress. *Environmental Science & Technology*, *50*, 6485–6494.

10 Resources Recovery Potential from Solid Waste and Bioeconomy Model toward Sustainable Production and Responsible Consumption

Abiola Folakemi Olaniran, Adekunbi Adetola Malomo, Bolanle Akinsanola, and Olubukola David Olaniran

10.1 INTRODUCTION

A million tons of solid waste is generated annually in rural areas. Solid waste management (SWM) has to be an integral part of an environmental management system with a significant global footprint and affordable budgets to manage it. As a global average, SWM costs had been projected to around USD 1000 per ton by 2025–2030. Waste generation is increasing at an unexpectedly rapid rate in populated countries as well as underdeveloped countries. The exponential increase in the human population, along with the swift growth of industries and cities, has resulted in a massive generation of trash. These are closely related to high living standards and per capita income, which are linked to high rates of refuse, leading to massive heaps of rubbish and lack of infrastructure for waste disposal, which prolongs the time that waste is in bins or on the roadside. According to a World Bank project, by the end of 2025, urban settlements should produce over two million tons of municipal solid waste (MSW), which is primarily made up of paper, plastic, and food and organic waste. The main issues facing SWM are related to the fundamental elements of the waste industry, such as waste production and insufficient procedures for trash collection, transportation, treatment, and disposal. The priority of a sustainable solid waste management program is to shift from traditional waste dumps that are cost intensive and hazardous to the environment toward proper waste management systems that retain useful resources within the economy (Das et al., 2019). The increase in waste generation is a global problem, and lack of responsibility from the waste generators and inadequate management of waste have worsened the challenge (da Vega et al., 2010; Vega & Bautista-Rodriguez, 2024). Solid waste is generated in schools, hospitals, stores, health facilities, hotels, beaches, and other centers. Solid waste management is a great global challenge which also causes economic loss to governments. According to the data generated by World Bank, municipal waste generation has reached 0.74 kg/day (Kaza et al. 2018). Solid waste causes pollution by generating greenhouse gases, a foul odour, contamination of water ways, floods, the spread of diseases, and can lead to a rise in temperature, thereby causing climate change. The estimated data on global generation of municipal solid waste is projected to be around 2.59 billion tonnes (BT) by 2030 and 3.4 BT by 2050 while emission of CO_2 equivalent is projected to be 2.38 BT by 2050 (Navia and Muñoz, 2015). Solid waste can be classified into non- combustible substances and combustible, high-moisture waste. Combustible waste consists of plastic, wood, paper, textiles, and other carbon

DOI: 10.1201/9781003668961-10

rich substances. This category of waste can be used in the generation of fuel, which reduces emission of CO_2 by 40% and also reduces waste in landfills by 50%. This kind of waste generates affordable and clean energy which aligns with the seventh United Nations Sustainable Development Goal (Yang et al., 2022).

10.2 SUSTAINABLE SOLID WASTE MANAGEMENT PRACTICES

The removal of materials and chemicals originating from fossil fuels as well as the combined energy and nutritional products of anaerobic digestion are linked to the largest environmental, social, and economic benefits of resource recovery from organic wastes. Multiple resources included in organic waste can be most effectively utilized by methods that complement one another. End markets, bioeconomy technology, and waste producers should foster an industrial symbiosis in order to create cascades of organic waste resources across the bioeconomy (Espinosa-Aquino et al., 2023). The concept of bioeconomy is substituting the use of biological resources that are known for being renewable for fossil resources in the form of processes, products, and services. These benefits capture nutrients which can replace frequently used organic fertilizers, biofuels (including natural gas and hydrogen), and organic precursors of chemicals obtained from fossil fuels (like pharmaceuticals or plastics) (Mihai et al., 2021). Making use of biological waste as a source of energy to power the bioeconomy can help guarantee a sustainable shift away from fossil fuels and toward bio-based resources. Additionally, the bioeconomy industry's residues and wastewaters contain potentially valuable organic and chemical resources that could be better utilized (Oguntuase et al., 2023). Through the monetization of waste and domestic biomass resources, a shift in the direction of a circular bioeconomy could improve security. Nevertheless, a lot of experts assert that the bioeconomy is inherently circular (Galanakis et al., 2022).

The "circularity" concepts must be emphasized if we are to stray from business as usual. Reducing waste and extending the useful life of goods, materials, and resources are essential components of a circular economy. According to Oguntuase et al. (2023), the circular bioeconomy places a strong emphasis on "maximizing the value of biomass over time through cascading and making use of residues and wastes in integrated, multi-output production chains (such as biorefineries) that prioritize the sustainable, resource-efficient valorization of biomass." A combination of measures, including multistakeholder collaboration and increased financial resources, must be taken in addition to the significant amounts of public and private funds that have already been provided globally through stimulus packages to promote the sustainable circular bioeconomy and ensure a swift and efficient transition (Fritsche et al., 2020). Furthermore, as small-scale local biorefineries adhere to rural development guidelines and take advantage of chances for resource-efficient payback chains and leverage, as well as unique capabilities within their various contexts, it should be a top priority to help them (Panoutsou & Chiaramonti, 2020). As reported by researchers, there should be sustenance of nature-based solutions like agroecology, reforestation, and interventions for low-carbon development in order to promote an effective green economic recovery. Enhancements in agricultural value chains that support biodiversity and sustainable food systems may be made easier by the extra incentive (Burger et al., 2018). These include, among other things, off-grid rural electrification, the development of low-energy buildings and the preservation of natural resources, and the integration of artificial intelligence (AI) and Internet and communication technologies (ICT) into industry. There could be an emergence of new economic models, production and consumption patterns, social norms, and governance structures. New developments may also benefit the manufacturing and food sectors in terms of processing (such as robotics-assisted food production) and production (such as carbon farming and climate-smart forestry) (Fritsche et al., 2020). Furthermore, the decentralization of food systems and biorefineries could safeguard smallholders, businesses, farmers, and consumers (for example, by using clever specialty finance schemes that support the concept of "biocities") (Fritsche et al., 2020).

10.3 THE SOCIOLOGICAL PERSPECTIVE OF SOLID WASTE RESOURCE RECOVERY AND ITS SUSTAINABILITY

Understanding and achieving sustainability is a critical issue in an era when our planet faces significant environmental, economic, social, and political challenges (Islam, 2017). Gould and Lewis emphasized the importance of this issue by stating, "The increasing urgency of the negative impacts of social systems on ecosystems created both the social space and social need for the emergence of environmental sociology." Islam (2017) argues that one of the key challenges in contemporary society is the correlation between social and environmental vulnerability. With the foregoing background, the sociological perspective to resource recovery from solid waste emphasizes the importance of considering waste not just as a nuisance or commodity but as a valuable resource that can benefit both the environment and society. This perspective highlights the social facets of waste management, advocating for an integrated, interdisciplinary approach that acknowledges the role of production, consumption, and lifestyle habits in waste generation (Concari et al., 2020). It also underscores the need for inclusive resource recovery strategies, such as informal but organized recycling initiatives, which not only contribute to environmental awareness but also have social benefits for waste collectors (Khan et al., 2022). Furthermore, the transition toward a circular economy is seen as essential for achieving sustainable resource management goals, requiring scientific and technological advancements, as well as changes in mentality, industry practices, and policies and regulations in the waste and resource management landscape (Raghu & Rodrigues, 2020).

Sanz-Hernández et al. (2019) identify significant gaps in the study of the bioeconomy and resource recovery from the perspective of the social sciences, highlighting several shortcomings in the existing literature. While previous research has predominantly concentrated on key areas such as energy-related issues, the agricultural and food security sector, health markets and human tissues, and waste resources and recycling, as well as strategies, policies, and governance of bioresources, it has overlooked crucial aspects. Specifically, there is a notable deficiency in studies examining the analytical role played by stakeholders in 'innovation niches' that drive bioeconomic dynamics, especially in relation to sustainability.

The concept of innovation niches refers to protected spaces where radical innovations can develop and mature, shielded from the prevailing market and regulatory pressures. These niches are crucial for the experimentation and development of new technologies, practices, and business models that can eventually transform mainstream markets and contribute to sustainable bioeconomic growth. Stakeholders, including entrepreneurs, researchers, policymakers, and civil society organizations, play a pivotal role in these niches by providing the necessary resources, knowledge, and support for innovation processes (Kędzia et al., 2022). The existing literature has extensively covered the technical and policy dimensions of the bioeconomy, exploring how bioresources can be harnessed for energy production, enhancing agricultural productivity, ensuring food security, and managing health and waste resources. These studies have provided valuable insights into the potential and challenges of bioeconomic strategies, emphasizing the importance of policy frameworks and governance structures in supporting bioeconomic initiatives. However, the social sciences have not adequately addressed the complex interactions among stakeholders that are essential for fostering innovation and promoting sustainable practices within the bioeconomy (Dunlap & Catton, 2022). Despite their importance, the role of these stakeholders in activating and sustaining bioeconomic dynamics remains underexplored. Understanding their interactions, motivations, and contributions is vital for creating effective strategies that foster innovation and ensure long-term sustainability of the bioeconomy and resource recovery. This gap in the literature calls for a more interdisciplinary approach that integrates social science perspectives into the study of the bioeconomy and resource recovery, focusing on the socio-economic and institutional factors that influence stakeholder engagement and innovation in this field. For example, in advocating for a transition to a circular economy to improve resource recovery and waste management, Gutberlet and de Carvalho Vallin (2024) emphasized the necessity of considering the socio-economic implications and their effects on societal sustainability.

The Resource Recovery from Waste program by United Nations as part of SDG 12 aims to develop both technical and social solutions for waste and resource management to maximize societal benefits. The program underscores the importance of stakeholder engagement in implementing sustainable solutions, highlighting the need for interventions that are high quality, legitimate, timely, and cost effective. This approach aligns with sustainability theory by emphasizing the integration of social, economic, and environmental considerations in developing sustainable practices (Tsai et al., 2021).

Sustainability theory encompasses the concept of meeting present needs without compromising future generations' ability to meet their own needs while maintaining ecological processes and biodiversity (Finlay, 2023; Zahedi, 2019). It emphasizes the importance of using natural resources judiciously to prevent environmental degradation and ensure long-term viability for human societies (Shakir Hanna & Cesaretti, 2019). The theory suggests a shift in conceptual frameworks toward an integrated approach with interconnected pillars focusing on political, economic, social, cultural, technological, ecological, and spiritual development within local, national, regional, and global contexts (Zahedi, 2019).

Sustainability theory incorporated diverse perspectives that emphasize and integrate social responses to environmental and cultural challenges. The theories strive to prioritize these responses, addressing the multifaceted problems faced by modern societies (Debrah et al., 2021). According to Robertson (2021), various models have been identified with different foci. An economic model, for instance, aims to sustain both natural and financial capital, recognizing the need for a balance between economic growth and environmental preservation. This approach focuses on maintaining the resources and financial stability necessary for long-term prosperity. An ecological model, on the other hand, prioritizes biological diversity and ecological integrity. It emphasizes the importance of preserving ecosystems and the myriad species they support, ensuring that natural processes continue to function effectively. This model underscores the need to protect the environment to maintain the planet's health and resilience. A political model of sustainability seeks to develop social systems that uphold human dignity. It focuses on creating equitable and just societies where individuals' rights and well-being are protected. This model highlights the importance of governance and policies that promote social equity, inclusion, and participation in decision-making processes. Religion is another model that has entered the sustainability discourse, offering symbolic, critical, and motivational resources for cultural change (Lestar & Böhm, 2020). Religious perspectives provide ethical frameworks and moral imperatives that can inspire individuals and communities to adopt more sustainable practices. These resources can motivate cultural shifts toward greater environmental stewardship and social responsibility. It was concluded that by integrating these various models, sustainability theory presents a holistic approach to addressing environmental and cultural challenges. It advocates for solutions that are not only ecologically sound and economically viable but also socially just and culturally meaningful (Brotosusilo et al., 2020). According to Longo et al. (2021), sustainability concerns are fundamentally sociological in nature since they stem from social structures, dynamics, and causes and necessitate modifications to social processes and human behavior. For instance, why do environmental issues like soil, water, and air pollution continue to exist if they are undesirable? Sociology provides helpful conceptual frameworks and analyses to assist in addressing this basic yet seemingly straightforward topic. Increasing the breadth and depth of our knowledge about the causes and mechanisms of environmental issues also increases the likelihood of finding successful solutions. Four overlapping conceptual and methodological aspects have also been identified within sociology that should be integrated into the core of sustainability theory analysis and research. These aspects include the role of social construction and critical realism, the dynamic between structure and agency, historical specificity, and the necessity of collective action.

Sociologists examine social relations within the framework of structures and institutions, where human interactions are guided by norms, rules, regulations, and values. Longo et al. (2021) further argues that although the field of sociology could improve in recognizing the natural foundations of social life, its insights into social relations, processes, structures, and institutions can be effectively extended to study society–environment dynamics. By doing so, we can deepen our understanding of the complex interplay between societal factors and environmental issues, ultimately fostering more holistic approaches to addressing these challenges. The integration of these sociological perspectives into sustainability research brings to light the complex interplay within and between various scales and levels of organization. It highlights how social contexts and fabrics influence knowledge and actions, showing that human lives are nested within and between social institutions. This perspective emphasizes the structural conditions that drive environmental degradation or restoration and explores the obstacles and opportunities for facilitating social change toward sustainability (Ma et al., 2024).

These insights underscore why humanity cannot resolve global environmental crises solely through technological solutions, individual behavior changes, or specific policy adjustments. Instead, they illustrate that viable sustainability solutions must address the intricate social relationships and structures that shape history. By grappling with these complex social dynamics, we can better understand and address the root causes of environmental challenges and develop more comprehensive and effective strategies for sustainable development. This comprehensive perspective is crucial for developing effective strategies to ensure a sustainable future for all. Researches have highlighted the significance of sustainability for societal development, urging for a co-evolutionary relationship between theories and society to accelerate progress toward more sustainable practices and policies (Fadhullah et al., 2022).

10.4 RESOURCE RECOVERY TECHNOLOGIES AND THE BIOECONOMY MODEL

Every year, solid waste is produced in both urban and rural locations without adequate consideration for locations in which there are few technological waste management techniques, particularly in the context of emerging and developing nations. The steady rise in human population, along with the swift pace of urbanization and industrialization, has resulted in a massive generation of trash. These activities are directly related to the deficiency of infrastructure for disposing of waste, which prolongs the time that waste remains in bins or by the side of the road. High rates of refuse are correlated with high living standards and per capita income, creating massive garbage piles. According to a World Bank project, urban settlements should produce about 1.3 million tons of MSW annually, and by the end of 2025, that amount should have doubled (Das et al., 2019). Solid recovery waste management is an essential component of any environmental management system with an international impact and substantial operating costs. The key elements of the waste sector are waste generation and processes for garbage collection, transportation, treatment, and disposal, which are the main obstacles facing waste recovery models. Traditional landfills, which can be costly and detrimental to the environment, should be replaced as a top priority by sustainable solid waste management programs with appropriate waste management techniques that preserve precious resources within budgetary constraints (Satpute et al., 2024). The waste industry needs to hire more qualified personnel and raise social awareness in order to overcome these disadvantages. Two crucial strategies for lowering waste and enhancing waste recovery include putting recycling equipment in buildings and teaching people how to generate less trash (Loan et al., 2023). Long-term disposal of waste in public areas ought to be regarded as a health issue. For sustainable waste management, solid waste must be properly categorized based on its constituents, including organic, glass, plastic, metallic, and flammable materials.

Because bioeconomy uses renewable biotechnological sources to produce energy, products, and materials across all viable economic sectors, it has been hailed as the miracle solid waste recovery paradigm. In order to improve the utilization and efficiency of waste materials, the bioeconomy model can offer new opportunities for lowering the production of solid waste and promoting the sustainable manufacturing of bioproducts from waste materials (Qin et al., 2021).

With a steady supply of environmentally friendly power, materials, and substrates, the bioeconomy is crucial for sustaining the circular economy. A wide range of thermochemical and biological processes can be used to produce waste-to-energy and several bioproducts through a chain of actions based on sustainability. An industrial process that is restorative and regenerative in nature, eliminating harmful generation while reusing and recycling waste materials and raw resources, is known as a circular economy (Guardia et al., 2019). In view of the hurdles humanity encounters from climate change, resource and energy depletion, and energy and food security, it is anticipated that the circular bioeconomy will grow sustainably. In order to optimize or maximize the social, environmental, and economic benefits by producing effective and sustainable products from solid waste, the waste recovery model primarily focuses on the generation of energy and biodegradable products by the combination of different processes in the utilization of solid waste (Duan et al., 2020).

In the midst of emerging and developing nations, the situation is even more concerning in rural areas where there are few scientific waste management solutions. Landfilling is the most often used method of disposing of waste worldwide because it is so easy to adopt. Many landfills remain neglected, with significant pollution, because of the hefty management costs involved. In solid waste management, landfilling is the process of systematically depositing waste products into specially designed landfills. It is frequently considered the most common method of garbage disposal. This strategy seeks to reduce the dangers to the public's health and the environment from waste accumulation while simultaneously addressing the urgent need for waste containment. Various forms of solid waste, such as construction, industrial, and municipal trash, are placed and compacted in controlled levels in landfills, which are designated locations. In order to guarantee containment, leachate management, and gas control, a number of engineering and operational procedures are used in landfill management. By taking these steps, possible environmental risks such soil erosion, contamination of groundwater, and pollution of the air are to be reduced. The first steps in the landfilling process involve choosing and preparing appropriate land locations, taking into account things like proximity to population centers, geological features, and legal requirements. Impermeable liners and leachate collection systems are commonly installed at landfills to stop pollutants from seeping into the nearby soil and groundwater. Moreover, vegetation cover and designed caps are used to reduce erosion and surface runoff while enhancing long-term stability. Over time, waste products dumped in landfills experience physical, chemical, and biological changes influenced by several factors including temperature, moisture content, and microbial activity. Organic matter decomposes as a result of these processes, producing landfill gas that is mostly made up of carbon dioxide and methane. Modern landfills use gas collection and recovery systems, which extract and treat landfill gas for beneficial use or flaring, to reduce greenhouse gas emissions and capture energy potential (Zeller et al., 2019). Even though landfilling is widely used, it has a number of environmental issues and obstacles. The most significant is the possibility of harmful materials from decaying waste contaminating groundwater. If the emissions of landfill gases are not effectively managed, they can represent a safety issue and contribute to climate change, especially methane emissions. Landfills can also have an impact on local communities' quality of life and property prices through their visual and olfactory effects. Alternate waste management techniques, such as waste reduction, recycling, composting, and waste-to-energy technologies, have been developed in response to these difficulties. Even while there are ways to reduce trash production and recover valuable resources using these approaches, landfilling is still an essential aspect of integrated solid waste management systems, especially for residual and nonrecyclable waste streams (van Praagh and Liebmann, 2021). Another important factor pertaining to solid waste is odor. The main volatile odoriferous substances released from landfills include benzoic acid, limonene, alpha-pinene, 2-butanone, beta-pinene, tetrachloroethylene, and dimethyldisulfide. Any waste management strategy will contain seven different categories of costs: back-end, remediation, upfront, social, operating, contingent, and environmental. The majority of waste management expenses are accounted for by upfront and operating expenditures, which include the out-of-pocket expenses for cars, waste collecting equipment, and personnel hires and maintenance to keep the system operational (Sonibare et al., 2019).

Knowledge on the physical composition is imperative to characterize and classify MSW for its proper management. SDG 12 Responsible Consumption and Production, where CBE (circular bioeconomy) can, for instance, drive the recovery of value from organic waste (e.g., wheat husk) into products that replace nonrenewable alternatives (e.g., fabrics, yarns, and threads) (Gonçalves & Maximo, 2023), or waste biomass from seafood production can be collected and processed into industrial compounds (e.g., mussel processing effluents into hyaluronic acid) on a commercial scale (Venugopal, 2022). Regardless of the location or nation, the first stage in every waste management strategy is monitoring the creation of waste. To improve the collection of waste from storage bins and vehicles, a number of advanced waste monitoring technologies have recently been developed, including geographic information systems, radio-frequency identification, ultrasonic sensors, and the international system for mobile/general radio packet service. Because it produces materials and soils that store carbon, biomass is a renewable carbon source that has the potential to replace carbon derived from fossil fuels in the economy when it comes to the production of food, fiber, chemicals, and fuels. For fossil fuel-dependent economies, biomass presents a local opportunity to sustainably contribute to global and national climate-related targets and transition to a decarbonized, more sustainable bioeconomy (Garvie et al., 2024).

Industrial production is still primarily dependent on nonrenewable resources and is excessively linear, even with the recent push for a circular economy. On a global scale, only a small percentage (12%) of the materials are derived from recycling (Mohan et al., 2019). In contrast, nonmetallic minerals such as sand or gravel account for around half of the extracted resources (Galanakis et al., 2022). To create resource-efficient, circular, and low-carbon solutions based on renewable energy and sustainably derived bio-based materials, adaptable innovations and workable technologies could be implemented. An automobile made of the biomaterial nanocellulose, which was produced in Japan in 2019, was five times lighter and stronger than steel. New biomaterials, including bioplastics, hold tremendous promise due to their lower carbon footprint and biodegradability than petrochemical products (Panoutsou & Chiaramonti, 2020). Products derived from wood, such as wood-based textiles, nanocellulose, and bioplastics, offer a source of sequestered carbon that can be utilised for construction, furniture, textiles, and fiber. An approach toward green recovery, climate change mitigation, and resilience in the post-pandemic world is valorizing woody biomass to produce a wide range of bio-based materials (Fritsche et al., 2020). New wood-based textiles have been reported to have a climate mitigation effect of 5 kg CO_2 per kg of product used compared to polyester. Moreover, a shift to biomaterials (based on engineering wood or bamboo) could substantially reduce the number of materials used and our cities' carbon footprint while creating durable carbon pools (Galanakis et al., 2022). Compared to concrete, the use of wood in construction has a climate mitigation effect of 2.4–2.9 kg CO_2 per kg of wood contained in products utilized, whereas each cubic meter of products stores one ton of carbon dioxide. Additionally, wood building is more resource-efficient, with the potential to cut construction materials usage by 50%. It is a top priority in the green recovery movement. However, there shouldn't be any more strain placed on natural resources as a result of the growing need for biomaterials. Cascading the use of biomass – which is a fundamental part of a circular bioeconomy – will contribute to reducing additional pressure on land for biomass (Fritsche et al., 2020).

The most meaningful way for waste reduction must be to educate citizens to produce less waste and also to install in-house recycling equipment to reduce the total waste load (Yang et al., 2022). Waste that is left in public spaces for longer periods of time should be considered a health risk (Loan et al., 2023). Although it is not commonly done, properly classifying waste according to its contents, that is, organic, plastics, glass, metal, and combustibles, is also crucial for sustainable waste management. The only ways to remedy these shortcomings are to increase social awareness and send skilled professionals into the trash business.

10.5 RECENT RESEARCH AND INNOVATION IN RESOURCE RECOVERY AND BIOECONOMY

Recent advances in resource recovery and bioeconomy include the development of medications, biofuels, renewable energy sources, green chemistry, and sustainable materials to boost economic activity and benefit the public. Nature-based inventions, such as biological control and helpful microbes, have recently advanced the bioeconomy by improving resource recovery and agricultural sustainability (Maciejczak, 2023). Solid waste recovery systems' resilience and mitigation strategies that allow adapting rapidly to inevitable crises should become a priority, ensuring that future shocks and extreme events will minimally affect food chains and vulnerable people (Mihai et al., 2021). The choice of technology to be used would rely upon the production and physical, concoction, and natural properties of the wastes (Chen et al., 2019). Thermochemical models are generally appropriate for low dampness content from municipal solid wastes, dry wastes, and lignocellulosic biomass; methods such as direct burning, gasification, and pyrolysis are recommended. Increasing circularity and integrating thermochemical and biochemical processes are also essential for increasing waste's value and producing biofuels and biobased products. Critical raw materials can be extracted by biorefinery methods, which can facilitate integration. In a more general view, it is prime time and an excellent opportunity to develop a transformative, circular, inclusive, and sustainable bioeconomy that includes all citizens, fosters innovation, and provides at least partial economic recovery solutions in the post–COVID-19 world (Javourez et al., 2024). It is vital for a paradigm shift from the well-known slogan of "no one left behind" to "leaving no one out." This change could be achieved in practice by promoting short- and long-term strategies and actual measures supporting communities, stake-holders, and operators to preserve and diversify economic activities, keep jobs, and ultimately build the required resilience to overcome the crisis (Garvie et al., 2024).

Coupled with decarbonization and sustainable production, these initiatives should encourage private sector investment in this area and track the effects of mitigation strategies. Bio-based developments are varied and spread across several disciplines. Some sample areas enumerated by Xu et al. (2021) include biomolecules, which involve molecule mapping, measurement, and engineering; biosystems, which involve engineering cells, tissues, and organs to create solutions like lab-grown meat, artificial organs for transplants, and advanced models for disease research; biomachines, which refer to the interface between biology and machines, such as biosensors for environmental monitoring, wearable health monitors, and robotic systems that mimic biological processes; agricultural innovations, of which typical examples include genetically modified crops, gene expression control methods, and biodegradable biosurfactants; material innovations, which include biocompatible proteins for tunable proton conduction and genetically modified fiber crops for waterproofing; and lastly, waste management, which involves utilizing synthetic organisms to break down and convert plastic waste into useful products (Pacheco-López et al., 2023).

A critical review by Huntington et al. (2022) on innovative resource recovery from industrial sites revealed that the adoption of green solvents offers both an eco-friendly alternative and the opportunity to rejuvenate waste from industrial sites. Also, green solvents play a crucial role in sustainable chemistry, emphasizing reduced environmental impact and resource efficiency. This approach aligns with circular economy principles and contributes to economic growth. Research has also been done on a number of devices to convert solid waste energy into clean water, bio-items, power, vaporous and fluid biofuel, and heat. Reusing solid waste entails collecting, repurposing, and recovering specific waste elements such as metal, paper, glass, and plastics to create novel substances or products. Recycling is a specific waste management technique that involves converting waste materials into new products, being one of the key strategies in a circular economy aimed at reducing the consumption of raw materials, thereby reducing energy usage and pollution (Tonini et al., 2021). Using recyclable materials to assemble reusable things, solid recovered waste such as damaged wood, paper, card, and plastics can be burned or converted to syngas, which can then be used to produce bioenergy or advanced biofuels. The decision of technology would rely upon the production and physical, concoction, and natural properties of the wastes (Chen et al., 2019). Burning reduces the volume, and the steam that is created can be used to generate energy. Appropriate separation and division of metropolitan waste into the recyclable part, leftover waste, and natural portion followed by their pretreatment is a key for feasible waste management practices. The circular economy model also extends the lifecycle of products through refurbishment, remanufacturing, and reuse in addition to recycling, which are focused on the model (Ashani et al., 2020).

Improved methods for the collection, transport, treatment, and disposal of waste, aimed at managing waste in a way that it becomes a resource rather than a burden, have been reported (Lin et al., 2022; Zan et al., 2022). These include adaptation of blockchain technologies, machine learning (ML), AI, internet of things, and industry 4.0 as technological enablers are not just supplementary tools but rather integral components that can significantly amplify the impact of circular economic models (Lei et al., 2023). Transparency and traceability are two of blockchain technology's main advantages (Frikha et al., 2024). The circular economy, which includes consumers, recyclers, and manufacturers among other stakeholders throughout a product's lifecycle, depends on innovations like this one to continue. The technology's capacity to track individual items in a transparent and safe manner enhances customer confidence and streamlines regulatory compliance, making the larger-scale adoption of solid waste recovery easier (Meier et al., 2023). AI and ML are two emerging technologies that are seen to be essential for automating and improving various operations inside the circular economy models. Particularly machine learning can adjust and improve over time, offering ever-more-efficient solutions for resource allocation, waste management, and even machinery predictive maintenance to prolong product lifecycles (Zan et al., 2022). Meanwhile, AI algorithms can examine vast datasets to spot resource-use inefficiencies, potentially opening the door to more environmentally friendly practices. ML may be used to automate waste sorting processes, predict waste generation patterns, and optimize collection routes, as reported by Prioux et al. (2023). As branches of ML and AI, big data, digitization, deep learning, artificial neural networks, and other emerging technologies are beginning to be recognized for their potential. Applications of artificial neural networks and deep learning for the circular economy model include better waste-sorting algorithms and predictive analytics for resource allocation. Large and complex datasets can be analyzed by these methods (Thaseen Ikram et al., 2023). Deep learning, with its ability to process vast amounts of unstructured data, can be used to identify and segregate recyclables from waste streams using image recognition; optimization techniques, when combined with neural networks, can further enhance these processes (Lin et al., 2022). Digitalization is an extensive process of integrating digital technologies into all aspects of economic activity from consumer engagement through apps that encourage recycling, to more complex tasks like automating waste sorting and resource recovery (de Oliveira et al., 2021; Seyyedi, et al., 2024). These digital platforms enable peer-to-peer sharing of resources, thereby extending product lifecycles and reducing waste, thus having the capacity to make circular practices more accessible, efficient, and scalable, so that while the deep learning can provide the analytical backbone for monitoring and optimizing resource flows in real time, digitalization refers to the broader integration of digital technologies into all aspects of the circular economy, pointing to a future where data-driven decision-making becomes the norm rather than the exception (Govindan, 2023). Big data can serve as the analytical engine that powers decision-making at various levels, since it has the ability to collect, analyze, and interpret large volumes of data in real time, allowing for more dynamic and responsive systems, for example, big data analytics can help in optimizing supply chains by predicting demand, thereby reducing overproduction and waste, as well as assist in resource allocation, ensuring that materials and energy are used as efficiently as possible, so that its importance lies in its ability to provide actionable insights that can drive sustainability at scale (Behl et al., 2023). In order to effectively perform predictive maintenance, optimize energy use, and manage solid waste, real-time data gathering and monitoring are critical to the adoption of the internet of things. This is achieved by emphasizing the interconnectedness of devices and systems. Interconnected sensors provide actionable insights that are imperative for immediate modifications, thus contributing to a sustainable model (Bibri et al., 2024). Industry 4.0 connotes the fourth industrial revolution, characterized by the integration of digital technologies into traditional industries showcasing the transformative potential of digital technologies to revolutionize manufacturing processes, supply chain management, and even consumer engagement, making them more efficient, transparent, and sustainable. Their importance in accelerating the transition to a circular and sustainable economic model offers unique capabilities when integrated for implementing and upscaling circular economic practices (Faisal, 2023). Studies have been reported on models for the bioeconomy and circular economy for synergies for biological resources maintenance, provision of renewable inputs for industries, and reduced dependence on solid and nonrenewable resources, while innovations in biotechnology have led to the development of bio-based chemicals and fuels, offering sustainable alternatives to their petrochemical counterparts (Laibach et al., 2019; Tsapekos et al., 2021). Research on extension of product lifecycles as key goals of the circular economy makes lifecycle assessment a methodological framework for evaluating the environmental impacts of a product or service throughout its entire lifecycle, from raw material extraction to production, use, and disposal (Antwi-Afari et al., 2023; Reslan et al., 2022). It provides valuable insights into where waste is generated and where efficiencies can be increased using principles of circularity that promote a lifecycle perspective to minimize waste and maximize resource efficiency (Duan et al., 2020; Toniolo et al., 2024). Smart cities signify an integrated approach to sustainability, where digital technologies are leveraged to improve urban services, reduce environmental impact, and enhance quality of life, serving as incubators for testing and implementing new technologies and practices, for example, sensor networks can monitor waste generation and recycling rates, while smart grids can optimize energy use across different sectors (Bibri et al., 2024). Acting as living labs for innovation, they can scale successful practices to larger populations (Nguyen et al., 2022). These technologies offer endless possibilities, for example, to optimize logistics, inventory management, and the use of scarce inputs, such as energy, water, or mineral resources; increase the efficiency, transparency, and accountability of the waste management process; create transparent and immutable records for sustainability metrics; develop decentralized platforms for sharing resources, which can be used by small firms to participate in the circular economy; provide consumers with verifiable information about the sustainability of their purchases, thereby encouraging more responsible consumption; smart cities; or the simplification of regulatory compliance (Khan et al., 2023). Models for product traceability are becoming increasingly important in a world concerned with ethical consumerism and sustainable supply chains, for example, these technologies can be utilized to analyze the lifecycle of products, from raw materials to end of life, and make this information transparent and accessible to consumers, which could revolutionize recycling efforts, as consumers and waste management entities would know exactly what materials are in a product and how to recycle them effectively; advanced encryption algorithms can be developed to secure this data, and ML algorithms could monitor these systems in real time for any security breaches or anomalies; secure and traceable data could provide verifiable proof of a product's environmental impact, which could be crucial for regulatory compliance and influence consumer behavior toward more sustainable choices (Frikha et al., 2024).

Studies on the use of new technologies with biorefinery model involve converting biomass into fuels, power, and value-added chemicals (Kiehbadroudinezhad et al., 2023). New technologies may be raised, for example, to explore digital technologies in monitoring, managing, and optimizing biorefinery operations, thus enhancing their efficiency and productivity; drive the feasibility and efficiency of using plastic waste in biorefinery processes; enhance the development of secure and transparent systems to trace and manage the lifecycle of plastic products; or allow the design of new products easily recyclable in biorefineries to create a closed-loop system where waste is minimized (Galanakis et al., 2022). Accordingly, some studies showed that the advancement of ecologically generous products and ventures could build up a practical, low-carbon environment that offers companies unprecedented opportunities to optimize resource use, improve process efficiency and reduce waste, although their true potential is realized when they are integrated into business models that advocate circularity, reuse, and regeneration (Chauhan et al., 2022; Erol et al., 2023; Esmaeilian et al., 2023; Turner et al., 2021).

10.6 FUTURE TRENDS AND OUTLOOK TOWARD TECHNOLOGICAL ADVANCEMENTS AND SUSTAINABILITY

In order to balance the financial, environmental, and cultural challenges brought on by the current direct use of resources, the circular economy has been employed to convert the value chain from a straight to a closed loop and improve the productivity of asset consumption. The bioeconomy is dependent on the conversion of renewable carbon stored in natural wastes and rural biomass into a wider range of products and commodities, such as feed, food, feed-based synthetic chemicals, bioenergy, and biopolymers. The core principles of the circular economy align with the bioeconomy and should promote the recycling and reuse of resources in an effort to lay the groundwork for integrated practical methods. A shift to a circular bioeconomy is also anticipated to establish a strong position on the creation of jobs and favorable financial improvement through social, mechanical, and authoritative innovation.

The adoption of the models will boost the productive cooperation of experts, academics, lawyers, financial analysts, and approach developers, all of whom are essential to establishing research partnerships for the reconstruction of solid waste, thus validating the circular economy as an effective asset that should be explored by companies to leverage novel technologies and generate value in a way that is both environmentally friendly and profitable. Becoming a responsible producer and consumer will lessen the load on collection services, treatment facilities, and final disposal facilities.

REFERENCES

Antwi-Afari, P., Ng, S. T., Chen, J., Oluleye, B. I., Antwi-Afari, M. F., & Ababio, B. K. (2023). Enhancing life cycle assessment for circular economy measurement of different case scenarios of modular steel slab. *Building and Environment*, *239*, 110411.

Ashani, P. N., Shafiei, M., & Karimi, K. (2020). Biobutanol production from municipal solid waste: Technical and economic analysis. *Bioresource Technology*, *308*, 123267.

Behl, A., Singh, R., Pereira, V., & Laker, B. (2023). Analysis of Industry 4.0 and circular economy enablers: A step towards resilient sustainable operations management. *Technological Forecasting and Social Change*, *189*, 122363.

Bibri, S. E., Huang, J., Jagatheesaperumal, S. K., & Krogstie, J. (2024). The synergistic interplay of artificial intelligence and digital twin in environmentally planning sustainable smart cities: A comprehensive systematic review. *Environmental Science and Ecotechnology*, *20*, 100433.

Brotosusilo, A., Nabila, S. H., Negoro, H. A., & Utari, D. (2020). The level of individual participation of community in implementing effective solid waste management policies. *Global Journal of Environmental Science and Management*, *6*(3), 341–354.

Burger, C., Kalverkamp, M., & Pehlken, A. (2018). Decision making and software solutions with regard to waste management. *Journal of cleaner production*, *205*, 210–225.

Chauhan, A., Sharma, N. K., Tayal, S., Kumar, V., & Kumar, M. (2022). A sustainable production model for waste management with uncertain scrap and recycled material. *Journal of Material Cycles and Waste Management*, *24*(5), 1797–1817.

Chen, J., Hua, C., & Liu, C. (2019). Considerations for better construction and demolition waste management: Identifying the decision behaviors of contractors and government departments through a game theory decision-making model. *Journal of Cleaner Production*, *212*, 190–199.

Concari, A., Kok, G., & Martens, P. (2020). A systematic literature review of concepts and factors related to pro-environmental consumer behaviour in relation to waste management through an interdisciplinary approach. *Sustainability*, *12*(11), 4452.

da Vega, C. A., Taboada-González, P., Aguilar-Virgen, Q., & Ojeda-Benítez, S. (2010). Solid waste generation and composition in Mexico: A comparison between rural and urban communities. *Journal of Environmental Science and Engineering*, *4*(11), 71.

Das, S., Lee, S. H., Kumar, P., Kim, K. H., Lee, S. S., & Bhattacharya, S. S. (2019). Solid waste management: Scope and the challenge of sustainability. *Journal of cleaner production*, *228*, 658–678.

Debrah, J. K., Vidal, D. G., & Dinis, M. A. P. (2021). Raising awareness on solid waste management through formal education for sustainability: A developing countries evidence review. *Recycling*, *6*(1), 6.

Duan, Y., Pandey, A., Zhang, Z., Awasthi, M. K., Bhatia, S. K., & Taherzadeh, M. J. (2020). Organic solid waste biorefinery: Sustainable strategy for emerging circular bioeconomy in China. *Industrial Crops and Products, 153*, 112568.

Dunlap, R. E., & Catton, W. R. (2022). Toward an ecological sociology: The development, current status, and probable future of environmental sociology. In William V. D'Antonio, Masamichi Sasaki, Yoshio Yonebayashi (Eds.), *Ecology, World Resources and the Quality of Social Life* (pp. 11–31). Routledge.

Erol, I., Peker, I., & Ar, I. M. (2023). Accelerating the transition to a circular economy: An investigation on the enablers of blockchain-based solar and wind energy supply chains. In Suhaib A. Bandh, Fayaz A. Malla, and Anh Tuan Hoang (Eds.), *Renewable Energy in Circular Economy* (pp. 219–237). Springer International Publishing.

Esmaeilian, S., Mohamadi, D., Esmaelian, M., & Ebrahimpour, M. (2023). A multi-objective model for sustainable closed-loop supply chain of perishable products under two carbon emission regulations. *Journal of Modelling in Management, 18*(2), 285–317.

Espinosa-Aquino, B., Gabarrell Durany, X., & Quirós Vargas, R. (2023). The role of informal waste management in urban metabolism: A review of eight Latin American countries. *Sustainability, 15*(3), 1826.

Fadhullah, W., Imran, N. I. N., Ismail, S. N. S., Jaafar, M. H., & Abdullah, H. (2022). Household solid waste management practices and perceptions among residents in the East Coast of Malaysia. *BMC Public Health, 22*, 1–20.

Faisal, M. N. (2023). Role of Industry 4.0 in circular supply chain management: A mixed-method analysis. *Journal of Enterprise Information Management, 36*(1), 303–322.

Finlay, E. (2023). Sustainability. *Development in Practice, 33*(1), 1–2.

Frikha, T., Ktari, J., Amor, N. B., Chaabane, F., Hamdi, M., Denguir, F., & Hamam, H. (2024). Low power blockchain in industry 4.0 case study: Water management in Tunisia. *Journal of Signal Processing Systems, 96*(3), 257–271.

Fritsche, U., Fritsche, U., Brunori, G., Chiaramonti, D., Galanakis, C.M., Hellweg, S, Matthews, R. & Panoutsou, C. (2020). *Future transitions for the Bioeconomy towards Sustainable Development and a Climate-Neutral Economy Knowledge Synthesis Final Report*. https://doi.org/10.2760/667966

Galanakis, C. M., Brunori, G., Chiaramonti, D., Matthews, R., Panoutsou, C., & Fritsche, U. R. (2022). Science of the total environment bioeconomy and green recovery in a post-COVID-19 era. *Science of the Total Environment, 808*, 152180. https://doi.org/10.1016/j.scitotenv.2021.152180; *Cultivation in Marginal Land for Bioenergy*. 1–22.

Garvie, L. C., Lee, D. J., & Kulišić, B. (2024). Towards a bioeconomy: Supplying forest residues for the Australian market. *Energies, 17*(2), 397

Gonçalves, M. L. M., & Maximo, G. J. (2023). Circular economy in the food chain: Production, processing and waste management. *Circular Economy and Sustainability, 3*(3), 1405–1423.

Govindan, K. (2023). How digitalization transforms the traditional circular economy to a smart circular economy for achieving SDGs and net zero. *Transportation Research Part E: Logistics and Transportation Review, 177*, 103147.

Guardia, L., Suárez, L., Querejeta, N., Rodriguez Madrera, R., Suárez, B., & Centeno, T. A. (2019). Apple waste: A sustainable source of carbon materials and valuable compounds. *ACS sustainable chemistry & engineering, 7*(20), 17335–17343.

Gutberlet, J., & de Carvalho Vallin, I. (2024). Waste picker rights and social inclusion: the creation of a university with knowledge democracy. *Detritus*, (26), 114.

Huntington, V. E., Coulon, F., & Wagland, S. T. (2022). Innovative resource recovery from industrial sites: a critical review. *Sustainability, 15*(1), 489.

Islam, M. S. (2017). Sustainability through the lens of environmental sociology: An introduction. *Sustainability, 9*(3), 474.

Javourez, U., Karan, S. K., & Hamelin, L. (2024). Setting the environmental performance threshold for diverting residual streams towards bioeconomy strategies. *Science of The Total Environment, 949*, 174481. https://doi.org/10.1016/j.scitotenv.2024.174481

Kaza, S., Yao, L., Bhada-Tata, P., & Van Woerden, F. (2018). *What a Waste 2.0: A Global Snapshot of Solid Waste Management to 2050*. World Bank Publications.

Kędzia, G., Ocicka, B., Pluta-Zaremba, A., Raźniewska, M., Turek, J., & Wieteska-Rosiak, B. (2022). Social innovations for improving compostable packaging waste management in CE: A multi-solution perspective. *Energies, 15*(23), 9119.

Khan, M. I., Khan, S., Khan, U., & Haleem, A. (2023). Modeling the Big Data challenges in context of smart cities–an integrated fuzzy ISM-DEMATEL approach. *International journal of building pathology and adaptation, 41*(2), 422–453.

Khan, S., Anjum, R., Raza, S. T., Bazai, N. A., & Ihtisham, M. (2022). Technologies for municipal solid waste management: Current status, challenges, and future perspectives. *Chemosphere*, *288*, 132403.

Kiehbadroudinezhad, M., Merabet, A., & Hosseinzadeh-Bandbafha, H. (2023). A life cycle assessment perspective on biodiesel production from fish wastes for green microgrids in a circular bioeconomy. *Bioresource Technology Reports*, *21*, 101303.

Laibach, N., Börner, J., & Bröring, S. (2019). Exploring the future of the bioeconomy: An expert-based scoping study examining key enabling technology fields with potential to foster the transition toward a bio-based economy. *Technology in Society*, *58*, 101118.

Lei, Z., Cai, S., Cui, L., Wu, L., & Liu, Y. (2023). How do different Industry 4.0 technologies support certain circular economy practices? *Industrial Management & Data Systems*, *123*(4), 1220–1251.

Lestar, T., & Böhm, S. (2020). Ecospirituality and sustainability transitions: Agency towards degrowth. *Religion, State & Society*, *48*(1), 56–73.

Lin, K., Zhao, Y., Kuo, J. H., Deng, H., Cui, F., Zhang, Z., Zhang, M., Zhao, C., Gao, X., Zhou, T. & Wang, T. (2022). Toward smarter management and recovery of municipal solid waste: A critical review on deep learning approaches. *Journal of Cleaner Production*, *346*, 130943.

Loan, L. T. T., Chung, D. K., Van Duy, L., Thiem, N. T., Thuy, P. T. T., Phong, N. T., Balanay, R.M. & Phi, D. X. (2023). Evaluation of knowledge, attitude and practice (KAP) toward waste separation at source: the case of Lao Cai City, Lao Cai Province, Vietnam. *Journal of Material Cycles and Waste Management*, *25*(5), 2984–2995.

Longo, S. B., Isgren, E., Clark, B., Jorgenson, A. K., Jerneck, A., Olsson, L., Kelly, O. M., Harnesk, D., & York, R. (2021). Sociology for sustainability science. *Discover Sustainability*, *2*, 1–14.

Ma, H., Ma, R., & Zhang, Q. (2024). Sustainable allocation of economy, society, environment and resource: a system dynamics approach. *Environmental Engineering & Management Journal (EEMJ)*, *23*(2), 425.

Maciejczak, M. (2023). Nature based innovations in the development of bioeconomy. *Roczniki (Annals)*, *2023*(2), 98–108.

Meier, O., Gruchmann, T., & Ivanov, D. (2023). Circular supply chain management with blockchain technology: A dynamic capabilities view. *Transportation Research Part E: Logistics and Transportation Review*, *176*, 103177.

Mihai, F. C., Gündoğdu, S., Markley, L. A., Olivelli, A., Khan, F. R., Gwinnett, C., Gutberlet, J., Reyna-Bensusan, N., Llanquileo-Melgarejo, P., Meidiana, C., Elagroudy, S, Ishchenko, V., Penney, S., Lenkiewicz, Z., & Molinos-Senante, M. (2021). Plastic pollution, waste management issues, and circular economy opportunities in rural communities. *Sustainability*, *14*(1), 20.

Mohan, S. V., Dahiya, S., Amulya, K., Katakojwala, R., & Vanitha, T. K. (2019). Can circular bioeconomy be fueled by waste biorefineries—A closer look. *Bioresource Technology Reports*, *7*, 100277.

Navia, R., & Muñoz, R. (2015). Biotechnologies for gaseous emissions and by-products management in waste treatment facilities. *Waste Management & Research*, *33*(11), 945–946.

Nguyen, H. T., Marques, P., & Benneworth, P. (2022). Living labs: Challenging and changing the smart city power relations? *Technological Forecasting and Social Change*, *183*, 121866.

Oguntuase, O. J., Adu, O. B., & Obayori, O. S. (2023). How do individual-level factors influence the adoption of low-carbon technology? Proposing and validating the bioeconomy technology acceptance model in the context of Africa. *Frontiers in Sustainability*, 4, 1148001. https://doi.org/10.3389/frsus.2023.1148001

Oliveira, M. D. P. S. L., de Oliveira, E. A., & Fonseca, A. M. (2021). Strategies to promote circular economy in the management of construction and demolition waste at the regional level: a case study in Manaus, Brazil. *Clean Technologies and Environmental Policy*, *23*, 2713–2725.

Pacheco-López, A., Gómez-Reyes, E., Graells, M., Espuña, A., & Somoza-Tornos, A. (2023). Integrated synthesis, modeling, and assessment (iSMA) of waste-to-resource alternatives towards a circular economy: The case of the chemical recycling of plastic waste management. *Computers & Chemical Engineering*, *175*, 108255.

Panoutsou, C., & Chiaramonti, D. (2020). *Socio-Economic Opportunities from Miscanthus. Energies*, *13*(11), 2741.

Prioux, N., Ouaret, R., Hetreux, G., & Belaud, J. P. (2023). Environmental assessment coupled with machine learning for circular economy. *Clean Technologies and Environmental Policy*, *25*(2), 689–702.

Qin, M., Chen, C., Song, B., Shen, M., Cao, W., Yang, H., Zeng, G. & Gong, J. (2021). A review of biodegradable plastics to biodegradable microplastics: Another ecological threat to soil environments? *Journal of Cleaner Production*, *312*, 127816.

Raghu, S. J., & Rodrigues, L. L. (2020). Behavioral aspects of solid waste management: A systematic review. *Journal of the Air & Waste Management Association*, *70*(12), 1268–1302.

Reslan, M., Last, N., Mathur, N., Morris, K. C., & Ferrero, V. (2022). Circular economy: a product life cycle perspective on engineering and manufacturing practices. *Procedia CIRP*, *105*, 851–858.

Robertson, M. (2021). *Sustainability Principles and Practice*. Routledge.

Sanz-Hernández, A., Esteban, E., & Garrido, P. (2019). Transition to a bioeconomy: Perspectives from social sciences. *Journal of cleaner production*, *224*, 107–119.

Satpute, P. A., Suryawanshi, S., & Kulkarni, S. (2024). Sustainable materials for the Indian industry: Now and ahead. *Materials Circular Economy*, *6*(1), 1–10.

Seyyedi, S. R., Kowsari, E., Gheibi, M., Chinnappan, A., & Ramakrishna, S. (2024). A comprehensive review integration of digitalization and circular economy in waste management by adopting artificial intelligence approaches: Towards a simulation model. *Journal of Cleaner Production*, 460, 142584.

Shakir Hanna, S. H., & Cesaretti, G. P. (2019). The theory of sustainable development: A review. *Rivista di studi sulla sostenibilità: IX*, 2, 13–28.

Sonibare, O. O., Adeniran, J. A., & Bello, I. S. (2019). Landfill air and odour emissions from an integrated waste management facility. *Journal of Environmental Health Science and Engineering*, *17*, 13–28.

Thaseen Ikram, S., Mohanraj, V., Ramachandran, S., & Balakrishnan, A. (2023). An intelligent waste management application using IoT and a genetic algorithm–fuzzy inference system. *Applied Sciences*, *13*(6), 3943.

Tonini, D., Schrijvers, D., Nessi, S., Garcia-Gutierrez, P., & Giuntoli, J. (2021). Carbon footprint of plastic from biomass and recycled feedstock: Methodological insights. *The International Journal of Life Cycle Assessment*, *26*, 221–237.

Toniolo, S., Russo, I., & Bravo, I. (2024). Integrating product-focused life cycle perspectives in the fresh food supply chain: Revealing intra-and inter-organizational views. *Sustainable Production and Consumption*, *48*, 46–61.

Tsai, F. M., Bui, T. D., Tseng, M. L., Lim, M. K., Wu, K. J., & Mashud, A. H. M. (2021). Assessing a hierarchical sustainable solid waste management structure with qualitative information: Policy and regulations drive social impacts and stakeholder participation. *Resources, Conservation and Recycling*, *168*, 105285.

Tsapekos, P., Khoshnevisan, B., Alvarado-Morales, M., Zhu, X., Pan, J., Tian, H., & Angelidaki, I. (2021). Upcycling the anaerobic digestion streams in a bioeconomy approach: a review. *Renewable and Sustainable Energy Reviews*, *151*, 111635.

Turner, J., Spencer, J., O'Brien, L., Palethorpe, S., Banford, A., & Taylor, R. (2021). Applying the waste hierarchy for future recycling of nuclear fuels. *Nuclear Future*, *17*(6), 43.

van Praagh, M., & Liebmann, B. (2021). Microplastics in landfill leachates in three Nordic countries. *Detritus*, *17*, 58–70.

Vega, D. I., & Bautista-Rodriguez, S. (2024). The impact of circular economy strategies on municipal waste management: A system dynamics approach. *Cleaner Engineering and Technology*, *21*, 100761.

Venugopal, V. (2022). Green processing of seafood waste biomass towards blue economy. *Current Research in Environmental Sustainability*, *4*, 100164.

Xu, A., Chang, H., Xu, Y., Li, R., Li, X., & Zhao, Y. (2021). Applying artificial neural networks (ANNs) to solve solid waste-related issues: A critical review. *Waste Management*, *124*, 385–402.

Yang, N., Li, F., Liu, Y., Dai, T., Wang, Q., Zhang, J., Dai, Z. & Yu, B. (2022). Environmental and economic life-cycle assessments of household food waste management systems: a comparative review of methodology and research progress. *Sustainability*, *14*(13), 7533.

Zahedi, S. (2019). Sustainable development theory: A critical perspective and an integrative model. *Journal of Economics and Sustainable Development*, *10*(21), 43–52.

Zan, F., Iqbal, A., Lu, X., Wu, X., & Chen, G. (2022). "Food waste-wastewater-energy/resource" nexus: integrating food waste management with wastewater treatment towards urban sustainability. *Water Research*, *211*, 118089.

Zeller, V., Towa, E., Degrez, M., & Achten, W. M. (2019). Urban waste flows and their potential for a circular economy model at city-region level. *Waste management*, *83*, 83–94.

11 Microbial Electrolysis
An Emerging Biotechnology for Energy Production and Wastewater Treatment

Sajjad Hussain, Hammad Khan, Arslan Maqbool, Muhammad Usama, and Saima Gul

11.1 INTRODUCTION

All energy sources, including renewable energy, have an environmental impact, much like any human endeavor (Sayed et al. 2021). The benefits of renewable energy sources over fossil fuels are evident, even though each has trade-offs of its own (Usman and Radulescu 2022). They lead to less use of water and land, less pollution of the air and water, minimal loss of wildlife and habitat, and either no or less greenhouse gas emissions. Furthermore, the combination of technological improvements and their local and decentralized nature produces notable benefits for the economy and society (Javid et al. 2021). The depletion of fossil fuel supplies and the rising demand for energy worldwide are the root causes of the current energy crisis. To meet this issue, scientists from all over the world are actively investigating alternate energy sources (Arutyunov and Lisichkin 2017). Furthermore, burning fossil fuels quickly releases greenhouse gases like CO_2, which exacerbates climate change worldwide. There are environmental problems because of the ever-increasing amount of wastewater produced by the population. Global sustainability depends on addressing these two urgent problems: wastewater management and energy shortages (Zhang et al. 2021). The complex mixture of organic substances found in wastewater usually needs to be treated before being disposed of. Combinations of biological, chemical, and physical processes are frequently used in this treatment. However, the energy-intensive machinery used in wastewater treatment facilities, such as pumps and blowers, raises the cost of operation. Anaerobic digestion has the ability to recover energy, but aerobic treatments aggravate the problem by producing large amounts of sludge, which raises expenses even more (Wainaina et al. 2020). Operational costs could be greatly decreased by combining value-added biochemical processes and wastewater treatment with the production of green energy. Because of its complicated composition, wastewater management requires flexible and reliable solutions (Okolie et al. 2022).

Hydrogen (H_2) is one of the most attractive bio-based chemicals and fuels that can be obtained from wastewater since it is a clean energy source that emits no CO_2 or other pollutants and is beneficial to the environment (Sazali 2020). Burning H_2 doesn't produce any ozone-depleting compounds, greenhouse gases, or acid rain. High-energy, calorific fuels like H_2 can be created biologically by dark fermentation, photo fermentation, and bio-photolysis. But in these activities, the metabolic pathways of microbes prevent full substrate utilization, which results in the creation of substitute by-products (Pagliano et al. 2017). Moreover, thermodynamic obstacles are the reason for the lower H_2 yields seen in traditional biological H_2 production techniques. By adding a tiny electrical input to

DOI: 10.1201/9781003668961-11

a bio-electrochemical system (BES) called a microbial electrolysis cell (MEC), these obstacles can be removed (Siwal et al. 2021). MEC is a cutting-edge, environmentally friendly technique that turns organic waste into hydrogen gas. Using MECs to produce H_2 from wastewater has potential applications in green energy production and wastewater treatment, providing a way around problems with traditional hydrogen production (Kadier et al. 2020). Using wastewater as a source of free electrons, MECs are a promising technique for combining energy recovery with wastewater treatment (Kadier et al. 2020). This chapter examines integrated techniques and conventional waste-to-energy (WTE) generation technologies, giving an overview of MEC, its operational principles, prospective electron transfers, and existing applications (Khan et al. 2022). Finally, it explores the benefits and potential challenges of MECs over traditional techniques. A comprehensive evaluation is given in this chapter, which also integrates and compares MEC with other traditional WTE technologies (Hosseinzadeh et al. 2022).

11.1.1 Overview of Microbiological Electrolysis

BESs present a feasible alternative because they can produce bio-chemicals without requiring external power input. Even when additional power is supplied to enhance production, the amount required is considerably lower than in conventional manufacturing processes. (Alawad et al. 2023). MECs are bio-electrochemical systems designed to extract biofuels from various substrates, particularly hydrogen (Kadier et al. 2020). MECs consist of an anodic compartment and a cathodic compartment, separated by a proton exchange membrane as depicted in Figure 11.1.

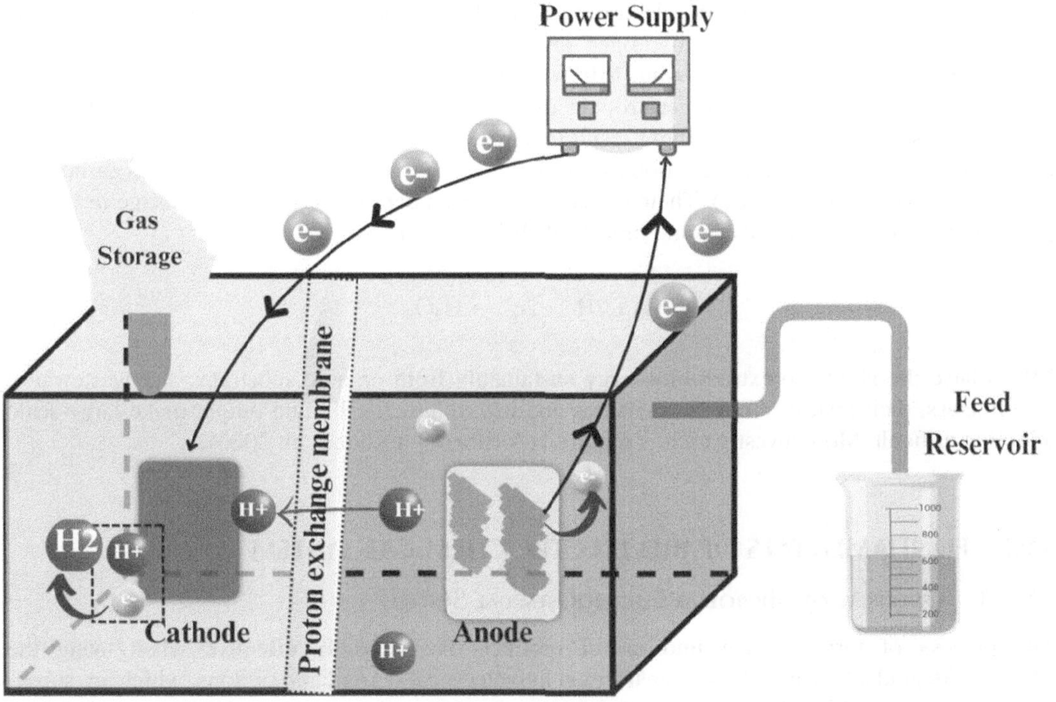

FIGURE 11.1 The diagram depicts a typical microbial electrolysis cell reactor schematically.

The anodic compartment operates in an anaerobic environment, while the cathodic compartment produces hydrogen gas (Zeppilli et al. 2019). The bacteria that are electrochemically active initiate the oxidation process of the organic fraction within the anolyte, leading to the release of electrons, protons, and CO_2 (as shown in Equation (1.1)). The MEC reactor uses electrons and protons to generate hydrogen gas (H_2) and methane (CH_4) (Hassanein et al. 2020). The single-chamber membrane-less setup is economically advantageous as it eliminates the need for a proton exchange membrane and shows superior performance efficiency due to reduced internal resistance compared to a dual-chamber setup (Masoudi et al. 2021). In theory, within MEC, the substrate's oxidation could potentially yield an anode potential (E^o) of up to −0.3 V compared to the normal hydrogen electrode. Anaerobic conditions do not allow spontaneous H_2 formation due to negative cell voltage (Kadier et al. 2016).

At anode:

$$C_6H_{12}O_6 + 6H_2O \rightarrow 6CO_2 + 24e^- + 24H^+ \tag{11.1}$$

$$CO_2 + 8H^+ + 8e^- \rightarrow CH_4 + 2H_2O \tag{11.2}$$

At cathode:

$$24H^+ + 24e^- \rightarrow 12H_2 \tag{11.3}$$

To produce H_2 at the cathode, an external voltage greater than 0.114 V is needed. Cell potential as high as 0.8 V may be necessary (Kadier et al. 2016). However, exceeding this threshold could rupture exoelectrogenic organisms, affecting biofuel yield in MECs. MECs are more efficient than typical water electrolysis processes and can utilize various organic substrates, making them a cleaner and more ecological alternative to electricity-driven H_2 production (Masoudi et al. 2021). MEC development initially focused on hydrogen production, but recent advancements have revealed diverse bio-recovery applications using MECs (Kadier et al. 2016). For instance, increasing the growth of acetoclastic and hydrogen-trophic methanogens (microbes that consume acetate and hydrogen, respectively) in MECs might result in a higher output of methane (CH_4). MECs produce hydrogen peroxide for oxygen reduction by employing carbon-based electrodes, such as carbon felt and fabric (Masoudi et al. 2021). Their versatility has been demonstrated by their usage in biofuel production, heavy metal recovery, and industrial chemical synthesis.

$$O_2 + 2H^+ + 2e^- \rightarrow H_2O_2 \tag{11.4}$$

MECs have the ability to extract bioenergy sustainably from organic substrates and wastewater; nevertheless, their restricted current densities and low volumetric biofuel output make large-scale adoption difficult. More investigation is required (Alcaraz-Gonzalez et al. 2023).

11.2 FUNDAMENTALS OF BIO ELECTROCHEMICAL SYSTEM

11.2.1 Overview of Microbial Electrochemical Systems

The process of turning trash into useful products is incredibly effective when using the MEC. Biodegradable materials are transformed into electrical current and protons, which are subsequently transformed into hydrogen gas, using exo-electronic microorganisms (Dörr et al. 2023). A small voltage, usually between 0.2 and 0.8 volts, is needed between the electrodes in this method, which has advantages for commercial use because of its improved efficiency and performance (Alcaraz-Gonzalez et al. 2023).

11.2.1.1 System of Double-Chamber MECs

A dual-chamber structure (anodic and cathodic chamber) in MEC with a membrane performing the role of a barrier can minimize the organic load and bacteria and consequently lower contaminants like CO_2, CH_4, and H_2S (Tong et al. 2023). Short-circuiting is avoided and internal resistance is reduced by this design. Neutral membranes, however, have the potential to obstruct the transit of protons and hydroxyl ions, creating a pH gradient and raising input resistance (Ramya and Senthil Kumar 2022). The change in pH across the membrane might slow down how fast hydrogen (H_2) is produced by stopping the activity of certain bacteria that play a role in generating electricity (called electrochemically active bacteria or EAB) and by increasing the strength of the input. It was shown that using a double-chamber microbial electrolysis cell was not a good method for getting H_2 from wastewater that doesn't have chemicals to control pH because the pH level dropped too much (Prathiba et al. 2022). Various types of double-chamber MECs have been developed. These include cruse type, barrel-shaped, circular like a disk, and concentric tube-shaped cells. They achieve this by bringing the electrodes closer together and using a porous geotextile membrane as the separator. Recently, a plane-type microbial electrolysis cell has been designed to decrease feed resistance. Figure 11.2 shows diagrammatically the double chamber MEC.

11.2.1.2 Single-Chamber MECs System

A single-chamber microbial electrolysis cell was elaborated by removing membranes from dual-chamber systems and combining both electrodes in a single chamber, potentially reducing membrane resistance losses (Figure 11.3) (Rozenfeld et al. 2021). It boasts simplicity in manufacturing and autoclaving processes while circumventing issues typically associated with membranes, such as fouling, degradation, and excessive cost. These single-chamber microbial electrolysis cells have been engineered in various configurations, utilizing materials ranging from glass cruets to plastic blocks as primary structures (Wang et al. 2020). Glass serum cruets aid in high-throughput bioelectrochemical investigations for selecting EAB or potential cathode materials in limited microbial electrolysis cells, as illustrated in Figure 11.3. Methanogenic bacteria's hydrogen utilization and re-oxidation pose challenges in single-chamber microbial electrolysis cells (CHRISPIM MC, 2021).

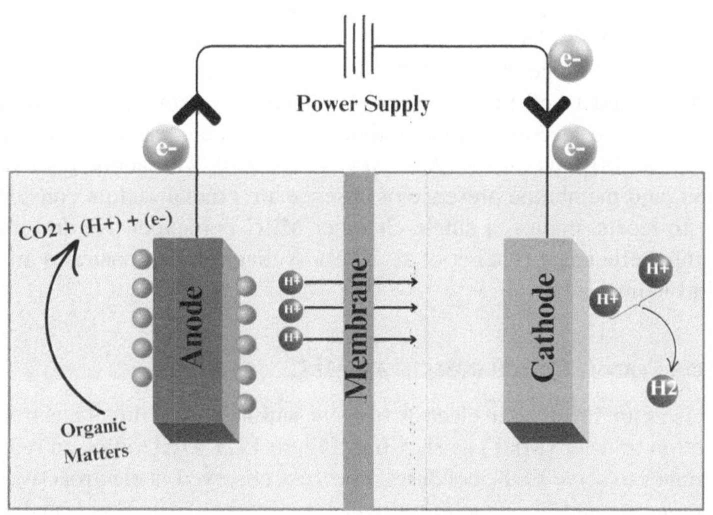

FIGURE 11.2 Diagrammatic representation of double chamber MEC.

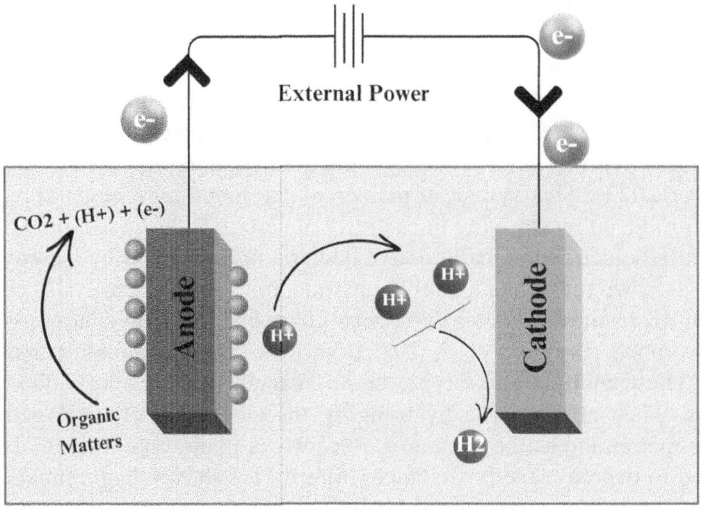

FIGURE 11.3 The diagrammatic representation of single chamber MEC.

UV irradiation and an active gas harvesting approach use a vacuum and gas-permeable membrane to prevent H_2 consumption, enhance hydrogen flow, enhance efficiency, and minimize burst risks (Kadier et al. 2019).

11.2.2 Principles and Mechanisms of Microbial Electrolysis

The MEC is a novel system that combines the functionality of microbial fuel cells (MFCs) without the need for oxygen, allowing for spontaneous oxidation and reduction at the cathode (Rousseau et al. 2020). In an MEC, the predominant populations at the anode are electrochemically active bacteria (EAB) or electrogenic bacteria, which decompose organic matter into protons, electrons, and CO_2 (Rout et al. 2023). Hydrogen (H_2) is produced at the cathode by reducing the generated protons and electrons. Consequently, obligatory anaerobes are typically observed in this environment (Koul et al. 2022). This technology's electrically aided operation expands the range of uses. A steady DC power source, usually in the range of 0.2 to 0.8 V, a microbially-enriched anode, a catalyzed cathode (Pt–catalyzed cathode), and a well-thought-out gas collection system are necessary for building and running an MEC. There are several architectural designs available, each of which is tailored to the functions for which the MEC is meant (Antevski et al. 2020). The anode-cathode distance, flow rates, configuration, and membrane presence or absence are crucial factors considered (Rout et al. 2023). According to recent studies, a single-chamber MEC is inexpensive and improves the process's total coulombic efficiency (Kadier et al. 2019). A diagram demonstrates an example of two chambers MEC in Figure 11.4.

11.2.3 Electron Transfer Mechanism in an MEC

Two main mechanisms are involved in electron transfer within MECs: direct electron transfer (DET) and mediated electron transfer (MET) as shown in Figure 11.5. DET relies on redox proteins situated on the membranes to serve as the conduits, a process observed in electroactive bacteria such as *Geobacter sulfurreducens* and *Shewanella putrefaciens* (Pyun 2020). This transfer occurs through direct contact with components such as outer c-type cytochromes and other heme proteins.

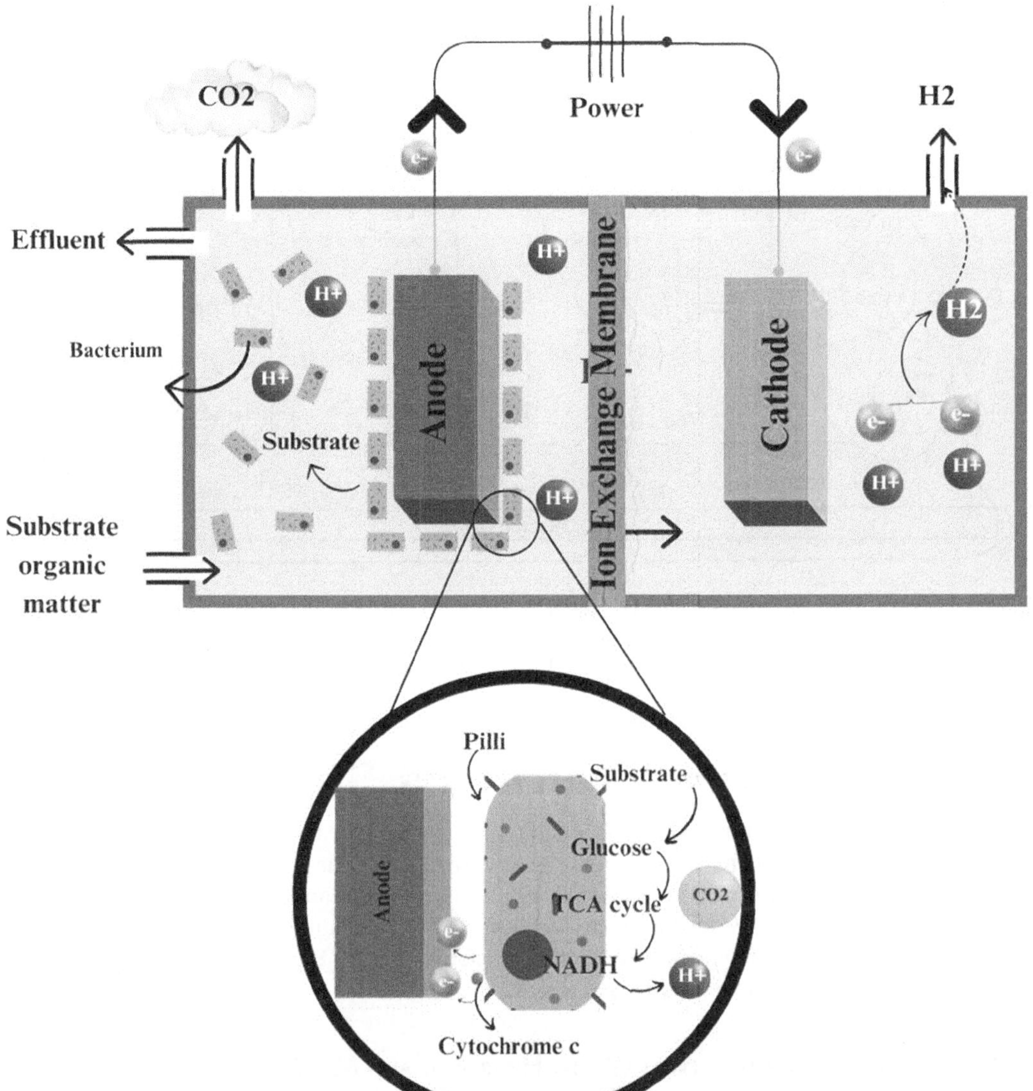

FIGURE 11.4 An illustration of a standard two-chamber MEC, with particular attention to electron transfer, cathode reduction, and anode oxidation.

Additionally, cell protrusions (pili) have been noted to facilitate electron transfer, particularly under anaerobic conditions (Ndayisenga et al. 2023). DET has also been observed with conductive nanowires, particularly when cytochromes MtrC and OmcA are involved. Micrographs from operational MFCs have revealed electron transfer occurring through cell surface protrusions (Brogi et al. 2018). On the other hand, MET involves the use of dissolved redox species as intermediaries, eliminating the need for direct contact between microbes and electrodes (Antevski et al. 2020). Instead, electron transfer is mediated by compounds that can undergo reversible oxidation at the cathode, which can be either endogenous or exogenous.

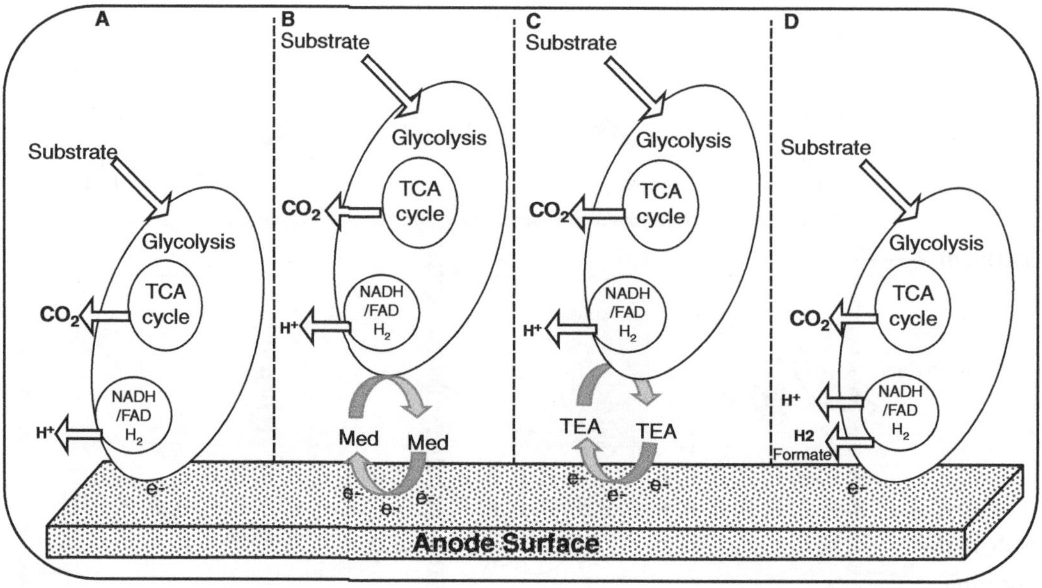

FIGURE 11.5 A schematic representation of the many electron transfer pathways that may occur in an MEC: (a) DET through membrane-bound c-type cytochromes or nanowires (b) MET through artificial mediators (c) MET via primary respiratory metabolites and (d) MET via secondary respiratory metabolites.

11.3 APPLICATIONS MICROBIAL ELECTROLYSIS

With advancements in reactor design and operational techniques, the applications of MECs are expanding into various domains. Broadly, we can discern at least four main categories for the potential use of microbial electrochemical systems: synthesis, waste treatment, resource recovery, and bio-sensing (Miller et al. 2019). Additionally, their versatile application extends to electrochemical research and integration with other electrochemical reactors. Figure 11.6 provides a schematic summary of the current applications of MECs across these categories.

11.3.1 Energy Production from Waste: Hydrogen and Methane

11.3.1.1 Hydrogen Production

MECs primarily harvest hydrogen (H_2) from organic substrates and the organic component of wastewater. The initial demonstration of bio-hydrogen retrieval through MECs was pioneered by Liu and Rozendal's research team (Liu et al. 2005) (Rozendal et al. 2006). Thermodynamic limitations prevent additional bacterial conversion of CH_3COO- to H_2, even though bacterial fermentation of one mole of glucose normally produces only four moles of H_2 and two moles of CH_3COO- as a by-product (Jungmeier 2017). Furthermore, many other by-products are produced while working with complicated substrates like wastewater, which hinders bacteria's ability to convert them into H_2 (Bengelsdorf et al. 2018). However, Liu et al. (2005) found that increasing an MFC's intrinsic potential with an extra external cathode potential of 0.25 V or more greatly increases H_2 recovery. Equation (11.5) states that the conversion of CH_3COO- to H_2 is an endothermic process. Therefore, the extra potential flow support overcomes this thermodynamical barrier, increasing the generation of H_2.

$$CH_3COO^- + 4H_2O \rightarrow 2HCO^{3-} + H^+ + 4H_2 \quad (11.5)$$

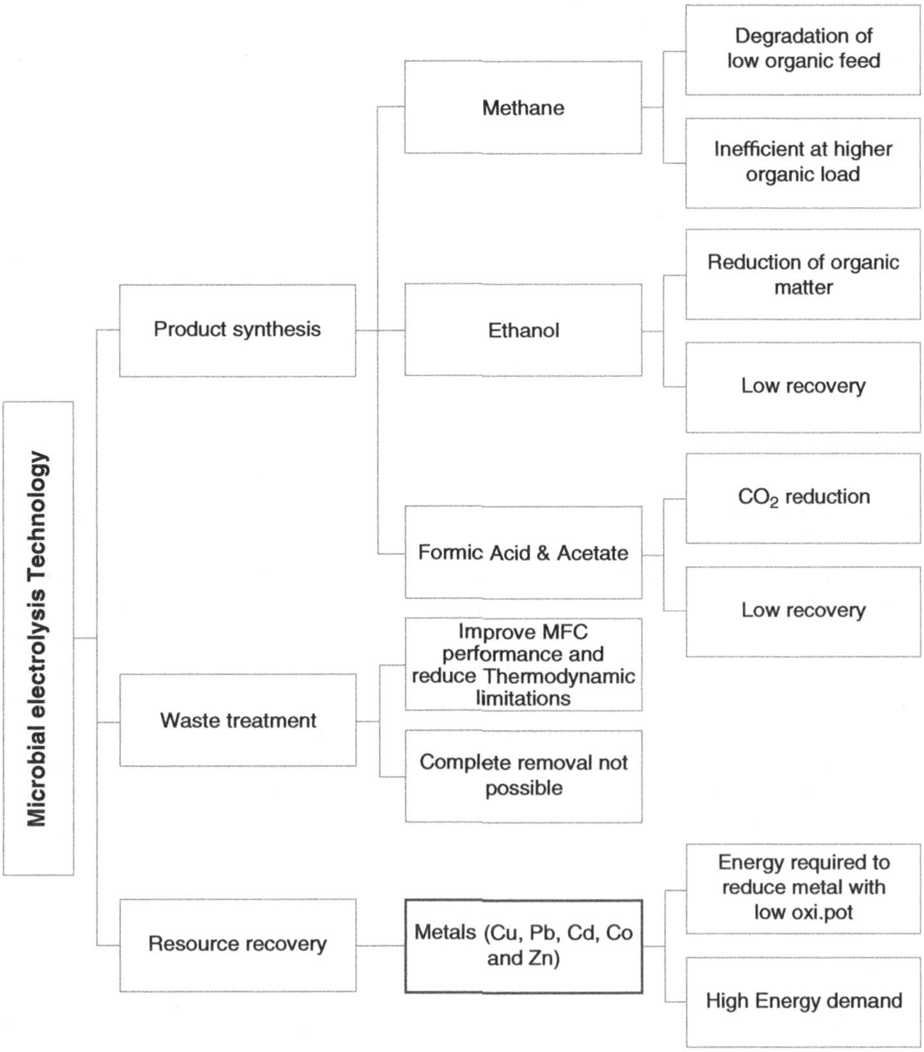

FIGURE 11.6 Current applications of MEC along with its drawbacks.

An MEC enhances the versatility of applications by facilitating electrical operation. It requires a constant DC power supply, a microbial-enriched anode, a catalyzed cathode, and a well-designed gas collection system (Kadier et al. 2016). Architectural designs are tailored to fit workflows or uses. MECs with a single chamber are more economical and have higher coulombic efficiency (Prathiba et al. 2022). The Cui research group demonstrated how methanogenesis could be efficiently repressed in a single-chamber MEC by maintaining an alkaline electrolytic environment with a pH of 11.2, which increased the production of hydrogen (H_2) (Cui et al. 2019). A decrease in methanogens within the cathodic biofilm may be the cause of the elevated H_2 production. Furthermore, the total drop in cathodic biomass at these high pH values led to a reduction in CH_3COO- consumption by cathodic species and an increase in CH_3COO- utilization by electroactive strains at the anode (Aboelela and Soliman 2022). Notwithstanding these developments, commercial H_2 extraction from MECs still faces difficulties, such as the unmodified cathodes slow hydrogen evolution reaction (HER) and the need for expensive catalysts to increase HER efficiency (Park et al. 2022).

11.3.1.2 Methane Production

The anaerobic decomposition coupled with a microbial electrolysis cell is expected to accelerate the production of methane from biomass. This process involves using both untreated waste-activated sludge and thermal-treated waste-activated sludge, which enhances the efficiency of methane production and the activity of microbes involved in the process (Figure 11.7). There are various obstacles in the way of MECs' production of bio-hydrogen. Homoacetogenic activity, in which cathodic hydrogen is changed into acetate, and hydrogen recycling, which entails reoxidation at the anode, both lower the efficiency of H_2 recovery (Saravanan et al. 2023). Consequently, since it is easier to collect and transport CH_4 than H_2, it could be preferable to increase rather than decrease CH_4 production in MECs (Pawar et al. 2022). Methane (CH_4) production in MECs is made possible via a process known as bio-electro methanogenesis, in which cathodic microorganisms use electrons from the anode to convert CO_2 to CH_4 (Eq. (11.6)).

$$CO_2 + 4H_2 \rightarrow CH_4 + 2H_2O \tag{11.6}$$

To enhance methane (CH_4) production, MECs are often combined with anaerobic digestion (AD) (Dange et al. 2021). The MEC-AD integrated system can function by either retrofitting the electrode assembly into an anaerobic digester directly or by recirculating digester feedstock between an MEC reactor and a traditional anaerobic digester (Wang, Liu, et al., 2022) (Figure 11.7). The kinetic and thermodynamic promotion of methanogenic activity by an environment in such integrated systems results in much higher bio-methane recovery. In hybrid MEC-AD systems, hydrogenotrophic methanogens enrich the cathodic biofilm, and there is direct interspecies electron transfer between exoelectrogens and the anode (Pelaz et al. 2023). This mechanism promotes acidogenesis, acetogenesis, and methanogenesis while lowering the partial pressure of H_2. Furthermore, a different wastewater, leachate, and municipal sludge can be treated by the MEC-AD setup. As a result, the MEC-AD hybrid is a potential technology for environmentally friendly biomethane conversion (Deutzmann et al. 2023).

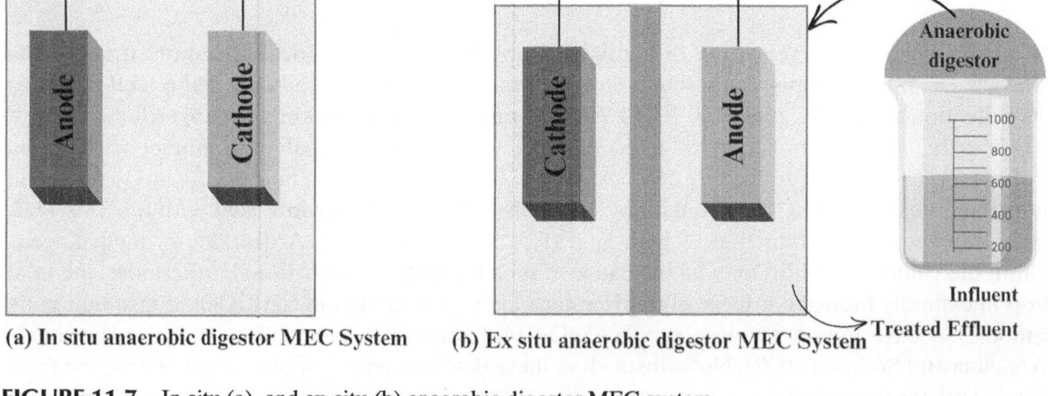

FIGURE 11.7 In situ (a), and ex-situ (b) anaerobic digester MEC system.

11.3.2 Wastewater Treatment and Resource Recovery

Most laboratory investigations have centered on the development of novel electrodes, experimenting with different cell configurations, pushing the boundaries of MEC efficiency, or evaluating the influence of diverse natural factors (Pawar et al. 2022). These investigations typically utilized artificial media to afford researchers control over mass structure. However, it has long been acknowledged that laboratory trials with authentic wastewater (WW) are indispensable for gauging the real-world viability of this technology (Savla et al. 2022).

11.3.2.1 Elimination of Complex Organic Contaminants

Sludge can be viewed as bio-solids rather than mere waste owing to its potential for energy generation (Casiano Flores et al. 2018). In China, the absence of anaerobic digesters in most WW treatment plants is primarily due to technical complexities and the high capital and operational costs involved. Typically, sewage sludge undergoes concentration and dewatering treatments. Recent study has illustrated the feasibility of employing ultrasound treatment (UT) in conjunction with an up-flow anaerobic sludge blanket (UASB) and an MEC system to enhance the sludge degradation process (Li et al. 2018). UT facilitates the dissolution of solid sludge into the liquid phase, followed by the removal of dissolved organic matter in an anaerobic digester combined with anodic oxidation (UASB-MEC) (Yang et al. 2021). After one-week anaerobic decomposition, the final soluble chemical oxygen demand (COD) and protein concentrations in the MEC effluent reached 569.5 and 373.1 ppm. This integrated reactor significantly enhances the efficiency of sludge anaerobic degradation and holds promise for widespread implementation in engineering applications (Yuan et al. 2022). The class of xenobiotic heterocyclic chemicals known as benzothiazole derivatives (BTHs) is distinguished by its resistance to natural degradation processes and toxicity (Li et al. 2018). Liu et al. looked into whether employing MECs could eliminate BTHs. Similarly, Miran et al. pointed to the treatment of 4-chlorophenol, a chlorinated phenol contaminant, in an MEC system that was enhanced with SRB (sulfate-reducing bacteria) (Miran et al. 2017). Their data reveal that the SRB–enriched MEC effectively removed chlorophenol and that the dechlorinated by-products were subsequently degraded. Furthermore, the successful reduction of halogenated aromatic compounds as 2,4-dinitrochlorobenzene, p-fluor nitrobenzene, and p-nitrophenol has been demonstrated by the integration of MEC with biological therapy. The utilization of MEC in conjunction with a UASB reactor is noteworthy because it promotes the conversion of p-chloronitrobenzene into p-chloroaniline, a substance that has a lower environmental effect (Xu et al. 2016). To sum up, MEC technology has a lot of potential as a cutting-edge wastewater treatment technique because of its excellent degrading efficiency, low maintenance requirements, and robust environmental sustainability (Pérez et al. 2023). These factors are crucial for achieving economic viability and efficient scalability (Bhattu and Singh 2023).

11.3.2.2 Inorganic Pollutants

Numerous inorganic pollutants, including phosphorus, perchlorate, nitrate, and sulfur can undergo degradation within MECs. This section primarily focuses on nitrates and sulfates. Nitrate contamination is a global groundwater issue, often addressed through physical, chemical, photocatalytic, or electrode ionization methods (Kundu et al. 2022). Biological denitrification, a heterotrophic process, is commonly utilized in wastewater treatment. In MECs, nitrate reduction at a bacteria-catalyzed cathode has been observed, with the application of external voltage enhancing electric current generation (Sayed et al. 2023). This method resulted in a significant decrease in nitrate levels, from 23.3 to 5.3 mg L^{-1} within 24 hours. However, there are still challenges in using MECs for nitrogen removal. These include issues with scaling up the technology, managing microbiological processes, ensuring long-term stability, and dealing with high costs. Additionally, both economic and environmental impacts need to be considered (Sayed et al. 2023). The enhancement of sulfate reduction through bio-electrochemical processes has been shown to significantly improve organic decomposition within MECs. Organic substrate oxidation occurs within the anode chamber, while sulfate acts as an electron acceptor at the bio-cathode (Manna and Sen 2023). Pérez et al. (2023) have demonstrated that MECs can enhance anaerobic acidogenesis when treating wastewater containing sulfate, yielding higher sulfate reduction rates even under heavy sulfate loads. Additionally, Jadhav et al. (2021) utilized a single-chamber MEC with an applied voltage of 0.7 V and 600 mg L^{-1} of potassium sulfide, achieving an 80.7% removal rate of sulfide, with the dominant bacteria identified as Geobacter (7.35%). To enhance the bio-electrocatalytic activity of MECs in sulfate removal, Wang et al. suggested utilizing neutral red as an electron transfer mediator, resulting in a 79% sulfate removal rate in the reactor (Wang, Zhang, et al., 2022). However, a key challenge in utilizing MECs for sulfate removal lies in determining the optimal operating conditions for real wastewater treatment (Singh et al. 2022). When given the right substrates, MECs exhibit a remarkable capacity for producing useful chemical compounds. In an MEC, for example, wastewater containing glycerol is decomposed to produce ethanol at anodic compartment, while cane molasses containing sugar is converted to succinic acid in the cathodic chamber (Rozendal et al. 2006; Park et al. 2022). To improve the generation of biofuel in MECs, methods including cathode modifications and catalyst coating are applied in addition to supplying suitable substrate. By electrodepositing Mn onto the carbon felt electrode, Anwer and colleagues, for example, increased the generation of acetic acid by decreasing CO_2 in a MEC by 43% (2.28 g/L) in comparison to unmodified carbon felt (2019).

Similarly, Luan et al. demonstrated how Methyl bacterium extorquens in an MEC can convert CO_2 without the need for other cofactors like nicotinamide adenine dinucleotide phosphate and H_2 (Luan et al. 2023). Under the correct operating circumstances, an MEC may produce a wide range of value-added goods. Nevertheless, the scalability of this inventive technology as a productive bio-refinery remains a subject of debate due to challenges such as inadequate biofuel recovery and a lack of clarity regarding the interaction among various microbiotas (Bengelsdorf et al. 2018). Hence, significant efforts are needed to address these issues and streamline MEC technology soon.

11.4 INTEGRATION WITH RENEWABLES

11.4.1 Using Solar and Wind Power for Microbial Electrolysis

Since solar energy is abundant and the Earth receives a significant amount of solar radiation (120,000 TW), it is considered a promising alternative to address the world's energy needs (Serale et al. 2018). Several research works have investigated the use of different solar cells, like dye-sensitized solar cells, to harvest sunlight and supply extra power for microbial electrochemical systems (MESs) (Absi Halabi et al. 2015). MESs can produce liquid or gaseous fuels like hydrogen (H_2), methane (CH_4), and ethanol for future use. They do this by converting solar energy into electricity using a natural, solar-powered MES system (Silvi 1995). As an intermittent power source that can operate day or night, solar light has been shown in recent studies to efficiently increase the production of biogas using polluted water and to modify the expression of electron transfer proteins like cytochrome c, which in turn affects the dynamics of microbial communities. Under mild conditions, a scalable and affordable photo-catalyst sheet powered only by sunshine, CO_2, and water has achieved a solar-to-acetate conversion efficiency of 0.7% thanks to the growth of the non-photosynthetic, CO_2–fixing acetogen *Sporomusa ovata* (Philibert 2005). The acetate that is produced is then employed in microbial electrochemical systems to produce electricity, demonstrating the possible combination of solar energy–based microbial electrochemistry. In further projects, the direct integration of dye-sensitized solar cells with the anode to create a

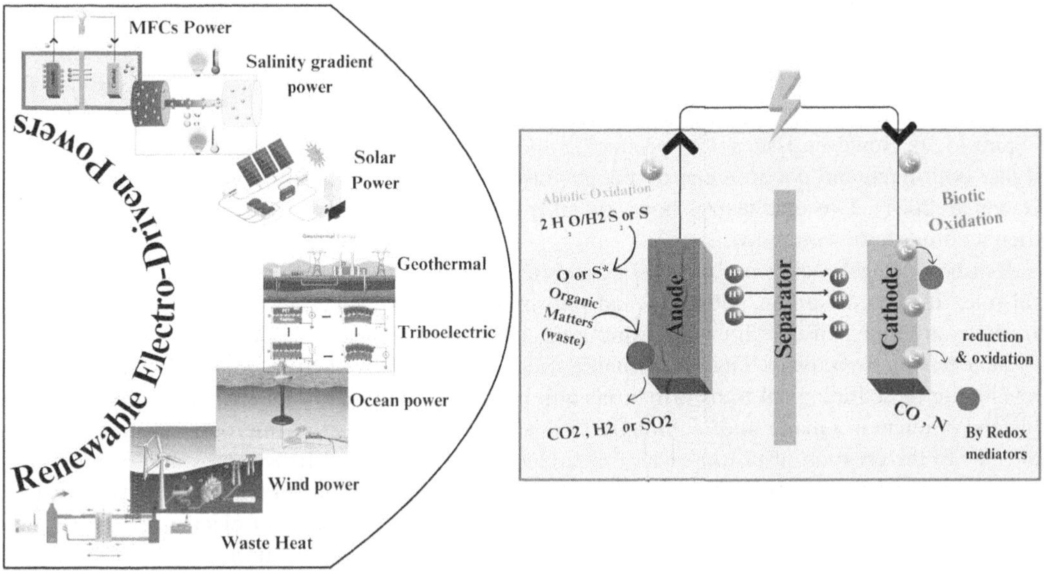

FIGURE 11.8 Alternative power supplies drive MESs.

photo-bioanode may save expenses and get around the mechanical difficulties of having separate units (Philibert 2005). Furthermore, the efficient conversion of solar light into energy could be achieved using nanotechnology to build easily scalable alternative electrode materials with enormous surface areas (Figure 11.8).

11.4.1.1 Underexploited Renewable Energy Sources

Besides established energy sources powering MESs, a variety of untapped renewable energy reservoirs are anticipated to broaden their applicability, encompassing waste heat, mechanical energy, and potential sources like wind, geothermal, and marine energy, particularly for field applications (Figure 11.8) (Ozin and Ghoussoub 2020). Wind power, the swiftest-growing renewable energy sector, has garnered significant interest in recent times, leading many nations to erect wind farms to bolster their energy production. These farms can span both onshore and offshore locations, reaching heights of hundreds of meters, catering to diverse settings from small residences to remote areas (Ozin and Ghoussoub 2020). In contrast, waste heat typically arises from machinery or industrial processes, offering an opportunity for environmental and economic gains through its recovery. Nevertheless, harnessing waste heat at low temperatures presents formidable challenges (2020). Furthermore, MESs could be powered by innovative self-sufficient triboelectric nano-generators, serving as novel mechanical energy harvesters (Crossley et al. 2014). These generators convert slight mechanical motions into electricity through contact electrification and electrostatic induction, exploiting intermittent contact-separation and shifting motions at differing triboelectric layers (Hamlehdar et al. 2019). For instance, the perpetually lost stir energy in MESs could be captured by triboelectric nano-generators to sustain their operation. The main difficulty is integrating and synchronizing these two systems in an efficient manner, especially when it comes to optimizing the use of real power output, like alternating current, without the need for further equipment (Naqvi et al. 2022).

11.4.2 Enhancing Sustainability Through Integration

Many countries have made the commitment to adopt carbon-neutral technologies by 2050 (Sun et al. 2022). In addition to enabling the absorption of CO_2 and fixation of nitrogen, MESs provide the capacity to convert surplus electricity from renewable sources into useful chemicals and fuels (Figure 11.9). However, issues with producing and distributing renewable electricity as well as satisfying both fixed and portable energy needs must be resolved before practical use can take place (Gao et al. 2021). This emphasizes how crucial it is to focus more on renewable electricity sources when it comes to powering MESs.

A carbon-neutral society embodies sustainable living practices, where microorganisms play pivotal roles in natural carbon, nitrogen, and phosphorus cycles (Zahed et al. 2021). These natural processes are indispensable for addressing long-standing environmental challenges such as waste (-water) resource recovery, environmentally friendly energy production, and pollutant removal. Developing these biological transformations quickly enough to create a system that is both sustainable and efficient is a major hurdle, though (Li et al. 2018). Electroactive microorganisms play critical roles in the creation of power and value-added products within MESs. The study of microbial electrical activity began more than a century ago with the discovery of microbial electrical contact with platinum electrodes (Cui et al. 2019). Later, more than 100 species of electroactive microbes were discovered, with Shewanella and Geobacter species acting as representative species for nanowires. Significantly, ten years ago, multicellular filamentous cable bacteria were discovered, demonstrating the capacity to carry electrons across significant distances and facilitating long-range electron transfer (Ptushenko 2020a). These cable bacteria, found at the anode of benthic MFCs, are poised for integration into MESs, either as pure cultures or within mixed batches. Their potential for soil and wastewater remediation is promising, given their capability to remediate oil-contaminated sites, fix nitrogen and CO_2, and potentially catalyze water splitting (Ptushenko 2020b). Furthermore, cable bacteria influence the environment of their sediment, function as barriers against H_2S in

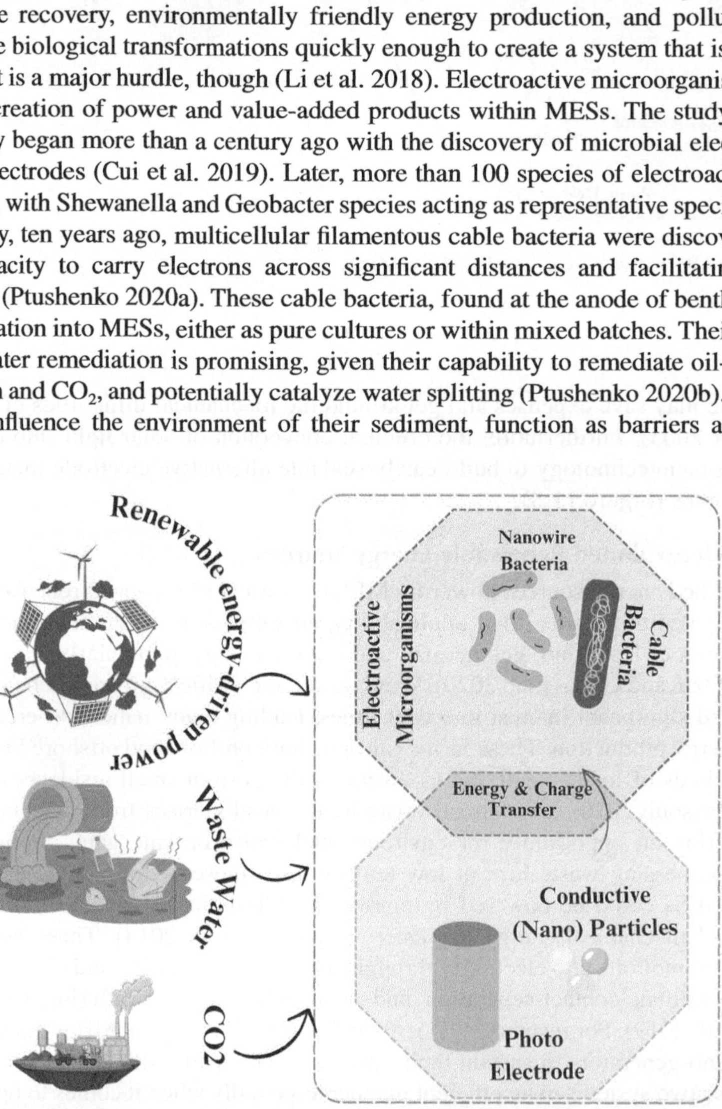

FIGURE 11.9 Diagrammatic illustration of the MESs powered by renewable electricity to create chemicals and fuels from CO_2.

seawater, and are speculated to establish electrical connections with other bacteria. However, the current limitation of existing only in natural sediment due to the lack of pure cultures hinders broader utilization (Ptushenko 2020a). It is anticipated that the scope for a variety of applications will expand with the integration of pure cultures of cable bacteria into mixed MESs as access to them becomes more practical (Ptushenko 2020a).

11.5 TECHNOLOGICAL ADVANCEMENTS

Using electroactive bacteria to transform organic matter or wastewater into valuable products like hydrogen gas or other useful compounds, MECs represent a promising technology (Kadier et al. 2016). Developments in this field have been directed toward improving MEC performance, effectiveness, and scalability by means of a variety of tactics, including new materials, microbial consortiums, and reactor architectures. The goal of electrode material research has been to increase the stability and efficiency of electron transport in MECs (Mier et al. 2021). For their conductivity and resistance to corrosion, this investigation includes metal-based materials like titanium, nickel, and stainless steel as well as carbon-based materials like carbon cloth, carbon nanotubes, graphene, and conductive polymers (Noreen et al. 2022). Membranes are essential for separating the cathode and anode chambers and promoting ion movement (Maqbool et al. 2023). Advances in membrane science have produced proton and ion exchange membranes with improved ion conductivity, durability, and selectivity, such as polymeric membranes and Nafion. To catalyze electrochemical reactions at the electrode contact, catalysts are essential (Wu et al. 2021). Nanostructured catalysts, such as platinum, palladium, and nickel nanoparticles, have been scrutinized to boost HER at the cathode and organic oxidation reactions at the anode. Additionally, bio-inspired catalysts derived from microbial enzymes or biofilms exhibit promise for MEC applications (Wu et al. 2021). To maximize MEC performance, researchers have customized metabolic pathways by designing microbial consortia. To successfully complete this project, a precise combination of electroactive microorganisms that can effectively use substrates and transmit electrons inside MEC systems must be chosen (Prathiba et al. 2022). Consortia engineering strategies are designed to increase product specificity, substrate utilization efficiency, and electron transfer rates. Benefits from utilizing mixed microbial cultures (MMC) include resilience, flexibility, and behavior akin to an ecosystem (Antevski et al. 2020). MMC–based MECs have improved resistance to changes in operating circumstances and can efficiently use a variety of organic feedstocks. Co-culturing, enrichment, and microbial community selection are some of the methods used to produce effective MMC for MEC applications (Cui et al. 2019).

Different electrode topologies, including flow-through, double-chamber, and single-chamber designs, have been studied to maximize mass and electron transport and reaction kinetics in MECs (Idris 2022). The goal of designs like three-dimensional structures and bipolar electrode arrays is to increase the active surface area, which improves electrode efficacy. Maintaining ideal substrate concentrations, pH values, and microbial activity in MEC reactors requires careful flow control (2022). To this end, mechanisms such as continuous flow reactors, recirculation setups, and membrane-based separators are employed to regulate substrate transport, mitigate mass transfer constraints, and elevate reactor performance. The transition of MEC systems from laboratory-scale to industrial-scale necessitates the resolution of engineering hurdles pertaining to reactor design, material selection, process refinement, and cost-effectiveness (Santoro et al. 2017). To facilitate the widespread adoption of MECs for sustainable bio production, exploration into modular reactor configurations, process intensification strategies, and synergies with other renewable energy technologies (e.g., solar panels) is underway (Wang, Liu, et al., 2022). In general, continuous research and innovation in novel materials, microbial consortia, and reactor designs are propelling the progress of microbial electrolysis technology, laying the groundwork for its commercialization and broad utilization in bioenergy and wastewater treatment applications (Khan et al. 2022).

11.6 CHALLENGES AND OUTLOOK

11.6.1 Scaling Up Challenges and Research Opportunities

MECs have become a hopeful route for generating sustainable hydrogen and other biofuel resources among the array of technologies aiming at recovering bio-resources (2020). Considerable work is being done to increase the technology's practical applications. Consequently, a great deal of research, both large scale and research scale, has been done to assess the viability of MECs in addition to wastewater treatment (Saravanan et al. 2023). In one such pilot scale, genuine urban wastewater was used at an applied cathodic voltage of 0.2 V vs. SHE to produce 4 L of hydrogen per day using a 130-liter cassette cell type MEC (Guerrero-Sodric et al. 2023). Throughout a five-month operational period, a consistent cathodic hydrogen recovery rate of 82% was maintained, highlighting the feasibility of scaling up MECs for practical applications (Guerrero-Sodric et al. 2023).

Since the intermittent and low-volume discharge of decentralized wastewater management systems aligns well with the operational cycle of MECs, there is potential for practical implementation of MECs in these systems. Heidrich et al. (2014) carried out an impressive pilot-scale study to demonstrate the applicability of this strategy (Heidrich et al. 2014). Over the course of a year, untreated household wastewater with influent COD values ranging from 147 to 1976 ppm was fed into a 100-liter MEC reactor for the study. Throughout the experiment, the MEC reactor ran at an applied cell potential of 1.1 V and reached a coulombic efficiency of 41.2%, producing an average of 0.6 liters of hydrogen each day (Heidrich et al. 2014). However, over the course of time, the optimum COD reduction varied, between 30% to 65%. Improper pumping from the grit compartment into the MEC digestor may have introduced indigestible materials, which could explain the fluctuation in COD removal rates (Heidrich et al. 2014). Most pilot MEC configurations have fallen short of expectations despite significant efforts, mainly because of low electrode current densities and insufficient energy recovery rates. For example, the conversion efficiency of H_2O_2 was 7.2% in a 110-liter double-chamber MEC (with a cathodic chamber capacity of 10 liters and an anodic chamber volume of 100 liters) at a current density of 0.96 mA/m^2 (Heidrich et al. 2014; Wijayasekera et al. 2022). The major challenges in scaling up MECs is the increased internal resistance that arises from larger component sizes and higher overpotential losses (Heidrich et al. 2014). Moreover, ohmic losses and unequal potential distribution make larger electrodes more vulnerable to higher potential drops, which reduces the efficiency of energy recovery even more. For larger MEC systems, recently developed methods for evaluating internal resistance and its fluctuations with design and operating circumstances can help optimize reactor architecture and performance (Jungmeier 2017; Pawar et al. 2022). Furthermore, techniques like coating with nanoparticles, altering electrodes with ammonia and plasma, and other physicochemical methods can significantly increase surface area, leading to faster reaction kinetics and higher current densities (Maqbool et al. 2023). As such, altered electrodes present a promising solution to important problems encountered while transferring MEC systems from lab to field use (Thapa et al. 2022).

Moreover, Cambrian Innovation appears to have brought MECs into the commercial sphere with the Eco Volt reactor, which is designed to treat wastewater with a high strength (Antevski et al. 2020). However, MECs must be assessed for their economic viability as bio-resource recovery systems to be viable with current wastewater treatment technologies. Using inexpensive materials, particularly for electrodes and current collectors, which is a major cost of total expenditure, may be necessary to streamline MEC design (Thapa et al. 2022). To reduce material costs, carbonaceous materials made from waste are being used more often as anodes in MFCs and MECs. Furthermore, one of the most important tools for demonstrating MECs' potential as carbon-neutral technologies is life-cycle assessment, or LCA (Prathiba et al. 2022). The greenhouse gases potential of the system was determined by an LCA conducted on a large scale MEC that was running at a wastewater stream rate of 65 liters per day and an applied cell potential of 0.5 V (Savla et al. 2021). This resulted in an estimation of 18.8 kg CO_2 equivalent per kg of H_2 produced. It is obvious that significant advancements in MEC design are necessary to promote environmental and technical maturity. The MECs' compatibility and incorporation with the current wastewater management foundation is another key factor to consider. The MEC system ought to have an adaptable architecture that facilitates its easy installation in wastewater treatment facilities, either as a side-stream process or downstream (Sheikh et al. 2023). MEC/membrane filtration and MEC-AD are two examples of hybrid configurations incorporating MECs that have been investigated to improve the functional capabilities of bio-electrochemical resource recovery systems. Furthermore, the integration of MEC technology into conventional units may enable its broader acceptance in residential and commercial wastewater treatment facilities (Sheikh et al. 2023). The system could become self-sustaining on a wider scale by combining MECs with innovative carbon-neutral energy sources like solar, osmotic, thermoelectric, and microbial power (Ekadewi et al. 2023). It is noteworthy that, when energy savings are converted into monetary terms, MEC operation may provide higher cost-effectiveness for treating natural home wastewater than conventional activated sludge procedures (Ekadewi et al. 2023). As of right now, MEC's technology readiness level is about a five, meaning that to raise it to higher levels, regulatory incentives as well as technological innovation are required (Rossi et al. 2023).

11.6.2 Potential for Decentralized Energy and Sustainable Wastewater Treatment

(MECs) hold potential as a sustainable wastewater treatment technology in the long term. Nevertheless, scaling up poses challenges its implementation (Radhika et al. 2022). Therefore, it is imperative to gather new experiences in scaling up to understand the key hurdles in MEC design. Despite the wide array of applications and promising prospects for MECs, their industrial feasibility remains a significant hurdle, necessitating the upscaling of systems to assess this technology (Rahimnejad 2023). It has been explored recently how MECs for hydrogen production can go from bench to pilot scale. Nevertheless, the hydrogen production rate and the Coulomb efficiency were below the maximum values reported in laboratory-scale research, indicating the necessity for additional optimization, even if the electric energy recovery exceeded 70% (Radhika et al. 2022). Moreover, several MEC applications, like metal recovery, chemical synthesis, and pollution removal, have not yet been scaled up from the laboratory level (Radhika et al. 2022).

Two major economic obstacles to scaling up are reactor design and electrode materials. The main goal should be to create affordable reactors that can rival more technology for treating wastewater and producing electricity (Dessì et al. 2021). This will help lower costs by increasing the number of production units while also making this technology more feasible and encouraging the creation of MEC reactors that are more effective. For example, inexpensive anode materials with a good record of electron transmission in MF Cs, such as carbon fiber brushes or activated carbon, could be good substitutes for MEC anodes. Like noble metals, bio-cathodes may present a viable substitute for cathode catalysts due to their low cost, superior stability, and favorable environmental characteristics. However, the efficacy of bio-cathodes in pilot-scale operations is still unknown (Borja-Maldonado and López Zavala 2022).

11.6.3 Conclusion

Traditional methods of treating wastewater seem inefficient due to their heavy reliance on energy, even though wastewater contains 18 kJ/g of COD, which might be effectively connected through MECs and other BESs. MECs have the capability to extract this stored chemical energy from wastewater, generating important chemicals such as H_2, CH_4, H_2O_2, alcohols, and various other industrial chemicals with minimal energy input. Additionally, they can reduce the organic load while recovering valuable resources. Although MECs have shown promise as efficient bio-refinery systems in both laboratory and field settings, their commercialization is impeded by challenges such as low bio-base fuel reclamation and current density. Biochemical catalysts have been used to improve the efficiency of MECs; however, for practical use, major improvements in reactor design and material configuration are still needed. Prioritizing future research should go toward determining if MECs are compatible with the current wastewater management infrastructure and whether they are techno-economically feasible. It is expected that MECs will become a viable alternative to offset the energy inefficiencies linked to traditional wastewater treatment techniques.

In addition to showing great promise in lowering wastewater pollution in industrial streams, MECs also make it easier to recover resources from industrial wastewater. This device produces power and hydrogen, a clean fuel with zero pollution, while also helping to reduce the amount of biodegradable material in the stream. MEC utilization has been shown to be effective for producing hydrogen, with an energy-positive and carbon-negative profile, in response to the growing worldwide need for hydrogen. Integrating newer technologies improves production and streamlines workflows. However, issues like structural constraints and microbial competition still exist, requiring a great deal of research to optimize yields and process efficiency. Optimizing setup efficiency can be achieved through improving substrate selection and reactor design. With its significant reduction in energy use, MEC technology is positioned to be a useful tool for resource recovery and energy generation.

11.6.4 Call for Further Research and Implementation

Since the inception of MECs, significant strides have been taken to enhance the technology through the application of diverse designs, membranes, electrode materials, microbes, and substrates. MECs not only excel in H_2 production from wastewater but also offer potential for recovering value-added products. When compared to traditional bio-hydrogen production techniques like photo-fermentation and DF, MECs show potential since they overcome the thermodynamic obstacles present in traditional methods. However, there are significant barriers to the widespread use of MECs due to their construction and operating expenses. A number of tactics have been used to reduce these expenses, including the use of less costly electrodes and membrane-less designs, which have the added benefit of high rates of H_2 production and recovery. MECs look better than other technologies for producing H_2 and treating wastewater, but they still need to be improved in order to overcome some operational issues. While using wastewater in MECs to produce H_2 is beneficial, uneven nutritional compositions generally necessitate nutrient supplementation, which makes MECs less competitive with competing technologies. MECs that can remove nutrients from wastewater have been made possible by recent advancements, but their high operating costs are still a source of worry. Although the manufacturing of fine chemicals for the food and pharmaceutical industries may pose health risks, the creation of important bio-chemicals such as acetate, ethanol, formic acid, or hydrogen peroxide in MECs using wastewater adds to the technology's appeal. Scaling up MECs entails the flow of large electric currents, with voltage fluctuations and losses directly dependent on current, necessitating careful consideration for the development of efficient MECs. Reports indicate that MEC efficiency is also influenced by various operating conditions, including inoculum type, applied voltage, ionic strength, temperature, pH, hydraulic retention time, and organic loading. Integrating H_2 and other value-added bio-chemicals production from wastewater treatment using MECs appears feasible, as the high operating costs could potentially be offset by other products. Therefore, it is hopeful that in the future, more efficient and cost-effective MECs will be developed, capable of operating at a commercial level. The outlook for MEC technology shows promise: Recently, various large-scale reactors have been designed, and initial industrial ventures are underway. However, significant challenges persist. High capital costs associated with MECs stand as the primary hurdle to their commercial viability. Additionally, effective management of hydrogen presents a technical challenge; inadequate control on the cathode side can lead to the proliferation of undesirable microorganisms, potentially hampering overall reactor performance. Methane-producing MECs could serve as a viable alternative to hydrogen-producing ones, expediting the commercial advancement of this technology. Streamlining the commercialization process of MECs could be facilitated through the establishment of a thorough strategy for data sorting, standardized laboratory methodologies, and characterizing MECs.

REFERENCES

Aboelela D, Soliman MA (2022) Hydrogen production from microbial electrolysis cells powered with microbial fuel cells. *Journal of King Saud University –Engineering Sciences*. https://doi.org/10.1016/j.jksues.2022.05.008

Absi Halabi M, Al-Qattan A, Al-Otaibi A (2015) Application of solar energy in the oil industry – Current status and future prospects. *Renewable and Sustainable Energy Reviews* 43: 296–314.

Alawad SM, Mansour RB, Al-Sulaiman FA, Rehman S (2023) Renewable energy systems for water desalination applications: A comprehensive review. *Energy Conversion and Management* 286: 117035.

Alcaraz-Gonzalez V, Flores-Estrella RA, Nolasco M, et al (2023) Microbial electrolysis cells for biohydrogen generation and wastewater treatment—A short review and current trends. In Debik E, Bahadir M, Haarstrick A (eds) *Wastewater Management and Technologies. Water and Wastewater Management.* Springer.

Antevski K, Bernardos CJ, Cominardi L, et al (2020) On the integration of NFV and MEC technologies: Architecture analysis and benefits for edge robotics. *Computer Networks* 175: 107274. https://doi.org/10.1016/j.comnet.2020.107274

Anwer AH, Khan MD, Khan N, et al (2019) Development of novel MnO2 coated carbon felt cathode for microbial electroreduction of CO2 to biofuels. *Journal of Environmental Management* 249: 109376. https://doi.org/10.1016/j.jenvman.2019.109376

Arutyunov VS, Lisichkin G V (2017) Energy resources of the 21st century: Problems and forecasts. Can renewable energy sources replace fossil fuels? *Russian Chemical Reviews* 86(9): 777–804. https://doi.org/10.1070/rcr4723

Bengelsdorf FR, Beck MH, Erz C, et al (2018) Bacterial anaerobic synthesis gas (syngas) and CO2 + H2 fermentation. In Sima Sariaslani and Geoffrey Michael Gadd (Eds.), *Advances in Applied Microbiology* (pp. 143–221). Academic Press.

Bhattu M, Singh J (2023) Recent advances in nanomaterials based sustainable approaches for mitigation of emerging organic pollutants. *Chemosphere* 321: 138072. https://doi.org/10.1016/j.chemosphere.2023.138072

Borja-Maldonado F, López Zavala MÁ (2022) Contribution of configurations, electrode and membrane materials, electron transfer mechanisms, and cost of components on the current and future development of microbial fuel cells. *Heliyon* 8(6): e09624.

Brogi A, Forti S, Ibrahim A, et al (2018) Survey high-performance modelling and simulation for selected results of the COST action IC1406 cHiPSet. *Future Generation Computer Systems* 29:325–336.

Casiano Flores C, Bressers H, Gutierrez C, de Boer C (2018) Towards circular economy – A wastewater treatment perspective, the Presa Guadalupe case. *Management Research Review* 41(5):554–575. https://doi.org/10.1108/MRR-02-2018-0056

CHRISPIM MC (2021) Resource recovery from wastewater treatment: Challenges, opportunities and guidance for planning and implementation. *Frontiers in Neuroscience* 14: 574619.

Crossley S, Whiter RA, Kar-Narayan S (2014) Polymer-based nanopiezoelectric generators for energy harvesting applications. *Energy Materials: Materials Science and Engineering for Energy Systems* 9(3): 135–141. https://doi.org/10.1179/1743284714Y.0000000605

Cui T, Hu Y, Shen B, Chen Q (2019) Task offloading based on lyapunov optimization for mec-assisted vehicular platooning networks. *Sensors (Switzerland)* 19(22): 4974. https://doi.org/10.3390/s19224974

Dange P, Pandit S, Jadhav D, et al (2021) Recent developments in microbial electrolysis cell-based biohydrogen production utilizing wastewater as a feedstock. *Sustainability (Switzerland)* 13(21):11752.

Dessì P, Rovira-Alsina L, Sánchez C, et al (2021) Microbial electrosynthesis: Towards sustainable biorefineries for production of green chemicals from CO2 emissions. *Biotechnology Advances* 46:107675.

Deutzmann JS, Kracke F, Spormann AM (2023) Microbial electromethanogenesis powered by curtailed renewable electricity. *Cell Reports Physical Science* 4(1):101515. https://doi.org/10.1016/j.xcrp.2023.101515

Dörr M, Dengner J, Maier CS, et al (2023) Short-term heat stress assays provide a standardized screening tool to assess impacts of microbiome manipulation on Aiptasia thermal stress tolerance. *bioRxiv* (preprint). https://doi.org/10.1101/2023.06.01.543236

Ekadewi P, Arbianti R, Gomez C, Utami TS (2023) Biohydrogen production in microbial electrolysis cell operating on designed consortium of denitrifying bacteria. *Food Technology and Biotechnology* 61(1):37–45. https://doi.org/10.17113/ftb.61.01.23.7496

Gao C, Ge H, Lu Y, et al (2021) Decoupling of provincial energy-related CO2 emissions from economic growth in China and its convergence from 1995 to 2017. *Journal of Cleaner Production* 297:126627. https://doi.org/10.1016/j.jclepro.2021.126627

Ghangrekar, M. M., Duteanu, N. M., Surampalli, R. Y., & Zhang, T. C. (Eds.). (2023). *Microbial Electrochemical Technologies: Fundamentals and Applications*. Wiley. https://doi.org/10.1002/9783527839001

Guerrero-Sodric O, Baeza JA, Guisasola A (2023) Exploring key operational factors for improving hydrogen production in a pilot-scale microbial electrolysis cell treating urban wastewater. *Chemical Engineering Journal* 469:144001. https://doi.org/10.1016/j.cej.2023.144001

Hamlehdar M, Kasaeian A, Safaei MR (2019) Energy harvesting from fluid flow using piezoelectrics: A critical review. *Renewable Energy* 143:1644–1665.

Hassanein A, Witarsa F, Lansing S, et al (2020) Bio-electrochemical enhancement of hydrogen and methane production in a combined anaerobic digester (AD) and microbial electrolysis cell (MEC) from dairy manure. *Sustainability (Switzerland)* 12(20):8491. https://doi.org/10.3390/su12208491

Heidrich ES, Edwards SR, Dolfing J, et al (2014) Performance of a pilot scale microbial electrolysis cell fed on domestic wastewater at ambient temperatures for a 12 month period. *Bioresource Technology* 173:87–95. https://doi.org/10.1016/j.biortech.2014.09.083

Hosseinzadeh A, Zhou JL, Li X, et al (2022) Techno-economic and environmental impact assessment of hydrogen production processes using bio-waste as renewable energy resource. *Renewable and Sustainable Energy Reviews* 156:111972.

Idris MO (2022) Review of bioelectrochemical studies of microbial fuel cells. In Mohamad Nasir Mohamad Ibrahim, Asim Ali Yaqoob and Akil Ahmad (Eds.), *Microbial Fuel Cells: Emerging Trends in Electrochemical Applications*. IOP Publishing, Ltd.

Jadhav DA, Mungray AK, Arkatkar A, Kumar SS (2021) Recent advancement in scaling-up applications of microbial fuel cells: From reality to practicability. *Sustainable Energy Technologies and Assessments* 45:101193.

Javid I, Chauhan A, Thappa S, et al (2021) Futuristic decentralized clean energy networks in view of inclusive-economic growth and sustainable society. *Journal of Cleaner Production* 309:127331.

Jungmeier G (2017) The biorefinery fact sheet. *International Journal of Life Cycle Assessment* 23:838–847.

Kadier A, Jain P, Lai B, et al (2020) Biorefinery perspectives of microbial electrolysis cells (MECs) for hydrogen and valuable chemicals production through wastewater treatment. *Biofuel Research Journal* 7(3):1163–1176.

Kadier A, Kalil MS, Abdeshahian P, et al (2016) Recent advances and emerging challenges in microbial electrolysis cells (MECs) for microbial production of hydrogen and value-added chemicals. *Renewable and Sustainable Energy Reviews* 61:501–525.

Kadier A, Kalil MS, Rai PK, et al (2019) Microbial electrolysis cells (mecs): A promising and green approach for bioenergy and biochemical production from waste resources. In R. Navanietha Krishnaraj, Rajesh K. Sani (Eds.), *Bioelectrochemical Interface Engineering* (pp. 209–234). Wiley.

Khan AH, López-Maldonado EA, Alam SS, et al (2022) Municipal solid waste generation and the current state of waste-to-energy potential: State of art review. *Energy Conversion and Management* 267:115974.

Koul Y, Devda V, Varjani S, et al (2022) Microbial electrolysis: A promising approach for treatment and resource recovery from industrial wastewater. *Bioengineered* 13(1):1481–1494.

Kundu D, Dutta D, Samanta P, et al (2022) Valorization of wastewater: A paradigm shift towards circular bioeconomy and sustainability. *Science of the Total Environment* 848:157740.

Li X, Chen L, Mei Q, et al (2018) Microplastics in sewage sludge from the wastewater treatment plants in China. *Water Research* 142:72–85. https://doi.org/10.1016/j.watres.2018.05.034

Liu H, Grot S, Logan BE (2005) Electrochemically assisted microbial production of hydrogen from acetate. *Environmental Science & Technology* 39(11):4317–4320. https://doi.org/10.1021/es050244p

Luan L, Ji X, Guo B, et al (2023) Bioelectrocatalysis for CO2 reduction: recent advances and challenges to develop a sustainable system for CO2 utilization. *Biotechnology Advances* 63:108130.

Manna M, Sen S (2023) Advanced oxidation process: A sustainable technology for treating refractory organic compounds present in industrial wastewater. *Environmental Science and Pollution Research* 30:59375–59391.

Maqbool A, Shahid A, Jahan Z, et al (2023) Development of ZnO-GO-NiO membrane for removal of lead and cadmium heavy metal ions from wastewater. *Chemosphere* 338:139622. https://doi.org/10.1016/j.chemosphere.2023.139622

Masoudi M, Rahimnejad M, Mashkour M (2021) Enhancing operating capacity of microbial fuel cells by using low-cost electrodes and multi anode-cathode connections in a membrane-less configuration. *International Journal of Hydrogen Energy* 46(5):4087–4098. https://doi.org/10.1016/j.ijhydene.2020.12.019

Mier AA, Olvera-Vargas H, Mejía-López M, et al (2021) A review of recent advances in electrode materials for emerging bioelectrochemical systems: From biofilm-bearing anodes to specialized cathodes. *Chemosphere* 283:131138. https://doi.org/10.1016/j.chemosphere.2021.131138

Miller A, Singh L, Wang L, Liu H (2019) Linking internal resistance with design and operation decisions in microbial electrolysis cells. *Environment International* 126:259–266. https://doi.org/10.1016/j.envint.2019.02.056

Miran W, Nawaz M, Jang J, Lee DS (2017) Chlorinated phenol treatment and in situ hydrogen peroxide production in a sulfate-reducing bacteria enriched bioelectrochemical system. *Water Research* 117:11–19. https://doi.org/10.1016/j.watres.2017.04.008

Naqvi A, Ali A, Altabey WA, Kouritem SA (2022) Energy harvesting from fluid flow using piezoelectric materials: A review. *Energies (Basel)* 15(12):4446.

Ndayisenga F, Yu Z, Wang B, Zhou D (2023) Effects of the applied voltage on electroactive microbial biofilm viability and hydrogen production in a recalcitrant organic waste-fed single-chamber membrane-free microbial electrolysis cell performance. *Chemical Engineering Journal* 469:144002. https://doi.org/10.1016/j.cej.2023.144002

Noreen S, Maqbool A, Maqbool I, et al (2022) Multifunctional mesoporous silica-based nanocomposites: Synthesis and biomedical applications. *Materials Chemistry and Physics* 285:126132. https://doi.org/10.1016/j.matchemphys.2022.126132

Okolie JA, Epelle EI, Tabat ME, et al (2022) Waste biomass valorization for the production of biofuels and value-added products: A comprehensive review of thermochemical, biological and integrated processes. *Process Safety and Environmental Protection* 159:475–502.

Ozin GA, Ghoussoub MF (2020) *The Story of CO2: Big Ideas for a Small Molecule*. University of Toronto Press.

Pagliano G, Ventorino V, Panico A, Pepe O (2017) Integrated systems for biopolymers and bioenergy production from organic waste and by-products: A review of microbial processes. *Biotechnology for Biofuels* 10:113.

Park SG, Rajesh PP, Sim YU, et al (2022) Addressing scale-up challenges and enhancement in performance of hydrogen-producing microbial electrolysis cell through electrode modifications. *Energy Reports* 8:12432–12446.

Pawar AA, Karthic A, Lee S, et al (2022) Microbial electrolysis cells for electromethanogenesis: Materials, configurations and operations. *Environmental Engineering Research* 27(6):210336.

Pelaz G, González R, Morán A, Escapa A (2023) Elucidating the impact of power interruptions on microbial electromethanogenesis. *Applied Energy* 331. https://doi.org/10.1016/j.apenergy.2022.120382

Pérez H, García OJQ, Amezcua-Allieri MA, Vázquez RR (2023) Nanotechnology as an efficient and effective alternative for wastewater treatment: An overview. *Water Science and Technology* 87(7):1607–1623. https://doi.org/10.2166/wst.2023.179

Philibert C (2005) *The Present and Future Use of Solar Thermal Energy as a Primary Source of Energy*. International Energy Agency, The internal Academy Council.

Prathiba S, Kumar PS, Vo DVN (2022) Recent advancements in microbial fuel cells: A review on its electron transfer mechanisms, microbial community, types of substrates and design for bio-electrochemical treatment. *Chemosphere* 286:131856. https://doi.org/10.1016/j.chemosphere.2021.131856

Ptushenko VV (2020a) Electric cables of living cells. I. Energy transfer along coupling membranes. *Biochemistry (Moscow)* 85(7):777–785.

Ptushenko VV (2020b) Electric cables of living cells. II. Bacterial electron conductors. *Biochemistry (Moscow)* 85(8):901–909.

Pyun SI (2020) Thermodynamic and electro-kinetic analyses of direct electron transfer (DET) and mediator-involved electron transfer (MET) with the help of a redox electron mediator. *Journal of Solid State Electrochemistry* 24:2123–2133. https://doi.org/10.1007/s10008-020-04780-2

Radhika D, Shivakumar A, Kasai DR, et al (2022) Microbial electrolysis cell as a diverse technology: Overview of prospective applications, advancements, and challenges. *Energies (Basel)* 15(22):8445.

Rahimnejad M (2023) Biohydrogen generation and MECs. In Mostafa Rahimnejad (Ed.), *Biological Fuel Cells: Fundamental to Applications* (pp. 321–349). Elsevier.

Ramya M, Senthil Kumar P (2022) A review on recent advancements in bioenergy production using microbial fuel cells. *Chemosphere* 288:132512. https://doi.org/10.1016/j.chemosphere.2021.132512

Rossi R, Nicolas J, Logan BE (2023) Using nickel-molybdenum cathode catalysts for efficient hydrogen gas production in microbial electrolysis cells. *Journal of Power Sources* 560:232594. https://doi.org/10.1016/j.jpowsour.2022.232594

Rousseau R, Etcheverry L, Roubaud E, et al (2020) Microbial electrolysis cell (MEC): Strengths, weaknesses and research needs from electrochemical engineering standpoint. *Applied Energy* 257:113938.

Rout PR, Goel M, Pandey DS, et al (2023) Technological advancements in valorisation of industrial effluents employing hydrothermal liquefaction of biomass: Strategic innovations, barriers and perspectives. *Environmental Pollution* 316:120601.

Rozendal RA, Hamelers HVM, Euverink GJW, et al (2006) Principle and perspectives of hydrogen production through biocatalyzed electrolysis. *International Journal of Hydrogen Energy* 31(12):1632–1640. https://doi.org/10.1016/j.ijhydene.2005.12.006

Rozenfeld S, Gandu B, Hirsch LO, et al (2021) Hydrogen production in a semi-single-chamber microbial electrolysis cell based on anode encapsulated in a dialysis bag. *International Journal of Energy Research* 45(6):8991–9001. https://doi.org/10.1002/er.7050

Santoro C, Arbizzani C, Erable B, Ieropoulos I (2017) Microbial fuel cells: From fundamentals to applications. A review. *Journal of Power Sources* 356:225–244. https://doi.org/10.1016/j.jpowsour.2017.03.109

Saravanan A, Karishma S, Kumar PS, et al (2023) Microbial electrolysis cells and microbial fuel cells for biohydrogen production: Current advances and emerging challenges. *Biomass Conversion and Biorefinery* 13:4021–4037.

Savla N, Guin M, Pandit S, et al (2022) Recent advancements in the cathodic catalyst for the hydrogen evolution reaction in microbial electrolytic cells. *International Journal of Hydrogen Energy* 47(15):9224–9240. https://doi.org/10.1016/j.ijhydene.2022.03.058

Savla N, Pandit S, et al (2021) Techno-economical evaluation and life cycle assessment of microbial electrochemical systems: A review. *Current Research in Green and Sustainable Chemistry* 4:100069.

Sayed ET, Olabi AG, Shehata N, et al (2023) Application of bio-based electrodes in emerging capacitive deionization technology for desalination and wastewater treatment. *Ain Shams Engineering Journal* 14(2):101918.

Sayed ET, Wilberforce T, Elsaid K, et al (2021) A critical review on environmental impacts of renewable energy systems and mitigation strategies: Wind, hydro, biomass and geothermal. *Science of the Total Environment* 766:144505. https://doi.org/10.1016/j.scitotenv.2020.144505

Sazali N (2020) Emerging technologies by hydrogen: A review. *Journal of Hydrogen Energy* 45(17)10010–10020.

Serale G, Fiorentini M, Capozzoli A, et al (2018) Formulation of a model predictive control algorithm to enhance the performance of a latent heat solar thermal system. *Energy Conversion and Management* 173:21–35. https://doi.org/10.1016/j.enconman.2018.07.099

Sheikh M, Harami HR, Rezakazemi M, et al (2023) Efficient NH3-N recovery from municipal wastewaters via membrane hybrid systems: Nutrient-Energy-Water (NEW) nexus in circular economy. *Chemical Engineering Journal* 465:142959.

Silvi C (1995) The future of solar energy. *Leonardo* 28(1):1–8. https://doi.org/10.2307/1576198

Singh D, Goswami RK, Agrawal K, et al (2022) Bio-inspired remediation of wastewater: A contemporary approach for environmental clean-up. *Current Research in Green and Sustainable Chemistry* 5:100262.

Siwal SS, Zhang Q, Saini AK, et al (2021) Recent advances in bio-electrochemical system analysis in biorefineries. *Journal of Environmental Chemical Engineering* 9(5):106176.

Sun LL, Cui HJ, Ge QS (2022) Will China achieve its 2060 carbon neutral commitment from the provincial perspective? *Advances in Climate Change Research* 13(3):370–383. https://doi.org/10.1016/j.accre.2022.02.002

Thapa BS, Kim T, Pandit S, et al (2022) Overview of electroactive microorganisms and electron transfer mechanisms in microbial electrochemistry. *Bioresource Technology* 347:126740.

Tong KTX, Tan IS, Foo HCY, et al (2023) Sustainable circular biorefinery approach for novel building blocks and bioenergy production from algae using microbial fuel cell. *Bioengineered* 14(1):2102376.

Usman M, Radulescu M (2022) Examining the role of nuclear and renewable energy in reducing carbon footprint: Does the role of technological innovation really create some difference? *Science of the Total Environment* 841:156662. https://doi.org/10.1016/j.scitotenv.2022.156662

Wainaina S, Awasthi MK, Sarsaiya S, et al (2020) Resource recovery and circular economy from organic solid waste using aerobic and anaerobic digestion technologies. *Bioresource Technology* 301:122778.

Wang B, Liu W, Zhang Y, Wang A (2022) Natural solar intermittent-powered electromethanogenesis towards green carbon reduction. *Chemical Engineering Journal* 432:134369. https://doi.org/10.1016/j.cej.2021.134369

Wang H, Song X, Zhang H, et al (2020) Removal of hexavalent chromium in dual-chamber microbial fuel cells separated by different ion exchange membranes. *Journal of Hazardous Materials* 384:121459. https://doi.org/10.1016/j.jhazmat.2019.121459

Wang XT, Zhang YF, Wang B, et al (2022) Enhancement of methane production from waste activated sludge using hybrid microbial electrolysis cells-anaerobic digestion (MEC-AD) process – A review. *Bioresource Technology* 346:126608.

Wijayasekera SC, Hewage K, Siddiqui O, et al (2022) Waste-to-hydrogen technologies: A critical review of techno-economic and socio-environmental sustainability. *International Journal of Hydrogen Energy* 47(41):17662–17685.

Wu Y, Liang Y, Wang H (2021) Heterogeneous molecular catalysts of metal phthalocyanines for electrochemical CO2Reduction reactions. *Accounts of Chemical Research* 54(19):3821–3831. https://doi.org/10.1021/acs.accounts.1c00200

Xu X, Shao J, Li M, et al (2016) Reductive transformation of p-chloronitrobenzene in the upflow anaerobic sludge blanket reactor coupled with microbial electrolysis cell: performance and microbial community. *Bioresource Technology* 218:740–746. https://doi.org/10.1016/j.biortech.2016.07.037

Yang Z, Li S, Ma S, et al (2021) Characteristics and removal efficiency of microplastics in sewage treatment plant of Xi'an City, northwest China. *Science of the Total Environment* 771:145377. https://doi.org/10.1016/j.scitotenv.2021.145377

Yuan F, Zhao H, Sun H, et al. (2022) Investigation of microplastics in sludge from five wastewater treatment plants in Nanjing, China. *Journal of Environmental Management* 301:113793. https://doi.org/10.1016/j.jenvman.2021.113793

Zahed MA, Movahed E, Khodayari A, et al (2021) Biotechnology for carbon capture and fixation: Critical review and future directions. *Journal of Environmental Management* 293:112841.

Zeppilli M, Paiano P, Villano M, Majone M (2019) Anodic vs cathodic potentiostatic control of a methane producing microbial electrolysis cell aimed at biogas upgrading. *Biochemical Engineering Journal* 152:107393. https://doi.org/10.1016/j.bej.2019.107393

Zhang D, Sial MS, Ahmad N, et al (2021) Water scarcity and sustainability in an emerging economy: A management perspective for future. *Sustainability (Switzerland)* 13(1):144. https://doi.org/10.3390/su13010144

12 Prediction of Heavy Metal Adsorption by Biomass Ash Using Artificial Neural Network

T.J. Kalanja, K. Chigayo, J. Kurasha, B. Mapani, H. Musiyarira, T.N. Sithole, and T. Falayi

List of Notations

C_e is the metal ion concentration in solution at equilibrium (mg/L)
C_o is the initial metal ion concentration in solution (mg/L)
q_e is the biomass ash adsorption capacity at equilibrium (mg/g)
q_m is the biomass ash maximum adsorption capacity (mg/g)
q_{ms} is the Sips model maximum adsorption capacity
q_t is the biomass ash adsorption capacity at time t (mg/g)
K_L is the Langmuir constant (L/g)
K_S is the Sips constant Ks (L^{ns}/mg^{ns})
N_s is the Sips constant
V is the volume of solution (L)
M is the mass of biomass ash (g)
K_F is the Freundlich adsorption capacity ($mg/g(L/mg)^{1/n}$)
n is the Freundlich adsorption intensity
k_1 is the pseudo first-order rate constant (min^{-1})
k_2 is the pseudo second-order rate constant (kg min^{-1})

12.1 INTRODUCTION

The mining industry is critical to the socio-economic growth of many resource-rich economies. The industry has been and continues to be the economic backbone of many nations, as evidenced by its average yearly economic growth, contribution to GDP, job creation, income production and role as a vital source of government fiscal receipts and foreign exchange earnings, among other things (Nambinga & Mubita, 2021). Mining activities provide resources in many daily activities (Carvalho, 2017); however, mining has had the most negative environmental effects and is associated with significant social repercussions and inequalities (Carvalho, 2017). Mine tailings, which are the waste materials left over after valuable metals or minerals have been extracted from an ore, are usually transported to a tailing storage facility (TSF) where they are deposited as a slurry. When the slurry settles in the TSF, the water percolates in the slurry and becomes mine leachate. Rainwater that falls on the TSF also contributes to the leachate problem. The metal-rich leachate produced has been demonstrated to have a negative impact on biological communities (living organism and non-living organism) (Riley, et al., 2018). Leachates contain significant levels of hazardous heavy metals such

DOI: 10.1201/9781003668961-12

as lead, chromium, mercury, manganese, zinc, arsenic, cadmium, aluminium, silver, copper and nickel. These heavy metals ions are recognised as poisonous, non-biodegradable and carcinogenic (Singh, et al., 2017). They are highly mobile and soluble in water, and they can weaken the immune system, lead to lung cancer, respiratory issues, stomach discomfort and ulcers, respiratory issues, skin allergies, kidney and liver diseases, and even result in death (Yahya, et al., 2022). Furthermore, other symptoms include weakness, tremors, a mask-like face and psychological disruption (Huang, et al., 2017).

There are numerous treatment techniques that have been used to treat mine leachates, including neutralisation, ion exchange and solvent extraction, ion exchange and precipitation, biosorption, adsorption and reverse osmosis (Ayala & Fernandez, 2019). These processes have some disadvantages, for example, hydroxide precipitation generates a huge amount of low-density sludge, which might pose dewatering and disposal challenges (Mthombeni, et al., 2018). The problem with sulphide precipitation is that in acidic settings, sulphide precipitates leading to production of poisonous H_2S vapours. Ion exchange cannot handle concentrated metal solutions; ion exchange resins must be renewed using chemical reagents as they are used up, and this procedure can produce a substantial amount of secondary pollution (Al-Asheh & Aidan, 2021). Membrane systems have the highest energy usage because of the high pressures involved (Marcelo, et al., 2018). Long-term metal hydroxide precipitation has the tendency to clog the limestone drains and lower the flow capacity of limestone drains used in neutralisation (O'Dwyer et al., 2022). Adsorption has received interest in the pursuit of alternative cheap methods to remove harmful metals ions from leachates due to its affordability and capacity to regenerate the adsorbent (Obayomi, et al., 2019). Additionally, adsorption is capable of extracting complex forms of metals from aqueous solutions at extremely low levels of contamination (Mthombeni, et al., 2018). There are numerous variables and these include surface area of adsorbent (Ogemdi, 2019), adsorbent particle size, adsorption time (Sun et al., 2019) and solubility of adsorbate, to mention but a few.

Agricultural wastes have been utilised as bio-sorbents for wastewater treatment since they are highly effective and show great promise (Yahya, et al., 2022). There are typically three types: (i) non-living biomass, such as bark, lignin, shrimp, krill, squid and crab shell; (ii) algal biomass; (iii) microbial biomass, such as algae, bacteria, fungi and yeast (Yahya, et al., 2022). Agricultural wastes such as ground nutshell (Bayuo, et al., 2019), cashew nut (Yahya, et al., 2022), sewage sludge (Seo, et al., 2020), shrimp shell (Nunez-Gomez, et al., 2019) and rice husk (Li, et al., 2017) are highly promising in their use as bio-sorbents. Prosopis tree is an invasive species found in Namibia. It is generally used as wood fuel to control its spread. A Namibian brewery uses the wood to fire its boilers as part of its sustainability plan. The brewery produces 20–30 tonnes per month of ash. To mitigate against the culture of landfilling, a possible use for the biomass ash as an adsorbent was investigated. The use of waste (biomass ash) to treat another waste (mine leachate) conforms to the dictates of sustainability, specifically Sustainable Development Goal 6 (SDG6).

An artificial neural network (ANN) algorithm is said to mimic the way the brain functions (Fertu et al., 2022). The ANN is made up of linked neurons which are connected via weights (Hafsa et al., 2020). The most common ANN is the feed forward multi-layer perceptron, which has at least an input and output node. The training of the ANN uses backward propagation algorithms (Kennedy et al., 2015). Each node is linked to each node in the next layer leading to prediction of outputs from inputs through non-linear functions. There are many advantages of using ANN and these include that (i) the algorithm is learnt from the data provided without the need to understand the science governing the process, and (ii) ANN are resilient to noise (Kennedy et al., 2015). ANN has been used successfully in the prediction of treatment of wastewater (Hafsa et al., 2020).

Therefore, the main purpose of this research was to characterise the biomass ash, determine the effect of solid loading, time and temperature on the adsorptive properties of the biomass ash, determine the adsorption isotherm and kinetic model and finally to use a trained ANN algorithm to predict the biomass adsorptive performance which can then be used in designing mine effluent plants.

12.2 MATERIALS AND METHODS

12.2.1 Materials

The leachate sample was obtained from a mine in Namibia and analysed to determine the concentration of heavy metal ions present.

The mine leachate was extremely acidic and contained concentrations of Cr (III), Fe (III), Cu (II), Zn (II) and Pb (II), which are above WHO statutory limits (WHO, 2017) (Table 12.1). The leachate can therefore be considered to be a pollutant and has to be treated before disposal into the environment. The biomass used in this study was supplied by the Namibian brewery and was used as received without any further modification.

12.2.2 Methodology

***Effect of solid loading*:** A 1–5 g sample of biomass ash was added to separate 100 ml solutions of leachate. The solutions were agitated at 200 rpm in a 250 ml conical flask at 298.15 K for 1 h. At the end of 1 h, the slurries were filtered with the assay kept for further analysis using the inductively coupled plasma spectrometer (ICP).

***Effect of residence time*:** A 2.5 g sample of biomass ash was added to a separate 100 ml of leachate in 250 ml conical flasks. The resulting slurries were agitated at 200 rpm at 298.15 K for 10, 20, 30, 40, 50 and 60 minutes, respectively. At the end of each specific time interval, the slurries were filtered with the filtrate retained for further ICP analysis. The above experiments were repeated at 308.15 K and 318.15 K to investigate the effect of temperature on the adsorption.

***Reuse of metal loaded biomass ash*:** Heavy metal loaded biomass ash (adsorbent that has been used to remove metals from leachate) was dried in an oven at 323.15K for 1 day. After drying, the heavy metal loaded biomass ash was crushed to a fine powder using a mortar and pestle. A 2.5 g sample of the heavy metal loaded adsorbent was then added to a fresh leachate solution and was agitated at 200 rpm at 298.15K for 1 h. At the end of 1 h, the slurries were filtered with the assay kept for further analysis using the ICP.

12.2.3 Analytical Techniques

The biomass ash was characterised for surface morphology using a scanning electron microscopy (SEM) (JEOL JSM-IT300) after the biomass ash was carbon coated. A Fourier transform infrared (FTIR) spectroscopy (PerkinElmer Spectrum Two) was used for the analysis of the functional groups present on the surface of biomass ash. A Voltcraft PH-100 ATC pH meter was used to measure the pH of the leachate. The inductively coupled plasma (ICP, PerkinElmer Optima 8000) was

TABLE 12.1
pH and Metal Parameters of Leachate

Parameter	Reading	(World-Health-Organisation, 2017) Statutory Limits
pH	0.89 ± 0.021602	-
Cr (III)	5.57	0.05
Fe (III)	1005.00	0.30
Cu (II)	16.70	1.50
Zn (II)	11.80	3.00
Pb (II)	2.32	0.05

used to analyse for metal ions in the leachate. The biomass ash elemental composition was determined using an X-ray fluorescence (XRF, Rigaku NEX CG). The particle size distribution of the biomass ash was determined using laser fraction technique (Malvern, MASTERSIZER 3000).

12.3 LITERATURE REVIEW

12.3.1 Adsorption Isotherms

Adsorption isotherms were employed to investigate the thermodynamic behaviour of the adsorption process. The experimental results were examined using the Langmuir, Freundlich and Sips isotherms to determine the metal adsorption capacities of the biomass ash.

The Freundlich isotherm is one of the adsorption empirical isotherms (Wang & Guo, 2020a). The non-linear equation for the Freundlich is given as

$$q_e = K_F C_e^{\frac{1}{n}} \tag{12.1}$$

Where $q_e\left(\text{mg}/\text{g}\right)$ and $C_e\left(\text{mg}/l\right)$ are adsorbed amount per mass of adsorbent and adsorbate concentrations at equilibrium, $K_F\left(\text{L}^{1/n}.\text{mg}^{1-1/n}.\text{g}^{-1}\right)$ and n are constants. The use of non-linear isotherms reduces errors associated with linearisation, which can be as high as 40% (Wang & Guo, 2020a). The Freundlich isotherms represent multi-layer adsorption on heterogenous adsorbents.

The Langmuir isotherm is one of the chemical adsorption isotherms (Wang & Guo, 2020a). The non-linear equation for the Langmuir equation is given as

$$q_e = \frac{q_m K_L C_e}{1 + K_L C_e} \tag{12.2}$$

where $K_L\left(\text{L}/\text{m}\right)$ is the ratio of the adsorption rate and desorption rate, $q_m\left(\text{mg}/\text{g}\right)$ is the maximum adsorption capacity of the adsorbent. The Langmuir isotherm is based on the premise of monolayer adsorption, homogenous distribution of adsorption, a constant adsorption energy and the interaction between adsorbate molecules is negligible.

The Sips model is a hybrid Langmuir and Freundlich isotherm model. The non-linear equation is given as

$$q_e = \frac{q_{ms} K_s C_e^{ns}}{1 + K_s C_e^{ns}} \tag{12.3}$$

where $q_{ms}\left(\text{mg}/\text{g}\right)$ is the maximum adsorbed amount, $K_s\left(\text{L}^{\text{ns}}/\text{mg}^{\text{ns}}\right)$ and ns is the Sips constant. This is a three-parameter isotherm which can be used for both homogeneous and heterogeneous adsorption. N_s is a measure of the heterogeneity of the system.

12.3.2 Adsorption Kinetics

The pseudo first order kinetic (PFO) is an empirical kinetic equation (Wang & Guo, 2020b). The non-linear equation for PFO is given as

$$q_t = q_e\left(1 - e^{-k_1 t}\right) \tag{12.4}$$

Where q_t is the amount of adsorbate adsorbed per mass of adsorbent after a time t. The PFO is associated with instances where external/internal diffusion is the rate controlling step thereby relating to physisorption (Wang & Guo, 2020b).

The pseudo second order kinetic (PSO) is an empirical kinetic equation. The non-linear equation for PSO is given as

$$q_t = \frac{q_e^2 k_2 t}{1 + q_e k_2 t} \tag{12.5}$$

The PSO is mostly associated with chemisorption (Zhang, 2019).

12.3.3 Adsorption Thermodynamics

The calculation of the changes in enthalpy, entropy and Gibbs free energy that occur during adsorption based on Equations (12.6) and (12.7).

$$\Delta G^{\circ} = -RT \ln K_D \tag{12.6}$$

Where is ΔG° the standard free energy change, J/mol, R is the universal gas constant, 8.314 J/mol K, T is the absolute temperature and K_D is the equilibrium constant $\left(q_e/c_e\right)$.

$$\Delta G^{\circ} = \Delta H^{\circ} - T\Delta S^{\circ} \tag{12.7}$$

A plot of ΔG° vs. temperature, T will be linear and the values of ΔH° and ΔS° can be determined from the slope and intercept of the plot.

12.3.4 Statistical Analysis

All experiments were done in triplicate with the average of the three tests being reported. Statistical significance was tested using a 2-tail t test at 95% confidence interval. Validation of the ANN can be done through mean square error (MSE) and R^2 correlation coefficient (Abbasi et al., 2022). The formulas for the two are shown in equation (12.8) and (12.9) respectively

$$R^2 = 1 - \frac{\sum_{i=1}^{N} \left(x_i - x_i^{ANN}\right)^2}{\sum_{i=1}^{N} \left(x_i - \bar{x}\right)^2} \tag{12.8}$$

$$\text{MSE} = \frac{\sum_{i=1}^{N} \left(x_i - x_i^{ANN}\right)^2}{N} \tag{12.9}$$

12.4 RESULTS AND DISCUSSIONS

12.4.1 Characterisation of the Biomass Ash

The XRF analysis showed that Ca and Mg were the major elements at 46.95% and 6.68% mass/mass respectively whilst the loss on ignition was 24%. The minor elements (1%–10%) were S, K, P, Si and Al. Laser diffraction analysis showed that the biomass ash had a specific external surface area of 420.7 m^2/kg with a surface area moment mean of 11.5 µm, and a volume moment mean was 34.9 µm. SEM (Figure 12.1) showed that the biomass ash was composed of irregularly shaped particles that have very fine particles spreading over the solid particles.

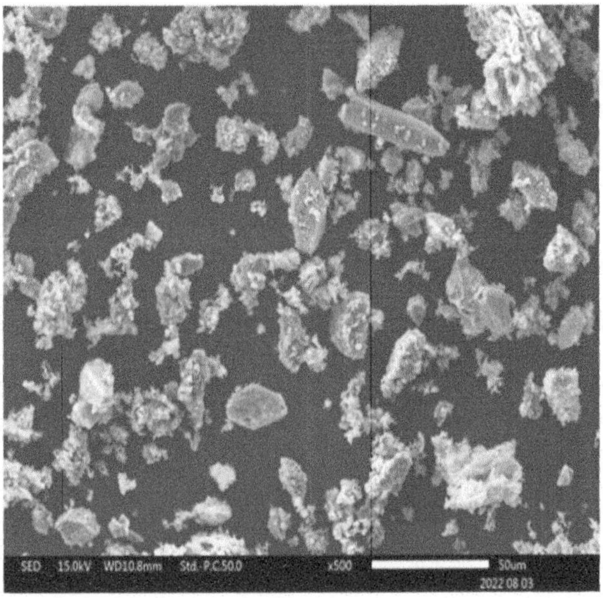

FIGURE 12.1 SEM of biomass ash at × 500 magnification.

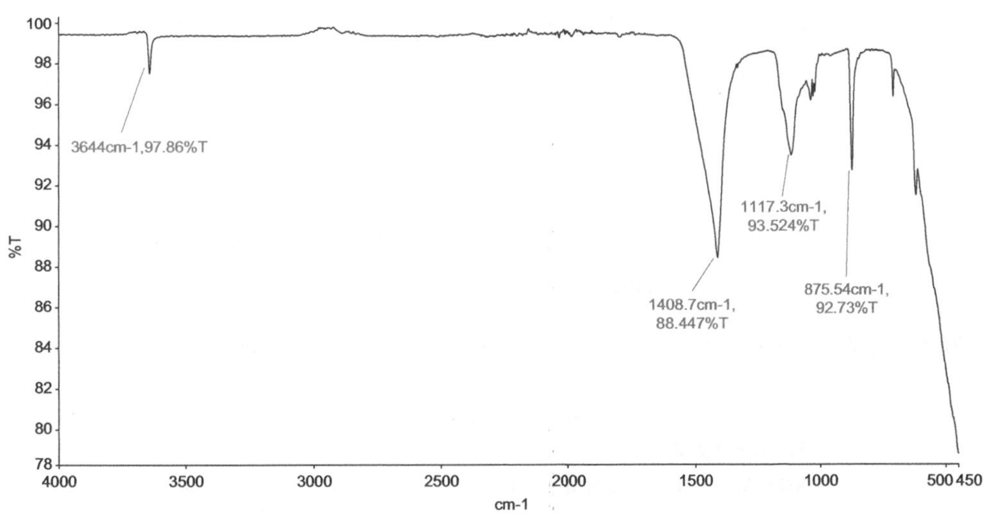

FIGURE 12.2 FTIR spectrum of the biomass ash.

FTIR (Figure 12.2) analysis showed the two distinctive regions, the fingerprint region (up to about 1450 cm^{-1}) and the functional group region above 1500 cm^{-1}. O – H stretching functional group was represented by the largest peak with wave number (3644 cm^{-1}). The spectral band at (1407.7 cm^{-1}) also indicated the O – H bending functional group. The band at 1117.3 cm^{-1} represents C – O stretching functional group. Furthermore, the band at 875.54 cm^{-1} represents C – H bending functional group (Mohrig, et al., 2010).

12.4.2 The Effect of Solid Loading on Heavy Metal Ion Removal Efficiency

There was an increase in % metal ion removal as biomass ash loading was increased from 1 to 5% m/v due to an increase in adsorption sites with increasing mass of the biomass ash (Sreedhar & Reddy, 2019) (Figure 12.3). The highest heavy metal ion removal was at 5% m/v biomass ash loading with 97% Cr (II), 100% Fe (II), 100% Zn (II), 100% Cu (II) and 100% Pb (II) removed, respectively. The pH of the leachate gradually increased during adsorption with the pH of the 2.5% solid loading experiment at 7.01 after1 h (Figure 12.3). This was due to the release of base metal ions (Ca (II) and Mg (II)) from the biomass into the leachate. This was evidenced by the XRF analysis of the adsorbent before and after adsorption (Table 12.2).

There was a significant statistical difference in metal ion removal between 1–5% m/v biomass ash loading, however from 2% to 5% m/v biomass ash loading there was no statistical significance ($p = 0.996 > 0.05$). This may have been due to metal ion exhaustion in the leachate. One of the aims of the study was to neutralise the pH of the leachate and hence a solid loading of 2.5% m/v (Figure 12.3) was chosen for the remainder of the experiments.

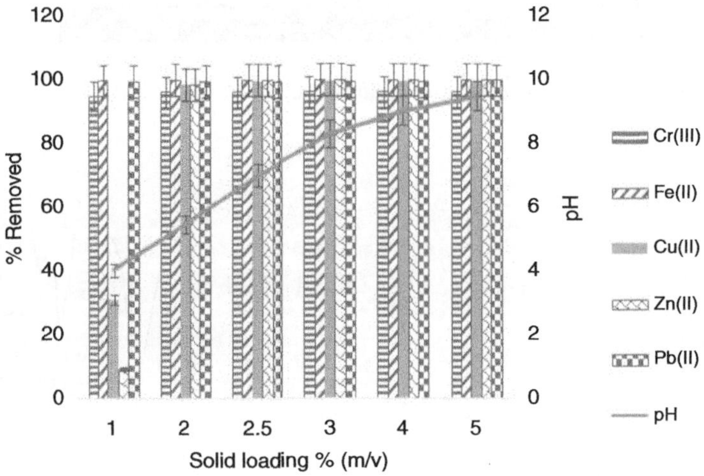

FIGURE 12.3 Variation in % removal of Cr, Fe, Cu, Zn & Pb and pH with adsorbent loading.

TABLE 12.2
XRF Analysis of Adsorbent

Components	Biomass Ash % Mass/Mass	Heavy Metal Loaded Biomass Ash (%m/m)
Mg	6.681	5.258
Al	1.157	1.700
Si	1.237	1.307
P	1.002	0.905
S	1.216	12.577
K	4.160	0.726
Ca	46.951	20.412

12.4.3 Effect of Time and Temperature

There was an increase in adsorption capacity of biomass ash with time as more metal ions were adsorbed onto the adsorbent (Figure 12.4). The adsorption capacity after 60 mins was 40.17, 0.66, 0.47 mg/g and 0.09 mg/g for Fe (II), Cu (II), Zn (II) and Pb (II) respectively. There was no statistical difference in adsorption capacity for all metal ions from 40 to 60 min. This may have been due to metal ion depletion in the solution. There was an increase in the adsorption capacity with temperature (from 25°C to 45°C) for all metal ions. The increase was however not statistically significant ($p > 0.05$) this may be due to the fact that at 25°C the rate of adsorption was over 96%.

12.4.4 Isotherms and Kinetics

The Sips isotherm could be used to model the adsorption of heavy metals onto biomass ash as the correlation coefficient was greater than 0.97 (Table 12.3). Freundlich and Langmuir parameters are not shown as the correlation coefficients for both isotherms were below 0.88 with large values for SSE. The Sips constant n_c was greater than unity for most metal ions showing how heterogenous the system was. The adsorption of the heavy metal ions onto biomass ash could be modelled well using PFO as the correlation coefficient was higher than 0.97 (Table 12.3) and the difference between experimental and calculated q_e was below 6% (Table 12.3). The PSO could not model the adsorption as it had a lower correlation coefficient $\left(R^2 < 0.54\right)$ and higher values for SSE. The PFO can be used to model physisorption, which agrees well with the thermodynamic calculations (Table 12.4).

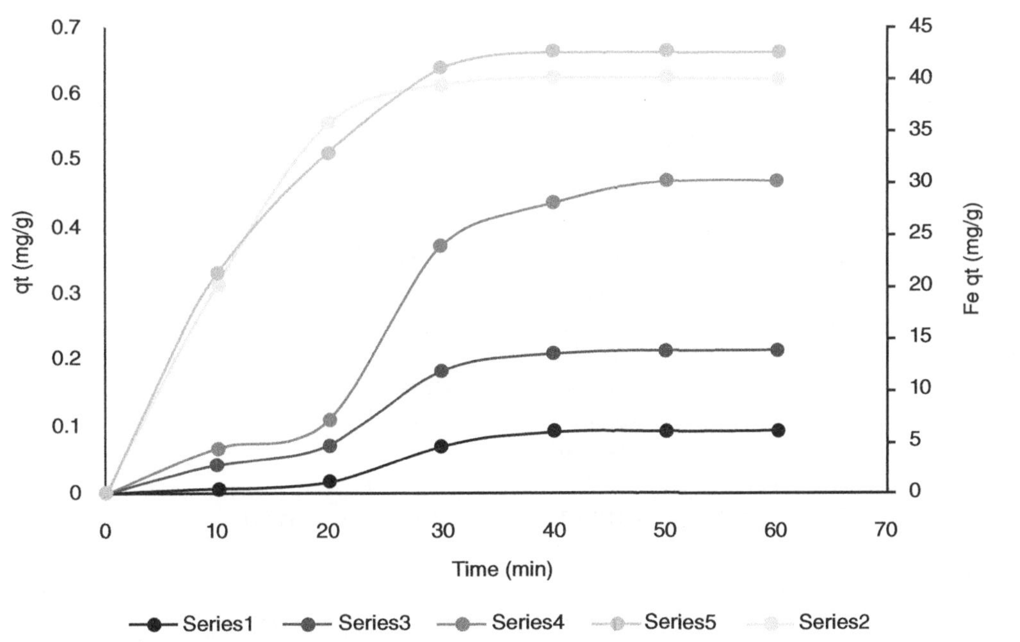

FIGURE 12.4 Variation of metal removal with time.

TABLE 12.3
Sips Isotherm and Pseudo First Order Kinetic Parameters

Metal Ion	Cr^{3+}	Fe^{2+}	Cu^{2+}	Zn^{2+}	Pb^{2+}
Sips Isotherm Parameters					
Q_{ms} (mg/g)	0.216	40.267	0.681	0.470	0.101
K_s $\left(L^{ns}/mg^{ns}\right)$	2.388	3.817	1.133	2.016	2.549
nc	3.842	2.162	0.811	1.472	1.967
SSE	1.864×10^{-4}	1.98×10^{-4}	0.001	1.284×10^{-4}	1.344×10^{-6}
R^2	0.978	0.993	0.996	0.970	0.999
Pseudo first order kinetics					
Q_e calculated (mg/g)	0.208	40.365	0.709	0.491	0.099
Q_e experimental (mg/g)	0.213	40.214	0.667	0.483	0.101
K_1 $\left(mol\,L^{-1}\right)^{1-n} s^{-1}$	0.0056	0.013	0.0364	0.0577	0.0032
SSE	1.432×10^{-4}	0.004	0.002	4.519×10^{-4}	1.107×10^{-5}
R^2	0.992	0.973	0.988	0.995	0.997

TABLE 12.4
Thermodynamic Data

T (K)	ΔG (J/mol)	ΔH (J/mol)	ΔS (J/mol)
Cr^{3+}			
298.15	−2157.7	18819	70.644
308.15	−3123.18		
318.15	−3570.59		
Fe^{2+}			
298.15	−3320.29	30244	112.560
308.15	−4428.71		
318.15	−5571.44		
Cu^{2+}			
298.15	−309.528	22733	77.315
308.15	−1109.53		
318.15	−1855.83		
Zn^{2+}			
298.15	−1737.94	15528	58.003
308.15	−2401.25		
318.15	−2897.99		
Pb^{2+}			
298.15	−2319.43	8068	34.851
308.15	−2675.99		
318.15	−3016.45		

12.4.5 Thermodynamics

The negative value of the Gibbs free energy increased for all metals as temperature increased (Table 12.4). This was an indication that the adsorption process of these metal ions onto biomass ash became more favourable at higher temperatures. The negative ΔG values indicated that the process is thermodynamically feasible. For all metal ions the ΔG was below −20 kJ/mol (Table 12.4), which supported the meaning of PFO results indicating that adsorption process was physisorption (Okoye & Obi, 2012; Abechi, 2018). The positive values for enthalpy showed that the adsorption process was endothermic. The values of entropy were also positive, indicating biomass ash's affinity for the metal ions.

12.4.6 Reuse of Adsorbent

Figure 12.5 shows the variation in metal removal with number of cycles the biomass ash could be used without regeneration.

Figure 12.5 showed that the removal of Cu (II) and Pb (II) remained unchanged after the first cycle of reuse of biomass ash. Chromium (III) removal decreased from 96% to 78.46% after the first cycle of reuse. Zinc (II) decreased from 100% to 89.83% after the first cycle of reuse. Lead (II) decreased from 100% to 84.91% after the first cycle of reuse. According to the findings, biomass ash could only be reused once, because after the second cycle of reusability, the pH of the solution was already below 7 (Figure 12.5) and the residual concentrations of Cr and Pb were above the World Health Organization statutory limits. Therefore, in total the biomass ash could be recycled without regeneration for two treatment cycles. Since the same biomass ash was utilised, the accessible adsorption sites remained unchanged, but the active sites decreased as there was more metal uptake every cycle, which is why there was a decrease in metal ions adsorption.

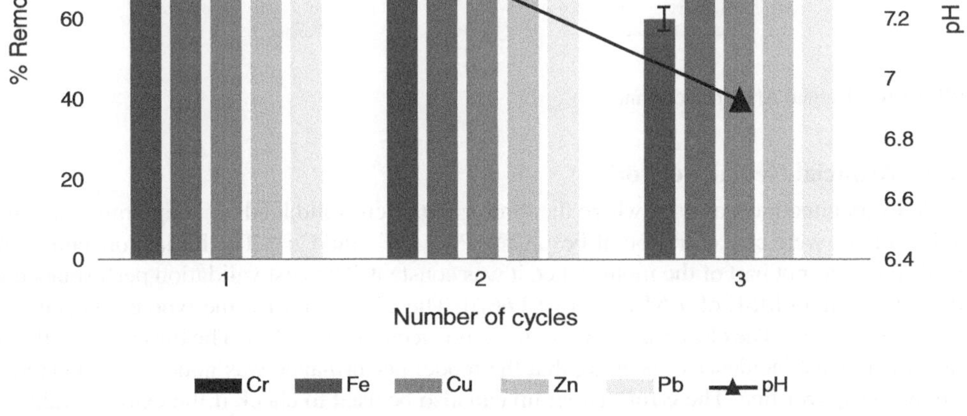

FIGURE 12.5 Variation in metal removal and pH with reuse of metal loaded biomass ash.

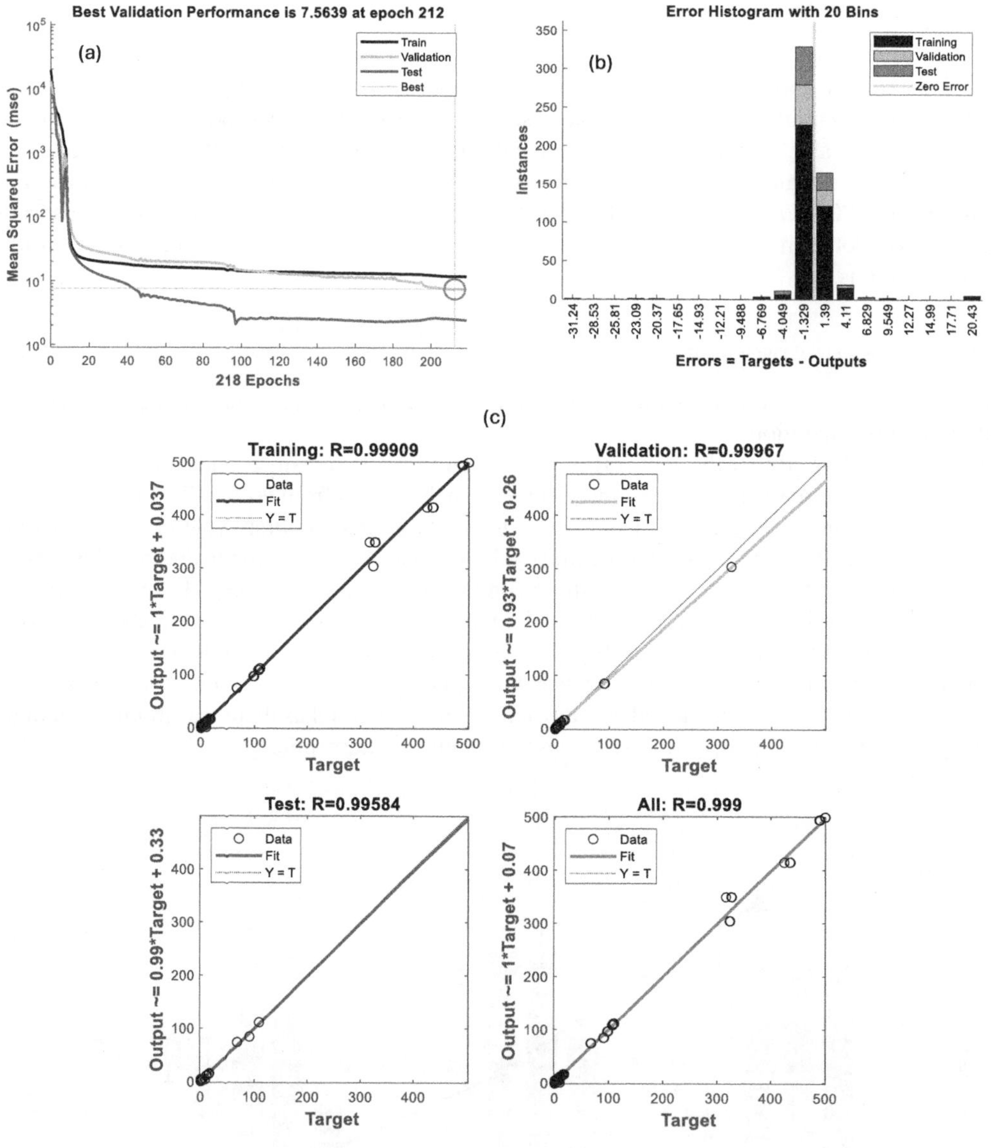

FIGURE 12.6 Trained ANN performance.

12.4.6.1 Artificial Neural Network

A 3,5,5 ANN architecture was used where the three inputs were solid loading, temperature and time, whilst the outputs were concentration of Fe^{2+}, Cr^{3+}, Pb^{2+}, Zn^{2+} and Cu^{2+}. The initial concentration of the metals ions was not part of the inputs since it was constant. The best validation performance was at epoch 212 with an MSE of 7.54 (Figure 12.6(a)). The correlation for the whole algorithm was 0.999 (Figure 12.6(c)). The closer R^2 to unity the more accurate the ANN. The bins close to the zero error had the highest incidences, meaning that the model performance was good as most errors are close to the zero-error line. The error histogram can also be used to check if the extreme values are outliers or not; for the training data the outliers did not affect the trained ANN since the histogram also shows that the bulk of the data falls within the −1.329 error (Figure 12.6(b) (Younis and Hakan, 2019).

Twenty blind samples from the same mine were used to test the robustness of the trained model. The adsorption was carried out using a 2.5% m/v solid loading at 25°C for 1 h. The trained showed an accuracy of 95.4%.

12.5 CONCLUSIONS

Prosopis tree biomass ash was capable of neutralising highly acidic mining effluent accompanied with over 96% removal of Cr (III), Fe (II), Cu (II), Zn (II) and Pb (II) using a solid loading of the adsorbent of 2.5% m/v. The sorption mechanism fitted physisorption and could be modelled well using the three parameter Sips isotherm and pseudo first order kinetic. The adsorption system was highly heterogeneous as evidenced by the non-unity of the Sips isotherm constant. The relatively high external surface area was responsible for the reusability of the metal loaded biomass ash. The adsorption of the metal cations onto the biomass ash was endothermic and retained a positive entropy value showing the high affinity for the metal cations by the biomass ash. Artificial neural networks could be used to predict the performance of the biomass ash as an adsorbent with a high correlation of 0.99. This study presented an opportunity for the use of biomass ash in the treatment of acidic leachate thereby satisfying the needs of a circular economy and sustainable development goal (SDG6).

REFERENCES

Abbasi E, Alavi Moghaddam MR and Kowsari E (2022) A systematic and critical review on development of machine learning based-ensemble models for prediction of adsorption process efficiency. *Journal of Cleaner Production*, 379(1): 134588, https://doi.org/10.1016/j.jclepro.2022.134588

Abechi SE (2018) Studies on the mechanism of adsorption of methylene blue onto activated carbon using thermodynamic tools. *Science World Journal*, 13(2): 17–19.

Al-Asheh S and Aidan A (2021) A comprehensive method of ion exchange resins regeneration and its optimization for water treatment. Promising techniques for wastewater treatment and water quality assessment. *IntechOpen*, 2021, http://doi.org/10.5772/intechopen.93429

Ayala J and Fernandez B (2019) Treatment from abondoned mine landfill leachates. Adsorption technology. *Journal of Materials Research and Technology*, 8(3): 2732–2740, https://doi.org/10.1016/j.jmrt.2019.04.009

Bayuo J, Peling-Ba KB and Abukasi MA (2019). Adsorptive removal of chromium (VI) from aqueous solution unto groundnut shell. *Applied Water Science* 9(4): 104, https://doi.org/10.1007/s13201-019-0987-8

Carvalho FP (2017). Mining industry and sustainable development: Time for change. *Food and Energy Security*, 6(2): 61–77, https://doi.org/10.1002/fes3.10

Fertu DI, Dragoi EN, Bulgariu L, Curteanu S and Gavrilescu M (2022). Modeling the biosorption process of heavy metal ions on soybean-based low-cost biosorbents using artificial neural networks. *Processes*, 10(3): 1–24, https://doi.org/10.3390/pr10030603

Hafsa N, Rushd S, Al-Yaari M and Rahman M (2020) A generalized method for modeling the adsorption of heavy metals with machine learning algorithms. *Water*, 12(12): 3490. https://doi.org/10.3390/w12123490

Huang H, Wang G, Shi Z, Li Z, Kang F, and Liu F (2017). Removal of hexavalent chromium in natural groundwater using activated carbon and cast iron combined system. *Journal of Cleaner Production*, 165(1): 667–676, https://doi.org/10.1016/j.jclepro.2017.07.152

Kennedy MJ, Gandomi AH and Miller CM (2015) Coagulation modeling using artificial neural networks to predict both turbidity and DOM-PARAFAC component removal. *Journal of Environmental Chemical Engineering*, 3(4): 2829–2838. https://doi.org/10.1016/j.jece.2015.10.010

Li W, Law F and Chan Y (2017) Biosorption studies on copper (II) and cadmium (II) using preheated rice straw and rice husk. *Environmental Science and Pollution*, 24(10): 8903–8915, https://doi.org/10.1007/s11356-015-5081-7

Marcelo L, Michelle GL, Gabriel M and Ricardo J (2018) State of art of landfill leachate treatment: Literature review and critical evaluation. *Ciência E Natura*, *40*(12): 78–87, https://doi.org/10.5902/2179460X35239

Mohrig JR, Hammond CN and Schatz, PF (2010) *Infrared Spectroscopy in Techniques in Organic Chemistry*. 3rd ed. New York: W. H. Freeman and Company.

Mthombeni NH, Mbakop S, Ray SC et al. (2018) Highly efficient removal of chromium (VI) through adsorption and reduction: A column dynamic study using magnetized natural zeolite–polypyrrole composite.

Journal of Environmental Chemical Engineering, 6(4): 4008–4017, https://doi.org/10.1016/j.jece.2018.05.038

Nambinga V and Mubita L (2021) *The Impact of Mining Sector to the Nambia Economy: Assessing Socio-economic and Environmental Effects*. Windhoek: Namibia National Planning Commission.

Nunez-Gomez D, Rodrigues C, Lapolli FR, and Lobo-Recio MA (2019) Adsorption of heavy metals from coal acid mine drainage by shrimp shell waste: Isotherm and continous-flow studies. *Journal of Environmental Chemical Engineering*, 7(1): 102787, https://doi.org/10.1016/j.jece.2018.11.032

O'Dwyer TF, Ghanim B, Courtney R, Hudson A, Tony Pembroke J and Murnane JG (2022) Sustainable treatment of acidic and alkaline leachates from mining and industrial activities: Current practice and future perspectives [Internet]. *Wastewater Treatment*. IntechOpen; 2022. Available from: http://doi.org/10.5772/intechopen.103889

Obayomi KS, Bello JO, Nnoruka JS, Adediran AA and Olajide PO (2019) Development of low-cost bio-adsorbent from agricultural waste composite for Pb(II) and As(III) sorption from aqueous solution. *Cogent Engineering*, 6(1), http://doi.org/10.1080/23311916.2019.1687274

Ogemdi IK (2019) Properties and uses of colloids: A review. *Colloid and Surface Science*, 4(2): 24–28, http://doi.org/10.11648/j.css.20190402.12

Okoye P and Obi C (2012) Thermodynamic and kinetic evaluations of some heavy metal ions on aluminum-pillared and unpillared bentonite clays. *International Archive of Applied Sciences and Technology*, 3(2): 58–67, http://doi.org/10.4314/jasem.v18i1.20

Riley AL, Pepper SE, Sexton EL (2018) Metal Removal from Synthetic Mine Tailing Leachates using Fixed-Bed Ion-Exchange Columns. Copper Cobalt Africa, incorporating the 9th Southern African Base Metals Conference. *Copper Cobalt Africa - 9th Southern African Base Metals Conference*, 10–12 July 2018, Livingstone, Zambia. Southern African Institute of Mining and Metallurgy.

Seo JH, Kim N, Park M, Lee S, Yeon S and Park D (2020) Evaluation of metal removal performance of rod-type biosorbent prepared from sewage-sludge. *Environmental Engineering Research*, 25(5): 700–706, https://doi.org/10.4491/eer.2019.201

Singh D, Kumar V, Mohan S, Bano D and Hasan SH (2017). Breakthrough curve modelling of graphene oxide aerogel packed fixed bed column for the removal of Cr(VI) from water. *Journal of Water Process Engineering*, 18(08): 150–158, https://doi.org/10.1016/j.jwpe.2017.06.011

Sreedhar I and Reddy IS (2019) Heavy metal removal from industrial effluent using bio-sorbent blends. *SN Applied Sciences*, 1(9): 1021. https://doi.org/10.1007/s42452-019-1057-4

Sun W, Roy A and Selim HM (2019). Residence time effects on molybdenum adsorption on soils: Elucidation by multi-reaction modeling and XANES analysis. *Soil Systems*, 3(3): 55, https://doi.org/10.3390/soilsystems3030055

Wang J and Guo X (2020a) Adsorption isotherm models: Classification, physical meaning, application and solving method. *Chemosphere*, 258(1): 127279, https://doi.org/10.1016/j.chemosphere.2020.127279

Wang J and Guo X (2020b) Adsorption kinetic models: Physical meanings, applications, and solving methods. *Journal of Hazardous Materials*, 390(1): 122156, https://doi.org/10.1016/j.jhazmat.2020.122156

World-Health-Organisation, 2017. *Guidelines for Drinking-water Quality*. 4th ed.

Yahya MD, Imam IA and Abdulkareem SA (2022) Column adsorption studies for the removal of chemical oxygen demand from fishpond wastewater using waste alum sludge. In *Advances in Remediation Techniques for Polluted Soils and Groundwater* (Gupta PK, Yadav B and Himanshu SK (eds)). Elsevier, Amsterdam, the Netherlands, pp. 21–48.

Younis MY and Hakan K (2019) Artificial neural network prediction of sulfur content of diesel fuel from its physical properties artificial neural network prediction of sulfur content of diesel fuel from its physical properties. *IOP Conference Series: Materials Science and Engineering*, 518(1): 1–8, https://doi.org/10.1088/1757-899X/518/6/062008

Zhang J (2019) Physical insights into kinetic models of adsorption. *Separation and Purification Technology 229*: 115832, https://doi.org/10.1016/j.seppur.2019.115832

13 Bio-Inspired Materials for Reducing Ecological Footprint and Life Cycle Assessment

Shubha Dwivedi, Naveen Dwivedi, and Aline Viancelli

13.1 INTRODUCTION

Global climate change is a major concern nowadays. Rapid industrialization and urbanization leads to altered climate patterns. The social environment (equal access to food and fresh water), the environment (weather contrasts with extremely low or high air temperatures, increased windiness, increased flooding in coastal areas, etc.) and annual economic losses are all at risk from climate change, which calls for mitigation measures. The European Union has established a goal to cut greenhouse gas emissions by 40% by 2030 in order to lessen the effects of climate change. To reduce the ecological footprint and increase production of sustainable biomaterial, it is required of every human being to reduce the pace of development. This is the prime time to develop biomaterials from natural sources for various applications. At every step of the production process, technology development or any type of product making, life-cycle assessment (LCA) is the primary requirement to check the impact of the process/product on the environment. For example, from the paper industry, it should be calculated how many trees were used for making one piece of paper. Contrary to this, what efforts are made by the individual to overcome the impact of their carbon footprint? This is how the entire world has to rethink, reevaluate, and restructure (3R) the process of development to meet the market demand and supply. This chapter will focus on the current statistics (year 2024) for ecological footprint and carbon footprints and also evaluate the process of biomaterial development.

13.2 ECOLOGICAL FOOTPRINT

Ecological footprints (EF) are the measure of the rate of consumption of natural resources and generation of waste. The assessment of ecological footprints is also associated with biocapacity or productive area. When a country's biocapacity exceeds its EF, it has an "ecological reserve" or "biocapacity reserve".

Reducing EF involves taking comprehensive measures to minimize the environmental impact of our activities. An EF measures the demand on Earth's ecosystems and compares this demand with the planet's capacity to regenerate resources and absorb wastes. The goal is to live within the means of our planet, ensuring that we do not deplete resources faster than they can be replenished. EF and biocapacity are expressed in globally comparable hectares with world average productivity.

If a population's EF exceeds the region's biocapacity (per capita consumption of natural resources), that region turns into biocapacity deficit. In more popular communications, it is also called "an ecological deficit". To reduce EF, eco-friendly measures need to be adopted.

A positive biocapacity reserve means, as shown in Figure 13.1, that the environment is producing or replenishing resources faster than humans are currently consuming them. A negative biocapacity reserve means that resource consumption by humans is faster than the environment can replenish them.

DOI: 10.1201/9781003668961-13

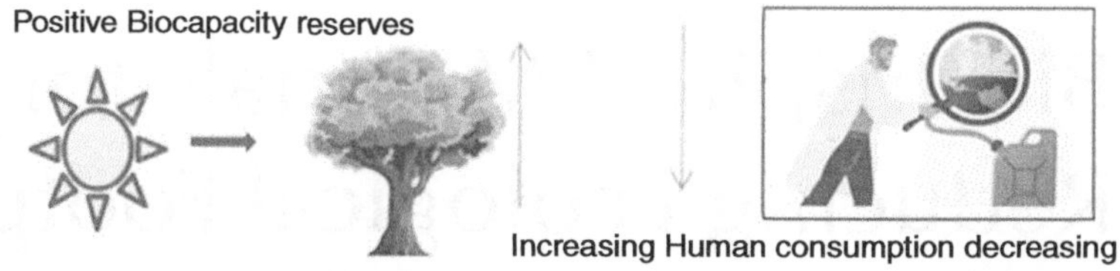

FIGURE 13.1 Positive biocapacity in which natural production increases contrary to human consumption.

The world-average ecological footprint was 2.77 global hectares per capita (12 billion total) in 2017, with an average biocapacity of 1.60 global hectares. This subtracts to a global deficit of 1.17 global hectares per person, or a biocapacity reserve of −1.17, meaning humanity's consumption of natural resources is currently outpacing the Earth's ability to replenish those resources. India's modest 1.19 clearly depicts the poverty situation in worldwide EF index. It reflects the country's widespread poverty and lower lifestyle, contrary to the rapid industrialization seen in recent era. On the other hand, the condition of the USA is quite different in the index. The US population density is low but the people's lifestyle is economically high, which creates a burden of exploitation of fossil fuel and carbon usage resulting in high carbon emissions.

Terminology related with ecological footprint is presented in Table 13.1.

Table 13.2 shows the comparative analysis of footprint in ecological perspectives of four major countries and Figure 13.2 shows the top 10 countries' ecological footprint analysis in a bar graph.

One can see from the statistics of the United Kingdom's EF that surprisingly the country is continuously making efforts to lower its carbon emissions and reduce consumption of fossil fuel. It was positioned at fifth place to drop down to ninth place on the EF index (dropped 27% of EF). There are

TABLE 13.1
Terminology Related with Ecological Footprint

Ecological Footprint (EF)	Its impact of carbon dioxide emissions, mining, waste generation and management, and deforestation on the environment.
Biocapacity	Its capacity to support life and accommodate its population's per capita EF.
Biocapacity Reserve	When a country's biocapacity exceeds its EF.
Ecological Deficit or Biocapacity Deficit	The ecological reserve becomes negative when a country's biocapacity is smaller than its EF.

TABLE 13.2
Comparative Analysis of Ecological Footprint of Countries

Particulars	China	USA	India	Japan
Ecological footprint per capita	3.71 hectares	8.04 gha	1.19 Lower footprint reflects the poverty of the country	4.65 hectare
Biocapacity	1.3 billion hectares,	3.45 hectares.	0.43 hectares	0.59 hectares per capita.
Ecological deficit	−4 billion gha	−1.49 billion hectares	−1 billion hectares	−517 million gha, the fourth-highest in the world

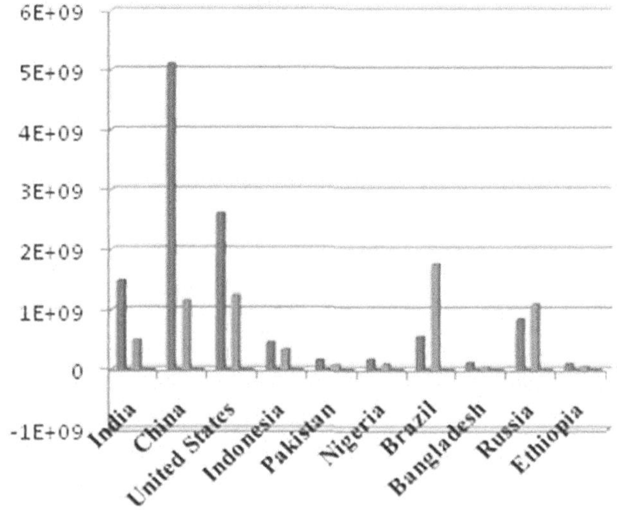

FIGURE 13.2 Ecological footprint analysis of top countries of the world in 2024.

several factors that affect the EF of a country, like level of policy framework, population size, rate of development, and so on. Highly developed countries have a high index of EF contrary to developing countries which have a small rate of development (Figure 13.2).

13.2.1 Types of Ecological Footprints

1. **Carbon Footprint:** Measures the amount of carbon dioxide emissions produced by human activities.
2. **Food Footprint:** Evaluates the environmental impact of producing, transporting, and consuming food.
3. **Water Footprint:** Assesses the total volume of fresh water used to produce goods and services consumed by an individual or community.
4. **Land Footprint:** Considers the amount of land required to produce the resources consumed and absorb the waste generated.
5. **Energy Footprint:** Accounts for the consumption of energy resources.
6. **Waste Footprint:** Examines the impact of waste production and disposal.

13.3 CARBON FOOTPRINT

A carbon footprint can be defined as the total amount of greenhouse gases (including carbon dioxide and methane) that are generated anthropogenically or by human influence. As per the report the highest carbon footprint for a person in the world was reported in Qatar, that is 37.6 metric tons per person (https://ourworldindata.org/co2/country/india).

The typical carbon footprint for a person in the United States is 16 tons, one of the highest rates in the world.

The average carbon footprint of a person in India is estimated to be 2.0 tons of carbon dioxide (CO_2) per year. Annual carbon dioxide emission of India in 2022 was 2.83 tons. Share of global CO_2 emission India in 2022 as 7.62%. Energy generation, steel and cement industries, fossil fuels, transportation, manufacturing, and agriculture are major drivers for carbon footprint. (https://climatetrade.com/the-worlds-most-polluting-industries)

Figures 13.3–13.6 present the statistics of carbon emission worldwide and India.

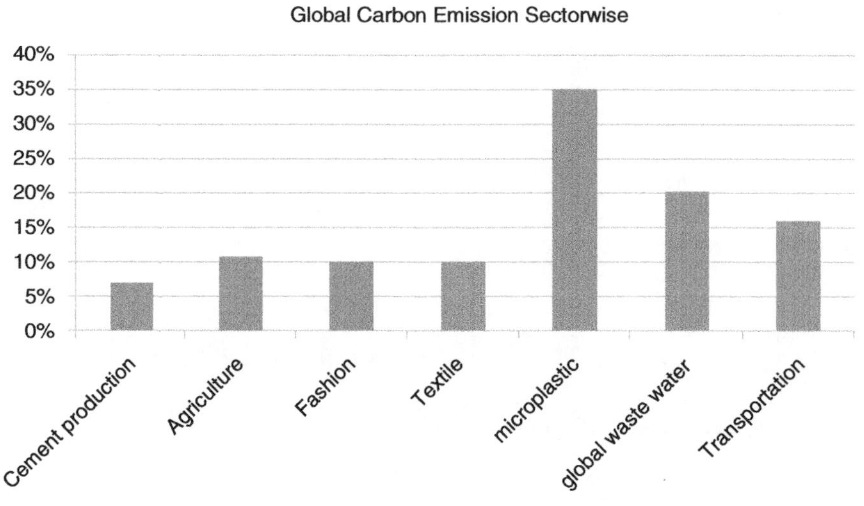

FIGURE 13.3 Global carbon emission.

Sector wise greenhouse gas emission (India).

Source (Sector-wise Assessment of Carbon Footprint across Major Cities in India (iisc.ac.in))

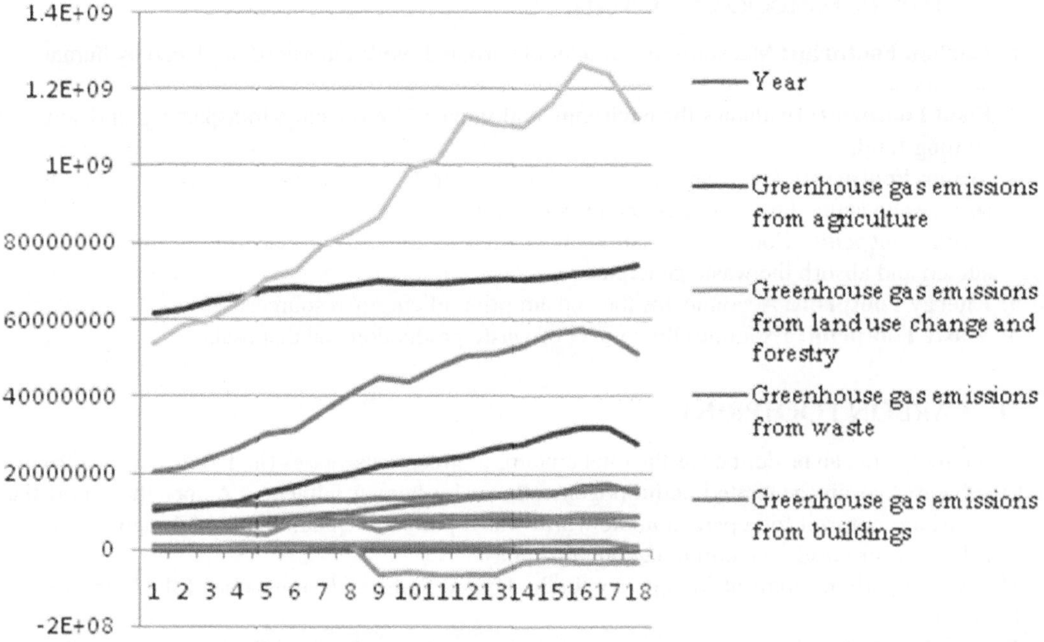

FIGURE 13.4 Sectorial carbon emission by India.

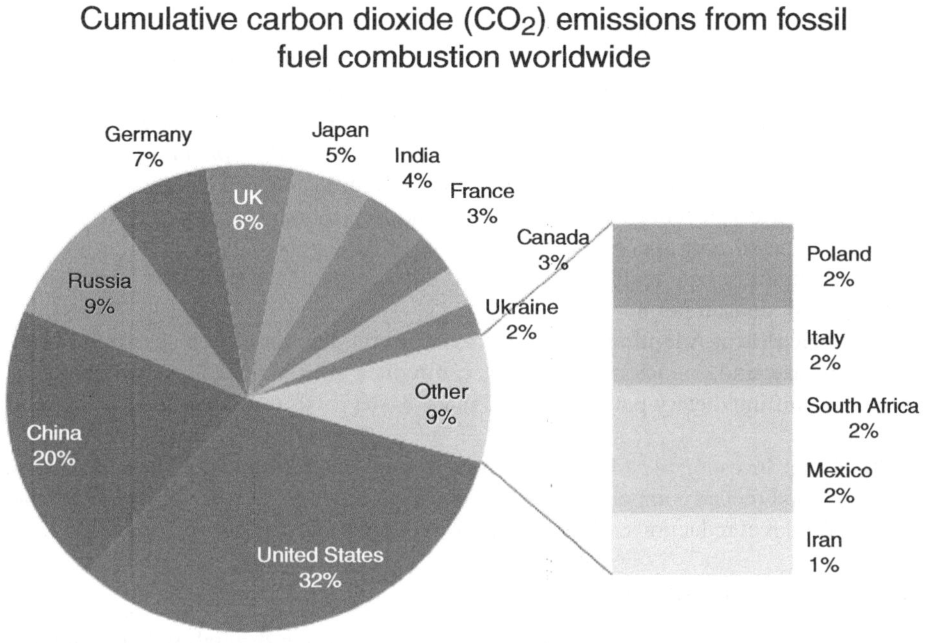

FIGURE 13.5 Cumulative CO_2 emission global.

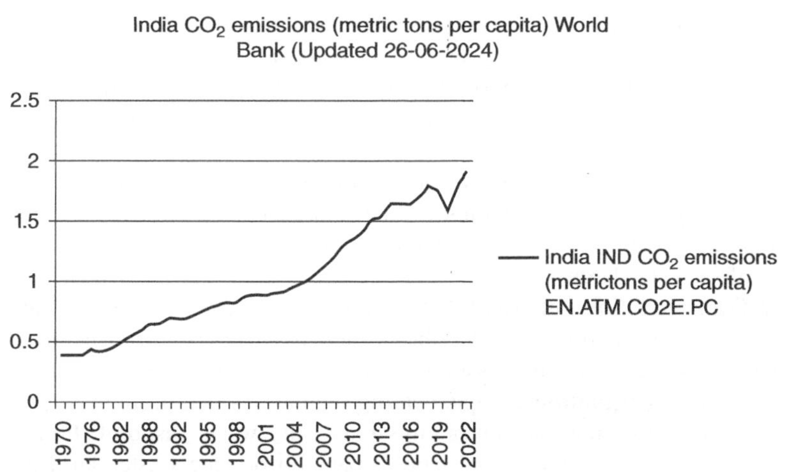

FIGURE 13.6 CO_2 emission India.

13.3.1 Strategies for Reducing Carbon Footprints

Reducing carbon footprints requires a multifaceted approach, involving both behavioral changes and technological innovations. Some effective strategies include

Energy Efficiency: Improving energy efficiency in buildings, transportation, and industrial processes can significantly reduce emissions. This can be achieved through measures such as insulation, energy-efficient appliances, and advanced manufacturing techniques.

Renewable Energy: Transitioning from fossil fuels to renewable energy sources, such as solar, wind, and hydroelectric power, can substantially lower carbon footprints. Investment in renewable energy infrastructure and technologies is crucial for this transition.

Sustainable Transportation: Promoting public transportation, cycling, walking, and electric vehicles can reduce emissions from the transportation sector. Urban planning that encourages compact, walkable communities can also play a significant role.

Waste Reduction and Recycling: Reducing waste and improving recycling rates can lower emissions associated with the production and disposal of goods. Implementing circular economy principles, where materials are reused and recycled, can minimize the environmental impact of consumption.

Sustainable Agriculture: Adopting sustainable agricultural practices, such as organic farming, agro forestry, and precision agriculture, can reduce emissions from the food sector. Additionally, shifting dietary patterns toward plant-based foods can lower the carbon footprint of diets.

Carbon Offsetting: Investing in carbon offset projects, such as reforestation, renewable energy, and methane capture, can compensate for emissions that are difficult to eliminate. While not a substitute for direct reductions, offsets can play a role in achieving net-zero emissions.

13.3.2 Policy and Regulatory Frameworks

Governments play a crucial role in driving carbon footprint reduction through policies and regulations. Some key approaches include

Carbon Pricing: Implementing carbon pricing mechanisms, such as carbon taxes or cap-and-trade systems, can provide economic incentives for reducing emissions. By putting a price on carbon, these policies encourage businesses and individuals to adopt low-carbon practices.

Renewable Energy Targets: Setting ambitious renewable energy targets and providing incentives for renewable energy development can accelerate the transition to a low-carbon energy system. Policies such as feed-in tariffs and renewable portfolio standards can support this transition.

Energy Efficiency Standards: Implementing energy efficiency standards for buildings, appliances, and vehicles can drive improvements in energy performance and reduce emissions. Building codes, fuel economy standards, and appliance efficiency regulations are examples of such standards.

Sustainable Transportation Policies: Promoting public transportation, active transportation, and electric vehicle adoption through policies such as subsidies, infrastructure investment, and urban planning can reduce emissions from the transportation sector.

Waste Management Regulations: Implementing regulations that promote waste reduction, recycling, and composting can lower emissions from waste disposal. Policies such as extended producer responsibility and landfill bans on organic waste can drive improvements in waste management.

Agricultural and Land Use Policies: Encouraging sustainable agricultural practices and protecting forests through policies such as subsidies, conservation programs, and land use planning can reduce emissions from the agriculture and land use sectors.

13.3.3 Future of Carbon Footprinting

As awareness of climate change grows, the importance of carbon footprinting is likely to increase. Future developments in this field may include

Advanced Technologies: Continued advancements in technologies such as AI, IoT, and blockchain can enhance the accuracy and transparency of carbon footprint assessments. These technologies can enable real-time monitoring, data integration, and verification of emissions data.

Integration with Sustainability Reporting: Carbon footprinting is increasingly being integrated into broader sustainability reporting frameworks, such as the Global Reporting Initiative and the Task Force on Climate-related Financial Disclosures. This integration can provide a more comprehensive view of an entity's environmental performance.

Consumer Awareness and Behavior Change: As consumers become more aware of their carbon footprints, demand for low-carbon products and services is likely to grow. This shift in consumer behavior can drive businesses to adopt more sustainable practices and innovate in low-carbon solutions.

Policy and Regulatory Developments: Governments are likely to continue strengthening policies and regulations to drive carbon footprint reduction. International agreements, such as the Paris Agreement, will play a critical role in shaping national and global climate policies.

Corporate Leadership and Collaboration: Businesses are expected to play an increasingly important role in addressing climate change through collaborative initiatives and commitments to carbon neutrality. Industry partnerships, such as the Science Based Targets initiative, can drive collective action toward reducing emissions.

13.4 BIOENGINEERING-INSPIRED MATERIALS

Many innovative approaches are available for designing products, systems, and processes, such as bionics, biomimetics, and biomimicry, which have drawn inspiration from the forms, behaviors, and processes found in nature which is truly green in process. There are various approaches mentioned in Figure 13.7 available to design biomaterials and their applicability in sustainability analysis.

Bioengineering-inspired materials offer a pathway toward reducing ecological footprints by leveraging renewable resources, enhancing biodegradability, improving energy efficiency, and enabling innovative solutions that align with sustainable life cycle assessments. These materials are

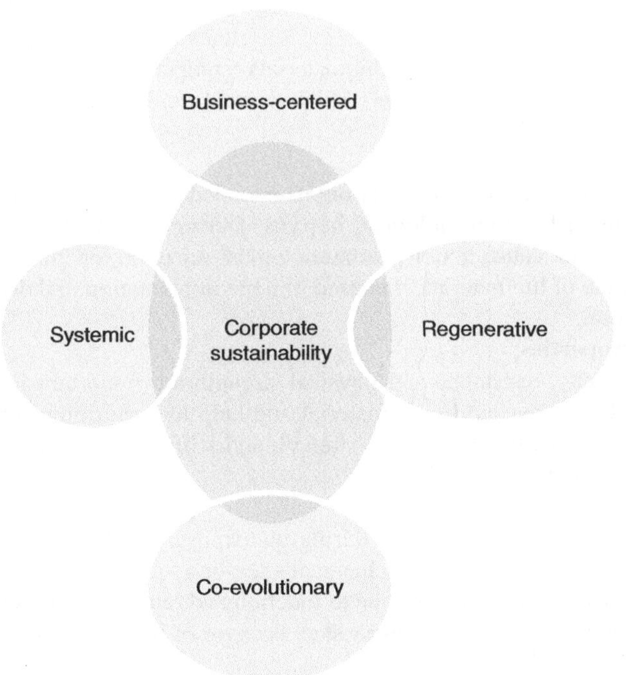

FIGURE 13.7 Approaches for designing of biomaterials.

crucial in advancing toward a more sustainable future across various novel polymeric substances featuring suitable mechanical qualities that are industries.

Bioinspired material can easily be degraded and available abandoned, most of the industries are focusing on the use of bioinspired material synthesis, and process of manufacturing is also being opted green (Pallas, G., et al., 2019).

Developing bio-based approaches for creating materials that degrade naturally due to the substantial environmental and climatic effects of producing both synthetic and natural composites, various industries are turning their attention to low carbon footprints. For example, in order to create synthetic textile fibers like nylon or polyester, a variety of nonrenewable, non-recyclable resources and petrochemicals are needed for the purpose of making textiles. In parallel, the synthesis of biodegradable polymer fibers like polylactic acid, which is derived from corn ethanol, and natural fibers like cotton that depend on industrial farming methods are getting the highlights.

13.5 CHARACTERISTICS OF BIOMATERIALS

Innovative technologies in the fields of bionics, biodesign, biomimetics, and biomimicry hold great potential for contributing to sustainability in a planned manner and provide various solutions to environment-related issues in a cost-effective way (Bjørn and Hauschild, 2018; Chappell et al., 2017).

In modern times, various new terminologies have emerged to define the flow of stimulus from nature to bioengineering solutions, including bio-inspiration, biomimicry, development of bioproducts, and biodesign. The objective of taking inspiration from nature is to provide conventional eco-friendly approaches for designing and construction of the biomaterials.

Sustainability is the functional and conventional key to unlock the hidden potential of nature. The whole world is trying to achieve the sustainable development goals by 2030. As per the agenda 2030, all nations of the world have already started their significant contribution to achieve global sustainability in all its magnitudes. Furthermore, with its foremost transformative aim, that is "leaving no one behind," based on the principle of including the most susceptible and weakest portions of the society. Consequently, it aims to eliminate poverty in all its forms and to combat discrimination and rising inequality within and between countries. Biomaterials are materials, either synthetic or natural, that are utilized to interact with biological systems in tissue engineering, implants, and medical devices. The intended use and the biomaterials' compatibility with the human body determine their design and choice. Key attributes of biomaterials are listed as follows:

1. **Biocompatibility:**
   The capacity of a biomaterial is to carry out its intended purpose without causing any unfavorable systemic or local physiological impacts. During implantation of biomaterial into the body, it must not cause toxicity, immunological rejection, or inflammation. Titanium is a good example of biomaterial; it is used in bone implantation and does not induce any immune response.
2. **Mechanical Properties:**
   Flexibility, elasticity, resistance and physical strength are important mechanical properties that should be conveyed by biomaterial and help in performing many physiological functions. Bone implants require high strength and stiffness, whereas skin grafts require flexibility and elasticity.
3. **Chemical Stability:**
   It is also an important characteristic; during the exposure of body fluids or tissue to the environment, it prevents any type of chemical alteration.
   Over time, the material should continue to function and retain its integrity without degrading. Surgical tools are made of stainless steel because of its ability to withstand oxidation and corrosion.

4. **Biodegradability:**
   After use, bioengineered materials usually dissolve into harmless components since they are inherently biodegradable. Conventional materials, such as plastics, have a centuries-long environmental history that contrasts sharply with this.
5. **Customizability and Functionality:**
   Bioengineering allows for the precise regulation of material qualities such as conductivity, strength, and flexibility. This enables tailored solutions that meet specific application needs, potentially reducing waste and resource usage.
6. **Surface Properties:**
   The properties of materials with the same chemical composition can vary greatly depending on whether the compound is applied as a bulk material or as a nanomaterial because of the massive reactive interface that nanoparticles provide between the particle and its local environment due to their large surface area to volume ratio (Bantz et al., 2014).
7. **Bioactivity:**
   Bioactivity is an important characteristic to design biomaterial. Any positive or negative impact that artificial matter have on biological things is referred to as bioactivity. A bioactive material is one that, when submerged in a solution similar to serum, selectively generates bone-like hydroxyapatite. Secondary metabolites produced by living things that are typically not necessary for their development and survival are known as bioactive chemicals. While researching on bioglass in 1969, Larry L. Hench came up with the idea of bioactivity. He then applied it to the subject of bioceramics, establishing a new category of bioactive materials. In the realm of biological materials, bioactivity has solidified its basis after more than half a century of development. In vivo degradation and absorption of β-tricalcium phosphate-absorbable bioceramics, for instance, can result in the replacement of these bioceramics by new tissues, which can trigger unique biological responses (Mousavi et al., 2022).
8. **Sterilizability:**
   Sterilizability is an important attribute of biomaterial. Natural biomaterials have biological, mechanical, and physical characteristics that work well together, as well as degradation rates that are suitable for cellular processes. Additionally, they have particular signals that facilitate efficient cell attachment, such as cell adhesion ligands.
9. **Toxicity:**
   The material shouldn't create negative bodily reactions or emit dangerous compounds. Numerous papers just address one toxicity metric, which might not be enough for a thorough assessment of the novel substance. The addition of enriching materials, such as ceramics based on hydroxyapatite, bioglasses, trace element oxides, and chemicals that promote proliferation or have antibacterial or anti-inflammatory qualities, is frequently linked to the enhancement of biocompatible materials. Important characteristics are absent from the majority of data on the assessment of the cytotoxicity of biomaterials supporting bone regeneration made of chitosan and bioglass that are reported in publications (Hosoyama et al., 2019; Islam et al., 2020; Chraniuk et al., 2022).
10. **Wear and Tear Resistance:**
    A material's resistance to wear, erosion, or degradation brought on by friction or mechanical stress. Hip implant materials must be able to endure frequent movement and friction without rapidly degrading.
11. **Renewable Resources:**
    Many elements used in bioengineering come from plants, algae, and even waste products (such as agricultural wastes). This teaches about dependence on limited fossil resources and the harm that extraction and processing do to the ecosystem.

12. **Carbon Footprint Reduction:**
 Bioengineering can produce products with a lower carbon footprint by either sequestering CO_2 during growth or using it as a feedstock. During their production process, bio-based polymers like PLA (polyacetic acid) can store carbon.

13.6 LIFE CYCLE ASSESSMENT (LCA)

The life cycle assessment (LCA) method for bionanotechnology and bio-nanoproducts can provide important details regarding the main advantages and environmental effects of this new technology. Experimental data on the properties and toxicity of bionanoparticles from research projects should be incorporated into LCA methodology, and potential LCA approaches are required. To get comprehensive data on the environmental performance of nanoproducts at every step of their life cycle, altered exposure and fate simulations are necessary. It is frequently denoted to as a "cradle-to-grave" analysis, although variations such as "cradle-to-gate" and "cradle-to-cradle" also exist.

Products based on nanomaterials and bionanotechnology are currently widely used in a variety of applications. The dearth of information regarding the possible effects of nanotoxicity on the environment and human health has led to debate over this tendency. Although LCA is thought to be the most effective method for evaluating the environmental behavior of new technologies, there are still a number of unknowns and data gaps, and the possible effects of released nanomaterials on human and environmental health have not yet been included in LCA databases. Using current models as a starting point, the primary goal of applying LCA techniques to goods based on nanotechnology is to determine characterization variables for bionanomaterials. Impact and inventory data could be incorporated into LCA (Szymanowski and Kostrzewa, 2019).

13.6.1 Key Phases of Life Cycle Assessment

LCA can deliver a more thoughtful approach for the major environmental issues and guarantee the environmental sustainability of nanomaterials (Nurul Umairah M. Nizam et al., 2021). LCA is an all-inclusive approach to evaluate environmental impacts of a product throughout its complete life cycle by recognizing the materials used and energy and emissions released to the environment, which is vital in evaluating the potential impacts of nanomaterial releases, as shown in Figure 13.8.

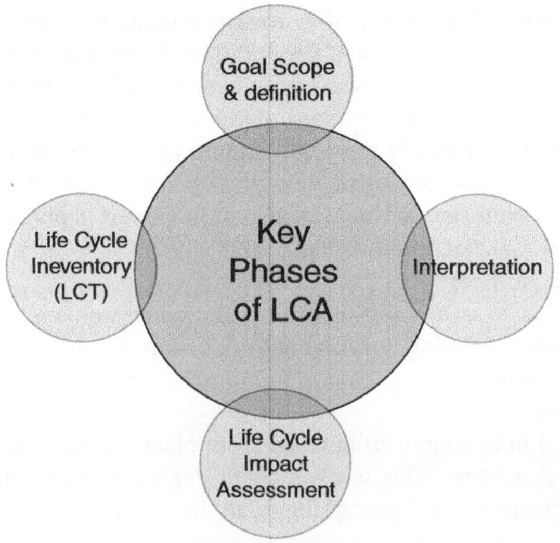

FIGURE 13.8 Key phases of LCA.

LCA is a world-wide standardized accepted methodology, based on the International Organization for Standardization (ISO) 14040 series (ISO 2006), encompassing four phases, as follows:

1. Goal and scope
2. Life cycle inventory
3. Life cycle impact assessment
4. Life cycle interpretation

13.6.2 Types of LCA

1. **Cradle-to-Grave:** From raw material extraction (cradle) to disposal or recycling (grave)
2. **Cradle-to-Gate:** Assesses the product life cycle from resource extraction to when the product leaves the manufacturing facility (before the use phase)
3. **Cradle-to-Cradle:** An extension of cradle-to-grave
4. **Gate-to-Gate:** EIA in the production phase to gate (Figure 13.9)

13.6.3 Applications of Life Cycle Assessment

1. **Product Design and Improvement**
    - LCA can inform designers and engineers about the environmental impacts of materials and processes, leading to more sustainable products.
    - **Example:** Switching to renewable materials or optimizing energy use in manufacturing.
2. **Policy Making**
    - Governments and regulatory agencies can use LCA results to develop environmental policies and regulations.
    - **Example**: Setting carbon emission limits based on the full life cycle impacts of products in specific industries.
3. **Environmental Labeling**
    - Companies use LCA to obtain eco-labels or certifications for their products, indicating lower environmental impacts.
    - **Example:** Labels like "Carbon Neutral" or "Ecolabel" rely on LCA to verify sustainability claims.
4. **Corporate Sustainability**
    - Companies use LCA to measure and improve the sustainability of their operations and supply chains.
    - **Example:** Identifying ways to reduce the carbon footprint of a company's product line.

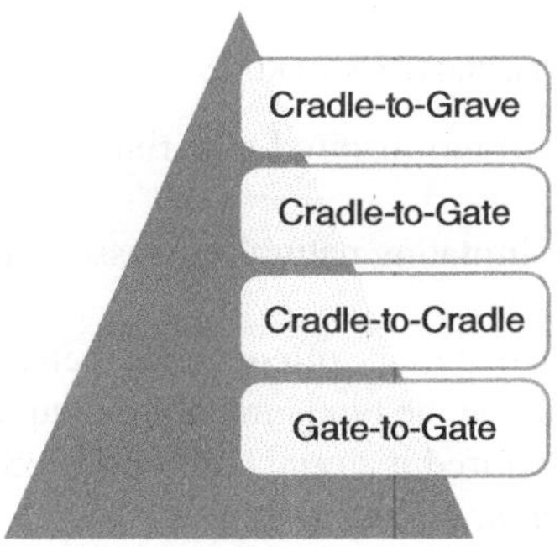

FIGURE 13.9 Types of LCA.

5. **Consumer Decision-Making**
   - LCA data can help consumers make more informed choices by understanding the full environmental impact of products.
   - **Example:** Choosing products with lower environmental impacts based on LCA–based labels.

13.6.4 Benefits of LCA

- **Comprehensive Environmental Assessment:** It looks at all stages of a product's life, providing a complete picture of its environmental impact.
- **Supports Decision-Making:** Helps in identifying trade-offs and making informed decisions that lead to more sustainable outcomes.
- **Encourages Innovation:** By revealing environmental "hotspots," LCA can stimulate innovations in product design, material selection, and manufacturing processes.

13.6.5 Challenges of LCA

- **Data Availability:** Accurate and comprehensive data can be difficult to obtain, especially for complex products.
- **Time and Resource Intensive:** Conducting a full LCA can be a lengthy and resource-demanding process.
- **Complexity:** Interpreting LCA results requires expertise, as impacts in one category (e.g., reduced emissions) might lead to increases in another (e.g., resource use).
- **Subjectivity:** Decisions on boundaries, impact categories, and assumptions can influence results, leading to variability in outcomes across different studies.

13.7 BIO-INSPIRED MATERIALS FOR REDUCING ECOLOGICAL FOOTPRINT AND LIFE CYCLE ASSESSMENT

Bio-inspired materials are designed by mimicking structures, functions, and processes found in nature. These materials draw inspiration from biological systems that have evolved over millions of years to be efficient, sustainable, and resilient. By applying nature's principles to material design, researchers and engineers aim to reduce the ecological footprint of products and improve their overall sustainability.

Incorporating **bio-inspired materials** into design and manufacturing can significantly reduce environmental impacts, as these materials are often renewable, biodegradable, or highly efficient. When evaluated through LCA, bio-inspired materials frequently show improved environmental performance compared to conventional materials.

13.7.1 Key Concepts of Bio-Inspired Materials

Definition of the four key concepts of bio-inspired material is given as follows: (Figure 13.10).

1. **Biomimicry:** The practice of imitating natural processes, structures, or systems to develop new materials and technologies.
2. **Sustainability:** Bio-inspired materials often utilize renewable resources, consume less energy during production, and can be biodegradable, reducing waste.
3. **Eco-Efficiency:** Many bio-inspired materials are optimized for efficiency, using minimal resources while providing maximum performance.
4. **Applicability:** Applicability of bio-material lies in various fields like environment, medical, pharmaceuticals etc.

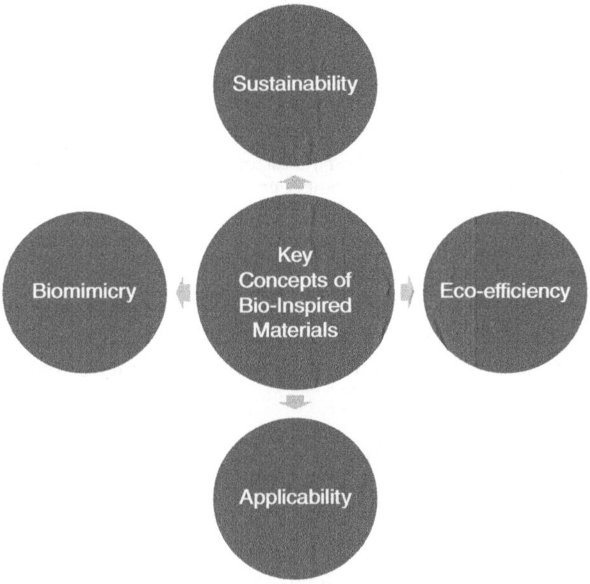

FIGURE 13.10 Key concept of bio-inspired material.

13.8 BIOENGINEERING-INSPIRED MATERIALS: CONCEPT AND PRINCIPLES

Bio-inspired materials offer significant potential for reducing the ecological footprint of products and processes. By mimicking nature's efficiency and sustainability, these materials can minimize resource use, lower energy consumption, and reduce waste. LCA serves as an essential tool for quantifying these benefits and guiding the development of bio-inspired innovations that align with environmental goals. Incorporating bio-inspired designs into product development not only helps industries move toward a circular economy but also reduces the overall life cycle environmental impacts, making these materials a key component in creating a more sustainable future.

Bioengineering-inspired materials represent a paradigm shift in sustainable material science, drawing inspiration from biological systems to create innovative solutions that minimize environmental impact across their lifecycles. This review explores the intersection of bioengineering, materials science, and environmental sustainability, focusing on the application of these materials in reducing ecological footprints through rigorous life cycle assessment (Wang X. et al., 2014).

Bioengineering-inspired materials, often referred to as biomimetic materials, emulate structural and functional properties found in nature. The concept stems from biomimicry, which involves observing and learning from natural systems to solve human challenges sustainably. These materials leverage biological principles such as self-assembly, hierarchical structures, and biocompatibility to enhance performance while minimizing environmental impact (Zhang, Y., et al., 2020).

13.8.1 Biomimetic Design and Synthesis

Biomimetic design principles guide the development of bioengineering-inspired materials. Key strategies include

- **Selection of Biological Models:** Identifying organisms or systems in nature with desirable properties (e.g., strength, flexibility, biodegradability).
- **Material Composition:** Using bio-based polymers, natural fibers, and additives to replicate biological structures and functions.

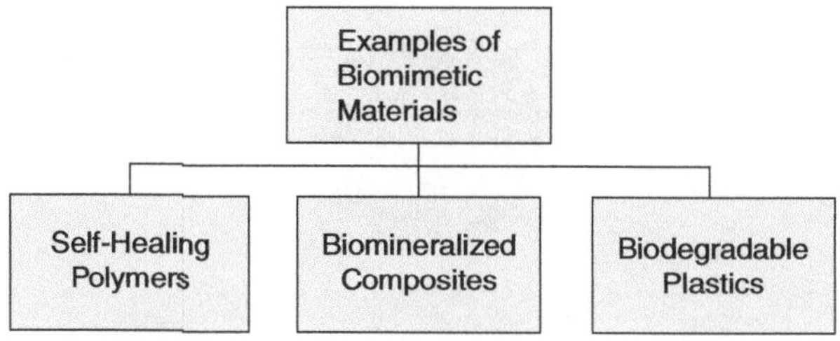

FIGURE 13.11 Examples of biomimetic material.

- **Fabrication Techniques:** Employing advanced manufacturing methods such as 3D printing, electrospinning, and nanotechnology to create biomimetic materials with precision and efficiency.

13.8.1.1 Examples of Biomimetic Materials

- **Self-Healing Polymers:** Inspired by the healing mechanisms found in biological tissues, self-healing polymers autonomously repair damage, extending material life span and reducing maintenance requirements.
- **Biomineralized Composites:** Mimicking the hierarchical structure of seashells and bones, biomineralized composites exhibit enhanced mechanical properties and biocompatibility; suitable for medical implants and structural applications.
- **Biodegradable Plastics:** Modeled after natural polymers that degrade in the environment, biodegradable plastics offer a sustainable alternative to traditional plastics, reducing landfill waste and marine pollution (Figure 13.11).

13.8.2 Challenges and Limitations

Despite their potential benefits, bioengineering-inspired materials face several challenges:

- **Performance Optimization:** Balancing biomimetic design principles with functional requirements (e.g., strength, durability) remains a challenge, requiring continuous innovation and material testing.
- **Scale-Up and Cost:** Scaling production processes to meet commercial demands while maintaining cost competitiveness poses challenges due to technological limitations and market dynamics.
- **Scale-Up and Cost:** Navigating regulatory frameworks and standards for bio-based and biodegradable materials can hinder widespread adoption and market penetration (Figure 13.12).

13.9 BENEFITS OF BIOENGINEERING-INSPIRED MATERIALS

- Reduced Ecological Footprint: Less use of resources and greenhouse gas emissions.
- Enhanced Sustainability: Raw materials that are available over the long term.
- Decreased Waste: Increased biodegradability and recycling potential.
- Enhanced Safety and Health: Nontoxic substances that are safer for human consumption.

In general, the bioinspired materials in various fields have a number of advantages. These benefits are elaborated as follows:

- Better performance
- Sustainability

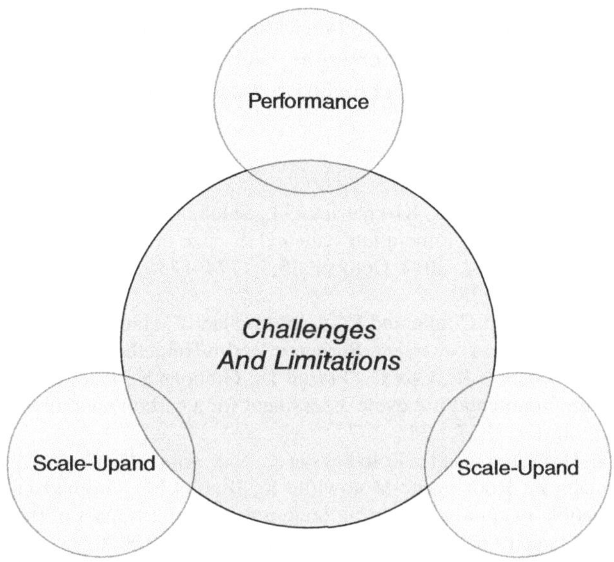

FIGURE 13.12 Challenges and limitations.

- Multifunctionality
- Adaptability
- Efficiency
- Novel solutions
- Biocompatibility
- Self-healing properties
- Finer utilization of resources
- Improved health
- Miniaturization
- Reduced toxicity
- Improved interface with biological systems
- Cost-effectiveness
- New functionalities

13.10 CONCLUSION

In the current scenario climate change and rapid alteration is a major problem. Industry should optimize their production aligned with demand and supply chain. Industries also have to check consumer interest and practices. Technologies aligned with sustainable development goals are more promising nowadays. Harnessing the nectar from the nature is the key towards sustainability. Policy framework to control the increasing population should also be an area of concern as population increase is directly proportional the exploitation of natural resources, resulting in an increased ecological footprint. Researchers, scientists, and industrialists should meet on a common platform and discuss the climate resilient/eco approaches or technologies to minimize the overexploitation of natural resources. "Go with green" or "sustainable lifestyle" is not only infectious motivational dialogue, it also consists of a lot of values which are to be needed for the protection of nature. Bioengineering-inspired materials are extant to ground-breaking methodology for mitigating the ecological impacts of recent industry and consumer practices. LCA plays an acute role in quantifying the environmental benefits, providing a comprehensive evaluation of the impacts associated with these materials across their entire life cycle. The state of the art defined so far provide indications to us to mark the conclusion that LCA is definitely a valuable, appreciable, and powerful tool for the assessment of sustainability parameters as emerging technologies, nevertheless several procedural issues prerequisite to be solved to deliver an inclusive assessment in the field of technologies.

REFERENCES

Bantz C., Koshkina O., Lang T., Galla H.J., Kirkpatrick C.J., Stauber R.H., Maskos M. The surface properties of nanoparticles determine the agglomeration state and the size of the particles under physiological conditions. *Beilstein J Nanotechnol.* 2014 October 15;5:1774–1786. doi: 10.3762/bjnano.5.188. PMID: 25383289; PMCID:PMC4222438.

Bjørn, A., Hauschild, M.Z. Cradle to Cradle and LCA. In Michael Z. Hauschild, Ralph K. Rosenbaum, Stig Irving Olsen (Eds.), *Life Cycle Assessment*; Springer: Berlin/Heidelberg, Germany, 2018; pp. 605–631.

Chappell M.A., Shih W.S., Bledsoe J.K., Cox C., Janzen D., Gibbons S., Patel R., Kennedy A.J., Brame J., Brondum M., et al. Environmental life cycle assessment for a carbon nanotube-based printed electronic sensor platform. *Adv. Mater.* 2017;1:345–347.

Chraniuk M., Panasiuk M., Hovhannisyan L., Zołędowska S., Nidzworski D., Ciołek L., Woźniak A., Kubiś A., Karska N., Jaegermann Z., Rodziewicz-Motowidło S., Biernat M., Gromadzka B., Assessment of the toxicity of biocompatible materials supporting bone regeneration: Impact of the type of assay and used controls. *Toxics.* 2022 January 6;10(1):20. doi: 10.3390/toxics10010020

Hosoyama K., Lazurko C., Muñoz M., McTiernan C.D., Alarcon E.I. Peptide-based functional biomaterials for soft-tissue repair. *Front. Bioeng. Biotechnol.* 2019;7:205. doi: 10.3389/fbioe.2019.00205

Islam M.M., Shahruzzaman M., Biswas S., NurusSakib M., Rashid T.U. Chitosan based bioactive materials in tissue engineering applications—A review. *Bioact. Mater.* 2020;5:164–183. doi: 10.1016/j.bioactmat.2020.01.012

ISO (2006). ISO 14040 International Standard. In: Environmental Management–Life Cycle Assessment–Principles and Framework. International Organisation for Standardization, Geneva, Switzerland. http://www.iso.org/iso/catalogue_detail?csnumber=37456

Mousavi S.M., Hashemi S.A., Kalashgrani M.Y., Omidifar N., Bahrani S., VijayakameswaraRao N., Babapoor A., Gholami A., Chiang W.H. Bioactive graphene quantum dots based polymer composite for biomedical applications. *Polymers*, 2022;14:617.

Nizam N.U.M., Hanafiah, M.M., Woon, K.S. A content review of life cycle assessment of nanomaterials: Current practices, challenges, and future prospects. *Nanomater.* 2021;11:3324.

Pallas G., Vijver M.G., Peijnenburg W.J.G.M., Guinée J. Life cycle assessment of emerging technologies at the lab scale: The case of nanowire-based solar cells. *J. Ind. Ecol.* 2019;24:193–204.

Szymanowski H., Kostrzewa, M. The importance of life cycle assessment in environmental evaluation of biomaterials. *J. Ecol. Eng.* 2019:20(9):1–11.

Wang, X., Liu, Z., Zhong, H., Guo, Z., Yuan, X. (2014, September). Mussel- inspired synthesis of polydopamine-functionalized graphene oxide hydrogel as broad- spectrum antimicrobial material. *Twelfth International Conference on Photonics and Imaging in Biology and Medicine (PIBM 2014)*, Wuhan, China (Vol. 9230, pp. 320–332). SPIE.

Zhang, Y., Mei, J., Yan, C., Liao, T., Bell, J., Sun, Z. Bioinspired 2D nanomaterials for sustainable applications. *Adv. Mater.* 2020;32(18):1902806.

14 Enzymes Based Technologies for Bioremediation of Environmental Pollution

Josimayra Almeida Medeiros, Carlos Eduardo de Farias Silva, Gabrielly Karla Silva Santos, João Victor Ferro, João Victor Oliveira Nascimento da Silva, Micaela Almeida Alves do Nascimentos, Yasmin Matos Cardoso dos Santos, and Brígida Maria Villar da Gama

14.1 INTRODUCTION

Unchecked population growth promotes the generation of large amounts of waste and also requires greater industrial development to meet the numerous needs of this increasingly demanding society. As a result, high volumes of pollutants are generated daily, becoming an environmental and social problem (Saravanan et al. 2021).

Among the solid or liquid wastes produced by various processes are dyes, pesticides, highly toxic compounds such as heavy metals, and materials that are difficult to degrade, such as hydrocarbons and plastics. This variety of pollutants and their high concentrations necessitate more robust and often more expensive treatment systems to comply with current environmental legislation (Mathew et al. 2017; Menger et al. 2021).

Bioremediation is a biological degradation mechanism that uses living organisms or enzymes derived from them to reduce pollutants to undetectable or non-toxic levels through a greener and more economical technology than the commonly used physico-chemical methods. Microorganisms such as fungi and bacteria are widely studied and used in waste treatment processes, whether liquid or solid, due to their proven efficiency in degrading different compounds and resistance to a variety of environmental conditions. Often, the high degradative capacity of these microorganisms is associated with the presence of intra- and/or primarily extracellular enzymes that act directly on the degradation of compounds, converting or transforming them into less toxic substances (Behbudi et al. 2021; Mousavi et al. 2021).

Thus, the enzymatic action in treatment processes has been mentioned by researchers in various studies over decades, and its isolated use is attracting interest in this sector due to the possibility of using waste/effluents for their production, reducing the initial cost, and the high efficiency and stability in the process, not requiring specific nutritional and environmental conditions for their development (Behbudi et al. 2021; Mousavi et al. 2021).

Enzymes are effective organic catalysts that act in degradation reactions by reducing the activation energy, thus making the process occur more quickly. Additionally, they act with the main characteristic of the active site and substrate interaction, making the catalysis process highly specific, reducing the formation of by-products and making them a potential remediation agent for application to different types of pollutant compounds (Antecka et al. 2018; Karthik et al. 2021).

Therefore, this chapter will address the enzyme as the protagonist of the remediation process, showing the main producing microorganisms, the mechanism of their interaction with different

DOI: 10.1201/9781003668961-14

types of pollutants, the optimal operational conditions, applications in different processes, and technological advances in this area.

14.2 MICROBIAL ENZYME PRODUCTION

Microorganisms such as bacteria and fungi can metabolize complex compounds due to their ability to produce enzymes that will break down these macromolecules into smaller molecules or convert them into substances that are easier to assimilate. These enzymes have high industrial applications and can be produced on various substrates through submerged and semi-solid state fermentation, utilizing conditions, and optimizing production to reduce operational costs, especially in the isolation and purification stages.

14.2.1 Bacterial Enzymes

Bacteria are microorganisms characterized by their ability to degrade a variety of pollutants and withstand extreme operational conditions. Along with these characteristics, they also possess the capability to produce a diversity of enzymes, which can be used as remediation agents for different environmental pollutants and have a faster growth rate than fungi. However, although these microorganisms metabolize various compounds when referring to enzymatic action, their individual performance may present limitations in degrading all present pollutants due to the specificity of the enzyme produced. Therefore, it is necessary to combine them with other enzymes or microorganisms, such as in the co-cultivation of different bacteria (Saravanan et al. 2021).

Bacterial enzymes act differently depending on the pollutant found in the environment. In the case of metals, for example, the most viable process is the transformation into a less toxic compound through mechanisms of methylation, oxidation, and reduction by enzymes such as reductases (*Streptomyces* sp.), which act in the reduction of copper, and ureases (*Sporosarcina Pasteurii*) which contribute to the biomineralization of metals like nickel, cobalt, zinc, and cadmium (Saravanan et al. 2021).

On the other hand, when in contact with organic or inorganic dyes, bacterial enzymes utilize the demineralization mechanism to perform both conversion and transformation processes. Notable enzymes include laccases (*Lysinibacillus sphaericus*), azoreductases (*Lysinibacillus sphaericus*), proteases (*Bacillus cereus* KM201428), and lipoxygenases (*Pseudomonas aeruginosa*) (Lu et al. 2020; Saravanan et al. 2021).

Additionally, the study by Sirajuddin et al. (2020) reveals that bacterial enzymes, such as carboxylesterase from *E. coli* IES-02, are effective in degrading malathion organophosphorus pesticides. The research showed that this enzyme achieved a degradation rate of 81% for the evaluated pollutant, transforming it into less toxic compounds like succinic acid and ethyl hydrogen formate, thus significantly contributing to environmental remediation.

14.2.2 Fungal Enzymes

Fungi are microorganisms with a high degradative potential, partly due to their ability to produce intra- and extracellular enzymes that promote the biotransformation of various compounds. For example, laccase and peroxidase enzymes are used to remove pollutants in various industries (Negi and Das, 2023). According to Deng, Zhao and Yang (2022), laccases are enzymes with low substrate specificity, capable of oxidizing a wide variety of xenobiotic compounds, including chlorinated phenolics. Supporting this, Cen et al. (2022) highlighted the high capacity for phenol removal by laccase produced by the fungus *Coriolopsis gallica*.

Similarly, versatile peroxidase (VP) is an enzyme produced by the fungus *Pleurotus eryngii* that contains heme with a wide variety of substrates, capable of degrading various pollutants, including azo dyes. Genetic engineering modifications can enhance its catalytic activity and expand its substrate spectrum. When incorporated into the cell wall, VP maintained 70% of its initial activity after 10 cycles of dye degradation, each cycle lasting 12 hours, making it an ideal platform for bioremediation of azo dye–contaminated environments (IlićĐurđić et al., 2020).

Another example is catalases (EC 1.11.1.6), produced by fungi, which are a class of oxidoreductases widely distributed in microorganisms, animals, and plants. They are heme tetrameric proteins shaped like dumbbells, with a molecular weight of 60 to 75 kDa. Due to their stability and robustness, catalase-derived structures are resistant to proteolysis, pH variations, and temperature-induced denaturation. They exhibit a broad working pH range (6.5–7.0) compared to the free enzyme, which functions optimally only at pH 6.5. The optimal working temperature after immobilization also improved to 40°C compared to 30°C for the free enzyme, reflecting its thermostability. Additionally, immobilized catalase was more tolerant of the inhibitory effects of metal ions than its soluble counterparts. Overall, the properties of immobilized catalase are attractive for various biotechnological applications (Gan et al. 2022).

14.3 OPERATIONAL FACTOR AFFECTING THE ENZYMATIC DEGRADATION PROCESS

Enzymes, like the microorganisms that produce them, have optimal conditions (temperature, pH) for their full activity and efficiency in the degradation process. Operational parameters such as pH and temperature are crucial for any biological process. Specifically for enzymes, temperature control is required since they are proteins that can undergo denaturation or have some of their properties altered when exposed to high temperatures due to changes in their surface charges (Saravanan et al. 2021).

However, there is no specific range that can be applied to all enzymes; thus, studying this variable during process development is essential. For example, some authors mention temperatures of 20 to 40°C for laccase (Navada and Kulal, 2021) and 15 to 30°C for lipases and peroxidases (Lopes et al. 2020). A similar behavior occurs with pH, where no optimal range exists. However, it is estimated that a neutral pH is most favorable for many enzymes. Nonetheless, enzymes like protease perform well in a basic medium, while peroxidase shows better performance in an acidic pH (Lopes et al. 2020).

Other factors to consider in the enzymatic degradation process are the concentration of pollutants and the enzyme dosage, as the good performance of the remediation system must directly relate to these two points. Although evidence shows that a higher amount of enzymes provides greater efficiency, if the pollutant concentration is low, enzymatic action will be limited. Conversely, if there is a high concentration of pollutants and a small amount of enzyme, the bioavailability of active sites will be low, compromising the system's efficiency and productivity (Saravanan et al. 2021).

Finally, enzyme immobilization can be an alternative that provides thermal stability to the protein and tolerance to harsh environments, such as low pH, in addition to prolonging the enzyme's lifespan and activity for multiple recycles. Therefore, considering the prior interaction between the support system and the enzyme when choosing immobilization is essential. Immobilization is generally key to maintaining stability, as it allows rapid kinetics and appreciable substrate conversion in harsh reaction environments (Karume, 2023). However, high recoverability of enzymatic activity after the immobilization process is necessary, along with stability on the chosen support for multiple cycles, increasing the process's profitability.

14.4 MECHANISMS OF ENZYMATIC POLLUTANT DEGRADATION

The enzymatic degradation of environmental pollutants aims to transform complex molecules into less toxic forms for the environment (Bilal et al. 2019). This process occurs in several specific chemical stages. Firstly, each enzyme has an active site with unique properties that precisely fit the substrate, the target pollutant. This fit is comparable to inserting a key into a lock (primarily the lock-and-key model, though there is also the induced-fit model, such as that present in lipases), ensuring reaction specificity.

Subsequently, the enzyme binds to the substrate, forming an enzyme-substrate complex, which induces a conformational change in the enzyme, further adjusting it to the substrate and facilitating the reaction. This structural adaptation of the substrate makes it more susceptible to degradation. In the next stage, the enzyme catalyzes the breakdown of the substrate into less toxic or more easily degradable final products. During this process, the chemical bonds of the substrate are broken or rearranged, transforming its molecular structure. After the reaction, the enzyme remains intact (provided there is no denaturation or inhibition) and available to catalyze another reaction with a new substrate molecule, restarting the degradation cycle. (Bala et al. 2022; Bilal et al. 2019; Saravanan et al. 2021).

Different types of enzymes act on various substrates. Some are specialized in organic pollutants, such as hydrocarbons, herbicides, pesticides, and plastics, while others are specific for inorganic substrates, like heavy metals. This diversity allows enzymatic degradation to be applied to a wide range of environmental contaminants.

Moreover, the enzymatic process is quick and easy to control, providing efficiency in degrading specific pollutants. Enzymes can operate under a wide variety of environmental conditions, which allows their application in diverse scenarios. Another positive aspect is that the process does not require an additional supply of nutrients, reducing costs. Enzymatic degradation also presents fewer mass transfer limitations, allowing more effective action. Enzymatic catalysis not only accelerates the speed of specific chemical reactions but also occurs efficiently without consuming the enzymes in the process.

This means that enzymes can be used repeatedly to degrade contaminants, making the process both economically and environmentally sustainable. Despite its advantages, enzymatic degradation faces challenges. Selecting the ideal microbial source to produce the desired enzymes can be complex, and enzyme activity and stability may vary with environmental conditions and the type of pollutant. Additionally, large-scale enzyme production and purification still need to be overcome, as well as the difficulty of reusing free enzymes (Saravanan et al. 2021; Thew et al. 2023; Zaaba and Jaafar 2020).

14.4.1 Enzymes on Organic Contaminants

Organic pollutants such as hydrocarbons, including petroleum and its derivatives, pesticides, herbicides, and plastics, are highly toxic and pose a significant environmental threat. Their stable structure prevents rapid natural degradation, resulting in long-term contamination of soils and waters. Enzymes play a crucial role in mitigating this environmental pollution by facilitating the degradation of these compounds.

Hydrolases, such as esterases and amidases, break chemical bonds including esters and amides, often found in pesticides and plastics. For example, esterases break the ester bonds present in organophosphorus pesticides, while amidases act on carbamate pesticides. Oxidoreductases, including oxidases, laccases, and cytochrome P450, catalyze oxidation-reduction reactions, transferring electrons between molecules, and are important in the degradation of hydrocarbons and aromatic compounds. Oxidases such as dioxygenases break C-H bonds in hydrocarbons, while laccases produced by fungi oxidize a variety of aromatic compounds, including pesticides. Lyases, on the contrary, break bonds without adding water, such as dehalogenases that remove halogen atoms from molecules like organochlorine pesticides. Transferases, such as glutathione S-transferases, move functional groups between molecules and can be important in detoxifying pesticides and heavy metals (Bhandari et al. 2021; Haripriyan et al., 2022; Mousavi et al. 2021; Rathore et al. 2022; Saravanan et al. 2021).

14.4.1.1 Plastic Residues

Plastic pollution is a global challenge, and finding solutions for its degradation is crucial. Hydrolases, enzymes belonging to the EC 3 class, play a vital role in this process as they specialize in breaking chemical bonds through hydrolysis. This process involves adding a water molecule to the chemical bond and breaking it into two parts. One of the most important hydrolases in plastic degradation is esterase (EC 3.1.1.1).

To break down the long polymer chains of plastics, esterase uses a two-step mechanism. Initially, in the adhesion phase, hydrolase enzymes, generally produced outside microbial cells, first bind to the plastic surface through hydrophobic interactions, because both the enzymes and the plastic are water-repellent. This is followed by hydrolytic cleavage, where the enzyme's active site breaks the chemical bonds in the polymer chain, dividing it into smaller molecules (monomers or dimers) (Thew et al., 2023).

Some microorganisms produce esterases and can contribute to plastic degradation, such as the yeast *Pseudozymaantarctica*, which produces an esterase that degrades biodegradable agricultural cover films (Yamashita et al., 2019), and *Aspergillus sp*., which mainly produces esterases, lipases, and ureases that can degrade LDPE (low-density polyethylene), HDPE (high-density polyethylene), PVC (polyvinyl chloride), PET (polyethylene terephthalate), and PU (polyurethane) (Nasrabadi et al., 2023). Moreover, various *Bacillus* species are known to produce esterases that degrade different types of plastics, including PET (polyethylene terephthalate) (Pantelic et al., 2023). Phosphotriesterases catalyze the hydrolysis of organophosphorus compounds found in some plastic additives, and amidases catalyze the breakdown of amide bonds found in some plastics.

Besides esterases, other enzymes also play an important role in polymer degradation, such as laccases (EC 1.10.3.2) among oxidoreductases, which oxidize phenolic compounds in some plastics.

14.4.1.2 Hydrocarbons

Hydrocarbon pollution, such as oil and its derivatives, causes considerable environmental damage. Oxygenases play a crucial role in the degradation of these compounds by using molecular oxygen (O_2) to add oxygen atoms to hydrocarbons, initiating their breakdown. Monooxygenases (EC 1.14.14), in particular, act on the degradation of hydrocarbons and other organic compounds by adding an oxygen atom to the substrate while the other oxygen atom is reduced to water during catalysis.

These enzymes are often found in bacteria and fungi that degrade hydrocarbons. For example, *Pseudomonas putida* is a bacterium widely studied for its ability to degrade aromatic and aliphatic hydrocarbons, including benzene, toluene, xylene, and naphthalene, through the production of various oxygenases, such as toluene/xylene monooxygenase (TXO) and naphthalene dioxygenase. Although laccases are more frequently found in fungi, some strains of *Pseudomonas putida* can produce these enzymes. Oxygenases catalyze the incorporation of oxygen into hydrocarbon molecules, while laccases oxidize the compounds, breaking them into smaller molecules (Choudhury et al., 2022). Additionally, species of *Mycobacterium* are known to degrade long-chain hydrocarbons, such as those found in crude oil, through the production of oxygenases. These enzymes, mainly PAH-RHDs (dioxygenases that hydroxylate polycyclic aromatic rings), can oxidize saturated and unsaturated hydrocarbons by introducing two oxygen atoms into the molecules. The presence of mycolic acids in the cell wall of *Mycobacterium* facilitates the degradation of hydrophobic compounds, such as hydrocarbons (Deng et al., 2023). Species of *Rhodococcus* are also versatile in the degradation of hydrocarbons, including crude oil, aromatic and aliphatic compounds, pesticides, and other organic compounds, due to their ability to produce a complex enzymatic system that involves oxygenases and other enzymes (Nazari et al. 2022).

Fungi such as *Aspergillus oryzae* and *Aspergillus flavipes* can degrade phenanthrene, a polycyclic aromatic hydrocarbon (PAH) considered toxic and pollutant. These species produce enzymes like manganese peroxidase (MnP) and lignin peroxidase (LiP), which may be involved in phenanthrene degradation. These fungi show potential for the bioremediation of soils contaminated by PAHs (de la Cruz-Izquierdo et al., 2021). Similarly, the fungal species *Trametes versicolor* produces laccase enzymes to degrade PAHs such as anthracene (ANT) and benzo[a]pyrene (B[a]P) (Deng, Wang, et al. 2022).

14.4.2 Enzymes on Inorganic Contaminants

Heavy metals are inorganic, nonbiodegradable, and highly toxic compounds, including arsenic (As), cadmium (Cd), chromium (Cr), lead (Pb), and mercury (Hg). By using the appropriate enzyme, it is possible to transform these ions into less toxic compounds that are easier to remove through mechanisms such as oxidation-reduction and methylation. In oxidation-reduction, ions are converted into less polluting and more easily removable forms through processes such as precipitation and volatilization. In methylation, heavy metals are converted into organic compounds that are easily absorbed and less harmful to the environment. Next, we will discuss some pollutants and the types of enzymes involved in their mitigation (Bharagava et al. 2020; Borah et al. 2020).

14.4.2.1 Arsenic

Arsenite oxidase (Aio/Aox) and arsenate reductase (ArsC) are enzymes capable of degrading arsenate ions (As). Arsenic can cause cardiovascular diseases, hemolysis, melanosis, encephalopathy, and severe issues in the nervous system and bone marrow. It can be found in mining wastewater, which, if discarded without proper treatment, contaminates soil and water bodies (Bharagava et al. 2020; Borah et al. 2020).

Arsenite oxidase can convert arsenite (As (III)) into arsenate (As (V)), a less toxic ion, through redox reactions. On the other hand, arsenate reductase can convert As (V) into As (III), which, although more toxic, is more soluble, allowing for its leaching. The more commonly used enzyme is arsenite oxidase, while arsenate reductase can be used in specific situations where removal by precipitation is more viable (Mousavi et al. 2021).

These enzymes are naturally present in the metabolism of some microorganisms, such as bacteria like *Escherichia coli*, which contains arsenate reductase in its plasmid. *Bacillus* sp., *Aeromonas* sp., *Acinetobacter* sp., and *Pseudomonas* sp. show high tolerance to environments contaminated with this metal and can degrade it, making them promising candidates for remediation. Bacteria can also degrade arsenic by performing methylation with the aid of the enzyme S-adenosylmethionine methyltransferase (ArsM), transforming it into a less toxic and more volatile organic compound (biovolatilization) (Kabiraj et al., 2022). Figure 14.1 illustrates how As remediation occurs in the bacterial cytoplasm.

These enzymes are also present in fungal species such as *Aspergillus* sp., *Penicillium* sp., *Rhizobium* sp., *Rhizopus* sp., *and Trichoderma* sp., which can be used to decontaminate environments containing arsenate. Additionally, the yeast *Saccharomyces cerevisiae* can also reduce As (III) using ArsC (Kabiraj et al., 2022; Mousavi et al., 2021). Tripathi et al. (2020) reported the presence of ArsM in the fungal strain *Humicola* sp., isolated from arsenic-contaminated soil, enabling its biovolatilization. Fungal strains of the *Ascomycetes* group and the *Trichocomaceae* family isolated from the same site also showed tolerance to As.

ArsC and ArsM are present in the metabolism of some microalgae species, allowing them to reduce and methylate As. They can perform the mitigation process in three steps: starting with the reduction of As (V), followed by the methylation of As (III), and finally their biosynthesis into arsenosugars and arsenolipids, and excretion. This mechanism is called biotransformation (Papry, Miah, and Hasegawa, 2020). Some examples of microalgae (or cyanobacteria) species are *Microcystis aeruginosa*, *Scenedesmus obliquus*, and *Chlamydomonas reinhardtii* (Wang et al., 2020).

14.4.2.2 Mercury

Mercury (Hg) is released into the environment during coal combustion, mining processes, and cement manufacturing, and is also found in crude oil. It is a highly volatile and toxic compound used

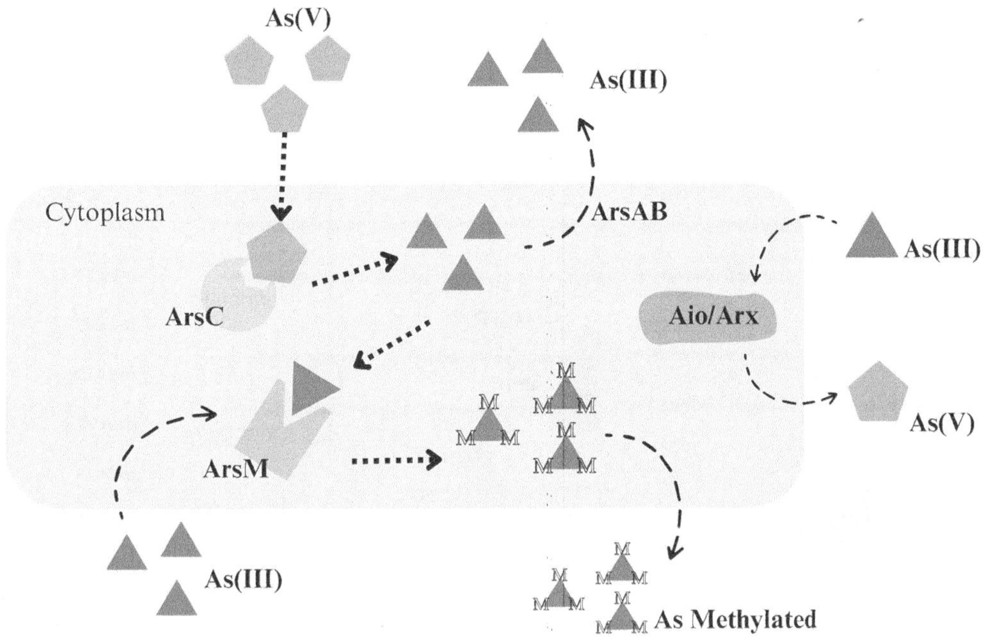

FIGURE 14.1 Remediation of arsenic in bacterial cytoplasm.

in thermometers, barometers, batteries, dental amalgams, and metal refining. At elevated levels, it causes harm to humans (dermatitis, bronchitis, nervous and immune system damage) and the environment (Pavithra et al. 2023).

Mercury reductase (MerA), using FAP and NADPH as electron sources, reduces Hg (II) to Hg (0), making the final ion gaseous and less toxic. Subsequently, methylation can be carried out with the help of the enzyme organomercurial lyase (MerB), which breaks carbon-mercury bonds, transforming them into organomercury compounds. The mer operon encompasses these and other enzymes/proteins that assist microorganisms in resisting Hg (Priyadarshanee et al., 2022). Figure 14.2 shows the action of these enzymes in mercury degradation. Bacteria such as *Pseudomonas, Bacillus*, *Enterobacter*, *Staphylococcus*, *Geobacter*, and *Acinetobacter* produce these types of enzymes, with the first three being the most used for Hg removal from soil and water. The fungi *Penicillium* and *Rhizopus* also have the capacity to remove Hg enzymatically (Mousavi et al., 2021; Narayanan, Ali, and El-Sheekh, 2023). Microalgae such as *Selenastrum minutum*, *Chlorella fusca* var. *fusca*, the diatom *Naviculapellicosa*, and the thermophilic alga *Galdiera sulphuraria* are also capable of reducing Hg (II) to Hg (0) and ÿ-HgS (Leong and Chang 2020).

14.4.2.3 Chromium

Chromium (Cr) is a heavy and carcinogenic metal found in wastewater from the metallurgical, textile, and electroplating industries. Through the action of specific enzymes, it is possible to transform the toxic ion Cr (VI) into Cr (III), which has lower toxicity and solubility. Bacterial species capable of producing enzymes that degrade chromium include *Pseudomonas, Bacillus, Escherichia, Shewanella*, and *Enterobacter*, with gram-positive bacteria being particularly effective at reducing chromium (VI) (Mousavi et al., 2021; Saxena et al., 2020). The removal of Cr has been observed using live and dead biomass of microalgae species such as *Chlorella* sp., *Phaeodactylum* sp., and *Navicula* sp. (Leong & Chang, 2020). Similarly, fungi like *Penicillium* sp., *T. pseudokoningii*, and *Aspergillus niger* are also capable of reducing Cr (VI) (Mala et al. 2020).

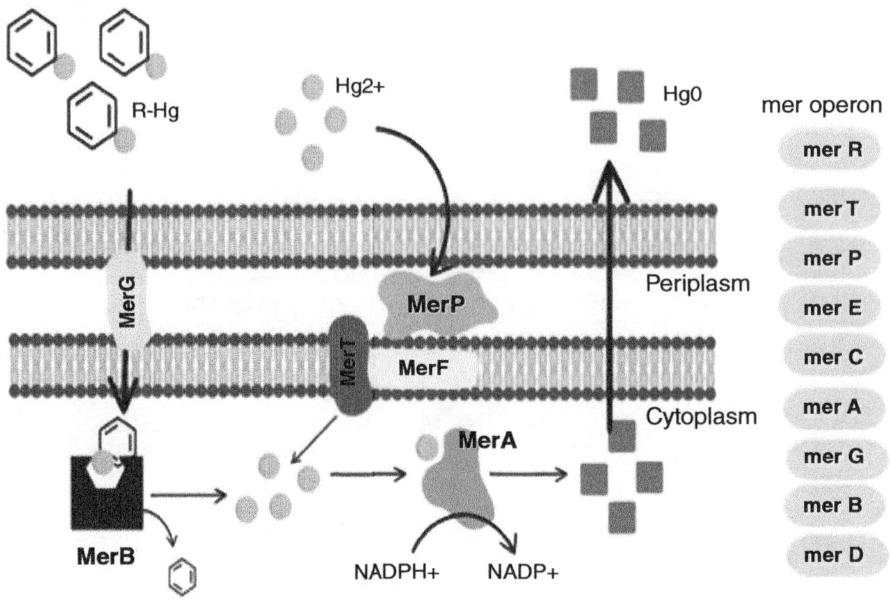

FIGURE 14.2 Enzymatic mechanism of mercury reduction.

Chromium reductase enzymes are known for reducing Cr (VI) and its intermediates simultaneously, and this reduction depends on NADH as the primary electron donor. The reduction process can occur intracellularly or extracellularly, either anaerobically or aerobically (Ramli et al., 2023). The anaerobic process occurs in two stages: first, Cr (VI) is reduced to Cr (V), then Cr (V) is further reduced to Cr (III). In the aerobic process, Cr (VI) acts as the terminal electron acceptor. Initially, an electron transfer reduces Cr (VI) to Cr (V), an unstable intermediate. Immediately after, a two-electron transfer mediated by NADH/NADPH or reserve electrons forms Cr (III) (Saxena et al. 2020). Figure 14.3 illustrates the aerobic and anaerobic pathways for chromium degradation.

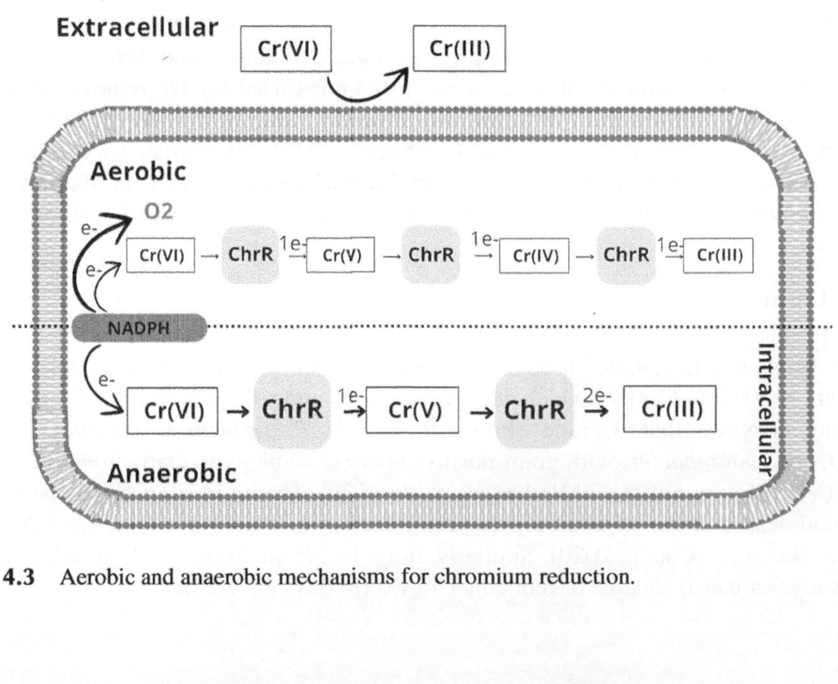

FIGURE 14.3 Aerobic and anaerobic mechanisms for chromium reduction.

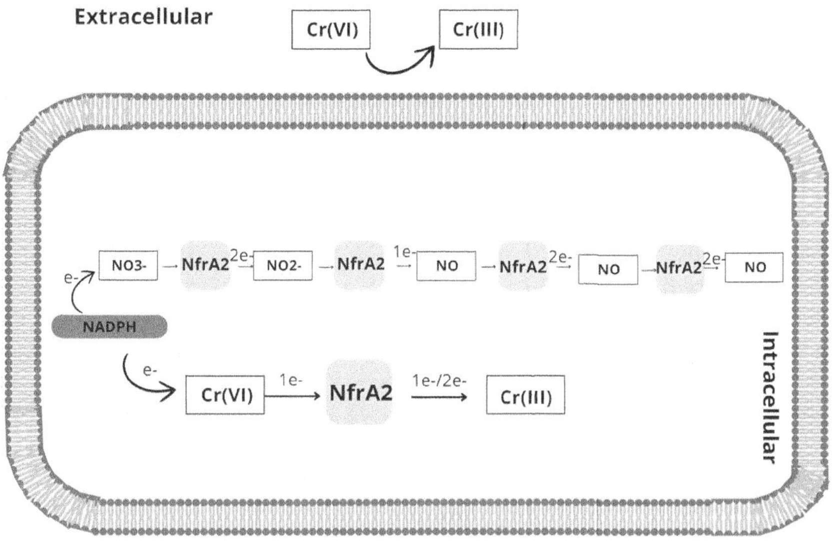

FIGURE 14.4 Mechanism of chromium reduction by nitroreductases.

When it comes to nitroreductase, its action is indirect. In this case, the enzymes involved in the reduction process are known as Type I, characterized by their ability to catalyze in the absence of O_2 and transfer two electrons during the reaction, with the most commonly known being NfsA, NfsB, and NfrA2. The reactions can occur intra- and extracellularly, requiring NADH and nitrate for ion reduction, as shown in Figure 14.4 (Ramli et al., 2023; Saxena et al. 2020).

14.4.2.4 Other Heavy Metals

The enzyme urease (Urea aminohydrolase EC 3.5.1.5) converts urea into carbonate, which can react with toxic metal ions to form insoluble compounds, a process known as biomineralization (Xia et al., 2021). Jalilvand et al. (2019) used ureolytic bacteria (*Stenotrophomonas rhizophila* and *Variovorax boronicumulans*) for the precipitation of Lead (Pb), Cadmium (Cd), and Zinc (Zn). Additionally, *Brucella intermedia*, another ureolytic bacterium, was used by Hu et al. (2024) for the biomineralization of Copper (Cu), Lead (Pb), and Zinc (Zn).

P-type ATPases can be used by the metabolism of some microorganisms as translocators of heavy metals. They also encode proteins for the transport and degradation of these metals, as noted by Xie et al. (2021) in the adsorption of Cu (II) and Pb (II) by the bacterium *Deinococcus wulumuqiensis*. Similarly, Sun et al. (2022) observed that enzymes from the ATPase family played an important role in the adsorption of Cu (II) and Cr (VI) by the bacterium *Stenotrophomonas rhizophila*.

14.5 BIOREMEDIATION THROUGH ENZYME ACTION

The active participation of enzymes in bioremediation processes has increased scientific investigations related to the efficiency of pollutant removal/degradation and enzyme-substrate interaction factors, be they hydrophobic, electrostatic, or hydrogen bonding (Liu et al., 2018). It is also worth noting the extensive applications of this method for the treatment of xenobiotics, that is, synthetic chemicals used in (agro)industrial sectors, such as alkanes, fuels, polyaromatic compounds, PAHs, herbicides, solvents, and industrial azo dyes, among others (Atashgahi et al. 2018; Goedtke et al. 2020).

The efficiency of applying an enzyme for effluent treatment depends mainly on the specific composition and origins of the effluent to be treated. For example, lipases can break down oils, fats, and greases, while proteases are efficient in decomposing proteins. Additionally, carboxylesterases can enhance the hydrolysis of ester bonds in carbamates, organophosphates, and other chlorinated organic compounds. Phosphodiesterases can hydrolyze a wide variety of organophosphates, while haloalkane dehalogenases can break the carbon-halogen bond in halogenated aliphatic compounds. Laccase and peroxidase exhibit significant enzymatic promiscuity, making them widely used due to their ability to act on various compounds (Azman et al. 2024; Feng et al., 2021; Mishra et al. 2020).

Furthermore, their functionalities as surface modifiers promote ideal reactivity conditions with multiple products and emulsifying capacity due to the presence of hydrophobic and hydrophilic amino acids with polypeptide chains. This makes enzymatic applications like amylase and cellulase highly soluble and easily biodegradable bioremediative mechanisms (Al-Maqdi et al. 2021; Pan et al. 2019; Saravanan et al. 2021).

More effective enzymatic bioremediation approaches must consider whether the enzyme employed in the process can perform its degradative functions sustainably and stably under adverse conditions. An example is the inadequacy of free laccases in treating wastewater effluents, a problem solved by using their immobilized form (Bilal et al. 2019; Lassouane et al. 2022; Wlizło et al. 2020).

Based on studies by Wu et al. (2024) on the behavior of enzymes obtained from the white-rot filamentous fungus *Trametes versicolor* in treating effluents containing 2,4,6-TCP (trichlorophenol), simulations were conducted in glass flasks containing 10 mL of 2,4,6-TCP solution (initial concentration varying from 20 to 160 mg L^{-1}), 1% acetonitrile, and 3 mM HBT, at 30°C with shaking at 150 rpm, reaction times ranging from 2 to 24 hours, and enzymes in both free and immobilized conditions. At lower 2,4,6-TCP concentrations, removal rates reached 100% with both free and immobilized enzymes. As the pollutant level increased, removal rates decreased to 20.93% and 60.92% at the maximum concentration (160 mg L^{-1}).

He et al. (2024) compared the performance in treating wastewater contaminated with nitrate using emulsified vegetable oils (EVO) in the immobilization process of α-amylase (α-EVO), cellulase (c-EVO), and protease (p-EVO) to obtain enzyme-EVO (1.0 mg mL^{-1}) (with a 1:50 ratio of enzymes to EVO based on soybean oil). To do this, 1500 mL bioreactors (with a wastewater volume of 1000 mL) were conditioned at 22 ± 2 °C in semi-continuous operation mode, ensuring the initial NO3–N concentration in each of the 3 ten-day cycles was equivalent to 50 mg L^{-1}. At the end of the tests, a higher degradation acceleration rate of NO3–N for c-EVO (32.80 ± 0.35 mg/Ld) and improved degradation in the last cycle for both c-EVO and p-EVO with respective percentages of 85.39% and 87.11% were observed.

In the treatment of PAHs, Guo et al. (2019) used lignin peroxidase (LiP) produced by the bacterium *Pichia methanolica* and subjected it to immobilization and purification processes with $Fe_3O_4SiO_2$polydopamine nanoparticles in glass flasks containing immobilized LiP varying from 0 to 1 g, 200 mL of sodium tartrate buffer, and 5 mg L^{-1} for each analyzed PAH (phenanthrene, fluoranthene, benzo(a)pyrene, and dibutyl phthalate) under 48-hour incubation. Total removals were obtained for tetracycline, dibutyl phthalate, 5-chlorophenol, and phenol when subjected to 0.75 g of LiP, while for phenanthrene, fluoranthene, and benzo(a)pyrene, 1 g of the enzyme was required, achieving maximum removals of 79.2%, 73.2%, and 64.8%, respectively.

Enzymatic remediation can be used for a wide variety of pollutants, as shown in Table 14.1, which lists various studies using enzymes to remove or modify different polluting compounds.

14.6 ENZYME-BASED TECHNOLOGIES

The characteristics of enzymes associated with their remedial capacity, as discussed in previous sections, make them promising agents in the bioremediation process. This is because, in addition to degrading pollutants, they can convert them into simpler and less toxic compounds quickly and efficiently. They can be used on a wide range of pollutants. For this reason, research is being conducted to optimize and scale the process from the laboratory to volumes close to those required in real processes, making this technology applicable to treatment processes (Behbudi et al., 2021;

TABLE 14.1
Studies on Enzyme Remediation of Pollutants

Enzymes	Source Microorganism	Pollutant	System Configuration	Main results	References
Laccase	*Ganoderma multipileum*	Chromium	The process was carried out by incubation in an Eppendorf for 120 min under stirring conditions of 200 rpm at 35°C containing 0.5 µg of partially purified laccase and initial chromium concentrations of 100, 150, 200, 250 and 300 mg L^{-1}.	The best chromium removal performances for concentrations of 100 and 150 mg L^{-1}were >94% (20 h incubation) and 72.18% (40 h incubation), respectively. For the initial concentration of 250 mg L^{-1}, with an incubation time of 120 h, 82.3% removal was achieved.	Alshiekheid et al. (2023)
Lipase	*Acinetobacter junii* VSJK-R6, *Pseudomonas composti* VSJK-R8e *Pseudomonas aeruginosa* VSJK-R9	Restaurant wastewater	The remediation process consisted of 250mL glass flasks containing 100mL of restaurant wastewater, supplemented with 0.5% NH_4Cl and 0,8% K_2HPO_4, and incubated for 5 days at 30°C under continuous *agitation at 100 rpm. The study* evaluated three different bioremediation configurations: (1) 5% V/V of the purified enzyme, (2) 1% V/V of the *A.junii, P. composti* and *P aeruginosa* consortium (1,5·10^8) cellsmL^{-1}) and (3) 5% V/V of the purified enzyme plus 1% V/V of the consortium.	For initial COD of 3200±80 mg L^{-1}, removals of 50%, 70.84% and 85.43%, respectively for configurations 1, 2 and 3, with the control reactor achieving only 22.93% removal. With initial oil *content equal to 890±27 mg* L^{-1}, removals of 63.82%, 76.29% and 90.11% respectively for configurations 1, 2 and 3, with the control reactor achieving only 28.98% removal.	Sutar et al. (2023)
Acetaldehydedehydrogenase, dioxygenase and acetyl COA acetyltransferase	*Pseudomonas proteolytica* GBPI_Hb61	Estriol	Glass flasks were used for optimization with a total of 41 tests for the parameters pH (5 to 9), temperature (5 to 25°C), inoculum volume (2 to 10% v/v), agitation (75 to 185 rpm) and estriol concentration (5 to 20 mg L^{-1}).	After optimizing the operating conditions, the best condition was pH 7, agitation of 130 rpm, inoculum concentration of 6% v/v and temperature of 25°C, in this way, the process achieved a degradation for estriol of 85.23% (initial concentration of 20 mg L^{-1}).	Thathola et al. (2024)

(Continued)

TABLE 14.1 *(Continued)*
Studies on Enzyme Remediation of PollutantsTreatment

Enzymes	Source Microorganism	Pollutant	System Configuration	Main results	References
Laccase	*Trametes hirsuta*	17α-ethinylestradiol (EE2)	The remediation process of EE2 present in effluents from a WWTP (initial concentration of 0.4 mg L^{-1}) had an incubation period of around 4.1d and took place in glass flasks at a temperature of 21°C using the constitutive laccase A0A6M5CX58 (Yg).	After bioremediation, a removal percentage of 80% was achieved for EE2	El Yagoubi et al. (2023)
Esterase, amidase and oxidoreductase	*Streptomyces* sp. PU10	Impranil (soluble polyester-PU)	The incubation process was carried out in an Eppendorf containing 10 mL of mineral salt medium, initial concentrations of Impranil (soluble polyester-PU) of 0.5%–1.5% (v/v) and 0.1% (v/v) of inoculum, for 72 hours at 37°C with 180 rpm agitation.	Impranil degradation was found to be >96% in 48 h of incubation for the initial concentrations of 0.5 and 1.0% (v/v), with stability until the end of the treatment time. For the 1.5% (v/v) concentration, around 80% degradation was obtained in 48 h and, at the end of the period, >96% was achieved	Pantelic et al. (2024)
Laccase	*Vibrio* sp. LA	Safranin 0, methylene blue, bromophenol blue, Eriochrome T black, crystal violet, malachite green and methyl orange	The remediation system consisted of glass flasks incubated for 6 and 24 hours at 28°C, pH 5, enzyme activity of 5 U mL^{-1} and initial concentrations of 60 mg L^{-1} Safranin 0, 10 mg L^{-1} methylene blue, 300 mg L^{-1} bromophenol blue, 300 mg L^{-1} Eriochrome T black, 10 mg L^{-1} crystal violet, 10 mg L^{-1} malachite green and 10 mg L^{-1} methyl orange.	The best degradation rates were obtained for the dyes Eriochrome black T (62.3 ± 3.1%), methyl orange (32.1 ± 1.3%) and bromophenol blue (46.3 ± 2.2%). This was followed by degradations of 23.6 ± 2.3%, 15.2 ± 1.2%, 15.2 ± 1.1% and 10.2 ± 0.4% respectively for methylene blue, malachite green, crystal violet and safranin 0.	Jiang et al. (2023)

Manganese peroxidase (MnP)	*Cerrena unicolor* BBP6	Remazol Brilliant Blue R (RBBR), Methyl Orange (MO), Crystal Violet (CV), Brilliant Blue R (BBR), and Bromophenol Blue (BPB)	Decolorization process given by incubation for 12 h at 30°C without stirring with 5 ml of 50 mM malonate buffer, 100 mg L^{-1} for each dye, 1 mM MnSO4, 0.1 mMH_2O_2 and enzyme activity of 100 U L^{-1}	Decolorization rate for RBBR, MO, BPB, CV and BBR dyes of 79.7%, 69.2%, 51.2%, 61.6% and 28.4%	Zhang, Zhang and Geng (2020)
Manganese peroxidase (MnP)	*Cerrena unicolor* BBP6	Fluorene (FLR) and Phenanthrene (PHE)	Degradation using 24 h incubation at 30°C without stirring in test tubes with 5 ml of 50 mM malonate buffer, 10 mg $L^{-1}$of FLR or PHE, 1 mM of MnSO4, 0.1 mM of$H_2O_2$and enzyme activity of 100 U L^{-1}	Degradation rates of polycyclic aromatic hydrocarbons FLR and PHE of 80.2% and 90.6% respectively	Zhang, Zhang and Geng (2020)
Laccase (Lac), Manganese peroxidase (MnP) and lignin peroxidase (LiP)	*Aspergillusiizukae*EAN605	Remazol Brilliant Blue R (RBBR)	Incubation for 4 h at 30°C in glass vials containing 6.15 mg of RBBR, 10 mL of sodium acetate buffer solution, 1 mg of crude enzyme with enzyme activity of 6.15, 2.58 and 127.99 U mg^{-1} respectively for Lac, MnP and LiP.	Decolorization of RBBR equivalent to 84.84% and confirmation of the high stability of the enzymes used at pH 3–9	Noman et al. (2020)
Laccase (Lac)	*Trametes versicolor*	Airport wastewater	Treatment in an airlift reactor with a reaction volume of 1L with effluent diluted 50% v/v with enzyme extract (1200 U mL^{-1}), aeration of 1.0 vvm, 30 °C with 9 days of incubation and initial COD concentration of 4544.44±78 mgL^{-1}	Removals of 43% for COD, degradability index equal to 1.36 and supernatant activity of 300 UmL^{-1}	Atilano-Camino et al. (2020)
Peroxidase	*Bacillus amyloliquefaciens*	Azo dyes (Congo red and Eriochrome black T) and triphenylmethane dyes (bromine cresol green and crystal violet)	Decolorization carried out in glass jars at 30 °C for 72 h containing 60 mg $L^{-1}$for each dye, 50 mM H_2O_2 in 4.5 ml of 0.05 M citrate buffer and enzyme activity of 0.672 U mL^{-1}	Removals 65.70%, 70.62%, 80.06% and 62.09% respectively for Congo red, bromine cresol green, Eriochrome black T and crystal violet	Yang et al. (2019)

Mousavi et al., 2021). It is important to highlight that direct enzymatic application is necessary mainly when the pollutants are very toxic or difficult to biodegrade and cannot undergo biological treatment with live cells in traditional processes (such as activated sludge and anaerobic digestion). In this way, enzymatic treatment can serve as a step to make the pollutants viable for further treatment by reducing their toxicity and recalcitrance.

In this context, Abejón et al. (2015) conducted a study on the feasibility of using large-scale enzymatic membrane reactors with microfiltration of 39 channels with an average pore diameter of 1.4 mm and 25 cm in length and membranes of 1.2 m in length, using immobilized laccase on ceramic membranes to degrade tetracycline. Based on kinetic parameters that evaluated hydraulic permeability, maximum reaction rate, and Michaelis-Menten constant, a mathematical model was proposed to predict the reactor's performance in different scenarios and configurations. Three situations were studied for implementing the treatment process, comprising effluents from wastewater treatment plants with characteristics typical of municipal, hospital, and industrial units. At the end of the study, the researchers concluded that from the simulated processes using enzymatic membrane reactors connected in series, this technology proved to be a viable technique, requiring only that the gelatin and enzyme layer be regenerated when enzymatic activity declines. This process can be resolved with a classic in situ cleaning treatment.

Another pilot-scale system, with a maximum treatment capacity of 2.4 tons/day, was proposed by Cho et al. (2008). In this case, the authors presented an efficient and innovative model in which peroxidase was immobilized on porous celite spheres and used in the treatment of wastewater released from a petrochemical plant, aiming to degrade toxic aromatic compounds in an integrated process that combines enzymatic catalysis and electrochemical oxidation. The COD removal efficiency using the integrated system was 85%–93%. The study concluded that the combined electro-enzymatic and electrochemical methods are twice as efficient in removing organic pollutants from petrochemical wastewater compared to methods used separately and also present lower energy consumption.

Additionally, to improve enzymatic remediation technology, a technical-scale study with a membrane reactor with an effective volume of 460 L was conducted by Gasser et al. (2014). In this study, researchers evaluated the immobilization of laccase on silica nanoparticles, resulting in a nanobiocatalyst that demonstrated a wide substrate spectrum in terms of degrading recalcitrant pollutants. The nanobiocatalysts were introduced into membrane reactors for tertiary wastewater treatment to eliminate phenolic compounds, particularly bisphenol A (BPA). The system efficiently removed BPA, and long-term field test results demonstrated the potential of silica nanoparticle/laccase composites in advanced wastewater treatment.

Some companies are already using enzymes in the bioremediation process. Notable among them is the Indian company Infinita Biotech Private Limited, recognized by the Department of Scientific and Industrial Research of the Ministry of Science and Technology of the Government of India. The company focuses on the creation, manufacture, and commercialization of enzymatic formulations for various industries. Infinita Biotech has developed a customized enzyme-based formulation obtained from fungi and bacteria, which can be mixed with microorganisms or applied directly to the effluent, transforming toxic compounds into biodegradable and easily collectable parts. The company highlights that specific enzymes catalyze specific pollutants. Inorganic pollutants such as phenols, benzidines, anilines, and other heterocyclic aromatic compounds are treated by oxidoreductases, peroxidases, and oxygenases, while organic pollutants are treated by protease, lipase, cellulase, amylase, urease, and xylanases (Ifinita Biotech Pvt Ltd, 2024).

Another notable company is Eurovix, an Italian company established in 1989, specializing in the manufacture of biointensifiers or biopromoters focused on sustainability. Eurovix has also developed products for effluent treatment through enzymatic remediation. For example, the Micropan bioactivator product line consists of approximately 50% high-performance microorganisms and 50% specific enzymes. The enzymes help break down the molecules of contaminant organic compounds, facilitating the action of the digestive microorganisms (Eurovix 2024).

14.7 CONCLUSION AND FUTURE PERSPECTIVES

Enzymes are biological catalysts required in various industrial processes as they accelerate reactions by lowering the activation energy. This feature has brought them increased visibility in the biotechnological sectors due to their high capacity to promote the degradation of various compounds, such as pollutants, in a stable, rapid, and highly specific manner. However, challenges remain regarding the high cost of applying these agents, especially in simpler processes or those with already established technologies, such as large-scale real effluent treatment.

Therefore, despite the numerous advantages provided by using enzymes, it is necessary to direct research efforts toward promoting the feasibility of this technology. This can be achieved mainly by reducing the costs associated with the production, isolation, and purification stages of these inputs. One effective strategy could be validating the use of the pollutant waste itself for enzymatic production, thereby associating the treatment process with enzymatic production. Then, promoting the recovery and reusability of enzymes after cycles is crucial.

REFERENCES

Abejón R, De Cazes M, Belleville MP, Sanchez-Marcano J. Large-scale enzymatic membrane reactors for tetracycline degradation in WWTP effluents. *Water Research* 2015; 73:118–131. https://doi.org/10.1016/j.watres.2015.01.012

Al-Maqdi KA, Bilal M, Alzamly A, Iqbal HMN, Shah I, Ashraf SS. Enzyme-loaded flower-shaped nanomaterials: A versatile platform with biosensing, biocatalytic, and environmental promise. *Nanomaterials* 2021; 11(6):1460. https://doi.org/10.3390/nano11061460

Alshiekheid MA, Umar A, Ameen F, Alyahya SA, Dufossé L. Biodegradation of chromium by laccase action of Ganoderma multipileum. *Journal of King Saud University – Science* 2023; 35:102948. https://doi.org/10.1016/j.jksus.2023.102948

Antecka K, Zdarta J, Siwińska-Stefańska K, Sztuk, G, Jankowska E, Oleskowicz-Popiel P, Jesionowski T. Synergistic degradation of dye wastewaters using binary or ternary oxide systems with immobilized laccase. *Catalysts* 2018; 8:402. https://doi.org/10.3390/catal8090402

Atashgahi S, Shetty SA, Smidt H, De Vos WM. Flux, impact, and fate of halogenated xenobiotic compounds in the gut. *Frontiers in Physiology* 2018; 9:888. https://doi.org/10.3389/fphys.2018.00888

Atilano-Camino MM, Álvarez-Valencia LH, García-González A, García-Reyes RB. Improving laccase production from Trametes versicolor using lignocellulosic residues as cosubstrates and evaluation of enzymes for blue wastewater biodegradation. *Journal of Environmental Management* 2020; 275:11231. https://doi.org/10.1016/j.jenvman.2020.111231

Azman AA, Muhd Noor ND, Leow ATC, Mohd Noor SA, Mohamad Ali MS. Identification and characterization of a promiscuous metallohydrolase in metallo-β-lactamase superfamily from a locally isolated organophosphate-degrading *Bacillus sp.* strain S3wahi. *International Journal of Biological Macromolecules* 2024; 271:132395. https://doi.org/10.1016/j.ijbiomac.2024.132395

Bala S, Garg D, Thirumalesh BV, Sharma M, Sridhar K, Inbaraj BS, Tripathi M. Recent strategies for bioremediation of emerging pollutants: A review for a green and sustainable environment. *Toxics* 2022; 10(8):484. https://doi.org/10.3390/toxics10080484

Behbudi G, Yousefi K, Sadeghipour Y. Microbial enzymes based technologies for bioremediation of pollutions. *Journal of Environmental Treatment Techniques* 2021; 9(2), 463–469. https://doi.org/10.47277/JETT/9(2)469

Bhandari S, Poudel DK, Marahatha R, Dawadi S, Khadayat K, Phuyal S, Shrestha S, Gaire S, Basnet K, Khadka U, Parajuli N. Microbial enzymes used in bioremediation. *Journal of Chemistry* 2021. https://doi.org/10.1155/2021/8849512

Bharagava RN, Saxena G, Mulla SI. Introduction to Industrial Wastes Containing Organic and Inorganic Pollutants and Bioremediation Approaches for Environmental Management. In: Saxena G, Bharagava RN (eds) *Bioremediation of Industrial Waste for Environmental Safety*. Singapore: Springer Singapore, pp 1–18, 2020.

Bilal M, Adeel M, Rasheed T, Zhao Y, Iqbal HM. Emerging contaminants of high concern and their enzyme-assisted biodegradation: A review. *Environment International* 2019; 124:336–353. https://doi.org/10.1016/j.envint.2019.01.011

Borah P, Kumar M, Devi P. Types of Inorganic Pollutants: Metals/Metalloids, Acids, and Organic Forms. In Pooja Devi, Pardeep Singh, and Sushil Kumar Kansal (Eds.): *Inorganic Pollutants in Water*. Elsevier, pp 17–31, 2020. ISBN 9780128189658. https://doi.org/10.1016/B978-0-12-818965-8.00002-0

Cen Q, Wu X, Cao L, Lu Y, Lu X, Chen J, et al. Green production of a yellow laccase by Coriolopsis gallica for phenolic pollutants removal. *AMB Express* 2022; 12:96. https://doi.org/10.1186/s13568-022-01434-6

Cho SCS, Lee HLH, Moon SMS. Integrated electroenzymatic and electrochemical treatment of petrochemical wastewater using a pilot scale membraneless system. *Process Biochemistry* 2008; 43(12):1371–1376. https://doi.org/10.1016/j.procbio.2008.08.006

Choudhury SP, Panda S, Haq I, Kalamdhad AS. Enhanced methane production and hydrocarbon removal from petroleum refinery sludge after Pseudomonas putida pretreatment and process scale-up. *Bioresource Technology* 2022.

de la Cruz-Izquierdo RI, Paz-González AD, Reyes-Espinosa F, Vazquez-Jimenez LK, Salinas-Sandoval M, González-Domínguez MI, Rivera G. Analysis of phenanthrene degradation by Ascomycota fungi isolated from contaminated soil from Reynosa, Mexico. *Letters in Applied Microbiology* 1 May 2021; 72(5):542–555. https://doi.org/10.1111/lam.13451

Deng J, Wang H, Zhan H, Wu C, Huang Y, Yang B, Mosa A, Ling W. Catalyzed degradation of polycyclic aromatic hydrocarbons by recoverable magnetic chitosan immobilized laccase from Trametes versicolor. *Chemosphere* 2022; 301:134753, ISSN 0045-6535, https://doi.org/10.1016/j.chemosphere.2022.134753

Deng W, Zhao W, Yang Y. Degradation and detoxification of Chlorophenols with different structure by LAC-4 laccase purified from white-rot fungus Ganoderma Lucidum. *International Journal of Environmental Research and Public Health* 2022; 19:8150. https://doi.org/10.3390/ijerph19138150

Deng Y, Mou T, Wang J, Su J, Yan Y, Zhang Y-Q. Characterization of three rapidly growing novel Mycobacterium species with significant polycyclic aromatic hydrocarbon bioremediation potential. *Frontiers in Microbiology* 2023; 14:1225746. https://doi.org/10.3389/fmicb.2023.1225746

El Yagoubi Y, Lemieux B, Segura P, Cabana H. Characterization of laccases from Trameteshirsuta in the context of bioremediation of wastewater treatment plant effluent. *Enzyme and Microbial Technology* 2023; 171:110308. https://doi.org/10.1016/j.enzmictec.2023.110308

Eurovix. Complex formulations designed to solve environmental problems. Entratico, Itália. Disponível em: https://eurovix.it/en/prodotti/enviro/. Acesso em 30 June 2024.

Feng S, Hao Ngo H, Guo W, Woong Chang S, Duc Nguyen D, Cheng D, et al. Roles and applications of enzymes for resistant pollutants removal in wastewater treatment. *Bioresource Technology* 2021; 335:125278. https://doi.org/10.1016/j.biortech.2021.125278

Gan J, Ashraf SS, Bilal M, Iqbal HMN. Biodegradation of environmental pollutants using catalase-based biocatalytic systems. *Environmental Research* 2022; 214:113914. https://doi.org/10.1016/j.envres.2022.113914

Gasser CA, Yu L, Svojitka J, Wintgens T, Ammann EM, Shahgaldian P, Corvini PF-X, Hommes G. Advanced enzymatic elimination of phenolic contaminants in wastewater: A nano approach at field scale. *Applied Microbiology and Biotechnology* 2014; 98:3305–3316. https://doi.org/10.1007/s00253-013-5414-8

Goedtke L, Sprenger H, Hofmann U, Schmidt FF, Hammer HS, Zanger UM, Poetz O, Seidel A, Braeuning A, Hessel-Pras S. Polycyclic aromatic hydrocarbons activate the aryl hydrocarbon receptor and the constitutive androstane receptor to regulate xenobiotic metabolism in human liver cells. *International Journal of Molecular Sciences* 2020; 22:372. https://doi.org/10.3390/ijms22010372

Guo J, Liu XC, Zhang XM, Wu J, Chai C, Ma D, Chen QH, Xiang D, Ge W. Immobilized lignin peroxidase on Fe3O4@SiO2@polydopamine nanoparticles for degradation of organic pollutants. *International Journal of Biological Macromolecules* 2019; 180:433–440. https://doi.org/10.1016/j.ijbiomac.2019.07.105

Haripriyan U, Gopinath KP, Arun J, Govarthanan M. Bioremediation of organic pollutants: A mini review on current and critical strategies for wastewater treatment. *Archives of Microbiology* 2022; 286. https://doi.org/10.1007/s00203-022-02907-9

He B, Li M, Zhao X, Zou H, Xu B, He J. Comparative study of the quick action effect of multiple enzyme-based nano-emulsified oils in enhancing nitrate contamination remediation in groundwater. *Environmental Research* 2024; 257:119297. https://doi.org/10.1016/j.envres.2024.119297

Hu X, Yu C, Shi J, He B, Wang X, Ma Z. Biomineralization mechanism and remediation of Cu, Pb and Zn by indigenous ureolytic bacteria B. intermedia TSBOI. *Journal of Cleaner Production* 2024; 436:140508. https://doi.org/10.1016/j.jclepro.2023.140508

Ifinita Biotech Pvt Ltd. Role of enzymes in wastewater treatment. Tamil Nadu, Ìndia. Available in: https://infinitabiotech.com/blog/role-of-enzymes-in-wastewater-treatment/. Acesso em 30 June 2024.

IlićĐurđić K, Ostafe R, ĐurđevićĐelmaš A, Popović N, Schillberg S, Fischer R, et al. Saturation mutagenesis to improve the degradation of azo dyes by versatile peroxidase and application in form of VP-coated yeast cell walls. *Enzyme and Microbial Technology* 2020; 136:109509. https://doi.org/10.1016/j.enzmictec.2020.109509

Jalilvand N, Akhgar A, Alikhani HA, Rahmani HA, Rejali F. Removal of heavy metals zinc, lead, and cadmium by biomineralization of urease-producing bacteria isolated from Iranian mine calcareous soils. *Journal of Soil Science and Plant Nutrition* 2019; 20:206–219. https://doi.org/10.1007/s42729-019-00121-z

Jiang J, Deng J-L, Wang Z-G, Chen X-Y, Wang S-J, Wang Y-C. Characterization of a New Laccase from Vibrio sp. with pH-stability, Salt-tolerance, and Decolorization Ability. *Molecules* 2023; 28:3037. https://doi.org/10.3390/molecules28073037

Kabiraj A, Biswas R, Halder U, Bandopadhyay R. Bacterial arsenic metabolism and its role in arsenic bioremediation. *Current Microbiology* 2022; 79:131. https://doi.org/10.1007/s00284-022-02810-y

Karthik V, Kumar P, Vo DVN, Selvakumar P, Gokulakrishnan M, Keerthana P, Audilakshmi V, Jeyanthi J. Enzyme-loaded nanoparticles for the degradation of wastewater contaminants: A review. *Environmental Chemistry Letters* 2021; 19:2331–2350. https://doi.org/10.1007/s10311-020-01158-8

Karume I. Nature of support system and enzyme as key factors in immobilized biocatalyzed processes. *Sustainable Chemistry for the Environment* 2023; 4:100048. https://doi.org/10.1016/j.scenv.2023.100048

Lassouane F, Aït-Amar H, Rodriguez-Couto S. High BPA removal by immobilized crude laccase in a batch fluidized bed bioreactor. *Biochemical Engineering Journal* 2022; 184:108489. https://doi.org/10.1016/j.bej.2022.108489

Leong, YK; Chang, JS. Bioremediation of heavy metals using microalgae: Recent advances and mechanisms. *Bioresource Technology* 2020; 303:122886. https://doi.org/10.1016/j.biortech.2020.122886

Liu Z, Liu Y, Zeng G, Shao B, Chen M, Li Z, Jiang Y, Liu Y, Zhang Y, Zhong H. Application of molecular docking for the degradation of organic pollutants in the environmental remediation: A review. *Chemosphere* 2018; 203:139–150. https://doi.org/10.1016/j.chemosphere.2018.03.179

Lopes LS, Vieira N, Da Luz JMR, Silva MDC, Cardoso WS, Kasuya MC. Production of Fungal Enzymes in Macaúba Coconut and Enzymatic Degradation of Textile Dye. *Biocatalysis and Agricultural Biotechnology* 2020, *26*, 101651, https://doi.org/10.1016/j.bcab.2020.101651

Lu J, Zhang C, Leong HY, Show PL, Lu F, Lu Z. Overproduction of lipoxygenase from Pseudomonas aeruginosa in Escherichia coli by auto-induction expression and its application in triphenylmethane dyes degradation. *Journal of Bioscience and Bioengineering* 2020; 129:327–332. https://doi.org/10.1016/j.jbiosc.2019.09.006

Mala JGS, Takeuchi S, Dhanasingh S, Mani U. Microbial chromate reductases: Novel and potent mediators in chromium bioremediation: A review. *Applied Microbiology: Theory & Technology* 2020; 12:32–44. https://doi.org/10.37256/amtt.112020222

Mathew AB, Sing, H, Biju VG, Krishnamurthy NB. Classification, source, and effect of environmental pollutants and their biodegradation. *Journal of Environmental Pathology, Toxicology and Oncology* 2017; 36(1):55–71. https://doi.org/10.1615/JEnvironPatholToxicolOncol.2017015804. PMID: 28605331

Menger F, Boström G, Jonsson O, Ahrens L, Wiberg K, Kreuger J, GagoFerrero, P. Identification of pesticide transformation products in surface water using suspect screening combined with national monitoring data. *Environmental Science & Technology* 2021; 55:10343–10353. https://doi.org/10.1021/acs.est.1c00466

Mishra B, Varjani S, Agrawal DC, Mandal SK, Ngo HH, Taherzadeh MJ, et al. Engineering biocatalytic material for the remediation of pollutants: A comprehensive review. *Environmental Technology & Innovation* 2020; 20:101063. https://doi.org/10.1016/j.eti.2020.101063

Mousavi SM, Hashemi SA, Iman Moezzi SM, Ravan N, Gholami A, Lai CW, Chiang WH, Omidifar N, Yousefi K, Behbudi G. Recent advances in enzymes for the bioremediation of pollutants. *Biochemistry Research International* 2021; 1–12. https://doi.org/10.1155/2021/5599204

Narayanan M, Ali SS, El-Sheekh M. A comprehensive review on the potential of microbial enzymes in multi-pollutant bioremediation: Mechanisms, challenges, and future prospects. *Journal of Environmental Management* 2023; 334:117532. https://doi.org/10.1016/j.jenvman.2023.117532

Nasrabadi AE, Ramavandi B, Bonyadi Z. Recent progress in biodegradation of microplastics by Aspergillus sp. in aquatic environments. *Colloid and Interface Science Communications* 2023. https://doi.org/10.1016/j.colcom.2023.100754

Navada KK, Kulal A. Kinetic characterization of purified laccase from *Trametes hirsuta*: a study on laccase catalyzed biotransformation of 1,4-dioxane. *Biotechnol Lett*, 2021, 43, 613–626. https://doi.org/10.1007/s10529-020-03038-1

Nazari MT, Simon V, Machado BS, Crestani L, Marchezi G, Concolato G, Ferrari V, Colla LM, Piccin JS. Rhodococcus: A promising genus of actinomycetes for the bioremediation of organic and inorganic contaminants, *Journal of Environmental Management* 2022; 323:116220, ISSN 0301-4797, https://doi.org/10.1016/j.jenvman.2022.116220

Negi BB, Das C. Mycoremediation of wastewater, challenges, and current status: A review. *Bioresource Technology Reports* 2023; 22:101409. https://doi.org/10.1016/j.biteb.2023.101409

Noman E, Al-Gheethi AA, Talip BA, Mohamed R, Kassim AH. Oxidative enzymes from newly local strain Aspergillus iizukae EAN605 using pumpkin peels as a production substrate: Optimized production, characterization, application and techno-economic analysis. *Journal of Hazardous Materials* 2020; 386:121954. https://doi.org/10.1016/j.jhazmat.2019.121954

Pan X, Fang Y, Wang L, Xie M, Hu B, Zhu Y, Zhao E, Pei F, Shen F, Li P, Shen X, Hu Q. Effect of enzyme types on the stability of oil-in-water emulsions formed with rice protein hydrolysates. *Journal of the Science of Food and Agriculture* 2019; 99(15):6731–6740. https://doi.org/10.1002/jsfa.9955

Pantelic B, Araujo JA, Jeremic S, Azeem M, Attallah OA, Siaperas R, Mojicevic M, Chen Y, Fournet MB, Topakas E, Nikodinovic-Runic J. A novel Bacillus subtilis BPM12 with high bis(2 hydroxyethyl)terephthalatehydrolyticactivityefficientlyinteractswithvirginandmechanicallyrecycledpolyethyleneterephthalate. *Environmental Technology & Innovation* 2023; 1864. https://doi.org/10.1016/j.eti.2023.103316

Pantelic B, Siaperas R, Budin C, De Boer T, Topakas E, Nikodinovic-Runic J. Proteomic examination of polyester-polyurethane degradation by Streptomyces sp. PU10: Diverting polyurethane intermediates to secondary metabolite production. *Microbial Biotechnology* 2024; 17:e14445.

Papry RI, Miah S, Hasegawa H. Integrated environmental factor-dependent growth and arsenic biotransformation by aquatic microalgae: A review. *Chemosphere* 2020; 303:135164. https://doi.org/10.1016/j.chemosphere.2022.135164

Pavithra KG, SundarRajan P, Kumar PS, Rangasamy G. Mercury sources, contaminations, mercury cycle, detection and treatment techniques: A review. *Chemosphere* 2023; 312:137314. https://doi.org/10.1016/j.chemosphere.2022.137314

Priyadarshanee M, Chatterjee S, Rath S, Dash H, Das S. Cellular and genetic mechanism of bacterial mercury resistance and their role in biogeochemistry and bioremediation. *Journal of Hazardous Materials* 2022; 423:126985. https://doi.org/10.1016/j.jhazmat.2021.126985

Ramli NN, Othman AR, Kurniawan SB, Abdullah SRS, Hasan HA. Metabolic pathway of Cr(VI) reduction by bacteria: A review. *Microbiological Research* 2023; 268:127288. https://doi.org/10.1016/j.micres.2022.127288

Rathore S, Varshneu A, Mohan S, Dahiya P. An innovative approach of bioremediation in enzymatic degradation of xenobiotics. *Biotechnology and Genetic Engineering Reviews* 2022; 38(1): 1–32. https://doi.org/10.1080/02648725.2022.2027628

Saravanan A, Kumar PS, Vo DVN, Jeevanantham S, Karishma S, Yaashikaa PR. A review on catalytic-enzyme degradation of toxic environmental pollutants: Microbial enzymes. *Journal of Hazardous Materials* 2021; 419:126451. https://doi.org/10.1016/j.jhazmat.2021.126451

Saxena G, Kishor R, Bharagava RN. Application of Microbial Enzymes in Degradation and Detoxification of Organic and Inorganic Pollutants. In: Saxena G, Bharagava RN (orgs.) *Bioremediation of Industrial Waste for Environmental Safety*. Singapore: Springer Singapore, pp 41–51, 2020. https://doi.org/10.1007/978-981-13-1891-7_3

Sirajuddin S, Khan MA, Qader SAU, Iqbal S, Sattar H, Ansari A. A comparative study on degradation of complex malathion organophosphate using of Escherichia coli IES-02 and a novel carboxylesterase. *International Journal of Biological Macromolecules* 2020; 145:445–455. https://doi.org/10.1016/j.ijbiomac.2019.12.192

Sun S, Zhang K, Wu Y, Zhu N, Wang Y, Chen J, Leng F. Transporter drives the biosorption of heavy metals by Stenotrophomonas rhizophila JC1. *Environmental Science and Pollution Research* 2022; 29:45380–45395. https://doi.org/10.1007/s11356-022-18900-0

Sutar VP, Mali GV, Upadhye V, Singh VK, Sinha RP. Purification of lipase from pseudomonas aeruginosa VSJK R-9 and its application in combination with the lipolytic consortium for bioremediation of restaurant wastewater. *Applied Biochemistry and Biotechnology* 2023; 195:1888–1903. https://doi.org/10.1007/s12010-022-04230-x

Thathola P, Agnihotri V, Pandey A, Upadhyay SK. Biodegradation of steroid hormone estriol by pseudomonas proteolytica GBPI_Hb61, a psychrotolerant himalayan bacteria. *Water, Air, & Soil Pollution* 2024; 235:289. https://doi.org/10.1007/s11270-024-07079-4

Thew CX, Lee ZS, Srinophakun P, Ooi CW. Recent advances and challenges in sustainable management of plastic waste using biodegradation approach. *Bioresource Technology* 2023; 374:128772. https://doi.org/10.1016/j.biortech.2023.128772

Tripathi P, Khare P, Barnawal D, Shanker K, Srivastava PK, Tripathi RD, Kalra A. Bioremediation of arsenic by soil methylating fungi: Role of Humicola sp. strain 2WS1 in amelioration of arsenic phytotoxicity in

Bacopa monnieri L. *Science of the Total Environment* 2020; 716:136758. https://doi.org/10.1016/j.scitotenv.2020.136758

Wang Z, Gui H, Luo Z, Sarakiotia I, Yan C, Laing G. Arsenic release: Insights into appropriate disposal of arsenic-loaded algae precipitated from arsenic contaminated water. *Journal of Hazardous Materials* 2020; 384:121249. https://doi.org/10.1016/j.jhazmat.2019.121249

Wlizło K, Polak J, Kapral-Piotrowska J, Grąz M, Paduch R, Jarosz-Wilkołazka A. Influence of carrier structure and physicochemical factors on immobilisation of fungal laccase in terms of Bisphenol a removal. *Catalysts* 2020; 10:951. https://doi.org/10.3390/catal10090951

Wu J, Gao D, Wang L, Du X, Zhang Z, Liang H. Bioremediation of 2,4,6-trichlorophenol by extracellular enzymes of white rot fungi immobilized with sodium alginate/hydroxyapatite/chitosan microspheres. *Environmental Research* 2024; 252:118937. https://doi.org/10.1016/j.envres.2024.118937

Xia X, Wu S, Zhou Z, Wang G. Microbial Cd(II) and Cr(VI) resistance mechanisms and application in bioremediation. *Journal of Hazardous Materials* 401:123685, 2021. https://doi.org/10.1016/j.jhazmat.2020.123685

Xie C, Guo S, Chen K, Yu X, Li Y, Xu X. Removal of copper and lead ions from water using the extremophile Deinococcuswulumuqiensis R12. *Desalination and Water Treatment* 2021; 228:276–285. https://doi.org/10.5004/dwt.2021.27338

Yamashita YS; Ueda H; Koitabashi M; Kitamoto H. Pretreatment with an Esterase from the Yeast Pseudozyma Antarctica Accelerates Biodegradation of Plastic Mulch Film in Soil under Laboratory Conditions. *Journal of Bioscience and Bioengineering* 2019, *127*, 93–98, https://doi.org/10.1016/j.jbiosc.2018.06.011

Yang J, Gao T, Zhang Y, Wang S, Li H, Li S, Wang S. Degradation of the phenolic β-ether lignin model dimer and dyes by dye-decolorizing peroxidase from Bacillus amyloliquefaciens. *Biotechnology Letters* 2019; 41:1015–1021. https://doi.org/10.1007/s10529-019-02696-0

Zaaba NF, Jaafar M. A review on degradation mechanisms of polylactic acid: Hydrolytic, photodegradative, microbial, and enzymatic degradation. *Polymer Engineering & Science* 2020; 60:2061–2075. https://doi.org/10.1002/pen.25511

Zhang H, Zhang X, Geng A. Expression of a novel manganese peroxidase from Cerrena unicolor BBP6 in Pichia pastoris and its application in dye decolorization and PAH degradation. *Biochemical Engineering Journal* 2020; 153:107402. https://doi.org/10.1016/j.bej.2019.107402

15 Bio-refinery Approach for a Sustainable Eco-system and Circular Bio-economy

Divya Bisen, Ashish Pratap Singh Chouhan, and Manish Pant

15.1 INTRODUCTION TO BIOREFINERY CONCEPTS

In the context of the diminishing availability of fossil fuel reservoirs and related environmental concerns, biorefineries are gaining popularity as sustainable positions that are comparable to those of fossil fuel–based refineries in the economy. A biorefinery system is a facility that facilitates the economically viable conversion of biomass and organic waste, which are non–fossil-based feedstocks, converted into an array of marketable items encompassing chemicals, materials, and energy through integrated, resource-efficient unit operations (Venkata Mohan et al., 2016). It is thought that using renewable feedstocks will help achieve carbon neutrality because biorefineries are crafted to extract value from diverse renewable sources like forestry, agriculture, aquaculture, and various forms of waste (like agricultural and biogenic municipal waste), transforming them into a spectrum of products. Each year, approximately 181.5 billion tons of lignocellulosic biomass (LCB) is generated, accounting for approximately 90% of biomass. However, only about 3% of this biomass is efficiently utilized and integrated into the circular economy (Kellock et al. 2019). LCB primarily consists of a linear aromatic polymer known as lignin, comprising coniferyl and sinapyl alcohol units. Cellulose, the crystalline polymer, comprises microfibrils that are interconnected through van der Waals forces and hydrogen bonds. Hemicellulose, on the other hand, is a composite of polysaccharides containing pentose and sugars, typically consisting of approximately 200 units (Agbor et al. 2011). A biorefinery is an environmentally sustainable facility that integrates several conversion technologies, including thermochemical, biochemical, and microorganism growth processes. Its main objective is to generate a variety of products, including biofuels, chemicals, energy, and valuable products through the efficient utilization of biomass resources (Thürer et al., 2018). Cellulose, hemicellulose, and lignin constitute the essential components of LCBs, which is predominantly derived from diverse sources like barley straw, coconut husk, wheat straw, corn stover, rice husk, sugarcane bagasse, empty fruit bunch, and wood. These overlooked resources hold promise for transformation into diverse products through valorization in an integrated biorefinery approach, aligning with the principles of a circular economy.

The global drive to advance the bioeconomy is clear as many nations and organizations actively implement specific policies and strategies to expand their bio-based activities. The primary objective is to actively contribute to the achievement of as many of the United Nations' 17 Sustainable Development Goals (SDGs) as possible (Heimann 2019). Biorefining converts biomass feedstocks into various marketable products, including food, animal feed, chemicals, materials, and a range of energy forms like biofuels, power, and heat. This process, employing eco-friendly and sustainable bioprocessing systems, aligns with SDGs (Tursi, 2019). As biomass is biogenic, and carbon dioxide (CO_2) emitted during various biochemical processes doesn't add to atmospheric CO_2 levels.

This chapter primarily delves into the concept of biorefinery and its pivotal role in waste management within the circular economy framework. It explores various origins of organic solid waste and

 DOI: 10.1201/9781003668961-15

the biorefinery technologies employed to craft bio-based products from this waste. Notably, it addresses a significant hurdle in biorefinery processes and emphasizes the necessity of techno-economic analysis for the success of any biorefinery operation, enriching the overall message. Consequently, this analysis furnishes comprehensive insights into the circular-economic dimensions of the biorefinery process. Additionally, it explores the concept of biorefinery and outlines potential avenues for future research.

Figure 15.1 shows that numerous countries worldwide are actively engaged in advancing biorefinery approaches, leveraging sustainable and innovative technologies to convert biomass into valuable products. Here are descriptions of a few notable countries for their contributions to biorefinery research and implementation.

India's involvement in biorefinery research reflects a commitment to sustainable development, energy security, and environmental stewardship. As the country continues to explore and implement biorefinery technologies, it contributes to the global transition toward a more sustainable and bio-based economy.

Germany stands at the forefront of biorefinery research with a strong emphasis on sustainability and green technologies. The country has established various research centers and industrial collaborations focusing on the efficient conversion of biomass into biofuels, biochemicals, and other high-value products. Germany's commitment to renewable energy and environmental stewardship has propelled its leadership in the biorefinery field.

The United States is a key player in biorefinery advancements, particularly in the development of biofuels. Various research institutions and universities are actively involved in exploring different biorefinery pathways, including the conversion of agricultural residues, algae, and woody biomass. The U.S. government has also provided support through funding initiatives to accelerate the commercialization of biorefinery technologies.

China has emerged as a major player in the field of biorefinery, driven by its commitment to reducing dependence on fossil fuels and addressing environmental concerns. Chinese research

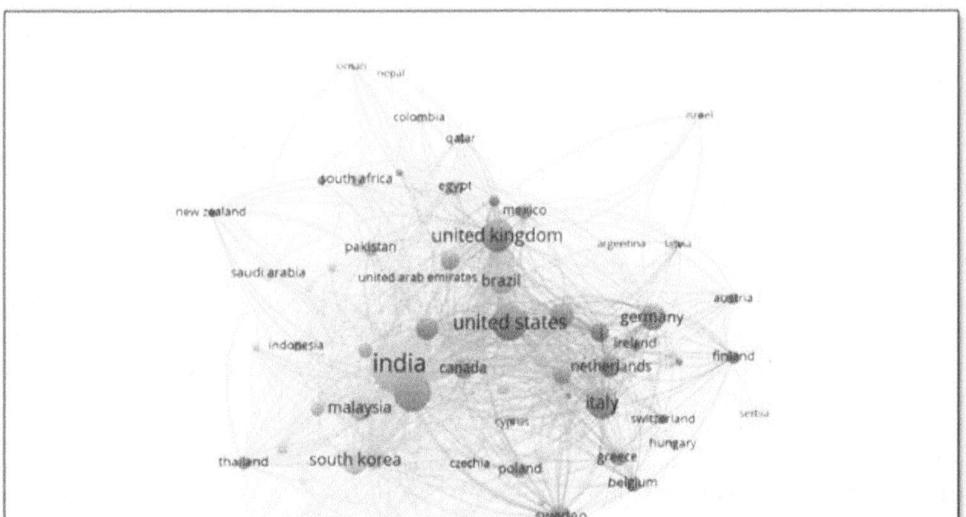

FIGURE 15.1 Bibliometric data of different countries on bio-refinery approach (VOS viewer software).

institutions and industries are actively involved in developing technologies for the transformation of various biomass feedstocks, including agricultural leftovers and municipal waste, into biofuels and chemicals. China's rapid growth in the biorefinery sector reflects its focus on sustainable development.

Finland has been proactive in promoting the economy, and biorefinery research is a key component of this strategy. The country is exploring advanced technologies to transform forest biomass into a range of bio-based products. Finland's emphasis on sustainability, utilizing its abundant forestry resources, aligns with the principles of circular economy and bio-based innovation.

15.2 ADVANCEMENTS IN THE RESEARCH AND ANALYSIS OF BIOREFINERY INITIATIVES

Biorefineries have progressed through four developmental stages. In the initial phase, first-generation biorefineries relied on a single feedstock and technology, producing a restricted array of products. Main feedstock sources comprised high-sugar crops, maize stover, or wheat straw (Karp and Shield, 2008). While some of these feedstocks have the potential to generate biofuels, the utilization of a significant quantity of cereals during production raises concerns about potential increases in food prices, which may have adverse effects on economic development. Moving to the second generation, biorefineries continued to employ a single feedstock and technology but demonstrated the capability to produce a broader range of products. Unlike the first generation, the feedstock expanded beyond food crops to include lignocellulosic sources. The third generation marked a substantial advancement, enabling the production of different products from diverse feedstocks and technologies, catering to the varied demands of the market. This phase fully harnessed the components present in diverse feedstocks; however, it faced challenges in enhancing production efficiency while simultaneously mitigating the environmental impacts of the production process. Additionally, the focus of this generation was on augmenting the economic value linked to the production of high-value products (Moretti et al. 2018). Third-generation biorefineries, employing comparatively cleaner treatment methods for biomass feedstocks, were categorized into centralized and distributed systems. In general, centralized systems exhibited lower environmental impacts across various categories compared to distributed systems (Kim and Dale, 2015). Advancing beyond the third-generation, fourth-generation biorefineries emphasizes the economic feasibility and environmental benefits linked to extensive microalgae cultivation. The objective is to attain an optimal equilibrium between economic efficiency and environmental advantages through microalgae cultivation (Nizami et al. 2017).

15.2.1 Classification of Feedstock

Thermochemical treatment employs heat to transform diverse feedstocks, including waste materials, into valuable commodities like energy, fuels, or chemicals. The classification of different feedstocks for thermochemical treatment depends on their composition, characteristics, and potential for conversion.

- **Biomass:** Biomass feedstocks are derived from organic materials of plant or animal origin. The concept of biomass encompasses all renewable plant-derived organic materials. This covers a broad spectrum of sources, including energy crops, trees, crops for both food and feed, agricultural leftovers, aquatic plants, animal wastes, municipal waste, and an array of other discarded materials (Tchapda and Pisupati, 2014, U.S. Energy Information Administration (EIA), 2013).
- **Municipal Solid Waste (MSW):** MSW is a heterogeneous mix encompassing kitchen and household items, plastics, rubber, metals, paper products, cardboard, inert materials, electronic waste, and more. The categorization of MSW involves separating it into

biodegradable and non-biodegradable fractions. Interestingly, only around 15% of waste generated in municipalities goes through recycling processes, while the majority finds its way to landfills or open disposal sites (Ouda et al. 2013).

- **Plastic waste:** Plastics have become integral to the modern world because they are lightweight, durable, energy-efficient, and have a rapid production rate and design flexibility. These materials are integral to a diverse range of industrial and domestic sectors, demonstrating their indispensability with continually broadening applications in various industries. However, the substantial volumes of plastic waste present a significant environmental challenge, primarily due to the challenges associated with their disposal (Guangyu et al., 2009; Xiangxue et al. 2012).
- **Tires and Rubber:** Used tires and rubber materials can be subjected to thermochemical processes to recover energy or produce valuable products like synthetic fuels. One crucial source of raw materials for the production of new rubber products is reclaimed rubber obtained from waste tires. Typically, powdered rubber is utilized in the manufacturing of various rubber products such as tires, tubes, furniture, and carpets. The reclamation of crumb rubber has garnered increased attention, particularly for its notable performance in civil engineering applications (Pacheco et al., 2012).
- **Waste Cooking Oil**: Used cooking oil from restaurants or households can be a feedstock for thermochemical processes. Waste cooking oil (WCO) is considered hazardous waste originating from the use of vegetable oils in cooking or frying processes. WCO has the potential for recycling into various valuable products, including biodiesel, lubricants, fuel oils, raw materials, and scrap metal.
- **Agricultural Residues:** Residual agricultural materials like straw, corn stover, and sugarcane bagasse are frequently utilized as raw materials for thermochemical treatments. Agricultural wastes encompass the by-products generated during the harvesting and processing of plant-based crops. These remnants are typically categorized into field residues and residues from agro-food processing activities. Notably, these agricultural remnants, being non-edible, avoid competition with food supplies or fertile arable lands. Despite the considerable environmental, societal, and economic advantages of harnessing woody biomass for diverse valuable products in household, industrial, and construction as well as power sectors, a substantial portion of forestry biomass remains largely unexplored (Singh et al. 2021).
- **Animal Manure:** Derived from livestock and poultry farming, animal manure consists of metabolic and waste by-products. Rich in organic matter and essential nutrients crucial for crop growth, this substance also presents potential ecological risks owing to potential pathogen content. Components of animal manure typically encompass metabolic waste, feces, waste feed, and residual feedwater. While serving as a crucial agricultural nutrient source, livestock manure can also lead to the release of greenhouse gases, like methane (CH4), via microbial decomposition (Nguyen et al. 2020).
- **Wastewater Sludge:** Sewage sludge, an unavoidable by-product produced by municipal wastewater treatment plants, poses a significant challenge in numerous countries due to its growing volume and the consequential impacts of its disposal. This semi-solid residual substance emerges from the settling of suspended solids during the wastewater treatment procedures. There are two main categories of sewage sludge: primary sludge, formed through gravitational settling that captures suspended solids and organic matter, and secondary sludge, which arises as microorganisms consume organic substances (Milieu Ltd, WRc, Risk and Policy Analysts Ltd (RPA), 2010).
- **Industrial Waste:** A substantial volume of industrial waste is generated globally each year. The disposal of such a significant quantity of waste on common land poses numerous environmental hazards. Utilizing these industrial wastes in water remediation offers an environmentally friendly, cost-effective, and efficient approach for their repurposing.

- **Algae:** There is a rising global interest in exploring the potential of algae for renewable fuel, food, and materials production. Algae, a diverse collection of single-celled and multicellular organisms that undergo photosynthesis, exhibit higher photosynthetic efficiency compared to other biomass sources. Notably, they can be cultivated in uncomplicated open saline ponds, eliminating competition for arable land in their production (Wang et al. 2008).
- **Coal and Lignite:** Traditional fossil fuels like coal can also be subjected to thermochemical processes. The global coal reserves remain extensive, characterized by low cost and high energy density. Coal is anticipated to play a significant role in meeting future energy demands in numerous nations, particularly in rapidly developing countries like China and India (Balat 2007).
- **Construction and Demolition Waste:** Materials from construction and demolition activities, such as wood, concrete, and asphalt, can be treated thermo-chemically to recover energy or valuable components.

The selection of feedstock relies on considerations like local accessibility, economic viability, and the desired final products. Thermochemical processes commonly used include pyrolysis, gasification, and liquefaction, each with its own set of advantages and limitations depending on the feedstock and intended output. The goal is often to maximize the energy recovery or the production of valuable chemicals while minimizing environmental impact.

15.3 CONVERSION TECHNOLOGY

Biorefineries employ a variety of conversion technologies to transform biomass into different products. The conversion of raw biomass into usable energy requires many processing steps, utilizing the three primary process technologies: bio-chemical, thermo-chemical, and physio-chemical. The two main methods of biochemical conversion are fermentation (which produces ethanol) and anaerobic digestion (which produces biogas). The four primary process choices for the thermo-chemical conversion methods that are discussed here are hydrothermal processing, gasification, combustion, and pyrolysis. In physio-chemical conversion, extraction stands as the key process, involving the crushing of oilseeds to extract oil and subsequent esterification.

15.3.1 Thermochemical Conversion

Biomass is a sustainable, clean, and green resource that can be used to make fuels to meet energy demands. Yet, there are many drawbacks to using biomass directly as biofuels, including low calorific value, unfavorable moisture content, anomalous composition, and characteristics (Bisen et al. 2025a). By enhancing the operating conditions, thermochemical techniques provide methods that significantly reduce undesirable by-products and mitigate these drawbacks. Thermochemical processes like combustion, gasification, pyrolysis, hydrothermal carbonization, and torrefaction help produce biofuels from both lignocellulosic and non-lignocellulosic materials (Tortosa-Masiá et al., 2007). Figure 15.2 shows the thermochemical conversion flow chart.

15.3.1.1 Torrefaction

Torrefaction, an intriguing thermochemical process commonly utilized as a thermal pre-treatment for biomass resources (Ong et al. 2020), improves the effectiveness of thermochemical conversion techniques like pyrolysis, liquefaction, and gasification by eliminating moisture and partially breaking down the biomass (Sarker et al., 2021). Through reducing oxygen levels and increasing carbon content, torrefaction enhances the biomass's capacity to generate heat. The objective of torrefaction is to produce materials with high hydrophobicity and solid fuel as a substitute for coal (Christoforou and Fokaides, 2018). Within this category, torrefaction is described as the thermal pre-treatment of biomass in an oxygen-free environment at temperatures ranging from 200 to 300°C, and it is

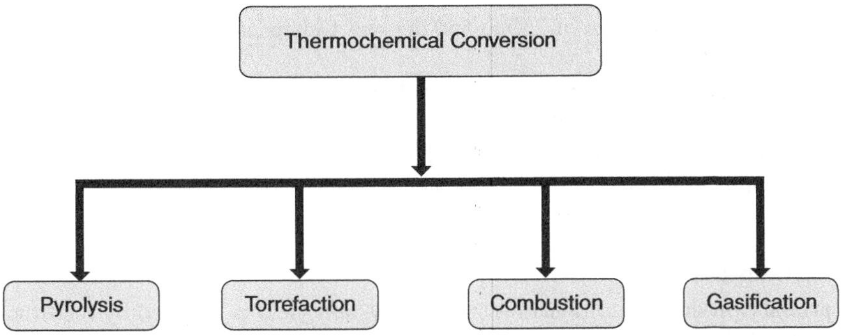

FIGURE 15.2 Flowchart of the thermochemical conversion process.

categorized into three severities: mild (235–275°C), moderate (255–275°C), and severe (275–300°C) torrefaction temperatures (Basu et al. 2014). Torrefaction plays a crucial role in thermochemical conversion by improving the energy content, stability, and handling characteristics of biomass. It aids in the effective harnessing of biomass resources for sustainable energy production and assists in tackling challenges related to the variability of biomass feedstocks in various conversion processes.

15.3.1.2 Pyrolysis

Pyrolysis involves breaking down biomass at high temperatures without oxygen, producing bio-oil, charcoal, and pyro gases like H_2, CH_4, CO, and CO_2. The bio-oil has two parts: an organic phase containing tar and dense hydrocarbons, which can be refined into cleaner fuels for transportation. However, to enhance its quality, bio-oil needs upgrading methods, both catalytic and non-catalytic, to eliminate oxygen, nitrogen, and sulfur compounds. These impurities not only decrease the fuel's energy value but can also lead to emissions like NOx and SOx during combustion (Chen et al. 2021). Pyrolysis methods are categorized into three primary groups: slow, rapid, and flash pyrolysis. Beyond these, additional techniques such as microwave pyrolysis, catalytic pyrolysis, catalytic hydro-pyrolysis, and hydrous pyrolysis, along with hydro-pyrolysis are also employed, expanding the spectrum of biomass conversion methods available for various applications (Bisen et al. 2025b).

15.3.1.3 Hydrothermal Liquefaction

Hydrothermal liquefaction (HTL) is a method where biomass undergoes liquefaction in high-temperature, high-pressure conditions, using water as a medium. Unlike pyrolysis, which breaks down organic materials without oxygen, HTL involves the presence of water, enabling the conversion of biomass into liquid form through this thermochemical process. In hydrothermal liquefaction, however, water is used as a reaction medium, distinguishing it from traditional pyrolysis. Usually, it

TABLE 15.1
Common Operational Parameters for Slow, Intermediate, and Fast Pyrolysis

Process	Temperature (°C)	Heating Rate (°C/min)	Vapour Residence Time	Bio-Oil (wt.%)	Bio-Char (wt.%)	Pyro Gas (wt.%)
Slow-pyrolysis	300–500	<50	>30 min	25–50	25–35	20–50
Intermediate pyrolysis	400–600	200–300	10 min	35–50	30–40	20–30
Fast-pyrolysis	400–900	10–1000	<2 s	55–75	10–25	10–30

(Sajdak and Słowik, 2014)

operates within a temperature span of 250 to 550 degrees Celsius and under pressures ranging from 5 to 25 megapascals (MPa).

During hydrothermal liquefaction, biomass is subjected to high temperature and pressure conditions in the presence of water, leading to the breakdown of complex organic compounds into simpler, liquid hydrocarbons. The resulting bio-oil serves multiple purposes; it can function as a renewable fuel source or serve as a raw material for various chemical processes (Das and Ganesh, 2003).

15.3.1.4 Combustion

Heat is the primary product of combustion technology, and yield is affected by feedstock, gasifier, and temperature (750–1500°C). Combustion can be used to remediate almost all solid wastes. In addition to producing energy, combustion is utilized in industry for drying, pre-heating, and steam generation. Carbon-based resources, such as coal, are widely employed in energy generation, releasing significant amounts of CO_2 into the atmosphere (Shahbaz et al. 2021). The process of completely converting carbon- and hydrogen-rich biomass into CO_2 and H_2O, generating heat, is termed full combustion. According to the International Energy Agency, coal is responsible for roughly 44% of CO_2 emissions, which fall within the range of 0.34 to 0.39 kilograms of CO_2 per kilowatt-hour ($kgCO_2/kWh$). The use of biomass in conjunction with coal could be a viable strategy for reducing total CO_2 emissions (Datar et al., 2004). More than 70% of all heat production comes from the burning of volatile gases. It occurs above the fuel bed, and yellow flames are typically present to indicate it. Little blue flames are indicative of char combustion in the fuel bed (Milne et al. 1998).

15.3.1.5 Gasification

Gasification involves the controlled, exothermic partial oxidation of biomass, optimizing process conditions to yield gaseous products abundant in CO, H_2, CH_4, and CO_2. This resulting gas can undergo purification for direct use as engine fuel or can be biologically fermented into liquid fuels or chemical building blocks. However, a challenge in gasification lies in handling higher molecular weight volatiles that condense into tars. These tars pose problems by fouling equipment and potentially creating long-term environmental hazards, such as polycyclic aromatic hydrocarbons (Schaleger et al. 1982). Additionally, by adjusting the gasification process's temperature, pressure, reaction time, reaction medium, feedstock concentration, and equivalence ratio along with catalyst loading, it is possible to increase the gasification process's efficiency as well as the rate at which carbon is transformed to hydrogen, as well as the selectivity of the hydrogen produced, and the calorific value of the syngas (Goudriaan et al. 2008).

15.3.2 Biological Conversion

Biological conversion technologies encompass a diverse set of processes that leverage biological agents, such as microorganisms or enzymes, to transform raw materials into valuable products. These technologies play a pivotal role in sustainable practices, offering eco-friendly alternatives to traditional industrial processes. One prominent aspect of biological conversion is microbial fermentation, a well-established method for converting organic materials into various products, including biofuels, chemicals, and pharmaceuticals (Brunori 2015). Microorganisms, such as bacteria, yeast, and fungi, serve as efficient biocatalysts in fermentation processes (Lynd et al. 2002). This technology not only diminishes dependence on fossil fuels but also aids in mitigating greenhouse gas emissions.

15.3.2.1 The Macroalgal Biorefinery Concept

Establishing a biorefinery that adheres to a comprehensive zero-waste approach is a crucial prerequisite for aligning with the key principles outlined in the European Commission's bio-economy framework (Lee and Kim, 2015). There is an urgent requirement to minimize waste and improve the economic viability of a bioprocess through the application of sustainable biotechnologies, enabling the extraction of multiple products from a single biomass source. Common principles in bioprocess design advocate prioritizing the extraction of higher-value products before those of lesser value. This involves recovering bioactive compounds, functional elements, sugars, and minerals. Any left-over residues should then be repurposed as feedstock for producing biofuels or platform chemicals, effectively minimizing waste (Cesario et al. 2018).

Despite the abundance of untapped biochemical constituents in macroalgae, several challenges must be addressed to realize the full potential of macroalgal biorefineries. These challenges encompass identifying the spectrum of high-value bio-products (in terms of quantities and values) obtainable from various macroalgal feedstocks. Equally crucial is assessing the technological and economic feasibility of the employed bioprocess and cascading approach (Prabhu et al. 2020).

In recent years, there has been a recognition among researchers of the need to move beyond singular product-oriented feedstock programs. They are now exploring multifaceted approaches to refine various types of macroalgae, aiming to generate multiple bio-products. An illustrative depiction of this diversified process is showcased in Figure 15.3, outlining a macroalgal biorefinery flowchart. This diagram encompasses both upstream and downstream processing sections, highlighting numerous biotechnological pathways utilized for enhancing the value of the feedstock.

The optimal bioprocessing technique for a specific product is largely contingent on the type of macroalgae utilized as feedstock and its compatibility with the selected bioprocessing technology. In the case of macroalgal biomass, upstream bioprocessing plays a crucial role in shaping the array of end-products downstream. Various cultivation parameters distinctly impact the biochemical composition of macroalgae (Khoo et al. 2019). Furthermore, choosing the right method or a blend of methods is pivotal based on the specific bioproduct being targeted. A notable instance is in the context of bioethanol production (Pablo et al. 2019).

15.3.2.2 Vermicomposting Process

This process involves transforming organic waste into a nutrient-enriched and stabilized form with the assistance of earthworms (Prasad et al. 2019). Conducted within a confined space, such as a bin, the process involves a mixture of solid waste and earthworms maintained under optimal environmental conditions. To enhance the nutrient content in the final compost and expedite the composting duration, various additives like cow manure and fly ash, along with microorganisms like potassium-fixing, nitrogen-fixing, and phosphorus-solubilizing bacteria, are introduced. These

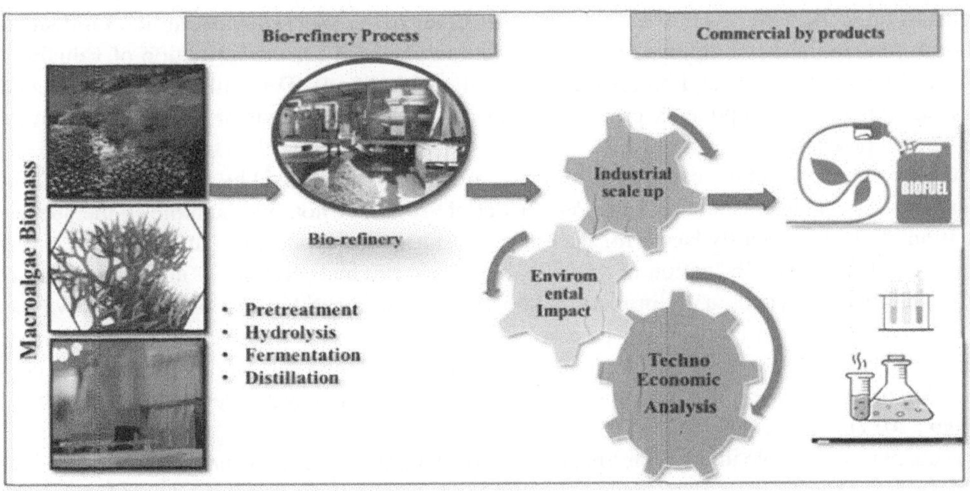

FIGURE 15.3 The process of converting macroalgae biomass into useful products involves several stages.

components serve to foster the growth of earthworms, ultimately leading to improved nutrient levels in the resulting compost, while simultaneously reducing the carbon-to-nitrogen (C:N) ratio and the overall composting time (Busato et al. 2012, Das et al. 2016, Malińska et al., 2017, Waqas et al. 2023).

Microbes and earthworms play a vital role in the degradation process by secreting a diverse array of digestive enzymes. These enzymes contribute to the breakdown of organic materials, facilitating the decomposition of the waste into simpler and more bioavailable components. The collaborative action of these microbial and earthworm activities helps in the efficient breakdown of complex organic substances, promoting the overall effectiveness of the composting or waste conversion process (Medina-sauza et al. 2019). Furthermore, the symbiotic relationship between earthworms and microbes fosters frequent interactions with the substrate, enhancing biochemical reactions. Through this collaboration, essential nutritive elements present in the waste materials are transformed into soluble forms, ultimately contributing to an improvement in the quality of vermicompost. The dynamic interplay between earthworms and microbes enhances the overall efficiency of the vermicomposting process by making key nutrients more readily available in soluble forms for plant uptake and soil enrichment (Garg and Gupta, 2009; Yuvaraj et al., 2021). This technique is cost-effective and environmentally friendly and has demonstrated successful applications in the decomposition of various waste types, including food waste, agricultural residues, and industrial waste. The process transforms these materials into biofertilizers, enhancing plant growth. Additionally, the technique exhibits the capability to remediate heavy metals from solid waste-contaminated sites. Its multifaceted benefits make it a sustainable and efficient approach to waste management while simultaneously contributing to environmental remediation and agricultural productivity (Alshehrei and Ameen, 2021). Figure 15.4 shows vermicomposting; organic waste is transformed into nutrient-rich compost through the activity of earthworms, illustrating a sustainable and efficient cycle of waste decomposition.

15.3.2.3 Insect-Based Biorefinery

This emerging technology utilizes insects to digest solid waste with varied organic compositions, aiming at recovering bio-products and managing waste effectively. Insect biorefinery is an innovative approach to waste management and resource generation, using the capacity of insects to decompose various organic waste materials. Insect-based feedstocks possess a short life cycle, need less land and water, emit fewer greenhouse emissions, and have superior feed conversion efficiency relative to traditional cattle (Kee et al. 2023).

Various insects, such as houseflies, grasshoppers, and black soldier flies, have been identified as efficient agents for the beneficial conversion of organic wastes into valuable sources of fats and proteins through the generation of larvae (Surendra et al. 2015). Nevertheless, agricultural and food wastes are predominantly employed as sustainable raw material for the formation of valuable bioproducts (like animal feeds and biofertilizers as well as biodiesel). This utilization is facilitated by the diverse chemical components present in these wastes, with insects serving as bioconversion agents in the process.

The growth and metabolic capacities of insects are heavily influenced by the nutritional elements present in the organic waste substrate (Peguero et al. 2022). A notable example is the successful utilization of black soldier fly larvae for the bioconversion of fruit waste, restaurant scraps, and dairy manure, resulting in the formation of valuable bio-based products (Leong et al. 2016, Wang et al. 2017). This process yields a by-product called frass from harvested larvae, which serves as an organic bio-fertilizer. In comparison to conventional methods, insect-based bioconversion offers advantages such as a shorter life cycle, lower greenhouse gas emissions, and enhanced feed conversion efficiency. Moreover, this technique demands less land and water resources (Van Huis and Oonincx, 2017).

Insects play a vital role in the biorefinery model by generating a diverse array of products, including biofuels, enzymes, polymers, resins, animal feed, and organic fertilizer. In addition to these

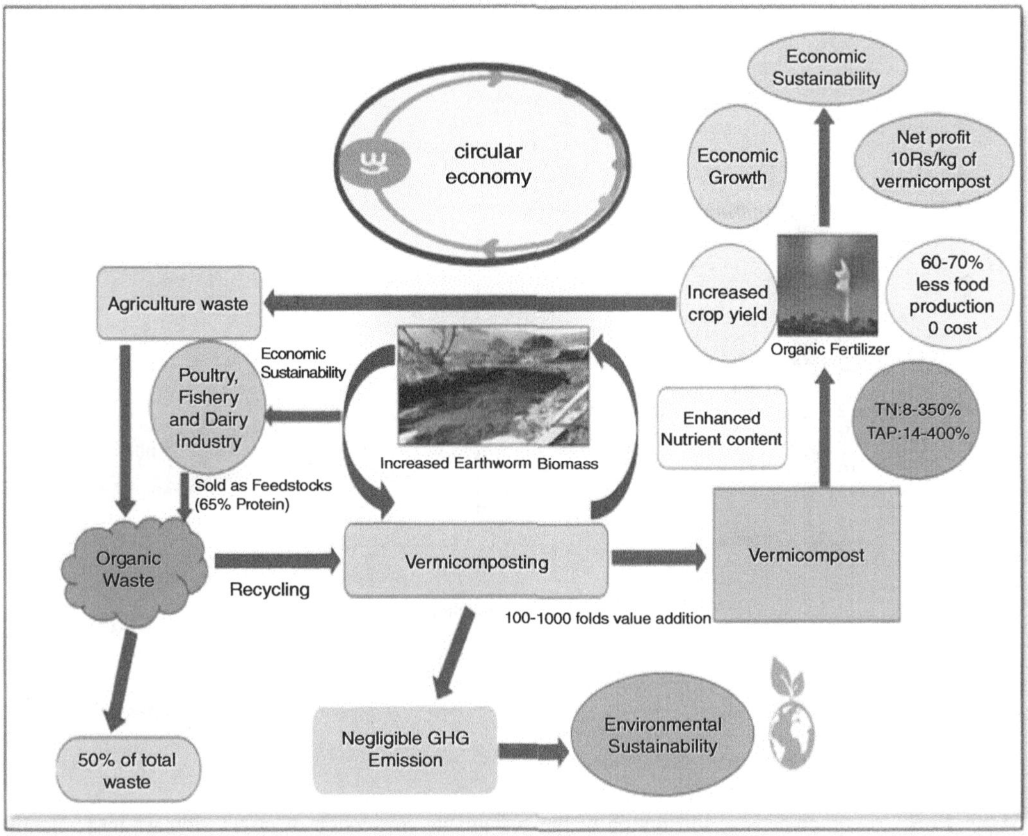

FIGURE 15.4 Organic waste transformed into nutrient-rich compost through the activity of earthworms in vermicomposting, illustrating a sustainable and efficient cycle of waste decomposition.

outputs, insects contribute significantly to the valorization of organic waste. Figure 15.5 provides a schematic representation of the concept of an insect-based biorefinery for the formation of biofuels and bio-based products. This section provides a summary of the various products that can be obtained through the adoption of an insect-based biorefinery (Bhaskar et al., 2018).

15.3.3 Anaerobic Digestion

This process entails the conversion of complex organic solid waste into carbon dioxide (CO_2) and biogas in the absence of oxygen, facilitated by a diverse community of microorganisms (Yasin et al. 2019). The optimal metabolism of these organisms relies on maintaining critical factors such as pH, retention time, volatile fatty acids, C:N ratio, temperature, total solids, particle size, and loading rate at their ideal levels (Varjani et al. 2021). The method is adaptable to various reactor configurations, including continuous, semi-continuous, or discontinuous, within both single-stage and two-stage anaerobic digestion systems (Franca and Bassin, 2020).

The single-stage digestion system comprises four phases: hydrolysis, acidogenesis, acetogenesis, and methanogenesis (Ren et al. 2018). In the initial hydrolysis phase, complex organic matter breaks down into simpler, soluble forms (such as amino acids, sugars, and fatty acids) facilitated by hydrolytic enzymes (Kumar and Samadder, 2020). After hydrolysis, acidogenesis involves further

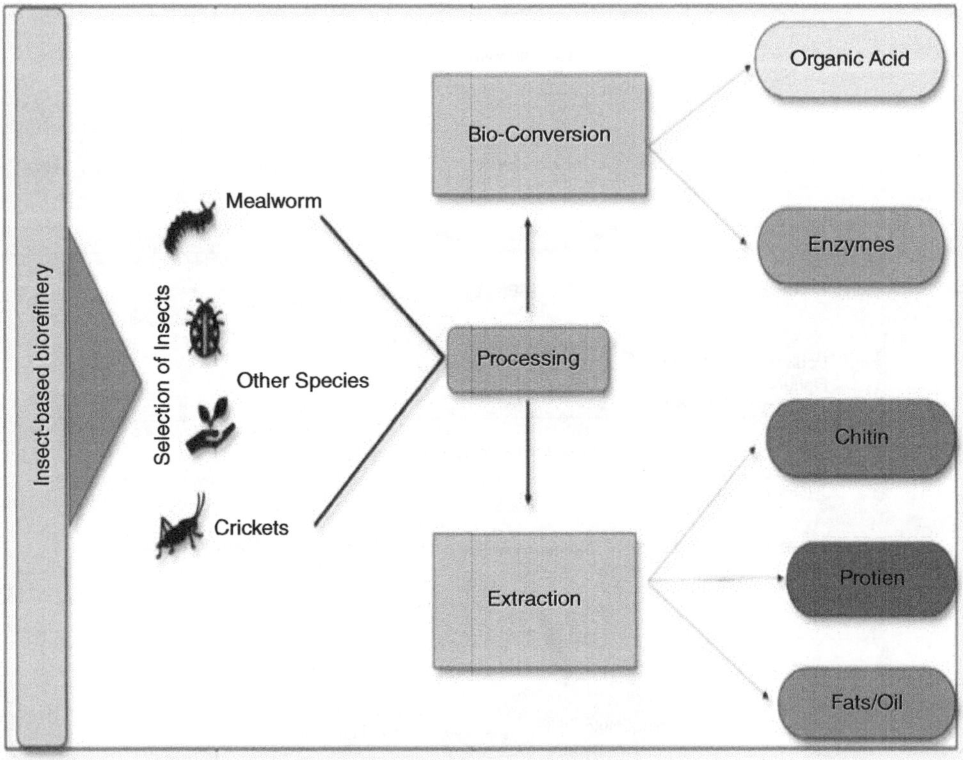

FIGURE 15.5 A comprehensive illustration of the sustainable conversion of insect biomass into valuable products such as protein, oil, and chitin.

degradation or fermentation of basic organic compounds, resulting in the production of volatile fatty acids, CO_2, hydrogen, and other bioproducts.

Anaerobic digestion is employed to extract value from diverse waste streams, encompassing household waste, livestock waste, food waste, and industrial waste (Dehkordi et al. 2020). This method efficiently diminishes waste materials and concurrently harnesses energy in the shape of biofuels derived from biomass. Noteworthy is the fact that agricultural and livestock wastes, renowned for their elevated organic matter content, are especially well-suited for the generation of biogas through anaerobic digestion (Fan et al. 2018).

The biogas composition is contingent on the specific biomass utilized. For instance, if plant food waste is chosen as the biomass source, biogas production typically comprises around 60% methane, complemented by carbon dioxide (CO_2) (Moustakas et al. 2020). This resulting biogas serves as a sustainable energy reservoir, catering to diverse needs like electricity generation, fuel, and heat production. Furthermore, the by-product (digestate) derived from the anaerobic digestion of organic solid waste is rich in nutrients and serves as a valuable biofertilizer (Rajendran et al., 2021; Slorach et al. 2019). Despite its simplicity, cost-effectiveness, and minimal emissions of gaseous pollutants, this method encounters notable drawbacks such as a relatively slow and limited processing capacity. Additionally, the necessity for substrate pretreatment and the challenge of maintaining a specific temperature range emerge as critical constraints that hinder its widespread applicability (Khan 2020). Figure 15.6 has shown a visual representation of the efficient transformation of organic waste into biogas and nutrient-rich digestate through microbial fermentation in an oxygen-free environment.

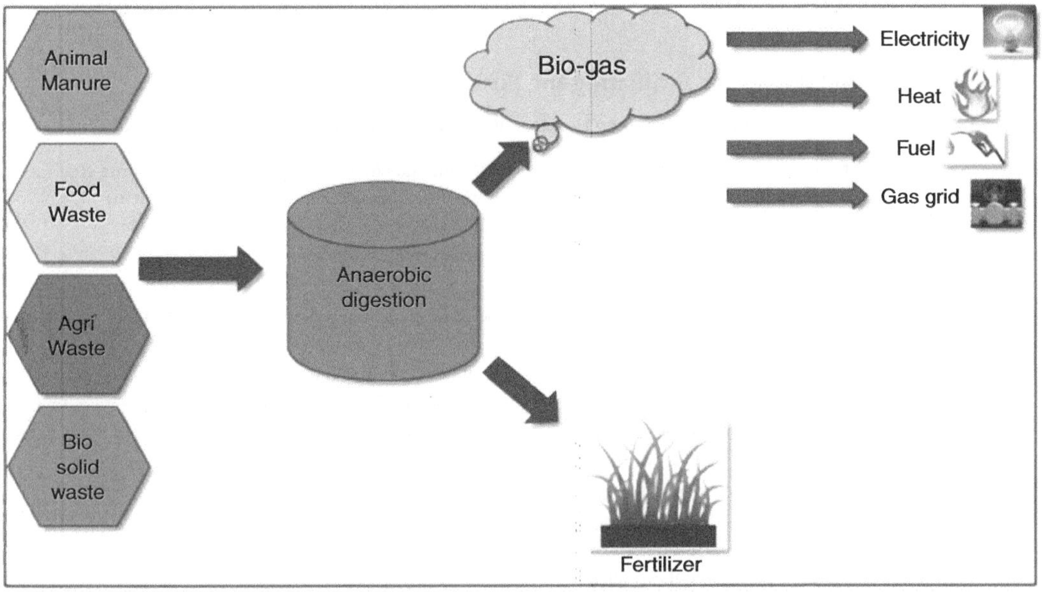

FIGURE 15.6 A visual representation of the efficient transformation of organic waste into biogas and nutrient-rich digestate through microbial fermentation in an oxygen-free environment.

15.4 SUSTAINABILITY ASSESSMENT OF BIOREFINERIES

Traditionally, evaluating the performance of a biorefinery has involved measuring its economic value through methods like net present value and other temporally adjusted approaches. Furthermore, environmental assessments have relied on life cycle assessment (LCA). However, a common trend in many studies is an exclusive focus on either economic or environmental criteria, often disregarding the potential synergies of multiple factors (Tuazon and Gnansounou, 2017). To fill this void, a thorough evaluation of biorefineries' sustainability can be conducted using Tuazon's triple-bottom-line framework. This framework systematically considers environmental, economic, and social dimensions, providing a holistic view of the overall sustainability of biorefinery operations (Liew et al. 2014). In a relevant study, the application of the triple bottom-line framework to assess the sustainable production of biofuel revealed its primary favorability, attributable to positive impacts on greenhouse gas (GHG) emissions and beneficial outcomes for rural communities. Despite these advantages, obstacles like operational and investment costs, as well as limitations in feedstock availability, pose challenges to the commercialization of biofuel production. Table 15.2 provides an overview of diverse methods employed for evaluating the sustainability of various biorefineries, along with accompanying observations on circular economy. The following sections will delve into a review of the methodologies applied in the sustainable assessment of biorefineries.

15.5 STRENGTHS AND WEAKNESSES OF BIOREFINERY

Biorefineries are essential enabling technologies for the successful and widespread implementation of the bioeconomy concept. IEA Task 42 establishes the definition of a biorefinery. Biorefinery is the sustainable conversion of biomass into a variety of energy sources and marketable products. A biorefinery employs a variety of processes, including mechanical, biological, chemical, and thermochemical, to convert biomass into various end products, including fertilizers, chemical building blocks, polymers, resins, food, animal feed, biomaterials, and energy (Gaffey et al. 2024).

TABLE 15.2
The Various Methodologies for Evaluating the Sustainability of a Biorefinery

Type of Biorefinery	Sustainability Performance Factor	Remarks on Circular Bioeconomy	References
Algal (Seaweed)	Considerations related to sustainability encompass aspects such as climate change impact, cumulative energy demand, marine eutrophication, and human toxicity.	Restoring the environment and mitigating climate change play pivotal roles in fostering a regenerative circular economy.	Seghetta et al. (2016)
Lignocellulosic (Dry wood)	Eco-efficiency indicator	The utilization of alternative resources has a positive influence on both the circular economy and environmental sustainability.	Bello et al. (2018)
Sugar	The examination of sustainability in various contexts involves analyzing factors such as fossil fuel consumption, carbon footprint, water usage, agro-industrial yield, human development, production costs, and the diversification of products.	The shift toward a circular economy signifies a paradigm change that inspires the reconfiguration of biorefineries, fostering the development and diversification of value-added green products.	García-Bustamante et al. (2018)
Waste (Food)	Key considerations in evaluating environmental impact include factors such as energy consumption, nutrient release (associated with eutrophication and acidification), and the payback period.	The economic viability of these processes is significantly dependent on the prevailing product prices.	Capson-Tojo et al. (2016)
Waste (Olive mill solid waste)	Net present value, internal rate of return, and payback period	In the circular economy, essential facets include the valuation of products and the reduction of pollution through waste management.	Serrano et al. (2017)
Waste (Wastewater)	Cost of production, climate change, resources, ecosystem quality, and human health are integral components in evaluating the overall sustainability and impact of various processes or systems.	The direct conversion of waste into fuels and chemicals contributes significantly to achieving a circular economy.	Shemfe et al. (2018)

Many studies focused solely on analyzing a single criterion, economic or environmental, neglecting the potential synergies of multiple criteria. The evaluation of biorefinery sustainability can be conducted by employing the triple bottom-line framework introduced by Tuazon (Tuazon et al. 2013). This framework encompasses the environmental, economic, and social aspects. Azapagic (2014) conducted a thorough review and analysis of the sustainability issues associated with integrated biorefineries. This review provided a comprehensive examination of both the strengths and weaknesses of integrated biorefineries by the triple bottom-line framework (Azapagic, 2014). Table 15.3 presents a compilation of diverse methods employed to evaluate the sustainability of various biorefineries. Additionally, it includes corresponding comments on their relevance to the circular economy. The following sections provide a review of the methodologies utilized in the sustainable assessment of biorefineries.

TABLE 15.3
The Strengths and Weaknesses of the Case Studies in Biorefineries within the Circular Economy

Type of Biorefinery	Sustainability Assessment Tool Used	Biorefinery Strengths	Biorefinery Weaknesses	References
Lignocellulosic (Dry wood)	Life cycle (eco-efficiency) assessment	An indicator measuring sustainability in terms of financial gains enables the blending of economic advancement and environmental safeguarding.	Challenges related to the pretreatment of biomass and the energy demands of the processing procedure.	Liew et al. (2014)
Lignocellulosic (Sugar)	Analytic hierarchy process	Shows favorable socio-economic effects through product diversification and increased profitability.	Requires more suitable approaches to assess the overall sustainability level of the entire system.	Seghetta et al. (2016)
Lignocellulosic/ Waste	Techno-economic analysis, and life cycle assessment	Integrating first and second-generation ethanol plants helps minimize investment risks and environmental impacts in comparison to standalone systems.	The economic viability relies on the pricing of co-products.	Ferreira et al. (2018)
Waste (Food)	Life cycle assessment	Sequentially employing resources through multiple stages of reuse, recycling, and bioenergy conversion.	Inconsistencies in material quality and insufficient exploration at the industrial scale.	Corrado and Sala (2018)
Waste (Municipal solid waste)	Economic value analysis and life cycle assessment	Generates wealth through the valorization of waste, with the potential to achieve sustainability independently.	Impurities in the feedstock.	Sadhukhan et al. (2016)
Waste (Slag)	Environmental and social sustainability assessment	An intentional investment in a biorefinery is expected to have a notably positive impact on the local and regional economy.	Approaches to managing and reporting sustainability, monitoring greenhouse gas emissions, adopting life-cycle thinking, and promoting awareness of circularity.	Husgafvel et al. (2017)
Waste (Wastewater)	Dynamic simulation, life cycle assessment, and techno-economic assessment	In addition to pollution control through wastewater treatment, it provides diverse advantages, including the synthesis of novel products and the implementation of processes free from fossil fuels.	These could involve complexities in accurately modeling dynamic system behavior, assessing the entire life cycle impact.	Shemfe et al. (2018)

15.6 ENVIRONMENTAL IMPACT AND FOOTPRINT ASSESSMENT

In biorefining, the objective is to extract maximum value from the biomass feedstock, minimizing waste and enhancing economic efficiency. This approach not only positions these processes as potential competitors to the petrochemical industry but also contributes to a decreased overall environmental footprint. The integration of various production methods is anticipated to streamline the overall production process, leading to notable reductions in both product costs and environmental impacts. Despite using non–fossil-based resources as feedstock, biorefineries still have difficulties with environmental sustainability. Although it is commonly believed that biorefineries have a beneficial environmental impact, this may not always be the case from a systems approach perspective (Saraiva, 2017). Relatively speaking, the goal of biorefineries is to create fewer carbon molecules than fossil fuel–based refineries. Developing novel, scalable, and flexible biobased production processes and technologies increases the complexity of the biorefinery process, which may ultimately lead to increased uncertainty (Byun and Han, 2020). This may result in a process that is no longer sustainable, which would have a greater financial and environmental cost. Choice of feedstock, manufacturing, process integration, and end-of-life use scenarios might potentially kickstart sustainable enterprises (Hottle et al. 2013). In this sense, biorefinery systems require a systematic process for identifying effects and evaluating their relative sustainability (Ögmundarson et al., 2020).

15.7 WASTE BIO-REFINERY PROMOTING A CIRCULAR ECONOMY

Petroleum or fossil fuel has long served as a crucial natural resource, serving as the primary feedstock for energy production (such as transportation fuels) and the synthesis of materials like plastics and chemicals. Despite their historical importance, these resources are limited and pose environmental hazards, actively contributing to climate change through greenhouse gas emissions, especially carbon dioxide (CO_2) into the atmosphere. The associated environmental concerns have heightened global awareness, leading to extensive research efforts focused on carbon mitigation and adaptation (NASA, 2019). Moving from a petroleum refining model to a waste biorefining model suggests significant work being done on GHG mitigation and carbon management. By transitioning from a linear economy based on the take, make, and dispose premise to a circular economy that is sustainable and circular, waste biorefinery helps to retain and recycle resources (Ferreira et al. 2019).

The use of waste materials, primarily municipal solid and liquid waste, in bioprocesses to create value-added bioproducts and metabolites considered waste-to-treasure has gained attention due to the renewable nature of the products and their environmentally benign biodegradability. The bioprocessing of waste biorefinery on the production of biopolymers and bioenergy not only solves environmental security and energy security issues but also indicates improved waste stream management. Due to the platform's cheap cost and sustainable manufacturing feedstock, it is both economically and environmentally sound (Mohan et al. 2016).

Numerous environmental issues, including global warming, water and environmental contamination, waste disposal, and natural resource depletion, can then be addressed. For example, the development of bioplastics or biopolymers that significantly replace traditional petrochemical plastics can assist in reducing plastic pollution, which has negative effects on soil and marine ecosystems (Chae and An, 2018; Derraik, 2002). Bioenergy and biofuels have been developed from microorganisms to achieve energy security, and their production is not dependent on petroleum feedstock. Biofuels for different applications (e.g., transportation fuels) can be used to reduce GHGs and control carbon. A cost-effective method of producing biofuels (or microbial lipid synthesis) can then be achieved by bioprocessing with a waste biorefinery (Brennan and Owende, 2010; Ragauskas et al., 2006). Figure 15.7 shows a schematic view on the outline of the structure of the biorefinery circular economy approach.

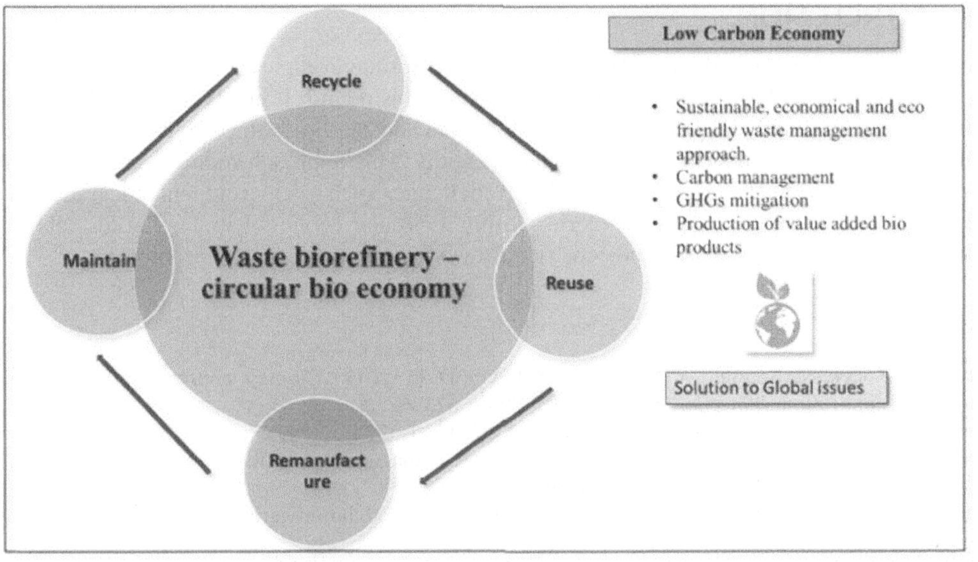

FIGURE 15.7 Schematic view on the outline of the structure of biorefinery circular economy approach.

15.8 CHALLENGES AND OPPORTUNITIES IN BIOREFINERY

The existing state of biorefineries faces several significant challenges that impact their integration into the current market dominated by fossil-based practices. Biorefineries face a multitude of challenges encompassing market acceptance, feedstock availability, and composition, ensuring volumes align with market demands, resource recovery efficiency, techno-economic feasibility, and environmental sustainability. The production chain of biorefineries is marked by an array of technical, ecological, economic, societal, and sustainable challenges. These challenges traverse various activities such as feedstock acquisition (including cultivation, production, harvest, and storage), considerations regarding feedstock costs, processing and pretreatment, valorization processes, implications on land use, impacts on food security, infrastructure and logistics, capital expenditure, the current state of research and development, product quality, market prices, and the progression of biorefining technologies.

Addressing these essential bottlenecks is crucial for the successful commercialization of biorefineries (Azapagic, 2014). Biorefineries categorize themselves based on the raw materials they employ (ranging from biomass and waste to gas), the biocatalysts in play (spanning bacteria, yeast, fungi, and algae), the chosen process or technology (encompassing fermentation, acidogenesis, methanogenesis, photosynthesis, thermochemical processes, and catalysis), or the specific output or iteration they aim for (first, second, third, and fourth generations). However, there's a recognized necessity to establish a clear, universally accepted framework linking the product range to existing markets. Such a system would effectively engage stakeholders and contribute to a deeper comprehension of biorefinery processes within the context of market dynamics (Brown, 2018).

To attain the sustainable exploitation of renewable resources, it is imperative to formulate strategies concentrating on transforming intricate biomolecules into more straightforward products (Alderton, 2020). This entails the strategic integration of unit operations, aiming to enhance resource recovery efficiency while reducing waste generation. The ultimate goal is to establish closed-loop systems that promote a circular and sustainable approach to resource utilization (Venkata Mohan et al., 2020).

15.9 CONCLUSION

Ensuring responsible waste disposal methods is crucial for environmental preservation and human well-being. The waste biorefinery concept stands out in this regard. It involves converting waste or by-products into valuable bioproducts like biopolymers and bio-lipids through bioprocessing, contributing significantly to a sustainable circular economy. This approach reduces carbon footprints and greenhouse gas emissions, addressing global challenges such as environmental degradation and food security. Embracing waste biorefineries within a circular economy strategy holds substantial promise in fostering a sustainable and eco-friendly world, warranting active promotion.

REFERENCES

Agbor, V. B., Cicek, N., Sparling, R., Berlin, A., & Levin, D. B. (2011). Biomass pretreatment: Fundamentals toward application. *Biotechnology Advances*, 29(6), 675–685.

Alderton, G. (2020). Clarifying diet studies. *Science*, 367, 1336–1338.

Alshehrei, F., & Ameen, F. (2021). Vermicomposting: A management tool to mitigate solid waste. *Saudi Journal of Biological Sciences*, 28(6), 3284–3293.

Azapagic, A. (2014). Sustainability considerations for integrated biorefineries. *Trends in Biotechnology*, 32(1), 1–4.

Balat, M. (2007). Influence of coal as an energy source on environmental pollution. *Energy Sources, Part A: Recovery, Utilization, and Environmental Effects*, 29(7), 581–589. doi: 10.1080/15567030701225260

Basu, P., Sadhukhan, A. K., Gupta, P., Rao, S., Dhungana, A., & Acharya, B. (2014). An experimental and theoretical investigation on torrefaction of a large wet wood particle. *Bioresource Technology*, 159, 215–222.

Bello, S., Ríos, C., Feijoo, G., Moreira, M. T. (2018). Comparative evaluation of lignocellulosic biorefinery scenarios under a life-cycle assessment approach. *Biofuels, Bioproducts and Biorefining*, 12(6), 1047–1064.

Bhaskar, T., Pandey, A., Mohan, S. V., Lee, D. J., & Khanal, S. K. (Eds.). (2018). *Waste biorefinery: potential and perspectives*. Elsevier.

Bisen, D., Chouhan, A. P. S., Pant, M., & Chakma, S. (2025b). Advancement of thermochemical conversion and the potential of biomasses for production of clean energy: A review. *Renewable and Sustainable Energy Reviews, 208*, 115016.

Bisen, D., Lanjewar, R., Chouhan, A. P. S., Pant, M., & Chakma, S. (2025a). Catalytic co-pyrolysis of rice husk and high-density polyethylene using dolomite for enhancement of bio-oil production and quality. *Environmental Science and Pollution Research*, 1–19.

Brennan, L., & Owende, P. (2010). Biofuels from microalgae—A review of technologies for production, processing, and extractions of biofuels and co-products. *Renewable and Sustainable Energy Reviews*, 14(2), 557–577.

Brown, T. R. (2018). Price uncertainty, policy, and the economic feasibility of cellulosic biorefineries. *Biofuels, Bioproducts and Biorefining*, 12(2018), 485–496.

Brunori, G. (2015). *Sustainable Agriculture, Forestry, and Fisheries in the Bioeconomy: A Challenge for Europe*. EU Commission.

Busato, J. G., Lima, L. S., Aguiar, N. O., Canellas, L. P., & Olivares, F. L. (2012). Changes in labile phosphorus form during maturation of vermicompost enriched with phosphorus-solubilizing and diazotrophic bacteria. *Bioresource Technology*, 110, 390–395.

Byun, J., & Han, J. (2020). Sustainable development of biorefineries: Integrated assessment method for co-production pathways. *Energy & Environmental Science*, 13(8), 2233–2242.

Capson-Tojo, G., Rouez, M., Crest, M., Steyer, J. P., Delgenès, J. P., Escudié, R. (2016). Food waste valorization via anaerobic processes: A review. *Reviews in Environmental Science and Biotechnology*, 15(3), 499–547.

Cesario, M. T., da Fonseca, M. M. R., Marques, M. M., & de Almeida, M. C. M. (2018). Marine algal carbohydrates as carbon sources for the production of biochemicals and biomaterials. *Biotechnology Advances*, 36, 798–817.

Chae, Y., & An, Y.-J. (2018). Current research trends on plastic pollution and ecological impacts on the soil ecosystem: A review. *Environmental Pollution*, 240, 387–395.

Chen, W. H., Lin, B. J., Lin, Y. Y., Chu, Y. S., Ubando, A. T., Show, P. L., Ong, H. C., Chang, J.-S.,Ho, S.-H., Culaba, A. B., Pétrissans, A., & Pétrissans, M. (2021). Progress in biomass torrefaction: Principles, applications and challenges. *Progress in Energy and Combustion Science*, 82, 100887.

Christoforou, E., & Fokaides, P. (2018). Recent advancements in torrefaction of solid biomass. *Current Sustainable/Renewable Energy Reports*, 5(2), 163–171.

Corrado, S., & Sala, S. (2018). *Designing Sustainable Technologies, Products, and Policies*. Springer International Publishing.

Das, D., Bhattacharyya, P., Ghosh, B. C., & Banik, P. (2016). Bioconversion and biodynamics of Eisenia foetida in different organic wastes through microbially enriched vermiconversion technologies. *Ecological Engineering*, 86, 154–161.

Das, P., & Ganesh, A. (2003). Bio-oil from pyrolysis of cashew nut shell: A near fuel. *Biomass and Bioenergy*, 25(1), 113–117.

Datar, R. P., Shenkman, R. M., Cateni, B. G., Huhnke, R. L., & Lewis, R. S. (2004). Fermentation of biomass-generated producer gas to ethanol. *Biotechnology and Bioengineering*, 86(5), 587–594. doi: 10.1002/bit.20071

Dehkordi, S. M. M. N., Jahromi, A. R. T., Ferdowsi, A., Shumal, M., & Dehnavi, A. (2020). Investigation of biogas production potential from mechanically separated municipal solid waste as an approach for developing countries (case study: Isfahan-Iran). *Renewable and Sustainable Energy Reviews*, 119, 109586.

Derraik, J. G. B. (2002). The pollution of the marine environment by plastic debris: A review. *Marine Pollution Bulletin*, 44(9), 842–852.

Fan, M., Zhang, S., Ye, G., Zhang, H., & Xie, J. (2018). Integrating sugarcane molasses into sequential cellulosic biofuel production based on the SSF process of high solid loading. *Biotechnology for Biofuels*, 11, 1–9.

Ferreira, J. A., Agnihotri, S., & Taherzadeh, M. J. (2019). Waste biorefinery. In M. J. Taherzadeh, K. Bolton, J. Wong, & A. Pandey (Eds.), *Sustainable Resource Recovery and Zero Waste Approaches*. The Netherlands: Elsevier, pp. 35–52.

Ferreira, J. A., Brancoli, P., Agnihotri, S., Bolton, K., Taherzadeh, M. J. (2018). A review of integration strategies of lignocelluloses and other wastes in 1st generation bioethanol processes. *Process Biochemistry*, 75, 173–186.

Franca, L. S., & Bassin, J. P. (2020). The role of dry anaerobic digestion in treating the organic fraction of municipal solid waste: A systematic review. *Biomass and Bioenergy*, 143, 105866.

García-Bustamante, C. A., Aguilar-Rivera, N., Zepeda-Pirrón, M., Armendáriz-Arnez, C. (2018). Development of indicators for the sustainability of the sugar industry. *Environmental and Socio-Economic Studies*, 6(4), 22–38.

Garg, V. K., & Gupta, R. (Eds.) (2009). Vermicomposting of agro-industrial processing waste. In *Biotechnology for agro-industrial residues utilisation: Utilisation of Agro-residues* (pp. 431–456). Dordrecht: Springer Netherlands.

Goudriaan, F., Naber, J. A., Van de Beld, B., Boerefijn, F. R., Van der Wal, S., Bos, G. M., & Zeevalkink, J. A. (2008). Thermal Efficiency of the HTU Process for Biomass Liquefaction. In A.V. Bridgwater (Ed.), *Progress in Thermochemical Biomass Conversion* (pp. 149–162). Netherlands: Wiley.

Gaffey, J., Collins, M. N., & Styles, D. (2024). Review of methodological decisions in life cycle assessment (LCA) of biorefinery systems across feedstock categories.

Guangyu, L., Jian, L., Xiaobo, M., et al. (2009). Pyrolysis of MSW plastics: Technologies and their reactors. *Environmental Engineering*, 27, 383–388.

Heimann, T. (2019). Bioeconomy and SDGs: Does the bioeconomy support the achievement of the SDGs? *Earth's Future*, 7, 43–57.

Hottle, T. A., Bilec, M. M., & Landis, A. E. (2013). Sustainability assessments of bio-based polymers. *Polymer Degradation and Stability*, 98(2013), 1898–1907.

Husgafvel, R., Poikela, K., Honkatukia, J., Dahl, O. (2017). Development and piloting of sustainability assessment metrics for the arctic process industry in Finland: The biorefinery investment and slag processing service cases. *Sustainability (Switzerland)*, 9(10), 1693.

Karp, A., & Shield, I. (2008). Bioenergy from plants and the sustainable yield challenge. *New Phytologist*, 179, 15–32.

Kee, P. E., Cheng, Y. S., Chang, J. S., Yim, H. S., Tan, J. C. Y., Lam, S. S., Lan, J. C., Ng, H. S., & Khoo, K. S. (2023). Insect biorefinery: A circular economy concept for biowaste conversion to value-added products. *Environmental Research*, 221, 115284.

Kellock, M., Maaheimo, H., Marjamaa, K., Rahikainen, J., Zhang, H., Holopainen-Mantila, U., & Kruus, K. (2019). Effect of hydrothermal pretreatment severity on lignin inhibition in enzymatic hydrolysis. *Bioresource Technology*, 280, 303–312.

Khan, I. (2020). Waste to biogas through anaerobic digestion: hydrogen production potential in the developing world case of Bangladesh. *International Journal of Hydrogen Energy*, 45(32), 15951–15962.

Khoo, C. G., Dasan, Y. K., Lam, M. K., & Lee, K. T. (2019). Algae biorefinery: review on a broad spectrum of downstream processes and products. *Bioresource Technology*, 292, 121964.

Kim, S., & Dale, B. E. (2015). Comparing alternative cellulosic biomass biorefining systems: Centralized versus distributed processing systems. *Biomass Bioenergy*, 74, 135–147.

Kumar, A., & Samadder, S. R. (2020). Performance evaluation of anaerobic digestion technology for energy recovery from organic fraction of municipal solid waste: A review. *Energy*, 197, 117253.

Lee, S. Y., & Kim, H. U. (2015). Systems strategies for developing industrial microbial strains. *Nature Biotechnology*, 33(10), 1061–1072.

Leong, S. Y., Kutty, S. R. M., Malakahmad, A., & Tan, C. K. (2016). Feasibility study of biodiesel production using lipids of Hermetia illucens larva fed with organic waste. *Waste Management*, 47, 84–90.

Liew, W. H., Hassim, M. H., Ng, D. K. S. (2014). Review of evolution, technology, and sustainability assessments of biofuel production. *Journal of Cleaner Production*, 71, 11–29.

Lynd, L. R., Weimer, P. J., van Zyl, W. H., & Pretorius, I. S. (2002). Microbial cellulose utilization: Fundamentals and biotechnology. *Microbiology and Molecular Biology Reviews*, 66(3), 506–577.

Malińska, K., Golańska, M., Caceres, R., Rorat, A., Weisser, P., & Ślęzak, E. (2017). Biochar amendment for integrated composting and vermicomposting of sewage sludge–the effect of biochar on the activity of Eisenia fetida and the obtained vermicompost. *Bioresource Technology*, 225, 206–214.

Medina-Sauza, R. M., Álvarez-Jiménez, M., Delhal, A., Reverchon, F., Blouin, M., Guerrero-Analco, J. A., Cerdán, C. R., Guevara, R., Villain, L., & Barois, I. (2019). Earthworms building up soil microbiota, a review. *Frontiers in Environmental Science*, 7, 81.

Milieu Ltd, WRc, Risk and Policy Analysts Ltd (RPA). (2010). Environmental, economic, and social impacts of the use of sewage sludge on land. Final Report, Part III: Project Interim Reports; DG ENV.G.4./ETU/2008/0076r.

Milne, T. A., Abatzoglou, N., & Evans, R. J. (1998). *Biomass Gasifier "Tars": Their Nature, Formation, and Conversion.* Golden, CO: National Renewable Energy Laboratory. doi: 10.2172/3726

Mohan, S. V., Butti, S. K., Amulya, K., Dahiya, S., & Modestra, J. A. (2016). Waste biorefinery: A new paradigm for a sustainable bioelectro economy. *Trends in Biotechnology*, 34(11), 852–855.

Moretti, M., Van Dael, M., Malina, R., & Van Passel, S. (2018). Environmental assessment of waste feedstock mono-dimensional and bio-refinery systems: Combining manure co-digestion and municipal waste anaerobic digestion. *Journal of Cleaner Production*, 171, 954–961.

Moustakas, K., Parmaxidou, P., & Vakalis, S. (2020). Anaerobic digestion for energy production from agricultural biomass waste in Greece: Capacity assessment for the region of Thessaly. *Energy*, 191, 116556.

NASA. (2019). Mitigation and adaptation I solutions—Responding to climate change. Retrieved from https://climate.nasa.gov/solutions/adaptation-mitigation/. Accessed 27 September 2019.

Nguyen, B. T., Trinh, N. N., & Bach, Q. V. (2020). Methane emissions and associated microbial activities from paddy salt-affected soil as influenced by biochar and cow manure addition. *Applied Soil Ecology*, 152, 103531.

Nizami, A. S., Rehan, M., Waqas, M., Naqvi, M., Ouda, O. K. M., Shahzad, K., et al. (2017). Waste biorefineries: Enabling circular economies in developing countries. *Bioresource Technology*, 241, 1101–1117.

Ögmundarson, Ó., Sukumara, S., Herrgård, M. J., & Fantke, P. (2020). Combining environmental and economic performance for bioprocess optimization. *Trends in Biotechnology*, 38(11), 1203–1214.

Ong, H. C., Chen, W. H., Singh, Y., Gan, Y. Y., Chen, C. Y., & Show, P. L. (2020). A state-of-the-art review on thermochemical conversion of biomass for biofuel production: A TG-FTIR approach. *Energy Conversion and Management*, 209, 112634.

Ouda, O. K., Cekirge, H. M., & Raza, S. A. (2013). An assessment of the potential contribution from waste-to-energy facilities to electricity demand in Saudi Arabia. *Energy Conversion and Management*, 75, 402–406.

Pablo, G., Domínguez, E., Domínguez V. D., Romaní, A., Domingues, L., & Garrote, G. (2019). Third-generation bioethanol from invasive macroalgae Sargassum muticum using autohydrolysis pretreatment as the first step of a biorefinery. *Renewable Energy*, 141, 728–735.

Pacheco, E., Visconte, L., Furtado, C., Neto, J. (2012). *Recycling of Rubber: Mechano-Chemical Regeneration.* New York: Advances in Materials Science Research, Nova Science Publishers.

Peguero, D. A., Gold, M., Vandeweyer, D., Zurbrügg, C., & Mathys, A. (2022). A review of pretreatment methods to improve agri-food waste bioconversion by black soldier fly larvae. *Frontiers in Sustainable Food Systems*, 5, 745894.

Prabhu, M. S., Israel, A., Palatnik, R. R., Zilberman, D., & Golberg, A. (2020). Integrated biorefinery process for sustainable fractionation of Ulva ohnoi (Chlorophyta): Process optimization and revenue analysis. *Journal of Applied Phycology*, 32, 2271–2282.

Ragauskas, A. J., Williams, C. K., Davison, B. H., et al. (2006). The path forward for biofuels and biomaterials. *Science*, 311(5760), 484–489.

Rajendran, N., Gurunathan, B., Han, J., Krishna, S., Ananth, A., Venugopal, K., & Priyanka, R. S. (2021). Recent advances in the valorization of organic municipal waste into energy using biorefinery approach, environment, and economic analysis. *Bioresource Technology*, 337, 125498.Ren, Y., Yu, M., Wu, C., Wang, Q., Gao, M., Huang, Q., & Liu, Y. (2018). A comprehensive review on food waste anaerobic digestion: Research updates and tendencies. *Bioresource Technology*, 247, 1069–1076.

Ren, Y., Yu, M., Wu, C., Wang, Q., Gao, M., Huang, Q., & Liu, Y. (2018). A comprehensive review on food waste anaerobic digestion: Research updates and tendencies. *Bioresource Technology*, 247, 1069–1076.

Prasad, M. N. V., de Campos Favas, P. J., Vithanage, M., & Mohan, S. V. (2019). *Industrial and municipal sludge: Emerging concerns and scope for resource recovery*. Elsevier.

Sadhukhan, J., Ng, K. S., Martinez-Hernandez, E. (2016). Novel integrated mechanical biological chemical treatment (MBCT) systems for the production of levulinic acid from a fraction of municipal solid waste: A comprehensive techno-economic analysis. *Bioresource Technology*, 215, 131–143.

Sajdak, M., & Słowik, K. (2014). Use of plastic waste as a fuel in the co-pyrolysis of biomass: Part II. Variance analysis of the co-pyrolysis process. *Journal of Analytical and Applied Pyrolysis*, 109, 152–158. doi: 10.1016/j.jaap.2014.07.001

Saraiva, A. B. (2017). System boundary setting in life cycle assessment of biorefineries: A review. *International Journal of Environmental Science and Technology*, 14(2017), 435–452.

Sarker, T. R., Nanda, S., Dalai, A. K., & Meda, V. (2021). A review of torrefaction technology for upgrading lignocellulosic biomass to solid biofuels. *BioEnergy Research*, 14, 645–669.

Schaleger, L. L., Figueroa, C., & Davis, H. G. (1982). Direct liquefaction of biomass: results from the operation of the continuous bench-scale unit in liquefaction of water slurries of Douglas fir wood. *Biotechnology for Biofuels and Bioproducts*, 12, 3–14.

Seghetta, M., Hou, X., Bastianoni, S., Bjerre, A. B., Thomsen, M. (2016). Life cycle assessment of macroalgal biorefinery for the production of ethanol, proteins, and fertilizers – A step towards a regenerative economy. *Journal of Cleaner Production*, 137, 1158–1169.

Serrano, A., Fermoso, F. G., Alonso-Fariñas, B., Rodríguez-Gutierrez, G., Fernandez Bolaños, J., Borja, R. (2017). Olive mill solid waste biorefinery: High-temperature thermal pre-treatment for phenol recovery and biomethanization. *Journal of Cleaner Production*, 148, 314–323.

Shahbaz, M., AlNouss, A., Ghiat, I., Mckay, G., Mackey, H., Elkhalifa, S., & AlAnsari, T. (2021). A comprehensive review of biomass-based thermochemical conversion technologies integrated with CO2 capture and utilization within BECCS networks. *Resources, Conservation and Recycling*, 173, 105734.

Shemfe, M., Gadkari, S., Yu, E., Rasul, S., Scott, K., Head, I. M., Gu, S., Sadhukhan, J. (2018). Life cycle, techno-economic and dynamic simulation assessment of bioelectrochemical systems: A case of formic acid synthesis. *Bioresource Technology*, 255, 39–49.

Singh, A., Nanda, S., Guayaquil-Sosa, J. F., & Berruti, F. (2021). Pyrolysis of Miscanthus and characterization of value-added bio-oil and biochar products. *Canadian Journal of Chemical Engineering*, 99, S55–S68.

Slorach, P. C., Jeswani, H. K., Cuéllar-Franca, R., & Azapagic, A. (2019). Environmental sustainability of anaerobic digestion of household food waste. *Journal of Environmental Management*, 236, 798–814.

Surendra, K. C., Sawatdeenarunat, C., Shrestha, S., Sung, S., & Khanal, S. K. (2015). Anaerobic digestion-based biorefinery for bioenergy and biobased products. *Industrial Biotechnology*, 11(2), 103–112.

Tchapda, A., & Pisupati, S. (2014). A review of thermal co-conversion of coal and biomass/waste. *Energies*, 7(3), 1098–1148.

Thürer, M., Tomašević, I., Stevenson, M., Qu, T., & Huisingh, D. (2018). A systematic review of the literature on integrating sustainability into engineering curricula. *Journal of Cleaner Production*, 181, 608–617.

Tortosa-Masiá, A. A., Buhre, B. J. P., Gupta, R. P., & Wall, T. F. (2007). Characterizing ash of biomass and waste. *Fuel Processing Technology*, 88(11–12), 1071–1081.

Tuazon, D., Corder, G., McLellan, B. (2013). Sustainable development: A review of theoretical contributions. *International Journal of Sustainable Future for Human Security*, 1(1), 40–48.

Tuazon, D., & Gnansounou, E. (2017). *Towards an Integrated Sustainability Assessment of Biorefineries*. Elsevier B.V.

Tursi, A. (2019). A review on biomass: Importance, chemistry, classification, and conversion. *Biofuel Research Journal*, 6, 962–979.

U.S. Energy Information Administration (EIA). (2013). *Annual Energy Outlook 2013 with Projections to 2040*. Washington, DC, USA.

Van Huis, A., & Oonincx, D. G. (2017). The environmental sustainability of insects as food and feed. A review. *Agronomy for Sustainable Development*, 37, 1–14.

Varjani, S., Shah, A. V., Vyas, S., & Srivastava, V. K. (2021). Processes and prospects on valorizing solid waste for producing valuable products employing bio-routes: A systematic review. *Chemosphere*, 282, 130954.

Venkata Mohan, S., Amulya, K., & Modestra, J. A. (2020). Urban biocycles-closing metabolic loops for resilient and regenerative ecosystem: A perspective. *Bioresource Technology*, 306, 123098.

Venkata Mohan, S., Nikhil, G. N., Chiranjeevi, P., Reddy, C. N., Rohit, M. V., Kumar, A. N., & Sarkar, O. (2016). Waste biorefinery models towards sustainable circular bioeconomy: Critical review and future perspectives. *Bioresource Technology*, 215, 2–12.

Wang, B., Li, Y., Wu, N., & Lan, C. Q. (2008). CO2 bio-mitigation using microalgae. *Applied Microbiology and Biotechnology*, 79(5), 707–718.

Wang, C., Qian, L., Wang, W., Wang, T., Deng, Z., Yang, F., Xiong, J., & Feng, W. (2017). Exploring the potential of lipids from black soldier fly: New paradigm for biodiesel production (I). *Renewable Energy*, 111, 749–756.

Waqas, M., Hashim, S., Humphries, U. W., Ahmad, S., Noor, R., Shoaib, M., Naseem, A., Hlaing, P. T., & Lin, H. A. (2023). Composting processes for agricultural waste management: A comprehensive review. *Processes*, 11(3), 731.

Xiangxue, Z., Jie, A., Yuzhong, W., et al. (2012). Progress of producing vehicle fuels from cracking waste plastics. *Chemical Industry and Engineering Progress*, 31, 389–401.

Yasin, M., Jang, N., Lee, M., Kang, H., Aslam, M., Bazmi, A. A., & Chang, I. S. (2019). Bioreactors, gas delivery systems, and supporting technologies for microbial synthesis gas conversion process. *Bioresource Technology Reports*, 7, 100207.

Yuvaraj, A., Thangaraj, R., Ravindran, B., Chang, S. W., & Karmegam, N. (2021). Centrality of cattle solid wastes in vermicomposting technology–A cleaner resource recovery and biowaste recycling option for agricultural and environmental sustainability. *Environmental Pollution*, 268, 115688.

16 Metabolic Engineering in Waste Prevention

Prashant Saxena and Madhvi Saxena

16.1 INTRODUCTION

The field of metabolic engineering makes a specialty of designing organic structures to Control flux in the route of preferred products. Creating forward engineering strategies that are based on associated principle and predictive fashions is one of the primary targets of metabolic engineering. The research of metabolic engineering has mainly focused on the change of a small range of genes with apparent connections to a centered pathway because such processes require both sufficient knowledge to broaden models and genetic engineering gear to construct and take a look at version predictions. Common alterations include the elimination of rival pathways, the insertion of heterologous genes, and/or the overexpression of the pathway's rate-proscribing steps. These sorts of efforts were successful in raising manufacturing titers from a whole lot of platform traces, which have a huge range of applications, consisting of bulk chemicals, biofuels, drugs, and food derivatives (Stalidzans & Dace, 2021).

One of the number one goals of metabolic engineering is to optimize present metabolic pathways to increase the yield of a target compound. This can contain overexpressing genes that encode key enzymes, knocking out competing pathways or introducing regulatory elements that facilitate flux through a pathway (Nielsen and Keasling, 2016). Metabolic engineers can introduce new pathways into an organism with the aim of moving genes from another species. This is not unusual in artificial biology, wherein pathways from flowers, microorganisms, or fungi are introduced into microbial hosts like *Escherichia coli* or *Saccharomyces cerevisiae* to create compounds which include biofuels, prescription drugs, or chemicals. Modern metabolic engineering relies heavily on computational fashions to monitor the effects of genetic modifications. Genome-scale metabolic models (GEMs) and other computational equipment permit engineers to simulate mobile metabolism and discover key targets for engineering (Stephanopoulos, 1998).

Metabolic engineering plays a key role in waste prevention by optimizing microorganisms to convert waste materials into valuable products. By redirecting metabolic pathways, artificial microbes can flip agricultural waste, commercial byproducts, or even municipal waste into biofuels, bioplastics, and different beneficial compounds. This technique not only reduces the environmental effect of waste disposal but also creates sustainable solutions for resource healing (Chen & Nielsen, 2019). One example is engineering *Escherichia coli* to convert glycerol, a byproduct of biodiesel manufacturing, into ethanol or different organic chemical compounds. Another process is the usage of synthetic microbes to break down plastic waste into biodegradable components that could then be upcycled into new materials (Keasling, 2010).

Waste prevention is an important problem for improved sustainability in contemporary industry. By minimizing waste at the source, agencies can reduce environmental impact, lower operational expenses, and improve useful resource performance. Waste prevention strategies encompass the adoption of cleaner production techniques, useful resource management, and the mixture of circular economy concepts, wherein waste is repurposed or recycled once more into the production.

DOI: 10.1201/9781003668961-16

Waste prevention consists of numerous advantages which include environmental effects, price savings, and regulatory compliances (Ghisellini et al., 2016).

Metabolic engineering plays a pivotal position in advancing sustainability and fostering the circular economic system by permitting the conversion of renewable belongings and waste materials into valuable product. By remodeling metabolic pathways in microorganisms, metabolic engineering lets in to produce primarily bio-based chemical compounds, fuels, and substances which can replace fossil-derived products (Figure 16.1). This no longer simply reduces dependency on nonrenewable resources, it also minimizes environmental effects via utilizing waste as a feedstock, promoting resource efficiency, and closing the loop in commercial procedures.

Metabolic engineering allows the transformation of various wastes consisting of agricultural residues, commercial by-product, and municipal waste into immoderate-cost products like biofuels, bioplastics, and place of information chemical substances. This manner helps the circular economy by decreasing waste and growing new revenue streams from otherwise discarded substances (Chen & Nielsen, 2019). With the assistance of metabolic engineering microbes to efficaciously convert renewable feedstocks (e.g., biomass, CO_2) into biofuels, chemicals, and materials, metabolic engineering offers a sustainable opportunity to petrochemical-based industries. This shift contributes to lowering greenhouse gas emissions and lowering the environmental footprint of business production (Tiso & Narancic, 2021) The circular economy system aims to create a closed-loop device in which products and materials are reused, repaired, or recycled, minimizing waste and aiding consumption. Metabolic engineering enables this by permitting the conversion of waste into new materials, thereby helping the recycling and upcycling of resources (Zhang & Show, 2020).

16.2 FUNDAMENTALS OF METABOLIC ENGINEERING

Metabolic engineering is the sector of study focused on the directed change of cell metabolism to acquire preferred outcomes, which include the improved manufacturing of a specific compound, the creation of recent metabolic pathways, or the optimization of present ones. It consists of the utility of genetic, biochemical, and computational equipment to regulate metabolic networks inside microorganisms, plants, or cells. Metabolic pathways, at the heart of metabolic engineering, are sequences of enzymatic reactions that convert substrates into product within a cell. Understanding and manipulating those pathways are essential to reach the desired outcomes in metabolic engineering. Metabolic engineering is implemented in diverse industries, along with the manufacturing of pharmaceuticals,

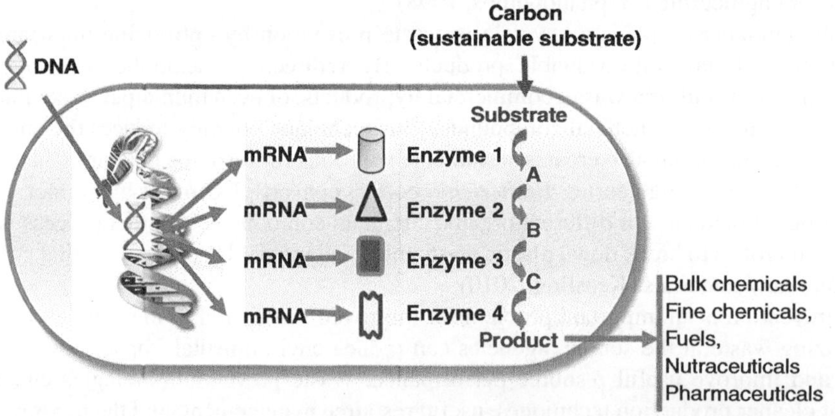

FIGURE 16.1 Process of metabolic engineering: conversion of biomolecules into useful products.

biofuels, food additives, and business enzymes. The ability to engineer organisms to provide particular compounds makes it a powerful tool for both research and industrial purposes (Stephanopoulos, 1998).

16.2.1 Tools and Techniques in Metabolic Engineering

Metabolic engineering is based on a range of modern-day tools and strategies to regulate and optimize the metabolic pathways within organisms to produce favored compounds. Methods span from genetic modification techniques to computational modeling, an important function in designing and imposing powerful metabolic engineering strategies. Metabolic engineering plays a big role in waste prevention by using optimizing microorganisms to transform waste materials into desirable products, as a consequence mitigating the environmental effect of waste and supporting a circular economy. Various equipment and strategies are used in this area to obtain efficient and sustainable waste control.

Synthetic biology methods permit the design and creation of novel metabolic pathways that can utilize waste materials as substrates. For example, artificial biology can be used to engineer pathways in bacteria or yeast to transform cellulose or lignin from plant waste into valuable chemical compounds. Designing and constructing biocatalysts that can degrade or convert waste materials into valuable products consists of engineering enzymes with optimal composition for breaking down complicated waste substances (Devi et al., 2022; Kawai & Morikawa, 2020).

MFA (metabolic flux analysis) is used to measure and examine the drift of metabolites through metabolic pathways. In waste prevention, MFA evaluates the best eco-friendly routes for converting waste substances into target product and detecting bottlenecks inside the manner (Yao et al., 2021). Another approach, flux balance assessment (FBA), is employed to simulate and optimize metabolic networks under different conditions. By modeling the metabolic strategies concerned in waste conversion, FBA can manually alter the engineering of microorganisms to maximize the efficiency of waste usage and product formation. Another approach for waste management is adaptive laboratory evolution, which includes evolving microorganisms under selective strains to increase their potential to system waste materials. This approach can enhance strain tolerance to toxins with the aid of by-products or boost performance in changing waste substrates (Jin & Nielsen, 2017).

16.2.2 Microorganisms and Plants in Metabolic Engineering

Metabolic engineering has emerged as a pivotal tool in waste prevention, wherein microorganisms and plants are genetically changed to facilitate their ability to transform waste substances into valuable products. This approach includes the manipulation of metabolic pathways within these organisms to optimize the utilization of waste substrates, thereby lowering environmental pollution and promoting sustainable practices. Microorganisms together with yeast, and fungi have been extensively engineered to degrade and assimilate organic waste, along with agricultural residues, industrial by-products, or even plastic waste, converting them into biofuels, bioplastics, and different valuable chemicals. For instance, genetically modified *Escherichia coli* strains have been engineered to convert lignocellulosic biomass into bioethanol, supplying a renewable alternative to fossil fuels (Lee et al., 2019) (Figure 16.2). Similarly, vegetation had been engineered to improve phytoremediation, the manner via which plants detoxify and dispose of pollution from the environment. Transgenic plants with better metabolic pathways can soak up heavy metals and different contaminants from the soil, transforming them into less toxic by-products or storing them in harvestable tissues for safe disposal. The integration of metabolic engineering in microorganisms and plants consequently holds considerable promise in mitigating waste technology and advancing circular economy concepts by turning waste into resources (Figure 16.3, Table 16.1).

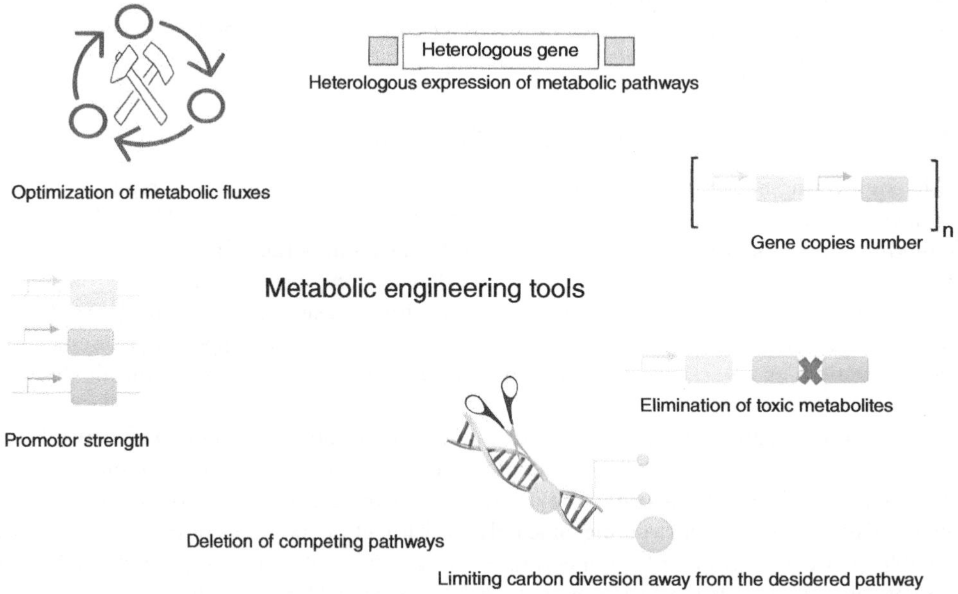

FIGURE 16.2 General tools for the metabolic engineering of microorganisms.

16.3 MICROBIAL CONVERSION OF WASTE

Microbial conversion of waste is an essential method of sustainable waste control, leveraging the natural metabolic strategies of microorganisms to transform diverse waste materials into valuable product. This biotechnological method ordinarily includes the usage of bacteria, fungi, and other microorganisms to degrade and metabolize natural waste, converting it into biofuels, bioplastics, and other biochemicals. For example, anaerobic digestion is a well-documented method wherein microbes degrade natural waste in the absence of oxygen, generating biogas, which is a renewable supply of energy, and digestate, which may be used as a nutrient-rich fertilizer (Tauseef et al., 2013).

Microbial conversion is mainly effective in treating agricultural waste, food waste, and sewage sludge, which are plentiful and often hard to manipulate through traditional methods. Advanced metabolic engineering has enabled the improvement of microbial traces with better skills, such as elevated resistance to toxic compounds, advanced substrate specificity, and higher product yields. For example, genetically engineered *Clostridium* species have been optimized to transform lignocellulosic biomass into butanol, a valuable biofuel, with better performance than wild-type traces (Lee et al., 2012).

Additionally, microbial conversion tactics are being included with other technologies, consisting of microbial fuel cells, to further facilitate energy restoration from waste. This technique not only reduces the environmental effect of waste but also contributes to the circular economy by means of converting waste into resources, thereby promoting sustainability and resource performance.

16.3.1 Agricultural and Food Waste Conversion

Agricultural and food waste conversion is an essential element of waste prevention techniques aimed toward mitigating the environmental effects of waste accumulation even while simultaneously creating valuable product. The great volume of waste generated from agricultural activities and food processing affords a giant problem; however, it additionally offers an opportunity for resource recuperation through modern conversion processes. These wastes, rich in organic nutrients and vitamins,

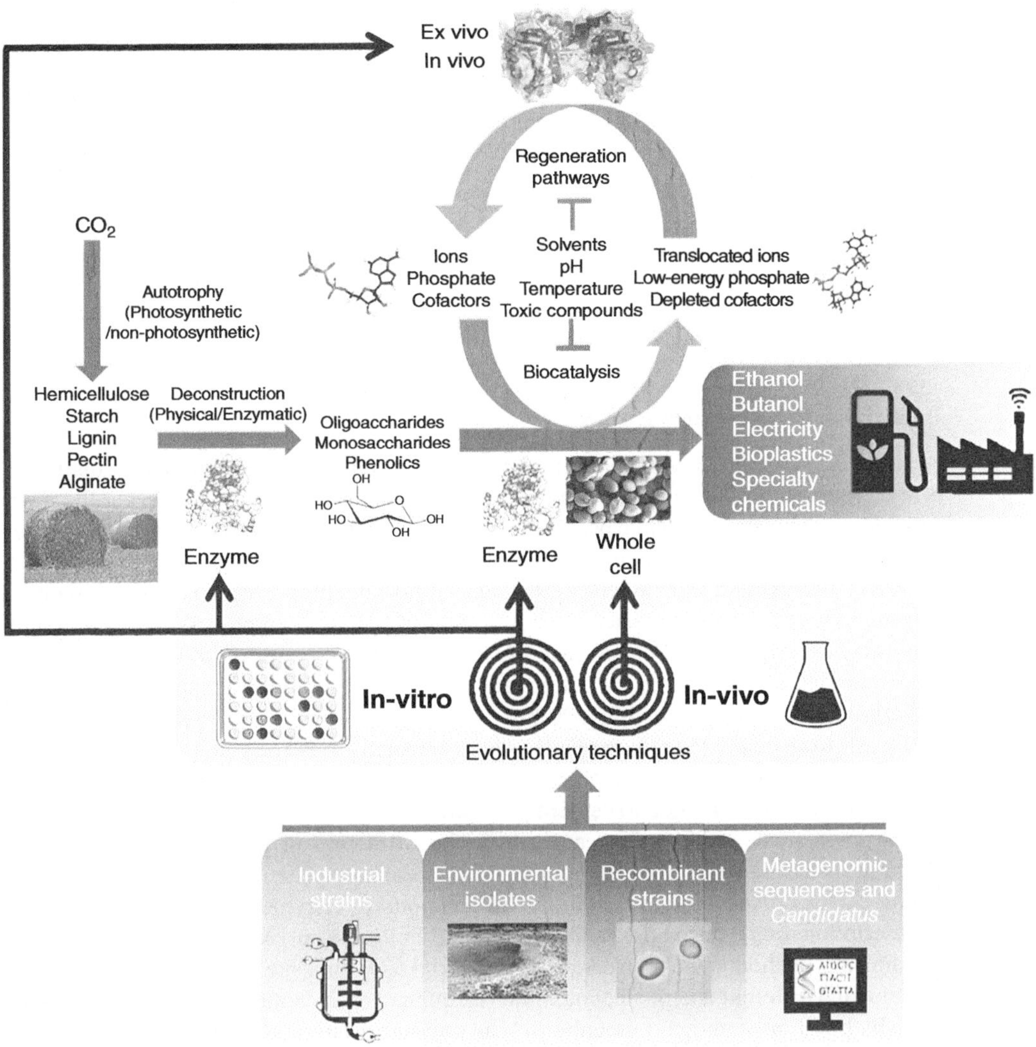

FIGURE 16.3 An overview of the current techniques available for modifying microbes' metabolic environments.

TABLE 16.1
List of Methods Applied on Plants for the Prevention of Biological Waste

S.No	Pollutant Treated	Method
1	Cadmium, Cobalt, Chromium, Nickel, Mercury, Lead, Selenium, Zinc	Phyto Extraction
2	Minerals deposit on lagoons; Phenolic and Chlorinated compounds	Phyto Stabilization
3	Hydrocarbons	Phyto Stimulation
4	Mercury, Methane, Selenium	Phytovolatilization
5	Nitrobenzene, DDT, RDX, Pesticides, Nitrites, and Phenols	Phyto Decomposition

may be repurposed through diverse organic, chemical, and thermochemical methods. For instance, anaerobic digestion is usually employed to transform food waste and agricultural residues into biogas, a renewable electricity source, and digestate, which can be used as a biofertilizer (Li et al., 2013).

Composting is another widely used technique where natural waste is biologically decomposed under managed aerobic situations, resulting in a nutrient-rich compost that enhances soil fitness and reduces the need for chemical fertilizers (Bernstad Saraiva Schott & Andersson, 2015). Additionally, superior technology, including pyrolysis and gasification, is being developed to thermochemically convert agricultural waste into biochar, bio-oil, and syngas, which have uses in energy production, soil modification, and as chemical feedstocks. The integration of these waste conversion methods into the agriculture and food manufacturing systems not only diverts waste from landfills but also contributes to the circular economy by transforming waste into valuable resources, reducing greenhouse gas emissions, and promoting sustainable agricultural practices.

16.3.2 Industrial Waste Valorization

Industrial waste valorization is a key method in waste prevention that includes the transformation of by-products and residues into valuable materials, energy, or chemicals, thereby lowering the environmental effects and enhancing resource performance. Traditional business techniques often generate considerable amounts of waste, which include stable waste, wastewater, and gaseous emissions, that can pose vast environmental and monetary challenges if now not properly managed. Valorization strategies purpose to transform these wastes into resources through modern processes, together with recycling, recovery, and upcycling. For instance, slag from metallic manufacturing may be repurposed as a production material or rather for natural aggregates in concrete, as a result decreasing the demand for virgin raw substances (Euroslag, 2020).

Similarly, fly ash from coal-fired power plants, which is usually a dangerous waste, may be converted into geopolymer cement, offering a more sustainable alternative to standard Portland cement and lowering carbon emissions related to cement production (Provis & Bernal, 2014).

Moreover, business waste valorization can involve the extraction of precious chemicals from waste streams. For instance, waste from the food and beverage enterprise, together with fruit peels and spent grains, can be processed to extract bioactive compounds, enzymes, and bioplastics, which have use in prescription drugs, cosmetics, and packaging (Dhillon, Kaur, & Brar, 2013).

Another promising method is the conversion of industrial CO_2 emissions into price-added products like methanol and different chemical compounds through carbon capture and utilization (CCU) technologies, which not only mitigate greenhouse gas emissions but also create financial benefits (Aresta, Dibenedetto, & Angelini, 2013). The integration of waste valorization approaches into business systems supports the circular economy by using turning waste into an resource, minimizing the need for landfills, decreasing raw fabric consumption, and fostering sustainable industrial practices.

16.3.3 Wastewater Treatment Using Engineered Microbes

Wastewater remediation using engineered microbes represents a cutting-edge technique for waste reclamation, where genetically modified microorganisms are deployed to facilitate the breakdown and removal of pollutants from wastewater. Traditional wastewater remediation methods depend upon natural microbial communities to degrade organic matter and get rid of contaminants, but these procedures may be limited in their performance, particularly whilst managing complicated or recalcitrant pollution. Engineered microbes, designed through metabolic engineering and synthetic biology, are tailor-made to target particular pollution, improve degradation fees, and even get better valuable resources from wastewater.

For example, genetically engineered strains of *Pseudomonas* and *Escherichia coli* have been advanced to break down toxic fragrant compounds, together with phenols and toluene, which are typically found in industrial wastewater (Wang et al., 2018). These microbes can release enzymes that are very efficient at catalyzing the breakdown of those pollutants, mainly through a quicker and more complete cleansing.

In addition to pollutant degradation, engineered microbes can be used to better obtain valuable vitamins and materials from wastewater. For example, modified Cyanobacteria were utilized to capture and convert extra nitrogen and phosphorus into biomass, which may be harvested and used as biofertilizer or biofuel feedstock (Nguyen et al., 2017). This now not only reduces nutrient pollution, which could lead to eutrophication in aquatic structures, but also creates a sustainable loop where waste is transformed into beneficial products.

Another innovative approach is the usage of engineered microbes in microbial gasoline cells, where they generate electricity from the natural compounds in wastewater, accordingly presenting a renewable energy source whilst treating the waste (Logan et al., 2019). Overall, the usage of engineered microbes in wastewater remediation holds great potential for enhancing the efficiency and sustainability of wastewater control, decreasing environmental pollutants, and contributing to a circular economy with the aim of converting waste into resources.

Current research is using genetically engineered *Escherichia coli* for the conversion of plastic waste, particularly polyethylene terephthalate (PET), into a cost-effective product. Researchers at the University of Edinburgh developed a strain of *E. Coli* able to break down PET into its monomers, which are then transformed into a biodegradable plastic called polyhydroxyalkanoate (PHA). This progressive technique not only effectively addresses the environmental challenge posed by plastic waste but also creates a sustainable loop through changing nondegradable waste into biodegradable materials (Tournier et al., 2020). This case highlights the capacity of microbial engineering in tackling one of the most persistent styles of environmental pollutants.

Another research study entails the usage of *Aspergillus Niger*, a filamentous fungus, in the valorization of citrus peel waste. A bio-conservation plant in Spain has applied this approach to convert citrus peels, a derivative of the juice enterprise, into citric acid via fermentation. This process not only provides a sustainable manner of controlling huge quantities of agricultural waste but also generates citric acid, which is extensively used in the food and beverage industry as a preservative and flavoring agent (Sanchez et al., 2020). The success of this microbial conversion procedure has led to the establishment of several comparable plants across Europe, contributing to the circular economy by transforming waste into valuable commodities.

16.4 BIODEGRADATION AND BIOREMEDIATION

Biodegradation and bioremediation are crucial strategies in waste prevention and environmental restoration. Biodegradation involves the breakdown of organic materials with the aid of microorganisms, which include microorganisms and fungi, into easier materials like water, carbon dioxide, and biomass. This process is essential for decreasing the accumulation of natural waste inside the surroundings, consequently stopping pollution, and promoting the recycling of nutrients returned to ecosystems (Das & Chandran, 2011).

On the other hand, bioremediation is the utilization of biodegradation methods in a managed way to ease cleanup of contaminated environments, such as soils, groundwater, and marine regions polluted with dangerous materials like oil spills, heavy metals, and insecticides. This approach harnesses the metabolic capabilities of precise microorganisms to degrade or transform pollution into less harmful compounds, thereby mitigating their effect on the environment. The effectiveness of bioremediation relies on elements such as the nature of the contaminant, environmental situation, and the presence of suitable microbial groups. Together, biodegradation and bioremediation play a pivotal function in waste prevention by way of facilitating the organic assisted elimination of pollution from the environment, contributing to the sustainability of ecosystems and the reduction of human impact on the planet (Vidali, 2001).

16.4.1 Principles of Biodegradation

The standards of biodegradation are foundational to effective waste prevention, as they rely on the herbal breakdown of organics through microbial action. Biodegradation operates on the principle that microorganisms, mainly bacteria and fungi, utilize organic waste as a source of energy and nutrients, converting complicated molecules into simpler, less harmful materials consisting of carbon dioxide, water, and biomass. This process takes place in diverse environmental settings, such as soil, water, and composting structures, and is conditional on factors including the chemical shape of the waste material, the supply of oxygen (aerobic versus anaerobic situations), temperature, pH, and the presence of specific microbial groups (Becker & Wittmann, 2019).

The price and volume of biodegradation are determined by the molecular complexity and toxicity of the waste fabric, with simpler, more biodegradable materials, which include food waste and paper, breaking down more swiftly than complicated artificial compounds like plastics. In waste prevention, the utility of biodegradation is strategically aimed toward minimizing the accumulation of natural waste in landfills and reducing the environmental footprint of waste disposal (Madigan et al., 2014). This is done with the aim of promoting conditions that enhance microbial activity, including composting, which hastens decomposition and the production of valuable by-products like compost, which can be used to complement soil. The principle of "biodegradability" also underpins the development of eco-friendly substances that can be designed to break down eventually, decreasing their long-term effect on the environment.

16.4.2 Bioremediation Techniques: In Situ and Ex Situ

In situ bioremediation refers to the remediation of contaminated soil, sediment, or groundwater directly at the site in which the contamination took place. This approach is especially popular in waste prevention as it removes the need for excavation, shipping, and disposal of affected materials, thereby reducing the possibility of secondary contamination and wasted energy. Common in situ bioremediation strategies encompass bioventing, biosparging, phytoremediation, and natural attenuation.

Bioventing consists of the controlled introduction of air or oxygen into the unsaturated zone of the tainted site to stimulate the microbial degradation of organic pollutants, specifically hydrocarbons. This method is powerful in treating soils with low permeability wherein oxygen exchange is constrained (Das & Dash, 2020). Biosparging, on the other hand, injects air or oxygen into the saturated region to facilitate the degradation of contaminants dissolved in groundwater, enhancing the performance of organic biodegradation approaches (Jain et al., 2020).

Phytoremediation makes use of vegetation to take in, concentrate, and metabolize contaminants, thereby cleaning up soils and groundwater. This technique is especially powerful for heavy metals and natural pollutants (Ali, Khan, & Sajad, 2013). Natural attenuation is predicated on the organic tactics of microbial degradation, dilution, sorption, and volatilization to lessen contaminant concentrations through the years without human intervention (Pereyra et al., 2020). These in situ strategies are frequently preferred because of their cost-effectiveness and minimal on-site disruption; but, they will require longer treatment times and are notably dependent on site-unique conditions inclusive of contaminant type and concentration, and environmental factors.

Ex situ bioremediation involves the removal of contaminated soil or water from the site for treatment elsewhere, allowing for more controlled and often faster remediation processes. This method is frequently selected when in situ methods aren't feasible due to site situations, which include high concentrations of contaminants, or when fast treatment is vital. Key ex situ strategies encompass biopiles, windrow composting, land farming, and bioreactors.

Biopiles contain the piling of contaminated soil in a managed environment where aeration and, if essential, nutrient amendments are carried out to improve microbial degradation (Vidali, 2001). Windrow composting mixes contaminated soil with natural material, and the aggregate is periodically cultivated to hold aerobic conditions, promoting microbial composition and contaminant breakdown (Singh et al., 2017). Landfarming entails spreading contaminated soil over a prepared bed and tilling it periodically to aerate and stimulate microbial degradation. This method is frequently used for the remediation of petroleum hydrocarbons (Das & Dash, 2020).

Bioreactors are specialized vessels where contaminated soil or water is treated below optimized conditions. These structures can be designed to manipulate temperature, pH, oxygen tiers, and nutrient delivery, making them especially eco-friendly for the rapid treatment of a extensive range of contaminants (Sharma, Singh, & Vaid, 2018). However, ex situ bioremediation can be greater high priced due to the charges associated with excavation, transportation, and infrastructure, however it is frequently greater effective in cases where infection levels are high and require more intensive treatment.

16.4.3 Engineered Microbes for Enhanced Biodegradation

The development of genetically engineered microbes (GEMs) represents a good sized development within the discipline of bioremediation and waste prevention. Traditional bioremediation is predicated on certainly happening microorganisms to degrade environmental contaminants, however the efficiency of those natural tactics can be confined by using elements such as the availability of appropriate microbes, the presence of a couple of contaminants, and environmental situations. GEMs are mainly designed to triumph over those obstacles by using improving the metabolic pathways accountable for the breakdown of pollution, enabling faster and extra efficient biodegradation.

One of the primary techniques for engineering microbes is the change of their metabolic pathways to optimize the degradation of particular contaminants. For instance, genes from one of kind organisms may be inserted right into a single microbial strain to create a "superbug" capable of degrading a broader variety of pollution. This technique has been used to enhance the degradation of petroleum hydrocarbons, chlorinated compounds, and heavy metals (Timmis & Pieper, 2020). For instance, engineered consortia were evolved to degrade combinations of natural pollutants such as polychlorinated biphenyls (PCBs) and polycyclic aromatic hydrocarbons (PAHs) (Sáez & Rojas, 2019). By combining the strengths of different microbes, those consortia can obtain extra complete waste breakdown, contributing to more powerful waste prevention.

The use of GEMs additionally extends to the enhancement of microbial resistance to environmental stresses, which includes severe pH, temperature, or the presence of poisonous materials. By incorporating stress response genes, engineered microbes can keep excessive biodegradation composition even in harsh conditions, making them extra dependable and flexible in a huge style of environments (Lee et al., 2021). This potential to feature correctly in sever conditions is important for the practical software program of GEMs in waste prevention, as inflamed websites often present difficult environments.

However, the usage of GEMs in environmental packages moreover raises issues associated with biosafety and regulatory troubles. The potential risks related to the discharge of engineered organisms into the environment, which includes accidental gene transfer or ecological disruption, require careful evaluation and management. Regulatory frameworks in many nations are being superior to ensure that the usage of GEMs in bioremediation is safe and effective (Haro-Moreno et al., 2021). Despite those worrying situations, the capability benefits of GEMs in waste prevention, particularly in improving the fee and scope of biodegradation strategies, cause them to a promising device for environmental manage.

16.4.4 Case Studies in Biodegradation of Plastics, Heavy Metals, and Other Pollutants

16.4.4.1 Biodegradation of Plastic

Plastics, particularly those derived from petroleum, present a considerable environmental mission because of their patience within the environment and resistance to natural degradation approaches. However, recent advances in microbial biodegradation have proven promise in addressing this problem. A super case look at includes Ideonella sakaiensis, a bacterium determined in 2016 that can degrade polyethylene terephthalate (PET), a commonplace plastic used in bottles and textiles. This bacterium produces two key enzymes, PETase and MHETase, that break down PET into its monomers, which could then be in addition degraded through the bacterium for power and boom (Yoshida et al., 2016). The discovery of I. Sakaiensis has unfolded new opportunities for biodegradation of plastics, as researchers work to optimize those enzymes to be used in business programs, probably lowering plastic waste on a massive scale.

Another sizable case examine is using wax moth larvae (*Galleria mellonella*) To biodegrade polyethylene, one of the most widely produced plastics. Research has proven that the larvae can digest polyethylene into ethylene glycol, a compound that is much less dangerous to the environment (Bombelli et al., 2017). This herbal biodegradation technique, facilitated by way of the intestine micro organism of the larvae, represents a singular method to plastic waste prevention. While nevertheless inside the experimental degrees, these findings advocate that biological organisms may want to play a vital role in mitigating plastic pollutants.

16.4.4.2 Biodegradation of Heavy Metals

Heavy metals which includes lead, mercury, and cadmium are persistent environmental pollutants that pose severe fitness dangers because of their toxicity and tendency to bioaccumulate. Bioremediation techniques utilising microbes for heavy steel degradation or transformation have been successfully applied in numerous case studies. One outstanding example is the usage of Pseudomonas putida to biotransform toxic mercury (Hg2) into tons less dangerous elemental mercury (Hg0) through the motion of the enzyme mercuric reductase (Barkay et al., 2003). Heavy metals which incorporates lead, mercury, and cadmium are continual environmental pollutants that pose severe health risks due to their toxicity and tendency to bioaccumulate. Bioremediation strategies using microbes for heavy metallic degradation or transformation were effectively carried out in numerous case research. One prominent instance is the usage of Pseudomonas putida to biotransform poisonous mercury (Hg^2) into plenty much less harmful elemental mercury (Hg0) through the action of the enzyme mercuric reductase (Khan et al., 2009). This method has been effectively carried out in mine websites, lowering the mobility and bioavailability of heavy metals and preventing their unfold through the surroundings. The use of SRBs in bioremediation has been especially a success in treating acid mine drainage, a chief environmental problem in mining areas.

16.4.4.3 Biodegradation of Other Pollutants

Polychlorinated biphenyls (PCBs) and polycyclic aromatic hydrocarbons (PAHs) are continual organic pollution (POPs) which are proof against natural degradation and pose giant environmental and fitness dangers. A hit case look at within the biodegradation of these pollutants entails the use of Sphingomonas species, bacteria known for their capacity to degrade a extensive variety of aromatic compounds, inclusive of PCBs and PAHs (Field & Sierra-Alvarez, 2008). Sphingomonas sp. Stress A1 has been proven to break down PCBs through a sequence of oxidative reactions, changing them into less poisonous compounds. This strain has been utilized in contaminated soils, in which it has notably decreased PCB degrees, demonstrating the potential of microbial bioremediation in managing POPs.

Similarly, Mycobacterium vanbaalenii PYR-1 has been correctly used within the degradation of PAHs in contaminated soil. This bacterium metabolizes PAHs via dioxygenation, main to the

TABLE 16.2
Comparison between Biodegradation and Bioremediation

Aspect	Biodegradation	Bioremediation
Definition	The breakdown of organic substances by natural biological processes.	The use of organism (usually microbes) to remove or neutralize contaminants from the environment
Scope	Natural Process that occurs in the environment without human intervention	Engineered process designed to accelerate the natural biodegradation process.
Application	Occurs in natural ecosystem (e.g., soil, water, composting).	Used in contaminated environments (e.g., oil spills, industrial sites).
Mechanism	Microorganisms like bacteria, fungi, and algae degraded organic materials.	Microorganism or plants are introduced or stimulated to degraded contaminants.
Effectiveness	Dependent on environmental conditions, type of substance, and microbial community.	More controlled and often more effective due to human intervention and optimization.
Environmental Impact	Minimal impact, natural and eco friendly	Can be ecofriendly, but potential risks if not properly managed (introduction to non-native species).
Key Differences	Biodegradation is a broader term encompassing all-natural degradation process.	Bioremediation is a specific application of biodegradation with human intervention.

formation of much less dangerous intermediates that may be similarly damaged down through other microbial groups (Kweon et al., 2011). The application of M. Vanbaalenii in field trials has ended in full-size discounts in PAH concentrations, highlighting its capability for huge-scale bioremediation efforts (Table 16.2).

16.5 METABOLIC ENGINEERING IN THE CIRCULAR ECONOMY

Metabolic engineering plays a crucial position in advancing the spherical economic system, mainly inside the context of waste prevention. The concept revolves around optimizing and remodeling the metabolic pathways of microorganisms or flora to convert waste materials into valuable products, thereby last the loop in useful resource use.

16.5.1 Designing Closed-loop Systems with Engineered Organisms

Designing closed-loop systems with engineered organisms for waste prevention is a transformative technique inside the spherical economic system, focusing at the seamless integration of biological techniques to recycle and repurpose waste. In such structures, waste from one system serves as a feedstock for each other, facilitated by using way of genetically engineered organism's tailor-made to perform specific bioconversions. These organisms, frequently microbes, are metabolically engineered to successfully degrade, assimilate, or convert waste substances into valuable product, which include biofuels, bioplastics, or enterprise chemical substances. For example, engineered Escherichia coli lines were advanced to transform lignocellulosic biomass, a not unusual agricultural waste, into ethanol and different chemicals (Kim et al., 2020). This approach now not excellent reduces the quantity of waste however additionally minimizes the want for virgin resources, thereby very last the materials loop.

A critical thing of this layout is the optimization of metabolic pathways inside the engineered organisms to facilitate yield, performance, and the form of substrates that can be processed. For instance, artificial biology techniques have enabled the improvement of microbial consortia in which wonderful strains concentrate on degrading high-quality additives of a waste movement, which incorporates plastics or herbal pollution, developing a synergistic and additional inexperienced method (Law et al., 2021). Moreover, those closed-loop systems can be incorporated into existing industrial procedures, making an allowance for the continuous reuse of substances inside the identical production cycle, appreciably lowering the environmental footprint and promoting sustainability.

The deployment of such systems requires a deep know-how of the interaction between the engineered organisms and the surroundings, making sure that they feature optimally below commercial situations at the same time as maintaining safety and balance. Regulatory frameworks and lifecycle tests also are crucial in comparing the long-term sustainability and effect of these technologies. Overall, the layout of closed-loop structures the use of engineered organisms represents a promising frontier in waste prevention and resource efficiency, aligning with the standards of the round economy to create sustainable, self-maintaining commercial ecosystems.

16.5.2 Carbon Capture and Utilization (CCU) in Waste Management

CCU is an emerging method in waste management that performs a pivotal role in waste prevention by using remodeling carbon dioxide (CO_2) emissions, a primary commercial waste product, into valuable resources. CCU technology attention on capturing CO_2 from commercial techniques or directly from the surroundings and in the end converting it into useful product along with fuels, chemical substances, constructing materials, and polymers. This technique not simplest mitigates greenhouse gas emissions however also diverts CO_2 from being an unused waste product right into a valuable feedstock, thereby integrating it into the round economy. For example, CO_2 can be transformed into methanol, a versatile chemical used in the manufacturing of plastics, fuels, and other substances, through catalytic procedures or microbial fermentation (Aresta, Dibenedetto and Angelini, 2013). This conversion not only prevents the release of CO_2 into the surroundings but also reduces the need for fossil fuels, contributing to resource efficiency and sustainability.

The success of CCU in waste prevention depends on the performance of the capture technique, the energy requirements for conversion, and the financial viability of the end products. Advances in catalytic technology and biotechnological strategies have appreciably improved the efficiency and scalability of CO_2 conversion processes. For example, engineered microorganisms were developed to convert CO_2 into various bio-based products, inclusive of biofuels and bioplastics, offering a sustainable opportunity to standard petrochemical tactics (Li et al., 2020). Furthermore, the mixing of CCU with renewable energy resources, together with sun or wind energy, can facilitate the sustainability of those methods by using imparting a low-carbon electricity input for the conversion steps.

CCU in waste control also aligns with broader waste prevention techniques via developing closed-loop systems where CO_2 is usually recycled rather than being emitted as waste. This technique not only contributes to reducing the carbon footprint of commercial sports but also helps the improvement of a circular economy in which waste is minimized and resources are continuously reused. However, for greater adoption, challenges consisting of the excessive preliminary expenses, electricity needs, and the need for supportive policy frameworks must be addressed. Overall, CCU represents a promising avenue for remodeling CO_2 from a waste product into a valuable resource, contributing to both waste prevention and climate change mitigation.

16.5.3 Production of Renewable Bio-Based Chemicals and Materials

The production of renewable bio-based chemical substances plays an important function in waste prevention by means of recycling biomass and different natural waste streams into valuable products, as a result decreasing reliance on fossil sources and minimizing environmental effects. Bio-based chemicals and materials are derived from renewable biological resources, including agricultural residues, forestry by-product, or even municipal waste, presenting a sustainable opportunity for petroleum-based products. This approach not only diverts organic waste from landfills but also converts it into expensive products that include bioplastics, biofuels, and biochemicals, contributing to a circular economy. For example, lignocellulosic biomass, which incorporates agricultural residues like corn stover and wheat straw, can be processed into primarily bio-based chemicals consisting of ethanol, butanol, and other platform chemical substances that function constructing blocks for a wide variety of product (Choi et al., 2015).

The use of renewable bio-based materials in place of traditional petrochemicals can substantially reduce greenhouse gas emissions and reduce the general environmental footprint of manufacturing techniques. Advanced biotechnological techniques, consisting of metabolic engineering and synthetic biology, have enabled the eco-friendly conversion of biomass into many chemical substances and materials. For instance, the engineered bacterium *Escherichia coli* has been optimized to produce primarily bio-based polymers like PHAs, which might be biodegradable and can replace conventional plastics in many situations (Lu, Tappel, & Nomura, 2009). These primarily bio-based materials now not only provide a sustainable opportunity but also lessen the accumulation of plastic waste within the environment, addressing one of the primary global waste challenges.

Moreover, the combination of primarily bio- based production with current waste control structures can facilitate the overall performance of resources use. By converting waste streams into valuable chemical compounds and materials, the circular economy model is bolstered, wherein waste is continuously recycled and reused rather than discarded. This method additionally helps waste prevention by way of lowering the need for virgin raw materials and minimizing the generation of waste at the supply. However, the great adoption of renewable bio-based chemical compounds and materials requires overcoming demanding situations along with excessive manufacturing fees, scalability, and the improvement of sturdy supply chains for biomass feedstocks. Continued studies and innovation in this area are essential to creating primarily bio-based products that are economically competitive and environmentally sustainable.

16.5.4 Potential for Scalability and Economic Viability

The capability for scalability and economic viability in waste prevention through innovative technology, along with waste-to-resource processes and sustainable materials production, hinges on numerous vital factors. Scalability refers to the capacity to expand those technologies from pilot tasks to business-scale operations, whilst monetary viability includes ensuring that these technologies are value-effective and competitive with traditional techniques. For instance, the scalability of bio-based chemical production from waste streams requires advancements in method optimization, feedstock management, and integration into present commercial systems. Technologies like anaerobic digestion and gasification, which convert organic waste into biogas and biochar, have verified full-size potential for large-scale application, provided that the systems are designed to handle numerous and variable waste streams successfully (Hammond et al., 2020).

Economic viability is encouraged by means of elements along with capital investment, operational prices, and market demand for bio-based products. While preliminary setup prices for superior waste-to-resource technologies can be high, economies of scale and technological improvements can reduce expenses over the years. For example, the manufacturing of biofuels from waste biomass has emerged as very economically feasible with upgrades in enzyme technologies and fermentation approaches, which facilitate yield and reduce manufacturing costs (Sadhukhan et al., 2014). Additionally, authority's guidelines, subsidies, and incentives play a crucial role in helping the adoption of these technologies through offsetting preliminary charges and fostering market improvement.

Moreover, a successful integration of these technologies into waste control structures calls for collaboration among stakeholders, including technology developers, policymakers, and enterprise players, to cope with regulatory and logistical challenges. The improvement of strong supply chains for biomass feedstocks and the established order of clean economic incentives can in addition enhance the scalability and monetary viability of waste prevention techniques. As research maintains to improve, the combined consequences of technological innovation, supportive policies, and market dynamics are predicted to enhance the feasibility and price-effectiveness of waste prevention technologies, making them increasingly possible on a larger scale.

16.6 CHALLENGES AND FUTURE DIRECTIONS

16.6.1 Technical Challenges in Waste Prevention

Technical challenges in waste prevention are multifaceted and frequently hinge on optimizing procedures, improving performance, and integrating new technology into current systems. One of the number one challenges is the eco-friendly conversion of various waste streams into valuable products. For instance, the heterogeneity of waste substances, such as mixed municipal stable waste or complicated commercial by means of-products, complicates their processing and calls for advanced sorting and preprocessing technologies to ensure powerful recuperation and utilization (Geyer et al., 2016). Additionally, scaling up laboratory or pilot-scale technologies to business tiers provides extensive hurdles, such as the want for robust and reliable systems which can handle large volumes and sundry waste types without compromising performance or protection (Zhao et al., 2020).

Another technical task is optimizing the performance of waste-to-resource technology, such as anaerobic digestion, gasification, and superior recycling techniques. These methods frequently require precise control over environmental situations, consisting of temperature, pressure, and chemical composition, to maximize yield and high-quality of the end product. Variability in feedstock characteristics can affect manner balance and efficiency, necessitating the improvement of adaptive and resilient technologies (Zhang et al., 2017). Furthermore, the economic viability of those technology is closely connected to their capacity to operate cost-correctly at scale, which involves minimizing operational costs and maximizing product cost.

Integration into present waste management infrastructures also poses demanding situations. This includes aligning new technology with set up systems for waste collection, transportation, and disposal, which may additionally require sizable adjustments or upgrades to house revolutionary processes (Dufour et al., 2020). Overcoming those technical demanding situations requires ongoing studies and improvement, collaboration throughout sectors, and funding in generation advancements to facilitate the effectiveness and feasibility of waste prevention techniques.

16.6.2 Economic and Regulatory Considerations

Economic and regulatory issues are vital within the successful implementation of waste prevention techniques. Economically, the viability of waste prevention technologies is heavily prompted by using initial capital investment, operational costs, and marketplace dynamics. High prematurely charges related to advanced waste processing technologies, such as anaerobic digestion and gasification, may be a barrier to full-size adoption (Makarov et al., 2020). However, economies of scale and technological improvements can reduce these expenses over the years. Additionally, the economic blessings of waste prevention, including the manufacturing of valuable through-products and savings from reduced waste management charges, can offset initial investments and provide lengthy-time period monetary incentives (Kumar et al., 2018).

Regulatory considerations play a pivotal function in shaping the landscape for waste prevention. Governments and regulatory our bodies establish tips and standards that have an effect on the design, implementation, and operation of waste management technology. Policies including extended manufacturer obligation (EPR) and waste diversion mandates can power the adoption of waste prevention practices via developing financial incentives and regulatory necessities for corporations (Morris & McIlveen, 2019). Conversely, complicated and restrictive policies can hinder innovation and slow down the deployment of latest technologies. Ensuring that guidelines are clear, supportive, and adaptable to technological improvements is essential for fostering a good environment for waste prevention (Frosch & Gallopoulos, 2020).

Collaboration among policymakers, industry stakeholders, and researchers is vital to address these monetary and regulatory demanding situations. Developing supportive policies that provide monetary incentives and create a market for recycled and bio-based products, alongside streamlining regulatory procedures, can enhance the feasibility and effectiveness of waste prevention techniques. Ultimately, balancing monetary and regulatory elements is key to reaching sustainable waste control and advancing the circular economy.

16.6.3 Future Prospects in Waste Prevention

The future potentialities in waste prevention are promising, driven by improvements in technology innovation in materials science, and evolving regulatory frameworks. One great location of development is the mixing of artificial intelligence (AI) and machine learning into waste management systems. AI can facilitate waste sorting performance, optimize recycling methods, and predict waste technology patterns, leading to more powerful waste prevention techniques (Kumar et al., 2021). Additionally, innovations in biodegradable and compostable substances provide a pathway to lowering plastic waste, with new polymers designed to break down more easily in organic environments and limit long-term environmental impact (Yuan et al., 2022).

The rise of the circular economy concept is likewise reshaping waste prevention efforts, emphasizing the remodel of products and techniques to facilitate reuse, recycling, and useful resource healing. Circular economic system practices, together with designing products for disassembly and the use of closed-loop production systems, are gaining traction and are expected to become more prevalent as companies and customers increasingly prioritize sustainability (Geissdoerfer et al., 2017). Moreover, improvements in waste-to-energy technologies, such as improved efficiency in converting waste to biofuels and chemical compounds, will probably play a critical function in closing the loop and minimizing waste (Krebs et al., 2020).

16.6.4 Ethical and Environmental Concerns

Ethical and environmental concerns are vital whilst enforcing waste prevention techniques and technologies. One primary concern is the capability for environmental justice disparities, where certain communities might also bear a disproportionate burden of waste management activities or susceptibility to pollutants from waste processing centers. Ensuring that waste prevention technologies are applied equitably and do not adversely affect vulnerable populations is important (Schroeder & Martin, 2020). Additionally, the lifecycle environmental effects of recent technology need to be carefully evaluated to avoid unintended results. For example, even as biodegradable plastics offer a promising opportunity to conventional plastics, their manufacturing and degradation approaches can now and again cause environmental alternatives, such as increased resource use or the release of microplastics (D'Amato et al., 2021).

Moreover, the shift in the direction of waste-to-energy technologies raises concerns about the sustainability of feedstock sources and the emissions produced at some point of conversion tactics. While these technologies can reduce waste volumes and produce better energy, they also can produce greenhouse gases and other pollution if no longer properly managed (Moss et al., 2020). The use of renewable resources and the implementation of advanced emission control technologies are crucial to mitigating these environmental impacts.

16.7 CONCLUSION

In this book chapter authors have discussed the metabolic engineering in waste prevention, highlighting its capacity to revolutionize how we manage and repurpose waste substances. Metabolic engineering, via the appropriate change of metabolic pathways in microorganisms and flora, offers revolutionary solutions for converting waste into useful material. This technique not only addresses the challenge of waste disposal but also contributes to the advent of sustainable systems. By harnessing the talents of engineered organisms, we will successfully transform agricultural residues, industrial waste-products, and different waste streams into useful biofuels and chemical and biodegradable substances. This technological development aligns with broader sustainability goals, offering a pathway to minimize environmental effects and decrease dependence on nonrenewable resources.

However, the successful implementation of metabolic engineering for waste prevention isn't always without challenges. The scalability of those technologies remains a substantial hurdle, requiring advancements in procedure optimization, feedstock control, and integration into current industrial structures. While pilot-scale successes show the potential of metabolic engineering, transitioning to business-scale operations entails overcoming technical barriers related to technique balance, performance, and value-effectiveness. Additionally, financial concerns, together with capital investment and operational fees, need to be addressed to make those technologies viable on a larger scale. Ongoing research and innovation are important to enhancing the efficiency and affordability of metabolic engineering solutions, exploring their practical approaches for waste control.

Regulatory and moral concerns also play a pivotal role within the deployment of metabolic engineering technologies. Clear and supportive regulatory frameworks are important to facilitate the adoption of those technologies, ensuring that they may be implemented competently and successfully. Additionally, ethical issues related to the environmental and social influences of these technologies need to be carefully controlled. Addressing problems such as environmental justice, transparency, and the lifecycle effect of engineered products is critical to retaining public trust and making sure that these technologies contribute positively to sustainable waste management.

Looking ahead, the mixing of metabolic engineering into waste prevention strategies holds amazing promise for advancing sustainability and useful resource efficiency. As technological advancements continue to adapt, and as regulatory and economic frameworks adapt to help these improvements, metabolic engineering is poised to play a key role in transforming waste control practices. By specializing in non-stop development, collaborative research, and comprehensive coverage, we can harness the full ability of metabolic engineering to create a greater sustainable and useful resource-efficient future, using significant development within the field of waste prevention.

ACKNOWLEDGEMENT

I am very grateful to the management of the Chandigarh University and the Department of Biotechnology Engineering and Food Technology Engineering for providing all the necessary facility to complete this book chapter.

REFERENCES

Ali, H., Khan, E., & Sajad, M. A. (2013). Phytoremediation of heavy metals—Concepts and applications. *Chemosphere*, 91(7), 869–881. https://doi.org/10.1016/j.chemosphere.2013.01.075

Aresta, M., Dibenedetto, A., & Angelini, A. (2013). The changing paradigm in CO2 utilization. *Journal of CO2 Utilization*, 3–4, 65–73. https://doi.org/10.1016/j.jcou.2013.08.001

Barkay, T., Miller, S. M., & Summers, A. O. (2003). Bacterial mercury resistance from atoms to ecosystems. *FEMS Microbiology Reviews*, 27(2–3), 355–384. https://doi.org/10.1016/S0168-6445(03)00046-9

Becker, J., & Wittmann, C. (2019). Microbial production of value-added bioproducts from glycerol. *Biotechnology*, 10(2), 553–557.

Bernstad Saraiva Schott, A., & Andersson, T. (2015). Food waste minimization from a life-cycle perspective. *Journal of Environmental Management*, 147, 219–226. https://doi.org/10.1016/j.jenvman.2014.07.048

Bombelli, P., Howe, C. J., & Bertocchini, F. (2017). Polyethylene biodegradation by caterpillars of the wax moth Galleria mellonella. *Current Biology*, 27(8), R292–R293. https://doi.org/10.1016/j.cub.2017.02.060

Chen, Y., & Nielsen, J. (2019). Advances in metabolic engineering. *Metabolic Engineering*, 52, 28–41. https://doi.org/10.1016/j.ymben.2018.11.004

Choi, S., Song, C. W., Shin, J. H., & Lee, S. Y. (2015). Biorefineries for the production of top building block chemicals and their derivatives. *Metabolic Engineering*, 28, 223–239. https://doi.org/10.1016/j.ymben.2014.12.007

D'Amato, D., Maffei, S., & De Gisi, S. (2021). Environmental and social sustainability of biodegradable plastics: A comprehensive review. *Journal of Cleaner Production*, 297, 126608. https://doi.org/10.1016/j.jclepro.2021.126608

Das, N., & Chandran, P. (2011). Microbial degradation of petroleum hydrocarbon contaminants: An overview. *Biotechnology Research International*, 2011, 1–13. https://doi.org/10.4061/2011/941810

Das, N., & Dash, H. R. (2020). *Microbial Biodegradation and Bioremediation*. Elsevier.

Devi, A., Bajar, S., Kour, H., Kothari, R., Pant, D., & Singh, A. (2022). Lignocellulosic biomass valorization for bioethanol production: A circular bioeconomy approach. *Bioenergy Research*, 15(4), 1820–1841. https://doi.org/10.1007/s12155-022-10401-9

Dhillon, G. S., Kaur, S., & Brar, S. K. (2013). Perspective of apple processing wastes as low-cost substrates for bioproduction of high value products: A review. *Renewable and Sustainable Energy Reviews*, 27, 789–805. https://doi.org/10.1016/j.rser.2013.07.045

Dufour, J., De Mestral, D., & Dufresne, J. (2020). Integrating advanced waste management technologies: Challenges and solutions. *Waste Management*, 104, 134–142. https://doi.org/10.1016/j.wasman.2020.02.001

Euroslag. (2020). *Position Paper on the Status of Ferrous Slag in European Legislation*. Retrieved from https://www.euroslag.org/publications/

Field, J. A., & Sierra-Alvarez, R. (2008). Microbial degradation of chlorinated dioxins. *Chemosphere*, 71(6), 1005–1018. https://doi.org/10.1016/j.chemosphere.2007.10.062

Frosch, R. A., & Gallopoulos, N. E. (2020). Strategies for sustainable waste management and the role of policy. *Journal of Cleaner Production*, 273, 122501. https://doi.org/10.1016/j.jclepro.2020.122501

Geissdoerfer, M., Savaget, P., & Bocken, N. M. P. (2017). The circular economy – A new sustainability paradigm? *Journal of Cleaner Production*, 143, 757–768. https://doi.org/10.1016/j.jclepro.2016.12.048

Geyer, R., Jambeck, J. R., & Law, K. L. (2016). Production, use, and fate of all plastics ever made. *Science*, 347(6223), 768–771. https://doi.org/10.1126/science.1260352

Ghisellini, P., Cialani, C., & Ulgiati, S. (2016). A review on circular economy: The expected transition to a balanced interplay of environmental and economic systems. *Journal of Cleaner Production*, 114, 11–32. https://doi.org/10.1016/j.jclepro.2015.09.007

Hammond, G. P., Jones, C. I., & Stokes, M. (2020). An introduction to the scalability of waste-to-energy technologies. *Renewable and Sustainable Energy Reviews*, 130, 109905.

Haro-Moreno, J. M., Rodríguez-Valera, F., López-García, P., & Moreira, D. (2021). Environmental biosafety of genetically engineered organisms for bioremediation: The challenges and prospects. *Frontiers in Microbiology*, 12, 676276.

Jain, R., Sharma, D. K., & Singh, D. (2020). In situ bioremediation of contaminated sites: A review on applications and perspectives. *Environmental Science and Pollution Research*, 27(21), 26424–26439.

Jin, Y. S., & Nielsen, J. (2017). Metabolic engineering of yeast for biofuel and biochemical production from lignocellulosic biomass. *Current Opinion in Biotechnology*, 46, 56–63. https://doi.org/10.1016/j.copbio.2017.02.005

Kawai, F., & Morikawa, M. (2020). Metabolic engineering for the valorization of lignocellulosic biomass into chemicals. *Frontiers in Bioengineering and Biotechnology*, 8, 594031.

Keasling, J. D. (2010). Manufacturing molecules through metabolic engineering. *Science*, 330(6009), 1355–1358. https://doi.org/10.1126/science.1193990

Khan, F. I., Husain, T., & Hejazi, R. (2009). An overview and analysis of site remediation technologies. *Journal of Environmental Management*, 71(2), 95–122. https://doi.org/10.1016/j.jenvman.2004.02.003

Kim, S. W., Kim, H. U., Lee, S. Y., & Kim, T. Y. (2020). Metabolic engineering of Escherichia coli for the production of short-chain fatty acids. *Nature Communications*, 11(1), 1–10.

Krebs, M. R., Moretto, M., & Zheng, H. (2020). Advances in waste-to-energy technologies: Progress and prospects. *Renewable and Sustainable Energy Reviews*, 131, 110042. https://doi.org/10.1016/j.rser.2020.110042

Kumar, A., Kumar, S., & Saini, A. (2021). Application of artificial intelligence in waste management: A review. *Waste Management*, 118, 283–294. https://doi.org/10.1016/j.wasman.2020.08.025

Kumar, S., Ranjan, R., & Yadav, P. (2018). Economic evaluation of waste-to-energy technologies: Case studies and current trends. *Renewable Energy*, 126, 698–711. https://doi.org/10.1016/j.renene.2018.04.055

Kweon, O., Kim, S. J., Holland, R. D., Chen, H., Kim, D., Gao, Y., Baek, S., Baek, D.-H., Ahn, H., & Cerniglia, C. E. (2011). Polycyclic aromatic hydrocarbon metabolism in Mycobacterium vanbaalenii PYR-1. *Journal of Bacteriology*, 193(10), 2571–2582.

Law, W. Y., Shams Yazdani, S., & Scrutton, N. S. (2021). Microbial consortia for the sustainable production of chemicals and fuels: Towards synthetic microbial ecosystems. *Current Opinion in Chemical Biology*, 59, 109–118.

Lee, J., Kim, S., & Hwang, K. (2021). Development of genetically engineered microorganisms for the bioremediation of contaminated soils. *Applied Microbiology and Biotechnology*, 105(9), 3655–3670.

Lee, S. Y., Kim, H. U., Park, J. H., Park, J. M., Kim, T. Y., & Lee, J. W. (2012). Metabolic engineering of microorganisms: General strategies and examples. *Nature Reviews Microbiology*, 10(6), 445–455. https://doi.org/10.1038/nrmicro2793

Lee, S. Y., Kim, H. U., Chae, T. U., Cho, J. S., Kim, J. W., Shin, J. H., Kim, D. I., Ko, Y. S., Jang, W. D., & Jang, Y. S. (2019). A comprehensive metabolic map for production of bio-based chemicals. *Nature Catalysis*, 2(1), 18–33. https://doi.org/10.1038/s41929-018-0212-4

Li, M., Zhou, S., Wang, B., Cheng, C., & Zhao, X. (2020). Biotechnological approaches for CO_2 utilization in sustainable biofuel production: Advances and perspectives. *Renewable and Sustainable Energy Reviews*, 130, 109936. https://doi.org/10.1016/j.rser.2020.109936

Li, Y., Park, S. Y., & Zhu, J. (2013). Solid-state anaerobic digestion for methane production from organic waste. *Renewable and Sustainable Energy Reviews*, 15(1), 821–826. https://doi.org/10.1016/j.rser.2010.07.042

Logan, B. E., Rossi, R., Ragab, A., & Saikaly, P. E. (2019). Electroactive microorganisms in bioelectrochemical systems. *Nature Reviews Microbiology*, 17(5), 307–319. https://doi.org/10.1038/s41579-019-0173-x

Lu, J., Tappel, R. C., & Nomura, C. T. (2009). Mini-review: Biosynthesis of polyhydroxyalkanoates. *Polymer Reviews*, 49(3), 226–248. https://doi.org/10.1080/15583720903048243

Madigan, M. T., Martinko, J. M., Bender, K. S., Buckley, D. H., & Stahl, D. A. (2014). *Brock Biology of Microorganisms* (14th ed.). Pearson.

Makarov, I., Kluge, J., & Zhao, W. (2020). Economic and technical analysis of waste-to-resource technologies. *Waste Management*, 104, 140–149. https://doi.org/10.1016/j.wasman.2020.02.003

Morris, J., & McIlveen, M. (2019). Policy frameworks for waste management: Implications for extended producer responsibility and recycling. *Waste Management & Research*, 37(5), 443–451.

Moss, T., Schaefer, D., & Evans, J. (2020). The environmental impact of waste-to-energy technologies: A review. *Renewable and Sustainable Energy Reviews*, 130, 109958. https://doi.org/10.1016/j.rser.2020.109958

Nguyen, H. T., Moser, M., Nguyen, D., & Eckstein, S. (2017). A review on biofuel production from microalgae and non-food crops. *ChemieIngenieur Technik*, 89(10), 1365–1374. https://doi.org/10.1002/cite.201700053

Nielsen, J., & Keasling, J. D. (2016). Engineering cellular metabolism. *Cell*, 164(6), 1185–1197. https://doi.org/10.1016/j.cell.2016.02.004

Pereyra, C. M., Luna, M. F., & de Salamone, I. G. (2020). Bioremediation of soil ecosystems from heavy metal contaminants through the combination of native microorganisms and legumes: A review. *Indian Journal of Microbiology*, 60(4), 401–409.

Provis, J. L., & Bernal, S. A. (2014). Geopolymers and related alkali-activated materials. *Annual Review of Materials Research*, 44(1), 299–327. https://doi.org/10.1146/annurev-matsci-070813-113515

Sadhukhan, J., Roy, S., & Wills, D. (2014). Process and economic optimization of biofuel production from lignocellulosic biomass. *Bioresource Technology*, 155, 80–88. https://doi.org/10.1016/j.biortech.2013.12.068

Sáez, J. M., & Rojas, L. A. (2019). Engineered microbial consortia for bioremediation of environmental pollutants. *Biotechnology Advances*, 37(5), 767–785. https://doi.org/10.1016/j.biotechadv.2019.03.005

Sanchez, A., Yánez, R., Belda-Galbis, C. M., & Monedero, M. A. (2020). Valorization of citrus waste through biotechnological approaches. *Current Opinion in Green and Sustainable Chemistry*, 26, 100382. https://doi.org/10.1016/j.cogsc.2020.100382

Schroeder, H., & Martin, J. (2020). Environmental justice and the ethics of waste management. *Journal of Environmental Management*, 268, 110702. https://doi.org/10.1016/j.jenvman.2020.110702

Sharma, B., Singh, S., & Vaid, R. (2018). A review on bioremediation of petroleum hydrocarbons in soil. *Journal of Environmental Biotechnology Research*, 1(1), 20–28.

Singh, J., Kaur, A., Singh, R., & Bajaj, P. (2017). Microbial degradation and bioremediation of persistent organic pollutants in soil. In: Environmental Microbial Biotechnology (pp. 1–24). Springer.

Stalidzans, E., & Dace, E. (2021). Sustainable metabolic engineering for sustainability optimisation of industrial biotechnology. *Computational and Structural Biotechnology Journal*, 19, 4770–4776. https://doi.org/10.1016/j.csbj.2021.08.034

Stephanopoulos, G. (1998). *Metabolic Engineering: Principles and Methodologies*. Academic Press.

Tauseef, S. M., Abbasi, T., & Abbasi, S. A. (2013). Anaerobic digestion of waste biomass for methane production: A review. *International Biodeterioration & Biodegradation*, 85, 41–51. https://doi.org/10.1016/j.ibiod.2013.06.019

Timmis, K. N., & Pieper, D. H. (2020). Bacteria designed for bioremediation. *FEMS Microbiology Letters*, 367(12), fnaa072. https://doi.org/10.1093/femsle/fnaa072

Tiso, T., & Narancic, T. (2021). The role of metabolic engineering in the circular economy. *Biotechnology Advances*, 53, 107794. https://doi.org/10.1016/j.biotechadv.2021.107794

Tournier, V., Topham, C. M., Gilles, A., David, B., & Marty, A. (2020). An engineered PET depolymerase to break down and recycle plastic waste. *Nature*, 580(7802), 216–219. https://doi.org/10.1038/s41586-020-2149-4

Vidali, M. (2001). Bioremediation: An overview. *Pure and Applied Chemistry*, 73(7), 1163–1172.

Wang, J., Chen, C., & Cui, K. (2018). Engineered bacteria for efficient bioremediation of heavy metal and organic pollutants. *Environmental Engineering Science*, 35(5), 540–553.

Yao, L., Zhang, X., & Liu, S. (2021). Engineering microbial systems for biotransformation of waste into value-added products: Progress and prospects. *Bioresource Technology*, 341, 125892. https://doi.org/10.1016/j.biortech.2021.125892

Yoshida, S., Hiraga, K., Takehana, T., Taniguchi, I., Yamaji, H., Maeda, Y., Toyohara, K., Miyamoto, K., Kimura, Y., & Oda, K. (2016). A bacterium that degrades and assimilates poly(ethylene terephthalate). *Science*, 351(6278), 1196–1199. https://doi.org/10.1126/science.aad6359

Yuan, J., Liu, L., & Zhang, C. (2022). Innovations in biodegradable plastics: Current state and future prospects. *Journal of Polymers and the Environment*, 30(2), 417–432.

Zhang, C., & Show, P. L. (2020). Metabolic engineering for the development of bio-based chemicals and fuels: A perspective on sustainability. *Bioengineered*, 11(1), 270–284.

Zhang, Y., Chen, H., & Chen, J. (2017). Technical challenges and innovations in waste-to-energy technologies: A review. *Renewable and Sustainable Energy Reviews*, 73, 156–172. https://doi.org/10.1016/j.rser.2017.01.097

Zhao, Y., Zhang, Y., & Liu, X. (2020). Scale-up of waste-to-resource technologies: Technical and economic considerations. *Bioresource Technology*, 313, 123678.

17 IoT/AI Based Solutions in Environmental Waste Management and Monitoring

Madhura Bodkhe, Gauri Chavan, Sunil Kulkarni, and Ajaygiri Goswami

17.1 INTRODUCTION

17.1.1 Waste Management Scenario

One inevitable consequence of human activity is waste. The amount and complexity of solid waste are increased by urbanization, economic expansion and growing levels of living in cities. It consists of materials that are not broken by biology, such as leather, plastics, rubbers, metals, glass, and electronic waste, as well as materials that are broken by biology, such as paper, textiles, food waste, and vegetable waste, as well as materials that are moderately degradable, like cardboard and wood. Most developing nations have highly degradable municipal solid waste (MSW), primarily made up of an organic percentage with a high moisture content. People in India are migrating from rural areas to cities owing to the nation's quick modernization and population increase, adding hundreds of tons of MSW every day. Growing urbanization and rapid population expansion have led to notable increase in the production of municipal solid waste as well as a wide range of waste compositions. India's megacities—Delhi, Mumbai, Kolkata and Chennai—produce the most MSW because of their high residential and floating population densities, which come from homes, workplaces, trade and commercial activity, industries, and health care facilities. Along with some non-governmental organizations (NGOs) with experience in this area of solid waste management, numerous local organizations have taken the initiative for effective waste collection. Waste collection has been found to be significantly higher in urban areas, such as metropolitan cities, than it is in rural areas. While states such as Arunachal Pradesh and Nagaland continue to follow non-scientific methods for waste collection transportation and are in violation of the 2000 MSW Rules, other states such as Gujarat, Maharashtra, Andhra Pradesh, Delhi, and Tripura have taken steps to improve collection efficiency (CPCB2011; Bhat et al., 2018).

17.1.2 Solid Waste Management Conventional Approach and Recent Development

There are many different types of materials and objects on the market as a result of economic growth and technological advancements. The amount of waste produced in all forms rises as population control gets more difficult and complex (Lin et al., 2022). The amount of garbage is continuously being controlled, but without public assistance, it becomes impossible to collect, manage, and treat it (Al Huraimel et al., 2022). Analyzing and implementing novel approaches and technologies that may be discovered will help us get better in this field. Important and necessary steps must be taken to check the waste that is not liquid, that is, to gather information about the solid waste produced across various localities in a large nation such as India. This in turn aids in determining appropriate methods and procedures for managing the waste produced. Both human health and environmental

DOI: 10.1201/9781003668961-17

safety will benefit from this action. Some of the most significant traditional techniques for managing solid waste are reviewed in this section, with an explanation of more recent methods provided later. The management of refuse from municipalities that is solid (MSW) is a crucial concern that requires international heed. The socioeconomic and demographic attributes of the population of a region influence the natural world, type, and sources of MSW (Ferronato et al., 2022). Gathering, moving, handling, and getting rid of MSW are the traditional strategies employed in its management. These methods are widely employed in both developed and developing nations, although their efficacy and efficiency differ based on the unique circumstances of each location. The first stage of managing MSW is collection, which entails obtaining waste from homes, businesses, and other establishments. In light of the area and population concentration, different collection techniques are employed. The most popular approach is door-to-door collection in urban areas; waste is gathered in countryside regions from specific central locations (Mahyari et al., 2022). The next step is to transport the waste, which entails moving waste from collection locations to facilities for treatment and disposal. When MSW is treated, its weight and volume are decreased using a variety of techniques, including incineration, recycling, and composting. These procedures assist in lowering the quantity of rubbish that must be abandoned and likewise recuperate important supplies, including materials and energy (Sliusar et al., 2022). The residual waste from MSW is disposed of by loading a confinement structure, such as a landfill. The choice of discarding technique is based on the land's accessibility as well as the geology and soil's suitability for waste containment. Generally, different approaches are used in the administration of solid waste through conventional techniques that differ based on the demographic and socioeconomic features of a given area. To ensure that these techniques are proficient and successful in handling the garbage and protecting the environment, proper planning, implementation, and monitoring are necessary for effective administration of solid waste (Sondh et al., 2022). Table 17.1 contains sustainable pathways for solid waste treatment (Radwan and Khan 2023).

TABLE 17.1
Sustainable Pathways for Solid Waste Treatment

Technique	Unit and Definition	References
Waste to energy Conversion- Incineration	This method of treatment entails burning the waste that is solid under controlled conditions at elevated temps. Heat is released in tandem with it.	(Patil et al., 2014)
Thermal Waste to Energy methods	Sturdy trash can be readily transformed using Waste to Energy technologies, particularly heating methods, into heat or fuel or oil. One popular technique for handling solid waste is incineration or thermal "Energy from Waste" technology.	(Muralidharan et al. 2022; Shah et al. 2021).
Bioeconomy and waste biorefinery	This method turns organic materials into useful products by using biological agents, like tiny organisms.	(Maurya et al. 2021)
Anaerobic Digestion	Sludge stabilization is most commonly achieved by anaerobic digestion. Additionally, biogas, which includes methane and can be used as an alternative energy source, is produced during this digestion process by a variety of anaerobic bacteria.	(Cao and Pawłowski, 2012)
Waste Valorization	Valorization as energy (waste-to-energy), which is carried out using technologiesValorization is the process of converting food waste into energy through the use of technologies that produce energy from that material. Technology—specifically, thermochemical technologies like gasification, pyrolysis, incineration, and plasma technologies—will be crucial.	(Bosmans et al., 2013; Gumisiriza et al., 2017)

Source: Compiled by the Author

17.1.2.1 Waste to Energy – Incineration

The most popular thermal treatment for MSW is incineration, which reduces the waste's volume and produces a sizable amount of energy from the collected waste. It entails burning MSW with oxygen at temperatures above a thousand degrees Celsius. Wastes are burned and turned into ash, CO_2, and H_2O vapor. This technique can be applied to recover energy for use in electricity generation or heating. In addition to providing energy, incineration technologies also have the benefit of lowering waste volume, making waste harmless, lowering transportation expenses, and lowering production of methane, a greenhouse gas. Regretfully, burning is not a good option in India. This might be caused by the MSW's elevated level of moisture content (40–60%), a high proportion of organic material (40–60%), reduced content of calories (800–1100 kcal/kg), and a high inert content (30–50%). In Timarpur, New Delhi, the first sizable MSW incinerator plant was constructed in 1987. Its installation cost was approximately 250 million Indian rupees (US$5.7 million) and it had a 300 t per day capacity. The purpose of the other incinerator plant, which was built in Trombay, Mumbai, is to burn exclusively institutional waste, primarily paper and other waste materials. Hospital waste is burned in small incinerators located in many cities (Mazzei and Specchia, 2023).

17.1.2.2 Thermal Waste to Energy Methods

Technology that converts waste into energy, especially thermal processing methods, makes it simple to convert solid wastes into heat or fuel/oil. Incineration, another name for thermal Waste to Energy technologies, is a popular technique for handling trash that is solid (Muralidharan et al., 2022; Shah et al., 2021). The procedure comprises producing heat and electricity by utilizing high temperatures to burn solid waste. This process is well-liked because it can produce energy and less waste, but it also releases harmful substances like furans and dioxins, which can be harmful to health. The goal of recent research has been to decrease the environmental impact and increase the efficiency of thermal Waste to Energy technologies.

17.1.2.3 Gasification

Thermal Waste to Energy methods, like gasification and combustion, are frequently used to manage MSW. With the use of heat in a controlled environment, solid waste can be transformed into valuable resources like fuel using these techniques. The process of combustion is thermochemical in nature, involving the complete oxidation of solid waste at temperatures that are usually in the range of 800 to 1200°C. Three systems make up the process: pretreatment, combustion, and post-treatment. Paper, cardboard, and plastic have all been processed using this method (Capoor and Parida 2021; Lin et al., 2022; Mazzei and Specchia 2023). An additional Waste to Energy (WTE) conversion by thermochemistry conversion method is gasification, which entails heating MSW feedstock to 500–900°C in a regulated atmosphere that includes air, steam, and oxygen. Syngas, primarily made up of CO and H_2, is the process's main by-product. Small amounts of ethylene, ethene, and CH_4 are also present. From this syngas, liquid fuels, specialty chemicals, and energy recovery can be produced. It's crucial to remember that additional study is required to maximize these technologies' effectiveness and reduce their detrimental effects on the environment.

17.1.2.4 Pyrolysis

A significant thermal method for handling solid waste. Pyrolysis is entirely different from gasification in that it doesn't require oxygen, whereas gasification uses less oxygen. This method's main flaw is that it requires complex and challenging wastewater treatment before discarding it. Furthermore, it produces toxic contaminants following the creation of the resource. Alaedini et al., (2023), Andeobu et al. (2022), Assef et al. (2022), Chen et al. (2023), and Fudge et al. (2021) are important studies on pyrolysis. MSW is combusted at temperatures between 572°F and 1832°F in the absence of air during the specialized thermal treatment process known as pyrolysis. Char, bio-oil, and syngas are among the products of this process. Particle size, composition of waste feedstock, residence time, heating rate, and temperature during pyrolysis are a few elements that affect yield and product quality.

17.1.2.5 Bioeconomy and Waste Biorefinery

There are various ways to manage solid waste in addition to Waste to Energy technologies, which comprise various thermal Waste to Energy technologies. In contrast to alternative approaches, the biorefinery for trash and bioeconomy exclusively employ wastes made of organic materials as raw materials, transforming them into important sources. Since waste biorefineries can create biofuels from biological refuse, particularly food scraps from solid trash, the concept has been gaining traction recently (Alam et al., 2022; Rather et al., 2022).

17.1.2.6 Anaerobic Digestion

Anaerobic digestion is thought to possess the ability to aid in the creation of biogas. From a theory perspective, anaerobic digestion is a long-standing technique, and research on the subject is still ongoing. Producing biogas using this method is time-consuming, which is a significant challenge. Finding a methodology to use in conjunction with alternative approaches to produce superior goods and lessen solid waste is a crucial aspect of anaerobic digestion. A method of anaerobic digestion can produce a CH_4 yield of solid waste per tonne that is 2–4 times higher than that of a landfill. Given its more advantageous uses, anaerobic digestion is clearly becoming more and more popular among refineries. The technique assists in resolving waste-related issues. Anaerobic digestion is not a novel technique; it has been widely applied in treatment of water facilities as well as currently a practical means of recovering resources from waste. It is going to be among the best technologies available for handling solid waste (He et al., 2021).

17.1.2.7 Waste Valorization

Waste valuation has the potential to be very successful in certain industries. For example, food waste's composition—more than 50% starch and 40% lipids—makes it an excellent source for waste valorization. Given the abundance of solid waste available, creative methods that combine various valuation techniques. Techniques will contribute to the production of useful output in the form of energy and various resources. Current Waste to Energy technologies are being combined in research to produce a favorable yield of reusable items. It is also greatly appreciated that power generation is a possibility. For example, combining anaerobic digestion with biorefinery enables them to function simultaneously, treating waste and transforming it into biofuels. This serves as a good substitute for the other approaches that are out there (Sajid et al., 2022). Because of its enormous value, the worth of every waste material is now well understood.

17.1.3 Liquid Waste Management Conventional Approach and Recent Development

In essence, wastewater is the city's leftover water supply after it has been put to use for other purposes. Generally speaking, there are three categories of sewage or wastewater: storm, industrial, and household. Additionally, sanitary sewage, which is wastewater from homes, is wastewater that comes from either greywater—which is water from sinks, bathtubs, washing machines, and other sources—or blackwater, which is toilet-related wastewater that contains paper towels, faeces, or urine. In order to fulfil its intended functions and serve the region under consideration, wastewater treatment can be broken down into a number of processes or units of operations that are selected and merged. It can generally be divided into three categories: primary, secondary, and tertiary. Primary treatment comprises initial physical and chemical purification processes, whereas secondary treatment concentrates on biological wastewater treatment. After being treated, wastewater that has low levels of organics and solids following both fundamental and advanced processes is suitable for reuse of the released wastewater. These include UV treatment, chlorination, and various filtration techniques that try to remove nutrients, change the pH, or disinfect after taking the effluent's receiving environment into account. The majority of pollutants are eliminated during tertiary treatment procedures, which transform wastewater into safe water that can be used for industrial, medicinal, and human consumption purposes (Altowayti et al., 2022).

17.1.3.1 Screening

The first step, or pre-process, in the removal of bigger solid waste from wastewater, like clothing, wood, kitchen trash, and other materials, is known as screening. In order to prevent possible harm to processing equipment, particularly a pump and procedures for handling smaller pollutants, screening also serves the purpose of separating the larger solid waste. Small treatment plants typically use hand-cleaned coarse screens, but mechanically cleaned screens are more effective and can be further divided into four categories: continuous belt screens, reciprocating rakes, chain-driven screens, and catenary screens (Pankratz, 2017).

17.1.3.2 Centrifugal Separation

This process produces sludge thickening, sludge dewatering, and liquid-solid separation using a decanter centrifuge. To separate and discharge solids or sludges based on their density, sewage is supplied to centrifugal machinery, which spins around at varying speeds (Lin and Wu, 2022). To meet disposal requirements, sludge discharge can be modified; it can be applied as a stackable cake without moisture (dried mud) to lower volume and weight prior to transporting or disposal, or it can be in a fluid state for land application. This method's drawback is that because high horsepower motors are used, it necessitates high capital costs and frequent maintenance.

17.1.3.3 Coagulation and Flocculation

The process of adding a coagulant to a charged particle that has stabilized is known as coagulation (Zakaria et al., 2022). On the other side, the mixing method known as flocculation facilitates particle settling and promotes agglomeration by adjusting the length and vigour of mixing. The added coagulant will result in the loss of charge of too-small colloidal particles in the wastewater to settle by gravity. (Amran and Mustapha, 2020). Introducing a coagulant into wastewater helps neutralize the electrical charges on fine colloidal particles, encouraging them to aggregate into larger, denser flocs that can be separated from the water through sedimentation by gravity. This is the fundamental principle behind coagulation and flocculation in wastewater treatment. Larger flocs form as a result of destabilization, which makes settling easier. The procedure's effectiveness is contingent upon several factors, including pH, temperature, and duration of contact. Frequently utilized inorganic coagulants include aluminium salt, starch, activated silica, ferric sulphate, ferric chloride, and alum. When these highly charged ions come into contact with water, they will neutralize the suspended particles. The generated inorganic hydroxides lead to the formation of short polymer chains, which facilitate the growth of micro-flocs. Inorganic coagulants can be used to remove the majority of suspended particles. Yet, some organic precursors might also have a reaction with chlorine to produce a lot of floc, which can collect and hold onto bacteria, along with by-products of disinfection. Because inorganic coagulants absorb alkalinity, they can change the pH of water. It is consequently necessary to use equipment for storage and feeding that is resistant to corrosion, and to dispose of the sedimentation floc in an environmentally responsible way.

17.1.3.4 Floatation

Air or gas can be added to the flotation process to eliminate oils, greases, and other substances suspended in the air, which condense into agglomerates at the water's surface (Badawi et al., 2022).

17.1.3.5 Sedimentation and Gravity-Based Separation

The process includes letting suspended particles fall under gravity to the bottom of tanks by keeping wastewater unaltered or somewhat unaltered for a variety of time lengths (Hamidi et al., 2022).

17.1.3.6 Aerobic Process

Lagoon, pond, and sludge-based aerobic wastewater treatment systems (Nguyen et al., 2021) efficiently eliminate organic, dissolved, and suspended contaminants, but they also generate bio-solids that need additional handling.

17.1.3.7 Anaerobic Process

When bacteria break down organic matter into simpler forms in wastewater, anaerobic decomposition takes place. This process produces gases such as methane, ammonia, hydrogen sulphide, and nitrogen. These gases are utilized to recover energy (Rout et al., 2021).

17.1.3.8 Encapsulation

Nitrogen gas, crucial for plant development, is present in the atmosphere and soils. Ammonia, which is used in fertilizers, is produced by the Haber-Bosch process from atmospheric nitrogen. However, it is primarily ejected as urea and NH^{4+} through the metabolism of humans. Traditional activated sludge treatment is effective for treating wastewater, but organisms with a slow growth rate, such as nitrates and creators of methane take time to be captured (Guerrero et al., 2019). Moreover, a greater concentration of biomass is necessary. Encapsulation and attachment are immobilization methods that enhance solid–liquid separation, cell concentration, and temperature sensitivity. Encapsulation uses ionotropic or thermal gelatine on porous polymeric materials such as polyvinyl alcohol (PVA), carrageenan, alginate, agar, and polyacrylamide (Guerrero et al., 2019). PVA is popular due to its stability and concentration.

17.1.3.9 Oxidation and Advanced Oxidation

Fenton's reagent, hydrogen peroxide, chlorine dioxide, ozone, and chlorine are some of the chemical oxidation methods used to treat and purify organic wastewater. Organic pollutants such as hydrocarbons, ammonia, phenols, and dyes are successfully eliminated by this approach. Large volumes of wastewater cannot be handled with advanced oxidation, even if it is more economical when paired with increased ultraviolet oxidation, peroxide and ozone, and photocatalysis (Khan et al., 2020). After main and secondary treatments, they ought to be utilized for disinfection instead.

17.1.3.10 Adsorption

By taking soluble molecules out of solid substrates by hand through the process of adsorption, organic contaminants are successfully removed from wastewater streams (Ayob et al., 2021). Adsorption is the process of transferring mass so that soluble molecules are taken out of a solid substrate (usually activated carbon). It removes a variety of organic substances, the most often of which is used is activated carbon, which is derived from biomass (De Gisi et al., 2016). For adsorption, activated carbon creates tiny, low-volume pores. Particle size, temperature, molecular weight, pH, iodine number, and adsorbate concentration are some of the variables that affect adsorption. Adsorption has an inverse relationship with how big the adsorbent is and a direct relationship with surface area. Another factor that affects adsorption is temperature. Because of surface charge, adsorption and pH are inversely related. When a solute that is either liquid or gas adsorbs on a surface made of a liquid or solid, an adsorbate is created. Chemical, physical, and ion exchange are examples of adsorption processes. Due to its dependence on temperature and local availability, adsorption is low-maintenance and cost-effective.

Current trends in liquid waste management make use of membrane technologies. This decade, membrane technologies have been widely used in the water treatment industry. Membrane-based systems are the most environmentally friendly and enduring for treating wastewater (Singh and Hankins (Eds.), 2016). This is because they don't require chemicals, are safe for the environment, and are simple to utilize across all industries. These membrane technologies have existed for some time due to their effectiveness and efficiency in producing superior effluent. A membrane is a selected fashion in which the component movement occurs in two phases. Membranes are classified as either isotropic or anisotropic (Cadee et al., 2011). The physical structure and composition of isotropic membranes are uniform. The non-uniform anisotropic state consists of many tiers, shapes, and patterns. The membrane's constituent materials can be categorized as organic and inorganic, respectively (Sales et al., 2006). Organic membranes are created using artificial organic polymer materials such as metals, ceramics, and zeolites are used to make inorganic membranes. The membrane can be categorized based on its shape, bulk arrangement, and utilization.

17.1.3.11 Membrane Technology

The following five membrane process types are frequently employed in non-equilibrium-based water and wastewater treatment:

- Ultra- and micro-filtration
- Membrane bioreactors
- Reverse osmosis
- Electrolysis
- Electrodialysis

By using these techniques, dissolved substances and finely divided particles can be extracted from liquids. All five technologies are based on polymeric membranes.

In ultra and micro-filtration, particles between 0.04 and 1 μm can be eliminated with this method as long as the total amount of suspended solids is less than 100 mg/L (Gupta et al., 2012). A pressure of one to three bars is applied to the filters, which are constructed from materials such as asbestos, nylon, fiberglass, cellulose, wool, rayon, and so on, and arranged in various designs. The life of the filter can only be extended by pre-removing suspended particulates.

A membrane bioreactor (MBR) combines membrane filtration devices—typically shallow pressured microfiltration or ultrafiltration membranes—with activated sludge. An MBR system typically has 10 or 11 subsystems, a post-disinfection process, and a good screen membrane zone. Membrane zone is the initial stage of a biological process when contaminants filtered by internal membranes are broken down by bacteria. During filtering, air is channelled through integrated diffusers that continuously massage the membrane's surface to help mix and encourage the introduction of oxygen to the biological process (Kim et al., 2015; Yin et al., 2020). The advantages of MBR include a smaller plant footprint, improved output quality, and a modular design that enables flexible assembly (Iorhemen et al., 2016).

Reverse osmosis has the potential to eliminate nearly 99% of organic dissolved substances, microorganisms, viruses, and other pathogens due to its narrower filter pore width compared to micro- and ultra-filtration (Gupta et al., 2012). Membranes consisting of cellulose, polyether, and polyamide, arranged in various shapes such as tube, disc, and spiral, are among those applied at pressures higher than the osmotic pressure. Nowadays, the most economical way to produce drinkable water from salted water and desalinated water is using reverse osmosis (RO). RO systems require very minimal maintenance because they are chemical-free, self-cleaning, and simply need power and high-pressure water. Pretreatment is also required to lower the concentration of dispersed and colloidal particles.

Electrolysis is a novel and differing approach from conventional therapy procedures. Soluble materials precipitate on the electrode surface during the electrolytic process, or they decay as a result of an electrochemical redox reaction. The majority of the organic materials are broken down into CO_2 and H_2O or other less hazardous or non-toxic substances, leaving ions of metal deposited on the surface of the electrode. This method has not yet been completely implemented for the processing of wastewater at the commercial level, mostly due to the present capital costs associated with electrodes (Fudge et al., 2021). Certain industrial wastewaters, such as those that are enriched in metal ions and organic materials, are treated with it.

High recovery rates and chemical-free operation are two benefits of electrodialysis (ED). The wastewater from slaughterhouses and desalination facilities and home sewage treatment centres can be used to produce water, as well as all other nutrients, salts, acids, and alkalis (Al-Amshawee et al., 2020). The concept underlying ED is that the membrane that exchanges ions acts as a kinetic energy to facilitate ion resolution separation through the use of electromechanical separation techniques. It is thought to be helpful in getting rid of organic debris, hardness, and electrolyte ions. The power source charges the electrode, which causes an electric current to flow through the ED stack. The nutritional solution's electrically charged anions travel to the anode, while the cation goes to the cathode. The membranes that exchange cations and anion block each other's cation passage, which results in a decrease in the concentration of salt in the dilution chamber and an upsurge in the concentration of salt in the concentrated portion (Mohammadi et al., 2021). The challenging recovery of nutrients, like sludge or gas that is discharged into the atmosphere, is the first step in the typical water treatment process. The benefit of ED is that, by means of the nutrient recycling process, it can simultaneously reduce the nutrient load on water resources and raise the nutrient load while concurrently boosting the amount of nitrogen and phosphorus available. Defilement with ED is the primary drawback of employing membranes to extract nutrients extracted from sewage. Anion exchange membranes alongside positive charges are more vulnerable to organic contamination because they strongly attract the majority of the negative charges of colloids and organic molecules (Mohammadi et al., 2021; Pankratz, 2017).

17.1.4 Artificial Intelligence

There is no broadly agreed, accepted phrase for artificial intelligence (AI) (Allen, 1998; Bhatnagar et al., 2018; Brachman, 2006; Kirsh, 1991; Monett and Lewis, 2018; Nilsson, 2009; Simon et al., 2000). It becomes harder for policy makers to forecast exactly what the AI platforms are going to accomplish in the near future as well as how the technology may get there, nonetheless due to the lack of an agreed-upon meaning of the term. Whether these types of AI algorithms are even desirable remained unaddressed (Bhatnagar et al., 2018). "The study and design of intelligent agents" is the contemporary definition of AI, whereas a system that recognizes its environment and acts to improve its chances of success is referred to as an intelligent agent. It is one of the newest disciplines, having emerged just after World War II. John McCarthy first proposed the name in 1956. Molecular biology and artificial science are two of the more popular scientific disciplines these days. As of right now, AI spans a wide range of subfields, from general-purpose domains like perceptual and educational processes to more specialized ones like chess playing, mathematical theorem proof, poetry composing, and disease diagnosis. It is theoretically applicable to all areas of human intellectual activity because it automates and systematizes intellectual tasks. AI is therefore a truly global field. It is focused on behaviour as well as mental processes and reasoning. Artificial intelligence can be evaluated against a standard definition of intelligence, which is typically rationality, or in terms of how closely it mimics human performance. If a system acts in accordance with its knowledge and accomplishes the "right thing," it is considered rational. Alan Turing created the Turing Test in 1950 in an attempt to give an adequate operational definition of intelligence. If a human interrogator asks some written questions and is unable to determine whether the written answers are from a person or not, the machine passes the test. Artificially intelligent systems operate inside a certain context and employ a particular rationality framework. An organization must be able to gather environmental data and utilize it as guidance when developing tasks. This is the agent in question. Furthermore, the agent needs to be able to solve this problem independently because tasks are needed to implement solutions. Usually, search-based issue solving is used to do this, in which choices are made after considering and looking through the surrounding circumstances. Everything boils down to logic (solving problems) and knowledge (obtaining information). Figure 17.1 depicts areas of application of AI.

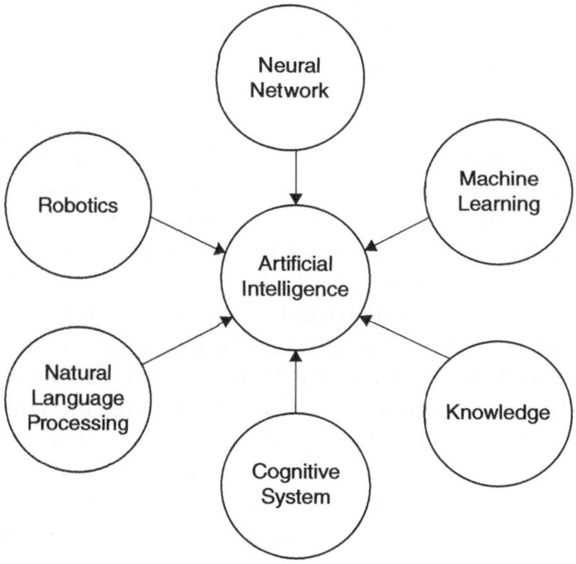

FIGURE 17.1 Areas of application of AI.

(Source: Compiled by the author)

17.1.5 Internet of Things

The Internet of Things (IoT) has been widely publicized on a national level and is widely regarded as the primary focus for advancements and innovations (Shi, 2022). A system of networked objects is called the IoT, while the internet itself is a network of communication that links individuals to knowledge (Verma, 2022). To enable the linking of objects with additional items and people at any point in time or location via any network, route, or service is, thus, the main objective of the IoT. Affordable cloud storage and improved statistics tools that let businesses handle and glean insights from massive amounts of data are two factors that have fuelled the growth of the IoT, particularly in workplace and industrial environments (Dubravac and Ratti, 2015). It's becoming widely acknowledged that the IoT represents the next evolution of the network. The IoT will make it possible to connect everyday objects to the internet for a variety of uses. Just 0.6% of potential IoT devices are thought to have been connected as of right now. In 2025, there are 5.56 billion internet users worldwide, representing 67.9% of the global population (Petrosyan, 2025). Also in 2025, there are 19.8 billion IoT devices worldwide (Vailshery, 2025). The IoT is a network of networks, or a collection of different "connected" items. As the internet grows, it has progressed beyond being a basic computer network. These days, a vast array of gadgets, including mobile phones, automobiles, industrial systems, cameras, toys, buildings, home appliances, and many more, may communicate with one another via the internet. No matter how big or little, these devices can perform process control, real-time monitoring, tracing, placement, control, and intelligent reorganizations. Wireless sensor networks possess crucial attributes and particular limitations when it comes to the IoT (Nazarov et al., 2022). To be able to guarantee the system's long-term viability with self-contained power supplies, wireless node hardware and network interaction protocols are also optimized for energy consumption (DeMedeiros et al., 2023). In recent years, there has been a notable surge in the quantity of gadgets with internet capabilities. Connecting people has become just one small part of a larger trend towards the merging of the digital and physical realms, even though its most notable commercial impact has been seen in the consumer electronics sector, specifically in the smartphone revolution and the growing popularity of wearable technology (watches, headsets, etc.). In light of all of this, it is projected that the range of devices and functionalities that the IoT can enable will continue to expand. This may be seen in the ambiguity around the name "Things," which makes it difficult to define the bounds of the Internet of Things, which are always expanding. As long as there is continued commercial success, the IoT will continue to offer an almost limitless array of options in both industry and research. The following study therefore talks about the many IoT domain applications that could be utilized as well as the associated research problems. Figure 17.1 depicts areas of application of IoT.

17.2 IOT FOR WASTE TREATMENT, MONITORING AND MANAGEMENT

Areas for application of IoT are shown in Figure 17.2. Waste sorting is very important for the proper disposal of large quantities of waste produced in our routine lives from numerous sources. Urban areas with growing economies that face impoverished waste collection services and fail to managee sanitary landfill collection exasperate the problem. The waste collection technique that was the execution of directives in many countries is an aggravation and the majority struggle due to the weak and fast-paced implementation. Currently, the volume of municipal solid waste is growing significantly with a rising population rate, economic upheavals, industrial evolution, changing utilization habits, and many other lifestyle factors of the urban population.

The proposed waste management system is based on a group of sensors and controllers. Wet and dry waste are separated and collected in different containers. This is done using the humidity sensor and IR sensor if the humidity sensor discovers moisture in the waste, only when the waste receptacle is open. Once the garbage receptacle is filled in a particular area, the garbage collector can collect waste (Kumar et al., 2016). The domain management company is notified to visit when a container is spilling over or approaching the saturation level so that the assigned sweeper could arrive in time and cancel it. IOT technology detects, activate, assembles, stores, and processes data by coupling devices to the internet (Zanella et al., 2014). Due to the IoT and AI, the waste management systems are shifting towards smarter and automized systems. There is a paradigm shift in challenges, resulting in a shift in management of wastewater treatment plants (Villarín & Merel, 2020).

The sweeper may not know the current scenario of a particular receptacle while it is simultaneously examining another receptacle; the previous receptacle may be full and possibly even overflowing. Thus, the proposed system would be favourable to the street sweeper as well as the person responsible for cleaning. Waste is classified into reusable waste, business waste, hospital waste,

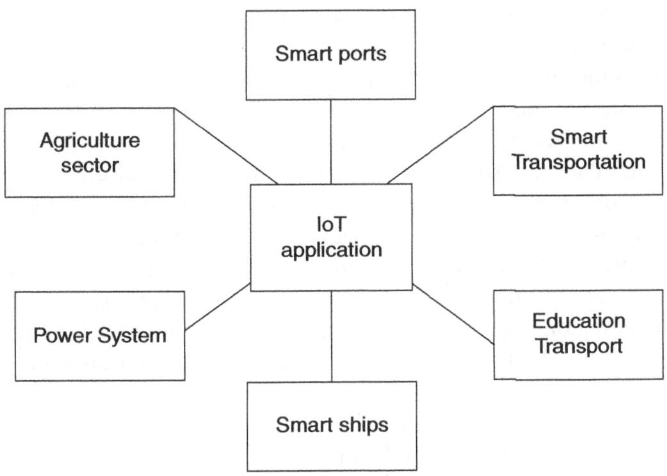

FIGURE 17.2 Areas of application of IoT.

(Source: Compiled by the author)

natural waste, virtual waste, nuclear waste, and plant waste (Balakrishna et al., 2020) according to the waste production supplier; for example business waste is classified according to its origin. Many studies predict that waste will reach massive quantities around 2025, with e-waste expected to make a major contribution due to its growth over time, and its use is expected to increase the cost of waste production. Real-time waste monitoring has been developed so that trash cans are exchanged as soon as they arrive (Shyam et al., 2017). IoT for waste management is being explored for researchers to make the waste management efficient in terms of time and handling (Haribabu et al., 2017, Dev et al., 2016, Anh Khoa et al., 2020).

Pollution may be prevented via removing chemical waste or other dangerous substances from landfills. Extracting the substances is expensive and risky, as it is not possible to differentiate among specific forms of waste and the capability of harm they could cause; in extreme cases, those methods can most effectively be performed via human methods. Many waste and recycling businesses have attempted to reduce separation and recycling costs, but to no avail. Resource restoration, including waste separation and recycling, often incurs costs with limited benefits. Resource refurbishment is a more intensive process involving separate waste collection (Corradini et al., 2022). To develop sustainable waste management strategies, Companies have used surveys and observations to assess individuals' intent to separate waste and willingness to pay for improved waste control services. Research has also examined environmental behavior using theories such as the theory of planned behavior, norm activation theory, and price discernment theory. In concept, gathering relevant information on recycled materials could cause higher recycling participation. This technique focuses on collecting and analyzing detailed data, making it customizable. Ideally, a device could predict recycling patterns based on disposal characteristics. This permits waste management corporations to monitor violations and recycling activities in real time (Kalirajan et al., 2021).

In this chapter, we explore how IoT technology can monitor wastewater released into water bodies like lakes, ponds, rivers, and so on. Sensors, which connect to the server whenever industrial waste is discharged, measure turbidity, temperature, and pH and transmit it to the server. When readings exceed set limits, waste is redirected to the recycling pit. This method repeats until safe levels are reached. Each time the system is triggered the user gets an alert. This real-time monitoring helps prevent the release of pollution into bodies of water, aids regulatory compliance, and allows environmental agencies to save resources and analyze disposal costs effectively (Krishnaraj et al., 2020). Water must be clean and sufficient for drinking, cleaning, and industrial use. Wastewater, which requires treatment before reuse, can harm animals, plants, and birds due to harmful chemicals and pathogens if left untreated. Effective wastewater management helps to recycle and conserve water rather than waste it. Proper treatment of effluent is vital to protect ecosystems and ensure safe reuse of water for various purposes (Deepnarain et al., 2020).

The effective utilization of wastewater helps reduce the environmental impact of pollution or industrial effluent. To prevent misuse, clear standards for water quality and safety measures must be established. Growing urbanization increases the risk of water scarcity, making clean water a basic human need (Jeong and Park, 2020). Recycling wastewater addressed this issue by storing treated sewage in an underground sump for reuse, reducing dependence on expensive groundwater and limiting the torrent of pollution into rivers and seas receiving wastewater discharge. Specific water applications may require additional treatment. Treating and recycling wastewater provides a cost-effective alternative to using freshwater sources such as surface water, rivers, and reservoirs. This is especially important in regions facing water scarcity and drought (Nayak et al., 2018). Untreated wastewater is often discharged into large bodies of water, causing environmental harm.

The reprocessing of wastewater is the best method to conserve water and limit contamination that harms the flora and fauna. Untreated wastewater does not naturally decompose (Rekha et al., 2020). Wastewater treatment removes pollutants and converts waste into usable water, potentially suitable for reuse or safe environmental discharge. In a basic treatment system, a Wi-Fi sensor network with a central hub is required. The data are collected, stored, and analyzed afterwards. The equipment uses a pump, effluent dumping chamber, and numerous sensor nodes to monitor fluid color changes.

Environmental factors like light and temperature can be regularly monitored by Wi-Fi sensors (Srivastava et al., 2019); however this Wi-Fi sensing was not very effective in managing water clarification treatments. Wi-Fi signals tend to experience interference and weakening when transmitted through water, making it challenging to collect precise and dependable data for monitoring water quality.

17.2.1 Application of IoT in Solid Waste Treatment, Monitoring, and Management

Figure 17.3 shows classification of solid waste. Many studies forecast that waste will reach a huge amount around 2025, while e-waste is expected to make a major benefaction due to its evolution over time, and its use is anticipate increasing the cost of waste production. The current tendency is that humans have the habit of throwing away their waste, usually each day, without enough insulation. They are unloading even if the trash can is already overflowing (Spirandelli et al., 2019). This results in poor environmental cleanliness which becomes a source of transportation. Authorities used to a sweeper to detach to remove easier waste once a day, but in some metro cities like Chennai, this is no longer enough. Additionally, the sweeper is no longer able to understand the current scenario of a positive field while simultaneously sweeping all other field, the preceding field will also be full and probably even overflowing. So sooner or later, during the rainy season, the overall work of the sweeper becomes boring and difficult (Memon et al., 2019). This is the time when the sweeper becomes boring and difficult. This is the time when the sweeper must take care of each thing. The proposed gadget should therefore be advantageous to both the street cleaner and the person accountable for cleaning process (Fallavi et al., 2017). Real time waste observing has been expanded so that trash cans are changed as soon as they arrive.

This device uses IOT program and sensors that ensure a welcoming environment and simplify the work loaded of waste collectors. The impartial of this study is to evolve a smart gadget that uses IOT devices and sensors to filter waste in real time more effectively and cost effectively. By using this device, we can reduce human efforts and ease the process of splitting wet and dry materials (Gupta et al., 2020). All real time facts can be updated in the mobile application connected to smart bin. This device offers an effective proceed towards hassle of waste control and a better environment. For example, a large-scale waste control device, using tool mastery necessity and different modelling description, can be designed to offer better usefulness than the proposed device based entirely on the internet of things. Some system developed using LoRa 32 TTL cento module at 433 MHz are superior hardware components used in conjunction with instrument mastery.

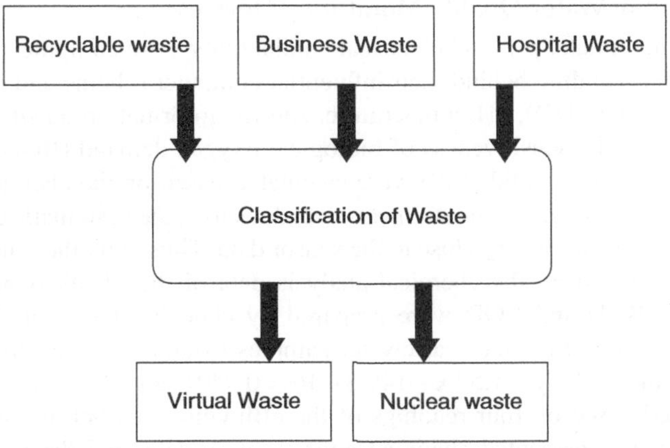

FIGURE 17.3 Classification of solid waste from IoT application point of view.

(Source: Compiled by the author)

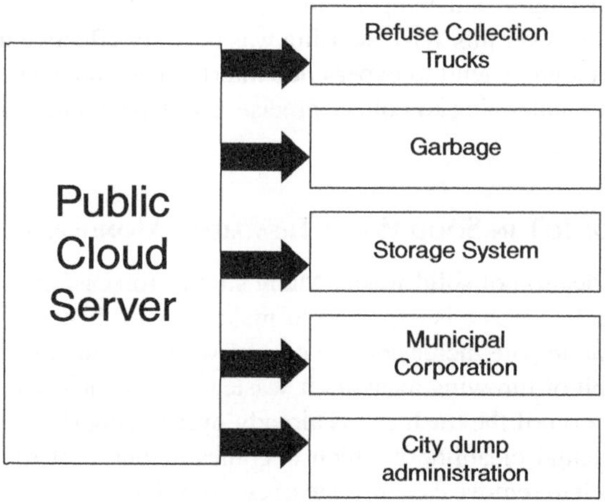

FIGURE 17.4 Architecture of SWC (Single wire Controller–area network) as a service.
(Source: Compiled by the author)

The fundamental principle for reducing waste produced daily is to apply the 3R principle or reduce, reuse and recycle (Nnaji et al., 2022). The justifiable development of the future will make wellness and groups increasingly crucial to ensure world class hygiene for humans in all areas of life. According to the government of India guidelines, emanate with input from the kingdom and major government, unconnected collection of waste according to organic criteria is predicted. Organic waste can be enriched in many different ways and has many interesting reprocess, including befoul and much more. Useful materials can be obtained from plant waste. Many governmental organizations and NGOs are working to reduce waste through various modern methods, such as the NGO called EFI, which launched an e-waste collection campaign in Chennai supported by training events and self-selected participants (Ojagh et al., 2021). The goal of the marketing campaign is to protect the environment and promote sustainable future collaborations. Figure 17.4 depicts architecture of a single-wire Controller Area Network.

17.2.2 Application of IOT in Liquid Waste Treatment, Monitoring, and Management

17.2.2.1 Evaluation of Water Quality Monitoring Devices

The dissolved oxygen, climate, pH, and electrical conductivity were computed by a senior consultant of the AQUAS Inc. standing behind their influential computer scheme and contrasting with the details of dissolved oxygen (DO), pH, temperature, and EC information on site (Capocasale et al., 2021). At the same time, the sensitiveness of biological oxygen demand (BOD), chemical oxygen demand(COD), and suspended solids (SS) were calculated based on the chemical analysis data of the identical wastewater sample. For the stainless-steel detectors, the systematic chemical data of the similar wastewater samples were very close to the sensor data. Thus, only the sensitivity of BOD and COD sensors was measured by the chemical-analysis data of the identical wastewater samples. Maturation curves of BOD and COD were prepared by choosing the systematic chemical data opposed to the sensor data of similar wastewater samples (Mahmoud and Mahmoud, 2021). The assessment curves of the COD ($y = 1.57x - 149.31$, $R^2 = 0.9401$) and BOD ($y = 0.24x + 29.10$, $R^2 = 0.9797$) were accepted based on four readings of the effluvium data before evaluating the sensor performance. The y-axis represented chemical concentration data, and the x-axis showed sensor readings (Kumar et al., 2016). The BOD and COD sensors were calibrated using periodic measurement curves and the connected wastewater sample with specialized equipment at AQUAS Inc.'s research facility.

17.2.2.2 Analysis of Water Quality

Sewage samples are analyzed for BOD, COD, and stainless steel valorization method. The wastewater sample is clarified and amended for anion and cations detection by using ion chromatography (or demineralization chromatography) (Şenol, 2021). Electrically charged conductance and the pH of the water sample were determined by pH/conductance meter (ExStik EC 500, EXTECH utensils, FLIR mercantile System, Goleta, CA, USA). Dissolved oxygen in the water sample was measured using a DO meter (Walk LAB, Trans implement PTE, Ltd, Singapore).

17.2.2.3 Statistical Analysis

A one-way ANOVA examination was performed using Origin 9.1 software (Original Lab, Northampton, MA, USA) followed by Tukeys test with a significance level of 0.05 (Sun et al., 2020).

17.3 AI FOR WASTE TREATMENT, MONITORING AND MANAGEMENT

Automatic technology is recognized as having three main levels: Artificial narrow intelligence (ANI), artificial general intelligence (AGI), and artificial superintelligence (ASI). According to Strelkova (2017) ANI, also known as weak AI, is designed to perform specific tasks within a single domain, such as Google Search. AGI, described by Wang (2019) as muscular machine learning, refers to AI that can reason. AGI is the evolution of AI that can reason and make decisions across various fields like a human, with autonomous vehicles being a common example. Culley et al. (2020) defined ASI as the stage where the AI surpasses human intelligence in nearly every aspect, including creativity, scientific reasoning, and interpersonal skills, making it more capable than the most intelligent human minds.

17.3.1 Application of AI in Solid Waste Treatment, Monitoring and Management

Waste classification plays a crucial role in the proper disposal of enormous quantities of waste generated daily existence from various sources. Built-up areas with growing economies face inadequate waste collection services and poor management of sanitary landfills, which worsens the problem. The waste collection method that was the execution of instruction in many countries is a provocation and the large part struggle due to weak and fast-stride recommendation (Kumari et al., 2023). The rise in municipal solid waste is driven by factors such as population growth, economic disruption, industrial development, changing consumption habits. To address this, an IoT-based waste management system has been proposed, using sensors and administrators to detect, sort, and monitor waste (Belenko et al., 2019). Moisture and IR sensors help separate wet and dry waste into appropriate bins. When bins approach capacity, the system alerts waste management teams so collectors can respond in time. Forecasts suggest that waste levels could peak around 2025, with e- waste becoming a major contributor due to increasing technology use. (Rahman et al., 2022). Despite these innovations, many people still routinely dispose of waste carelessly.

17.3.2 Application of AI in Liquid Waste Treatment, Monitoring, and Management

17.3.2.1 Water Standard Monitoring for Data Accession

Monitoring water quality is essential for collecting accurate data used in machine learning applications within Waste Water Treatment (WWT). Continuous observation of influent and effluent water using sensors generates valuable datasets (Singh et al., 2023). The probability of data used to train and test AI models essentially stays constant, Despite varying dataset sizes across studies, most use 60 to 80% of the data for testing and then retain the data for model training. Researchers have developed sensors capable of real-time detection, data analysis, and automated control (Carreres-Prieto et al., 2023; Emaminejad et al., 2022; Zhang et al., 2020, 2022). Key water quality metrics, like

BOD, COD, pH, and DO, are monitored for influent water. Advanced machine learning models in highly accurate prediction and monitoring of water quality, supporting public health and informing government decision-making about water quality management. Haimi et al. (2013) conducted a literature review to highlight the uses of data-driven soft sensors for the remote prediction and monitoring of organic wastewater treatment plant (WWTP). A comprehensive evaluation of developments in soft sensors for remote WWTP monitoring was also carried out by Ching et al. (2021). Additionally, Schneider et al. (2019) demonstrated that photoelectric sensors can effectively monitor on-site WWTP independently for a year without maintenance, demonstrating that powerful soft detectors have potential for long-term, real-time water quality monitoring. Classification of AI models is depicted in Figure 17.5.

17.3.2.2 Laboratory Scale Search

Due to its low cost, short experimental duration, security controllability, ease of use, and high reproducibility, laboratory-scale research has proven to be a valuable instrument in the development of novel procedure pattern and WWT innovations. wraps up our ordinary uses of artificial technology representation for lab-based experimentation. Design and bioreactor optimization are the primary areas of use of AI models in WWT in laboratory-scale research (Jawad et al., 2021; Kamali et al., 2021; Narayanan and Narayan, 2019). Using film procedures or membrane cultivation vessels (MBRs), some experiments have conducted thorough assessments of the applications of Artificial technology models for WWT in recent publications. laminate filtration (MFD) combined with biological (microbial) treatment is a practical and effective wastewater treatment technique. It essentially combines a traditional biotic behaviour system with vital corporeal liquid-solid integration, creating a hybrid system.

17.3.2.3 Process Design

AI representation has demonstrated significant benefits in mitigating or lessening the complication of the WWT process, which entails intricate process design and operational circumstances. A composite association between the different input and product adaptability can be established using AI models quickly and efficiently. Time, temperature, pH, the starting application of contaminants, and characteristics related to tributary water standard are frequently employed as input variables. The primary proposition variables are the removal and surface assimilation orderliness of pollutants, or the framework related to sewage water standard. AI models have been effectively used in a number of WWT–related areas, including energy consumption and operational parameter optimization, effluent water quality prediction, and WWT presentation (Shi & Xu, 2018). From the standpoint of

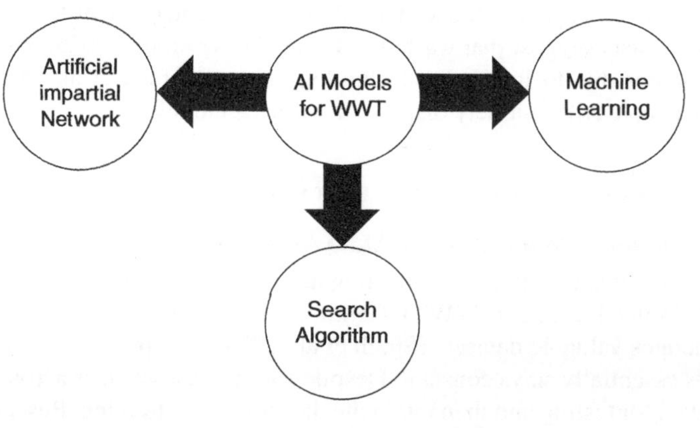

FIGURE 17.5 Classification of AI models.
(Source: Compiled by the author)

WWT process design, the applications, process parameter optimization and process performance prediction are the core foci of AI models (Bein et al., 2021).

17.3.2.4 Process Parameter Optimization

Process parameter optimization's primary goals are cost reduction and improved WWT efficiency. The use of AI models to optimize process parameters is compiled. In order to forecast the ideal operating parameters for increasing the green microphysics biomass in an algal biomaterial, Nayak et al. (2018) employed a composite ANN-GA prototype. The researchers discovered that the 4-12-1 cartography is the best network construction, with a maximum correlation coefficient of $R = 0.9947$ and a minimal mean square error. These variables increased the productivity of algal biomass by approximately 57%, and they had a CO_2 segregation rate of 578.1 $^{-1}$ d^{-1} and a COD depletion of 95.9 –2.4%.

17.3.2.5 Process Performance and Prediction

Researchers' attention has been drawn to process presentation, which is a crucial component of the WWT process. Pollutant removal efficiency, WWTP performance, ideal process conditions, and effluent quality are examples of common procedure performance forecasts. A concise overview of the ways AI models are being used to forecast process execution was provided by Nourani et al. (2018), who analyzed Nicosia WWTP performance using individual AI and ensemble models in order to forecast WWTP performance. The outcomes demonstrate that the NN troupe prototype has the best prognosis presentation among the troupe models, while the ANFIS model outperforms other single AI models. Additionally, to estimate the effluent water quality in real time Xie et al. (2022) paired GA with an enhanced feed forward neural network.

17.4 DISCUSSION

Using an array of studies involving the filling and emptying of garbage cans in real experimental surroundings, we created the framework that was suggested. The smart waste management mechanism using the Net Logo platform is created by utilizing numerous simulations in time T (minutes). Twenty to 25 trash cans are first allocated at random to a smart city with TBL 0% and TBC green. Throughout the test run, a single test bin (TB) is chosen, and several rounds are carried out on that particular bin, taking into consideration its waste or filling level. For a TB, there are eight different rounds that exactly match how much is put in the bin. These records contain a variety of readings concerning the act of tossing trash towards a bin. Bin-unloading is the process that happens when a bin is completely full. The weight level recorded in a testing bin over time is described. The weight of the solid waste in a bin is shown, expressed in kilograms per time T. An AI model that includes an ANN, or artificial neural network, can describe intricate nonlinear interactions between numerous inputs and outputs, yet at its fundamental level, it is a data-driven, black-box model. In other words, the AI model is able to offer an illustration of the relationship between the inputs and outcomes; it is unable to offer mechanism-level information about the topic of matter being studied. Even though more research needs to be done in the field of science, AI models have proven to be a powerful tool with exciting potential for engineering in WWT domains. The main reason is the persistent confusion that surrounds the fundamental technology of a significant amount of field-related WWT problems, although the AI models used in the previously mentioned research usually work well when confronting specific problems. Cloud and mobile applications exchange the distance determined by the ultrasonic sensor and the processing command transmitted by the controller. The node MCU, temperature sensor, moisture sensor, ultrasonic sensor, and Wi-Fi module make up the system. Numerous sensors, including weight sensors and infrared sensors, can be utilized to discover the presence of waste in the bin. On the other hand, the suggested approach uses ultrasonic sensors, which provide a precise interpretation of the waste level. A sensor (garbage application) uses ultrasonic sound waves to bounce off the garbage in the bin to determine how far it still has to be filled. Once that's done, it sends a notification to the mobile application that reads "RANGE EXCEEDED" along with a hardcoded bin ID.

17.5 CONCLUSION

Waste management becomes essential for attaining sustainability, enhancing urban mobility, cutting down on waste, and protecting natural resources, all of which are necessary for turning traditional cities into smart cities. The literature from recent years has been updated to investigate different features and attributes of intelligent waste management systems with artificial intelligence and IoT integration. Waste production is rising in direct proportion to the growth in the global population, urbanization, and industrialization. To safeguard the environment, it is crucial to have a comprehensive, efficient waste treatment system. As previously mentioned, to design a treatment facility that is distinct, effective, economical, and ecologically friendly, a variety of techniques and technologies, each with advantages and disadvantages, must be considered. A thorough literature analysis was conducted to describe solutions to the problems that were identified. This included considering data transmission, detection, analysis, and processing to arrive at an effective waste-handling solution. By tracking waste containers and the amount of garbage deposited, identifying high-demand locations in order to optimize the collection of solid waste, advising the shortest path and in fact engaging with the public to promote disposal when waste can be placed in the container, AI and the IoT can be used to foster good citizenship and prevent serious issues that arise from garbage piling up outside garbage collectors' driveways. Waste is efficiently collected and treated by the system that is suggested in this review article. It also saves time and fuel expenses. In light of this, it is possible to enhance the current waste management system and move society in the direction of a greener and healthier way of living.

REFERENCES

Al Huraimel, K., Alhosani, M., Gopalani, H., Kunhabdulla, S., & Stietiya, M. H. (2022). Elucidating the role of environmental management of forests, air quality, solid waste and wastewater on the dissemination of SARS-CoV-2. *Hygiene and Environmental Health Advances*, *3*, 100006.

Alaedini, A. H., Tourani, H. K., & Saidi, M. (2023). A review of waste-to-hydrogen conversion technologies for solid oxide fuel cell (SOFC) applications: Aspect of gasification process and catalyst development. *Journal of Environmental Management*, *329*, 117077.

Alam, P., Sharholy, M., Khan, A. H., Ahmad, K., Alomayri, T., Radwan, N., & Aziz, A. (2022). Energy generation and revenue potential from municipal solid waste using system dynamic approach. *Chemosphere*, *299*, 134351.

Al-Amshawee, S., Yunus, M. Y. B. M., Azoddein, A. A. M., Hassell, D. G., Dakhil, I. H., & Hasan, H. A. (2020). Electrodialysis desalination for water and wastewater: A review. *Chemical Engineering Journal*, *380*, 122231.

Allen, J. F. (1998). AI growing up: The changes and opportunities. *AI Magazine*, *19*(4), 13–13.

Altowayti, W. A. H., Shahir, S., Othman, N., Eisa, T. A. E., Yafooz, W. M., Al-Dhaqm, A., Soon, C. Y., Yahya, I. B., Che Rahim, N. A. N. B., Abaker, M., & Ali, A. (2022). The role of conventional methods and artificial intelligence in the wastewater treatment: A comprehensive review. *Processes*, *10*(9), 1832. https://doi.org/10.3390/pr10091832

Amran, N. A., & Mustapha, S. N. A. (2020). Oil–water separation techniques for bilge water treatment. In Chandrasekaran, P.T., Javaid, M.S., Aftab Sadiq, A. (Eds), *Resources of Water*. Intech Open: London, UK. https://doi.org/10.5772/intechopen.91409

Andeobu, L., Wibowo, S., & Grandhi, S. (2022). Artificial intelligence applications for sustainable solid waste management practices in Australia: A systematic review. *Science of the Total Environment*, *834*, 155389.

Anh Khoa, T., Phuc, C. H., Lam, P. D., Nhu, L. M. B., Trong, N. M., Phuong, N. T. H., Van Dung, N., Tan-Y, N., Nguyen, H. N., & Duc, D. N. M. (2020). Waste management system using IoT-based machine learning in university. *Wireless Communications and Mobile Computing*, *2020*, 1–13.

Assef, F. M., Steiner, M. T. A., & de Lima, E. P. (2022). A review of clustering techniques for waste management. *Heliyon*.

Ayob, S., Othman, N., Altowayti, W. A. H., Khalid, F. S., Bakar, N. A., Tahir, M., & Soedjono, E. S. (2021). A review on adsorption of heavy metals from wood-industrial wastewater by oil palm waste. *Journal of Ecological Engineering*, *22*(3), 249–265.

Badawi, A. K., Ismail, B., Baaloudj, O., & Abdalla, K. Z. (2022). Advanced wastewater treatment process using algal photo-bioreactor associated with dissolved-air flotation system: A pilot-scale demonstration. *Journal of Water Process Engineering*, *46*, 102565.

Balakrishna, S., Solanki, V. K., Gunjan, V. K., & Thirumaran, M. (2020). Performance analysis of linked stream big data processing mechanisms for unifying IoT smart data. In: Gunjan, V., Garcia Diaz, V., Cardona, M., Solanki, V., Sunitha, K. (Eds) *ICICCT 2019 – System Reliability, Quality Control, Safety, Maintenance and Management. ICICCT 2019*. Springer, Singapore. (pp. 680–688). https://doi.org/10.1007/978-981-13-8461-5_78

Bein, E., Zucker, I., Drewes, J. E., & Huebner, U. (2021). Ozone membrane contactors for water and wastewater treatment: A critical review on materials selection, mass transfer and process design. *Chemical Engineering Journal*, *413*, 127393

Belenko, V., Krundyshev, V., & Kalinin, M. (2019, July). Intrusion detection for Internet of Things applying metagenome fast analysis. In *2019 Third World Conference on Smart Trends in Systems Security and Sustainability (WorldS4)*, London, UK, 2019 (pp. 129–135). IEEE Explore.

Bhat, R. A., Dar, S. A., Dar, D. A., & Dar, G. H. (2018). Municipal solid waste generation and current scenario of its management in India. *International Journal of Advance Research in Science and Engineering*, *7*(2), 419–431.

Bhatnagar, S., Alexandrova, A., Avin, S., Cave, S., Cheke, L., Crosby, M., …, Hernández-Orallo, J. (2018). Mapping intelligence: Requirements and possibilities. In Müller, V. (Ed) *Philosophy and Theory of Artificial Intelligence 2017* (pp. 117–135). Springer, Cham. International Publishing.

Bosmans, A., Vanderreydt, I., Geysen, D., & Helsen, L. (2013). The crucial role of Waste-to-Energy technologies in enhanced landfill mining: A technology review. *Journal of Cleaner Production*, *55*, 10–23.

Brachman, R. J. (2006). AI more than the sum of its parts. *AI Magazine*, *27*(4), 19–19.

Cadee, K., O'Leary, B., Smith, P., Slunjski, M., & Bourke, M. (2011). World's first magnetic ion exchange (MIEX®) water treatment plant to be installed in Western Australia. In *American Water Works Association Conference, Washington, D.C. Proceedings* (pp. 11–15).

Cao, Y., & Pawłowski, A. (2012). Sewage sludge-to-energy approaches based on anaerobic digestion and pyrolysis: Brief overview and energy efficiency assessment. *Renewable and Sustainable Energy Reviews*, *16*(3), 1657–1665.

Capocasale, V., Gotta, D., Musso, S., & Perboli, G. (12–16 July 2021). A Blockchain, 5G and IoT-based transaction management system for SmartLogistics: An Hyperledger framework. In *Proceedings of the 2021 IEEE 45th Annual Computers, Software, and Applications Conference (COMPSAC)* (pp. 1285–1290), Madrid, Spain.

Capoor, M. R., & Parida, A. (2021). Current perspectives of biomedical waste management in context of COVID-19. *Indian Journal of Medical Microbiology*, *39*(2), 171–178.

Carreres-Prieto, D., García, J. T., Carrillo, J. M., & Vigueras-Rodríguez, A. (2023). Towards highly economical and accurate wastewater sensors by reduced parts of the LED-visible spectrum. *Science of The Total Environment*, *871*, 162082.

Chen, W., Ye, M., Li, M., Xi, B., Hou, J., Qi, X., Zhang, J., Wei, Y., & Meng, F. (2023). Characteristics, kinetics and product distribution on pyrolysis process for waste wind turbine blades. *Journal of Analytical and Applied Pyrolysis*, *169*, 105859.

Ching, P. M., So, R. H., & Morck, T. (2021). Advances in soft sensors for wastewater treatment plants: A systematic review. *Journal of Water Process Engineering*, *44*, 102367.

Corradini, E., Nicolazzo, S., Nocera, A., Ursino, D., & Virgili, L. (2022). A two-tier Blockchain framework to increase protection and autonomy of smart objects in the IoT. *Computer Communications*, *181*, 338–356.

Culley, C., Vijayakumar, S., Zampieri, G., & Angione, C. (2020). A mechanism-aware and multiomic machine-learning pipeline characterizes yeast cell growth. *Proceedings of the National Academy of Sciences*, *117*(31), 18869–18879.

De Gisi, S., Lofrano, G., Grassi, M., & Notarnicola, M. (2016). Characteristics and adsorption capacities of low-cost sorbents for wastewater treatment: A review. *Sustainable Materials and Technologies*, *9*, 10–40.

Deepnarain, N., Nasr, M., Kumari, S., Stenström, T.A., Reddy, P., Pillay, K., Bux, F. (2020). Artificial intelligence and multivariate statistics for comprehensive assessment of filamentous bacteria in wastewater treatment plants experiencing sludge bulking. *Environmental Technology & Innovation*, 19, 100853.

DeMedeiros, K., Hendawi, A., & Alvarez, M. (2023). A survey of AI-based anomaly detection in IoT and sensor networks. *Sensors*, *23*(3), 1352.

Dev, A., Jasrotia, M., Nadaf, M., & Shah, R. (2016). IoT based smart garbage detection system. *International Research Journal of Engineering and Technology*, *3*, 12.

DuBravac, S., & Ratti, C. (2015). The internet of things: evolution or revolution? *American International Group*.

Emaminejad, S. A., Morgan, V. L., Kumar, K., Kavathekar, A., Ragush, C., Shuai, W., Jia, Z., Huffaker, R., Wells, G., & Cusick, R. D. (2022). Statistical and microbial analysis of bio-electrochemical sensors used for carbon monitoring at water resource recovery facilities. *Environmental Science: Water Research & Technology*, *8*(10), 2052–2064.

Fallavi, K. N., Kumar, V. R., & Chaithra, B. M. (2017, February). Smart waste management using Internet of Things: A survey. In *2017 International Conference on I-SMAC (IoT in Social, Mobile, Analytics and Cloud) (I-SMAC) Palladam, India* (pp. 60–64). IEEE.

Ferronato, N., Mendoza, I. J. C., Portillo, M. A. G., Conti, F., & Torretta, V. (2022). Are waste-based briquettes alternative fuels in developing countries? A critical review. *Energy for Sustainable Development*, *68*, 220–241.

Fudge, T., Bulmer, I., Bowman, K., Pathmakanthan, S., Gambier, W., Dehouche, Z., Al Salem, A.M., & Constantinou, A. (2021). Microbial electrolysis cells for decentralised wastewater treatment: The next steps. *Water*, *13*(4), 445.

Guerrero, J. A., Almeida-Naranjo, C. E., & Villamar Ayala, C. A. (2019). Improvement of nutrients removal from domestic wastewater by activated-sludge encapsulation with polyvinyl alcohol (PVA). *Journal of Environmental Science and Health, Part A*, *54*(8), 721–727.

Gumisiriza, R., Hawumba, J. F., Okure, M., & Hensel, O. (2017). Biomass waste-to-energy valorisation technologies: A review case for banana processing in Uganda. *Biotechnology for Biofuels*, *10*, 1–29.

Gupta, O., Goyal, N., Anand, D., Kadry, S., Nam, Y., & Singh, A. (2020). Underwater networked wireless sensor data collection for computational intelligence techniques: Issues, challenges, and approaches. *IEEE Access*, 8, 122959–122974.

Gupta, V. K., Ali, I., Saleh, T. A., Nayak, A., & Agarwal, S. (2012). Chemical treatment technologies for wastewater recycling—An overview. *RSC Advances*, *2*(16), 6380–6388.

Haimi, H., Mulas, M., Corona, F., & Vahala, R. (2013). Data-derived soft-sensors for biological wastewater treatment plants: An overview. *Environmental Modelling & Software*, *47*, 88–107.

Hamidi, S., Banaee, M., Pourkhabbaz, H. R., Sureda, A., Khodadoust, S., & Pourkhabbaz, A. R. (2022). Effect of petroleum wastewater treated with gravity separation and magnetite nanoparticles adsorption methods on the blood biochemical response of mrigal fish (Cirrhinuscirrhosus). *Environmental Science and Pollution Research*, *29*, 3718–3732.

Haribabu, P., Kassa, S. R., Nagaraju, J., Karthik, R., Shirisha, N., & Anila, M. (2017, December). Implementation of a smart waste management system using IoT. In *2017 International Conference on Intelligent Sustainable Systems (ICISS)*, Palladam, India (pp. 1155–1156). IEEE.

He, Z. W., Yang, W. J., Ren, Y. X., Jin, H. Y., Tang, C. C., Liu, W. Z., Yang, C. X., Zhou, A. J., & Wang, A. J. (2021). Occurrence, effect, and fate of residual microplastics in anaerobic digestion of waste activated sludge: A state-of-the-art review. *Bioresource Technology*, *331*, 125035.

Iorhemen, O. T., Hamza, R. A., & Tay, J. H. (2016). Membrane bioreactor (MBR) technology for wastewater treatment and reclamation: Membrane fouling. *Membranes*, *6*(2), 33.

Jawad, J., Hawari, A. H., & Zaidi, S. J. (2021). Artificial neural network modeling of wastewater treatment and desalination using membrane processes: A review. *Chemical Engineering Journal*, *419*, 129540.

Jeong, S., & Park, J. (2020). Evaluating urban water management using a water metabolism framework: A comparative analysis of three regions in Korea. *Resources, Conservation and Recycling*, *155*, 104597.

Kalirajan, K., Nambiar, M. V., Vinodhini, K., & Ramya, E. (2021, May). IOT based industrial waste water monitoring and recycling. In *International Conference on Computing, Communication, Electrical and Biomedical Systems (ICCCEBS) 2021* 25-26 March 2021, Coimbatore, India. *J. Phys.: Conf. Ser.* 1916 012119.

Kamali, M., Appels, L., Yu, X., Aminabhavi, T. M., & Dewil, R. (2021). Artificial intelligence as a sustainable tool in wastewater treatment using membrane bioreactors. *Chemical Engineering Journal*, *417*, 128070.

Khan, A. H., Khan, N. A., Ahmed, S., Dhingra, A., Singh, C. P., Khan, S. U., Mohammadi, A. A., Changani, F., Yousefi, M., Alam, S., Vambol, S., Vambol, V., Khurshid, A., & Ali, I. (2020). Application of advanced oxidation processes followed by different treatment technologies for hospital wastewater treatment. *Journal of Cleaner Production*, *269*, 122411.

Kim, I., Choi, D. C., Lee, J., Chae, H. R., Jang, J. H., Lee, C. H., Park, P.-K., & Won, Y. J. (2015). Preparation and application of patterned hollow-fiber membranes to membrane bioreactor for wastewater treatment. *Journal of Membrane Science*, *490*, 190–196.

Kirsh, D. (1991). Foundations of AI: The big issues. *Artificial Intelligence*, *47*(1–3), 3–30.

Krishnaraj, N., Elhoseny, M., Thenmozhi, M., Selim, M. M., & Shankar, K. (2020). Deep learning model for real-time image compression in Internet of Underwater Things (IoUT). *Journal of Real-Time Image Processing*, *17*, 2097–2111.

Kumar, N. S., Vuayalakshmi, B., Prarthana, R. J., & Shankar, A. (2016, November). IOT based smart garbage alert system using Arduino UNO. In *2016 IEEE Region 10 Conference (TENCON)*, Singapore (pp. 1028–1034). IEEE.

Kumari, N., Pandey, S., Pandey, A. K., & Banerjee, M. (2023). Role of artificial intelligence in municipal solid waste management. *British Journal of Multidisciplinary and Advanced Studies*, *4*(3), 5–13.

Lin, C. C., & Wu, J. M. (2022). A novel centrifugal filtration device. *Separations*, *9*(5), 129.

Lin, K., Zhao, Y., Kuo, J. H., Deng, H., Cui, F., Zhang, Z., Zhang, M., Zhao, C., Gao, X., Zhou, T., & Wang, T. (2022). Toward smarter management and recovery of municipal solid waste: A critical review on deep learning approaches. *Journal of Cleaner Production*, *346*, 130943.

Mahmoud, M. S., & Mahmoud, A. S. (2021). Wastewater treatment using nano bimetallic iron/copper, adsorption isotherm, kinetic studies, and artificial intelligence neural networks. *Emergent Materials*, *4*, 1455–1463.

Mahyari, K. F., Sun, Q., Klemeš, J. J., Aghbashlo, M., Tabatabaei, M., Khoshnevisan, B., & Birkved, M. (2022). To what extent do waste management strategies need adaptation to post-COVID-19? *Science of the Total Environment*, *837*, 155829.

Maurya, D. K., Kumar, A., Chaurasiya, U., Hussain, T., & Singh, S. K. (2021). Modern era of microbial biotechnology: opportunities and future prospects. In Manoj Kumar Solanki, Prem Lal Kashyap, Rizwan Ali Ansari, Baby Kumari (Eds) *Microbiomes and Plant Health* (pp. 317–343). Academic Press, Florida.

Mazzei, H. G., & Specchia, S. (2023). Latest insights on technologies for the treatment of solid medical waste: A review. *Journal of Environmental Chemical Engineering*, *11*, 109309.

Memon, S. K., Shaikh, F. K., Mahoto, N. A., & Memon, A. A. (2019). IoT based smart garbage monitoring & collection system using WeMos & Ultrasonic sensors. *In 2019 2nd International Conference on Computing, Mathematics and Engineering Technologies (iCoMET)* (pp. 1–6), Sukkur, Pakistan. DOI: 10.1109/ICOMET.2019.8673526

Mohammadi, R., Tang, W., & Sillanpää, M. (2021). A systematic review and statistical analysis of nutrient recovery from municipal wastewater by electrodialysis. *Desalination*, *498*, 114626.

Monett, D., & Lewis, C. W. (2018). Getting clarity by defining artificial intelligence—A survey. In Müller, V. (eds) *Philosophy and Theory of Artificial Intelligence 2017. PT-AI 2017. Studies in Applied Philosophy, Epistemology and Rational Ethics*, vol 44 (pp. 212–214). Springer, Cham. https://doi.org/10.1007/978-3-319-96448-5_21

Muralidharan, V., Palanivel, S., & Balaraman, M. (2022). Turning problem into possibility: A comprehensive review on leather solid waste intra-valorization attempts for leather processing. *Journal of Cleaner Production*, *367*, 133021.

Narayanan, C. M., & Narayan, V. (2019). Biological wastewater treatment and bioreactor design: A review. *Sustainable Environment Research*, *29*(1), 1–17.

Nayak, M., Dhanarajan, G., Dineshkumar, R., & Sen, R. (2018). Artificial intelligence driven process optimization for cleaner production of biomass with co-valorization of wastewater and flue gas in an algal biorefinery. *Journal of Cleaner Production*, *201*, 1092–1100.

Nazarov, D., Nazarov, A., & Talu, S. (2022, August). Big data technologies in energy. In *AIP Conference Proceedings*, Almaty, Kazakhstan (Vol. 2656, No. 1). AIP Publishing.

Nguyen, P. Y., Carvalho, G., Reis, M. A., & Oehmen, A. (2021). A review of the biotransformations of priority pharmaceuticals in biological wastewater treatment processes. *Water Research*, *188*, 116446.

Nilsson, N. J. (2009). *The Quest for Artificial Intelligence*. Cambridge University Press, Cambridge, UK.

Nnaji, P. C., Anadebe, V. C., Onukwuli, O. D., Okoye, C. C., & Ude, C. J. (2022). Multifactor optimization for treatment of textile wastewater using complex salt–Luffa cylindrica seed extract (CS-LCSE) as coagulant: response surface methodology (RSM) and artificial intelligence algorithm (ANN–ANFIS). *Chemical Papers*, *76*(4), 2125–2144.

Nourani, V., Elkiran, G., & Abba, S. I. (2018). Wastewater treatment plant performance analysis using artificial intelligence–an ensemble approach. *Water Science and Technology*, *78*(10), 2064–2076.

Ojagh, S., Cauteruccio, F., Terracina, G., & Liang, S. H. (2021). Enhanced air quality prediction by edge-based spatiotemporal data preprocessing. *Computers & Electrical Engineering*, *96*, 107572.

Pankratz, T. M. (2017). *Screening Equipment Handbook*. CRC Press.

Patil, A. A., Kulkarni, A. A., & Patil, B. B. (2014). Waste to energy by incineration. *Journal of Computing Technologies*, 3, 12–15.

Petrosyan, A. (2025). *Number of Internet and social media users worldwide as of July 2024* [Data set]. Statista. Retrieved July 9, 2025, from https://www.statista.com/statistics/617136/digital-population-worldwide/

Radwan, N., & Khan, N. A. (2023). A systematic review of solid waste management (SWM) and artificial intelligence approach. 27 February 2023, PREPRINT (Version 1) available at *Research Square* https://doi.org/10.21203/rs.3.rs-2575251/v1

Rahman, M. W., Islam, R., Hasan, A., Bithi, N. I., Hasan, M. M., & Rahman, M. M. (2022). Intelligent waste management system using deep learning with IoT. *Journal of King Saud University-Computer and Information Sciences*, *34*(5), 2072–2087.

Rather, R. A., Wani, A. W., Mumtaz, S., Padder, S. A., Khan, A. H., Almohana, A. I., Almojil, S. F., Alam, S. S., & Baba, T. R. (2022). Bioenergy: A foundation to environmental sustainability in a changing global climate scenario. *Journal of King Saud University-Science*, *34*(1), 101734.

Rekha, P., Sumathi, K., Samyuktha, S., Saranya, A., Tharunya, G., & Prabha, R. (2020, March). Sensor based waste water monitoring for agriculture using IoT. In *2020 6th International Conference on Advanced Computing and Communication Systems (ICACCS)*, Coimbatore, India. (pp. 436–439). IEEE.

Rout, P. R., Zhang, T. C., Bhunia, P., & Surampalli, R. Y. (2021). Treatment technologies for emerging contaminants in wastewater treatment plants: A review. *Science of the Total Environment*, *753*, 141990.

Sajid, M., Raheem, A., Ullah, N., Asim, M., Rehman, M. S. U., & Ali, N. (2022). Gasification of municipal solid waste: Progress, challenges, and prospects. *Renewable and Sustainable Energy Reviews*, *168*, 112815.

Sales, M. G. F., Delerue-Matos, C., Martins, I. B., Serra, I., Silva, M. R., & Morais, S. A. C. M. (2006). A waste management school approach towards sustainability. *Resources, Conservation and Recycling*, *48*(2), 197–207.

Schneider, M. Y., Carbajal, J. P., Furrer, V., Sterkele, B., Maurer, M., & Villez, K. (2019). Beyond signal quality: The value of unmaintained pH, dissolved oxygen, and oxidation-reduction potential sensors for remote performance monitoring of on-site sequencing batch reactors. *Water Research*, *161*, 639–651.

Şenol, H. (2021). Methane yield prediction of ultrasonic pretreated sewage sludge by means of an artificial neural network. *Energy*, *215*, 119173.

Shah, A.V., Srivastava, V.K., Mohanty, S.S., & Varjani, S. (2021). Municipal solid waste as a sustainable resource for energy production: State-of-the-art review. *Journal of Environmental Chemical Engineering*, *9*, 105717.

Shi, H. (2022, December). Research on the application of the internet of things based on A-IoT. In *2022 6th International Seminar on Education, Management and Social Sciences (ISEMSS 2022)*, Chongqing, China. (pp. 1496–1502). Atlantis Press.

Shi, S., & Xu, G. (2018). Novel performance prediction model of a biofilm system treating domestic wastewater based on stacked denoising auto-encoders deep learning network. *Chemical Engineering Journal*, *347*, 280–290.

Shyam, G. K., Manvi, S. S., & Bharti, P. (2017, February). Smart waste management using Internet-of-Things (IoT). In *2017 2nd International Conference on Computing and Communications Technologies (ICCCT)*, Chennai, India (pp. 199–203). IEEE.

Simon, H. A., Bibel, W., Bundy, A., Berliner, H., Feigenbaum, E. A., Buchanan, B. G., Selfridge, O., Michie, D., Nilsson, N., Sloman, A., Waltz, D., Brooks, R., Davis, R., Shrobe, H., Boden, M. A., Michalski, R., Feldman, J., Dreyfus, H. L., Schank, R. C., Amarel, S., Hayes-Roth, B., Pearl, J., & McCarthy, J. (2000). AI's greatest trends and controversies. *IEEE Intelligent Systems and Their Applications*, *15*(1), 8–17.

Singh, N. K., Yadav, M., Singh, V., Padhiyar, H., Kumar, V., Bhatia, S. K., & Show, P. L. (2023). Artificial intelligence and machine learning-based monitoring and design of biological wastewater treatment systems. *Bioresource Technology*, *369*, 128486.

Singh, R., & Hankins, N. (Eds.). (2016). *Emerging Membrane Technology for Sustainable Water Treatment*. Elsevier Science, Amsterdam, Netherlands.

Sliusar, N., Filkin, T., Huber-Humer, M., & Ritzkowski, M. (2022). Drone technology in municipal solid waste management and landfilling: A comprehensive review. *Waste Management*, *139*, 1–16.

Sondh, S., Upadhyay, D. S., Patel, S., & Patel, R. N. (2022). A strategic review on Municipal Solid Waste (living solid waste) management system focusing on policies, selection criteria and techniques for waste-to-value. *Journal of Cleaner Production*, *356*, 131908.

Spirandelli, D., Dean, T., Babcock Jr, R., & Braich, E. (2019). Policy gap analysis of decentralized wastewater management on a developed pacific island. *Journal of Environmental Planning and Management*, *62*(14), 2506–2528.

Srivastava, A., Grotjahn, R., Ullrich, P. A., & Risser, M. (2019). A unified approach to evaluating precipitation frequency estimates with uncertainty quantification: application to Florida and California watersheds. *Journal of Hydrology*, *578*, 124095.

Strelkova, O. (2017). Three types of artificial intelligence. https://eztuir.ztu.edu.ua/bitstream/handle/123456789/6479/142.pdf?sequence=1&i

Sun, C., Puig, V., & Cembrano, G. (2020). Real-time control of urban water cycle under cyber-physical systems framework. *Water*, *12*(2), 406.

Vailshery L.S. (2025). *Number of IoT connected devices worldwide* [Graph]. Statista. Retrieved July 9, 2025, from https://www.statista.com/statistics/1183457/iot-connected-devices-worldwide/

Verma, S. B. (2022). Emerging Trends in IoT and Computing Technologies. In *International Conference of Emerging Trends in IoT and Computing Technologies (ICEICT 2022)*. Goel Institute of Technology and Management, Lucknow, India.

Villarín, M. C., & Merel, S. (2020). Paradigm shifts and current challenges in wastewater management. *Journal of Hazardous Materials*, *390*, 122139.

Wang, P. (2019). On defining artificial intelligence. *Journal of Artificial General Intelligence*, *10*(2), 1–37.

Xie, Y., Chen, Y., Lian, Q., Yin, H., Peng, J., Sheng, M., & Wang, Y. (2022). Enhancing real-time prediction of effluent water quality of wastewater treatment plant based on improved feedforward neural network coupled with optimization algorithm. *Water*, *14*(7), 1053.

Yin, X., Li, J., Li, X., Hua, Z., Wang, X., & Ren, Y. (2020). Self-generated electric field to suppress sludge production and fouling development in a membrane bioreactor for wastewater treatment. *Chemosphere*, *261*, 128046.

Zakaria, S. N. F., Abdul Aziz, H., & Mohamad, M. (2022). Comparison performance of coagulation flocculation process and combination with ozonation process of stabilized landfill leachate treatment. *Water Environment Research*, *94*(8), e10770.

Zanella, A., Bui, N., Castellani, A., Vangelista, L., & Zorzi, M. (2014). Internet of things for smart cities. *IEEE Internet of Things journal*, *1*(1), 22–32.

Zhang, W., Tooker, N. B., & Mueller, A. V. (2020). Enabling wastewater treatment process automation: leveraging innovations in real-time sensing, data analysis, and online controls. *Environmental Science: Water Research & Technology*, *6*(11), 2973–2992.

Zhang, W., Wei, S. P., Winkler, M. K., & Mueller, A. V. (2022). Design of a soft sensor for monitoring phosphorous uptake in an EBPR process. *ACS ES&T Engineering*, *2*(10), 1847–1856.

18 Nano-Biosensors for Detecting Environmental Pollutants

Bhaskar Garg

18.1 INTRODUCTION

Environmental pollution remains a pervasive challenge, driven largely by industrialization, urbanization, and agricultural activities. The release of hazardous substances, the environmental pollutants, into air, water, and soil has dire consequences for ecosystem and human health. The environmental pollutants are quite diverse, encompassing a wide array of substances such as heavy metals, pesticides, organic pollutants, and pathogenic microorganisms. Each type of pollutant among these prime categories has its own properties and impact on the surrounding environment as well as individuals as detailed in this chapter.

The expression heavy metals basically describes those metals that are relatively dense, exhibit high molecular weight or atomic mass, and are toxic even at minimal absorption. Though uncommon in the Earth's crust, heavy metals can enter the environment through industrial discharges, mining activities, agricultural runoff, and improper disposal of electronic waste. Owing to their nondegradability and tendency of bioaccumulation, heavy metals spawn a growing threat to individuals as well as to the environment. Among a variety of metals, mercury, lead, cadmium, and arsenic are considered as the most toxic. For example, lead contamination is of significant concern due to its widespread use in batteries, paints, and gasoline. Lead exposure can cause severe neurological and developmental issues, especially in children, leading to cognitive impairments and behavioral problems (Flora et al. 2012). Mercury is released into the environment through coal combustion, mining, and industrial processes. It can bioaccumulate in aquatic food chains, leading to high levels in fish and seafood. Chronic mercury exposure can damage the nervous, digestive, and immune systems (Rice et al. 2014). Similarly, cadmium is commonly associated with industrial emissions, phosphate fertilizers, and battery manufacturing. It accumulates in the human body, affecting the kidneys, bones, and respiratory system. On the other hand, arsenic contamination is prevalent in groundwater, particularly in regions with natural mineral deposits. Long-term exposure to arsenic-contaminated water can cause skin lesions, cancer, cardiovascular diseases, and diabetes in the individuals.

In line with heavy metals, pesticides from agricultural runoff can contaminate water bodies and affect nontarget species. The organochlorines, organophosphates, and neonicotinoids are some common pesticides that could persist in soil and water, entering the food chain. Owing to their long-term environmental impact and bioaccumulation in wildlife, the use of highly toxic organochlorines such as DDT and chlordane is restricted in many countries. Organophosphates such as malathion and parathion, on the other hand, are less persistent but highly toxic and inhibit acetylcholinesterase, causing neurological damage. Neonicotinoids are systemic pesticides that have been linked to declines in bee populations and other beneficial insects (Wood and Goulson 2017). In modern agro-settings, neonicotinoids are widely used.

DOI: 10.1201/9781003668961-18

Polycyclic aromatic hydrocarbons (PAHs) are well-known carcinogens generated from incomplete combustion of biomass and fossil fuels. These organic pollutants can cause various health problems upon long-term exposure (Kim et al. 2013). In contrast to PAHs, the organic pollutants from pharmaceuticals and personal care products (PPCPs) enter the environment through wastewater effluents and can affect aquatic organisms even at low concentrations. In general, PPCPs include a wide range of substances, ranging from antibiotics to endocrine-disrupting chemicals (EDCs) (Kümmerer 2009). Among persistent organic pollutants, bisphenol A (BPA) is considered as one of the most toxic industrial pollutants found in nearly all types of plastics and resins. BPA has been shown to disrupt endocrine function and is associated with various health issues, including cancer and reproductive disorders (Rochester 2013).

Waterborne pathogenic microorganisms such as viruses and bacteria serve yet other examples of pollutants that can contaminate water bodies in regions having inadequate water treatment infrastructure, posing serious health hazards. For instance, certain strains of *Escherichia coli (E. coli)* such as *E. coli* O157:H7 can cause severe gastrointestinal illness. Likewise, *Salmonella* species (*S. bongori* and *Salmonella enterica*) and *Vibrio cholerae*, the pathogens associated with waterborne outbreaks or contaminated water sources, can cause systemic infections and severe dehydration, respectively.

In the light of the preceding discussion, it is reasonable to state that a robust eco-monitoring system is a global priority due to the strong relationship between human health and environmental pollution. Consequently, the research practices focusing on the design and development of sensitive and selective detection methods for environmental pollutants are quite significant. Though effective and accurate, the conventional detection methods, for example, chromatography and mass spectrometry, are often cumbersome, expensive, unsustainable, and require sophisticated laboratory settings. This scenario necessitates the development of more efficient, sensitive, sustainable, and cost-effective detection technologies that could also function in urgent situations or real-time problems such as COVID-19 pandemic.

Among the existing methods for environmental pollutants detection, nano-biosensors have emerged as a promising alternative owing to their high sensitivity, specificity, rapid response times, portability, miniaturization, and many other useful properties. A critical analysis pertaining to the advantages of using nano-biosensors over conventional methods for detection of environmental pollutants is summarized in Table 18.1.

TABLE 18.1
An Overview on the Advantages of Using Nano-biosensors over Traditional Methods for Detecting Environmental Pollutants

Aspects	Advantages of Nano-biosensors	Traditional Methods
Sensitivity	Detect pollutants at very low concentrations due to high sensitivity.	Often require higher concentrations of pollutants.
Selectivity	High selectivity due to specific interactions with target pollutants.	Suffer from lower selectivity and interference from nontarget substances.
Response time	Rapid response time enables real-time monitoring.	Time-consuming, requiring lengthy sample preparation and analysis.
Miniaturization	Small size allows for deployment in various and remote environments.	Often require bulky equipment and are less suitable for in situ or portable applications.
Cost-effectiveness	Lower costs due to small material usage and mass production capabilities.	Involve higher costs for equipment, reagents, and labor-intensive processes.

(*Continued*)

TABLE 18.1 *(Continued)*
An Overview on the Advantages of Using Nano-biosensors over Traditional Methods for Detecting Environmental Pollutants

Aspects	Advantages of Nano-biosensors	Traditional Methods
Power consumption	Low power consumption, beneficial for portable and remote sensing applications.	Typically require significant power for operation and maintenance.
Versatility	Can be designed to detect a wide range of pollutants.	Typically require different instruments and procedures for various types of pollutants.
Data accuracy	High data accuracy due to precise detection capabilities of nanomaterials.	High and variable but may require extensive calibration and validation.
Real-time monitoring	Facilitates continuous and real-time monitoring of environmental conditions.	Generally provide delayed results due to offline analysis.
Environmental impact	Low impact due to minimal use of materials and small sensor size.	Typically require large quantities of chemicals and produce more waste.
Integration with IoT	Easy integration for continuous data monitoring and transmission.	Traditional methods are less adaptable to IoT integration.
User-friendly	Can be designed for easy use even by non-specialists.	Often require specialized training and expertise for accurate operation.

Overall, this chapter aims to provide a critical overview on nano-biosensors, including their working principles, types, and applications in the detection of various environmental pollutants. The chapter will also cover an adequate discussion on future prospects and challenges in the field. To the best of our knowledge, the content(s) presented herein are not covered previously within the proposed framework.

18.2 NANO-BIOSENSORS

18.2.1 An Overview of Major Components of Nano-Biosensors

Nano-biosensors are advanced analytical devices that integrate the principles of nanotechnology with biological recognition mechanisms to detect environmental pollutants with high sensitivity and specificity. These sensors are designed to recognize and quantify trace amounts of contaminants, making them invaluable in environmental monitoring (Dhahi et al. 2024). In general, nano-biosensors can be broadly classified into two types as detailed in the following.

18.2.1.1 Biological Recognition Element

The biological recognition element or bio-receptor is the core component responsible for the specific interaction with the target analyte, that is pollutant under study. Varying types of bio-receptors can be used such as an enzyme or protein, antibody, nucleic acid, and whole cells. Enzymes are extensively utilized in the development of biosensors, with one key advantage being the ease of scaling up their production (Sarkar et al. 2019). Mechanistically, the enzymes catalyze specific reactions with target pollutants, providing high specificity. However, the activity of enzymes is largely dependent on temperature and pH conditions. These issues can be mitigated by immobilization of active enzymes on a variety of substrates having high surface areas, providing long-term stability and reproducibility, an important aspect in the development of enzyme-based biosensors.

On the other hand, antibodies, the proteins generated by the immune system to recognize and combat infections, play a crucial role in biosensors. When an antigen binds to an antibody within a biosensor, a concentration-dependent signal is generated and arrested by a transducer (Zhang, Zeng, and Liu, 2017), allowing for highly selective detection. These antibodies can be tailored to target a wide range of molecules, including those with low molecular weights. However, the process of immobilization can alter an antibody's affinity and specificity, thereby affecting the performance of the biosensor.

Genosensors, the DNA biosensors, utilize DNA or RNA as their biological recognition elements. These sensors exhibit high selectivity by interacting specifically with their complementary polymerase chains, and their sensitivity can be enhanced by increasing the chain length of the gene (Pohanka 2018). Genosensors are capable of detecting DNA/RNA fragments as well as various other biomolecules or chemical compounds. However, due to certain limitations, the applications of most of the genosensors are restricted to clinical settings.

Aside from those of enzymes, antibodies, and genosensors, the whole-cell biosensors employ entire cells as recognition elements. Although this approach can reduce specificity with relatively longer response times, the whole-cell biosensors are highly versatile and can interact with a diverse array of target compounds, including, but not limited to, metal ions, hormones, EDCs, and pathogens (Sciuto et al. 2019).

Over the past several years, the whole-cell biosensors have appeared as a cost-effective and highly specific and sensitive substitute for traditional bio-receptors. These bioreporters are living microorganisms, having chromosomes and containing regulatory promoter(s). When the regulatory promoter is activated or repressed by a specific chemical or environmental stress, the reporter gene produces detectable signals (Jiang et al. 2021). In general, the whole-cell biosensors can be categorized into three classes based on the parameters they detect and how these parameters are transformed into quantifiable signals: whereas class-I bioreporters respond to precise objects by mounting the output signal, class-II bioreporters react toward strain situations by boosting the output signal. In contrast, class-III bioreporters respond to the compounds by decreasing the output signal. These bioreporters are particularly suitable for use as water monitoring biosensors. While class-I whole-cell biosensors are commonly used in detecting water contaminants, class-II and -III are also employed for similar applications but to a lesser extent (Zhu, Elcin, *et al.*, 2022). Overall, the whole-cell biosensors can respond to the presence of pollutants by changing their metabolic activity, which can be measured with ease.

Molecularly imprinted polymers (MIPs) represent yet another class of synthetic receptors that are created by incorporating a target template during the synthesis process, resulting in the generation of particular detection sites (Haupt and Mosbach 2000). These sites enable the MIPs to selectively detect the target analyte. MIPs are known to address the restrictions associated with the stability of enzymes and the high production cost of antibodies. MIPs can exhibit high affinity and specificity, offering greater stability compared to natural receptors (Shrivastav et al. 2015). Their primary applications include the detection of pharmaceuticals, with detection limits comparable to those of antibodies. Although the efficiency and robustness of MIPs are still emerging, they hold potential as feasible alternate materials as far as bulk synthesis and profitability are concerned in the regime of nano-biosensors.

18.2.1.2 Transducer

The transducer converts the biological interaction into a measurable physical signal. The type of signal generated depends on the transduction mechanism employed. In analogy to the biological detection element, transducers can also be of different types. For instance, transducers that detect changes in the absorption of light, fluorescence of light or phosphorescence of light are termed as optical transducers. Similarly, the electrochemical transducers measure changes in the electrical properties such as voltage, current, and so on. The thermal transducers measure changes in temperature as a result of biological interaction. Apart from these, the detectable changes in either mechanical or mass properties on the surface of any sensor can be attributed to the piezoelectric transducers.

At this juncture, it would be worthwhile to shed light on different components and the processing of the signal. Basically, there are three major components, namely, amplifiers, filters, and analog-to-digital converters (ADC) in a signal processor. The amplifiers increase the strength of the signal for

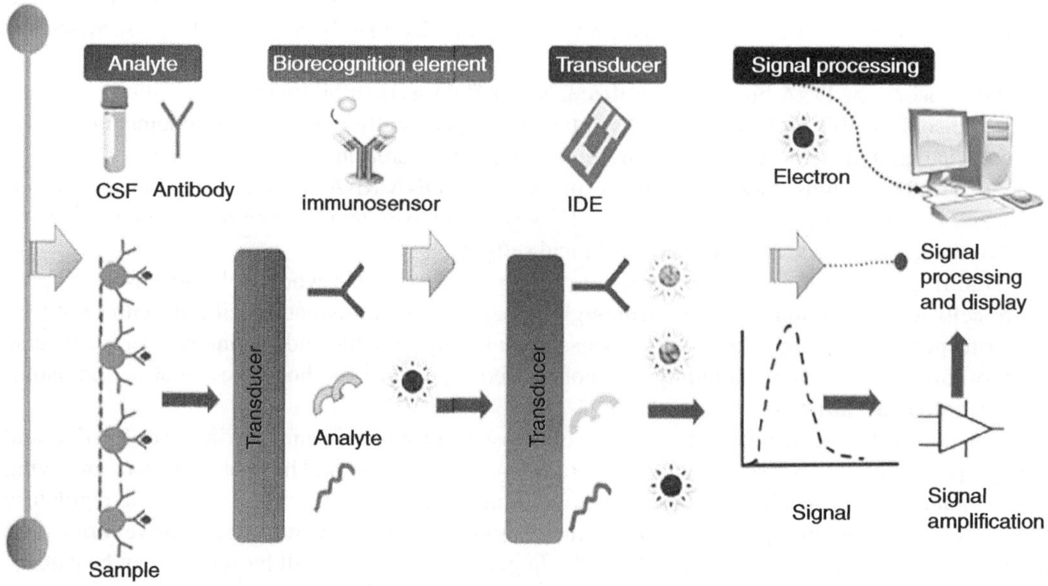

FIGURE 18.1 Schematic representation of a nano-biosensor, highlighting its three main components: the biological recognition element, the transducer, and the signal processor. It also illustrates the interaction between the target analyte and the biological recognition element, followed by the transduction of this interaction into a measurable signal, and finally, the processing of the signal for output.

Reproduced from ref. (Dhahi et al. 2024)

better detection. The filters remove background noise and interference, and ADC convert the analog signal into a digital form for further analysis and display. In brief, the signal processor amplifies and processes the raw signal from the transducer into a form that is easily interpretable. This step is crucial for enhancing the signal-to-noise ratio and ensuring accurate measurements. The fundamentals associated with the working action of nano-biosensors are represented in Figure 18.1.

18.2.2 Advancements in Nano-Biosensor Technologies

Over the past decades, nanotechnology has sparked significant curiosity and importance in the study of extremely controlled materials at nanoscale. The hierarchical arrangement of nanomaterials (NMs) with sophisticated nano-architectures offers substantial potential for wide technological applications including biosensing. Typically, the interaction between surface organizations, morphology, and physico-chemical properties leads to the creation of excellent biosensing materials. Among such materials, gold nanoparticles (Au NPs), carbon nanotubes (CNTs), graphene and related materials, and nanodots have opened up numerous avenues in the regime of nano-biosensor technologies. A summary of these intriguing materials are given in the following subsections.

18.2.2.1 Carbon Nanotubes

CNTs are remarkable NMs characterized by their high electrical conductivity, exceptional mechanical strength, and extensive surface area, making them particularly suitable for enhancing the sensitivity of biosensors. These properties enable CNTs to serve as excellent transducers in biosensor devices. They can be functionalized with a variety of bioreceptors such as enzymes, antibodies, or DNA, which allows them to detect specific pollutants with high precision.

Recent advancements have highlighted the versatility of CNTs in detecting a wide range of contaminants. For instance, a multi-walled (MW) CNT–based biosensor has been effectively utilized to detect heavy metals like cadmium with high sensitivity and selectivity. Differential pulse voltammetry (DPV) studies have shown that functionalizing MWCNTs with ethyl green (EG), an indicator dye, can detect Cd (II) ions as low as 2 nM detection limit, with a sensitivity of ca. $5nAnM^{-1}$ (Sreekanth et al. 2021). Besides heavy metals, functionalized CNT biosensors have also been used to identify pesticides, pharmaceuticals, and other environmental pollutants through π-π stacking, hydrophobic, and other non-covalent interactions. Furthermore, CNTs have shown great promise in pathogen detection too. By attaching specific antibodies or DNA probes to the CNT surface, these biosensors can detect bacteria, viruses, and other microorganisms with high specificity and rapid response times. This application is particularly valuable in medical diagnostics and environmental monitoring, where timely detection of pathogens is crucial. As such, the continuous development of CNT–based nano-biosensors underscores their potential in detection and quantification of environmental pollutants (Ferrier and Honeychurch 2021; Meskher et al. 2023).

18.2.2.2 Graphene

Graphene, one of the newest carbon materials, is renowned for its extraordinarily high electrical conductivity, mechanical strength, tunable bandwidth, amendable optical properties, and a relatively large specific surface area. Since its inception in 2004, researchers worldwide are continually making new breakthroughs in the study and applications of graphene and related materials across various scientific and technological domains, including biosensing (Fadeel et al. 2018; Garg et al. 2014, 2016; Garg, Bisht, and Ling, 2015; Garg and Ling 2013, 2018; Garg, Sung, and Ling, 2015; Kulakova and Lisichkin 2021). For instance, many gas sensors based on graphene domain are notable for their extreme sensitivity, enabling single-molecule detection. On the other hand, recent advancement of biomedical devices focuses on reducing the size of sensor element(s) while retaining or increasing the selectivity and sensitivity, a feature directly linked with the high surface area and an excellent electron mobility of graphene materials. The similar features in conjunction with ease of functionalization or modification of graphene surface have also been exploited in the reliable detection, in particular, electrochemical detection of a variety of pesticides such as organophosphates, organochlorine, nitro pesticides, and pesticides having non-aromatic rings (Kumaran et al. 2022; Zhu, Feng, and Lian, 2022). For example, in an elegant work, Tan et al. used a graphene-based nanocomposite electrode in place of conventional glassy carbon electrode (Tan et al. 2015). The nanocomposite was developed by making use of gold, a molecularly imprinted polymer (synthesized from monomeric methyl acrylic acid), and reduced graphene oxide (rGO). This modification led to significant improvements in the linear range and lower detection limit when carbofuran, a highly toxic pesticide, was used as an analyte. Specifically, the graphene-modified nanocomposite electrode exhibited a noticeable decrease in the charge-transfer resistance relative to that of conventional electrode. Impressive as these results on their own are, this study did not elucidate the fundamental mechanisms accountable for these improvements.

Given the noteworthy detrimental effects of glyphosate, an herbicide, on both public health and environment, a screen-printed gold electrode was used to anchor graphene-based composite material for the detection of glyphosate in different samples of river water (Thanh et al. 2021). The as-developed nano-biosensor could successfully detect glyphosate with a detection limit as low as 0.08 ppb. In yet other work, an array platform comprising gold electrode on graphene paper was found to be effective in recognizing glyphosate at nanomolar scale (Scandurra et al. 2022). Aside from the use of electrochemical methods, graphene has also been used to develop fluorescent detection techniques. In this context, for instance, a graphene-based nano-biosensor was developed to detect BPA following Förster resonance energy transfer (FRET) (Gupta and Wood 2017). The detection limit of graphene-biosensor was found to be 0.1 ng/mL. These works underscore the capability of graphene and related materials in developing highly sensitive yet selective nano-biosensors for comprehensive environmental monitoring, where a variety of contaminants may be present.

18.2.2.3 Quantum Dots (QDs)

A quantum dot (QD) is an inorganic core having organic molecules attached to its surface, representing an innovative type of highly fluorescent materials of size 1–10 nm, in general. As such, quantum dots are semiconductor nanocrystals, typically derived from the groups II, III, V, and VI of the periodic table, that exhibit unique optoelectronic properties due to quantum confinement effects. For instance, the larger QDs (5–6 nm in size) produce red or orange emission while green or blue emission is generated from small-sized QDs (2–4 nm in size). Nevertheless, these specific emission colors are directly related to the precise composition of QDs. Unlike traditional organic dyes, QDs exhibit high resistance to photobleaching, making them stable under prolonged exposure to light. In this context, they can absorb light over a wide range of wavelengths but emit light at a relatively narrow specific wavelength, enhancing the signal-to-noise ratio. Based on their structure, in general QDs can be categorized into several types such as doped QDs, core-shell QDs, and others. Their exceptional ability to tune the emission colors provides a handle to use them for the simultaneous detection of multiple targets. Furthermore, their facile modifications, preferably encapsulation of their surfaces with a variety of ligands (for instance, polyethylene glycol and mercaptopropionic acid) or synthetic reactions like salinization, escalate their potential as nano-biosensors (Zhang et al. 2022). Though effective, the inherent toxicity of metal-based QDs impedes their sustainability for practical applications. This drawback can be circumvented using zero-dimensional carbon nanodots (CNDs) or CQDs.

CNDs can be considered as quasi-spherical carbon-based nanoparticles (CNPs) having either amorphous or crystalline core. Based on the nuclear magnetic resonance spectroscopy, it was hypothesized that the internal core of CNDs is largely sp^2-hybridized while the outer core contains carbon atoms with sp^3-hybridization. The major components of CNDs are C, H, O, and N elements in a varying ratio subjected to their synthesis protocol. The UV-Visible absorption band of CNDs typically occurs in the range from 260 to 320 nm, which may experience a bathochromic shift after functionalization of the CNDs' surface with silicon and amine compounds. The as-resulted surface passivation is a key factor for exceptional photoluminescence properties of CNDs, manipulating the photo-induced electron-transfer mechanism in a variety of practical applications, including biosensing and enzyme mimetics (Garg and Bisht 2016; Ke *et al.* 2014; Liu et al. 2007). Furthermore, the end-results can be enhanced by integrating CNDs with a variety of metal NPs and dopants or developing CNDs from other nanomaterials of superior properties such as graphene and graphitic carbon nitride. In such cases, the synthesized NDs are commonly referred to as GQDs or N-doped GQDs. For instance, by intelligent use of co-precipitation method, GQDs were integrated with Fe-NPs to develop composite materials, namely, GQDs-Fe_3O_4. Owing to the synergistic interactions between Fe-NPs and GQDs, the as-developed GQDs-Fe_3O_4 composite materials were evaluated to be the exceptional catalysts in the detection and removal of a variety of phenolic compounds from water, providing a way to mimic the natural peroxidase enzymes (Wu et al. 2014). Analogous with carbon, water soluble SiQDs have also been utilized as effective biosensors in the fluorescent detection of environmental pollutants (Zhang et al. 2023), for instance, metal ions and bacteria, following a variety of mechanisms such as FRET, metal-enhanced fluorescence, inner filter effect, and surface energy transfer.

In the light of aforesaid discussion, it is reasonable to state that the advancements in nanofabrication techniques have significantly raised the performance of diverse nano-biosensors in the realm of environmental pollutants. At this juncture, however, it is worth commenting here that the different classes of nanomaterials, 0D QDs, 1D CNTs, and 2D graphene, reasonably overlap or intersect each other, and should be taken as a flexible guide to define their types as summarized in Figure 18.2.

18.3 OPTICAL NANO-BIOSENSORS

Optical nano-biosensors rely upon optical transduction due to the interaction between electromagnetic radiation and a specific analyte of interest. These specific interactions in conjunction with kinetic studies can be accurately studied through different analytical tools such as UV-Visible, fluorescence, and Raman spectroscopy, avoiding any magnetic or electrical interference. Despite the difficulties associated with their miniaturization, the optical nano-biosensors are quite sensitive and

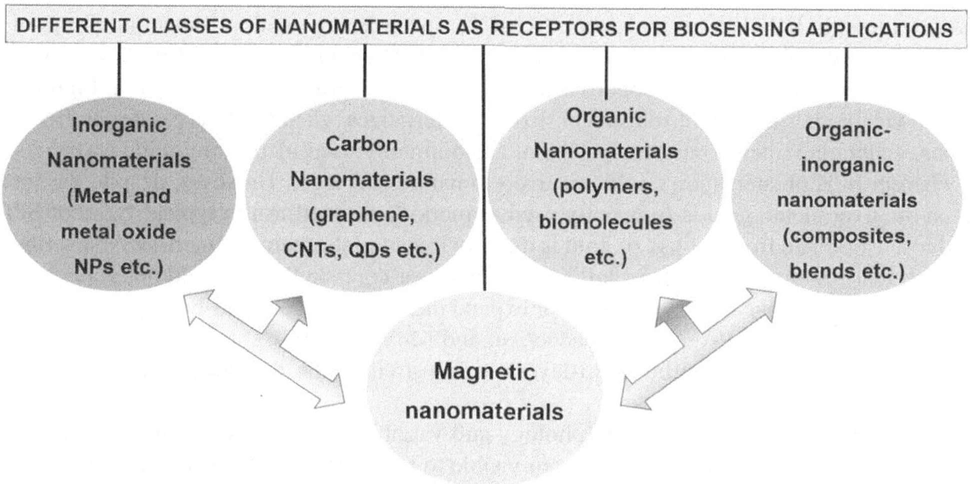

FIGURE 18.2 A schematic showcasing the different classes of nanomaterials used as nano-biosensors for environmental pollutants.

capable of detecting a variety of pollutants at nanomolar or even higher levels. In analogy to optical sensors, in recent years, optical biosensors have been designed to leverage the optical components of mobile phones for capturing signals and converting them into quantifiable values. For instance, high-resolution cameras facilitate data acquisition, and the phones' built-in lights serve as sources for excitation of light. Consequently, the optical features (like no. of pixels, fluorescence, color etc.) relevant to a particular image can be examined to extract the appropriate data. However, the complexity of this data demands the use of deep-learning and machine-learning tools. In an elegant work, recently, labeled antibodies that could emit bright fluorescence have been used to detect and measure toxic BPA in different water bodies through the use of mobile platform as mentioned previously (Cheng et al. 2022). This work holds great promise for on-site detection of water pollutants in the near future. At this juncture, it would be worthwhile to shed light on different parameters that could be required to evaluate for biosensing applications. For example, fluorescence quenching is one of the most important parameters that simply points to a bimolecular-type interaction. This type of interaction usually results in decreasing the fluorescence intensity of the fluorophore. This aspect becomes hard to evaluate sometimes as a significant quenching in fluorescence can also be observed due to environmental factors. Nevertheless, this issue can be circumvented efficiently if the analyte itself serves as a fluorophore. FRET–based quenching is yet other parameter, which typically occurs when the fluorescence spectrum of one species, the donor, overlaps with the excitation spectrum of another species, the acceptor. This results in the non-radiative transferal of energy from donor to acceptor, quenching the fluorescence intensity of donor (Gupta and Wood 2017). The other two fluorescence parameters include life-time measurement and the intensity measurements of the spontaneous fluorescence emission.

Aside from that use of fluorescence-based optical transducer, a number of other transducers such as surface plasmon resonance (SPR) including localized SPR (LSPR), interferometry, fiber optics, gratings, and optical resonators have also been employed in the detection of water pollutants. In detail, interferometry utilizes different optical paths as waveguides, producing controllable interferences. Interferometry provides a way to regulate the refractive index (RI) or by measuring the intensity of light beams. Currently, a variety of interference spectroscopies such as reflectometric interference (RIf) are available for the detection of different water pollutants, in particular, pathogens. For example, in an elegant work, mesoporous silicon having a diameter of 10–40 nm and inexpensive lectins were integrated to develop a biosensor that could successfully detect *E. coli* and

S. aureus through RIf-Fourier transform (RIf-FT) spectroscopy with a detection limit of ca. 10^3 cells/mL, and a linear response of 10^3–10^5 cells/mL (Yaghoubi et al. 2020). A schematic for overall surface modification of mesoporous silicon substrate as transducer is shown in Figure 18.3.

SPR is yet another important transducer, which materializes when photons excite the free metal electrons, generating a temporary wave. The most commonly used metals are gold (Au) and silver (Ag) wherein SPR phenomenon could occur up to nanometer level. However, despite the intense plasmon of silver metal, gold is frequently used as mono-layer platform in typical-types of SPR or LSPR biosensors. The frequent use of gold is directly linked to its chemical inertness. Nevertheless, it has been realized that the use of metallic multi-layers or core-shell NPs could not only provide a broader quantifying range but also increase or expand the sensitivity of nano-biosensor. The hybrids of Au/Ag/Au, titanium (Ti)/Ag/Au, graphene/Au, and MoS_2/Al/MoS_2-graphene, for instance, constitute some examples of metallic multi-layers, being used in the development(s) of biosensors (Kamaruddin et al. 2016).

It is well-known that the tunable morphology and variable size of NPs significantly affect their optical properties that could be switched from visible to near infra-red (NIR) wavelengths, improving LSPR. For example, in particular the SPR can be altered effectively from visible to NIR if one switches from spherical-shaped NPs to rod-shaped NPs. Furthermore, the presence of loud tips like stars on NPs further contributes toward the sensitivity, affording outstanding outcomes. A contemporary application of plasmons rely upon SP–coupled emission (SPCE), enhancing the fluorescence emission. Mechanistically, SPCE occurs when the surface plasmon of metal interacts with near-field fluorophore, exhibiting intriguing optical properties, including greater directivity, distinctive polarization, an excellent wavelength determination, and reduction in the background signals. For instance, SPCE–based fluorescent sensors have been used to detect tannic acid in aqueous solution up to picomolar level.

The RI can also be changed by making the use of grating structures on a desired surface, exhibiting certain patterns periodically. When the period of the index difference exceeds the light wavelength (within the grating), diffraction effects dominate. Conversely, if the grating period is shorter, light broadcast will progressively resemble through a constant medium as the period drops. For biosensing applications, the gratings are typically employed by determining the diffraction competence of analytes in a solution or by calculating the angle-resolved diffraction productivity. One of the most appealing features of optical gratings is that they can be easily integrated with other devices to enhance their functions. For instance, in an elegant work, a grating-coupled SPR has been employed to detect estrogens in environmental water samples (Liu et al. (2021). Despite this, it is

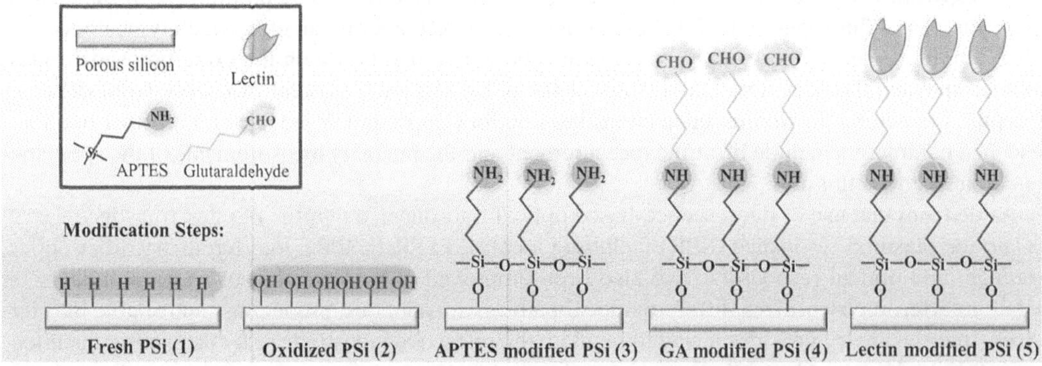

FIGURE 18.3 A schematic for the functionalization of porous silicon substrate having lectin and N-containing groups.

Reproduced from ref. (Yaghoubi *et al.* 2020)

reasonable to state that the use of optical grating couplers is still in its infancy as far as the nano-biosensors–based environmental monitoring is concerned.

Optical resonators (ORs) are another device wherein photons are arranged to a confined space, amplifying the signal intensities. One of the important advantages of these resonators is that they can exhibit a relatively high detection limit as the interaction between light and the sample (under study) can occur multiple times. A factor designated as 'Q' is often applied to determine the efficacy of a resonator in terms of shifts in the resonance and thus, the capacity of biosensor. Ring resonators such as microspheres are the most common optical resonators. The opto-fluidic ring resonator (OFRR) and silicon-on isolator (SOI) constitute some typical examples of ring resonators. Whereas a silicon ring is used in SOI, an internally functionalized microtube is used in OFRR having an external light source. The SOI is quite efficient in mitigating the surface disturbances due to the high RI of silicon. On the other hand, OFRR offers many advantages: 1) a small amount of sample is required, 2) simultaneous detection of multi-analytes, and 3) high precision in the results. Aside from those of SOI and OFRR, other ring resonators, for instance, sub-wavelength grating (SWG) can be used to incur high sensitivity in various aspects. Collectively, however, the use of optical resonators is quite limited in the detection of environmental pollutants, specifically, water.

In recent years, fiber optics (FOs) are at the forefront of modern science, engineering, and technology, displaying excellent optical properties both in bare, and when integrated with ORs, SPRs, and RIfs. FOs are also highly resistant toward corrosion, and electro-magnetic and electro-static interferences while making their operation and installation more facile for environmental applications.

As effective as these optical transducers (as described earlier) on their own are, relatively more attention has been paid in developing the fluorescence-based nano-biosensors for detection of a variety of pollutants in different water bodies with high detection limits (parts per billion (ppb)). Some of these selected works are summarized in Table 18.2. In order to get an authoritative details on either fluorescence or other types of optical nano-biosensors, the readers are advised to consult a recent review article in the field (Herrera-Domínguez et al. 2023).

TABLE 18.2
Fluorescent Detection of Environmental Pollutants in Different Water Samples

Environmental Pollutant	Bio-Recognition Element	Types of Water Samples	LOD[a] (ppb)	References
Mercury (Hg)	DNA	Surface Water	0.24	Long et al. (2013)
Atrazine (A chlorinated herbicide)	MIP[b]	Lake water	0.77	Liu et al. (2011)
Atrazine	Whole cell	Tap water	80	Scognamiglio et al. (2019)
Atrazine Isoproturon (An urea herbicide) Diuron (An urea herbicide)	Whole cell	DI water	10 (each)	Gosset et al. (2019)
2,4-Dichlorophenoxyacetic acid (herbicide)	Antibody	Drinking water	7.53	Liu et al. (2017)
BPA[c] (A toxic organic compound)	DNA	Bottled water Tap water River water	0.001 (each)	Gupta and Wood (2017)
BPA	Antibody	Drinking water	0.03	Liu et al. (2017)
BPA	Antibody	Lake water Tap water	0.076	Cheng et al. (2022)

(Continued)

TABLE 18.2 ***(Continued)***
Fluorescent Detection of Environmental Pollutants in Different Water Samples

Environmental Pollutant	Bio-Recognition Element	Types of Water Samples	LOD[a] (ppb)	References
1,2-Dibromoethane	Enzyme	River water	2400	Bidmanova et al. (2016)
1,2-Dichloroethane			2700	
1,2,3-Trichloropropane			1400	
3-Chloro-2-(chloromethyl)-1-propene			1400	
γ-Hexachlorocyclohexane			12100	
(Organic compounds)				
Sulfadimidine (An antibacterial drug)	Antibody	Waste water Lake water Bottled water	0.06 (each)	Liu et al. (2014)
Diclofenac (An anti-inflammatory drug)	Whole cell	Waste water	2900	Schirmer et al. (2019)
Ciprofloxacin (An antibiotic)	MIP	River water	1900	Huang et al. (2021)
17β-estradiol (a steroid hormone)	DNA	Waste water	0.14	Yildirim et al. (2012)
Microcystin-LR (A toxin)	Antibody	Drinking water	0.67	Liu et al. (2017)

[a] Limit of detection.

[b] Molecularly imprinted polymer.

[c] Bis-phenol A.

18.4 ELECTROCHEMICAL NANO-BIOSENSORS

Electrochemical nano-biosensors are advanced analytical devices that integrate nanotechnology with electrochemical sensing principles to detect and quantify biological molecules with high sensitivity and specificity. These sensors utilize nanomaterials, such as nanoparticles, nanowires, carbon nanotubes, and graphene, to enhance their performance, making them pivotal in environmental monitoring. Electrochemical biosensing mechanisms involve the conversion of a biological interaction into an electrical signal that can be measured and analyzed. The primary electrochemical sensing mechanisms are amperometry, potentiometry, conductometry, and impedimetry, which also serve as the basis for classification of electrochemical biosensors as described in the following.

18.4.1 Amperometric Nano-Biosensors

Amperometric nano-biosensors are a type of electrochemical biosensor that measures the current generated by the oxidation or reduction of an electroactive species at the electrode surface under a controlled potential. In detail, in the first step, the bio-receptors interact with the target analyte. Often, this interaction either leads to an electrochemical reaction of the analyte or a product of the biochemical reaction. As a result, the electron transfer occurs and current is generated, which is directly proportional to the concentration of the analyte. The major components of amperometric biosensors include the working electrode, reference electrode, and counter electrode. The working electrode relies upon common materials such as platinum, gold, and carbon-based materials. However, the electrode surface can be further modified using mediators or nanomaterials to enhance the electron transfer as well as sensitivity. The second electrode, that is the reference electrode, provides a stable reference potential against which the working electrode potential is measured. The saturated calomel electrode (SCE) and Ag/AgCl are frequently used as reference electrodes. The counter electrode completes the electrical circuit and current flows. Mostly, inert materials like graphite and platinum are used to make the counter electrode. The amperometric biosensors offer high sensitivity, dynamic and fast linear response, and real-time monitoring of specific analyte of interest, qualifying amperometry as a method of choice for bulk fabrication and commercialization. Nevertheless, owing to the high interferences and overlapping reactivity with other electroactive compounds, the amperometric biosensors may exhibit poor selectivity.

The amperometric biosensors can be further classified into three types. First generation biosensors rely on the natural substrate and product of the enzymatic reaction that are electroactive. Second-generation biosensors utilize small redox-active molecules or mediators to shuttle electrons between the enzyme and the electrode, enhancing the efficiency of electron transfer. Ferrocene, quinones, and a variety of metal complexes constitute some common examples of the mediators. In contrast to second generation biosensors, third generation biosensors aim to achieve the direct electron transfer between the enzyme's active site and the electrode, avoiding any mediators. This type of biosensors usually rely upon nanomaterials and are at the forefront of modern science, technology, and engineering. In this book chapter, mainly the third generation amperometric nano-biosensors have been highlighted.

18.4.2 Potentiometric Nano-Biosensors

Potentiometric nano-biosensors are a class of biosensors that measure the potential difference between two electrodes to detect biological or chemical substances. Typically, a reference electrode and a working electrode are used. The reference electrode maintains a constant potential, whereas the working electrode's potential changes in response to the analyte concentration. When the target analyte binds to the recognition element, it induces a change in the charge distribution, which in turn affects the potential. This change is measured and correlated to the concentration of the analyte. A variety of NMs such as CNTs, graphene, metallic NPs, and silicon nanowires can be used, displaying high sensitivity, fast response time, and miniaturization in various settings. Nevertheless, stability, an effective integration with existing systems, and a relatively high cost remain a challenge in widespread adoption of these sensors. Going back to COVID-19 pandemic, an MIP–integrated potentiometric nano-biosensor was formulated that could efficiently discriminate between corona virus and other types of viruses (Lee et al. 2022). In particular, the nano-biosensor was found able to achieve respective detection limits of 200 PFU/mL and 100 pg/mL for uncontaminated suspension of virus spots and spike proteins.

18.4.3 Conductometric Nano-Biosensors

Conductometric nano-biosensors are a type of biosensor that measure changes in the electrical conductivity of a solution or material in response to the presence of a specific analyte. In conductometric biosensors, the transducer converts the bio-recognition event into a measurable change in electrical conductivity. This is often facilitated by nanomaterials. The conductometric nano-biosensors offer numerous benefits. Firstly, they do not require any reference electrode. Secondly, the transducers used in these types of biosensors are generally not sensitive toward light while they typically consume low power. Thirdly, low-cost materials can be used for production in bulk as well as for the miniaturization purposes. This can be done with ease just by using a thin film of electrode. In general, many conductometric nano-biosensors have been developed whereby electrode surface was used as a substrate to form thin metallic layers of nickel, gold, copper, and so on for immobilization purposes.

18.4.4 Impedimetric Nano-Biosensors

Based on the most commanding electrochemical sensing method, the impedimetric biosensors measure the impedance (resistance and reactance) of a system, which changes in response to the interaction between the bio-recognition element and the target analyte. Impedance is a complex quantity that combines both the resistance (real part) and the reactance (imaginary part) of the system. As the basic components, two or more electrodes are used to apply an alternating current (AC) signal and measure the resulting impedance. In line with other types of biosensors, bio-recognition elements such as enzymes and nucleic acids are generally used to bind the target analyte in a buffer solution, the medium in which the impedance is measured. When the target analyte binds to the bio-recognition element on the electrode surface, it causes changes in the charge distribution, ionic concentration, or dielectric properties of the medium. These changes affect the impedance, which can be correlated to the analyte concentration. The impedimetric biosensors usually offer high sensitivity for the determination of interfacial properties, operating onto the functionalized surfaces. At this juncture, it would be fruitful to shed light on the technical differences between an impedimetric immunosensor and aptasensor. Whereas the former relies upon the formation of an immune-complex at the surface of the sensor, the latter displays integration of tiny single-stranded DNA/RNA (the aptamers) with robust binding affinity and stability, increasing the resistance (charge/electron-transfer) (Mohammad-Razdari et al. 2019). As such, the label-free recognition can be realized. Apart from biosensing, the usefulness of the technique can be extended to characterize novel materials based on appropriate information of electrochemical processes such as corrosion inhibition, fuel-cell electrochemistry, and charging-discharging of batteries. Nevertheless, the technique requires careful control of experimental conditions.

18.4.5 Voltammetric Nano-Biosensors

Voltammetry, one of the most common and routinely used techniques covers a variety of electrochemical methods, for instance, cyclic voltammetry (CV), square wave voltammetry (SWV), and differential pulse voltammetry (DPV). Collectively, all these techniques rely upon quantifying the current response when a pulse is applied to the working electrode. Interestingly, the differences in the current can deliver appreciable statistics about the existence of a particular analyte as well as its concentration while implementing voltammetric biosensors. The major benefit of these biosensors is that they can determine multiple analytes simultaneously. In this regime, it is apparent to define different voltammetric techniques adequately as given in the subsequent paragraphs.

CV is an electrochemical technique widely used for studying redox processes, characterizing electrode materials, and investigating reaction mechanisms. It involves sweeping the potential of a working electrode linearly with time, reversing the direction at a predetermined limit, and measuring the resulting current to create a cyclic voltammogram. In CV, the potential applied to the working electrode is ramped linearly from an initial value to a vertex potential and then reversed back to the initial potential. This sweep is often repeated multiple times. The current response is plotted as a function of the applied potential. This plot typically shows peaks corresponding to the oxidation and reduction reactions of the analyte. The key parameters of CV are peak current, peak potential, and scan rate. Researchers use CV to characterize new electrode materials, assess their electrochemical properties, and evaluate their suitability for various applications, such as energy storage and sensors. CV also helps in elucidating the reaction mechanisms by providing information on intermediate species, reaction pathways, and the influence of various factors on the redox processes. Despite this, CV also has some limitations like electrode fouling wherein the surface can affect reproducibility and accuracy of results. The experimental conditions such as scan rate and electrolyte composition significantly influence the results.

DPV is an advanced electrochemical technique used to measure the current response of an analyte as a function of an applied potential with enhanced sensitivity and resolution. In DPV, a series of potential pulses is superimposed on a linear potential sweep, and the current is measured just before and at the end of each pulse. The difference in current (hence "differential") is plotted against the applied potential, resulting in a differential voltammogram. This technique effectively reduces background noise and enhances the detection of small, closely spaced redox events, making it particularly useful for detecting low concentrations of analytes and for applications requiring high sensitivity, such as trace analysis in environmental monitoring, pharmaceuticals, and biochemical assays.

SWV is an electrochemical technique that combines features of cyclic and pulse voltammetry to provide high sensitivity and resolution in analyte detection. In SWV, a symmetrical square wave potential is superimposed on a staircase waveform, and the current is measured at the end of each forward and reverse pulse. The difference in current is plotted against the potential to produce a voltammogram. This method allows for rapid and sensitive analysis, minimizing background noise and capacitive currents. SWV is particularly useful for studying fast electron transfer processes, detecting low concentrations of electroactive species, and applications in environmental monitoring.

Collectively, owing to their inherent advantages, for instance, ability to provide wide information in electron-transfer kinetics, diffusion rate, adsorption rate, redox reaction's thermodynamics, a variety of voltammetric biosensors have been designed and applied for the determination of environmental pollutants, in particular, heavy metals. Some of these works have been highlighted below in the regime of three-electrode system.

Dai and Compton (2006) augmented CV by making the use of Pt NPs onto GCE through electrodeposition. This electrochemical set up was found to be highly sensitive in the unswerving detection of As^{3+} (arsenic ion) with a detection limit of ca. 35 μg/L. In other work, selective and sensitive impedimetric recognition of cadmium ions (Cd^{2+}) was achieved by modifications of screen-printed electrodes (one-use) with nanomaterials or composites of metallic oxides, CNTs, and cellulose (Magar et al. 2022). The detection limit of this assay was estimated to be above the picomolar concentration.

Polyurethanes (PUs) are versatile polymers that are used in a wide range of applications due to their flexibility, durability, and resistance to abrasion. The unique properties of polyurethanes arise from the chemical reactions between polyols and diisocyanates, which can be tailored to meet specific performance requirements. On another front, while essential in trace amounts for biological processes, copper ions can become significant environmental pollutants when their concentrations rise due to industrial activities. These ions can contaminate water sources, leading to toxic effects on aquatic life and potentially entering the human food chain. Taking this into account, Aragay and Merkoçi (2012) developed a nanocomposite, utilizing PUs and Pt NPs as dopants, with a size similar to a QD (≤ 5 nm). The as-developed nanocomposite could be used as voltammetric biosensor in the detection of copper ions (Cu^{2+}), exhibiting linear range and detection limits of 100–1000 and > 16.5 ng/mL, respectively. In other works, a variety of functionalized PUs, for example, poly(ester-urethane) urea, have been integrated with diverse NPs like Au NPs onto different electrodes to determine and quantify highly toxic Hg^{2+}, electrochemically, in fish tissue with high sensitivity and selectivity (March et al. 2015).

Cr(VI) ions are highly toxic, often released from industrial processes such as electroplating, leather tanning, and manufacturing. These ions can contaminate water and soil, posing serious health risks to humans and wildlife. Exposure to Cr(VI) can cause severe respiratory issues, skin irritation, and an increased risk of cancer. Its persistence in the environment and potential for bioaccumulation make it crucial to implement stringent regulations to minimize its impact. In a sophisticated study (Wyantuti et al. 2015), Au NPs were used to develop an electrochemical assay for the ultra-sensitive detection (2.4 ng/L) of Cr(VI).

Robinson et al. (2016) conducted a study where they synthesized and characterized carbon nanofibers (CNFs) for use in the electrochemical detection of Pb^{2+}. They made up nano-electrode-arrays from vertically aligned CNFs (VACNFs). These nano-electrodes significantly improved the performance of anodic stripping voltammetry for detecting Pb^{2+}, achieving a detection limit of 1.73 nM. As such, the significance of nanomaterials in detecting heavy metals using electrochemical sensors is evident. Metal NPs and carbon nanomaterials have been widely studied as electrode transformers and have been reported in numerous analyses of heavy metals in environmental samples.

Aside from heavy metals, much emphasis has been placed on voltammetric detection of pesticides, utilizing carbon nanomaterials, in particular, graphene. Some of these important works, wherein graphene is used in the modification of electrodes for the development of nano-biosensors, are mentioned next (Table 18.3).

To identify methyl parathion by SWV, Gong et al. (2011) developed a highly sensitive detection method by modifying an electrode with a gold-graphene composite. Their research reported an impressively low LOD at 0.002 μM, showcasing the effectiveness of their method. The exceptionally low LOD can be credited to the synergistic properties of the gold-graphene composite. Whereas the Au NPs provided a high surface area and enhanced the electron transfer processes, graphene offered remarkable mechanical strength, a large surface area, and superior electron mobility. When combined, these materials created a highly conducive environment for the immobilization of nitroaromatic organophosphates, like methyl parathion. This combination of properties makes the gold-graphene composite an ideal material for the sensitive and precise electrochemical detection of methyl parathion.

In 2014, Li et al. (2014) concentrated on developing a nano-biosensor for methyl parathion detection, utilizing a graphene-gadolinium-Prussian blue nanocomposite. This innovative sensor

TABLE 18.3
Electrochemical Detection of Pesticide Pollutants Using Functionalized Electrodes

Pesticide Pollutant	Electrochemical Method	Modified Electrode Material	LOD[a] (μM)	References
Carbofuran	DPV[b]	MIP-rGO-Au/GCE[c]	0.02	Tan et al. (2015)
Methyl parathion	SWV[d]	Au NPs-chi-GNs[e]/GCE[c]	0.77	Gong *et al.* (2011)
Methyl parathion	DPV[b]	Graphene-gadolinium-prussian blue/GCE[c]	0.001	Li et al. (2014)
Methyl parathion	Amperometry	MoS_2-graphene/GCE[c]	0.003	Govindasamy et al. (2017)
Glyphosate	Amperometry	Pt-laser-induced graphene	3.03	Johnson et al. (2022)
Dimethoate	DPV[b]	N-doped graphene/AuNPs/GCE[c]	8.7	Zhang, Fa, et al. (2017)
Carbaryl and fenamiphos	CV[f] and DPV[b]	rGO-Cu/CuO-Ag/GCE[c]	0.005 & 0.003	Hashemi et al. (2019)
Carbaryl and paraquat	DPV[b]	Ag-graphene/B-doped diamond/GCE[c]	> 0.001 &~ 0.002	Pop et al. (2018)
Diuron	LSV[g]	rGO-Au NPs/SPE[h]	39	Shams et al. (2016)
Diuron	DPV[b]	Graphene/GCE[c]	5	Zafar et al. (2023)

[a] Limit of detection.

[b] Differential pulse voltammetry.

[c] Glassy carbon electrode.

[d] Square wave voltammetry.

[e] Graphene nanosheets.

[f] Cyclic voltammetry.

[g] Linear sweep voltammetry.

[h] Screen-printed electrode.

demonstrated a remarkably low LOD of 1 nM, with a linear response range extending from 0.008 to 10 mM. The amalgamation of gadolinium-Prussian blue NPs onto the graphene surface significantly boosted the sensor's electro-catalytic activity, resulting in an enhanced detection performance. The nanocomposite biosensor operates by leveraging the unique properties of its components. Graphene provides a large surface area, excellent electrical conductivity, and mechanical strength, while gadolinium-Prussian blue NPs offer high electro-catalytic activity and stability. This combination enhances the electron transfer rate, leading to more sensitive and accurate detection of methyl parathion. Despite the promising results of this biosensor for methyl parathion, it is crucial to conduct comprehensive studies to evaluate its selectivity and stability over extended periods. Understanding these parameters is essential for assessing the practical applicability of the biosensor in real-world conditions.

In 2017, Govindasamy et al. (2017) introduced an advanced nano-biosensor for detecting methyl parathion using a graphene-molybdenum disulfide (MoS_2) nanocomposite, paired with amperometry technique. Their research demonstrated that the sensor exhibited a linear detection range spanning from 10 nM to 1.9 mM, indicating its capability to accurately measure methyl parathion across a broad concentration spectrum. The sensor's LOD was found to be quite impressive (0.0032 μM), underscoring its high sensitivity. The graphene-MoS_2 nanocomposite leverages the complementary properties of both materials. Graphene, known for its exceptional electrical conductivity, large surface area, and mechanical strength, provides a robust platform for the sensor. MoS_2, with its excellent electro-catalytic activity and layered structure, enhances the sensor's electron transfer efficiency and surface reactivity. Together, these materials create a synergistic effect that significantly improves the biosensor's performance.

Recently, Johnson et al. (2022) have introduced an innovative and selective detection method for glyphosate, utilizing Pt-decorated laser-induced graphene (LIG) combined with an enzyme, glycine oxidase. This approach provides a scalable and highly selective tool for glyphosate detection in complex matrices. The Pt NPs deposited on the LIG enhance the electrode's catalytic activity and conductivity, facilitating improved electron-transfer processes. Glycine oxidase, an enzyme specific to glyphosate, ensures high selectivity by catalyzing the oxidation of glyphosate, leading to measurable electrochemical signals. This combination allows for reliable detection of glyphosate even in intricate samples, showcasing the method's potential for practical applications in environmental monitoring as well as food safety analysis.

Zhang, Fa, *et al.* (2017) engineered an N-doped graphene-gold nanocomposite biosensor for dimethoate pesticide. This biosensor demonstrated a remarkably wide linear detection range, spanning from 1×10^{-12} to 4×10^{-8} M, showcasing its ability to accurately quantify dimethoate across varying concentrations. The sensor's detection limit was determined to be 8.7×10^{-13} M, underscoring its high sensitivity and effectiveness. The N-doping in graphene contributes to increased active sites and enhanced electron transfer rates, while the Au NPs improve the conductivity and catalytic activity of biosensor. However, despite the promising performance metrics, the study did not thoroughly investigate the biosensor's selectivity in the presence of other potential contaminants. Such studies would be crucial to validate the practical applicability of this N-doped graphene-gold nanocomposite in diverse environmental monitoring.

Carbaryl is a widely used carbamate insecticide known for its effectiveness against a broad spectrum of pests in agriculture and home gardens. Fenamiphos is an organophosphate nematicide used primarily to control nematodes in various crops, known for its systemic activity and persistence in soil. However, both carbaryl and fenamiphos serve as environmental pollutants. Whereas carbaryl poses risks to aquatic life and beneficial insects due to its toxicity and persistence in water and soil, fenamiphos can lead to soil and groundwater contamination, negatively affecting nontarget organisms and potentially disrupting ecosystems. Given the detrimental effects of these pollutants, Hashemi et al. (2019) developed an rGO–based nanocomposite, namely, rGO-Cu/CuO-Ag, to modify GCE for the concurrent electrochemical analysis (CV and DPV) of both carbaryl and fenamiphos. Under standardized experimental conditions, the nanocomposite-modified GCE exhibited respective LOD values of 0.005 and 0.003 μM for carbaryl and fenamiphos with appreciable range of linearity.

Aside from that of N-doping (as discussed previously), B-doping has also been utilized in the fabrication of electrochemical biosensor. Specifically, Pop et al. (2018) constructed an Ag-graphene-modified B-doped diamond electrodes for the concurrent detection of carbaryl and paraquat with LOD values of > 0.001 and ~0.002 μM, respectively.

Shams et al. (2016) developed a screen-printed electrode modified with rGO-Au nanocomposite for detection of diuron, an herbicide. Using LSV, the biosensor exhibited a linear response with LOD of ca. 0.004 nM. This synergy resulted in a highly sensitive and accurate sensor for diuron, highlighting the potential of rGO-Au nanocomposites in environmental monitoring.

Effective as these nanocomposite-modified electrodes are on their own, recently, Zafar et al. (2023) have provided compelling evidence for the effectiveness of bare graphene-coated electrodes in detecting diuron through DPV. Remarkably, the graphene-coated electrodes achieved a low LOD of 5 μM, showcasing their high sensitivity. These findings emphasize the capability of graphene-based electrodes to provide reliable and sensitive detection of diuron, offering a straightforward approach that avoids the complexity of additional modifications.

18.5 FIELD EFFECT TRANSISTORS

At this juncture, it would be worthwhile to shed light on field effect transistors (FETs), which are simply represented as electrochemical sensors in the existing literature. However, the working principle may be different. In general, an FET is composed of a semiconductor source, drain, and gate regions. Traditional FETs usually rely upon p-n junctions (source and drain) in which a source supplies the electrons (or holes) while a drain receives them. The earlier designed FETs were primarily based on silicon-containing metal oxides semiconductors. Recently, a diverse range of FETs, namely, thin-film transistors, chemically sensitized FETs (ChemFETs), and ion-selective EFTs are configured, avoiding the use of silicon substrate. Nevertheless, metal-oxide-based FETs are still the preferred choice in the detection of environmental pollutants, especially pesticides. However, FETs–based biosensing is a topic of its own, and have not been covered in this chapter, similar to those of other types of biosensors, including, enzyme-based biosensors, which have shown a burst in the research activities after the discovery of graphene.

18.6 CONCLUSIVE REMARKS AND FUTURE DIRECTIONS

This chapter underscores the pivotal role of advanced nano-biosensors in addressing the pressing issue of environmental pollution. Through an in-depth exploration of optical and electrochemical nano-biosensors, the chapter highlights their extraordinary capabilities in detecting a wide range of environmental contaminants, including heavy metals, pesticides, and organic pollutants. The deployment of various nanoarchitectures, in particular, carbon-based nanomaterials, has demonstrated significant potential in enhancing the sensitivity and specificity of these sensors, making them invaluable tools for environmental monitoring. The versatility of nanomaterials has enabled the development of a wide range of optical and electrochemical nano-biosensors, addressing the diverse and complex nature of environmental pollutants. Additionally, the ability of optical nano-biosensors to provide real-time monitoring is a significant advancement, which is crucial for mitigating potential harm to public health and ecosystems. The integration of these developed biosensors with other advanced technologies such as Internet of Things and artificial intelligence promises to further enhance their functionality and effectiveness in environmental monitoring.

However, several challenges remain that need to be addressed to fully realize the potential of nano-biosensors. Firstly, ensuring long-term stability is one of the key challenges as biosensors must maintain their performance over extended periods and under varying environmental conditions. Secondly, enhancing selectivity is another critical area of focus; though it has been achieved by few research groups, efforts should be directed toward designing biosensors that can distinguish between similar pollutants and reduce cross-reactivity, thereby improving the accuracy of pollutant detection. In line with this, the cost-related barriers also need to be addressed to facilitate widespread adoption. Future developments should aim toward this direction and making these biosensors more affordable for large-scale environmental monitoring applications. Transitioning from laboratory-scale research to large-scale production is essential for commercialization, necessitating the establishment of scalable manufacturing processes. Furthermore, the development of regulatory frameworks and standardization protocols for the use of nano-biosensors in environmental monitoring are essential. Advancements in nano-biosensor technology will benefit from interdisciplinary collaboration among chemists, physicists, biologists, material scientists, engineers, and policymakers. Such collaboration can drive innovation, address technical challenges, and support the development of comprehensive strategies for environmental protection. On a final note, there is always immense potential that the imaginative concepts we see in today's science fiction could inspire the breakthroughs in nano-biosensor technology that we will rely on for environmental monitoring in the future.

ACKNOWLEDGEMENT

This work did not receive any funding. B. G. conveys sincere thanks to his institution for endorsing research and development initiatives.

REFERENCES

Aragay, G. and Merkoçi, A. 2012. Nanomaterials application in electrochemical detection of heavy metals. *Electrochim. Acta* 84: 49–61.

Bidmanova, S., Kotlanova, M., Rataj, T., Damborsky, J., Trtilek, M. and Prokop, Z. 2016. Fluorescence-based biosensor for monitoring of environmental pollutants: from concept to field application. *Biosens. Bioelectron.* 84: 97–105.

Cheng, Y., Wang, H., Zhuo, Y., Song, D., Li, C., Zhu, A. and Long F. 2022. Reusable smartphone-facilitated mobile fluorescence biosensor for rapid and sensitive on-site quantitative detection of trace pollutants. *Biosens. Bioelectron.* 199: 113863.

Dai, X. and Compton, R. G. 2006. Detection of as (III) via oxidation to As (V) using platinum nanoparticle modified glassy carbon electrodes: Arsenic detection without interference from copper. *Analyst* 131: 516–521.

Dhahi, T. S., Dafhalla, A. K. Y., Tayfour, O. E., Mubarakali, A., Alqahtani, A. S., Ahmed, A. E. T., Elobaid, M. E., Adam, T. and Gopinath, S. C. B. 2024. Advances in nano sensors for monitoring and optimal performance enhancement in photovoltaic cells. *iScience* 27: 109347.

Fadeel, B., Bussy, C., Merino, S., Vázquez, E., Flahaut, E., Mouchet, F., Evariste, L., Gauthier, L., Koivisto, A. J., Vogel, U., Martín, C., Delogu, L. G., Buerki-Thurnherr, T., Wick, P., Beloin-Saint-Pierre, D., Hischier, R., Pelin, M., Carniel, F. C., Tretiach, M., Cesca, F., Benfenati, F., Scaini, D., Ballerini, L., Kostarelos, K., Prato, M. and Bianco, A. 2018. Safety assessment of graphene-based materials: Focus on human health and the environment. *ACS Nano* 12: 10582–10620.

Ferrier, D. C. and Honeychurch, K. C. 2021. Carbon nanotube (CNT)-based biosensors. *Biosensors* 11: 486.

Flora, G., Gupta, D. and Tiwari, A. 2012. Toxicity of lead: A review with recent updates. *Interdiscip. Toxicol.* 5: 47–58.

Garg, B. and Bisht, T. 2016. Carbon nanodots as peroxidase nanozymes for biosensing, *Molecules* 21: 1653.

Garg, B., Bisht, T. and Ling, Y.-C. 2014. Graphene-based nanomaterials as heterogeneous acid catalysts: A comprehensive perspective. *Molecules* 19: 14582–14614.

Garg, B., Bisht, T. and Ling, Y.-C. 2015. Graphene-based nanomaterials as efficient peroxidase mimetic catalysts for biosensing applications: An Overview. *Molecules* 20: 14155–14190.

Garg, B., Bisht, T. and Ling, Y.-C. 2016. Graphene-based nanomaterials: Versatile catalysts for carbon-carbon bond forming reactions. *Curr. Org. Chem.* 20: 1547–1566.

Garg, B. and Ling, Y.-C. 2013. Versatilities of graphene-based catalysts in organic transformations. *Green Mater.* 1: 47–61.

Garg, B. and Ling, Y.-C. *In Carbon Nanomaterials Sourcebook*, ed. K. D. Sattler, CRC Press, Boca Raton: 2018, 1st ed., Chapter 14, pp. 309–333.

Garg, B., Sung, C.-H. and Ling, Y.-C. 2015. Graphene-based nanomaterials as molecular imaging agents. *WIREs Nanomed. Nanobiotechnol.* 7: 737–758.

Gong, J., Miao, X., Zhou, T. and Zhang, L. 2011. An enzymeless organophosphate pesticide sensor using Au nanoparticle-decorated graphene hybrid nanosheet as solid-phase extraction. *Talanta* 85: 1344–1349.

Gosset, A., Oestreicher, V., Perullini, M., Bilmes, S. A., Jobbágy, M., Dulhoste, S., Bayard, R. and Durrieu C. 2019. Optimization of sensors based on encapsulated algae for pesticide detection in water. *Anal. Methods*. 11: 6193–6203.

Govindasamy, M., Chen, S. M., Mani, V., Akilarasan, M., Kogularasu, S. and Subramani, B. 2017. Nanocomposites composed of layered molybdenum disulfide and graphene for highly sensitive amperometric determination of methyl parathion. *Microchim. Acta* 184: 725–733.

Gupta, S. and Wood, R. 2017. Development of FRET biosensor based on aptamer/functionalized graphene for ultrasensitive detection of bisphenol A and discrimination from analogs. *Nano-Struct. Nano-Objects* 10: 131–140.

Hashemi, P., Karimian, N., Khoshsafar, H., Arduini, F., Mesri, M., Afkhami, A. and Bagheri, H. 2019. Reduced graphene oxide decorated on Cu/CuO-Ag nanocomposite as a high-performance material for the construction of a non-enzymatic sensor: Application to the determination of carbaryl and fenamiphos pesticides. *Mater. Sci. Eng. C* 102: 764–772.

Haupt, K. and Mosbach, K. 2000. Molecularly imprinted polymers and their use in biomimetic sensors. *Chem. Rev.* 100: 2495–2504.

Herrera-Domínguez, M., Morales-Luna, G., Mahlknecht, J., Cheng, Q., Aguilar-Hernández, I. and Ornelas-Soto, N. 2023. Optical biosensors and their applications for the detection of water pollutants. *Biosensors* 13: 370.

Huang, Q. D., Lv, C. H., Yuan, X. L., He, M., Lai, J. P. and Sun, H. 2021. A novel fluorescent optical fiber sensor for highly selective detection of antibiotic ciprofloxacin based on replaceable molecularly imprinted nanoparticles composite hydrogel detector. *Sens. Actuators B Chem.* 328: 129000.

Jiang, B., Song, Y., Liu, Z., Huang, W. E., Li, G., Deng, S., Xing, Y. and Zhang, D. 2021. Whole-cell bioreporters for evaluating petroleum hydrocarbon contamination. *Crit. Rev. Environ. Sci. Technol.* 51: 272–322.

Johnson, Z. T., Jared, N., Peterson, J. K., Li, J., Smith, E. A., Walper, S. A., Hooe, S. L., Breger, J. C., Medintz, I. L., Gomes, C. and Claussen, J. C. 2022. Enzymatic laser-induced graphene biosensor for electrochemical sensing of the herbicide glyphosate. *Glob. Challenges* 6: 2200057.

Kamaruddin, N. H., Bakar, A. A. A., Yaacob, M. H., Mahdi, M. A., Zan, M.S.D. and Shaari S. 2016. Enhancement of chitosan-graphene oxide SPR sensor with a multi-metallic layers of Au–Ag–Au nanostructure for lead(II) ion detection. *Appl. Surf. Sci.* 361: 177–184.

Ke, Y., Garg, B. and Ling, Y.-C. 2014. Waste chicken eggshell as low-cost precursor for efficient synthesis of nitrogen-doped fluorescent carbon nanodots and their multi-functional applications. *RSC Adv.* 4: 58329–58336.

Kim, K.-H., Jahan, S. A., Kabir, E. and Brown, R. J. 2013. A review of airborne polycyclic aromatic hydrocarbons (PAHs) and their human health effects. *Environ. Int.* 60: 71–80.

Kulakova, I. I. and Lisichkin, G. V. 2021. Prospects for using graphene nanomaterials: Sorbents, membranes, and gas sensors. *Russ. J. Appl. Chem.* 94: 1177–1188.

Kumaran, A., Vashishth, R., Singh, S., Surendran, U., James, A. and Chellam, P. V. 2022. Biosensors for detection of organophosphate pesticides: Current technologies and future directives, *Microchem. J.* 178: 107420.

Kümmerer, K. 2009. Pharmaceuticals in the environment. *Ann. Rev. Environ. Resour.* 34: 57–75.

Lee, W. I., Subramanian, A., Mueller, S., Levon, K., Nam, C. Y. and Rafailovich, M. H. 2022. Potentiometric biosensors based on molecular-imprinted self-assembled monolayer films for rapid detection of influenza A virus and SARS-CoV-2 spike protein. *ACS Appl. Nano Mater.* 5: 5045–5055.

Li, Y., Xu, M., Li, P., Dong, J. and Ai, S. 2014. Nonenzymatic sensing of methyl parathion based on graphene/gadolinium Prussian Blue analogue nanocomposite modified glassy carbon electrode. *Anal. Methods* 6:2157–2162.

Liu, H., Ye, T. and Mao, C. 2007. Fluorescent carbon nanoparticles derived from candle soot. *Angew. Chem. Int. Ed.* 46: 6473–6475.

Liu, L., Zhou, X., Lu, M., Zhang, M., Yang, C., Ma, R., Memon, A. G., Shi, H. and Qian, Y. 2017. An array fluorescent biosensor based on planar waveguide for multi-analyte determination in water samples. *Sens. Actuators B Chem.* 240: 107–113.

Liu, L. H., Zhou, X. H., Xu, W. Q., Song, B. D. and Shi, H. C. 2014. Highly sensitive detection of sulfadimidine inwater and dairy products by means of an evanescent wave optical biosensor. *RSC Adv.* 4: 60227–60233.

Liu, R., Guan, G., Wang, S. and Zhang Z. 2011. Core-shell nanostructured molecular imprinting fluorescent chemosensor for selective detection of atrazine herbicide. *Analyst* 136: 184–190.

Liu, L., Zhang, X., Zhu, Q., Li, K., Lu, Y., Zhou, X. and Guo, T. 2021. Ultrasensitive detection of endocrine disruptors via superfine plasmonic spectral combs. *Light Sci. Appl.* 10: 181.

Long, F., Zhu, A., Shi, H., Wang, H. and Liu J. 2013. Rapid on-site/in-situ detection of heavy metal ions in environmental water using a structure-switching DNA optical biosensor. *Sci. Rep.* 3: 1175.

Magar, H. S., Magd, E. E. A.-E., Hassan, R. Y. A. and Fahim, A. M. 2022. Rapid impedimetric detection of cadmium ions using Nanocellulose/ligand/nanocomposite (CNT/Co3O4). *Microchem. J.* 182: 107885.

March, G., Nguyen, T. D. and Piro, B. 2015. Modified electrodes used for electrochemical detection of metal ions in environmental analysis. *Biosensors* 5: 241–275.

Meskher, H., Mustansar, H. C., Thakur, A. K., Sathyamurthy, R., Lynch, I., Singh, P., Han, T. K. and Saidur, R. 2023. Recent trends in carbon nanotube (CNT)-based biosensors for the fast and sensitive detection of human viruses: a critical review. *Nanoscale Adv.* 5: 992–1010.

Mohammad-Razdari, A., Ghasemi-Varnamkhasti, M., Izadi, Z., Ensafi, A. A., Rostami, S. and Siadat, M. 2019. An impedimetric aptasensor for ultrasensitive detection of Penicillin G based on the use of reduced graphene oxide and gold nanoparticles. *Microchim. Acta* 186: 372.

Pohanka, M. 2018. Overview of piezoelectric biosensors, immunosensors and DNA sensors and their applications. *Materials* 11: 448.

Pop, A., Lung, S., Orha, C. and Manea, F. 2018. Silver/graphene-modified boron doped diamond electrode for selective detection of carbaryl and paraquat from water. *Int. J. Electrochem. Sci.* 13: 2651–2660.

Rice, K. M., Walker, E. M. Jr., Wu, M., Gillette, C. and Blough, E. R. 2014. Environmental mercury and its toxic effects. *J. Prev. Med. Public Health* 47: 74–83.

Robinson, J. E., Heineman, W. R., Sagle, L. B., Meyyappan, M. and Koehne, J. E. 2016. Carbon nanofiber electrode array for the detection of lead. *Electrochem. Commun.* 73: 89–93.

Rochester, J. R. 2013. Bisphenol A and human health: A review of the literature. *Reprod. Toxicol.* 42: 132–155.

Sarkar, A., Sarkar, K. D., Amrutha, V. and Dutta, K. Tools, Techniques and Protocols for Monitoring Environmental Contaminants, Elsevier, Amsterdam, The Netherlands: 2019, 1st ed. In Satinder Kaur Brar, Krishnamoorthy Hegde, and Vinayak Laxman Pachapur (Eds.), *An Overview of Enzyme-Based Biosensors for Environmental Monitoring*, pp. 307–329.

Scandurra, A., Censabella, M., Gulino, A., Grimaldi, M. G. and Ruffino, F. 2022. Gold nanoelectrode arrays dewetted onto graphene paper for selective and direct electrochemical determination of glyphosate in drinking water. *Sens. Bio-Sensing Res.* 36: 100496.

Schirmer, C., Posseckardt, J., Schröder, M., Gläser, M., Howitz, S., Scharff, W. and Mertig, M. 2019. Portable and low-cost biosensor towards on-site detection of diclofenac in wastewater. *Talanta* 203: 242–247.

Sciuto, E. L., Coniglio, M. A., Corso, D., van der Meer, J. R., Acerbi, F., Gola, A. and Libertino, S. 2019. Biosensors in monitoring water quality and safety: an example of a miniaturizable whole-cell based sensor for Hg^{2+} optical detection in water. *Water* 11: 1986.

Scognamiglio, V., Antonacci, A., Arduini, F., Moscone, D., Campos, E. V. R., Fraceto, L. F. and Palleschi G. 2019. An eco-designed paper-based algal biosensor for nanoformulated herbicide optical detection. *J. Hazard. Mater.* 373: 483–492.

Shams, N., Lim, H. N., Hajian, R., Yusof, N. A., Abdullah, J., Sulaiman, J., Ibrahim, I., Huang, N. M. and Pandikumar, A. 2016. A promising electrochemical sensor based on Au nanoparticles decorated reduced graphene oxide for selective detection of herbicide diuron in natural waters. *J. Appl. Electrochem.* 46: 655–666.

Shrivastav A. M., Mishra, S. K. and Gupta, B. D. 2015. Localized and propagating surface plasmon resonance based fiber optic sensor for the detection of tetracycline using molecular imprinting. *Mater. Res. Express.* 2: 35007.

Sreekanth, S. P., Alodhayb, A., Assaifan, A. K., Alzahrani, K. E., Muthuramamoorthy, M., Alkhammash, H. I., Pandiaraj, S., Alswieleh, A. M., Le, Q. V., Mangaiyarkarasi, R., Grace, A. N. and Raghavan, V. 2021. Multi-walled carbon nanotube-based nanobiosensor for the detection of cadmium in water. *Environ. Res.* 197: 111148.

Tan, X., Hu, Q., Wu, J., Li, X., Li, P., Yu, H., Li, X. and Lei, F. 2015. Electrochemical sensor based on molecularly imprinted polymer reduced graphene oxide and gold nanoparticles modified electrode for detection of carbofuran. *Sens. Actuators B Chem.* 220: 216–221.

Thanh, C. T., Binh, N. H., Duoc, P. N. D., Thu, V. T., Trinh, P. V., Anh, N. N., Tu, N. V., Tuyen, N. V., Quynh, N. V., Tu, V. C., Thao, B. T. P., Thang, P. D., Abe, H. and Chuc, N. V. 2021. Electrochemical sensor based on reduced graphene oxide/double-walled carbon nanotubes/octahedral Fe_3O_4/chitosan composite for glyphosate detection. *Bull. Environ. Contam. Toxicol.* 106: 1017–1023.

Wood, T. J. and Goulson, D. 2017. The environmental risks of neonicotinoid pesticides: A review of the evidence post 2013. *Environ. Sci. Pollut. Res.* 24: 17285–17325.

Wu, X., Zhang, Y., Han, T., Wu, H., Guo, S. and Zhang, J. 2014. Composite of graphene quantum dots and Fe_3O_4 nanoparticles: Peroxidase activity and application in phenolic compound removal. *RSC Adv*. 4: 3299–3305.

Wyantuti, S., Ishmayana, S. and Hartati, Y. W. 2015. Voltammetric determination of Cr (VI) using gold nanoparticles-modified glassy carbon electrode. *Procedia Chem*. 16: 15–23.

Yaghoubi, M., Rahimi, F., Negahdari, B., Rezayan, A. H. and Shafiekhani, A. 2020. A lectin-coupled porous silicon-based biosensor: Label-free optical detection of bacteria in a real-time mode. *Sci. Rep*. 10: 16017.

Yildirim, N., Long, F., Gao, C., He, M., Shi, H.-C. and Gu, A. Z. 2012. Aptamer-based optical biosensor for rapid and sensitive detection of 17β-estradiol in water samples. *Environ. Sci. Technol*. 46: 3288–3294.

Zafar, M. A., Liu, Y., Hernandez, F. C. R., Varghese, O. K. and Jacob, M. V. 2023. Plasma-based synthesis of freestanding graphene from a natural resource for sensing application. *Adv. Mater. Interfaces* 10: 2202399.

Zhang, Q., Zhang, X., Ma, F. and Zhang, C.-Y. 2022. Advances in quantum dot-based biosensors for DNA-modifying enzymes assay. *Coord. Chem. Rev*. 469: 214674.

Zhang, Y., Cai, N. and Chan, V. 2023, Recent advances in silicon quantum dot-based fluorescent biosensors. *Biosensors* 13: 311.

Zhang, Y., Fa, H., He, B., Hou, C.-J., Hou, D.-Q., Xia, T.-C. and Yin, W. 2017. Electrochemical biomimetic sensor based on oxime group-functionalized gold nanoparticles and nitrogen-doped graphene composites for highly selective and sensitive dimethoate determination. *J. Solid State Electrochem*. 21: 2117–2128.

Zhang, Z., Zeng, K. and Liu, J. 2017. Immunochemical detection of emerging organic contaminants in environmental waters. *TrAC Trends Anal. Chem*. 87: 49–57.

Zhu, J., Feng, M. and Lian, G. 2022. Graphene based FET biosensor for organic-phosphorous sample detection and the enzymatic analysis. *Crystals 12*: 1327.

Zhu, Y., Elcin, E., Jiang, M., Li, B., Wang, H., Zhang, X. and Wang, Z. 2022. Use of whole-cell bioreporters to assess bioavailability of contaminants in aquatic systems. *Front. Chem*. 10: 1018124.

19 Industrial Framework for Net Zero Emission Transition in Emerging Economies

Riti Kumari, Naveen Dwivedi, and Harmanpreet Meehnian

19.1 INTRODUCTION

With the ever-increasing threat of climate change and the present need to keep the increase of the global average temperature to 1.5°C above pre-industrial levels, achieving net zero emissions has become one of the most strategic global goals. This commitment, which has been made clear in the Paris Agreement, necessitates a radical shift in the way societies produce and consume energy, extract and manage resources, and pursue industrialization. There is a particular role of fast-urbanizing economies that are going through industrialization in making this shift. These countries are faced with the inverse relationship of promoting business and developing the economy with the increasing greenhouse gasses they emit. Unlike the developed nations that contributed more to carbon emissions in the past, such emerging economies have room for best practices as they grow, employing low-carbon technologies, and including low emission discharge into their strategies from the start. This however is accompanied by elongated stress brought about by the objectives that dominate this approach. Unlike the situation in advanced economies, many of the emerging economies continue to achieve economic progress dependent on fossil fuels, which are not only a source of energy, but also an important engine of economic activities as well as job creation. Moving away from these traditional energy sources to renewable alternatives requires substantial investment, technological innovation, and capacity building, all of which can be particularly daunting for countries with limited financial resources. There are social and political contexts as well, which need consideration such as balancing the needs of other parties, namely the industrial, governmental, and non-governmental sectors, while ensuring that the transition does not reinforce already existing disparities, or create additional ones. Nevertheless, satisfying these requirements to obtain net zero emissions gives an exceptional opportunity to developing countries as well. Investing in clean energy, energy efficiency, and clean production processes will not only shrink the carbon footprints of these nations, but increase energy security, generate jobs, and stimulate the economy as well. Also, the transition to a low-carbon economy may enable developing countries to capture the low-carbon and environmentally friendly growing opportunities in the international markets. There is need to develop policies and strategies that encourage clean energy and industrial measures without jeopardizing economic growth and social justice. Provided it is equipped with the appropriate strategies and support, the developing world would enhance measurable impact in global efforts to reduce emissions levels while attaining positive development objectives both to its people and the world. The path to net zero is challenging, but it is also an opportunity for emerging economies to redefine their development trajectories in a way that is both economically viable and environmentally sustainable.

DOI: 10.1201/9781003668961-19

19.1.1 Global Climate Imperative

19.1.1.1 Understanding the Net Zero Mandate

The world is in a critical situation given the state of the global climate change phenomenon, it is evident that there is a need for drastic change for all sectors and all the countries. A greater proportion of the world's population has begun appreciating the need for "net zero" emissions. The more a country manages to sequester, the further away it is from the chance of improving global warming since this will ensure that no more greenhouse gases will be further added to the atmosphere in such a country where net zero is tenable. This set goal is at the heart of climate change endeavors, as showcased by the Paris Agreement and the follow-up climate talks (Climate Analytics, 2023; IPCC, 2021). The rationale for the net and carbon neutrality concept is crystal clear. The reports of the Intergovernmental Panel on Climate Change, for example, declared more than once that it is necessary to keep the rise of the world's temperature not exceeding 1.5°C above the level of the preindustrial period in order to avoid the most negative climate impacts (IPCC, 2021). In order for emissions around the globe to reach net zero target year, global warming should be confined to temperature rise, which is quite clear to understand. This time frame explains the gravity of the problem and stresses the need for attention and rapid systemic change of energy, industrial, and land use within a few geopolitical spans (Climate Analytics, 2023).

19.1.1.2 The Unique Role of Emerging Economies in Global Emission Reduction

These emerging economies, which are undergoing fast-paced industrialization and urbanization, sit at the core of the global emissions challenge and are quite crucial in meeting the net zero target. Such countries/regions in this category include China, India, Brazil, and South Africa, all grouped by high growth, hunger for energy, and an increasing middle class (World Bank, 2022). Hence, they also contribute a lot to the global warming load. For instance, China and India, both growing developing countries, feature on the top three lists of the world's most prolific emitters of greenhouse gases (Climate Action Tracker, 2023). However, emerging economies also present an interesting scenario where considerable emission cuts can be achieved. As opposed to developed countries that have already grown large structures that are carbon heavy, most emerging economies have the opportunity to directly advance to low-carbon practices and technologies (World Resources Institute, 2021). This transition is not only practicable but also brings returning economic benefits to counteracting and eliminating unsustainable energy system lock-in expenses and decreasing the risks associated with dependence on carbon taxes and strict environmental policies in the future. Furthermore, many emerging economies are already experiencing the severe impacts of climate change, such as extreme weather events, sea-level rise, and water scarcity, making the case for climate action both an environmental and economic imperative (IEA, 2022; UNEP, 2021).

19.1.2 Balancing Economic Growth and Environmental Sustainability

19.1.2.1 Challenges and Opportunities in the Transition

The November transition in particular leads to a net zero economy in emerging economies, which leads to both a social and a structural tension. The tension lies on the one hand in the imperative now facing the emerging economies: the achievement both of economic growth and environmental sustainability. Because of the situation of poverty and inequality in the regions, economic development is still the concern as developing economies also have this importance for industrialization because industrialization is what spurs growth and creates employment. But the standard methods of industrialization entrenched in fossil fuel energy-based systems that defeat the purpose of net zero called for (IRENA, 2022).

The problem lies in separating environmental protection on the one hand and economic growth on the other. Sufficient policy frameworks, large amounts of money, and the coordination of countries are necessary for that goal (World Bank, 2022). Energy, construction, transport and agribusiness are among such processes that should go through radical changes. For instance, the change of energy sources to renewable energy such as solar, wind, and hydropower will reduce greenhouse gases while creating jobs and securing energy. It follows that such investments as the use of energy conservation technologies, modern principles of the circular economy, and rational distribution of land will contribute not only to a gradual reduction of emissions and consumption of resources but also to long-term economic competitiveness (IEA, 2022).

19.1.2.2 Industrial Development and Climate Responsibility

Developing countries are said to be held back by a dual contradiction as they strive to attain net zero emission targets, one which involves industrialization and retaining carbon-intensive sectors that contribute to economic growth but also emit greenhouse gases. Economic activities still hold relevance to industrialization history as these give avenues for better living standards, poverty alleviation, and overall communities' well-being. However, such development has to be consistent with the international climate change objectives if people are to live tomorrow. This involves changing course and making people understand that environmental protection must be a core part of industrial policies and practices (UNEP, 2021). Emerging economies are, however, already taking leading initiatives in this regard. For instance, China has accounted for nearly half of the deployment of wind and solar capacities of the world, becoming the global leader in the renewable energy industry (IRENA, 2022). Furthermore, India has ambitious renewable energy goals and various energy efficiency and eco-friendly urban development programs in place. Also, Brazil has made encouraging progress in enhancing biofuels use and decreasing deforestation, and South Africa has developed national frameworks for renewable energy and energy efficiency measures targeting development (World Bank, 2022). But these initiatives have to be intensified and duplicated across all sectors. By prioritizing green technologies, sustainable resource management, and low-carbon innovations, emerging economies can achieve their development goals while contributing to global climate mitigation efforts (World Resources Institute, 2021).

19.2 GLOBAL NET ZERO EMISSION GOALS AND INDUSTRIAL RESPONSIBILITY

Climate change requires a global response. Therefore, international agreements on this topic are critical. For example, the Kyoto Protocol was one of the first documents that sought to regulate emissions of developed countries (UNFCCC, 2023). However, the realization that such a narrow approach did not bring progress resulted in the emergence of the Paris Agreement in 2015, which provided an overarching legal framework where the world practically aspires to keep temperature increase well below 2°C warming with the encouragement of targeting 1.5°C (UNFCCC, 2015). In this agreement, countries are free to determine how deeper cuts can be made without mandating absolute reductions in emission of pollutants, while more respectful of the cooperation obligation of developing countries through investment and technology transferring (UNFCCC, 2015). Ongoing climate conferences such as the COP are also an important means of checking and enhancing these obligations. As an example, COP28 is expected to devote resources to raising ambition levels, mobilizing funds for adaptation and mitigation purposes, and promoting the deployment of technologies that are essential for the reduction of carbon emissions, especially in the industry sector (IEA, 2021a). For the industries, such obligations create both risks and new opportunities. A paradigm is expected because of the new climate goals with respect to the global agenda at corporate level. Such agreements stimulate the new kind of inventiveness and partnerships, giving the industry a task to support solving the change problem (UNFCCC, 2015).

19.2.1 International Climate Agreements and Policies

19.2.1.1 Paris Agreement and COP28 Outcomes

The Paris Agreement, signed by 196 parties in December 2015, represents a historic turning point in the global effort to address climate change (UNFCCC, 2015). The agreement's central aim is to hold the increase in global average temperature to well below 2°C above preindustrial levels, while pursuing efforts to limit the temperature rise to 1.5°C (IPCC, 2021). This temperature threshold is imperative, as research suggests that going beyond 1.5°C could result in catastrophic and irreversible effects on the environment, economies, and human society (IPCC, 2021). Countries are required under the provisions of the Paris agreement to come up with and submit within a duration of five years Nationally Determined Contributions (NDCs) which are supposed to provide ways in which such countries will manage to abate greenhouse gases and cope with the impacts of climate change (UNFCCC, 2015). As of 2023, more than 80% of global emissions are covered through these NDCs with countries vowing to more ambitious objectives (Climate Action Tracker, 2023). Even so, it is estimated that the current NDCs, if all are implemented as planned, will only constrain the increase of global temperatures to about 2.4°C by the year 2070, which could not be able to achieve the 1.5 degrees (Climate Action Tracker, 2023). This gap explains why there is a demand for greater resolve and faster progress on climate change, particularly in the industrial sector that produces around 23% of the world's CO2 emissions (IEA, 2021b). The Coalition for Climate Action: Final Report Element ~ The 28th Conference of the Parties (COP28), expected to convene in November 2023, is expected to support the efforts that will help to enhance action on climate globally (IEA, 2021c). Cognizant of the achievements that were made during past COPs, it is envisaged that COP28 will enhance the level of global action by keeping the correct focus on Paris Agreement implementation and increasing willingness to change for climate (UNFCCC, 2023). Advanced Outcomes from COP28 further include the Global Stocktake, which is one of the deliverables of COP28, and aims to undertake a detailed evaluation of climate actions that have been undertaken in relation to the set system of the Paris Agreement (UNFCCC, 2023). This process will highlight gaps in current efforts, particularly in high-emission sectors like industry, and drive countries to submit more ambitious NDCs by 2025 (IEA, 2021b). Additionally, COP28 is likely to address the need for increased climate finance, which is crucial for enabling developing countries to meet their climate goals (UNFCCC, 2023). The industrial sector, in particular, will require significant investments in clean technologies, estimated at around $4 trillion annually by 2030, to transition towards net zero (IEA, 2021c).

19.2.1.2 Sustainable Development Goals (SDGs) and Industry's Role

The United Nations Sustainable Development Goals (SDGs), which were adopted in 2015 and are part of the 2030 Agenda for Sustainable Development, gives global outreach to challenges such as poverty, inequalities, environmental degradation, climate change, peace, and justice (United Nations, 2015). Among the 17 SDGs, climate action and the accountability of industries in driving the net zero framework can be traced in some of the goals.

SDG 13: This particular SDG refers to climate change action and runs concurrent with the Paris Agreement calling for action to climate change. The industrial sector here is one of the major emission contributors and hence one of the targets on net zero achievements (IEA, 2021b). This entails that industries have to change their structure and management by utilizing energy-saving technologies, using alternative energies, and better resource management. As an illustration, the coal and cement industry accounts for 14% of CO2 emissions globally and needs to incorporate low-carbon working technologies to carbon capture and storage in order to offer global emission reductions of 90% (IEA, 2021b).

SDG 7: Affordable and Clean Energy emphasizes the necessity of energy access that is dependable and environmentally friendly. Because industry constitutes 42% of the total energy consumption, a complete transformation toward environmentally sound energy systems such as solar and wind is the type of land use that is necessary to abate pollution (IEA, 2021b). According to the International Energy Agency (IEA), renewable sources of energy have to account for 87% of all electricity power generated by 2050, which opens these sectors toward making the shift into energy and achieving the related climate objectives (IEA, 2021c).

SDG 9: Industry, Innovation, and Infrastructure aims at enhancing industrialization in a sustainable manner. The so-called circular economy and its business models, which are based on reclaiming and regenerating materials, have the potential to mitigate emissions and enhance resource efficiency (Ellen MacArthur Foundation, 2020). If circular practices were adopted on a global scale, it is estimated that greenhouse gas emissions would be reduced by 39% and resource extraction by 28% by 2032 (Ellen MacArthur Foundation, 2020).

SDG 12: Responsible Consumption and Production views production processes from a sustainability standpoint and urges industries to abate waste and pollution in the processes, and design processes that are resource efficient. Green supply chains that were first employed by companies such as Unilever and Patagonia offer an example of how business sustainability relates to corporate strategies and performance to produce better social and economic outcomes (Purton, 2024).

19.2.2 Industrial Sectors and Their Emission Contributions

In many countries and regions across the world, including the United States, the industrial sector is among the most noticeable sectors contributing to economic growth. However, it is worthwhile to note that they are also still one of the most significant sectoral sources of greenhouse gas (GHG) emissions (IEA, 2021b). From energy to food production, every type of industrial activity contributes a fair share towards carbon dioxide (CO2), methane (CH4), and other greenhouse gas emissions (IPCC, 2021). It is important to also address these emissions in order to meet net zero targets and to avoid the nonreversible consequences of climate change (IPCC, 2021).

19.2.2.1 Energy and Power Generation

The energy and power generation sector ranks among the largest contributors to global warming, with GHG emissions constituting over 73% (IEA, 2021b). This sector covers both the system of electricity and heat production, which is responsible for about 42% of global CO_2 emissions (IEA, 2021b). The reason for the supremacy of this sector in emissions is the consumption of fossil fuels like coal, oil, and gas in the generation of electric power for industrial and heating purposes.

As of today, carbon dioxide emissions related to the operation of coal-fired power plants, which individually contribute to a staggering amount of power generation; single-handedly accounts for nearly 820 grams of CO_2 emissions considering the output level of a kilowatt hour (kWh) of electricity. This is significantly lower than gas power plants developed, which emit around 450 g CO2/kWh. Even though the share of coal in the global energy mix is decreasing, specifically in more developed countries, it still supplies 36% of global electricity generation. This trend of relying on coal has been more apparent in the developing countries like India and China where urbanization and industrialization are still in demand for coal energy.

There is a clear trend of a shift to a less carbon-intensive energy mix. This trend is further augmented by renewable energy devices becoming even cheaper still. Solar PV (Solar Photovoltaic), wind, and hydro are at the forefront of this deployment as solar energy systems costs have decreased 85% in the last 10 years (IEA, 2021b). The share of renewable electricity was 29% in 2021, most of which was contributed by wind and solar energy, which grew measurably. According to the projections by the IEA, nearly 90% of the world's electricity generation could be fueled by renewable energy sources by 2050 in line with the net zero emissions scenario (IEA, 2021b). In this regard, Germany and Denmark have made remarkable developments as their renewables share within the electrical portfolio grew about 40% electricity mix, while China tops the global rank in terms of wind and solar installed capacity (World Energy Council, 2022).

Nevertheless, a mere shift to renewables does not cut it when it comes to decarbonizing the energy sector. To achieve more radical fades in emissions, other factors such as grid rehabilitation, energy storage systems, and carbon capture and storage technologies are necessary. The IEA claims that in the year 2030, a $4 trillion investment in clean energy infrastructure would be required every year to attain the global net zero aspiration by the year 2050 (IEA, 2021c).

19.2.2.2 Manufacturing and Heavy Industry

The manufacturing and heavy industry sector is another significant contributor of global GHG emissions, responsible for about 23% of the total emissions of CO2. Such industries include the production of iron, cement, aluminum, chemicals, and plastics, and other construction materials that play an important role in people's lives but also douse the atmosphere with pollutants. Nevertheless, processes used in the production pattern within these industries are mainly fossil fuel-based and energy-demanding (IEA, 2022).

Cement making accounts for about 8% of global CO_2 emissions stemming from human activity. This is mainly due to the use of quicklime in cement production, manufactured from heating limestone, which releases vast amounts of CO (Lehne & Preston, 2018). Although this industry is crucial for building and infrastructure, the cement industry lags in its commitment to minimize emissions as encumbered by the construction activity benefiting from its industrial empires (Lehne & Preston, 2018). Innovative technologies, such as carbon capture and storage (CCS), alternative cement materials, and low-carbon production methods, are being developed to mitigate these emissions. For instance, the use of carbon capture and utilization (CCU) technologies in cement plants could potentially reduce emissions by up to 90% (IEA, 2021c).

Another major contributor to CO_2 emissions is the steel industry, which accounts for around 7–9% of these emissions. The bulk of steel is made in blast furnaces using iron ore and coke when smelting. Transitioning the steel industry to low-carbon substitutes is quite essential toward the attainment of the net zero goals (IEA, 2022). One such technology is green hydrogen, which is hydrogen produced using renewable energy. That's huge because green hydrogen can substitute coke when making steel and thus result in lower greenhouse gas emissions (Algers, 2024). In Sweden, fossil-free steelmaking is being developed by the Swedish steelmaker SSAB, mining giant LKAB, and energy corporation Vattenfall under the HYBRIT project, which has already released the first batch of green steel in the world.

The chemical industry, specifically focusing on petrochemicals, fertilizers, and plastics production, is responsible for approximately 5%–6% of global GHG emissions (IEA, 2022). The manufacturing of chemicals, including that of ammonia and ethylene, is characterized by extremely energy-demanding processes. The primary methods of decarbonizing the chemical industry will be a combination of energy efficiency, electrification, and the utilization of renewable or bio-based feedstock materials (IEA, 2022).

It is estimated that over $2 trillion will have to be invested to decarbonize these industries over the next 30 years or so because this collaboration highly concentrates on increasing the application of clean technologies in the industry (Purton, 2024).

19.2.2.3 Transportation and Logistics

The transportation sector generates roughly 15% of the total worldwide GHG emissions, mainly due to emissions associated with the fossil fuel combustion of vehicles, ships, planes, and trains. As per the distribution of emissions within the transportation sector, road transport comprises roughly 75% of the emissions with the dominant transport modes being gasoline and diesel-powered passenger vehicles and using trucks and buses (International Transport Forum [ITF], 2021). In transportation, this is the most critical sector with emissions classified approximately at 7.3 GtCO2e. Global transport emissions in the year 2020 were 7.3 gigatons of CO_2 equivalent, $GtCO_2e$, making transport one of the sectors that received much-needed attention from climate mitigation agendas (ITF, 2021).

Indeed, the increase in the popularity of electric vehicles (EVs) is one of the lowest-emission solutions for combating pollution from road transport. By the end of 2022, the number of EVs exceeded 16.5 million, which is a dramatic rise from only 3.4 million EVs available in 2019 (IEA, 2023). In nations such as Norway, which leads the way for EV adoption, electric vehicles accounted for more than eight out of ten sales of new cars, indicating scope for a worldwide shift (IEA, 2023). However, this transition is limited and will only succeed to the extent that adequate charging availability is provided and electricity grids are decarbonized so that EVs indeed become a reduction in emissions (IEA, 2023).

Aviation is only responsible for approximately 2%–5% of the carbon emitted by both international and national logging activities. As the market for air travel around the world increases, then emissions in aviation will also increase even with the creations of emission standards and regulatory instruments. According to the Interpersonal Air Transport Association, they are achieving net zero emissions by 2050 with some measures such as increasing efficiency in the use of fuel, the introduction of sustainable aviation fuels (SAFs), and the use of electric or hybrid airplanes. It is estimated that SAFs, which have been shown to lower carbon emissions by 80% as compared to regular jet oil, will aid in the Green Mission of the aviation industry although a few more hurdles in production and cost-effectiveness still exist (International Air Transport Association, 2021).

While around 3% of total emissions on a global scale, seaborne trade, also known as maritime shipping is responsible for the carriage of about 90% of the trade. The description says that the International Maritime Organization has set a target of a 50 % reduction of emissions from shipping and pollution associated with it by 2050 (International Maritime Organization [IMO], 2020). More cleaner fuels such as LNG, biofuel, and hydrogen fuel cells will be promoted. Other cleaner shipping methods, such as wind-assisted propulsion and development of zero-emission vessels, are also under progress to lower the carbon footprint in sea transport (IMO, 2020).

19.2.2.4 Agriculture and Land Use

Employing SLCPs and policies, agriculture and land use change amounts to about 24% of the global GHG emissions. Emissions are a result of deforestation and conversion of lands to agricultural lands, enteric fermentation in livestock, rice paddies, and poor agricultural practices (IPCC, 2019; Smith et al., 2014). The sector is responsible for releasing huge amounts of CH_4 and N_2O, which have a far more greenhouse effect compared to CO_2 (Myhre et al., 2013).

Especially methane formed during enteric fermentation of various animals, such as beef cattle, sheep,and goats contributes to the global warming potential. Emissions from livestock farming contribute roughly 14.5% of global GHG emissions (Gerber et al., 2013). New methods and approaches such as methane-reducing feed additives and better breeding of livestock are underway to further reduce emissions from methane by ruminants (Hristov et al., 2013).

Deforestation and land-use change, particularly in tropical regions, are major contributors to CO_2 emissions. The loss of forests in regions like the Amazon, often for agricultural expansion, releases stored carbon into the atmosphere (Baccini et al., 2017). Between 2001 and 2020, the world lost 411 million hectares of forest, contributing to significant CO_2 emissions (Global Forest Watch, 2021). Protecting forests, promoting reforestation, and improving land management practices are critical to reducing emissions from land-use change (Griscom et al., 2017). In addition to mitigation, the agriculture sector can also serve as a carbon sink through sustainable practices like agroforestry, regenerative agriculture, and carbon sequestration in soils (Lal, 2020).

According to Figure 19.1a, Energy and Power Generation is the largest source of global greenhouse gas emissions, contributing 54.1% of the total. This is followed by Agriculture and Land Use at 17.8%, Manufacturing and Heavy Industry at 17.0%, and Transportation at 11.1%. Figure 19.1b highlights that within industrial sectors, Livestock Farming is the highest emitter, accounting for

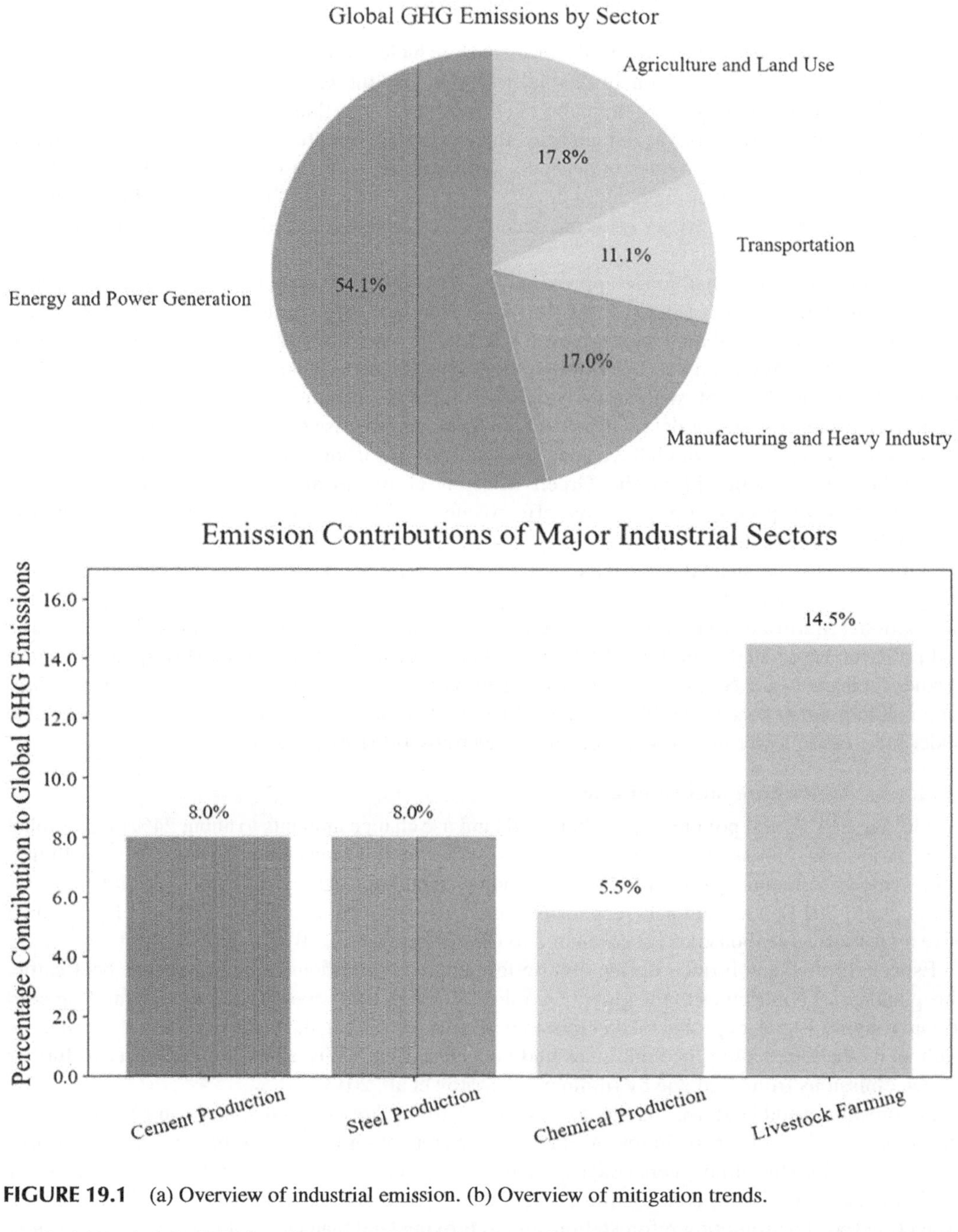

FIGURE 19.1 (a) Overview of industrial emission. (b) Overview of mitigation trends.

nearly 14% of global GHG emissions, while Cement and Steel Production contribute around 7%–8% each, and Chemical Production slightly over 4%.

19.3 FRAMEWORK FOR NET ZERO TRANSITION IN INDUSTRY

Organized by factors such as structural shifts, "decarbonization," which aims at reducing the processes that lead to high levels of greenhouse gases, is still gradual thanks to a plethora of loopholes within the policies. According to Table 19.1, the industrial sector's mission of moving toward net zero emissions is a complex task and requires a broad strategy of policy, regulation, and economics,

TABLE 19.1
Key Policy and Technological Initiatives for Industrial Net Zero Transition

Country	Policy/Regulation	Technological Initiative	Emission Reduction Target	Challenges
India	Renewable Purchase Obligation (RPO); National Action Plan on Climate Change (NAPCC)	Investment in Solar and Wind Energy; National Hydrogen Mission	Net Zero by 2070	High dependence on coal; Infrastructure and funding gaps
China	Ultra-Low Emission Standards for Steel and Cement Industries	Carbon Capture Utilization and Storage (CCUS); Hydrogen Energy	Peak emissions by 2030, Net Zero by 2060	High carbon intensity in industrial sectors
Brazil	National Policy on Climate Change; Incentives for biofuels (RenovaBio)	Use of bioenergy in industrial sectors; Expansion of Wind and Solar	Net Zero by 2050	Deforestation and land-use conflicts
South Africa	Carbon Tax; Just Energy Transition Partnership (JETP)	Development of Green Hydrogen Economy	Net Zero by 2050	Energy sector dominated by coal
Indonesia	Low Carbon Development Initiative (LCDI); Energy Conservation Policy	Electrification of Industrial Processes; Biomass Energy	Net Zero by 2060	Deforestation, reliance on coal
Mexico	Energy Transition Law; Carbon Pricing Mechanisms	Investment in Renewable Energy, Energy Efficiency Technologies	Net Zero by 2050	Limited investment in clean energy infrastructure

as well as technology. This approach is particularly important for developing countries, in particular as it compromises any prospects of maintaining the industrial development necessary for economic growth (McKinsey & Company, 2023). Additional substantiation follows on the necessity components supporting the provision of such transition.

19.3.1 Policy and Regulatory Framework

Cost-effective policy frameworks are fundamental cornerstones in guiding the industrial sector towards net zero emissions. Local governments theory agrees that the central government indeed comes in with strong policies and regulations and sets the pace for the transition at the speeds that are considered desirable (IEA, 2022).

19.3.1.1 Government Policies and Regulations

1. **Emission Reduction Targets:** Setting legally binding national emission reduction targets aligned with the Paris Agreement. For instance, as of 2023, over 130 countries have pledged to achieve net zero emissions by 2050, including major emerging economies like China (by 2060) and India (by 2070) (World Bank, 2023).
2. **Environmental Standards:** Taking effective measures to prevent pollution during industrial operations and avoid processes with large adverse effects. The European Union's Industrial Emissions Directive aims to control pollution from discharges from large industrial facilities, moving them to the use of the so-called Best Available Techniques. In the same way, China's Ultra-Low Emission Standards for the steel industry target controlling emissions of particulate matter (PM), sulphur dioxide (SO_2), and nitrogen oxides (NO_x) by a very wide margin (IIGCC, 2023).

3. **Renewable Energy Mandates:** Mandating a certain percentage of energy consumption to come from renewable sources. India's Renewable Purchase Obligation requires specific industries to source a portion of their power from renewable energy, aiming for 21% by 2022. This policy has driven substantial growth in India's renewable energy capacity, reaching over 100 GW by 2021 (IEA, 2022).
4. **Energy Efficiency Standards:** Prescribing basic energy requirements for industrial production methods and goods. China's Top 1,000 Enterprises Energy Conservation Program, initiated in 2006, aimed to improve energy efficiency by reducing energy consumption in key industries. The program helped save the equivalent of approximately 150 million tons of coal by 2010. Additionally, it promoted the reduction of electric vehicle and electric bicycle use in specific sectors, achieving a 20% reduction in energy consumption in civilian and industrial operations (World Bank, 2023).
5. **Reporting and Disclosure Requirements:** Requiring industries to monitor and report their GHG emissions. the U.S. Environmental Protection Agency's Greenhouse Gas Reporting Program mandates large emitters to report their emissions data annually, covering facilities that emit 25,000 metric tons or more of CO_2 equivalent per year (McKinsey & Company, 2023).

19.3.1.2 Carbon Pricing Mechanisms

The two types of Carbon Pricing Mechanisms are:

1. **Carbon Taxes:** A carbon tax directly sets a price on carbon by defining a tax rate on GHG emissions or the carbon content of fossil fuels. Examples include
   - Sweden introduced a carbon tax in 1991, which has gradually increased to approximately $137 per ton of CO_2 in 2023, one of the highest globally. This policy has contributed to Sweden reducing its GHG emissions by 29% from 1990 levels while maintaining economic growth (IEA, 2022).
   - South Africa implemented a carbon tax in 2019, starting at approximately $8 per ton of CO_2, aiming to reduce emissions and encourage investment in low-carbon technologies (IIGCC, 2023).
2. **Emissions Trading Systems (ETS):** An ETS, or cap-and-trade system, sets a cap on total emissions and allows industries to buy and sell emission allowances. Notable examples include:
   - European Union Emissions Trading System (EU ETS): The world's largest carbon market, covering around 40% of the EU's GHG emissions. Since its inception in 2005, emissions from installations covered by the EU ETS have declined by about 35% by 2020 (IEA, 2022).
   - China's National ETS: Launched in 2021, it initially covers the power generation sector, responsible for over 40% of China's emissions, and is expected to expand to other industrial sectors (McKinsey & Company, 2023). As of 2023, there are 68 carbon pricing initiatives implemented or scheduled for implementation, covering about 23% of global GHG emissions, according to the World Bank (2023). Carbon pricing revenues reached about $53 billion in 2020, providing funds that can be reinvested into low-carbon technologies and supporting vulnerable communities.

19.3.1.3 Incentives for Green Technology Adoption

1. **Tax Incentives and Rebates:** Offering tax credits, deductions, or rebates for investments in energy efficiency, renewable energy installations, or research and development (R&D) in sustainable technologies.
   - United States: The Investment Tax Credit provides a 26% tax credit for solar systems on residential and commercial properties. The Production Tax Credit offers incentives for

wind energy production, supporting the growth of renewable energy sectors (World Bank, 2023).
- India: Provides accelerated depreciation benefits and tax holidays for renewable energy projects, encouraging investment in solar and wind energy sectors, contributing to India's renewable capacity reaching over 100 GW as of 2021 (IIGCC, 2023).

2. **Subsidies and Grants:** Direct financial support for adopting clean technologies or conducting R&D in sustainability.
   - European Union: The Horizon Europe program allocates €95.5 billion (2021–2027) for research and innovation, with significant funding directed toward climate action and clean technologies (IEA, 2022).
   - China: Offers substantial subsidies for EV manufacturing and purchases, contributing to China becoming the world's largest EV market, with over 6 million EVs sold by 2022 (McKinsey & Company, 2023).
3. **Low-Interest Loans and Financing Mechanisms:** Providing favorable financing options to reduce the cost of capital for green investments.
   - Green Bonds: Governments and financial institutions issue green bonds to raise capital for projects with environmental benefits. The global green bond market surpassed $1 trillion in cumulative issuance by 2020 (World Bank, 2023).
   - Japan: Offers low-interest loans through the Japan Finance Corporation for SMEs investing in energy-efficient equipment, promoting energy conservation across industries (IIGCC, 2023).
4. **Public-Private Partnerships:** Collaborations between government entities and private sector companies to develop and deploy sustainable technologies and infrastructure.
   - Australia's Clean Energy Finance Corporation: A government-owned institution investing over AU$10 billion in clean energy projects, leveraging private sector capital to drive the clean energy transition (McKinsey & Company, 2023).
   - India's National Investment and Infrastructure Fund: Partners with private investors to fund sustainable infrastructure projects, including renewable energy and green transportation initiatives (World Bank, 2023).
5. **Feed-in Tariffs (FiTs) and Net Metering:** Policies that guarantee a fixed, premium price for renewable energy producers, encouraging investment in renewable energy generation.
   - Germany: Pioneered FiTs with the Renewable Energy Sources Act, significantly boosting renewable energy capacity, which accounted for over 46% of electricity consumption in 2020 (IEA, 2022).
   - Malaysia: Implemented FiTs to promote renewable energy, achieving a cumulative installed capacity of over 1,600 MW by 2020 (World Bank, 2023).

19.3.2 Corporate Strategies for Emission Reduction

1. **Setting Corporate Net Zero Goals:** Targets of net zero emissions have increased rapidly, especially among corporations, extending to the requirement to take action on climate change issues and to comply with international agreements, such as the Paris Agreement. By 2023, almost one out of every five of the world's largest corporations, which represent nearly 68% of world GDP, had already set net zero aims, a remarkable improvement from merely 16 in 2019 (GreenBiz, 2022). These targets are aimed at companies' public announcement pledging to bring down their emissions to zero or very close to it by the year in mid-century in line with the scientists' guidance of containing temperature rise to 1.5°C I.

Often to enhance the chances of these commitments being adhered to, a number of corporations are embracing Science Based Targets. These targets, approved by the Science Based Targets initiative, help to establish a timeline with date ranges of between 5 and 10 years that are consistent with the emissions reductions that must be achieved. For example, Microsoft has set a target not only to achieve carbon net zero emissions by 2030 but also to go further back in achieving net neutrality by 2050 (GreenBiz, 2023).

2. **Integrating Sustainability into Corporate Governance:** Making sustainability an integral part of corporate governance is very important, especially with the current growth of climate change risks and opportunities. Organizations are creating a new position called a chief sustainability officer, and this person is responsible for all activities related to sustainability and protecting the climate across the organization. One benefit of this trend is an increased number of the board's committees concentrating on environmental issues where climate change risk management is already overseen in more than 70% of S&P 500 companies (MSCI, 2023).

Moreover, in many corporations, attention to Environmental, Social, and Governance (ESG) factors and metrics is stitched into corporate governance with executive pay based on non-financial results. Unilever is one such example and ties a portion of its executive bonuses to such climate and social targets as emissions reductions and resource use efficacy. Such practices not only promote sustainability but also help to bring about corporate governance change needed for climate change. Climate change has fostered the corporate social responsibility movement and consequently the promotion of climate change-related corporate activities. Climate change is now considered an integral part of the company's activities and stakeholder evaluations (MSCI, 2023).

3. **Transparency and Reporting: Transparency and Reporting:** GHG Protocol and Task Force on Climate-related Financial Disclosures (TCFD)
   Effective corporate emission reduction cannot ignore transparency. Companies, aided by sustainable mechanisms like the GHG Protocol, can record and control emissions attributed to every segment of a value chain, summarized in three scopes:
   - **Scope 1:** The first scope covers direct emissions from sources owned or controlled by the legal body (for instance, industrial plants).
   - **Scope 2:** The second scope derives most emissions virtually from the third countries regarding the purchased power.
   - **Scope 3:** The third scope covers emissions that are not controlled by company operational activities; for example, emissions from the production process for suppliers and emissions during the use of the company's products.

This is not the only case in which the GHG Protocol is being adopted together with the TCFD guidance, as in the case of companies in their majority. Over 2,600 corporations in the world with a total market value of $25 trillion, are committed as of 2021 to making financial disclosures in line with these standards aimed at climate change (MSCI, 2023).

19.3.3 Technological Innovations

Gearing up for a forecast of net zero emissions in most industries would require a turn to a mix of old and new devices to make several core processes carbon neutral. Approximately 25% of global greenhouse gas emissions come from industrial activity, according to Figure 19.2, so new technologies should be adopted to fulfil the climatic goals (Purton, 2024). The following section describes how the chemical industries fit in their efforts to achieve this transformation.

1. **Renewable Energy Adoption in Industrial Processes**
   Embracing renewable energy in industrial practices is changing how industries produce emissions. Solar, wind, biogas, and other renewable resources are increasingly used in manufacturing, production lines, and operating heavy machinery.
   - **Renewable adoption by the world:** In 2021, electricity production from renewable energy added 260 GW of capacity, with new photovoltaic capacity installations reaching

175 GW (IEA, 2021b). By 2030, an increase in the use of renewables is expected to meet at least 60% of the electricity required by industries to achieve global energy decarbonization goals (IRENA, 2022).

- **Specific case in the sector:** The aluminum industry is a key example. Alcoa has reduced its energy emissions by 70%–80% by switching to hydropower instead of conventional fossil fuel sources (Smith and Taylor, 2022).
- **Utilizing biomass with cement:** The cement sector accounts for 7% of global emissions, and biomass usage in cement industries offers growth potential due to biomass fuels emitting up to 40% less CO_2 when scaled (IEA, 2021b).

2. **Carbon Capture, Utilization, and Storage (CCUS)**
3. CCUS technology is crucial in decarbonizing industries with zero emissions since it is impractical to electrify or switch entirely to renewable energy. The technology captures and stores CO_2 from industrial processes, either storing it underground or recycling it to aid in processes like enhanced oil recovery or synthetic fuel creation (IEA, 2021c).
   - **Deployment status:** As of 2022, around 30 commercial CCUS projects are operating worldwide, capturing about 40 million tons of CO_2 annually. By 2030, this figure is expected to increase tenfold, allowing industries like cement, steel, and chemicals to significantly reduce emissions (Global CCS Institute, 2022).
   - **Sector impact:** The cement sector, for instance, could reduce over 50% of its CO_2 emissions through widespread CCUS implementation (IEA, 2021c).
4. **Electrification and Energy Efficiency Improvements**
   Electrification is vital for various sectors, allowing them to convert carbon-intensive processes to electricity or climate-friendly heat energy (IRENA, 2022).
   - **Electrification of steelmaking:** The industry is shifting from emission-intensive, coal-based blast furnaces to electric arc furnaces using electricity to melt steel scrap. This shift has reduced emissions by 60% with conventional electricity sources and up to 90% with clean energy (Smith and Taylor, 2022).

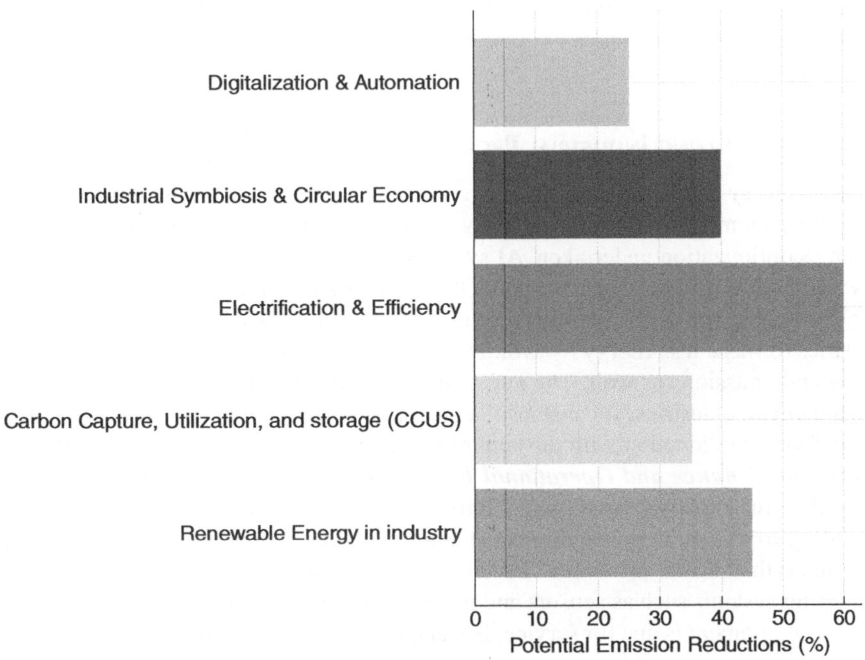

FIGURE 19.2 Technological innovations and their impact on emission reductions in industry.

- **Smart energy systems:** In factories, smart energy management systems powered by artificial intelligence are being deployed to improve energy efficiency, leading to 15%–25% energy savings (Purton, 2024).

5. **Digitalization and Automation**
   The use of digital technologies like the Internet of Things (IoT), machine learning, and automation improves industrial efficiency and reduces emissions. Digitalization enhances energy and carbon management using real-time data (McKinsey & Company, 2023).
   - **Projected impact of digitalization:** By 2030, digital technologies are expected to reduce industrial emissions by at least 12% while improving energy efficiency (IRENA, 2022).
   - **Case example:** Chinese manufacturers have implemented a digital supply chain energy management system that monitors energy use across supply chains, reducing emission concentrations by 20% and energy use by 15% (McKinsey & Company, 2024).

19.4 ARTIFICIAL INTELLIGENCE AND DIGITAL INNOVATIONS FOR NET ZERO

The journey to net zero emissions is being radically transformed by the integration of artificial intelligence (AI) and digital innovations, which are playing a pivotal role in reshaping industrial processes. These technologies offer unparalleled efficiency by leveraging real-time data analytics, predictive models, and automation to optimize energy use, reduce waste, and lower emissions. AI–powered systems monitor energy consumption across industries, dynamically adjusting operations to minimize excess energy use while maximizing the incorporation of renewable energy sources (Derrick, 2023). Predictive maintenance, another critical AI application, ensures that industrial equipment operates at peak efficiency by forecasting potential failures before they occur, reducing downtime and unnecessary energy loss (Butcher, 2023). Additionally, digital twins—virtual replicas of physical assets—allow companies to simulate and optimize industrial operations without disrupting actual production, leading to more energy-efficient and sustainable practices (ABB & Industrial, 2022). As industries adopt AI to streamline workflows, enhance decision-making, and reduce their carbon footprints, these innovations are accelerating the global push toward net zero emissions, offering scalable solutions that balance economic growth with environmental sustainability (ABB & Industrial, 2022).

19.4.1 AI in Optimizing Industrial Processes

***AI-Driven Energy Management Systems*:** Energy management is mainly done through energy management systems powered by AI whereby energy consumption is tracked and resources optimization undertaken. AI systems have the capability to forecast demand and those demand changes, especially with the deployments and uptake of more renewable energy sources, and subsequently manage energy flows (Derrick, 2023). It also enables industries to waste less energy by avoiding energy imbalance, which cuts down the costs of energy and emissions as well. The latest advanced AI technologies have begun to change energy-hungry industries, for instance, a steel plant or even a chemical plant, in terms of carbon footprint reduction with the help of energy management (Butcher, 2023)

***Predictive Maintenance and Operational Efficiency*:** Predictive maintenance has got to be one of the greatest advantages gained through the adoption of AI in the industrial sphere. According to this strategy, equipment is maintained periodically, leading to stretches in downtimes that waste resources. Unlike this, AI–enabled predictive maintenance uses monitoring systems such as sensors and machine learning-assisted algorithms that indicate when the equipment is due for service. By analyzing prior records of sensor data, historical performance analysis, and patterns of usage, AI can forecast some machine failures. This enhances machinery operation, decreases maintenance-related expenditure, and enhances the working lifespan of industrial equipment too (Butcher, 2023) This also improves productivity in such industries as they are able to work at maximum productivity with very minimal interruptions.

19.4.1.1 Digital Twins and Smart Manufacturing

***Real-Time Monitoring and Decision-Making*:** Digital twins connect the physical and virtual worlds to enhance operations through timely interventions and decision-making. In construction, a perfect "digital twin" of building assets can be created, allowing for constant monitoring of industrial processes, performance evaluations, and quick corrections. This feature is particularly important for production industries, where delicate processes must be carefully managed to achieve optimal output while minimizing energy consumption. Moreover, in cases where continuous feedback is provided through sensors, digital twins enable manufacturers to quickly identify inefficiencies and adjust parameters in real time, improving production and reducing power consumption (Segovia and Garcia-Alfaro, 2022).

***Reducing Emissions Through Digital Simulations*:** The advance of digital twins offers enormous flexibility as it allows for the playing out of various scenarios without affecting the actual operations. Within the context of digital twin technology, manufacturers have the capacity to perform an energy and cost analysis of various configurations and operational tactics within the virtual sphere and settle for the best possible configuration. They simulate alterations to production methods in front of the computer and runtime estimate their influencing factors on emissions and come up with a lower carbon emission practice. For instance, in the automobile industry, the framework of digital twins is implemented in designing and manufacturing electric cars by reducing emissions from all the processes in the EV manufacturing supply chain (Segovia and Garcia-Alfaro, 2022).

***AI for Emission Reduction and Carbon Capture*:** Smart technologies, which include carbon capture and storage and emission monitoring systems, coupled with AI, are also being employed in the effort to cut down industrial emissions. Environmental data can be processed by AI and extrapolated to trends on possible emissions, whose turnaround activities can be recommended to be taken up, paving the way for immediate reducing actions (Derrick, 2023). In the carbon capture segment, AI techniques will enhance carbon capture technology due to the predictive ability showing the peak emission time and place, which increases the performance and application of carbon capture (Segovia and Garcia-Alfaro, 2022).

19.5 CHALLENGES FOR EMERGING ECONOMIES

As soon as the developing countries come up with plans to further progress economically, they are faced with a number of challenges that hinder them from attaining all their purposes. Such challenges are somehow comprehensive, going through economic, infrastructural, and socio-economic spheres. Economically, these countries are under unbearable limits; for instance, the procedures associated with transitioning to greener technologies require heavy impositions of investments (Zhironkin, 2022). Further, very few domestic resources and adverse international situations limit opportunities to access green funds (UNEP, 2022). According to Table 19.2, these problems are aggravated by infrastructural deficits, where old systems are not able to incorporate renewable energy systems, while modern and robust infrastructure is not yet designed and built (Gurara et al., 2018). Moreover, factors such as low R&D capabilities constrain local innovation, which is essential for customizing global best practices to fit the local environment (Bhattacharyya, 2020).

On the other hand, these countries have to confront the integration of the economy and the environment as they aspire for rapid development in a recently developing industry, the "green economy," to ensure that there are no massive rationalizations or increasing current inequalities (UNCTAD, 2022). Such a change not only needs a fair shift from previous occupations for employees in conventional sectors but also better integration of inexpensive clean energy for disadvantaged

TABLE 19.2
Challenge for Emerging Economies

Category	Challenge	Details	Data/Statistics	References
Economic Constraints	**Capital Investment Needs**	Significant capital needed for clean technology transitions.	EMDEs: $1.8–$2.8 trillion annually by 2030s. 2022: $773 billion.	IEA, IFC, World Bank Reports
	Access to Green Financing	Challenges include regulatory frameworks, market access, and capital shortages.	$900 billion–$1.1 trillion needed annually by 2030s. Higher capital costs in EMDEs.	IEA, World Bank, IMF Reports
Infrastructure & Technological Gaps	**Outdated Infrastructure**	Outdated grids and transport networks limit renewable energy adoption.	Significant portions of EMDEs lack reliable electricity and clean cooking solutions.	IFC, IEA Reports
	R&D and Local Innovation	Limited R&D and innovation hinder clean technology development.	More R&D investment needed for affordable, scalable, and local technologies.	International Energy & Development Agencies
Socio-Economic Factors	**Employment and Just Transition**	Green transition impacts traditional industries, requiring job creation and reskilling.	Potential job losses; need for retraining and social safety nets. Opportunities in green industries.	Global Economic and Labor Reports
	Energy Access and Equity	Balancing sustainability with improving energy access is crucial, especially in regions with unreliable services.	Millions lack electricity and clean cooking solutions. Decentralized systems like mini-grids offer solutions.	IFC, Various International Energy Reports
	Balancing Industrial Growth with Sustainability	Integrating environmental policies with industrial growth strategies is essential.	Environmental considerations must be included in growth strategies. International support is crucial.	World Bank, International Development Organizations

people (IEA, 2021c). To address these multifaceted challenges, emerging economies must pursue innovative, context-specific solutions, backed by strong international cooperation and support (World Bank, 2021).

19.5.1 Economic Constraints and Investment Needs

19.5.1.1 Capital Investment Requirements for Clean Technology

As noted, developing economies require significant levels of capital for them to be able to make the switch to clean energy. It is assumed in the IEA's *Energy Technology Perspective* that achieving sustained economic development and climate change target is going to require that clean energy investment in Emerging Markets and Developing Economies (EMDEs) increase to between US$1.8 and 2.8 trillion annually, and this has to be by the 2030s (IEA, 2023). As seen, the need is projected to surge from the US$773b that was put to use in 2022 (IEA, 2023). These investments are important to enhance the deployment of these clean technologies in terms of lowering the emissions from electricity generation, energy conservation, and transportation. Such strategies do not abolish the existing risk of funds being sourced outside EMDEs since in EMDEs the cost of capital for these investments is still much higher than in advanced economies (IEA, 2023).

19.5.1.2 Access to Green Financing and International Support

Acquiring green financing is almost impossible for most emerging economies. They have challenges like poor regulatory environments, barriers to foreign funders, and very little internal financing. All of these are requirements in investments. However, public resources are not enough and a lot of private investment is required. Private investment is projected to contribute about $900 billion to $1.1 trillion a year by early 2030 (Purton, 2024).

19.5.2 Infrastructure and Technological Gaps

19.5.2.1 Existing Infrastructure Limitations

The current infrastructure in many emerging economies appears inadequate for the efficient deployment of clean technologies. For example, weathering of these regions' electrical grids, lack of storage, or impractical transportation options is an impedance for the use of excess energy created by renewable sources. Moreover, these regions also use fossil fuels for energy and therefore any change to green infrastructure is expensive and tedious (IEA, 2023). In this regard, it becomes a matter of great importance to modernize the current structures as these primary need of the population is electricity and clean cooking fuel which most of the developing world still lacks.

19.5.2.2 Need for R&D and Innovation in Local Contexts

To overcome that, there is an urgent requirement for indigenous innovations which will respond to the societal needs and the environmental issues in such regions. Allocation of resources for R&D will allow the creation of low-cost and large-scale clean technology that would be relevant to the communities where such technologies will be applied. It is understood that development of local innovative strategies to overcome the local constraints is equally important as international collaboration plus support building in the local organizations (IEA, 2023).

19.5.3 Socio-Economic Factors

19.5.3.1 Employment and Just Transition

One should not forget that transitioning to a green economy is not only a technical process but also entails socio-economic impacts that are particularly important with regard to the employment sector. For example, coal mining and manufacturing, among other industries, employ a bulk of the workers in many newly industrialized countries. These industries are bound to be irrelevant due to the increasing emphasis on green modes of production. Involvement of every section of the society in the green economy development requires identification of new sources of employment in green sectors, providing skills reskilling to the displaced employee and offering assistance to young people at risk of becoming unemployed due to changes in the economy (Purton, 2024).

This problem is particularly acute in a number of the emerging economies, where energy access is still not so easy for some of the population. There must be a middle ground where the two goals can be met—energy access should not come at the expense of environmental favors but promote them. Mini-grids and off-grid solar may not only expand energy access but may do it with carbon and other emissions reduction. Even still, these must complement policies that would ensure that access remains affordable and eludes exclusion of the marginalized groups (IEA, 2023).

19.5.3.2 Balancing Industrial Growth with Sustainability

It is well understood that most of the emerging economies have commonly adopted an industrialization strategy in the realization of their economic growth and this is at times in contention with the sustainability objective. This scenario heretofore requires a more strategic way of balancing developmental and environmental objectives by embedding environmental concerns within the industrial development policies. This could be providing cleaner production technologies, strict compliance with environmental laws, and coercive yet transformative policies for the industries which make sure that they go green. Furthermore, there is also the need for international collaboration to give the necessary technical and financial assistance in relation to industrial development to sustainability (Purton, 2024)

To sum up, the emerging economies are confronted by major stumbling blocks when attempting to provide solutions to the problems of sustainable development. Solving such problems requires a complex set of activities such as significant financial resources, improvement of certain infrastructural elements and certain socially just economic strategies. The role of the state, private business and the international community is necessary in order to solve such problems and provide a safe tomorrow.

19.6 CASE STUDIES OF EMERGING ECONOMIES IN TRANSITION

19.6.1 India—Adopting Digital Technologies in Energy-Intensive Sectors

19.6.1.1 Introduction

The economy of India is developing at a high rate, which causes strain on energy-demanding sectors of the country. To embrace globalization and growth within the sector, there is a growing trend of using advanced technologies in energy consumption and carbon reduction within this entity, particularly in steel, cement, and power generation. This case study investigates the practices of digital transformation to achieve these goals and the effects on industry and sustainability.

19.6.1.2 Context and Challenges

India's energy-intensive industries, such as steel, cement, and power generation, are essential for its growth, but they also consume a significant amount of energy and emit large volumes of greenhouse gases. The major constraints include:

- **High Energy Consumption:** Manufacturing sectors are responsible for nearly 40% of the industrial energy use in India, contributing significantly to the nation's carbon emissions (IEA, 2021a).
- **Operational Inefficiencies:** Many industries have not adopted newer systems, leading to inefficiencies and excessive energy consumption (Shinkevich et al., 2020).
- **Environmental Impact:** The emissions from these industries counteract India's climate goals, affecting the atmosphere (IEA, 2021b).

19.6.1.3 Digital Technologies in Action

India is implementing a range of digital technologies to address these challenges:

1. **Smart Manufacturing and Industrial IoT**
   - **Steel Industry:** Companies like Tata Steel and JSW Steel are incorporating IoT solutions to optimize energy consumption. Tata Steel, for example, uses smart sensors and real-time monitoring systems, which have led to a 10% improvement in energy efficiency and a 15% reduction in downtime (Seth et al., 2021; Tata Steel, 2024).
   - **Cement Production:** UltraTech Cement employs advanced control systems and data analytics to optimize kiln operations, resulting in a 7% reduction in energy consumption per ton of cement produced (McKinsey & Company, 2024; UltraTech Cement, 2023).
2. **Energy Management Systems**
   - **Power Generation:** NTPC Limited, India's largest power utility, uses digital EMS to monitor and manage energy flows across its plants. The integration of these systems has

improved grid stability and operational efficiency, leading to a 5% reduction in energy losses (IEA, 2021b; NTPC, 2023).

3. **Data Analytics and Artificial Intelligence**
   - **Predictive Analytics:** Predictive maintenance powered by data analytics is being used to foresee equipment failures and optimize maintenance schedules in the cement and steel industries, minimizing energy waste and extending equipment lifespans (Shinkevich et al., 2020).
   - **AI–Driven Optimization:** AI algorithms are refining energy use across processes. In steel manufacturing, AI optimizes energy consumption in real time, resulting in a 12% decrease in energy usage (IBM, 2023).

19.6.1.4 Results and Impact

The integration of digital technologies has produced significant benefits across India's energy-intensive sectors:

- **Increased Efficiency:** Energy efficiency has improved by 7% to 15%, depending on the sector and technology used (Tata Steel, 2024; UltraTech Cement, 2023).
- **Cost Savings:** Tata Steel reported annual cost savings of approximately $50 million due to enhanced energy management practices (Tata Steel, 2024).
- **Reduced Emissions:** UltraTech Cement's energy efficiency measures have reduced CO2 emissions by 1.2 million tons annually (UltraTech Cement, 2023).

19.6.2 China—Leveraging AI for Smart Manufacturing

19.6.2.1 Introduction

China, the largest manufacturing country in the world, is incorporating AI into manufacturing processes. Businesses are leveraging AI technologies to improve operational efficiency, lower costs, and attain their sustainable development objectives. This case study outlines how China is applying AI to make breakthroughs in smart manufacturing, with examples from the electronics, automotive, and heavy machinery industries.

19.6.2.2 Context and Challenges

Manufacturing remains key to China's economic power, but challenges remain:

- **High Resource Consumption:** Conventional production methods are energy-intensive, generating high levels of waste (McKinsey & Company, 2024).
- **Labor Costs and Skills Gap:** Rising wages and a shortage of semi-skilled workers push manufacturers to adopt automation and other advanced technologies (IBM, 2023).
- **Quality Assurance:** Ensuring mass production quality and reducing defect rates continue to be major hurdles (Huawei, 2022).

19.6.2.3 AI Technologies in Action

1. **Predictive Maintenance and Process Optimization**
   - **Electronics Industry:** Huawei and Lenovo use AI–enabled predictive maintenance to track equipment and preempt failures with minimal intervention. This approach has reduced unplanned downtime by 15% (Huawei, 2022).
   - **Automotive Sector:** BYD and Geely leverage AI to optimize supply chains and assembly line processes. Geely's AI–driven supply chain optimizations have improved inventory turnover rates by 25% (McKinsey & Company, 2024).
2. **Quality Control and Defect Detection**
   - **Heavy Machinery:** Companies like SANY and XCMG use AI–powered visual inspection systems to detect defects in real time. This has reduced defect rates by about 30%, improving product reliability (SANY, 2024; XCMG, 2023).

3. **Smart Robotics and Automation**
   - **General Manufacturing:** AI–based robotics are increasingly deployed in manufacturing, with Foxconn and DJI using collaborative robots to assist human operators in automation processes. This has boosted output from the assembly line by 20% (Foxconn, 2023).

19.6.2.4 Results and Impact

The integration of AI into China's manufacturing sector has yielded notable outcomes:

- **Enhanced Efficiency:** AI optimizations have improved efficiency by up to 20% (Huawei, 2022).
- **Cost Savings:** Predictive maintenance has reduced maintenance costs by 18% (McKinsey & Company, 2024).
- **Improved Quality:** AI–driven quality control systems have reduced defect rates by as much as 30% (SANY, 2024).

According to Figure 19.3a, it illustrates the growth of the Global Green Bond Market from 2015 to 2023. The chart shows a steady increase in green bond issuance, which represents financial instruments used to fund environmentally beneficial projects. Starting at approximately US$100 billion in 2015, the market has grown consistently each year, reaching around US$500 billion by 2023. Notable acceleration in growth is observed between 2020 and 2021, continuing its upward trajectory. This trend underscores the expanding role of green bonds in promoting sustainable investments and addressing climate change globally.

According to Figure 19.3b, it highlights the distribution of jobs created by sector through reskilling programs. The Renewable Energy sector leads with the highest number of jobs created, approaching 140,000. This is followed by Energy Efficiency, which creates around 120,000 jobs. Green

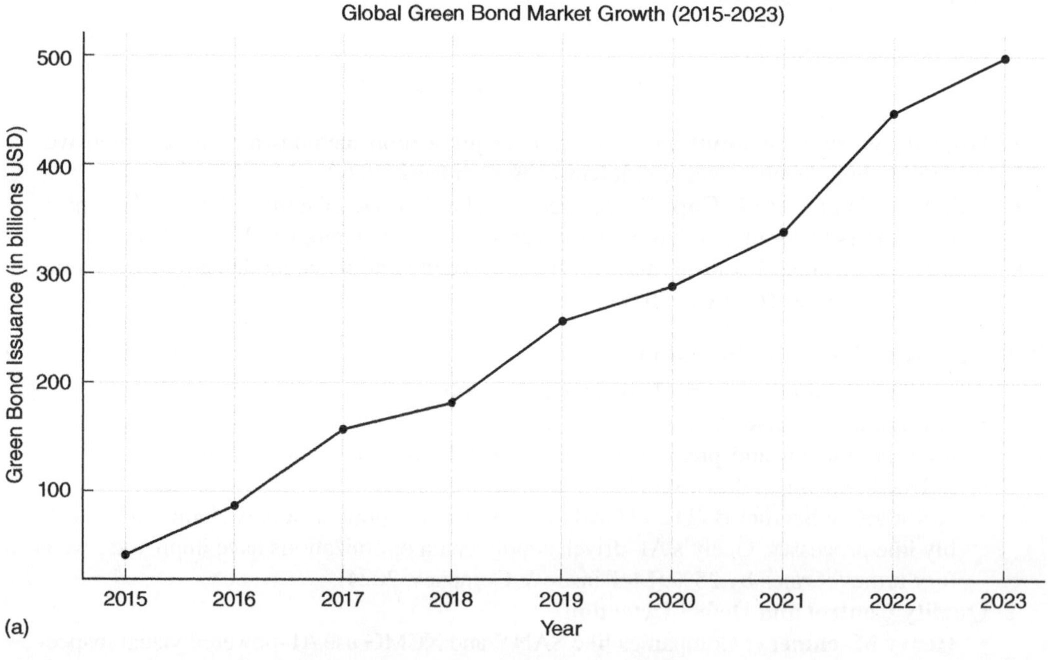

FIGURE 19.3 (a) Global green bond market growth.

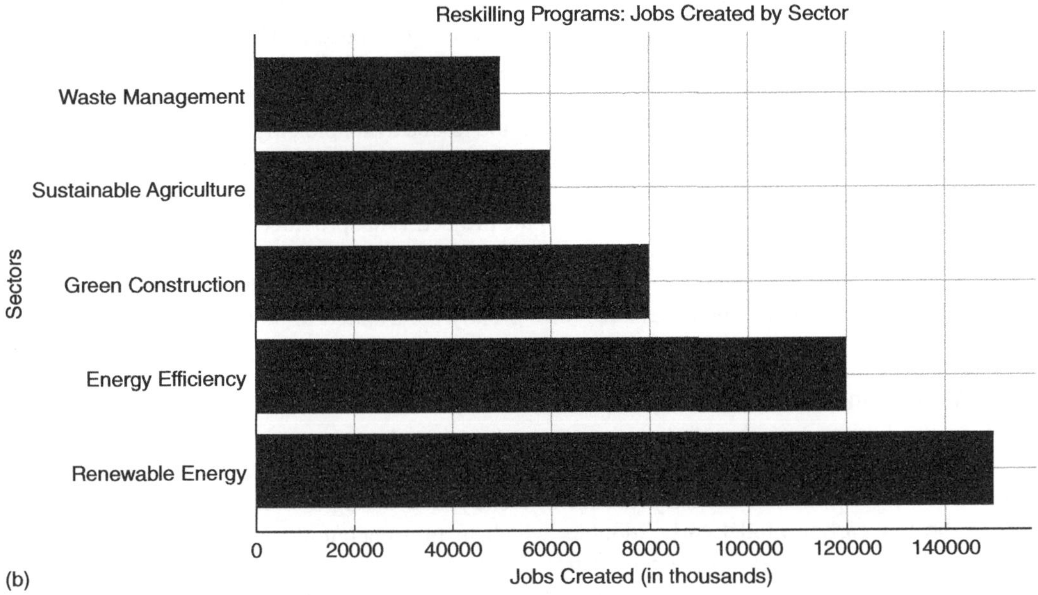

FIGURE 19.3 (Continued) (b) Jobs created by sector.

Construction comes next, contributing to roughly 80,000 new jobs. Sustainable Agriculture and Waste Management generate fewer jobs, with approximately 60,000 and 40,000 jobs, respectively. The data emphasizes the significant job creation potential in sectors associated with sustainability and green initiatives.

19.7 FUTURE DIRECTIONS AND MULTILATERAL COOPERATION

The acceleration in the journey toward net zero emissions rests with replicating the best practices in different sectors, promoting their uptake in developing countries and enhancing the cooperation in technology and knowledge transfer. It has been shown that such innovations can be profitable and environmentally friendly by the early designers, yet any expansion to new geographical areas poses likely more complexities that require effective financial as well as other forms of support. International financial institutions, the World Bank and Green Climate Fund in particular, are among the main stakeholders who provide the necessary funding and insurance against the risks (World Bank, 2022). Multilateral treaties, especially the Paris Agreement, establish the principles for international cooperation so that the technological, financial, and human resources are transferred to the countries that need them (UNFCCC, 2015). Future efforts will first focus on reconstitution and widening of these agreements to cover how countries will adjust and enhance resilience to climate change while insisting on punishment for breach of contract to ensure net zero for all countries remains a reality (World Bank, 2022). Through collective action, global cooperation can steer the world to its sustainable low carbon economy (IEA, 2023).

19.7.1 Scaling Best Practices Across Industries

Attaining net zero emissions in the world calls for adequate technology and dissemination of best practices. Environmental innovations have been initiated by highly competitive companies and veterans who have convinced them of the advantages of better environmental practices. For example, the renewable energy industry, most notably wind and solar energy in Denmark and Germany, has made large wind farm and solar photovoltaic deployments an economic reality (IEA, 2021b). Equally, leaders within the green manufacturing industries have illustrated how energy conservation and reducing emissions-based technologies could be both a strategic differentiation and increase returns of investment (World Bank, 2022).

However, the problem is that these practices should not be available only to the franchise leaders but should be propagated in geography and industries. They point as well to those characteristics that should be emulated in order to prevent some risks or pitfalls, those that are related to raising finance or those pertaining to the management of the project. What provided the initial success to the so-called early movers, if one may call them as such, was the policy thrust in terms of establishing subsidies and incentives, commitment to blue sky research, and a combination of public-private partnerships (UNEP, 2022). Such measures have greatly contributed to advancing green transition processes in the mature economies (World Bank, 2022).

19.7.1.1 Lessons from Early Adopters and Industry Leaders

Reaching the net zero points calls for a complete change in the business model of companies and governments in all sectors—incorporating sustainability as a main business operation. Some industries such as the automotive industry, the tech industry, and the agricultural industry have served well as the low-carbon model early adopters which in turn encourages other industries through spillover benefits (UNEP, 2022). To demonstrate, Tesla and Nissan have demonstrated the automobile industry's move to electric vehicles from internal combustion engine vehicles toward a new era. These companies not only minimized their footprint by investing in electric vehicle technologies and infrastructure but also defined what the future of mobility should look like (IEA, 2021b).

Along similar lines, in the agricultural sector, precision farming guided by AI and IoT–based sensing devices has maximized resource efficiency by minimizing water and fertilizer usage and increasing yields. These may be good in explaining how technology and pursuing sustainability objectives do not negate one another but rather complement and provide for optimal results (UNEP, 2022). It will be important to scale up this practice by encouraging more sharing of best practices through global industry forums and knowledge exchange to encourage sustainability across industries (World Bank, 2022).

19.7.2 Global Cooperation for Technology and Knowledge Transfer

The robust engagement of countries in the process of reaching net zero emissions is especially crucial for the so-called "east moves" where there is a direct incorporation of the know-how and technology from the advanced countries into the less-advanced countries. Formed bodies in it will be sufficient, using management, resources, and funds available only in advanced countries or innovative markets. Nevertheless, these developments will not have a broad effect without proper solutions regarding the dissemination of this knowledge to developing countries. The main barrier in technology transfer, however, seems to be the price tag on the advanced green technologies, including CCS system, electric vehicles, and renewable energy facilities (IEA, 2023). Due to the lack of funds, developing countries that already have large debts or are steeped in poverty usually find it difficult to participate in such investment without aid. Global and regional development banks as well as other institutions such as the World Bank and the Green Climate Fund are key to these undertakings since they help subsidize the process through concessional loans and grants to reduce the burden of technology cost for developing nations (World Bank, 2022).

19.7.2.1 Strengthening International Collaboration for Net Zero Goals

To strengthen international collaboration, multilateral initiatives such as the Paris Agreement and Mission Innovation have been designed to foster cooperation between countries (UNFCCC, 2015). Mission Innovation, for example, brings together 23 countries and the European Union to accelerate public and private clean energy innovation (IEA, 2023). This initiative promotes joint R&D efforts in key areas such as energy storage, hydrogen fuel, and next-generation renewables (UNEP, 2022).

Countries can also work together to create regional technology hubs, where expertise in green technologies can be shared. The establishment of centers of excellence for green technology in regions like Africa, Asia, and Latin America would enable local engineers, scientists, and policymakers to gain firsthand experience with advanced technologies (UNFCCC, 2015). These hubs can also promote the development of localized solutions that are economically viable and suited to local conditions (IEA, 2023).

19.8 CONCLUSION

Transitioning toward net zero emissions in developing nations is not only something that must be done for the sake of the environment but it also becomes a route for advancing industrialization in a sustained manner. These economies are central to fighting climate change as demonstrated in this chapter, helping in achieving global climate targets while providing for economic growth. But with the right policies, adoption of new technologies, good corporate governance, and global collaboration, the industries in these regions can help spearhead such change.

On the other hand, the barriers to change such as financial resource base availability, infrastructural deficits, and socio-economic factors are huge, and so also are the change triggers. Employing AI, Internet of Things, and blockchain technologies as well as green financial tools can hasten the achievement of net zero emissions in difficult to abate sectors. Furthermore, the source countries that have been discussed in this chapter, the emerging economies, are already changing and adopting technologies that can be replicated anywhere around the world.

In the end, for this transition to be efficient, there is a need for all players, including governments, industries, financial institutions, and civil society organizations, to work in unison. Such reforms may be more progressive than previously believed, and great hope should be placed in the enhancement of international collaboration and cooperation.

REFERENCES

ABB & Industrial: Pioneering AI-Powered Energy Management. (2022). *Energy Magazine*. Retrieved from Energy Digital. pp. 19–22.

Algers, J. (2024, February 12). *Leading with industrial policy: Lessons for decarbonization from Swedish green steel*. Roosevelt Institute. https://rooseveltinstitute.org/publications/leading-with-industrial-policy/

Baccini, A., Walker, W. S., Carvalho, L., Farina, M., Sulla-Menashe, D., & Houghton, R. A. (2017). Tropical forests are a net carbon source based on aboveground measurements of gain and loss. *Science*, *358*(6360), 230–234. https://doi.org/10.1126/science.aam5962

Bhattacharyya, S. C. (2020). Energy access in developing countries: Challenges and ways forward. *Elsevier*, pp. 35–72.

Butcher, A. (2023, September 20). *The energy dilemma: AI's double-edged sword. Wärtsilä Insights*. Wärtsilä. Retrieved from https://www.wartsila.com/insights/article/the-energy-dilemma-ai-s-double-edged-sword

Climate Action Tracker. (2023). Paris agreement 1.5°C global warming limit still within reach, but emissions must peak by 2025. *Climate Action Tracker*. pp. 1–5.

Climate Analytics. (2023). Understanding the Paris agreement's long-term temperature goal. *Climate Analytics*. pp. 1–3.

Derrick, M. (2023). Artificial intelligence in energy management. *Energy Digital*. pp. 1–2.

Ellen MacArthur Foundation. (2020). The circular economy: A transformative approach. pp. 1–4.

Foxconn. (2023). AI-enabled Robotics and manufacturing efficiency. *Foxconn Technology Group*. pp. 1–2.

Gerber, P. J., Steinfeld, H., Henderson, B., Mottet, A., Opio, C., Dijkman, J., & Tempio, G. (2013). Tackling climate change through livestock: A global assessment of emissions and mitigation opportunities. *FAO*. pp. 120–140.

Global CCS Institute. (2022). CCUS projects: Deployment and future impact. Retrieved from https://www.globalccsinstitute.com. pp. 1–3.

Global Forest Watch. (2021). Forest loss. *Global Forest Watch*. https://www.globalforestwatch.org. pp. 1–3.

GreenBiz. (2023, April 24). *Over 239 companies are projected to miss net-zero targets, reports GreenBiz*. Aquila. https://aquila.is/knowledge-hub-article/over-239-companies-are-projected-to-miss-net-zero-targets-reports-greenbiz/

GreenBiz Group. (2022). *State of Green Business 2022: Climate and economic transformation*. GreenBiz. Retrieved from https://info.greenbiz.com/rs/211-NJY-165/images/State-of-Green-Business-2022-Report.pdf

Griscom, B. W., Adams, J., Ellis, P. W., Houghton, R. A., Lomax, G., Miteva, D. A., & Woodbury, P. (2017). Natural climate solutions. *Proceedings of the National Academy of Sciences*, *114*(44), 11645–11650.

Gurara, D., Klyuev, V., Mwase, N., & Presbitero, A. F. (2018). Trends and challenges in infrastructure investment in developing countries. *International Development Policy | Revue internationale de politique de développement*, *10*(1). https://doi.org/10.4000/poldev.2802

Hristov, A. N., Oh, J., Giallongo, F., Frederick, T. W., Harper, M. T., Weeks, H. L., & Bravo, D. M. (2013). Mitigation of methane and nitrous oxide emissions from animal operations: A review of enteric and manure sources. *Journal of Animal Science*, *91*(11), 5045–5069.

Huawei. (2022). AI and predictive maintenance in electronics manufacturing. *Huawei Technologies*. pp. 1–3.

IBM. (2023). Artificial intelligence for optimized manufacturing. *IBM*. pp. 1–2.

IEA. (2021a). Energy efficiency in India: Opportunities and challenges. *International Energy Agency*. pp. 1–4. Retrieved from https://www.iea.org/articles/e4-country-profile-energy-efficiency-in-india

IEA. (2021b). Global energy review 2021. Retrieved from https://www.iea.org/reports/global-energy-review-2021. pp. 1–6.

IEA. (2021c). World energy outlook 2021. *International Energy Agency*. Retrieved from https://iea.blob.core.windows.net/assets/4ed140c1-c3f3-4fd9-acae-789a4e14a23c/WorldEnergyOutlook2021.pdf. pp. 1–3.

IIGCC. (2023). Decarbonizing industry report. pp. 1–5. Retrieved from https://igcc.org.au/wp-content/uploads/2023/10/IGCC-Sector-Pathways-Report_2023.pdf

Intergovernmental Panel on Climate Change (IPCC). (2021). *Climate Change 2021: The Physical Science Basis. Contribution of Working Group I to the Sixth Assessment Report of the Intergovernmental Panel on Climate Change* (pp. 1–7). Cambridge University Press. Retrieved from https://www.ipcc.ch/report/ar6/wg1/downloads/report/IPCC_AR6_WGI_SummaryVolume.pdf

International Air Transport Association. (2021). *Climate change: Fact sheet* (pp. 1–3). IATA. Retrieved from https://www.iata.org/contentassets/d13875e9ed784f75bac90f000760e998/fact_sheet_on_climate_change.pdf

International Energy Agency. (2022). *CO_2 emissions in 2022*. International Energy Agency. https://www.iea.org/reports/co2-emissions-in-2022. pp. 1–6.

International Energy Agency (IEA). (2023). Net zero by 2050: A roadmap for the global energy sector. Retrieved from https://www.iea.org/reports/net-zero-by-2050. pp. 1–5.

International Maritime Organization. (2020). *Fourth IMO GHG Study 2020: Executive summary* (pp. 1–3). IMO. Retrieved from https://wwwcdn.imo.org/localresources/en/OurWork/Environment/Documents/Fourth%20IMO%20GHG%20Study%202020%20Executive-Summary.pdf

International Renewable Energy Agency. (2022). *World Energy Transitions Outlook 2022: 1.5 °C pathway* (pp. 1–4). IRENA. Retrieved from https://www.irena.org/publications/2022/Mar/World-Energy-Transitions-Outlook-2022

International Transport Forum. (2021, May 17). *ITF Transport Outlook 2021* (pp. 1–4). OECD Publishing. https://doi.org/10.1787/16826a30-en

IPCC. (2019). Special report on climate change and land: Summary for policymakers. Retrieved from https://www.ipcc.ch/srccl/chapter/summary-for-policymakers/. pp. 1–7.

Lal, R. (2020). Regenerative agriculture for food and climate. *Journal of Soil and Water Conservation*, *75*(5), 123A–124A. https://doi.org/10.2489/jswc.2020.0620A

Lehne, J., & Preston, F. (2018, June). *Making concrete change: Innovation in low-carbon cement and concrete*. Chatham House. https://www.chathamhouse.org/2018/06/making-concrete-change-innovation-low-carbon-cement-and-concrete

McKinsey & Company. (2023). AI and digital transformation in heavy industry. *McKinsey Global Institute*. pp. 1–4.

McKinsey & Company. (2024, February 21). *Adopting AI at speed and scale: The 4IR push to stay competitive*. McKinsey & Company. https://www.mckinsey.com/capabilities/operations/our-insights/adopting-ai-at-speed-and-scale-the-4ir-push-to-stay-competitive

MSCI Sustainability Institute. (2023, May). *MSCI Net-Zero Tracker: May 2023 update*. MSCI. https://www.msci.com/documents/1296102/38217127/NetZero-Tracker-May.pdf

Myhre, G., Shindell, D., Bréon, F.-M., Collins, W., Fuglestvedt, J., Huang, J., & Zhang, H. (2013). Anthropogenic and natural radiative forcing. In T. F. Stocker, D. Qin, G.-K. Plattner, M. Tignor, S. K. Allen, J. Boschung, A. Nauels, Y. Xia, V. Bex, & P. M. Midgley (Eds.), *Climate Change 2013: The Physical Science Basis. Contribution of Working Group I to the Fifth Assessment Report of the Intergovernmental Panel on Climate Change* (pp. 659–740). Cambridge University Press. https://www.ipcc.ch/site/assets/uploads/2018/02/WG1AR5_Chapter08_FINAL.pdf

NTPC Limited. (2023). *Sustainability data trends 2023* (pp. 1–5). NTPC Limited. Retrieved from https://ntpc.co.in/sites/default/files/inline-files/Sustainability%20Data%20Trends%202023_0.pdf

Purton, M. (2024, August 28). Clean energy investment is set to double that of fossil fuels – here's how developing nations can also benefit. *World Economic Forum*. Retrieved from https://www.weforum.org/stories/clean-energy-investment-just-transition/

SANY Heavy Industry. (2024, May 13). *Sustainable Development Report 2023* (pp. 1–3). SANY Heavy Industry. Retrieved from https://www.sanyglobal.com/press_releases/2645/

Segovia, M., & Garcia-Alfaro, J. (2022). Design, modeling and implementation of digital twins. *Sensors*, *22*(14), 5396. https://doi.org/10.3390/s22145396

Seth, R., Bhushan, B., Sharma, N., Kumar, R., & Kaushik, I. (2021). Applicability of Industrial IoT in Diversified Sectors: Evolution, Applications and Challenges. In R. Kumar, R. Sharma, & P. K. Pattnaik (Eds.), *Multimedia Technologies in the Internet of Things Environment*. Springer. https://doi.org/10.1007/978-981-15-7965-3_4

Shinkevich, M. V., Vertakova, Y. V., & Galimulina, F. F. (2020). The synergy of digitalization within the framework of increasing energy efficiency in the manufacturing industry. *International Journal of Energy Economics and Policy*, *10*(3), 456–464. https://www.econjournals.com/index.php/ijeep/article/view/10203

Smith, J., & Taylor, P. (2022). Unlocking the net-zero transformation of industries. In *Fostering Effective Energy Transition 2022*. World Economic Forum. https://www3.weforum.org/docs/WEF_Energy_Transition_Index_2022.pdf

Smith, P., Bustamante, M., Ahammad, H., Clark, H., Dong, H., Elsiddig, E. A., & Tubiello, F. (2014). Agriculture, forestry and other land use (AFOLU). In O. Edenhofer, R. Pichs-Madruga, Y. Sokona, E. Farahani, S. Kadner, K. Seyboth, A. Adler, I. Baum, S. Brunner, P. Eickemeier, B. Kriemann, J. Savolainen, S. Schlömer, C. von Stechow, T. Zwickel, & J. C. Minx (Eds.), *Climate Change 2014: Mitigation of Climate Change. Contribution of Working Group III to the Fifth Assessment Report of the Intergovernmental Panel on Climate Change* (pp. 811–922). Cambridge University Press. https://www.ipcc.ch/site/assets/uploads/2018/02/ipcc_wg3_ar5_chapter11.pdf

Tata Steel Limited. (2024). *Driving technology transformation* (pp. 1–3). In *Integrated Report 2023–24*. Tata Steel Limited. Retrieved from https://www.tatasteel.com/investors/integrated-report-2023-24/driving-technology-transformation.html (tatasteel.com)

UltraTech Cement Limited. (2023). *Integrated and Sustainability Report 2022–23* (pp. 11–15). UltraTech Cement Ltd. Retrieved from https://www.grasim.com/Upload/PDF/ultraTech-cement-limited-annual-report-23.pdf

UNCTAD. (2022). *A green industrial policy for the post-pandemic recovery: Building back better through investment and technology transfer* (pp. 46–50). United Nations Conference on Trade and Development. https://unctad.org/system/files/official-document/diae2022d1_en.pdf

UNEP. (2022). Financing for development: Green fund challenges. *United Nations Environment Programme*. pp. 1–3.

United Nations. (2015). *Transforming our world: The 2030 Agenda for Sustainable Development* (pp. 31–37). United Nations. Retrieved from https://sustainabledevelopment.un.org/content/documents/212522030%20Agenda%20for%20Sustainable%20Development%20web.pdf

United Nations Environment Programme. (2021). *Emissions gap report 2021: The heat is on* (pp. 1–6). Nairobi: UNEP. Retrieved from https://www.unep.org/resources/emissions-gap-report

United Nations Framework Convention on Climate Change. (2023, December 13). *COP28 agreement signals "beginning of the end" of the fossil fuel era* (Press release, pp. 1–3). UNFCCC. Retrieved from https://unfccc.int/cop28/5-key-takeaways

United Nations Framework Convention on Climate Change (UNFCCC). (2015). The Paris agreement. Retrieved from https://unfccc.int/process-and-meetings/the-paris-agreement/the-paris-agreement. pp. 21–25.

World Bank. (2022). World Bank climate change action plan 2021–2025. Retrieved from https://www.worldbank.org/en/news/press-release/2021/06/22/world-bank-group-climate-change-action-plan-2021-2025. pp. 1–6.

World Bank. (2023). *State and Trends of Carbon Pricing 2023* (pp. 8–84). World Bank. Retrieved from https://openknowledge.worldbank.org/server/api/core/bitstreams/58f2a409-9bb7-4ee6-899d-be47835c838f/content

World Bank Group. (2021). *From crisis response to resilient recovery: Moving toward green, resilient, and inclusive development* (GRID) (pp. 46–55). World Bank. Retrieved from https://documents1.worldbank.org/curated/en/120541633011500775/pdf/The-World-Bank-Annual-Report-2021-From-Crisis-to-Green-Resilient-and-Inclusive-Recovery.pdf

World Energy Council. (2022). *World Energy Issues Monitor 2022: Global report* (13th ed.). World Energy Council. https://www.worldenergy.org/assets/downloads/World_Energy_Issues_Monitor_2022_-_Global_Report.pdf

World Resources Institute. (2021). *State of Climate Action 2021: Systems transformations required to limit global warming to 1.5 °C* (pp. 14–18). World Resources Institute. Retrieved from https://www.wri.org/publications/state-climate-action-2021

Xuzhou Construction Machinery Group (XCMG). (2023, July 14). *XCMG Machinery releases HANYUN OS 2023, empowering new industrialization with intelligent and digital transformation* [Press release]. XCMG Machinery. Retrieved from https://www.prnewswire.com/news-releases/xcmg-machinery-releases-hanyun-os-2023-empowering-new-industrialization-with-intelligent-and-digital-transformation-301877317.html

Zhironkin, S. (2022). *Advances in sustainable development: Technological and economic overview. Energies, 15*(17), 6451. https://doi.org/10.3390/en15176451

20 A Vision on the Development of Carbon Industry, Enactment of Adaptation Policies, Guidelines, Ethical Considerations and Public Perception

Kah Yee Lim and Keng Yuen Foo

20.1 INTRODUCTION

Since the 19th century, the emergence of new carbon materials has created great excitement and expectations among the scientific community. Carbon, the most versatile and abundant element on earth, serves as the promising precursor to a large number of novel allotropic forms, ranging from the amorphous carbon, graphite, diamond, carbon fiber and graphene, to the nanoscale carbon nanotube and fullerene, both from the fundamental viewpoint and for future applications (Bianco et al., 2020). From the history of materials perspective, the unique characteristics of carbon depend mainly on the structural design and arrangement of atomic structure. The specific properties of carbon materials, including the chemical stability, high-specific strength and modulus, high electrical conductivity, low density and thermal expansion, and good resistance to heat and resistance, have inspired important new industrial material or impetus for researchers in developing high-performance applications. However, the grand challenge of carbon-based materials research is to produce valuable, novel and high-performance carbon materials from sustainable, scalable, economically viable large-scale production process (Li *et al.*, 2022).

Establishment of new policies related to carbon technology innovation research is a daunting quest. Given the rapid growth and novel innovation of carbon-derived technologies, the enforcement of the governing policy, the scale-up of research and the marketing of final products are still in the rudimentary explorative states. In the United States, the burst of technology supporting policy, especially on graphene, is still at the relatively early phase, ahead of other policies. The understanding on the ethical issues of nanotechnology, or known as "nano-ethics" and environmental or planetary virtue ethics, is a consensus protocol for realistic merging, in promoting shared interests and experiences of nanotechnologists, researchers and environmentalists to the general public. These policy- and market-driven approaches would not only underline the wise deployment of the emerging technology, but also represent a viable protection to the environment and human health.

Centralised or integrated nanotechnology and health risk management framework have always been contentious scopes within the European Union (EU). According to Justo-Hanani and Dayan (2015), the respective regulatory law and policy have shown significant and non-significant impacts on commercial markets outside the EU's borders, putting European countries at the forefront status of law and legal framework developments within this area. However, with the rising popularity of

DOI: 10.1201/9781003668961-20

the regulatory policies in many countries, the dynamics of the societal implications and humanities in engendering the public acceptance of carbon technology development and commercialisation remain insufficiently explored. In this sense, this chapter reports the development trends of carbon technology, from the perspectives of overcapacity problems, mineral resources, product prices, factor costs, ethical issues, health hazards and environmental risks. The enactment of industrial policies/guidelines and precautionary principles together with the key strategies for supporting ethical decision making are briefly elucidated.

20.2 RECENT DEVELOPMENT OF THE CARBON INDUSTRY

Carbon, denoted as $^{6}_{12}C$ in Group 14 of Mendeleev's periodic table, is the fourth (4th) most abundant and indispensable element in nature by mass, after helium (He), hydrogen (H) and oxygen (O). It is a non-metallic and tetravalent chemical element, and the four valence electrons of carbon atom provide an ideal building block for different elements to form stable multi-atomic configuration with different lengths or rings, due to their unique characteristics of valence states, surface bonding and hybridisation. Amorphous carbon, graphite and diamond are the three well-known naturally occurring carbon allotropes. Thanks to the various combinations of carbon atom hybridisation, *sp* (carbyne or linear acetylenic carbon), sp^2 (graphite, graphene, fullerenes, nanotube and nanofiber), sp^3 (diamond and lonsdaleite), sp-sp^2 and sp^2-sp^3 (Q-carbon and amorphous carbon) can be constructed by changing the periodic motifs within the hybridised bonding networks. Figure 20.1 demonstrates the historical timeline of carbon materials development.

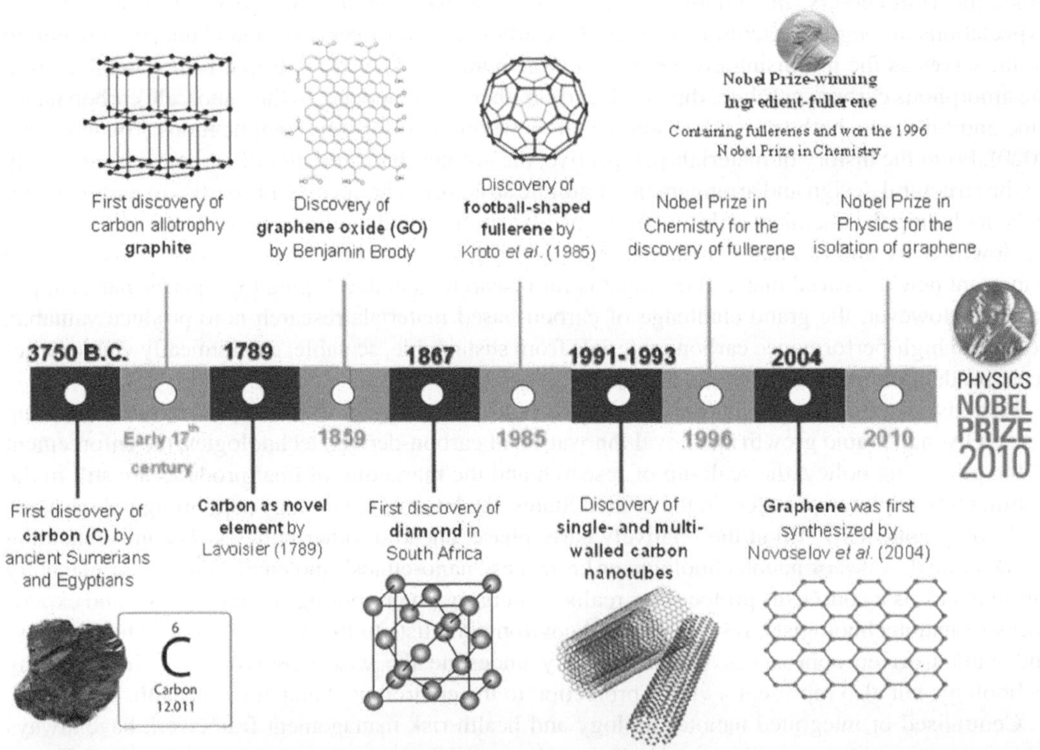

FIGURE 20.1 Historical timeline of carbon materials development.
(Modified from Porto et al. 2020)

From Figure 20.1, the discovery of carbon can be traced back to prehistoric period, predominantly in the form of coal and charcoal. Carbon was first discovered by ancient Sumerians and Egyptians (3750 B.C.), and has been defined as a novel element by Antoine Lavoisier in 1789 (Lavoisier, 1789). Another carbon allotropy, graphite, has been verified by Vandermonde et al. (1788) via oxidisation technique, similar to diamond as proposed by Lavoisier (1772). In 1985, a football-shaped carbon (fullerene), originally named as "Buckminsterfullerene", was initially introduced by Kroto et al. (1985). It represented a memorable "eureka moment" within the research community, to be awarded the 1996 Nobel Prize in Chemistry. Iijima (1991) was credited with the discovery of hollow and nanoscale diameter carbon filaments, particularly single-walled and multi-walled carbon nanotubes in 1991–1993. Graphene, a single-layer of hexagonal close-packed carbon atoms, was first unambiguously synthesised by Novoselov et al. (2004), which was credited from the idea of Boehm (1962), who awarded Nobel Prize in 2010. The development of carbon-based materials with high degrees of *sp* hybridisation and small-sized structures has been growing rapidly since the 19th century, and experienced a continuous surge of interest in its research field.

20.2.1 Mineral Resources Endowment and Development

Carbon (C) is an essential element, and its geologic origin in ancient sediment/sedimentary system is central to the evolution of Earth and life. The mystery of carbon mineralogy began >4 billion years ago, with only approximately 0.2% of total carbon present on Earth's crust in its native forms as coal, diamond and graphite (Hazen et al., 2008; Zhang and Yin, 2012). Coal (anthracite and bituminous) is the cheapest and most abundant mineral carbon source, accounting for 60%–98% of carbon content, as compared to diamond and graphite. According to BP Statistical Review of World Energy (2021), Australia, China, India, Russia and the United States are the top five coal mining countries globally, with the cumulative production at 815 billion tonnes. Most of the coal reserves are made up of anthracite and bituminous (70%), and hold significant roles in electric power generation, metallurgy/steel manufacturing, and capacitive carbon and activated carbon production industries (Zhang et al., 2022).

Diamond is a solid and crystalline allotrope of carbon, with a single element carbon (99.95%) in a tetrahedron form. Natural diamond is a rare mantle mineral formed over a million years, and mostly found in the metamorphic rocks or from the alluvial deposits. There have been different descriptions of the historical discovery of diamond mining, from its first inception to today, establishing India, Brazil, South Africa and Russia as the largest mining areas of diamond (Kjarsgaard et al., 2022). The rarity of natural diamond has captured much attention by virtue of its unique and extreme properties as the most expensive, precious and esteemed components of jewellery (Nemanich et al., 2014). However, there is limited success pertaining to the application of diamond in specific perspectives.

Within the mineral kingdom, graphite, made up of hexagonal layers of carbon honeycomb lattice, has a pre-eminent position in the crystalline carbon allotropes. It is the cheapest source of pure carbon deduced mainly from the metamorphic and igneous rocks. The resulting crystalline structure of graphite forms one of the softest minerals, which can be characterised into three principal types, namely flake graphite, lump graphite and amorphous graphite. These different types of graphite demonstrate a variety of physical properties, surface appearance, structural stability, chemical composition (carbon content) and flake size (Jara et al., 2019). Since the discovery of high-grade lump graphite during the 16th century, graphite has been of great value in new technological innovations.

According to the updated database from US Geological Survey (USGS, 2023), the distribution of graphite ore is extremely uneven throughout the regions and continents, with respect to the ore types, grades and characteristics. Among the top natural graphite producing countries, China holds the leading position, accounting for more than 65% or 850,000 metric tons of the world's graphite production in 2022, which were about 40% amorphous and 60% flake graphite. Graphite mineral resources have been subjected to intensive exploration and exploitation to meet the rapidly rising global supply and demand. It is identified as an important raw material in the European Commission's list (EC, 2014). Parallel with its development to support different high-technological applications, graphite is prelisted to be a "hot' priority commodity, and the fundamental technological, operational and economical parameters of the individual graphite deposits should be taken into consideration when prioritising the targets of exploration and development.

20.2.2 Product Price and Cost Factor

In recent years, the global carbon material market has experienced massive expansion, owing to the availability of raw precursors or processing technologies in supporting high quality carbon-derived products. Of increasing importance is the development of carbon particles with new morphologies and superior physiochemical properties, notably the carbon nanotubes, buckyballs, carbon fibers, graphene and other derivatives. The feasibility of these carbon-derived products is attributed to the significant price reduction, to offer avenues, opportunities and research directions for possible future applications, as compared to several years earlier. Table 20.1 documents the carbon markets,

TABLE 20.1
Carbon Markets, Products' Pricing and Potential Applications of Carbon Derivatives

Carbon Material	Carbon Content (%)	Price Range (USD/ton)	Global Market Size	Application	Reference
Carbon black	95	1,000–4,000	16.4 million metric tons/year	Plastics, electronic, tire, paint and coating	Fortune Business Insight (2022)
Graphite					
Amorphous graphite	70–85	80–835	1.3 million metric tons/year	Refractories, steel, batteries, fuel cells, water purification, electronic and electrical products, and spintronic application	Statista Research Department (2023)
Vein (lump and chip) graphite	90–99	500–2,500			
Flake graphite	85–97	500–10,000			
Graphene					
Super-jumbo flake (> 500 μm)	97–99	4,000–6,000	330 million metric tons/year	Nuclear reactor, aerospace system, and other specialised niche applications	Jara et al. (2019); Statista Research Department (2023)
Jumbo flake (300–500 μm)	95–99	2,500–3,000		Expandable graphite, composites and lithium-ion battery applications	
Battery-grade flake (180–500 μm)	99.9	5,000–20,000			
Large flake (150–300 μm)	>99	800–3,000		Spherical graphite	
Medium flake (106–150 μm)	>99	500–3,000			
Small flake (75–106 μm)	95–98	900–1,200		Industrial uses	
Amorphous (< 106 μm)	94–97	300–500			
Carbon fiber	40–99	20–28,000	160 thousand metric tons/year	Robotics, marine, aircraft and aerospace production	Chemanalyst (2020)

(Continued)

TABLE 20.1 *(Continued)*
Carbon Markets, Products' Pricing and Potential Applications of Carbon Derivatives

Carbon Material	Carbon Content (%)	Price Range (USD/ton)	Global Market Size	Application	Reference
Carbon nanotube (Single-walled, double-walled and multi-walled)	≥90%	3,000–200,000	95.4 thousand metric tons/year	Energy storage, molecular electronic, thermal materials, fabrics and fibers, catalyst supports, biomedical, air & water filtration, and conductive plastics/ adhesives	DeMordaunt et al. (2022)
Fullerenes	99.95%	35×10^6 –$8{,}835 \times 10^6$	10 thousand metric tons/year	Solar cell, electronic battery, fire retardant paint, antioxidant/ antibacterial materials, polymer reinforcement, medical and pharmaceuticals application	Dos Santos et al. (2023)

products' pricing and potential applications of carbon derivatives. From Table 20.1, the carbon market size ranged from 14.4 thousand metric tons/year (carbon nanotubes) to 300 million metric tons/ year (graphene), which are the major market outlets. From the presented data, the product pricing for the carbon derivatives can vary enormously and affected by the product quality and purity, flake size and grade, features, functionalities and performances. In order to make a better approximation of the manufacturers' costs, the economies of scale need to be further considered.

Large-scale production, long-term cycling stability, and low-cost and rapid manufacturing techniques are the dominant cost factors for any product to move toward commercialisation and industrial applications; however, they are rarely reported in the synthesis/literature review (Ma et al., 2017; Yang et al., 2019). As a rule of thumb, there are five parameters, including (i) precursor and manufacturing costs, (ii) up-scaling potential, (iii) product yield and quality, (iv) process complexity and safety, and (v) toxicological effects on environment and human health, that would contribute to the success, influence and constraints of the development. Researchers have reported that raw material remains the major production cost factors, responsible for more than 50% of the manufacturing cost of carbon (Choi et al., 2019; Liu et al., 2022; Nunna et al., 2019; Peijs et al., 2022). To date, the development of new, scalable, simple and cost-effective synthesis method for mature applications has been continued, forecasting the targeted price at 50 cents per gram, to advance the future growth in both innovative and revolutionary manners (Wang, 2018).

20.3 OVERCAPACITY ISSUES IN THE CARBON INDUSTRY

The overcapacity issue in the carbon industry has become an inevitable challenge on the economic development of large manufacturing countries. Carbon overcapacity crisis has reached an alarming new height, in which the production volume is persistently greater than the actual demand. According to the data from the Statista Research Department (2023), the global market value of carbon materials was recorded at approximately 348 million tons in 2022, while the apparent consumption was merely 16 million tons. This unprecedented level of carbon overcapacity has severely distorted the world market, and threatened the viability of the carbon producers.

In the global carbon materials market, graphene remains the most stable emerging technology. However, its large-scale commercialisation, industrialisation and marketisation or the mature industrial supply chain have not been fully established. According to the Global Graphene Market Research Report in 2016, there has been a significant outpaced imbalance between high production rate and low commercial market demand of graphene, of which less than 5% of graphene production is used for high-tech applications. The huge gateway and demand for graphene originated from the scientific-research requirements, and a small fraction of manufacturers, due to the unstable business model, low profitability, product positioning, limited competitive growth regime and minimum development (Liu and Gao, 2019). In addition, the overcapacity issues of the carbon industry would aggravate existing environmental pollution, global warming, climate change and additional public health hazards.

Driven by the world's largest basic research and industrial forces of graphene, the United States is at the forefront in the field. As the main exporters of graphene-related products, the United States (USA) has dominated the North American market since 2019, and it is predicted to ascend to top position with the collaborations of a number of manufacturing industries and research centres/institutes on the research and development (R & D) activities within this emerging technology. Similar to USA, the graphene market in the European region showed a tremendous growth rate, with 33.1% of shared revenue in 2019 (Trache *et al.*, 2020). European Research Council has invested USD 1 billion in potential and new emerging projects that may develop the national and regional markets, and this rising application in different industrial sectors, specifically biomedical, defence, automobile, marine and aerospace/aircraft, are driving the regional product supply and demand over the forecasting period. Therefore, the establishment of industrial-level strategic development plans, integrating the main goals and key research areas for the improvement of propulsion control systems and supporting plans, and introduction of a favourable industrial new growth point has been promoted.

Meanwhile, the graphene market in the Asia Pacific region have grown sharply, and in China, this industrial chain is in its embryonic stage, connecting the raw materials, synthesis methods, product development and commercialisation to the down-stream applications. Although China has set up graphene-related governing policies, administrative regulations and departmental rules, challenges related to the strategic layout, insufficient momentum of its driving locomotive, weak industrial foundation and the scattered innovation system remain critical challenges to be overcome. In this perspective, the Ministry of Industry and Information Technology, Ministry of Science and Technology, and National Development and Reform Commission of the Chinese government, have jointly published "Several Opinions on Accelerating the Innovation and Development of Graphene Industry" for the steady development of the graphene industry and stimulation of the innovation of technology in China. The major weaknesses of the carbon industry within the Asia Pacific Region are illustrated in Figure 20.2.

Addressing the carbon industry's overcapacity issues is filled with continuing challenges, and warrants more research/scholarly contributions from other perspectives. The common approach to controlling the overcapacity in carbon industry relies on the free markets, in which price falls significantly or stays below current market cost, and forces the small or inefficient suppliers to leave the market (Dong et al., 2021). Moreover, the implementation of environmental regulations/policies and technological innovations may not alleviate the overcapacity problems, while sustainable development requires collective and collaborative actions/efforts across different departments (governments, enterprises and the market) to achieve a sustainable equilibrium (Fan et al., 2021). As a whole, the changing policy intensity may significantly reduce the overcapacity of the carbon industry, which can differ according to the alteration of policies, enterprises and countries.

20.4 ENACTMENT OF ADAPTATION POLICY AND GUIDELINES

The potential environmental and health risks of complex design strategies or applications of carbon derivatives are hard to predict and justify. During the last few years, a number of national and international policies, laws, rules and regulations supporting the manufacturing, upgrading or

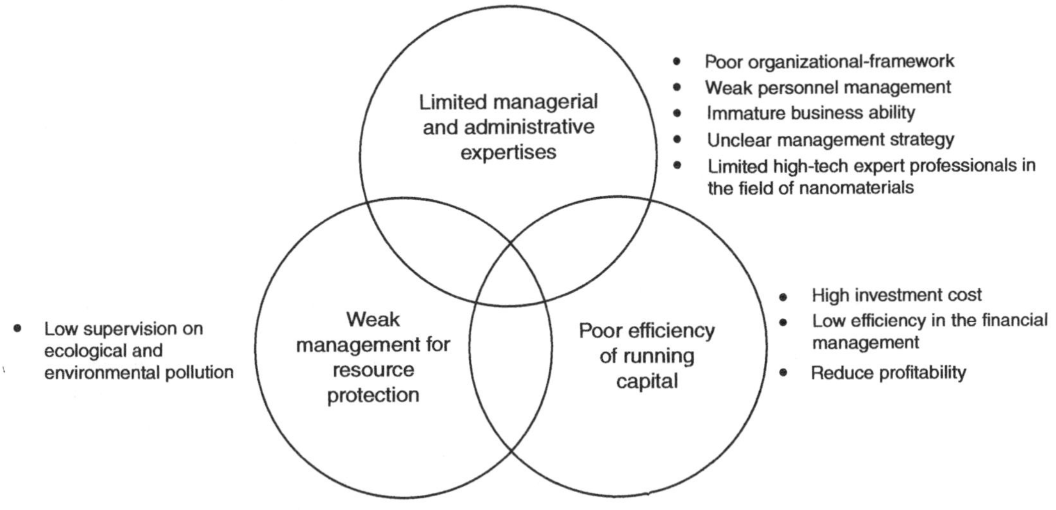

FIGURE 20.2 The major weakness of graphene sector in the Asia Pacific Region (Liu and Gao, 2019).

establishment of carbon-related downstream industries have been introduced. Parallel with the development, the commercialisation standards related to the production cost, risk management, and accessibility to the local and international trade markets have been analysed. Recognising the needs and aspirations of industrial partners to work on carbon materials in a trusted environment, with high indexes of consistency and reliability, European Commission (EC), European Carbon and Graphite Association, National Graphene Association and the Graphene Council have set up a series of guidelines for identification of gaps in the existing national and international standards, explorations of key areas and issues, and formulation of a technological road-map for carbon industry's development.

The Materials System Analysis, initiated by the EC in 2012, was designed to gather the crucial data on sustainable resource management, for twenty-eight (28) materials relevant to the European Union's Economy (EC, 2020a). The EU Critical Raw Materials List, EU trade negotiation and the underlying methodology/framework are the key instruments within the context/concept of the EU's raw materials policy (EC, 2014). The report is related to the five major battery manufacturing materials (cobalt, lithium, manganese, nickel and graphite), according to the selection, quality, application in the power- or battery-based systems, essential for clean energy and transportation requirements (EC, 2020b).

In addition, carbon fiber-related policies, in terms of trade policies, manufacturing support system, development, deployment and end-use market supporting policies, have been implemented by several key countries, mainly China, EU countries, Japan, Mexico, South Korea and USA, as direct or indirect support to the industry. Integration of funded collaborative research-and-development programs, together with the public–private partnerships and direct government procurement practice, would signify significant benefit and support to further spur the innovation clusters, economic development, employment opportunities and end-use applications. The carbon fiber manufacturing advancements in the USA are the establishment of Manufacturing Technology (ManTech) program by US Department of Defence, and Consortium for Accelerated Innovation and Insertion of Advanced Composites 2014 by the National Institute of Standards and Technology, to (a) develop new insight of the research roadmap for carbon fiber, (b) validate emerging cross-cutting concepts, processes and technologies, and (c) establish innovation manufacturing ecosystem, to the commercial market (Cook and Booth, 2017).

As the fourth-largest carbon fiber manufacturer in the world, China has made great achievement in the carbon fiber production and application, with the support of a series of strategic industrial policies and action plans (Table 20.2). These policies may reflect the research roadmap for carbon

TABLE 20.2
Supporting Policies for Carbon Materials Development (Liu and Gao, 2019; Teng et al., 2023)

Carbon Material	Policy	Remark
Carbon fiber	"13th Five-Year Plan" for national science and technology innovation	This policy focuses on the development of carbon fiber and carbon fiber composites, and made great breakthroughs in the key techniques for preparation, assessment, modification and application.
	Special policy of 15 items of carbon fiber	Guide and support high-quality development of the carbon fiber industry.
	Guidance Catalogue for Industrial Development and Transfer	Shanghai, Jiangsu and other places have accelerated the development of carbon fibers, high-performance carbon-derived functional composites and other derivatives.
	Three-Year Action Plan	It is demonstrated that the (a) development of high-performance and high-quality carbon fiber derivatives, and (b) application and production process for the advanced composite manufacturing should be emphasised and improved.
Graphene	"12th Five-Year Plan" for New Material Industry (2011–2015)	China's new material industry construction goal is to position itself as the leading industry in the national economy, and initially realise the strategic transformation from Material Power to Material Great Power.
	Special Plan for Nano Research in National Major Scientific Research Planning during the 12th Five Year Period	The development of graphene-based materials is regarded as a key support for the nano-materials program.
	Market Access Conditions for Graphite Industry	China has implemented market admittance management within the graphite industry.
	Implementation Plan of Key Material Upgrading Projects	China has promoted 20 new key materials (graphene) to achieve stable production volume and large-scale practical applications.
	Made in China 2025	It (a) clarifies the key position of graphene at the forefront of materials research, (b) emphasises the strategic layout planning and development, and (c) points out the pathway for enhancing China's innovation and manufacturing.
	Several Opinions on Accelerating the Innovation and Development of Graphene Industry	It proposes graphene as a key pilot industry, and lists out the emphases, goals and road map of graphene industrial development by the end of 2020.
	Outline of the National Strategy of Innovation-driven Development	It proposes the development of disruptive technologies and continues to spawn new industries. It also advocates that nanotechnology, graphene and other technologies should play the leading role in the new materials industry.
	Technology & Product Catalogues for the Promotion and Application in Building Materials Industry (2016–2017)	Graphene-powder and graphene-based heavy anti-corrosion coating development.
	"13th Five-Year Plan" for New Material Industry	Parallel with the development of strategic emerging industries and development of graphene to be a key frontier material.
Graphene	Guiding Opinions on Accelerating the Innovation and Development of New Material Industry	China has actively developed frontier materials, not limited to fundamental research, accumulative development of graphene, smart manufacturing products, and additive materials.

fiber research and development, and clearly point out the standpoint of China on the worldwide carbon industrial development. In 2013, a "Three-year Action Plan" has been developed to (a) boost carbon technology innovation, (b) promote industrial transformation processes, and (c) enhance the upstream and downstream of the carbon fiber industrial chains. China's Carbon Valley in Jilin province is an integrated industrial chain from raw material to final product, demonstrating the fulfilment of the action plan. Accordingly, China's research and policy portfolio have been formed in an orderly manner, to generate a more solid foundation for the 14th Five-Year Plan (2021–2025), to (a) propel new type of industrialisation, industrial upgrading and infrastructure improvement, and (b) narrow the gap between the supply and demand (Cook and Booth, 2017).

In general, EU and its member countries have been more flexible on the policies regulating carbon fiber exports and end-use markets than the USA and China. Much of this carbon research and development has historically been conducted via the EU Framework Programme for Research and Innovation, with the amount exceeding US$6 million to the project entitled "Functionalized Innovative Carbon Fibers Developed from Novel Precursors with Cost Efficiency and Tailored Properties (FIBRALSPEC)" (EC, 2016). Moreover, EU's Horizon 2020 is the EU Framework Programme for Research and Innovation (2014–2020), with a research budget at US$80 billion (EC, 2016). Graphene Flagship, the EU's largest research initiative, has been funded in the 7th EU Framework Programme; with a total budget of US$1 billion, focusing mainly on the scientific research and industrial applications of graphene, and the safety assessment of graphene with regards to the environment and public health (Arvidsson et al., 2018). The legislation of Registration, Evaluation, Authorisation and Restriction of Chemicals (REACH), dating from 2008, is a regulation adopted in EU to strengthen the knowledge of graphene nanomaterial safety and protection of human health (Pavlicek et al., 2021). In 2019, a free web tool, eREACHNano, was launched and introduced as a good online resource to outline the concept of nanomaterials, risks and EU regulation (European Food Safety Authority, 2020).

Similarly, in the UK, British Standards Institution (BSI) represents the authority for all international standard organisations, specifically as the European Committee for Standardization, European Committee for Electrotechnical Standardization, European Telecommunications Standards Institute, International Electrotechnical Commission and International Organization for Standardization (ISO), to set up expert panels and infrastructures for more than 11,000 specialists for national, international and Publicly Available Specification standards development paradigm in carbon-related fields, especially graphene nanomaterials. In 2016, China International Graphene Industry Union set up a platform for the international collaboration and cooperation between business partners, researchers and financial agencies, while the assessment, and quality standards and new platforms for research, entrepreneurship and business development were established for commercialisation of research findings (Industry Research Information Bank, 2017). Additionally, the UK-China Joint Working Group on graphene standardisation process has recently been formed as a modified partnership between the BSI and Standardisation Administration of China, with the primary aims to promote and strengthen the collaboration between two countries on graphene standardisation, and standard proposals related to graphene information, fabrication, characterisations, properties, drying, handling, storage, transportation, safety testing and applications (BSI, 2018). Table 20.2 summarises the supporting policies for the carbon industry's development.

20.5 ETHICAL CONSIDERATIONS AND PUBLIC PERCEPTION

20.5.1 Misconceptions About Ethics and Emerging Technologies

The concept of emerging technology has constantly evolved for a wide range of potential applications, leaving behind the complexity of ethical and societal considerations (Figure 20.3). The misconceptions and uncertainties of the issues are incredibly important in affecting the development and application of different carbon derivatives. Table 20.3 summarises the common misconceptions related to the social and ethical concerns about the emerging technologies.

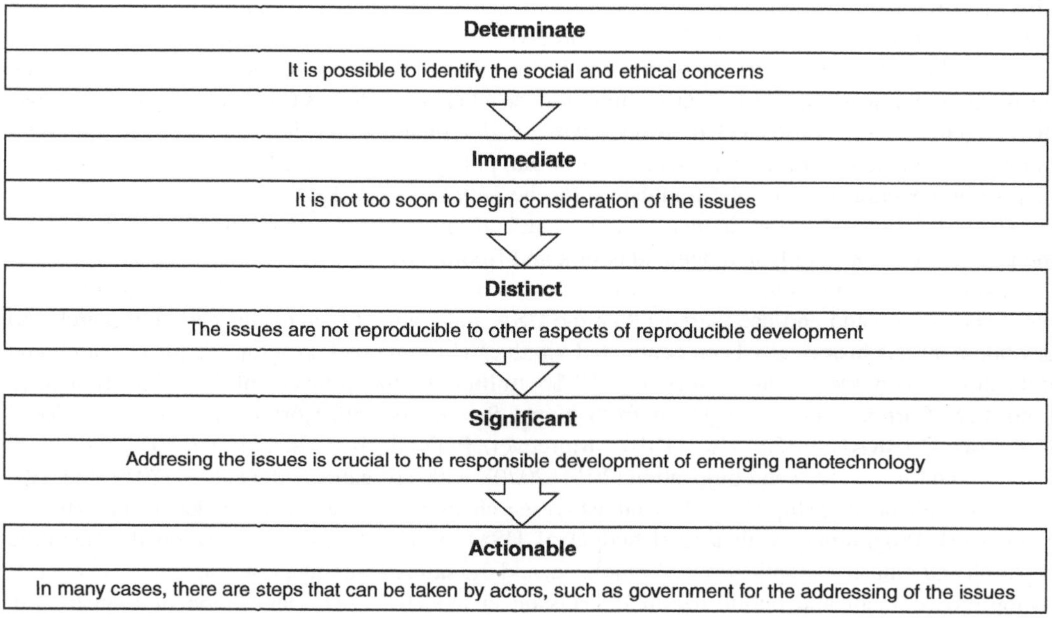

FIGURE 20.3 The ethical and social issues of emerging technologies (Sandler, 2009).

TABLE 20.3
Three Most Common Misconceptions Related to the Social and Ethical Issues (Sandler, 2009)

Misconception	Remark
Too soon to tell	• This misconception arises from a primary focus mainly on the technology itself, with various contextual factors, notably the limited access to technology, information/data insecurity, and insufficient oversight of biological defence (bio-defence) research. • It becomes apparent that it is not too early to identify, and to begin to respond to the associated ethical and social considerations.
The inevitable goodness of the emerging technology revolution	• This misconception resulted from a preoccupation that the technology contributes to comfort, safety, health and longevity of industrialised community. • It requires novel laws and regulations (environmental and workplace safety laws), government agencies (Department of Labor and Environmental Protection Agency), new frontiers of ethics (business, environmental and medical ethics), and a robust social movement for reforming.
Social and ethical issues focus on public acceptance	• This misconception stems from the desire to promote the commercialisation potential of emerging technologies, and public opposition has largely resulted from misunderstandings or baseless concerns. • Ethics plays a robust role in the responsible development of technology to ensure public acceptance.

Within the literature, the first-hand personal experience and interaction between the emerging technology and society are low and minimal, and therefore, the reliances on personal experience/anecdotes (individual opinion on its use), media coverage, social network structure and cultural predispositions for information collection and decision making are essential (Nelson and Gorichanaz, 2019). Recent discussions have highlighted the crucial role of US National Nanotechnology

Initiative, a long-term research and development initiative, to stimulate broader, unique and constructive perspectives on the relationship between technology, environment, policies and society, in addressing the salience of ethical and social issues within the ongoing responsible technological development.

20.5.2 A Typology of Ethical and Moral Issues

The typology of ethical issues is the intention to emphasise the social, moral and ethical problems associated with carbon related technologies, illuminating the full range of considerations (inclusiveness), distinguished by specific features (timeline, familiarity, definiteness, regulatory relevance and reliability); or categories that are neither too little nor too much to be taken into account. The ethical challenges associated with emerging technology are mutually compatible, and the specific issue might fall into more than a category, while some perspectives might overlap across these categories, from the perspective of risk evaluation, relations of power, and societal capacity approaches.

20.5.2.1 Effects of Social Context

The effects of social context occur from the interaction of technology with the under-represented characteristics of social-institutional system into which the technology is emerging. The new and emerging science and technology would not be responsible for this particular issue, as it is not the main cause of the problematic features of social context that give rise to them. In most cases, it is reasonably expected that technology will accelerate the deterioration of the social context scenario, due to the distinct physical-chemical characteristics of carbon materials, the functionalities of existing products, or the rate, size and number of innovations associated with the rapid science or engineering advancement. However, such problematic behaviors could sometimes offer new opportunities, and lead to the addressing of these social context issues. The cases of this social context issues include unequal access to technology and healthcare, educational inequalities, limited data protection and privacy preservation, inefficiency in intellectual property/patent systems, and accessibility to environmental risks.

20.5.2.2 Morality and Techno Culture Issues

The issue of morality arises from the interaction between emerging technology and morally contentious activity/practice, in supporting research and engineering practices (engineering processes, tools and techniques), uses and applications. The issue of techno culture is derived from the misuse of emerging technology in social systems. The arguments of techno culture issues include (a) over-reliance on the promise of technological fixes to manage the problematic impacts and underlying causes, (b) overestimation of the available technical capabilities in the prediction and control of biological systems, and (c) technological mediation in view of the relationship and experience for the conservation of nature. It prioritises the elite-controlled quantitative risk assessment model toward a more inclusive, transparent and precautionary approach for the determination of responsiveness to uncertainties and problematic side effects associated with an emerging technological innovation system. In contrast to the social and contested moral contexts, techno culture problems are less important to technology development. However, techno culture issues are distinct and susceptible to nanotechnology as it consists of the ability to precisely characterise, design and control the behavior of matters at the atomic-molecular level.

20.5.2.3 Form-of-Life Issues

Form-of-life issues arise from the synergistic effects of technology from different aspects of human situation, including the human person, the relationship between one another, and the inter-relationship to the natural environment, on which laws, social norms/standards, obligation (duty), social responsibility practices and legitimate institutions are predicated. Technology development demonstrates significant impacts on the forms/concepts of social interaction. These form-of-life issues are likely to occur within the environmental realms, in which artificial alternatives have been adopted into material science, or sources of substitutions against other resources, or commodities previously extracted and exploited from natural resource systems. This trend may speed up the capacity and performance of emerging science and technology for materials characterisation, designation and construction at the atomic, molecular and macromolecular levels. Although the disruption and reconfiguring of social practices can be beneficial, the accomplishment in the form of life issues requires technological adaptation and technical response to the development.

20.5.2.4 Transformational Issues

The questions of transformation might arise from the exploitation and deployment of technology, especially as the new concept of hybrid technologies, notably in biotechnology, information technology system, computer science and mechanical engineering, robotics and cognitive science, to transform and support different aspects of human well-being. This might be accomplished by the significant alteration of the available kind of creatures, reconstituting the relationship to the natural environment, or self-awareness of the autonomous AI system. Within this framework, the configuration of an ethical terrain and ethical landscape would raise substantial challenges to the social context, form-of-life, techno culture developments and contested moral problems.

20.5.3 Identify the Current State of Hazards and Risks

20.5.3.1 Health and Safety of the Public

The health benefits and safety concerns of engineered carbon materials have been investigated for the past few decades, but specific toxicological investigations, health and safety regulations related to carbon nanomaterial, particularly graphene, have not been well developed. In 2013, the first early research was carried out (Donaldson et al., 2013) to raise the warning flag on graphene, at least for the nanoscopic platelets of the tested substances. Similar verification has been conducted on the carbon nanotubes and graphene-related health risks, even at a trace quantity (in the order of g, mg or μg), either by inhalation, ingestion or dermal penetration (Bussy et al., 2013). The extent and uptake mechanisms of graphene via inhalation are the most likely route of unintended occupational exposure or the changing environmental conditions. From the prolonged perspective, respiratory complications or other serious diseases, specifically inflammation disorder, interstitial fibrosis, mutagenic response, tumours and cancers may take place. An overview of the toxicological studies on carbon derivatives on human health is summarised in Table 20.4.

Within this framework, a set of three generalised guidelines has been proposed and implemented for the health risk prevention of researchers, workers, engineers, technicians and chemists, to serve as a reference to the working teams, and to raise public awareness related to the carbon materials' potential health and safety effects (Bussy et al., 2013), including

1. Application of individual carbon sheets, that are accessible by the immune cells to be engulfed and removed from site, in which they were found in the human body.
2. Application of stable and individual carbon sheets, that could be easily dispersed in the aqueous medium, and to minimise their clumping and aggregation in the human body.
3. Application of carbon materials, or chemically modified carbon nanomaterials, which could be easily removed from or biodegraded in the human body, and to prevent damages from chronic accumulation in the living cells and tissues.

Preventive measures and safety guidelines ought to be adopted among researchers, chemists, engineers, managers and workers, that will involve in multiple stages of the manufacturing process, with different degrees of exposure. A clearer definition and understanding of different exposure levels of carbon-derived materials, and routes of process are particularly necessary (Sousa et al., 2020).

TABLE 20.4
An Overview of the Toxicological and Human Health Risk Studies on Carbon Derivatives

Carbon Derivative	Cell Line	Health Effect	Mechanism	Reference
Carbon black	Human umbilical vein endothelial cells (HUVECs) and human monocytic cells (THP-1)	• Endothelial dysfunction	ROS generation	Cao et al. (2014)
Single- and multi-walled carbon nanotube	Human alveolar carcinoma epithelial cells (A549) and human monocytic cells (THP-1)	• Lung cancer and tumours	DNA damage and inflammatory response	Yamashita et al. (2010)
Single-walled carbon nanotube	Human lung epithelial cells (SAEC, BEAS-2B)	• Lung fibrosis and cancer	Genomic DNA damage and apoptosis	Luanpitpong et al. (2014)
	Human epidermal keratinocytes (HaCaT cells)	• Skin disease, dermatitis and hyperkeratosis	Cellular toxicity	Shvedova et al. (2003)
	Human intestinal carcinoma cell line (Caco-2)	• Gastrointestinal toxicity	Induce inflammatory responses and oxidative stress	Jos et al. (2009); Pichardo et al. (2012)
	Human gingival/oral fibroblasts	• Geno-cytotoxicity	Apoptosis	Cicchetti et al. (2011)
Multi-walled carbon nanotube	Human lung epithelial cells (A549, BEAS-2B)	• Lung tumor	Direct-oxidative DNA damage	Cavallo et al. (2012); Haniu et al. (2013); Hirano et al. (2010)
	Human embryonic kidney cell (HEK293)	• Renal toxicity	ROS, protein expression, genotoxicity and apoptosis	Reddy et al. (2010)
Multi-walled carbon nanotube, carbon nanofiber, carbon black	Human lung-tumor cell lines (H596, H446 and Calu-1)	• Asbestos-induced disease	Cell membrane damage	Magrez et al. (2006)
Graphene	Airway epithelial cell line BEAS-2B	• Chronic obstructive pulmonary disease (COPD) • Formation of granulomas and lung fibrosis	Apoptosis	Tsai et al. (2017)
Graphene oxide	Human neutrophils	• Platelet aggregation • Pulmonary thromboembolism • Granuloma formation	Enzymatic degradation	Kurapati et al. (2015)
Graphene nanoparticle	Human monocytic cell line (THP-1)	• Lung and pleural inflammation • Frustrated phagocytosis • Activation of pro-inflammatory cytokine (NAPL3)	Cell membrane damage and ROS generation	Schinwald et al. (2012)

(*Continued*)

TABLE 20.4 *(Continued)*

An Overview of the Toxicological and Human Health Risk Studies on Carbon Derivatives

Carbon Derivative	Cell Line	Health Effect	Mechanism	Reference
Graphene and single-walled carbon nanotube	Neuronal PC12 cells	• Necrosis	Oxidative stress and apoptosis	Zhang et al. (2010)
Graphene nanoplatelets	Human lung epithelial cells (A549)	• Lung cancer	Oxidative stress, apoptosis, LDH, superoxide dismutase and autophagy	Nasirzadeh et al. (2019)
Graphene oxide	Human fibroblast cell	• Lung inflammation and subsequent granulomas	Cell floating and apoptosis	Wang et al. (2011)
Fullerenes	Human umbilical vein ECs (HUVECs)	• Atherosclerosis • Ischemic heart disease	Ubiquitin-autophagic death pathways	Yamawaki and Iwai (2006)
	Human lens epithelial cells (HLE B-3)	• Retinal damage	Apoptosis and necrosis	Roberts et al. (2008); Wielgus et al. (2010); Zhao et al. (2009)
	Renal proximal tubule cells (LLC-PK1)	• Renal toxicity	Oxidative stress, cytoskeleton and mitochondrial disruption	Johnson-Lyles et al. (2010)

Accordingly, the obligation of regular risk assessments (Directives 89/391/EEC and 98/24/EC), in every aspect of the operation, particularly related to carbon nanomaterials, has been synergised. The communication between the industries, organisations, researchers and related stakeholders is facilitated by the ISO risk management terminology standard (ISO/TS 80004-13), for elucidation of the terms, types, properties, production and characterisation techniques (Pollard and Clifford, 2017).

20.5.3.2 Environmental Protection

The overwhelming development of the carbon industry is raising concern on the possible direct or indirect effects to the natural environment and ecosystem. Although the biocompatibility and toxicity evaluation of carbon derivatives are the heart of concern, the hazardous risk to the natural ecosystem remains undeveloped. According to He et al. (2017), the adverse impacts of ecological exposure to carbon materials are complex and uncontrolled, and largely affected by its physicochemical behaviors, sources, environmental fates and ecological conditions. Understanding the mechanisms of microbial toxicity is meaningful and essential for the assessment of ecological risks of carbon materials. To date, the potential toxicological effects of carbon derivatives on bacteria, fungi, algae, plants and animals have been reported. Comprehensive information on the ecological risks of carbon materials within the biological systems have been documented in Table 20.5.

20.5.4 The Precautionary Principle

The precautionary principle (PP), a key regulatory instrument by the EU, is a new guideline for the reliable management of environmental science and public health. Article 15 of the Rio Declaration

TABLE 20.5
Overview of the Ecological Risks of Carbon Materials

Carbon Material	Microorganism	Remark	Reference
Graphene nanosheet	*Escherichia coli, Salmonella typhimurium, Enterococcus faecalis* and *Bacillus subtilis*	• The MIC values of *E. coli, S. typhimurium, B. subtilis* and *E. faecalis* were 1, 1, 4 and 8µg/mL, respectively, much lower than the MIC value of standard antibiotic (kanamycin).	Krishnamoorthy et al. (2012)
Graphene oxide (GO), multi-layer graphene (MG) and reduced graphene oxide (rGO)	*Chlorella pyrenoidosa (Freshwater algae)*	• The EC_{50} of GO, MG and rGO were 37, 62 and 34 mg/L, respectively, which induced membrane damage by direct physical penetration/extraction.	Zhao et al. (2017)
GO and rGO	*Pseudomonas aeruginosa*	• The maximum loss of *P. aeruginosa* viability was detected at the GO and rGO concentrations of 150 µg/ml. • The possible mechanism could result in physical damage of cellular membranes, with the deterioration of cell integrity and DNA/RNA leakage.	Gurunathan et al. (2012)
GO	• *Artemia salina (brine shrimp)* • *Chironomus sancticaroli (chironomidae)* • *Daphnia magna (planktonic microcrustacean)* • *Hydra attenuata (freshwater polyp)* • *Lemna minor (aquatic plant)* • *Raphidocelis Subcapitata (green algae)* • *Caenorhabditis elegans and Panagrolaimus sp. (nematodes)*	The EC_{50} of these organisms was detected as > 100 mg/L, and arranged in the decreasing pattern: *D. magna>Panagrolaimus sp.>L. minor.*	Castro et al. (2018)
rGO	• *Aspergillus niger* • *Aspergillus oryzae* • *Fusarium oxysporum*	The IC_{50}findings of rGO towards *A. niger*, *A. oryzae* and *F. oxysporum* have been identified to be 50, 100 and 100 µg/mL, respectively.	Sawangphruk *et al.* (2012)
Carbon nanofiber and multi-walled (MW) carbon nanotube	• *Chlorella vulgaris*	• The algal growth inhibition rate occurred at the concentration at 1 µg/mL • The possible mechanisms are induced by oxidation, ATP production and physical stress.	Pereira *et al.* (2014)
Single-walled (SW) carbon nanotube	*Saccharomyces cerevisiae*	Single-walled carbon nanotube (376.4 and 188.2 mg/L) induces oxidative stress and apoptosis, and oxidative stress has achieved the mitochondria-dependent apoptotic pathway.	Zhu et al. (2018)

(Continued)

TABLE 20.5 *(Continued)*
Overview of the Ecological Risks of Carbon Materials

Carbon Material	Microorganism	Remark	Reference
Single-walled (SW) carbon nanotube	• *Daphnia magna (microcrustacean)* • *Oryzias latipes (fish)* • *Raphidocelissubcapitata and Chlorella vulgaris (microalgae)*	• Single-walled carbon nanotube inhibits the growth rates of *R. subcapitata* and *C. vulgaris*, with EC_{50} (effective concentrations) at 29.99 and 30.96 mg/L, respectively. • It did not induce short-term toxic effects against *D. magna* and *O. latipes* up to the concentration at 100 mg/L.	Sohn et al. (2015)
Multi-walled (MW) carbon nanotube	*Escherichia coli* and *Staphylococcus aureus*	The best concentrations of multi-walled carbon nanotube for the maximum inhibition have been identified at 80 and 60 µg/mL for *E. coli* and *S. aureus*, with inhibition rates of 75.92% and 63.88%, respectively.	AboNeima et al. (2020)
	Chlorella sp. (green algae)	The IC_{50} value for multi-walled carbon nanotube was 12.4 mg/L, mainly could be explained by the hybrid implications of aggregation/agglomeration, shading effects, oxidative stress and physical interactions with cells.	Long et al. (2012)
	• *Chironomidae (larvae)* • *Ceriodaphniaaffinis (crustaceans)* • *Scenedesmus quadricauda (algae)* • *Escherichia coli (bacteria)*	Safe levels of the carbon-nanomaterial in the aquatic ecosystem are < 2 mg/L.	Gusev et al. (2012)
Fullerene	*Enterococcus faecalis, Escherichia coli, Klebsiella pneumoniae, Pseudomonas aeruginosa, Salmonella typhi, Staphylococcus* sp., *Streptococcus sp.*	• Gram-positive bacteria were inhibited at 50 mg/L, while gram-negative organisms were not inhibited even at concentrations up to 500 mg/L (fullerene). • The action of fullerene on gram-positive species was driven by the attachment to the cell wall, and destruction of membrane permeability and integrity.	Tsao et al. (2002)
	Pseudomonas aeruginosa	• The LC_{50} of fullerene on *P. aeruginosa* was recorded at 1,336 mg/L. • Mechanism analysis of cell damage on the tested species by fullerenes indicated apoptotic and necrotic responses.	Indeglia et al. (2018)
	Embryonic zebrafish	• Exposure to 200 µg/L of fullerene induced significant increases of malformations, genetic damage, pericardial edema syndromes and mortality.	Usenko et al. (2007)

on Environment and Development states that "PP should be extensively applied according to their capabilities for the protection of natural environment" (United Nations Environment Programme, 1992). PP has also been designed as a "principle of risk optimization or action" with six central supports: (a) preventive or precautionary action to reduce uncertainties; (b) shifting the heavy burden of proof and financial responsibility to the proponents of an activity; (c) exploration of new actions against harmful activities; (d) goal setting; (e) acceptance of a duty to monitor, understand, investigate, inform and act; and (f) enhanced public participation in democratic decision making, modify conventional legal approaches, and limit the cases of "threats of serious or irrevocable damages" under the conditions of scientific uncertainty. Table 20.6 lists the two major types of PP under different jurisdictions.

Within the PP systematisation, the traditional risk management approaches, notably the (a) tolerable risk, (b) cost–benefit analysis and (c) feasibility principle, are insufficient to fulfill the requirement for risk management and plans. From the perspective of an advocate of precautionary action, the principle of acceptable risk relies on the risk assessment model describing the possible risks of an agent, and seeking solutions or strategies to reduce the risk level to a threshold of socially acceptable. Different forms and types of carbon-derived materials present strikingly different toxicity risks for human health and environment according to various aspects, including raw sources, synthesis and manufacturing processes, and unique physio-chemical characteristics of products (size, shape, surface area, chemical compositions and coating). Until now, however, there are limited available scientifically validated testing methods or data for the quantitative estimation of risk (Farrow, 2004).

Cost-benefit balancing is the second pillar of the traditional risk management model, serving both the risk assessment and management paradigms. Apart from the previously discussed acceptable risk model, both of the benefits and risks of technology, in most circumstances, are equivalent to the proposed risk management options, especially the preservation of environmental and public health. However, the cost–benefit studies to manage the benefits, risks, properties, uncertainties and future scenario/direction of carbon-based technology are ill-equipped, due to the diversity and enormous number of potential applications. Current understanding of the cost–benefit balancing tools for global nanotechnology industry would obscure the cost-variance significance between different applications. On the contrary, implementing individual cost–benefit analysis for specific applications and uses would overwhelm the available risk-assessment and resource management plans (Marchant et al., 2008).

Feasibility is the third principle of a traditional risk management system. It is an economically feasible technology that could reduce risk to a minimum level, without the requirement of additional information. This approach has attracted aesthetic concern and popularity among the researchers, stakeholders and policy makers, as it allows circumvention over risk management and moves forward to the heart of the matter. This approach, however, may under-regulate or over-regulate the existing risks, according to the efficacy of the available technique in minimising the uncertainty and

TABLE 20.6
Types of Precautionary Principle (PP) (Ahteensuu and Sandin, 2012)

Jurisdictions	Weak PP	Strong PP
Extent of threat	Precautionary measures are carried out when major and irreversible risks exist	Precautionary measures are carried out when minor and reversible risks exist
Degree of uncertainty	Scientific proof is needed	Only speculative evidence of risk is needed
Extent of actions	Only low-cost precautionary measures may be taken	Precautionary measures irrespective of their cost are taken
Approach	• Non-preclusion • Margin-of-safety	• Best available technology • Prohibitory

unacceptable risks. Emerging nanotechnology is not applicable to this traditional form of risk management (Marchant et al., 2008).

In summary, none of the three traditional models could assess and manage the technology of risk of new emerging technology effectively, given the tremendous uncertainties under the real practical conditions. Nano-ethicists consider that cost–benefit analysis is not viable for the requirement of environmental health and safety (EHS) due to the difficulties in quantifying the element, composition and configuration for long-term sustainability, while the available life-cycle theory is neither widely implementable nor easily enforceable. However, PP is widely implied as a pragmatic strategy of the EHS risk management toolbox when limited knowledge has prevented risk assessment to be conducted.

At the present stage of the development of new emerging nanotechnology, the responsibility for taking precautionary/preventive actions, to a larger extent, could be placed on the researcher and the research communities, to modify the conventional legal approach to the responsibility for serious and unforeseeable damages. However, some philosophers reserve sharp criticism on PP, as this principle may paradoxically fail, due to the broad-scale impacts from different levels of risk tolerance and societal interests. This ambiguity leads to significant difference and inconsistency of its application. Weckert (2007) has defended that PP is a perfect complement of the traditional risk management approach to weaken anthropocentrism's long-term resource maximising belief and creed, by integrating both negative and positive forms of responsibilities, and credible threats to the recognised confirmed cases. PP is both a holistic feature (the rights of commercial corporations are nullified if the aspects have negatively impacted the whole community) and consequentiality behavior (consider the overall consequences), and it remains valid when it is practically applied. For the last few years, it has been adopted into a wide variety of treaties and protocols, including the EU in a communication to the Parliament (EC, 2000), that worked mainly on the environmental and health risks; Maastricht Treaty (Maastricht, 1992) and the Rio Declaration on Environment and Development (UNEP, 1992).

20.5.5 Strategies Supporting Ethical Decision Making

Ethical decision-making is a critical process for professional work in all fields; particularly it has been directed to the science and academia sectors. Most of the researchers in the field of nanoscience/nanotechnology have focused on the biases, which may be reacted to undermine ethical decision-making. Earlier research studies, especially in ethical decision-making, were matched with Rest's four-stage model (1986), specifically (1) recognition and judgment of ethical issues and criteria; (2) weighting the importance of ethics and moral components above all other considerations; and (3) acting within the framework of ethical principles. Rest's model has provided a unified or consistent framework for ethical decision-making since its introduction. Treviño (1986) has also proposed a model, after recognising the potential ethical and moral considerations in nanotechnology development, and both individual difference and situational preparations that could moderate its judgment, which has been expanded to the second stage of Rest's model (Rest, 1986). Similarly discussed by Hunt and Vitell (1986), Jones (1991) has further improved the model as an elaboration elucidating the principal underlying processes in the third stage of this model. This general theory was similarly supported by Tenbrunsel and Smith-Crowe (2008), who adopted meta-analytic methods for ethical decision-making in both sets of moral and amoral tracks. The four-stage model of ethical-decision making study is illustrated in Figure 20.4.

20.6 CONCLUSION

The continuously growing interest in the carbon-derived materials, featuring well-distinct physical and chemical properties, has signified a variety of breakthroughs of the high-performance energy-related technologies. Crucially, there are a great deal of uncertainties about the trajectory and

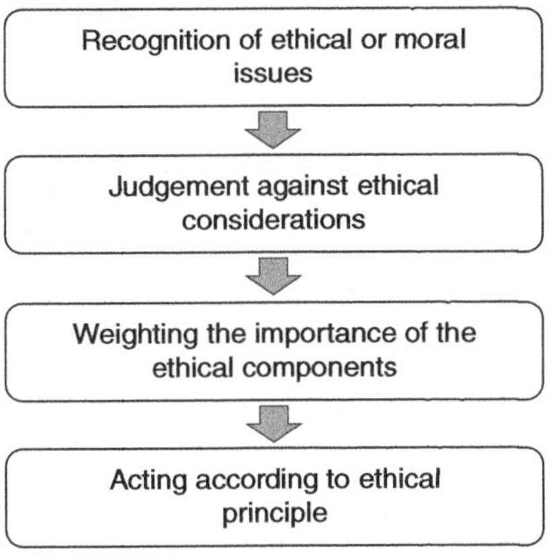

FIGURE 20.4 Four-stage model in ethical-decision making (Haines et al., 2008).

characteristics of carbon industry development and technology transfer surrounding the environment, human health and safety implications. Accordingly, numerous new policies and initiatives have been launched from multiple angles to promote emergent carbon research, commercialisation, balance or to bridge the interests of both environment and human beneficiaries. Therefore, this chapter has deliberately elucidated deeper perspectives on a set of anticipatory practices shaping the expectations of carbon technology across its development, related policy and ethics, and highlighted the relevant practice and performative effects. Still at the infancy stage of development, it is an important scope for future researchers and policy makers to examine the framework and scrutinise the unique impacts of the anticipatory practice from a normative perspective for the sustainable future.

REFERENCES

AboNeima, S. E., Motaweh, H. A., Elsehly, E. M. 2020. Antimicrobial activity of functionalised carbon nanotubes against pathogenic microorganisms. *IET Nanobiotechnol.*, 14: 457–464. doi:10.1049/iet-nbt.2019.0342

Ahteensuu, M., Sandin, P. 2012. The Precautionary Principle. In: Roeser, S., Hillerbrand, R., Sandin, P., Peterson, M. (eds) *Handbook of Risk Theory*. Springer, Dordrecht. doi:10.1007/978-94-007-1433-5_38

Arvidsson, R., Boholm, M., Johansson, M., de Montoya, M. L. 2018. "Just Carbon": Ideas about graphene risks by graphene researchers and innovation advisors. *Nanoethics*, 12: 199–210. doi:10.1007/s11569-018-0324-y

Bianco, A., Chen, Y., Frackowiak, E., Holzinger, M., Koratkar, N., Meunier, V., Mikhailovsky, S., Strano, M., Tascon, J. M., Terrones, M. 2020. Carbon science perspective in 2020: Current research and future challenges. *Carbon*, 161: 373–391. doi:10.1016/j.carbon.2020.01.055

Boehm, H. P., Clauss, A., Fischer, G. O., Hofmann, U. 1962. Dünnstekohlenstoff-folien. *Zeitschrift Für Naturforschung B*, 17: 150–153. doi:10.1515/znb-1962-0302

BP Statistical Review of World Energy 2021. 70th ed., Available online: https://www.bp.com/content/dam/bp/business-sites/en/global/corporate/pdfs/energy-economics/statistical-review/bp-stats-review-2021-full-report.pdf (Accessed 20 December 2023).

British Standards Institution (BSI). 2018. *Developing a UK Standards Strategy for Graphene*. Available at: https://www.bsigroup.com/globalassets/localfiles/en-gb/about-bsi/nsb/innovation/uk-standards-strategy-for-graphene-report.pdf (Accessed 8 January 2024)

Bussy, C., AliBoucetta, H., Kostarelos, K. 2013. Safety considerations for graphene: Lessons learnt from carbon nanotubes. *Acc. Chem. Res.*, 46: 692–701. doi:10.1021/ar300199e

Cao, Y., Roursgaard, M., Danielsen, P. H., Møller, P., Loft, S. 2014. Carbon black nanoparticles promote endothelial activation and lipid accumulation in macrophages independently of intracellular ROS production. *PLoS One*, 9: e106711. doi:10.1371/journal.pone.0106711

Castro, V. L., Clemente, Z., Jonsson, C., Silva, M., Vallim, J. H., de Medeiros, A. M. Z., Martinez, D. S. T. 2018. Nanoecotoxicity assessment of graphene oxide and its relationship with humic acid. *Environ. Toxicol. Chem.*, 37: 1998–2012. doi:10.1002/etc.4145

Cavallo, D., Fanizza, C., Ursini, C. L., Casciardi, S., Paba, E., Ciervo, A., Fresegna, A. M., Maiello, R., Marcelloni, A. M., Buresti, G., Tombolini, F., Bellucci, S., Iavicoli, S. 2012. Multi-walled carbon nanotubes induce cytotoxicity and genotoxicity in human lung epithelial cells. *J. Appl. Toxicol.*, 32: 454–464. doi:10.1002/jat.2711

Chemanalyst. 2020. *Carbon Fiber Market Analysis: Industry Market Size 2015–2035*. Available at: https://www.chemanalyst.com/industry-report/carbon-fiber-market-676 (Accessed 9 January 2024).

Choi, D., Kil, H. S., Lee, S. 2019. Fabrication of low-cost carbon fibers using economical precursors and advanced processing technologies. *Carbon*, 142: 610–649. doi:10.1016/j.carbon.2018.10.028

Cicchetti, R., Divizia, M., Valentini, F., Argentin, G. 2011. Effects of single-wall carbon nanotubes in human cells of the oral cavity: Geno-cytotoxic risk. *Toxicol. in Vitro*, 25: 1811–1819. doi:10.1016/j.tiv.2011.09.017

Cook, J. J., Booth, S. 2017. Carbon fiber manufacturing facility siting and policy considerations: International comparison (No. NREL/TP-6A20-66875). National Renewable Energy Lab (NREL), Golden, CO (United States). doi:10.2172/1365711

DeMordaunt, A., Sharma, S., Stypula, M., Charlton, C., Lin, S. 2022. Market Analysis: Upcycling Natural Gas into Solid Carbon Products (No. DOE/NETL-2021/3211). National Energy Technology Laboratory (NETL), Pittsburgh, PA, Morgantown, WV and Albany, OR (United States). doi:10.2172/1838316

Donaldson, K., Schinwald, A., Murphy, F., Cho, W. S., Duffin, R., Tran, L., Poland, C. 2013. The biologically effective dose in inhalation nanotoxicology. *Acc. Chem. Res.*, 46: 723–732. doi:10.1021/ar300092y

Dong, C., Qi, Y., Nemet, G. (2021). A government approach to address coal overcapacity in China. *J. Cleaner Prod.*, 278, 123417. doi:10.1016/j.jclepro.2020.123417

Dos Santos, F. N., Nascimento, M. M., da Rocha, G. O., de Andrade, J. B. 2023. The occurrence of pristine and functionalized fullerenes as constituents of airborne aerosols. *Sci. Rep.*, 13: 4248. doi:10.1038/s41598-023-31119-4

European Commission (EC). 2000. *Communication from the Commission on the Precautionary Principle*. European Commission, Brussels. Available at: https://op.europa.eu/en/publication-detail/-/publication/21676661-a79f-4153-b984-aeb28f07c80a/language-en (Accessed 11 January 2024)

European Commission (EC). 2014. *Report on Critical Raw Materials for the EU. Report of the Ad-hoc Working Group on Defining Critical Raw Materials*. European Commission (EC), Brussels, Belgium. Available at: http://ec.europa.eu/enterprise/policies/raw-materials/files/docs/crm-report-on-critical-raw-materials_en.pdf (Accessed 10 January 2024)

European Commission (EC). 2016. *What is Horizon 2020? Horizon 2020*. Available at: https://ec.europa.eu/programmes/horizon2020/en/what-horizon-2020 (Accessed 12 January 2024)

European Commission (EC). 2020a. *Critical Raw Materials for Strategic Technologies and Sectors in the EU—A Fore Sight Study*. Publications Office of the European Union, Luxembourg. doi:10.2873/865242

European Commission (EC). 2020b. *Study on the EU's List of Critical Raw Materials. Final Report*. Publications Office of the European Union, Luxembourg. doi:10.2873/11619

European Food Safety Authority (EFSA). 2020. Annual report of the EFSA scientific network of risk assessment of nanotechnologies in food and feed for 2019. *EFSA Supporting Publications*, 17: 1784E. doi:10.2903/sp.efsa.2020.EN-1784

Fan, B., Guo, T., Xu, R., Dong, W. (2021). Evolutionary game research on the impact of environmental regulation on overcapacity in coal industry. *Math. Probl. Eng.*, 2021: 5558112. doi:10.1155/2021/5558112

Farrow, S. 2004. Using risk assessment, benefit-cost analysis, and real options to implement a precautionary principle. *Risk Anal.*, 24: 727–735. doi:10.1111/j.0272-4332.2004.00471.x

Fortune Business Insight. 2022. *Carbon Black Market Size, Share & COVID-19 Impact Analysis, by Type (Furnace Black, Channel Black, Thermal Black, Acetylene Black, Others), by Grade (Standard Grade, Specialty Grade), by Application (Tire, Non-tire Rubber, Inks & Coatings, Plastics, Other)*. Available at: https://www.fortunebusinessinsights.com/industry-reports/carbon-black-market-101718 (Accessed 2 January 2024)

Gurunathan, S., Han, J. W., Dayem, A. A., Eppakayala, V., Kim, J. H. 2012. Oxidative stress-mediated antibacterial activity of graphene oxide and reduced graphene oxide in Pseudomonas aeruginosa. *Int. J. Nanomed.*, 7: 5901–5914. doi:10.2147/IJN.S37397

Gusev, A. A., Fedorova, I. A., Tkachev, A. G., Godymchuk, A. Y., Kuznetsov, D. V., Polyakova, I. A. 2012. Acute toxic and cytogenetic effects of carbon nanotubes on aquatic organisms and bacteria. *Nanotechnol. Russ.*, 7, 509–516. doi:10.1134/S1995078012050060

Haines, R., Street, M. D., Haines, D. 2008. The influence of perceived importance of an ethical issue on moral judgment, moral obligation, and moral intent. *J. Bus. Ethics*, 81: 387–399. doi:10.1007/s10551-007-9502-5

Haniu, H., Saito, N., Matsuda, Y., Tsukahara, T., Maruyama, K., Usui, Y., Aoki, K., Takanashi, S., Kobayashi, S., Nomura, H. and Okamoto, M., Shimizu, M., Kato, H. 2013. Culture medium type affects endocytosis of multi-walled carbon nanotubes in BEAS-2B cells and subsequent biological response. *Toxicol. in Vitro*, 27: 1679–1685. doi:10.1016/j.tiv.2013.04.012

Hazen, R. M., Papineau, D., Bleeker, W., Downs, R. T., Ferry, J. M., McCoy, T. J., Sverjensky, D. A., Yang, H. 2008. Mineral evolution. *Am. Mineral.*, 93: 1693–1720. doi:10.2138/am.2008.2955

He, K., Chen, G., Zeng, G., Peng, M., Huang, Z., Shi, J., Huang, T. 2017. Stability, transport and ecosystem effects of graphene in water and soil environments. *Nanoscale*, 9, 5370–5388. doi:10.1039/C6NR09931A

Hirano, S., Fujitani, Y., Furuyama, A., Kanno, S. 2010. Uptake and cytotoxic effects of multi-walled carbon nanotubes in human bronchial epithelial cells. *Toxicol. Appl. Pharmacol.*, 249: 8–15. doi:10.1016/j.taap.2010.08.019

Hunt, S. D., Vitell, S. 1986. A general theory of marketing ethics. *J. Macromark.*, 6: 5–16. doi:10.1177/027614678600600103

Iijima, S. 1991. Helical microtubules of graphitic carbon. *Nat.*, 354: 56–58. doi:10.1038/354056a0

Indeglia, P. A., Georgieva, A. T., Krishna, V. B., Martyniuk, C. J., Bonzongo, J. C. J. 2018. Toxicity of functionalized fullerene and fullerene synthesis chemicals. *Chemosphere*, 207, 1–9. doi:10.1016/j.chemosphere.2018.05.023

Industry Research Information Bank. 2017. *Present Situation Analysis and Future Development Suggestion of Graphene Industry [EB/OL]*. Available at: https://wenku.baidu.com/view/4e4161416bec0975f465e2dd.html (Accessed 9 January 2024)

Jara, A. D., Betemariam, A., Woldetinsae, G., Kim, J. Y. 2019. Purification, application and current market trend of natural graphite: A review. *Int. J. Min. Sci. Technol.*, 29, 671–689. doi:10.1016/j.ijmst.2019.04.003

Johnson-Lyles, D. N., Peifley, K., Lockett, S., Neun, B. W., Hansen, M., Clogston, J., Stern, S. T., McNeil, S. E. 2010. Fullerenol cytotoxicity in kidney cells is associated with cytoskeleton disruption, autophagic vacuole accumulation, and mitochondrial dysfunction. *Toxicol. Appl. Pharmacol.*, 248: 249–258. doi:10.1016/j.taap.2010.08.008

Jones, T. M. 1991. Ethical decision making by individuals in organizations: An issue-contingent model. *Acad. Manage. Rev.*, 16: 366–395. doi:10.5465/amr.1991.4278958

Jos, A., Pichardo, S., Puerto, M., Sánchez, E., Grilo, A., Cameán, A. M. 2009. Cytotoxicity of carboxylic acid functionalized single wall carbon nanotubes on the human intestinal cell line Caco-2. *Toxicol. in Vitro*, 23: 1491–1496. doi:10.1016/j.tiv.2009.07.001

Justo-Hanani, R., Dayan, T. 2015. European risk governance of nanotechnology: Explaining the emerging regulatory policy. *Res. Policy*, 44: 1527–1536. doi:10.1016/j.respol.2015.05.001

Kjarsgaard, B. A., de Wit, M., Heaman, L. M., Pearson, D. G., Stiefenhofer, J., Janusczcak, N., Shirey, S. B. 2022. A review of the geology of global diamond mines and deposits. *Rev. Mineral. Geochem.*, 88: 1–117. doi:10.2138/rmg.2022.88.01

Krishnamoorthy, K., Veerapandian, M., Zhang, L. H., Yun, K., Kim, S. J. 2012. Antibacterial efficiency of graphene nanosheets against pathogenic bacteria via lipid peroxidation. *J. Phys. Chem. C*, 116: 17280–17287. doi:10.1021/jp3047054

Kroto, H. W., Heath, J. R., O'Brien, S. C., Curl, R. F., Smalley, R. E. 1985. C60: Buckminsterfullerene. *Nat.*, 318: 162–163. doi:10.1038/318162a0

Kurapati, R., Russier, J., Squillaci, M. A., Treossi, E., Ménard-Moyon, C., Del Rio-Castillo, A. E., Vazquez, E., Samorì, P., Palermo, V., Bianco, A. 2015. Dispersibility-dependent biodegradation of graphene oxide by myeloperoxidase. *Small*, 11: 3985–3994. doi:10.1002/smll.201500038

Lavoisier, A. L. 1772. Second Mémoire sur la Destruction du Diamant au Grand Verre Brillant de Tschirnhausen, connu sur le Nom de Lentille du Palais Royal. *Mémoires Académie des Sciences*, pages 564–591.

Lavoisier, A. L. 1789. *Traitéélémentaire de chimie. Œuvres de Lavoisier*, Volume 1, 1864, pages 1–433, Imprimerieimpériale, Paris.

Li, H., He, X., Wu, T., Jin, B., Yang, L., Qiu, J. 2022. Synthesis, modification strategies and applications of coal-based carbon materials. *Fuel Process. Technol.*, 230: 107203. doi:10.1016/j.fuproc.2022.107203

Liu, F., Wang, Q., Zhai, G., Xiang, H., Zhou, J., Jia, C., Zhu, L., Wu, Q., Zhu, M. 2022. Continuously processing waste lignin into high-value carbon nanotube fibers. *Nat. Commun.*, 13: 5755. doi:10.1038/s41467-022-33496-2

Liu, P., Gao, Y. 2019. Graphene's potential in the future industrial development of China. *Resour. Policy*, 61: 118–127. doi:10.1016/j.resourpol.2019.02.007

Long, Z., Ji, J., Yang, K., Lin, D., Wu, F. 2012. Systematic and quantitative investigation of the mechanism of carbon nanotubes' toxicity toward algae. *Environ. Sci. Technol.*, 46: 8458–8466. doi:10.1021/es301802g

Luanpitpong, S., Wang, L., Rojanasakul, Y. 2014. The effects of carbon nanotubes on lung and dermal cellular behaviors. *Nanomed.*, 9, 895–912. doi:10.2217/nnm.14.42

Ma, J., Zheng, X., Lei, H., Ke, W., Chen, C., Chen, Z., Yang, G., Fang, G. 2017. Highly efficient and stable planar perovskite solar cells with large-scale manufacture of e-beam evaporated SnO_2 toward commercialization. *Sol. RRL*, 1: 1700118. doi:10.1002/solr.201700118

Maastricht. 1992. *The Maastricht Treaty. Provisions amending the treaty establishing the European Economic Community with a view to establishing the European Community*. Available at: http://www.eurotreaties.com/maastrichtec.pdf (Accessed 11 January 2024)

Magrez, A., Kasas, S., Salicio, V., Pasquier, N., Seo, J. W., Celio, M., Catsicas, S., Schwaller, B., Forró, L. 2006. Cellular toxicity of carbon-based nanomaterials. *Nano Lett.*, 6: 1121–1125. doi:10.1021/nl060162e

Marchant, G. E., Sylvester, D. J., Abbott, K. W. 2008. Risk management principles for nanotechnology. *Nanoethics*, 2: 43–60. doi:10.1007/s11569-008-0028-9

Nasirzadeh, N., Azari, M. R., Rasoulzadeh, Y., Mohammadian, Y. 2019. An assessment of the cytotoxic effects of graphene nanoparticles on the epithelial cells of the human lung. *Toxicol. Ind. Health*, 35: 79–87. doi:10.1177/0748233718817180

Nelson, J., Gorichanaz, T. 2019. Trust as an ethical value in emerging technology governance: The case of drone regulation. *Technol. in Soc.*, 59: 101131. doi:10.1016/j.techsoc.2019.04.007

Nemanich, R. J., Carlisle, J. A., Hirata, A., Haenen, K. 2014. CVD diamond—Research, applications, and challenges. *MRS Bull.*, 39, 490–494. doi:10.1557/mrs.2014.97

Novoselov, K. S., Geim, A. K., Morozov, S. V., Jiang, D. E., Zhang, Y., Dubonos, S. V., Grigorieva, I. V., Firsov, A. A. 2004. Electric field effect in atomically thin carbon films. *Sci.*, 306: 666–669. doi:10.1126/science.1102896

Nunna, S., Blanchard, P., Buckmaster, D., Davis, S., Naebe, M. 2019. Development of a cost model for the production of carbon fibres. *Heliyon*, 5: e02698. doi:10.1016/j.heliyon.2019.e02698

Pavlicek, A., Part, F., Rose, G., Praetorius, A., Miernicki, M., Gazsó, A., Huber-Humer, M. 2021. A European nano-registry as a reliable database for quantitative risk assessment of nanomaterials? A comparison of national approaches. *NanoImpact*, 21: 100276. doi:10.1016/j.impact.2020.100276

Peijs, T., Kirschbaum, R., Lemstra, P. J. 2022. A critical review of carbon fiber and related products from an industrial perspective. *Adv. Ind. Eng. Polym. Res.*, 5: 90–106. doi:10.1016/j.aiepr.2022.03.008

Pereira, M. M., Mouton, L., Yéprémian, C., Couté, A., Lo, J., Marconcini, J. M., Ladeira, L. O., Raposo, N. R., Brandão, H. M., Brayner, R. 2014. Ecotoxicological effects of carbon nanotubes and cellulose nanofibers in Chlorella vulgaris. *J. Nanobiotechnol.*, 12: 15. doi:10.1186/1477-3155-12-15

Pichardo, S., Gutiérrez-Praena, D., Puerto, M., Sánchez, E., Grilo, A., Cameán, A. M., Jos, Á. 2012. Oxidative stress responses to carboxylic acid functionalized single wall carbon nanotubes on the human intestinal cell line Caco-2. *Toxicol. in Vitro*, 26: 672–677. doi:10.1016/j.tiv.2012.03.007

Pollard, A. J., Clifford, C. A. 2017. Terminology: The first step towards international standardisation of graphene and related 2D materials. *J. Mater. Sci.*, 52, 13685–13688. doi:10.1007/s10853-017-1567-7

Porto, L. S., Silva, D. N., de Oliveira, A. E. F., Pereira, A. C., Borges, K. B. 2020. Carbon nanomaterials: Synthesis and applications to development of electrochemical sensors in determination of drugs and compounds of clinical interest. *Rev. Anal. Chem.*, 38: 20190017. doi:10.1515/revac-2019-0017

Reddy, A. R. N., Reddy, Y. N., Krishna, D. R., Himabindu, V. 2010. Multi wall carbon nanotubes induce oxidative stress and cytotoxicity in human embryonic kidney (HEK293) cells. *Toxicol.*, 272: 11–16. doi:10.1016/j.tox.2010.03.017

Rest, J. R. 1986. *Moral Development: Advances in Research and Theory*, page 224. Praeger, New York. http://hdl.handle.net/10822/811393

Roberts, J. E., Wielgus, A. R., Boyes, W. K., Andley, U., Chignell, C. F. 2008. Phototoxicity and cytotoxicity of fullerol in human lens epithelial cells. *Toxicol. Appl. Pharmacol.*, 228: 49–58. doi:10.1016/j.taap.2007.12.010

Sandler, R. 2009. Nanomedicine and nanomedical ethics. *Am. J. Bioethics*, 9: 16–17. doi:10.1080/15265160902995117

Sawangphruk, M., Srimuk, P., Chiochan, P., Sangsri, T., Siwayaprahm, P. 2012. Synthesis and antifungal activity of reduced graphene oxide nanosheets. *Carbon*, 50: 5156–5161. doi:10.1016/j.carbon.2012.06.056

Schinwald, A., Murphy, F. A., Jones, A., MacNee, W., Donaldson, K. 2012. Graphene-based nanoplatelets: a new risk to the respiratory system as a consequence of their unusual aerodynamic properties. *ACS Nano*, 6: 736–746. doi:10.1021/nn204229f

Shvedova, A., Castranova, V., Kisin, E., Schwegler-Berry, D., Murray, A., Gandelsman, V., Maynard, A., Baron, P. 2003. Exposure to carbon nanotube material: Assessment of nanotube cytotoxicity using human keratinocyte cells. *J. Toxicol. Environ. Health Part A*, 66: 1909–1926. doi:10.1080/713853956

Sohn, E. K., Chung, Y. S., Johari, S. A., Kim, T. G., Kim, J. K., Lee, J. H., Lee, Y. H., Kang, S. W., Yu, I. J. 2015. Acute toxicity comparison of single-walled carbon nanotubes in various freshwater organisms. *Biomed Res. Int.*, 2015: 323090. doi:10.1155/2015/323090

Sousa, S. P., Peixoto, T., Santos, R. M., Lopes, A., Paiva, M. D. C., Marques, A. T. 2020. Health and safety concerns related to CNT and graphene products, and related composites. *J. Compos. Sci.*, 4, 106. doi:10.3390/jcs4030106

Statista Research Department. 2023. *Graphene Industry Worldwide*. Available at: https://www.statista.com/topics/8769/graphene-industry-worldwide/#topicOverview (Accessed 5 January 2024)

Tenbrunsel, A., Smith-Crowe, K. 2008. Ethical decision making: Where we've been and where we're going. *The Acad. Manage. Ann.*, 2: 545–607. doi:10.1080/19416520802211677

Teng, H., Li, S., Cao, Z., Li, S., Li, C., Ko, T. J. 2023. Carbon fiber composites for large-scale wind turbine blades: Applicability study and comprehensive evaluation in China. *J. Mar. Sci. Eng.*, 11: 624. doi:10.3390/jmse11030624

Trache, D., Thakur, V. K., Boukherroub, R. 2020. Cellulose nanocrystals/graphene hybrids—A promising new class of materials for advanced applications. *Nanomater.*, 10: 1523. doi:10.3390/nano10081523

Treviño, L. K. 1986. Ethical decision making in organizations: A person-situation interactionist model. *Acad. Manage. Rev.*, 11: 601–617. doi:10.5465/amr.1986.4306235

Tsai, S. M., Bangalore, P., Chen, E. Y., Lu, D., Chiu, M. H., Suh, A., Gehring, M., Cangco, J. P., Garcia, S. G., Chin, W. C. 2017. Graphene-induced apoptosis in lung epithelial cells through EGFR. *J. Nanopart. Res.*, 19, 262. doi:10.1007/s11051-017-3957-9

Tsao, N., Luh, T. Y., Chou, C. K., Chang, T. Y., Wu, J. J., Liu, C. C., Lei, H. Y. 2002. In vitro action of carboxyfullerene. *J. Antimicrob. Chemother.*, 49: 641–649. doi:10.1093/jac/49.4.641

U.S. Geological Survey (USGS). 2023. *Mineral Commodity Summaries 2023 - GRAPHITE (NATURAL) Data Release*. Available at: https://pubs.usgs.gov/periodicals/mcs2023/mcs2023-graphite.pdf (Accessed 2 January 2024)

United Nations Environment Programme (UNEP). 1992. *Rio Declaration on Environment and Development*. Available at: http://www.unep.org/ (Accessed 11 January 2024)

Usenko, C. Y., Harper, S. L., Tanguay, R. L. 2007. In vivo evaluation of carbon fullerene toxicity using embryonic zebrafish. *Carbon*, 45: 1891–1898. doi:10.1016/j.carbon.2007.04.021

Vandermonde, C. A., Monge, G., Berthollet, C. L. 1788. Mémoire sur le fer considéré dans ses différens états métalliques. *Mém. Acad. Roy. Sci.* 132–200.

Wang, B. 2018. The future of manufacturing: A new perspective. *Eng.*, 4, 722–728. doi: 10.1016/j.eng.2018.07.020

Wang, Y., Liu, J., Liu, L., Sun, D. D. 2011. High-quality reduced graphene oxide-nanocrystalline platinum hybrid materials prepared by simultaneous co-reduction of graphene oxide and chloroplatinic acid. *Nanoscale Res. Lett.*, 6, 241. doi:10.1186/1556-276X-6-241

Weckert, J. 2007. Technological determinism, values and the precautionary principle in nanotechnology. *Eubios J. Asian int. Bioeth.*, 17, 72. http://hdl.handle.net/10822/964683

Wielgus, A. R., Zhao, B., Chignell, C. F., Hu, D. N., Roberts, J. E. 2010. Phototoxicity and cytotoxicity of fullerol in human retinal pigment epithelial cells. *Toxicol. Appl. Pharmacol.*, 242: 79–90. doi:10.1016/j.taap.2009.09.021

Yamashita, K., Yoshioka, Y., Higashisaka, K., Morishita, Y., Yoshida, T., Fujimura, M., Kayamuro, H., Nabeshi, H., Yamashita, T., Nagano, K., Abe, Y. 2010. Carbon nanotubes elicit DNA damage and inflammatory response relative to their size and shape. *Inflammation*, 33: 276–280. doi:10.1007/s10753-010-9182-7

Yamawaki, H., Iwai, N. 2006. Cytotoxicity of water-soluble fullerene in vascular endothelial cells. *Am. J. Physiol. Cell Physiol.*, 290: C1495–C1502. doi:10.1152/ajpcell.00481.2005

Yang, D., Chen, W., Zhang, X., Mi, L., Liu, C., Chen, L., Guan, X., Cao, Y., Shen, C. 2019. Facile and scalable synthesis of low-cost FeS@ C as long-cycle anodes for sodium-ion batteries. *J. Mater. Chem. A*, 7: 19709–19718. doi:10.1039/C9TA05664E

Zhang, X., Zhu, S., Song, W., Wang, X., Zhu, J., Chen, R., Ding, H., Hui, J., Lyu, Q. 2022. Experimental study on conversion characteristics of anthracite and bituminous coal during preheating-gasification. *Fuel*, 324, 124712. doi:10.1016/j.fuel.2022.124712

Zhang, Y., Ali, S. F., Dervishi, E., Xu, Y., Li, Z., Casciano, D., Biris, A. S. 2010. Cytotoxicity effects of graphene and single-wall carbon nanotubes in neural phaeochromocytoma-derived PC12 cells. *ACS Nano*, 4: 3181–3186. doi:10.1021/nn1007176

Zhang, Y., Yin, Q. Z. 2012. Carbon and other light element contents in the Earth's core based on first-principles molecular dynamics. *Proc. Natl. Acad. Sci.*, 109: 19579–19583. doi:10.1073/pnas.1203826109

Zhao, B., He, Y. Y., Chignell, C. F., Yin, J. J., Andley, U., Roberts, J. E. 2009. Difference in phototoxicity of cyclodextrin complexed fullerene [(γ-CyD) 2/C60] and its aggregated derivatives toward human lens epithelial cells. *Chem. Res. Toxicol.*, 22: 660–667. doi:10.1021/tx800478u

Zhao, J., Cao, X., Wang, Z., Dai, Y., Xing, B. 2017. Mechanistic understanding toward the toxicity of graphene-family materials to freshwater algae. *Water. Res.*, 111: 18–27. doi:10.1016/j.watres.2016.12.037

Zhu, S., Luo, F., Li, J., Zhu, B., Wang, G. X. 2018. Biocompatibility assessment of single-walled carbon nanotubes using Saccharomyces cerevisiae as a model organism. *J. Nanobiotechnol.*, 16: 44. doi:10.1186/s12951-018-0370-1

Index

Pages in **bold** refer to tables.

A

activated sludge process, 6
adaptation policy and guidelines (carbon industry), 418; *see also* carbon industry; governance and policy
adsorption, 82, 88, 256–267
  hazardous organics, 88–91
  heavy metals, 88
  modeling, ANN, 256–267
  nano-biomaterials, 89–91
  regeneration and scale-up, 21–25
advanced oxidation (biotechnological context), 287–299; *see also* enzymes; laccases; peroxidases
agricultural runoff, pollution impacts, 2–4; *see also* sustainable agriculture
AI (artificial intelligence) for waste treatment/monitoring, 357–360
  integration with IoT, 353–357, *353*; *see also* digital innovations; IoT
air biofiltration, 5–7
  VOC control, 6–7
algal biosorption, 6–8
anaerobic digestion (biogas), 8
ANN (artificial neural networks) for adsorption prediction, 256–267; *see also* machine learning; modeling

B

bioaugmentation, 5
bioeconomy (circular), resources recovery, 220–230
  future trends, 230
bioengineering-inspired materials, 275–283
biofilters and bioscrubbers, 5–7
bioreactors (air/water interfaces), 6–7
biorefinery, 304–320
  concepts, 304
  conversion technology, 308
  environmental footprint, 318
  circular economy, 318–320
  sustainability assessment, 315
bioremediation, 5–8, 39–55
  case studies, 10–12
  concept and approach, 40
  challenges, 50–51
  planning & execution, 51–55; *see also* enzymes; microbial remediation; phytoremediation
biostimulation, 5
biosensors, nano-biosensors, 366–383
  challenges and outlook, 382–383
  electrochemical, 376–382
  FET, 382
  optical, 372–376
biosorption, 6–9
  algae, 8
  fungi/bacteria, 9; *see also* adsorption
bio-waste to energy/products, *see* anaerobic digestion, biorefinery, resource recovery

C

carbon footprint, 269–275
  links to LCA, 278–283, *278*
carbon industry
  developments and overcapacity, 414–418
  ethics and public perception, 421–430
case studies (bioremediation)
  Exxon Valdez, 10–11
  Kuwaiti oil fields, 11–12
  Love Canal, 11
circular economy, 318–320
  links to metabolic engineering, 335–339
clean water—policies/acts (reference), 5
climate change—pollution interplay, 4–5
cometabolism, 9
composting, 8
constructed wetlands (engineered), **3**
  in water resources, 60–76
contaminants in water, 82–96
  heavy metals, 88
  hazardous organics, 88–91
  pharmaceuticals, **3**, 287–299
conventional technologies, global mitigation, 121–126
cross-media integration (bio + conventional methods), 27–29

D

denitrification (groundwater), 9–10
digital innovations for net-zero, 400; *see also* AI; IoT
dyes, enzymatic degradation, 287–299; *see also* laccases; peroxidases

E

ecological footprint, 269–275; *see also* LCA
education & community engagement (pollution solutions), 4–5
electrochemical nano-biosensors, 376–382
emerging technologies (enviro-biotech), 29–33
  nanobiotechnology, 30–31
  synthetic biology, 31–33
enzymatic technologies, 285–299
  applications, 294–299
  mechanisms, 287–294
  microbial enzyme production, 286
  operational factors, 287
enzymes, 5–6, 9, 14, 17–20, 22–26, 29, 40–42, **48–49**, 54, 81, 126, 138, 146–147, 154–157, **158**, 159–160, 191, 197, 207–209, 211–212, 247, 285–299, 310, 312–313, 325, 327, 330–331, 334, 368–372
ethics and public perception (carbon industry), 421–430

F

field-effect transistor (FET) sensors, 382
food chain & water quality, 2–4

G

genetic engineering for remediation, 6
governance & policy (pollution control), 4–5
  for NBS, 72

H

hazardous waste & e-waste: bio-recovery/recycling, **3**
heavy metals
  adsorption, 88
  biosorption, 6–9
  phytoremediation, 196–205, *205*, 213
hydrocarbons (soil/air), microbial degradation, 9–11
  bioventing/biosparging, 7

I

industrial framework for net-zero (emerging economies), 394–409
  case studies, 404–407
  challenges, 401–404
IoT for waste management/monitoring, 353–357
  integration with AI, 357–360
IP & ethics, *see* ethics and public perception

L

laccases, 6, 17–18, **48**, 156–157, 286, 288–289, 294
LCA (life cycle assessment), 278–283, *278*
  concepts and links to footprint, 269–275
  materials perspective, 280–283
lignocellulosic residues → high-SSA carbon, 101–113
Love Canal (case), 11

M

machine learning, *see* ANN
materials, bio-inspired (features, performance), 275–283
metabolic engineering in waste prevention, 325–339
  circular economy, 335–339
  fundamentals, 326
microbial electrolysis (MEC), 234–251
  applications, 240–247
  fundamentals, 236–240
  renewables integration, 244–247
  scale-up/challenges and outlook, 248–251
microbial fuel cells (MFC), *see* microbial electrolysis, wastewater energy recovery
microbial remediation (mechanisms, applications), 9–10; *see also* bioremediation
microplastics (context in water pollution), **3**
modeling for adsorption (ANN), 256–267
multi-disciplinary frameworks (environment + materials + policy), 117–132, 269–283

N

nano-biomaterials for remediation, 80–96
  enhanced nano-biomaterials, 89–91
  environmental cleanup, 91–96
nano-biosensors, *see* biosensors
nanobiotechnology in pollution treatment, 30–31
nature-based solutions (NBS), 60–76
  benefits, 70–71
  challenges, 71
  future directions, 74–75
  policy & governance, 72
  principles & types, 67
net-zero transition (industry; emerging economies), 387–409

O

omics in phytoremediation (heavy metals), 200–213
organics (hazardous) removal, 88–91
  enzymatic, 287–299

P

peroxidases (lignin, HRP etc.) in degradation, 287–299; *see also* dyes; enzymes
pharmaceutical pollution: wetlands & enzymatic routes, **3**, 287–299
phytoremediation, 196–205, *205*, 213
  plant species and mechanisms, 200–205
  omics, 204–213
plastic pollution; bioplastics & enzymes, **3**, 7
policy and regulation (pollution control; NBS; water), 4–5, 72
pollutants by class
  dyes, 287–299
  heavy metals, 88, 196–205, 256–267
  hydrocarbons, 9–11
  pesticides, **3**, 287–299
  pharmaceuticals, **3**, 287–299
pollution—air, water, soil (overviews, sources, impacts), 1–5
  urgency, 4–5
process integration (biotech + conventional), 27–29

R

radioactive pollution (bio options), **3**
regeneration & scale-up (biotech processes), 21–25
renewables integration with MEC, 244–247
resource recovery from solid waste, 220–230
  recent research, 224–230
  technologies, 224–227
risk & perception, *see* ethics and public perception

S

scale-up challenges (biotech processes), 24–25
sensors (environmental, nano-biosensors), 366–383
  electrochemical, 376–382
  FET, 382
  optical, 372–376

soil pollution & microbial rejuvenation, 136–173
limitations/advancements, 170–172
mechanisms, 148–171
sustainable agriculture (precision/organic), 5
sustainable solutions—regulatory, technological, community, 4–5
synthetic biology in remediation, 31–33
systems integration (bio + conventional), 27–29

T

thermal/energy considerations (process efficiency), 8
toluene/VOC biodegradation, 6, 10
toxic organics, *see* enzymes, pollutants

U

urban/industrial emissions (bio-abatement), 6–7

V

value from waste (bioeconomy; resource recovery; biorefinery), 220–230, 304–320
VOC control by biofiltration, 6–7

W

waste management—IoT/AI, 353–360
wastewater treatment
activated sludge, 6
enzymatic, 287–299
MEC, 234–251
water resources management via NBS, 60–76
waste water energy recovery, 235

Z

zero-emission, *see* net-zero transition

£ - 150